Memristors, Dynamics and Complexity

Historical Development and Recent Advancement

WORLD SCIENTIFIC SERIES ON NONLINEAR SCIENCE

Editor: Leon O. Chua
University of California, Berkeley

Series B. SPECIAL THEME ISSUES AND PROCEEDINGS

Volume 1: Chua's Circuit: A Paradigm for Chaos
Edited by *R. N. Madan*

Volume 2: Complexity and Chaos
Edited by *N. B. Abraham, A. M. Albano, A. Passamante, P. E. Rapp and R. Gilmore*

Volume 3: New Trends in Pattern Formation in Active Nonlinear Media
Edited by *V. Perez-Villar, V. Perez-Munuzuri, C. Perez Garcia and V. I. Krinsky*

Volume 4: Chaos and Nonlinear Mechanics
Edited by *T. Kapitaniak and J. Brindley*

Volume 5: Fluid Physics — Lecture Notes of Summer Schools
Edited by *M. G. Velarde and C. I. Christov*

Volume 6: Dynamics of Nonlinear and Disordered Systems
Edited by *G. Martínez-Mekler and T. H. Seligman*

Volume 7: Chaos in Mesoscopic Systems
Edited by *H. A. Cerdeira and G. Casati*

Volume 8: Thirty Years After Sharkovskiĭ's Theorem: New Perspectives
Edited by *L. Alsedà, F. Balibrea, J. Llibre and M. Misiurewicz*

Volume 9: Discretely-Coupled Dynamical Systems
Edited by *V. Pérez-Muñuzuri, V. Pérez-Villar, L. O. Chua and M. Markus*

Volume 10: Nonlinear Dynamics & Chaos
Edited by *S. Kim, R. P. Behringer, H.-T. Moon and Y. Kuramoto*

Volume 11: Chaos in Circuits and Systems
Edited by *G. Chen and T. Ueta*

Volume 12: Dynamics and Bifurcation of Patterns in Dissipative Systems
Edited by *G. Dangelmayr and I. Oprea*

Volume 13: Modeling and Computations in Dynamical Systems: In Commemoration of the 100th Anniversary of the Birth of John von Neumann
Edited by *E. J. Doedel, G. Domokos and I. G. Kevrekidis*

Volume 14: Dynamics and Control of Hybrid Mechanical Systems
Edited by *G. Leonov, H. Nijmeijer, A. Pogromsky and A. Fradkov*

Volume 15: From Physics to Control Through an Emergent View
Edited by *L. Fortuna, A. Fradkov and M. Frasca*

Volume 16: Frontiers in the Study of Chaotic Dynamical Systems with Open Problems
Edited by *Z. Elhadj and J. C. Sprott*

Volume 17: Advances on Nonlinear Dynamics of Electronic Systems
Edited by *A. Buscarino, L. Fortuna and R. Stoop*

Volume 18: Generalized Synchronization and Generalized Consensus of System Arrays
Edited by *L. Min and G. Chen*

Volume 19: Recent Trends in Chaotic, Nonlinear and Complex Dynamics
Edited by *J. Awrejcewicz, R. Shanmuganathan and M. Ragulskis*

Volume 20: Dynamics, Control, and Synchronization of Nonlinear Systems
Edited by *Nijmeijer and J. P. Ramírez*

Volume 21: Memristors, Dynamics and Complexity: Historical Development and Recent Advancement
Edited by *Guanrong Chen*

To view the complete list of the published volumes in the series, please visit:
http://www.worldscientific.com/series/wssnsb

World Scientific Series on Nonlinear Science Series B Vol. 21
Series Editor: Leon O. Chua

Memristors, Dynamics and Complexity

Historical Development and Recent Advancement

edited by

Guanrong Chen
City University of Hong Kong, Hong Kong, China

World Scientific

NEW JERSEY • LONDON • SINGAPORE • BEIJING • SHANGHAI • HONG KONG • TAIPEI • CHENNAI • TOKYO

Published by

World Scientific Publishing Co. Pte. Ltd.
5 Toh Tuck Link, Singapore 596224
USA office: 27 Warren Street, Suite 401-402, Hackensack, NJ 07601
UK office: 57 Shelton Street, Covent Garden, London WC2H 9HE

British Library Cataloguing-in-Publication Data
A catalogue record for this book is available from the British Library.

World Scientific Series on Nonlinear Science Series B — Vol. 21
MEMRISTORS, DYNAMICS AND COMPLEXITY
Historical Development and Recent Advancement

ISBN 978-981-98-3062-6 (hardcover)
ISBN 978-981-98-3063-3 (ebook for institutions)
ISBN 978-981-98-3064-0 (ebook for individuals)

For any available supplementary material, please visit
https://www.worldscientific.com/worldscibooks/10.1142/14805#t=suppl

Desk Editor: Rhaimie Wahap

Typeset by Stallion Press
Email: enquiries@stallionpress.com

https://doi.org/10.1142/9789819830633_fmatter

Preface

The year 2025 marks the 35th anniversary of the *International Journal of Bifurcation and Chaos*, founded by Professor Leon O. Chua at the University of California, Berkeley.

To commemorate this milestone, the November 2025 issue was specially curated under the title *A Panorama of Nonlinear Dynamics and Complex Systems*. It presents twelve contributions from leading experts, addressing core themes of chaos theory and bifurcation analysis central to the journal's mission. Professor Chua authored the keynote article, offering a sweeping overview of two pivotal topics — memristors and the edge of chaos — and illuminating their deep interconnections within the broader landscape of nonlinear dynamics.

This book brings together the twelve articles originally published in that celebratory issue, presenting their state-of-the-art insights in a single collection. It is designed to be both authoritative and accessible, engaging scientists, professionals, and graduate students across disciplines including statistical physics, applied mathematics, electrical and mechanical engineering, computer science, brain science, and the wider study of complex systems. The volume is envisioned as a significant resource and accessible reference, supporting research and development throughout the diverse landscape of nonlinear dynamics.

Guanrong Chen
Editor
City University of Hong Kong
Hong Kong, China

https://doi.org/10.1142/9789819830633_fmatter

Contents

https://doi.org/10.1142/9789819830633_0001

Chapter 1

MEMRISTORS on Edge of Chaos: Quo Vadis?#

Leon O. Chua
Department of Electrical Engineering and Computer Sciences,
University of California, Berkeley, CA 94720, USA
chua@berkeley.edu

Editorial Revisited[1]

Never in the annals of science and engineering has there been a phenomenon so ubiquitous, a paradigm so universal, or a discipline so multi-disciplinary as that of chaos. Yet chaos represents only the tip of an awesome iceberg, for beneath it lies a much finer structure of immense complexity, a geometric labyrinth of endless convolutions, and a surreal landscape of enchanting beauty. The bedrock which anchors these local and global bifurcation terrains is the omnipresent nonlinearity that was once randomly linearized by the engineers and applied scientists of yore, thereby forfeiting their only chance to grapple with reality.

We now know that bifurcation and chaos are rather mundane phenomena that could occur in many natural and man-made systems where nonlinearity is present. The robustness of such counterintuitive, if not bizarre, phenomena has not only overthrown the age-old dogma of a deterministic world but also unloosed the underpinning of many long-held diagnostic and design methodologies from engineering and the applied sciences, which were based on linear approximations. Even the normally benign subcritical Hopf bifurcation phenomena could lead to catastrophic consequences if a system imbued with such dynamics were inadvertently kicked out of its unsuspected failure boundary. On the other hand, recent laboratory observations suggesting that chaotic regimes may in fact represent the ground state of the central nervous system point to the intriguing possibility of exploiting and controlling chaos for future scientific and engineering applications.

Be that as it may, the subject of bifurcation and chaos has already established itself as a fundamental discipline cutting across virtually all scientific fields. This discipline is truly multidisciplinary and unique because it shares no territorial precedents. For what discipline could one conjure up an analogue situation where an electronic engineer could engage in an in-depth dialogue with a biologist on the dynamics of irregular heartbeat rhythms, with a mathematician on the dimension of the Cantor set, with an economist on the chaotic patterns of the stock market, with a psychologist on the brain wave series registered during dreams, with a social scientist on the catastrophe of prison riots, or even with a musician on the technique of playing a clarinet to induce period doubling?

In this 35th anniversary issue, we focus on two, among many, exciting and flourishing research areas, namely, ***memristors*** [Chua, 1971, 2018a, 2018b; Strukov *et al.*, 2008; Tour & He, 2008; Williams, 2008] and *edge of chaos* [Chua, 2024]. The fundamental theory which gives birth to the powerful "Edge of Chaos criteria" is the "*principle of local activity*" [Chua, 2005], which can be interpreted as the complement of the "*Second Law of Thermodynamics*".

#To cite this article, please refer to its earlier version published in the *International Journal of Bifurcation and Chaos*, Vol. 35, No. 14 (2025) 2540001 (67 pages), DOI: 10.1142/S0218127425400012.

[1]Extracted from Vol. 1, No. 1.

But the highlight of this 30th anniversary exposition is a colorful *panorama* for resolving the *Hodgkin–Huxley blunder* of identifying the sodium and potassium ion channels erroneously as time-dependent *resistances* [Chua *et al.*, 2012b]. The climax of the *panorama* is our definitive solution of the heretofore unresolved *Turing Instability* and *the Smale Paradox* [Chua, 2024].

Keywords: Memristor; edge of chaos; local activity principle; Hodgkin–Huxley circuit model; anomalous impedance; Kenneth Cole; Aplysia; Eric Kandel; Brian Josephson; Ivan Pavlov; synapse; action potential; Turing instability; Smale paradox; complexity theory.

1. Genesis of the Memristor

The dawn of nonlinear electronics was ushered in during the 1960s by a Cambrian-esque explosion of newly minted two-terminal electronic devices, bearing such intimidating monikers as Esaki diode, Josephson junction, varactor diode, thyristor, impact ionization avalanche transit-time (IMPATT) diode, Gunn diode, and ovonic threshold switch, etc. Trained in the old school of *linear* circuit analysis, hordes of electronics engineers were awed and shocked upon witnessing a parade of such strongly *nonlinear* and *dynamical* electronic devices unfolding at such a breathless rate. To many, the surreal proliferation of exotic devices would conjure the opening scene of Dickens's tale of yore: "*It was the best of times, it was the worst of times, it was the age of wisdom, it was the age of foolishness, it was the epoch of belief, it was the epoch of incredulity, it was the season of Light, it was the season of Darkness, it was the spring of hope, it was the winter of despair, we had everything before us, we had nothing before...*".

To analyze circuits made of *strongly nonlinear* and *dynamical* electronic devices [Chua, 1969, 2003], it is necessary to have realistic device models made of well-defined *nonlinear* circuit elements as building blocks, which did not exist then, because the electrical engineers from that bygone epoch were taught to overcome nonlinearities by expanding them in a Taylor series and then retaining only the linear term that neatly maps into a linear circuit model. But just like the ancient parable about the "*blind man and the elephant*", such models invariably gave rise to grossly inaccurate and misleading results. Unfortunately, the "*linearize then analyze*" culture endowed upon the electronic engineers of the day had made it impossible to devise such generalizations, due in part to the lack of a circuit-theoretic foundation for basic *nonlinear* circuit elements. Even Feynman *et al.* [1963] would sloppily use the old-fashioned name *Condenser* — instead of capacitor — in his 1963 classic, "The Feynman Lectures on Physics", when he wrote on pp. 6–12, Vol. II, "The coefficient of proportionality is called the *Capacity*, and such a system of two conductors is called a *Condenser*". Electronic circuit theory is concerned only with the prediction of the voltage $v(t)$ and current $i(t)$ associated with the external terminals sticking out of an enclosing cocoon (henceforth called a *black box*) where its interior may contain some newly minted nonlinear electronic device, or an interconnection of various run-of-the-mill electronic devices, batteries, and so on. And thus, basic *nonlinear* circuit elements must be defined from a *black-box perspective*, independent of their internal composition, material, geometry, and architecture.

From the circuit-theoretic point of view, the three basic two-terminal circuit elements are defined in terms of a relationship between two of the four fundamental circuit variables, namely, the *current* $\boldsymbol{i}$, the *voltage* v, the *charge* q, and the *flux* φ. Out of the *six* possible combinations of these four variables, five have led to well-known relationships. Two of these relationships are already given by $\boldsymbol{q(t) = \int_{-\infty}^{t} i(\tau)d\tau}$ and $\boldsymbol{\varphi(t) = \int_{-\infty}^{t} v(\tau)d\tau}$. Three other relationships are given, respectively, by the *axiomatic* definition of the three classical circuit elements, namely, the *resistor* (defined by a relationship between v and $\boldsymbol{i}$), the *inductor* (defined by a relationship between φ and $\boldsymbol{i}$), and the *capacitor* (defined by a relationship between q and v). Only one relationship remains undefined: the relationship between φ and q. From the *logical* as well as *axiomatic* points of view, it is necessary for the sake of completeness to postulate the existence of a fourth basic two-terminal circuit element, which is characterized by a φ–q relationship. This element is christened the ***memristor*** because it behaves like a nonlinear resistor with memory, *remembering things past* [Chua, 2011, 2014, 2015a; Chua *et al.*, 2012a; Volkov, 2018].

2. Genesis of the Edge of Chaos

According to several prominent authors, including Stephen Hawking, a main part of 21st-century science will involve ***complexity*** research. The intuitive idea is that global patterns and structures emerge from locally interacting elements like atoms in laser beams, molecules in chemical reactions, proteins in cells, cells in organs, neurons in brains, agents in markets, and so on, by self-organization [Mainzer, 2007]. However, what is the cause of *self-organization*? **Complexity** phenomena have been reported from many disciplines (e.g. biology, chemistry, ecology, physics, sociology, economy, etc.) and analyzed from various perspectives, such as *Schrödinger's order from disorder* [Schrödinger, 1948], *Prigogine's dissipative* structure [Prigogine, 1980], *Haken's synergetics* [Haken, 1988], Packard's edge of chaos [Packard, 1988; Langton, 1990], etc. But concepts of ***complexity*** are often based on examples or metaphors only. We argue for a mathematically precise and rigorous definition of **local activity** [Chua, 2005] as the cause of ***self-organizing complexity,*** which can be tested in an explicit and constructive manner [Mainzer & Chua, 2013].

Boltzmann's struggle in understanding the physical principles distinguishing between living and nonliving matter, Schrödinger's negative entropy in metabolism [Schrödinger, 1948], Turing's basis of morphogenesis [Turing, 1952], Prigogine's intuition of the instability of the homogeneous [Prigogine, 1980], Haken's synergetics [Haken, 1988], Gell-Mann's amplification of fluctuation [Gell-Mann, 1994, 1995] are in fact all direct manifestations of a *fundamental principle of locality.* The ***principle of local activity*** can be considered the complement of the *second law of thermodynamics*, explaining the emergence of order from disorder instead of disorder from order, in a *quantitative* way, at least for reaction–diffusion systems.

The ***principle of local activity*** is precisely the missing concept to explain the emergence of complex patterns in a homogeneous medium. Leon Chua defined this principle in the theory of nonlinear electronic circuits in a mathematically rigorous way [Chua, 1998]. ***The local activity principle*** can be generalized and proven for the class of nonlinear reaction–diffusion systems in physics, chemistry, biology, and brain research. The ***principle of local activity*** is the cause of symmetry breaking in *homogeneous* media [Chua, 2015b].

A completely new application of ***local activity*** is the so-called ***edge of chaos*** [Chua, 2013, 2022], where most complex phenomena emerge. In this particular case, a *hidden excitability* allows a unit to be destabilized when interacting with *dissipative* environments. Although a diffusion process has a tendency to equalize differences, an originally *dead* or inactive cell becomes *alive* or active upon coupling with other cells by diffusion [Smale, 1974]. This phenomenon seems counterintuitive, but can be mathematically and rigorously proven and confirmed with different applications in reality.

In the parameter spaces of *reaction–diffusion systems*, the domains of *local activity* can be visualized by computer simulations. The ***edges of chaos*** are very small regions with the ability of *creation*. For example, in a new Hodgkin–Huxley axon circuit model [Chua *et al.*, 2012a], the ***edge of chaos*** domain of the brain can be determined with a spectacular accuracy, namely, with less than 1 mV and 2 μA **[Chua *et al.*, 2012b]**. In the brain, tiny domains are the origin of *action potentials* and attractors, which are correlated with emerging mental and intellectual abilities. Thus, the ***edge of chaos*** domains seem to be hidden in the domains of activity like pearls in shells on the sea floor [Chua, 2025a, 2025b; Zhang *et al.*, 2020]. Actually, in research, the phrase "edge of chaos" was often used as a metaphor, but not as a mathematically precise concept. In particular, the ***Edge of Chaos*** has only been simulated via discrete maps [Packard, 1988; Langton, 1990]. Its discovery in the parameter spaces of *continuous-time dynamical systems* is, to our best knowledge, completely new in *complexity* research. The *local activity principle* and its pearl, the ***edge of chaos***, are couched in rigorous mathematics. Above all, they are characterized by constructive procedures to compute and visualize their complexity. Therefore, researchers from other disciplines who can describe their dynamical systems via *differential equations*, such as reaction–diffusion equations, can actually easily calculate, with a computer, the *parameter values* where *complexity* and *creativity* can occur.

We argue that the ***principles of local activity*** and ***edge of chaos*** are really fundamental in science. In some applications of physics, chemistry, biology, brain research, and technology, the domains of ***local activity*** and ***edge of chaos*** have been computed and visualized [Mainzer & Chua, 2013;

Williams, 2017]. In other disciplines of natural science, for example, in *quantum cosmology*, ***local activity*** can even be identified as *symmetry breaking* of *local gauge symmetries* generating the *complexity* of matter and forces in our universe. *Supramolecular chemistry* and *nanoscience* are also examples of 21st-century science exploring *self-organizing complex* molecular systems.

The life sciences are sometimes defined as the sciences of *complexity*, because cellular organisms are typical examples of *complex systems* with emerging effects that cannot be predicted in the long run. For physicians, the *complexity* and vulnerability of human organisms are everyday experiences. Finally, brain research and neuropsychology try to explain the origin of our mental and emotional abilities with nonlinear dynamics of complex neural systems. Therefore, it would be very helpful to know the origin of nonlinear effects in all of these *complex systems* of nature by applying the principle of ***local activity*** and its crown jewel, *the* ***edge of chaos***.

Twenty-first-century engineering sciences are more and more inspired by the life sciences [Mainzer & Chua, 2013]. In *systems biology*, cells, organs, and organisms are considered *complex* systems that can be modeled by *complex* networks with great similarity to electronic circuits. In the tradition of engineering sciences, *synthetic biology* uses schemes of systems biology to construct new artificial organisms for practical applications (e.g. new bacteria for cleaning polluted water). Robots become more and more autonomous and ***self-organizing*** systems. The question arises whether the principle of ***local activity*** can be applied in these systems. Recently, local activity and the ***edge of chaos*** were computed and visualized in parameter spaces of ***memristors***, which can be applied as nanoelectronic devices in artificial brains. In general, the ***emergence of complex behavioral patterns can also be explained by symmetry breaking in homogeneous networks***.

In the age of globalization, the Earth system grows together with human civilization. The climate and ecological systems can no longer be separated from one another, but depend on industrial growth and energy policies. Global communication networks and infrastructures, as well as financial dependencies of banks and states, are driven by the *nonlinear dynamics* of ***complex systems***. One of the main insights of nonlinear dynamics is the emergence of systemic risks, which are caused by the interactions of many factors and players in the whole system. The emergence of systemic risks from ***complex systems*** *dynamics* is a challenge for control tasks in engineering sciences, as well as regulation and governance in social systems. We need early warning systems in the technical and natural sciences, as well as economics and politics. Therefore, we must consider applications of cross-over disciplines in economic, financial, and social systems with the *emergence* of equilibrium states, *symmetry breaking* at critical points of *phase transitions*, and risky action at the *edge of chaos*. In any case, the driving causes of *symmetry* breaking and the *emergence of complexity* are ***locally active*** elements, cells, units, or agents in dynamical systems.

Thus, the demand for ***local activity*** in *societal systems* leads us to a plea for *locally active* ("entrepreneurial") agents. Entrepreneurial activities are not restricted to industry, but mean any kind of technical, economic, and social engagement leading to new innovations. With respect to sustainability, we need innovations considering benefits for the whole Earth system, not only narrow-minded profits. In short: Sustainable innovations must be the goal of entrepreneurial creativity. Moreover, according to the mathematics and philosophy of ***local activity***, interaction of agents in complex systems is not sufficient for innovations. Innovative creativity of local individuals is the driving force of entrepreneurship. However, of course, ***local activity*** of entrepreneurship needs interaction with other agents in order to achieve benefits for the whole system. Obviously, the ***principle of local activity*** involves deep challenges of ethics and responsibility in society [Mainzer & Chua, 2013].

3. Visions of Memristors on Edge of Chaos

Edge computing is a distributed information technology architecture in which data are processed in proximity to the sources, for example, sensors or Internet-of-Things devices [Krestinskaya *et al.*, 2019; Premsankar *et al.*, 2018], that generate them. In the traditional computing approach, data are preliminarily sent to a remote center before elaboration. This unavoidably slows down the extraction of information, as the transfer procedure is susceptible to bandwidth limitations, latency issues, and potential disruptions over the Internet. Another major issue is the huge amount of associated energy consumption. The energy consumption for sending data to a data center is more than 10 times that for processing it locally on the edge using memristive systems, because sending data involves converting analog data into digital, storing them, and transferring them over long distances.

A distributed signal processing paradigm is of vital necessity for a time-efficient and energy-efficient solution of demanding computational challenges, emerging frequently in the Big Data era we are currently living in. Technical systems operating within an edge-computing information technology environment are endowed with data storage and processing blocks, which, together with power and transceiver units, allow local real-time control over critical processes in various domains, including the military sector, the medical field, and industry. Only the main results of the local computations may then be transferred, upon request, across the Internet to some remote center for further elaboration, which resolves the data communication issues affecting the traditional computing approach. In the past three years, there has been an increased emphasis on solving practical design challenges for implementing neuro-memristive solutions for energy-efficient computing on the edge. The complexity of problems at the edge keeps on increasing, in problems such as autonomous driving, surveillance and tracking of multiple objects, and real-time tracking of humans, all of which demand increased computational density and, at the same time, need to meet low energy requirements. The dot-product computations that are at the heart of neural computations in a crossbar architecture align well with the energy-efficient requirements of edge devices [Rao *et al.*, 2023; Song *et al.*, 2024; Huang *et al.*, 2025].

Design optimization and tools that can help to automate the design are an emerging area of interest in designing edge-computing hardware. More recently, there have been efforts to make use of large language models to generate the chip designs and automate the time-consuming verification tasks. The use of generative AI and automating chip design for edge AI computing is futuristic, yet quickly becoming a reality due to the emergence of foundry-supported open-source chip design tools. Open chip movements such as RISC-V initiatives will further speed up the design-to-test times for neuro-memristive chips for edge computing, as the designers have a well-developed hardware and software stack that can be readily integrated to implement a complete ecosystem for building applications with edge AI chips.

4. A Panorama on Complexity

Our final section is a testimony to the age-old Chinese proverb, "***A picture is worth a thousand words***". Instead of echoing time-weary anecdotes and theories on **complexity** from scientific organizations and the popular press, we opted to exhibit a *panorama* of well-posed pictures for illuminating our theory on **complexity** that is based on our **Principle of *Local Activity*** [Chua, 2005] and its crown jewel, the edge of chaos [Chua, 1998], both of which were rigorously conceived and proved mathematically, where the fundamental nonlinear dynamics were based on a new basic circuit element dubbed the memristor (Exhibits 29, 30, 32). An example of a first-order memristor has a small-signal equivalent circuit consisting of an *infinite* resistance, aka an *open circuit*, in parallel with a -2 Ohm resistance in series with a -2 Henry inductance [Chua, 2025a], shown in Exhibit 117. This unique circuit is called Chua's Riddle, because adding an external resistance R^*, where $0 < R^* < 2$ Ohms, in series with the memristor will produce a benign damped exponential waveform current response $i(t)$, for any initial current $i(0)$. But if $R^* > 2$ Ohms, the response current $i(t)$ blows up! The memristor defined in Chua's Riddle is on the edge of chaos [Chua, 2025a].

Only recently, the well-known **Turing Instability** [Chua, 2024] was shown to be triggered by the Edge of Chaos (Exhibits 89–94, 96–97, 100–103). Likewise, the well-known **Smale Paradox** was also resolved via the Edge of Chaos (Exhibits 105–108).

Last, but not least, we call attention to the following Nobel laureates who won their accolades without realizing their discoveries were all direct consequences of the nonlinear dynamics of memristors.

- **Pavlov**, 1904 (Exhibits 49–54).
- **Hodgkin and Huxley**, 1961 (Exhibit 12).
- **Josephson**, 1973 (Exhibits 47–48).
- **Kandel**, 2000 (Exhibits 41–46).

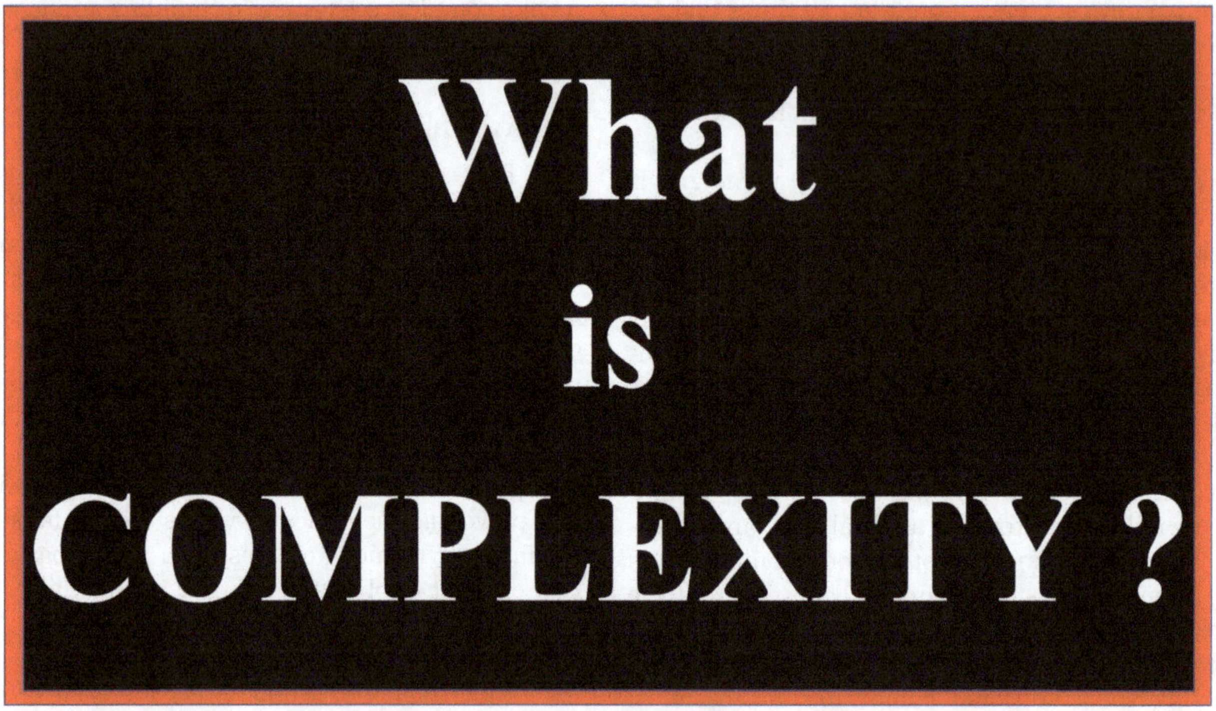

Exhibit 0

Exhibit 1

Exhibit 2

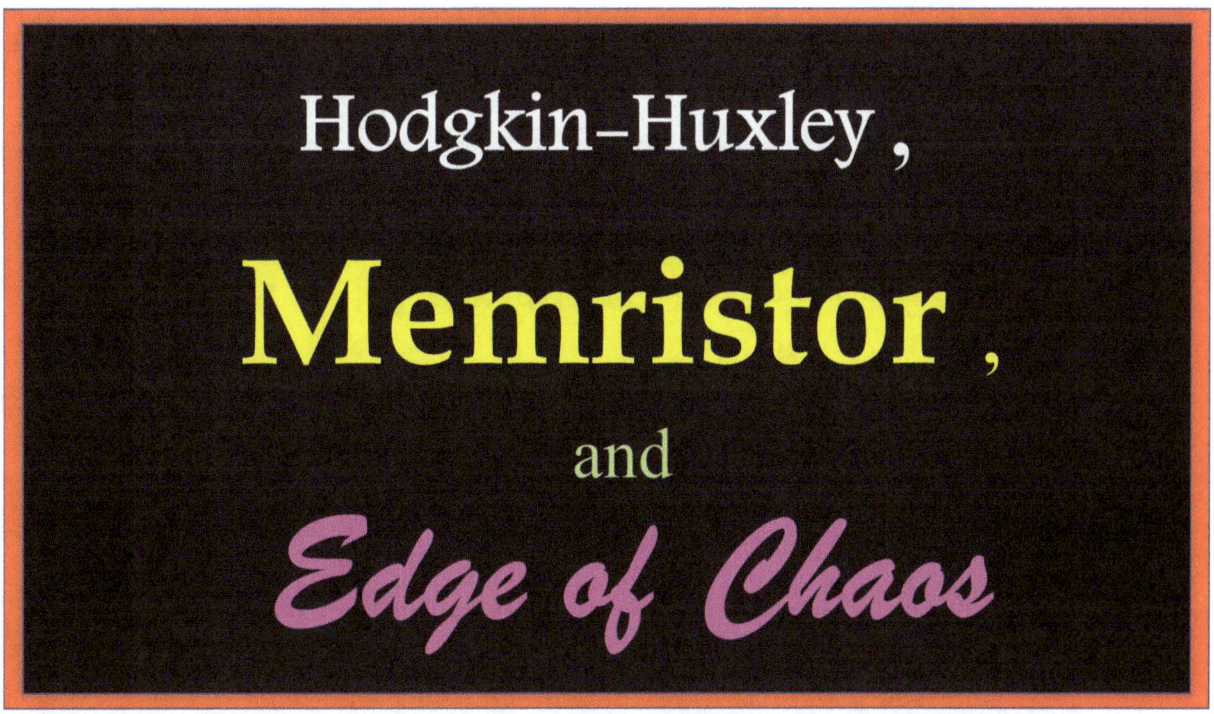

Exhibit 3

Exhibit 4

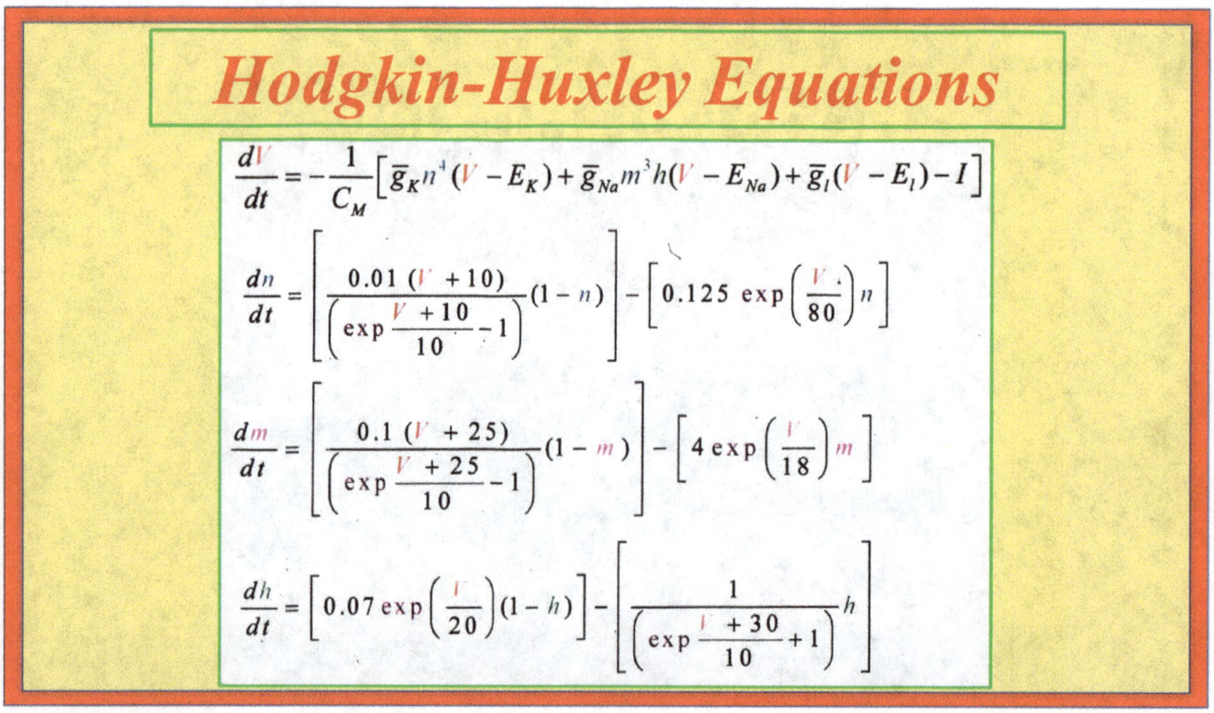

Exhibit 5

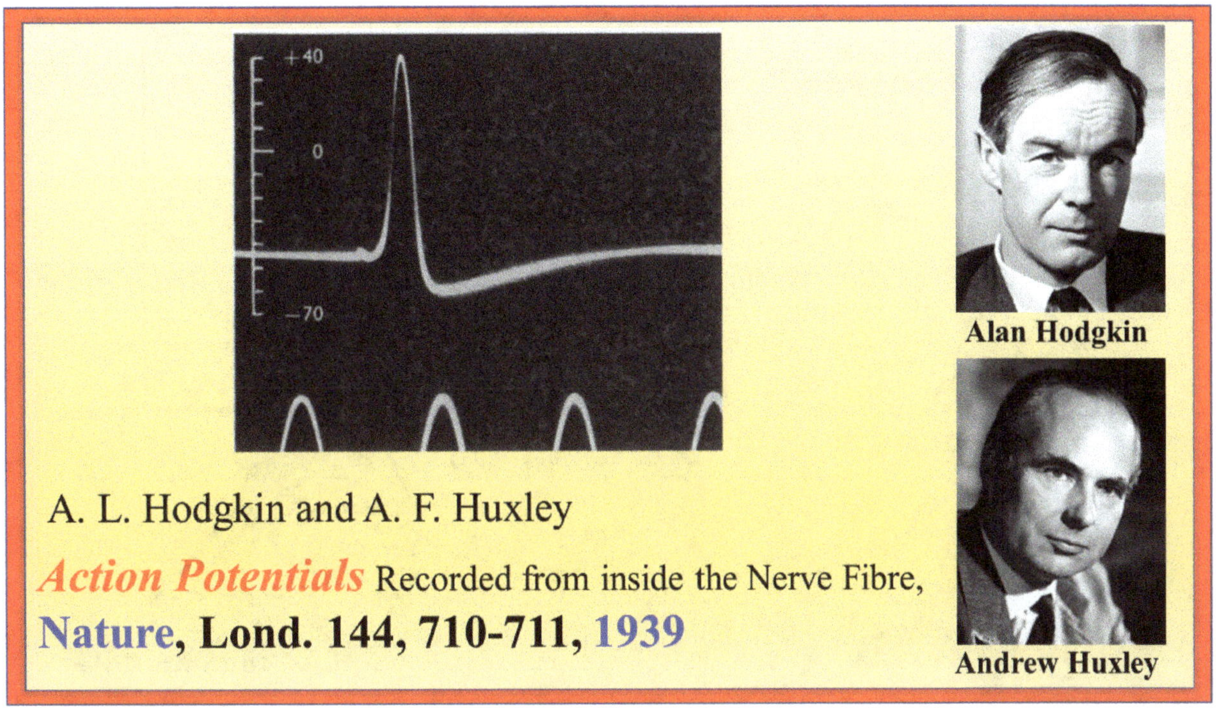

Exhibit 6

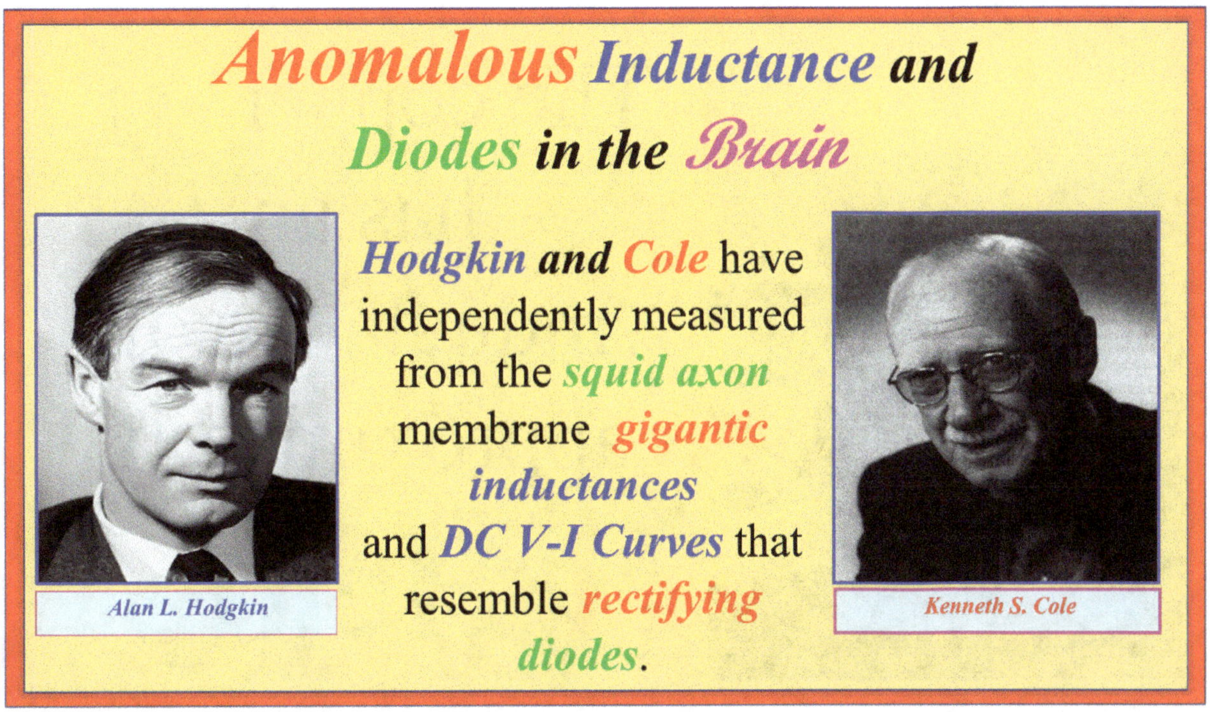

Exhibit 7

Exhibit 8

Exhibit 9

The suggestion of an inductive reactance anywhere in the system was shocking to the point of being unbelievable.

Kenneth Cole

Exhibit 10

Hodgkin-Huxley Mystery

Although the Hodgkin-Huxley Equations have been around for more than 70 years, we still do not understand the complex nonlinear dynamics responsible for the generation of the ***action potentials*** (***spikes***).

Exhibit 11

Hodgkin's Blunder

Hodgkin had struggled in vain searching for a *physical* interpretation of the *squid axon inductance*. He failed because he had mistaken the *axon* for a *time-varying conductance*, when in fact it has a simple explanation if the *Potassium* and *Sodium ion channels* are identified as *memristors*.

A. L. Hodgkin,
"The ionic basis of electrical activity in nerve and muscle,"
Biological Review, Vol. 26, pp. 339-409, 1951

Exhibit 12

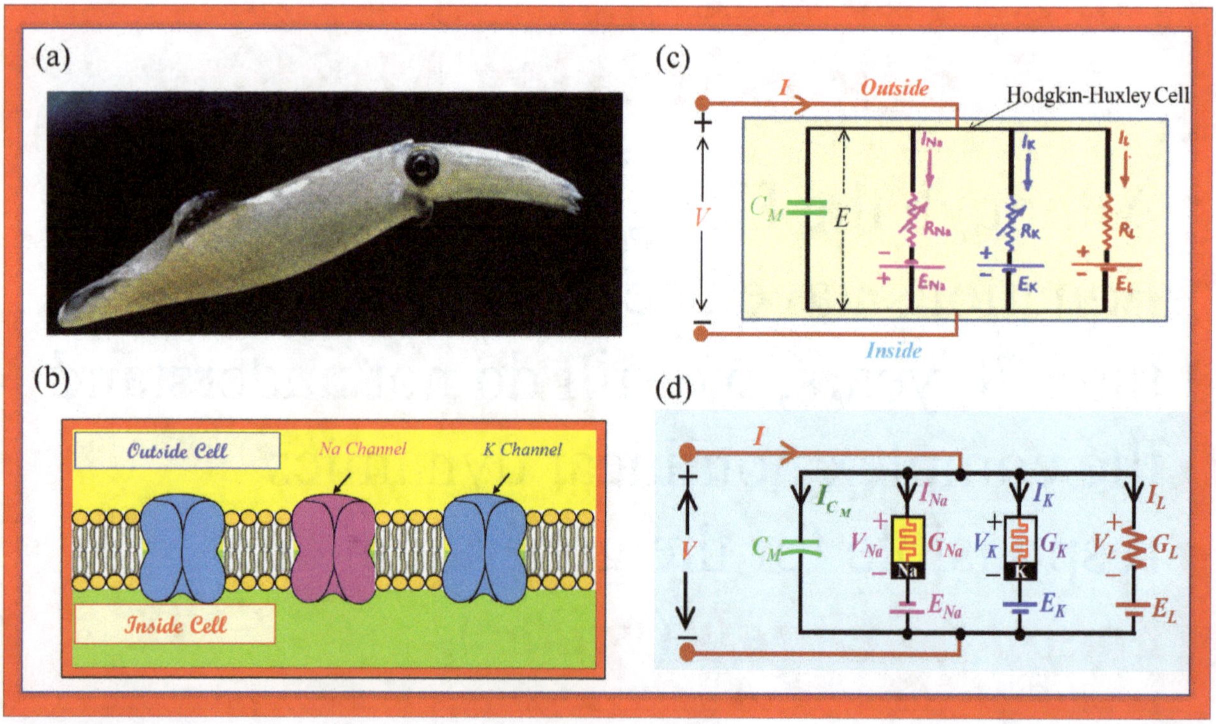

Exhibit 13

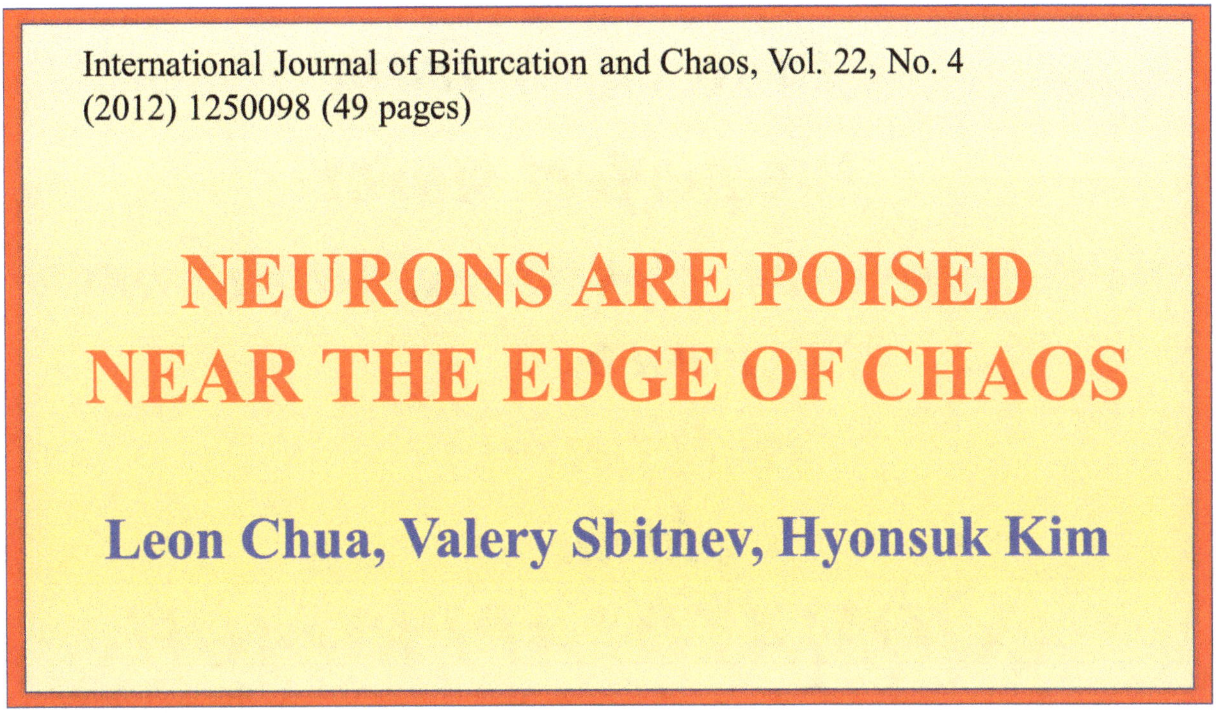

Exhibit 14

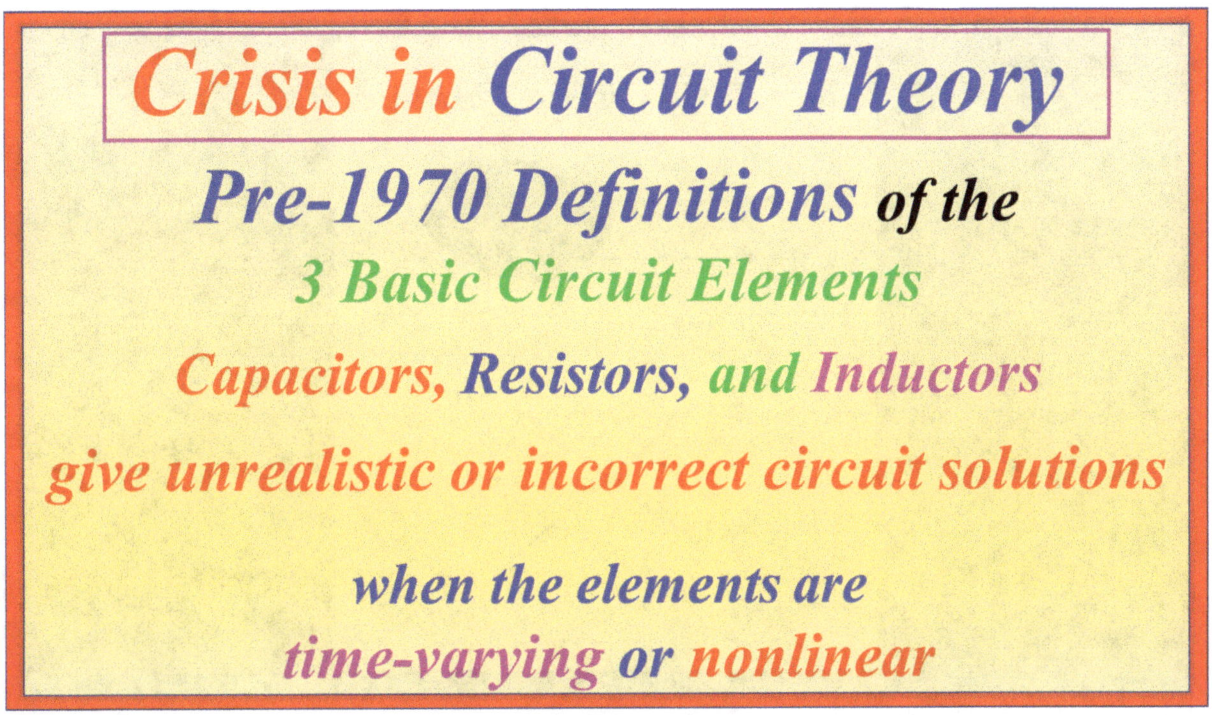

Exhibit 15

Exhibit 16

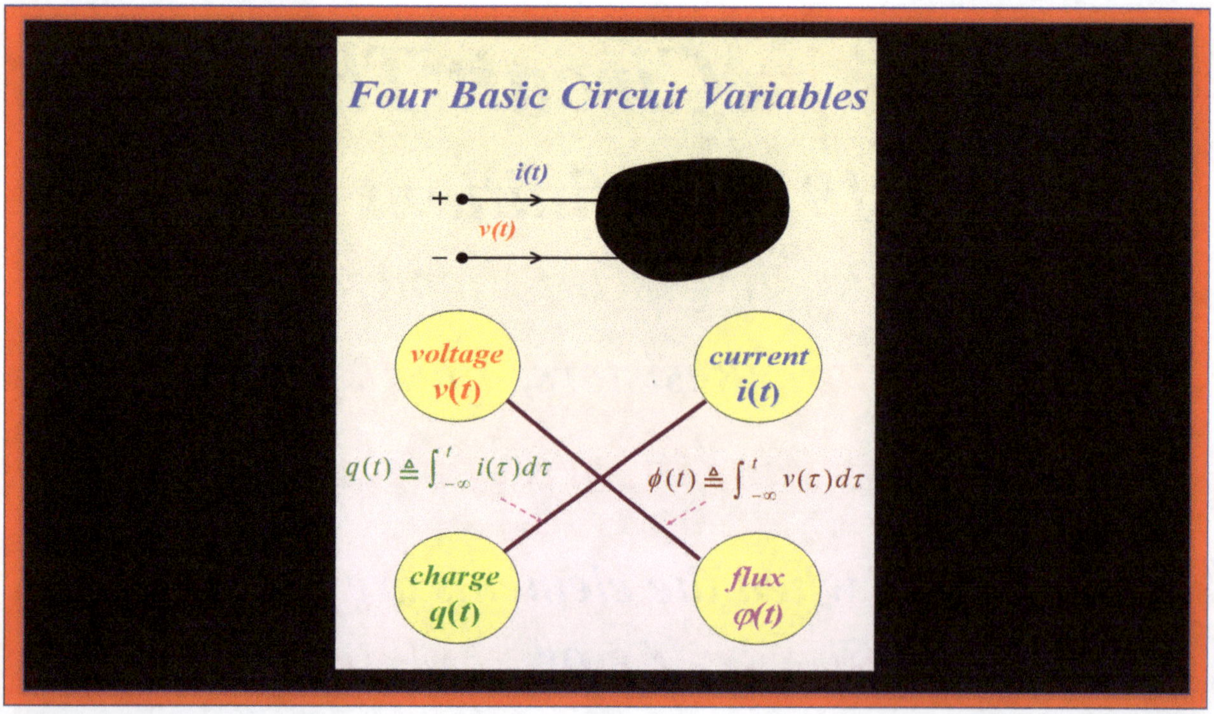

Exhibit 17

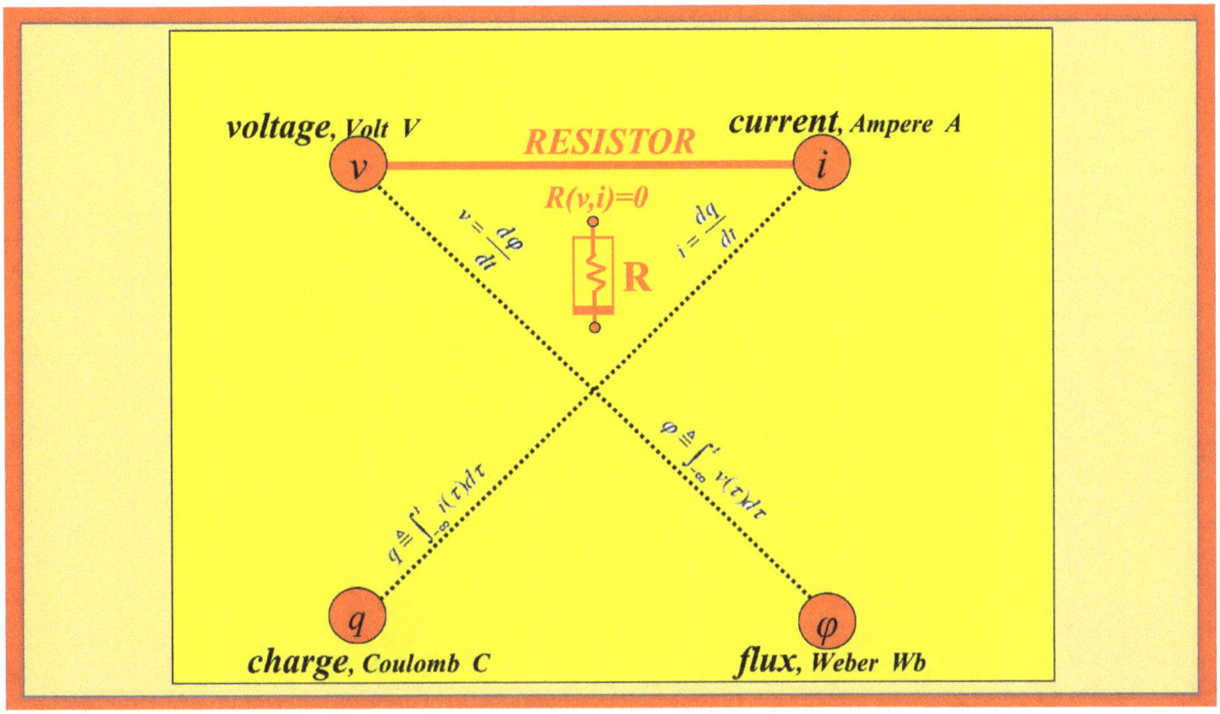

Exhibit 18

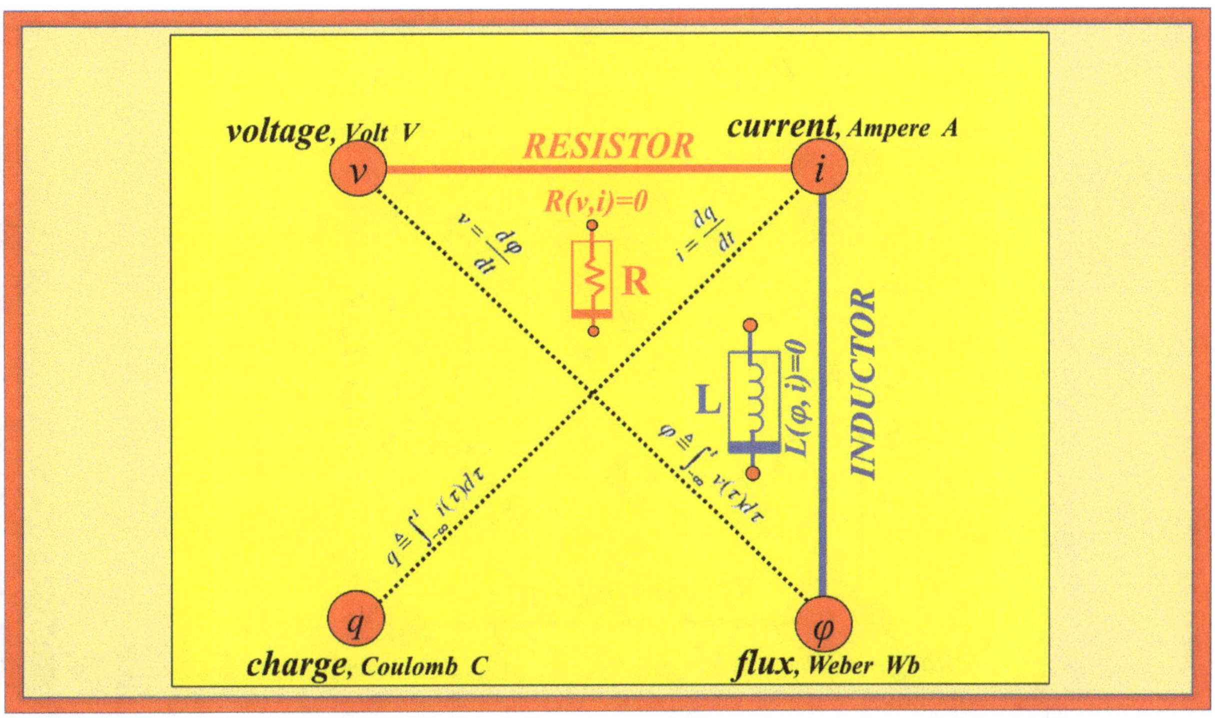

Exhibit 19

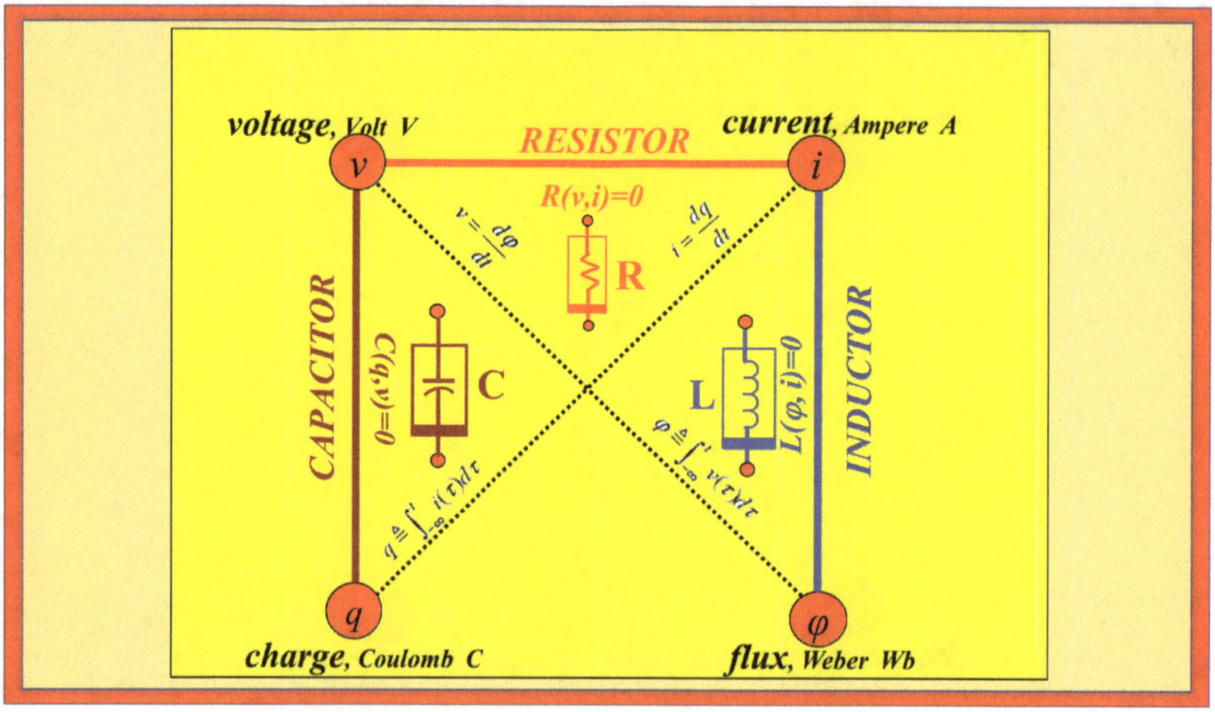

Exhibit 20

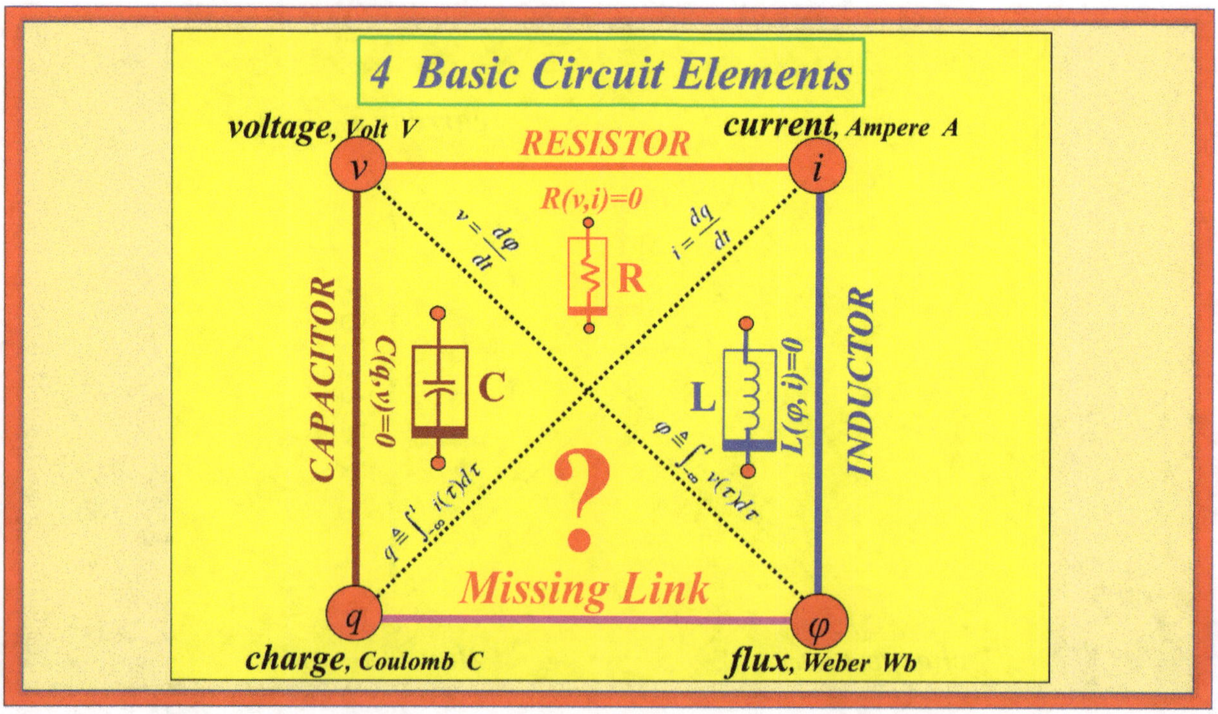

Exhibit 21

Exhibit 22

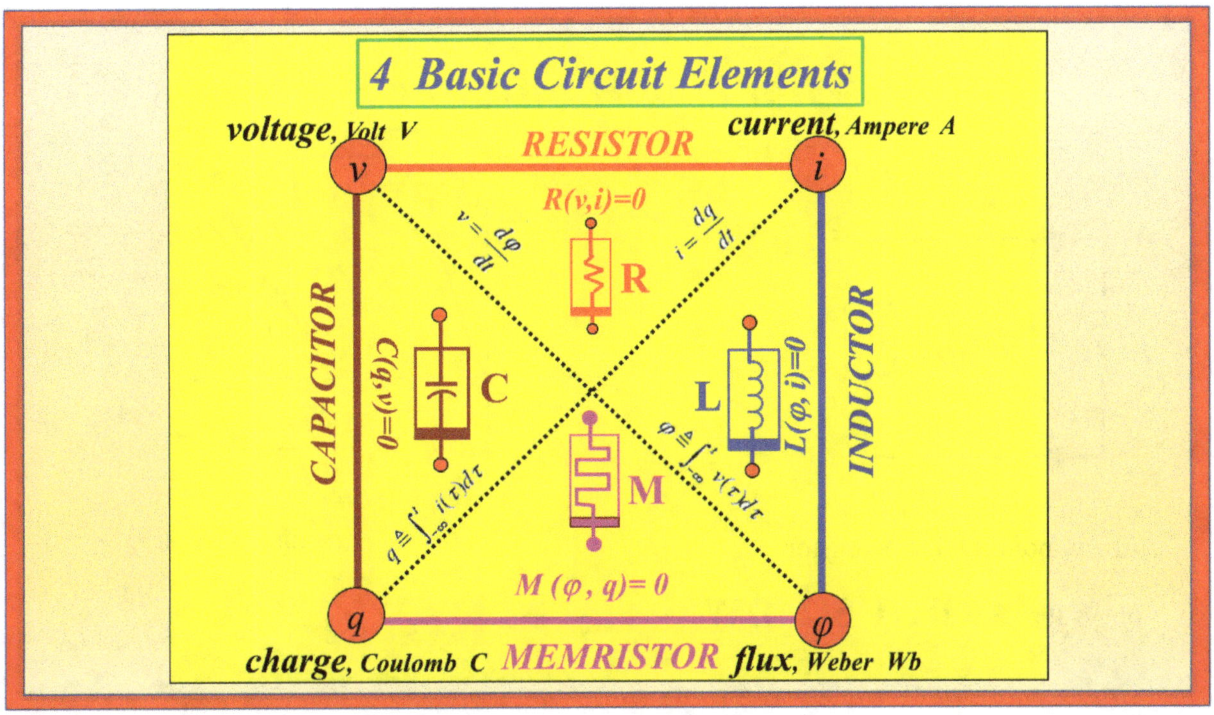

Exhibit 23

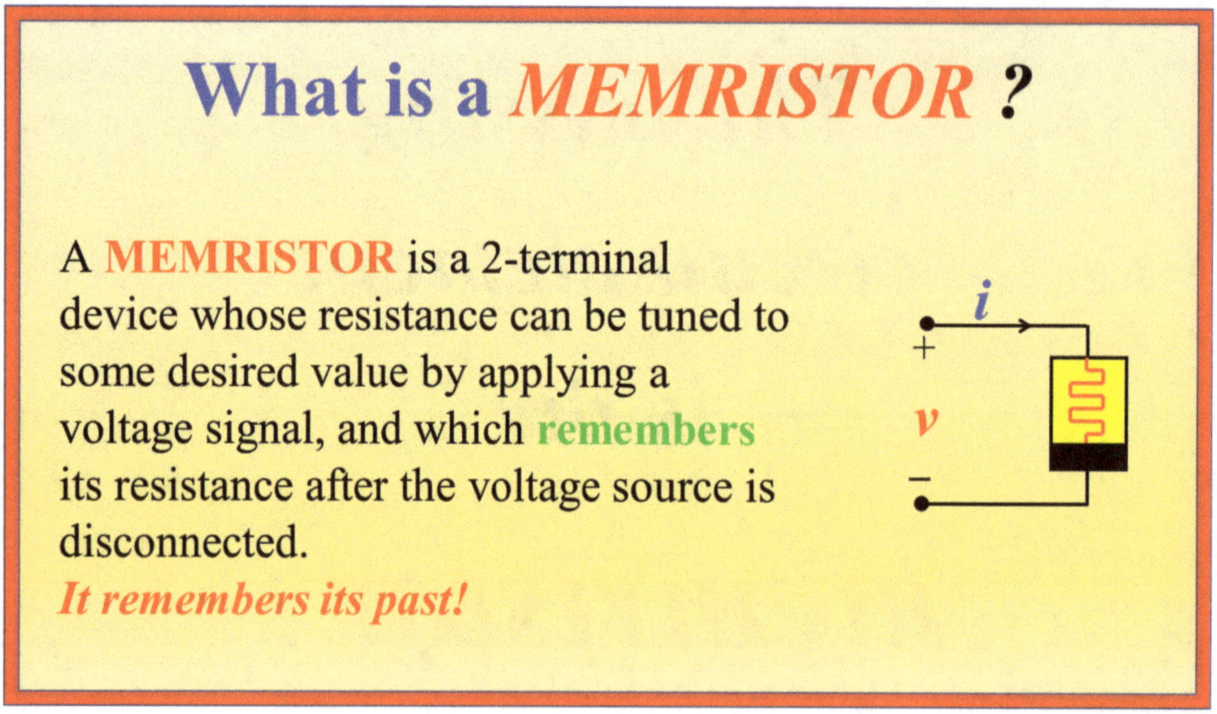

Exhibit 24

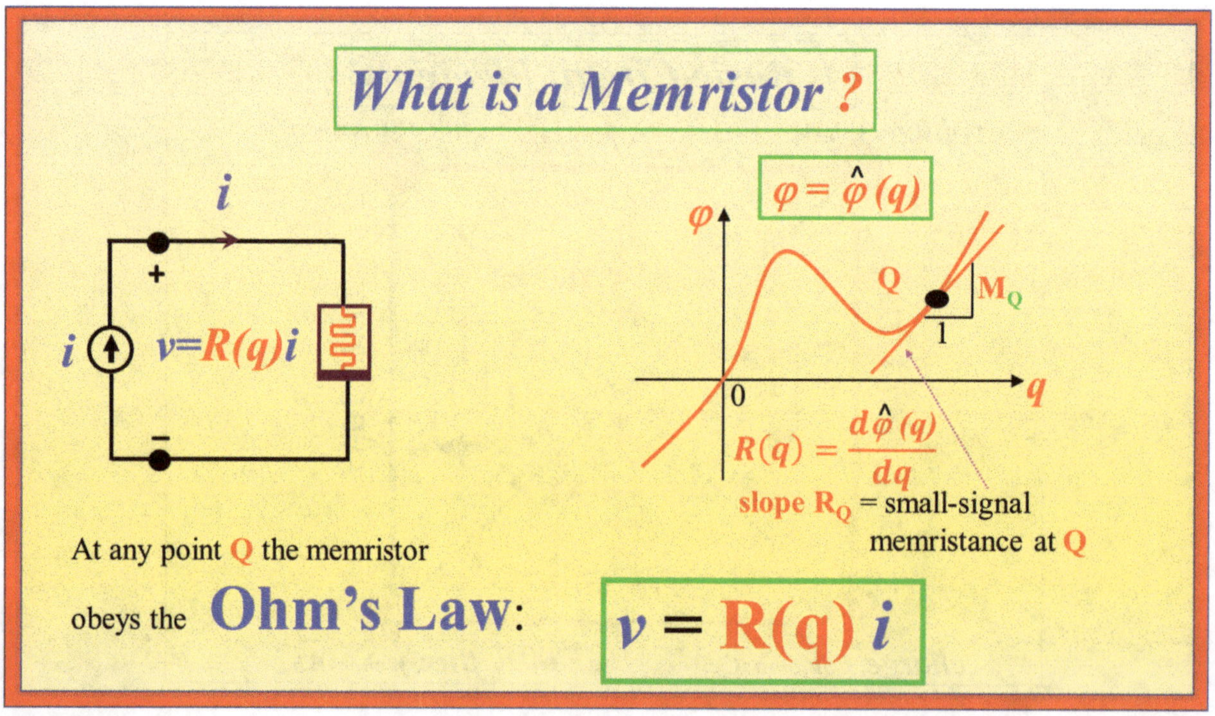

Exhibit 25

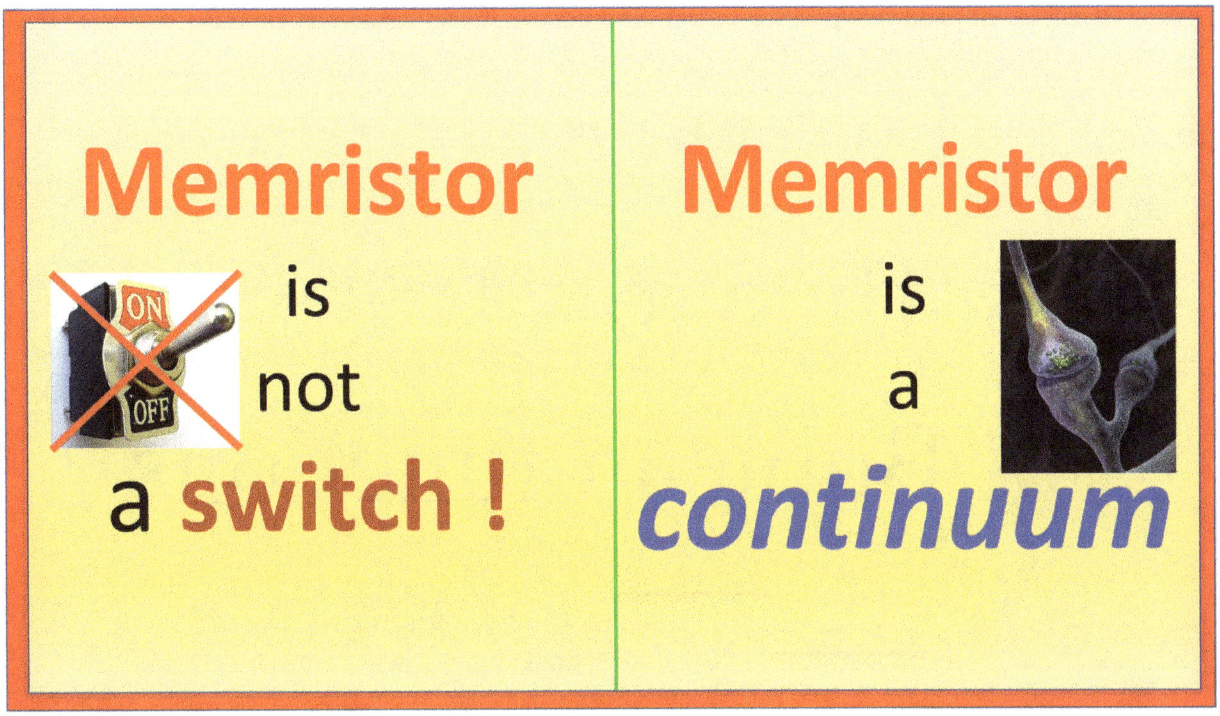

Exhibit 26

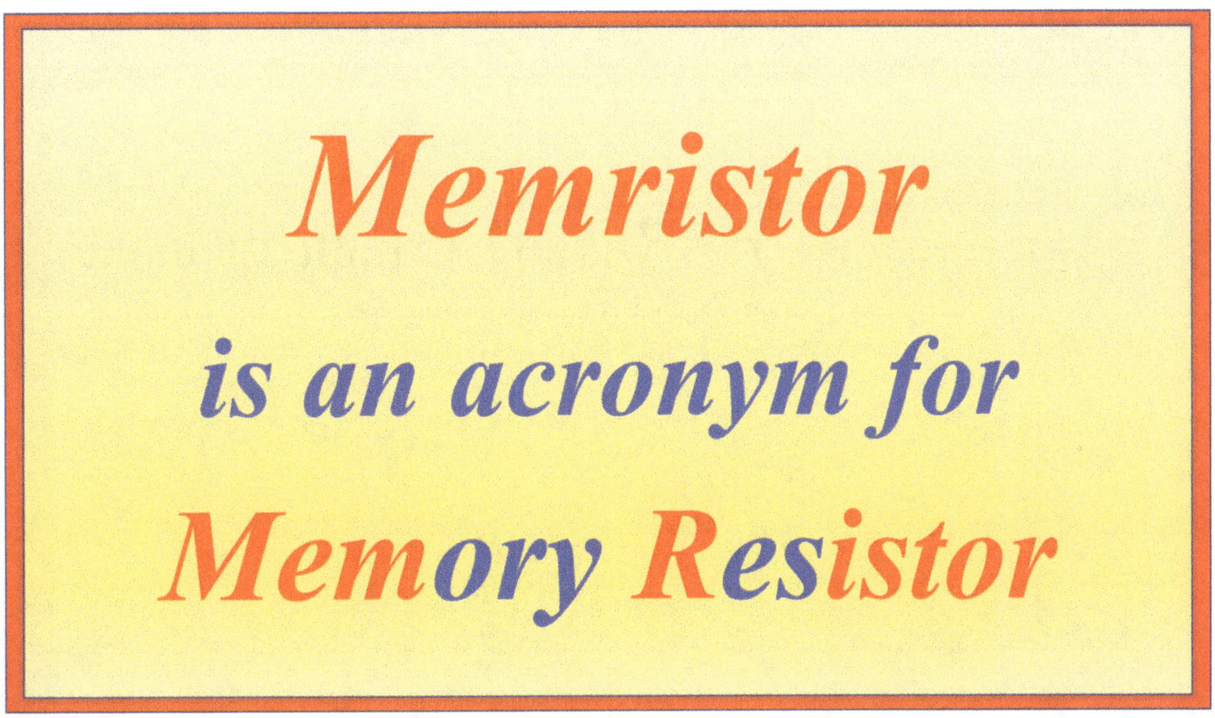

Exhibit 27

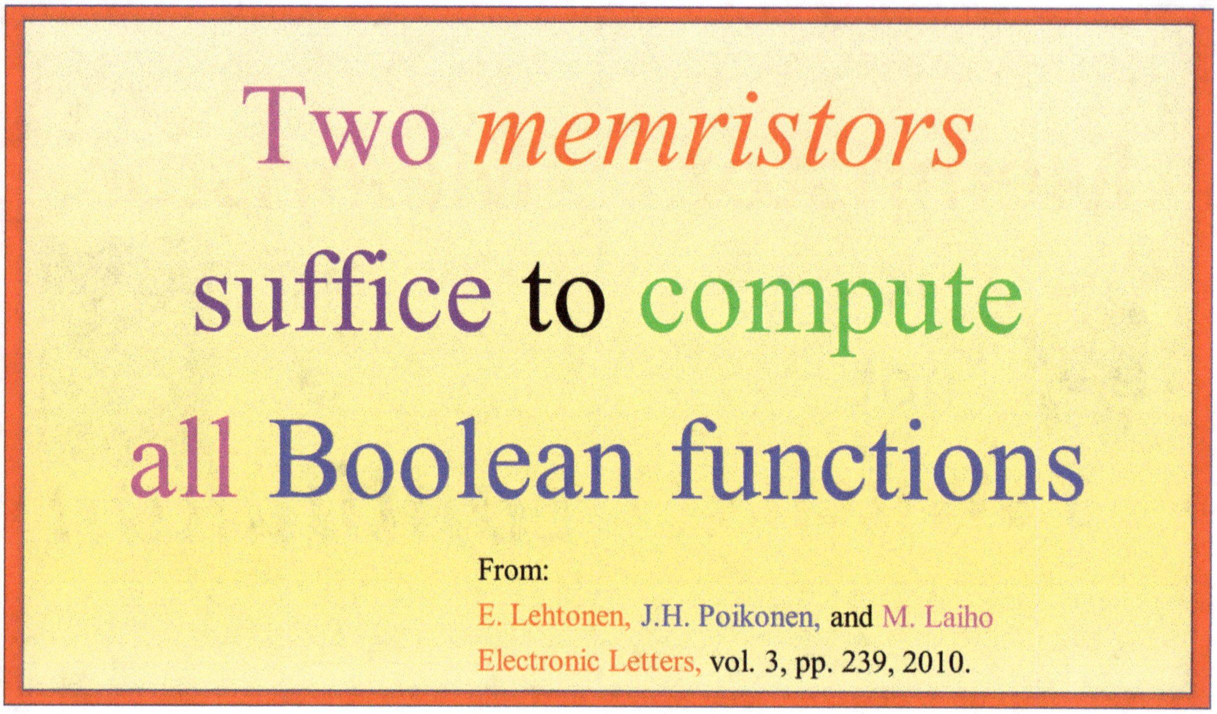

Exhibit 28

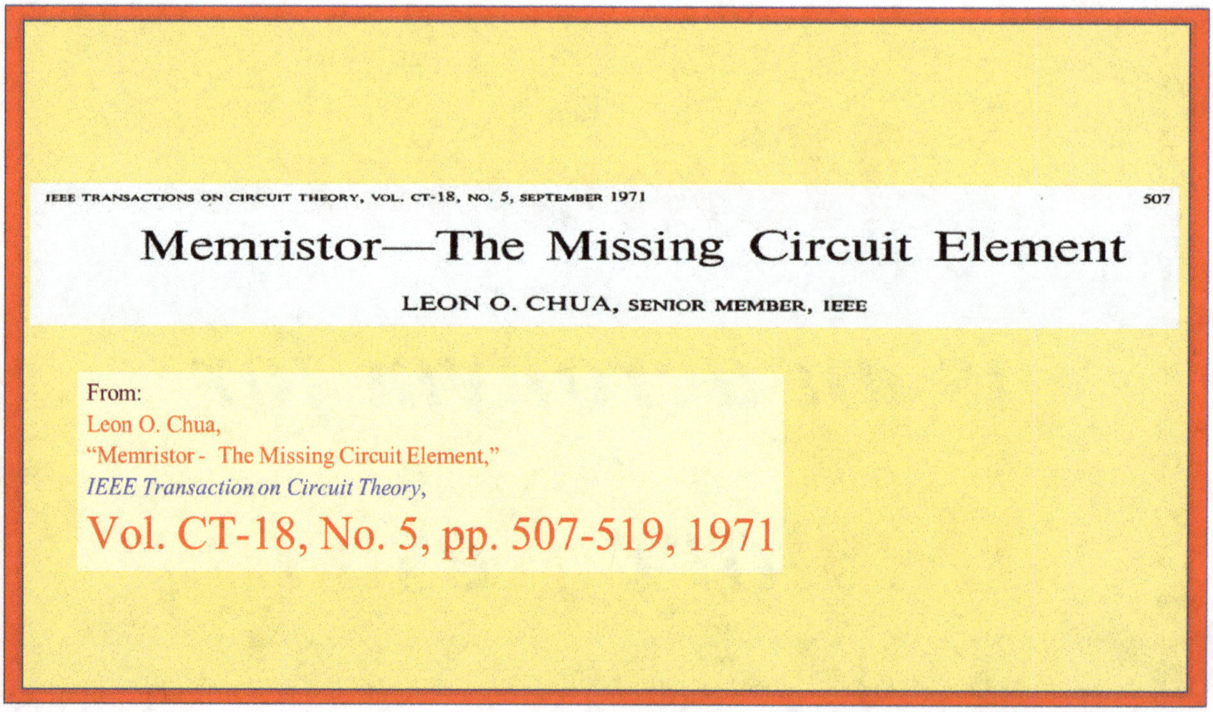

Exhibit 29

Exhibit 30

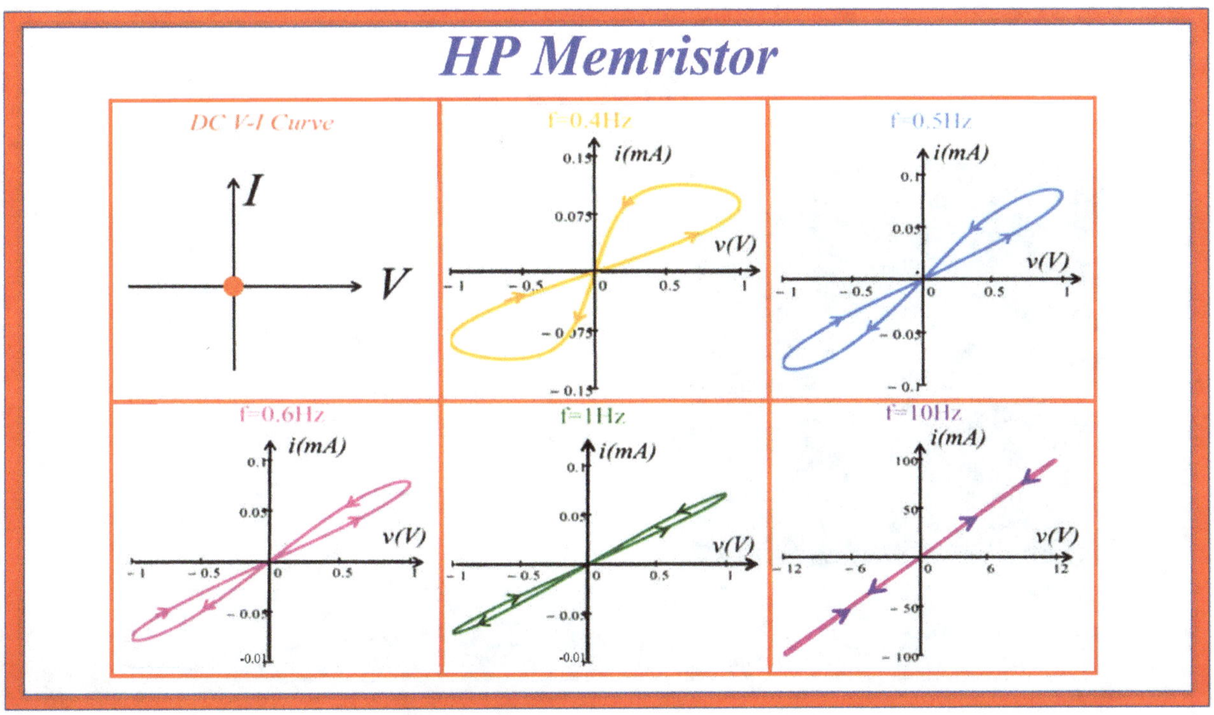

Exhibit 31

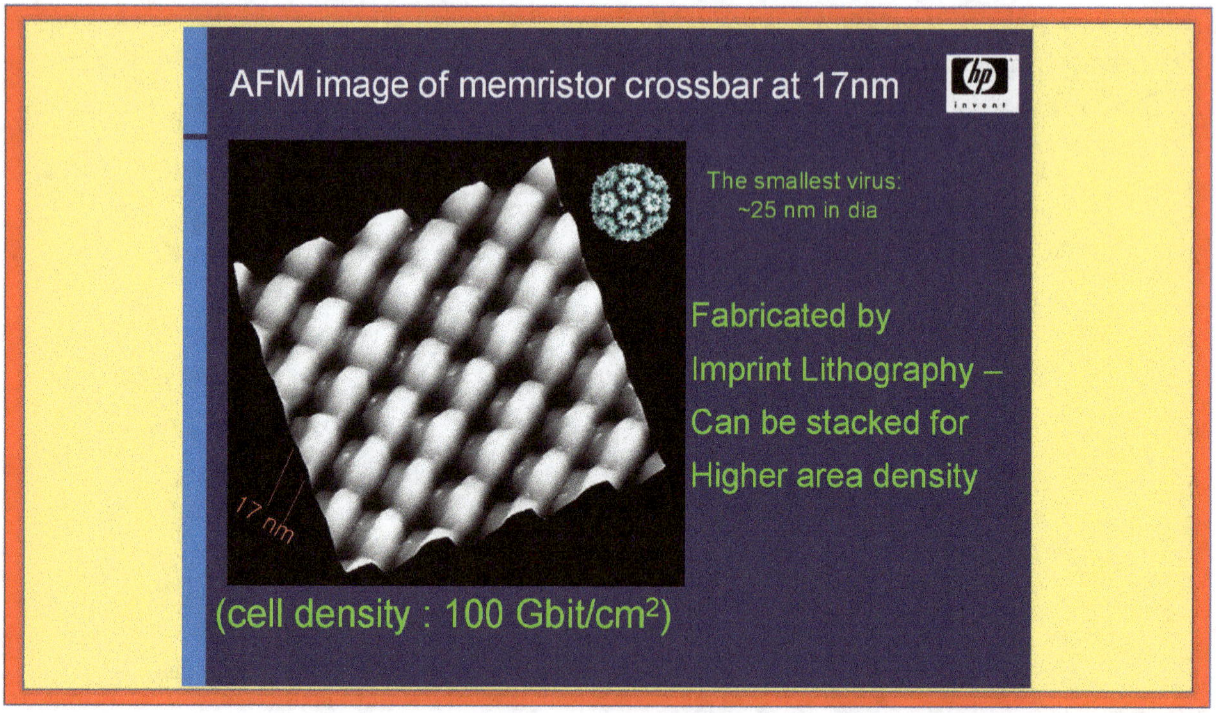

Exhibit 32

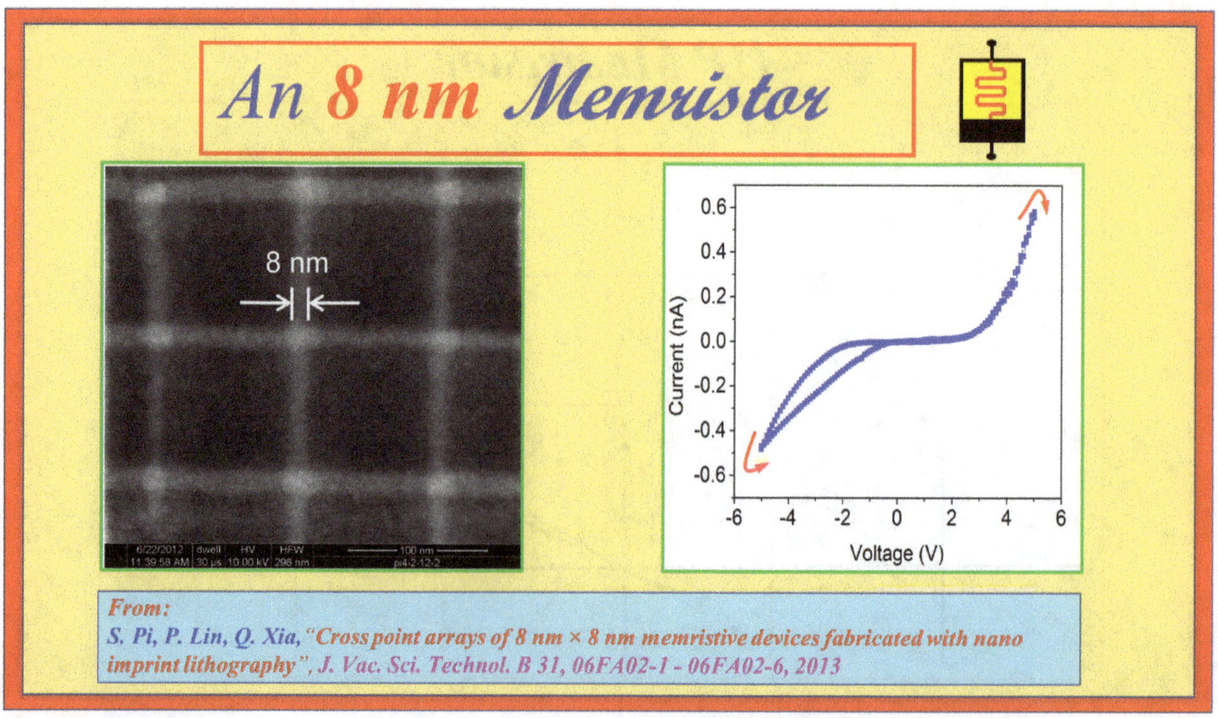

Exhibit 33

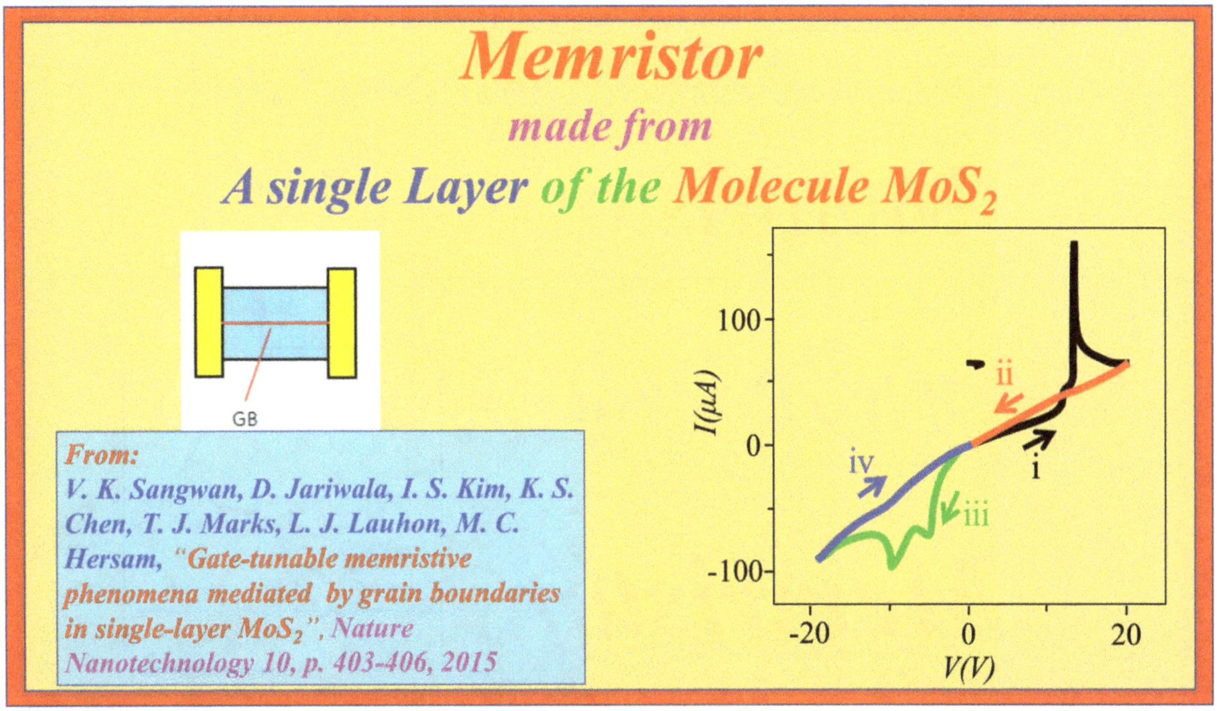

Exhibit 34

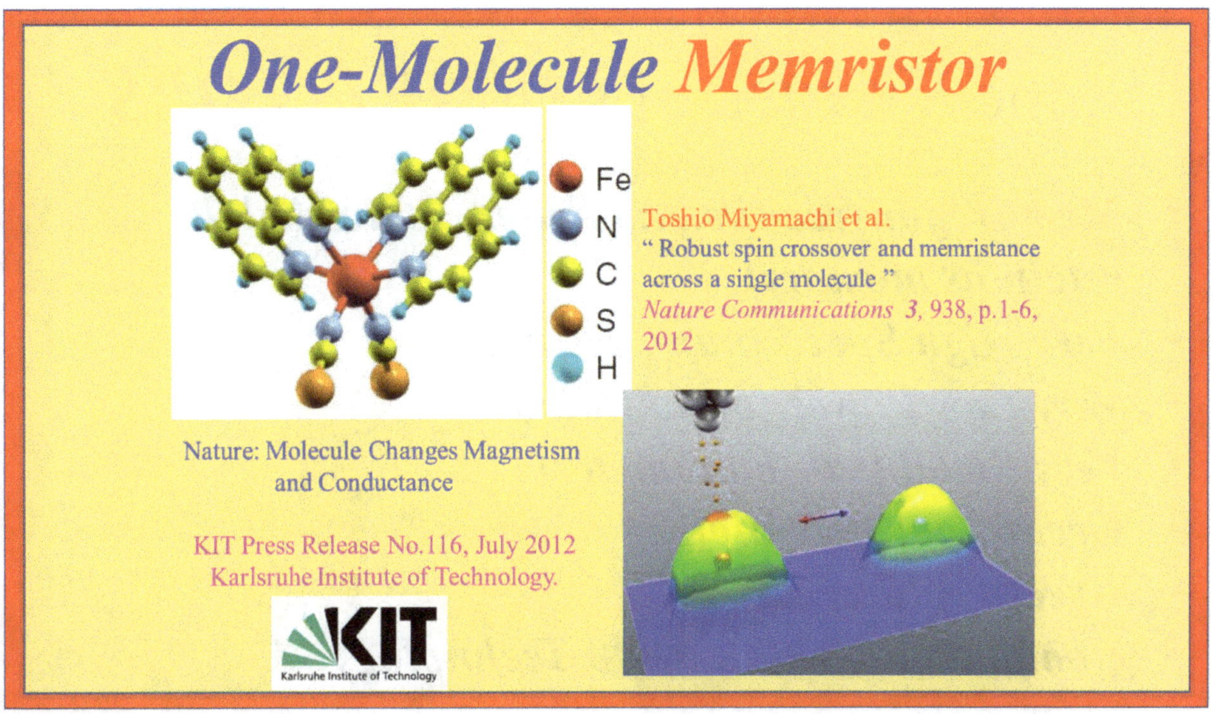

Exhibit 35

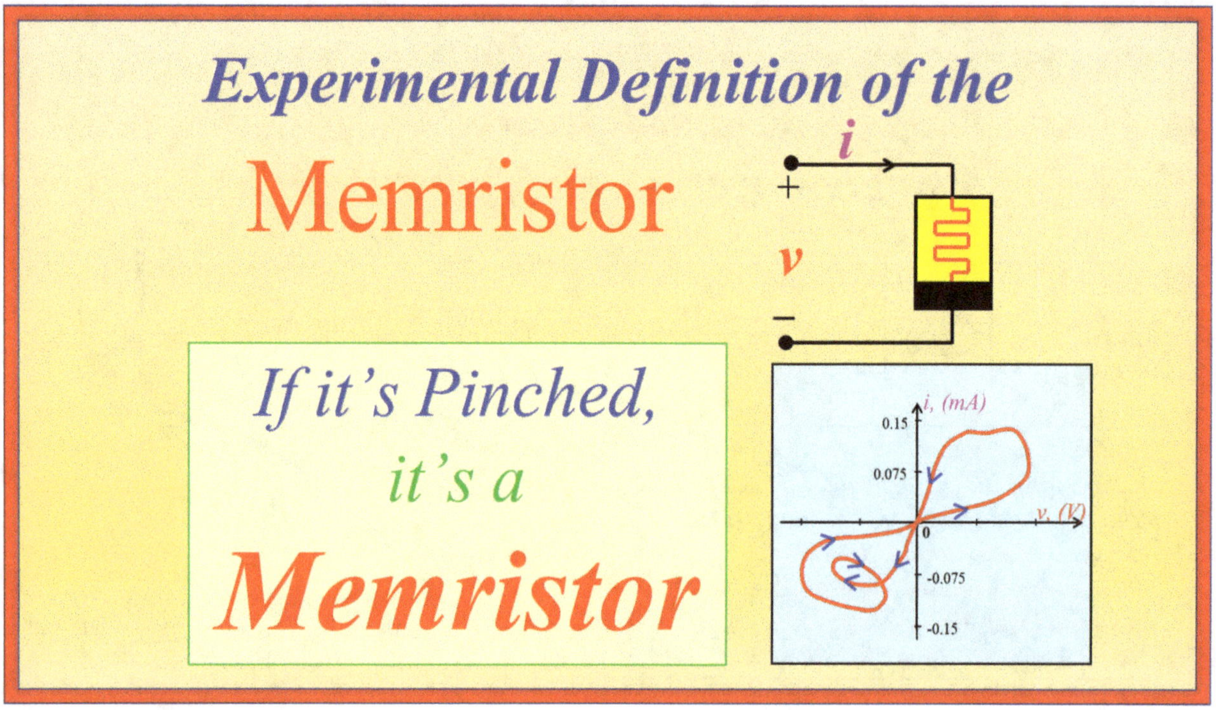

Exhibit 36

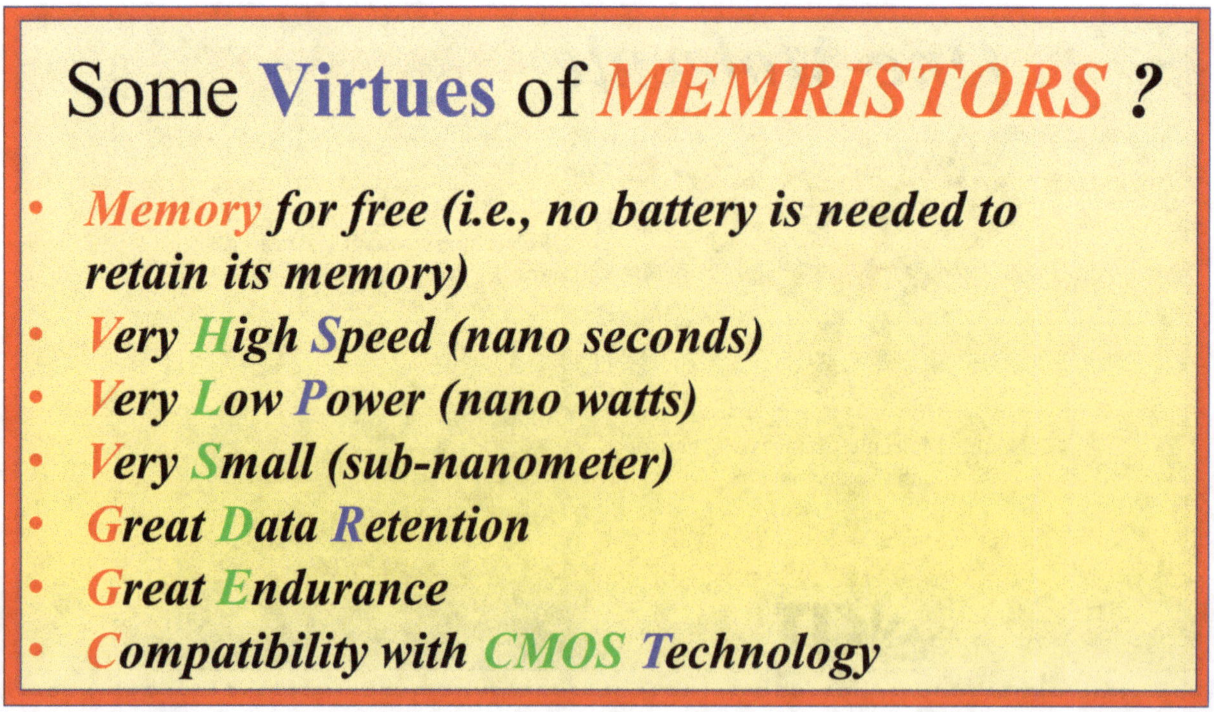

Exhibit 37

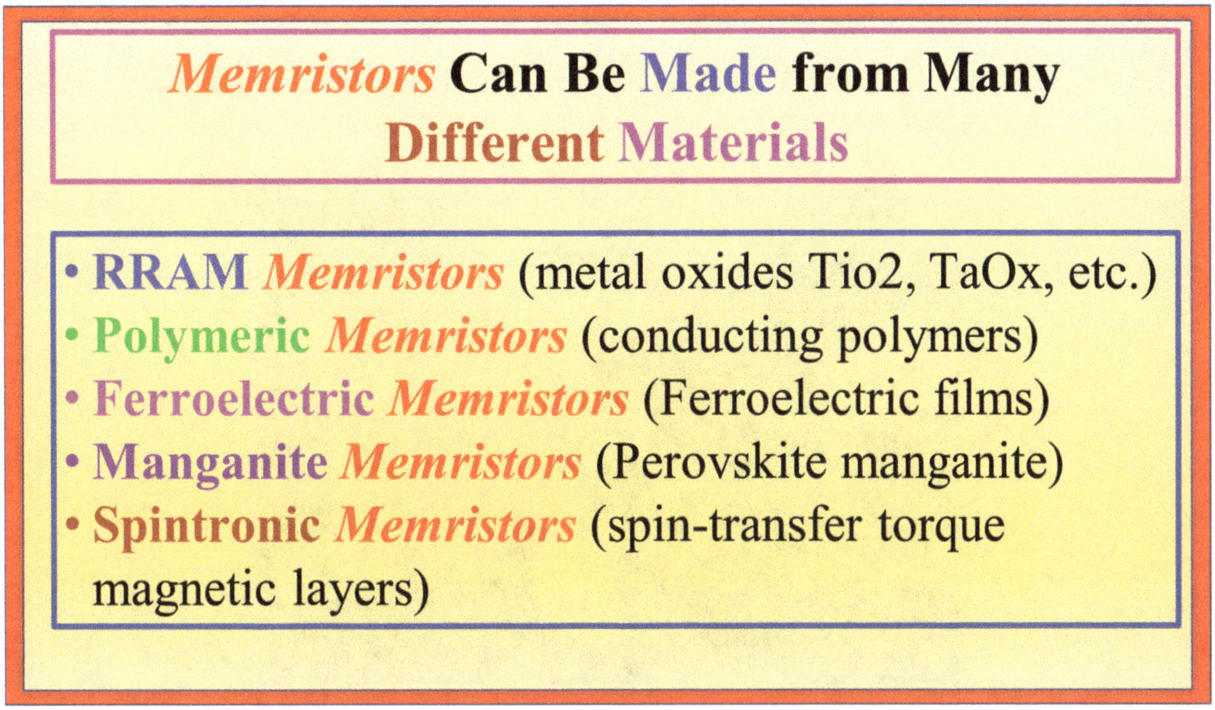

Exhibit 38

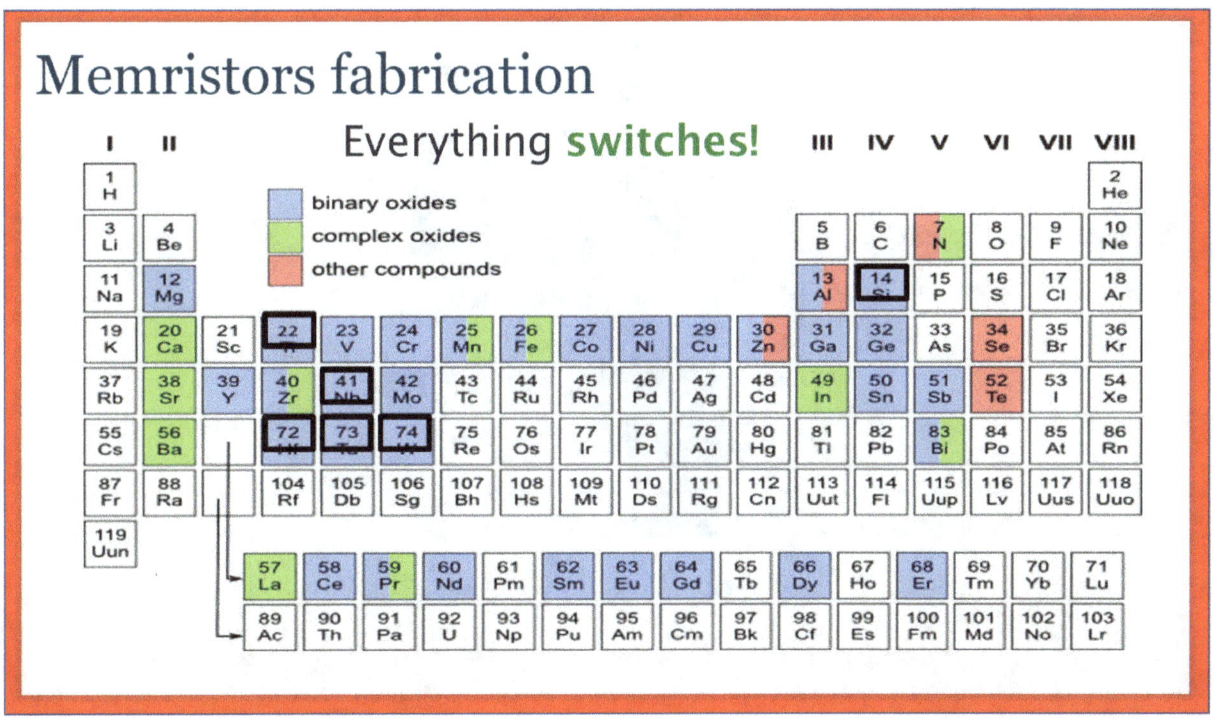

Exhibit 39

Exhibit 40

Exhibit 41

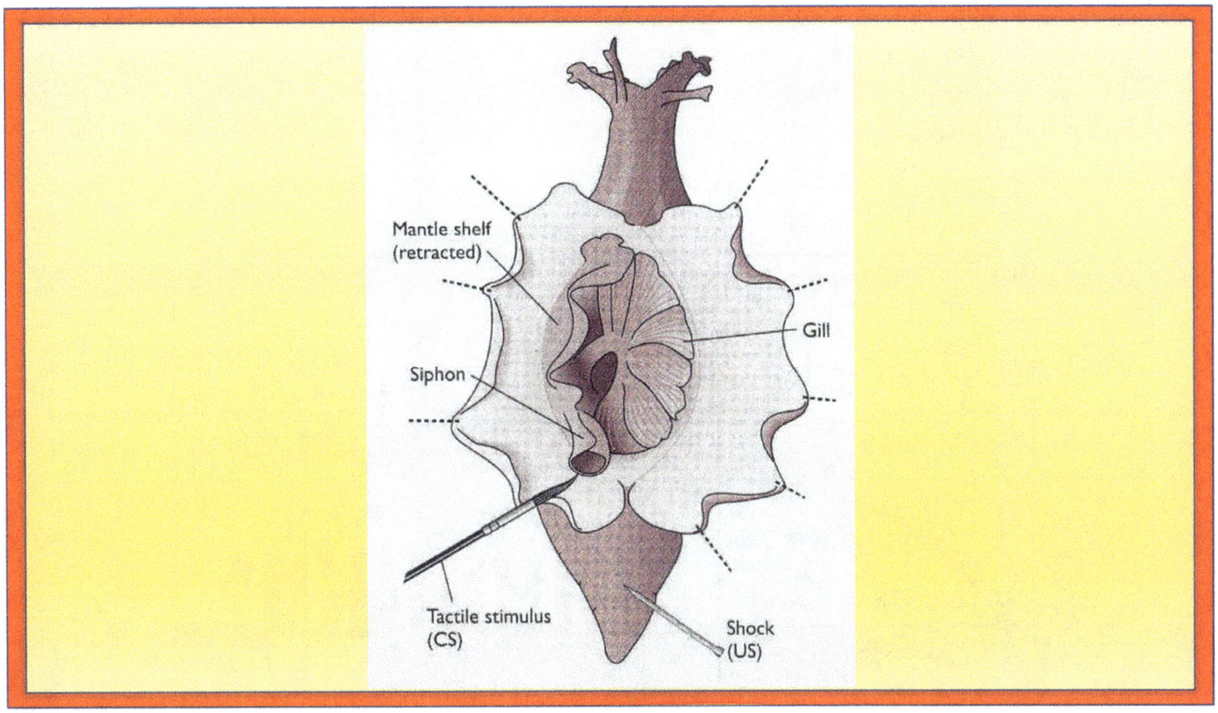

Exhibit 42

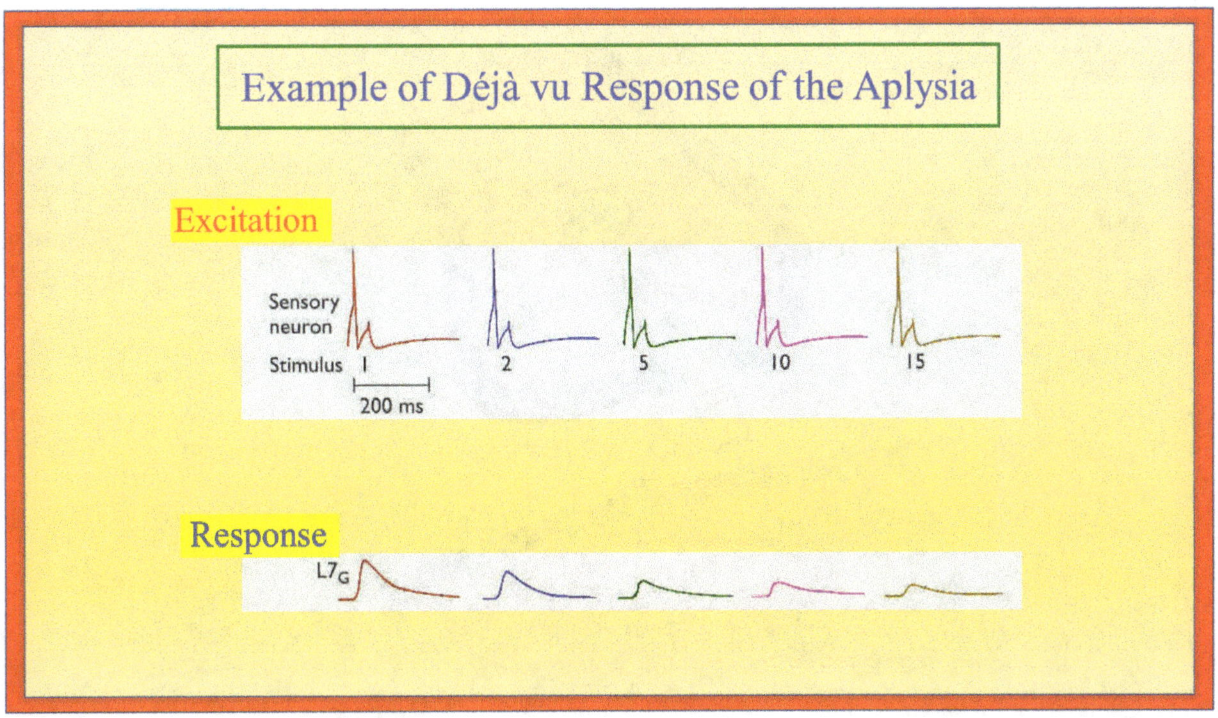

Exhibit 43

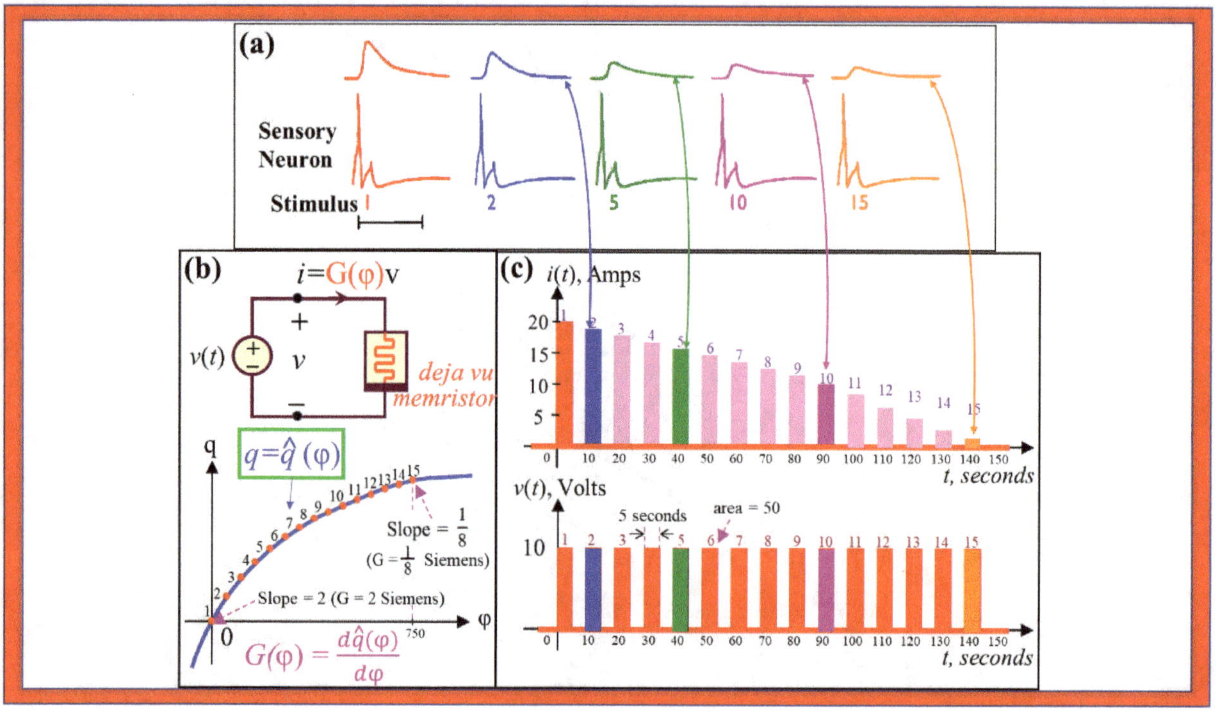

Exhibit 44

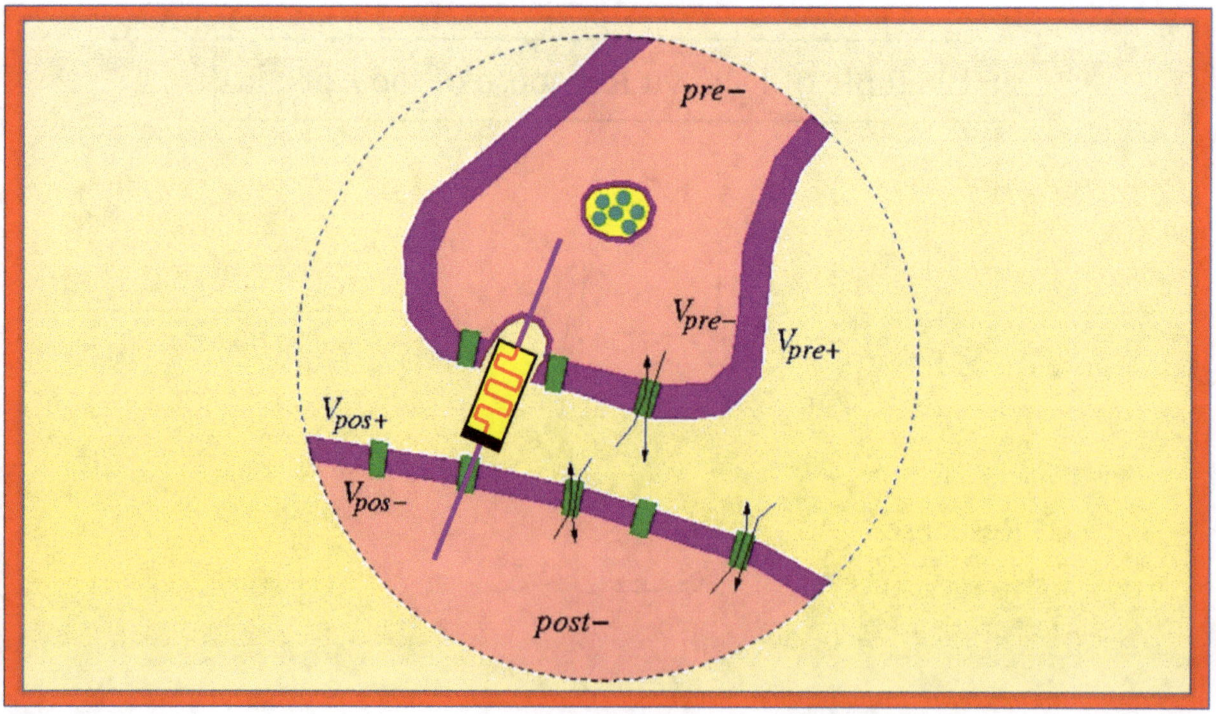

Exhibit 45

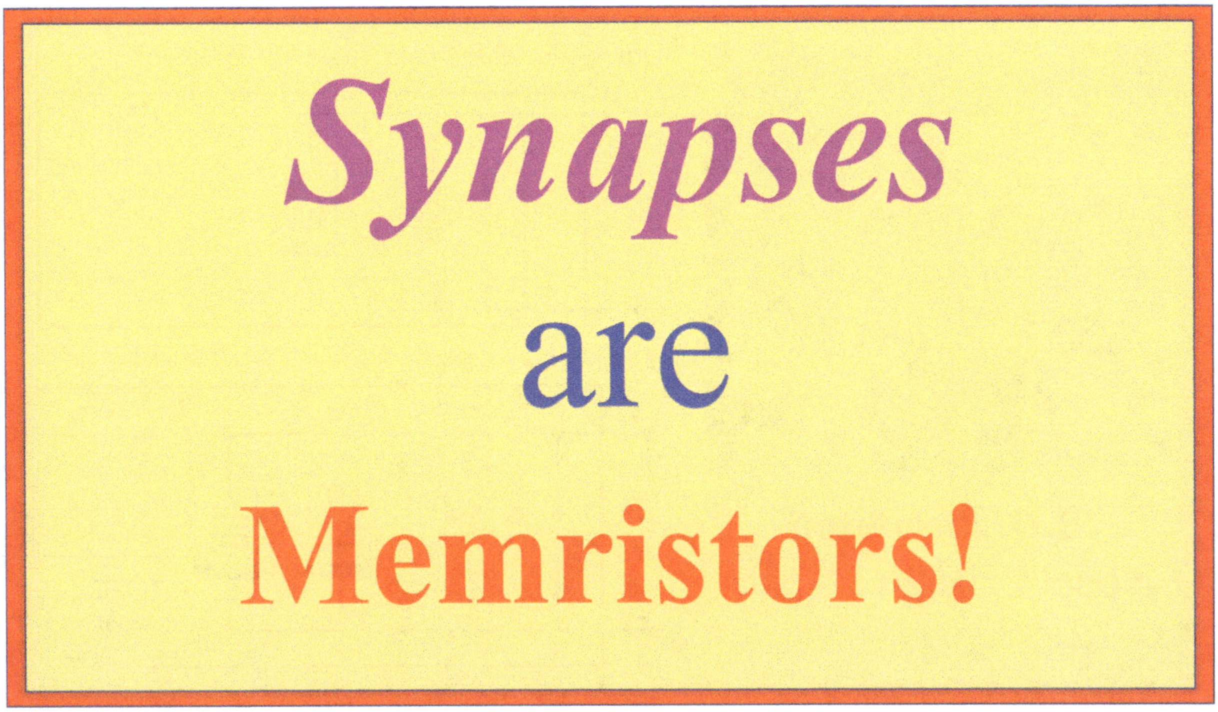

Exhibit 46

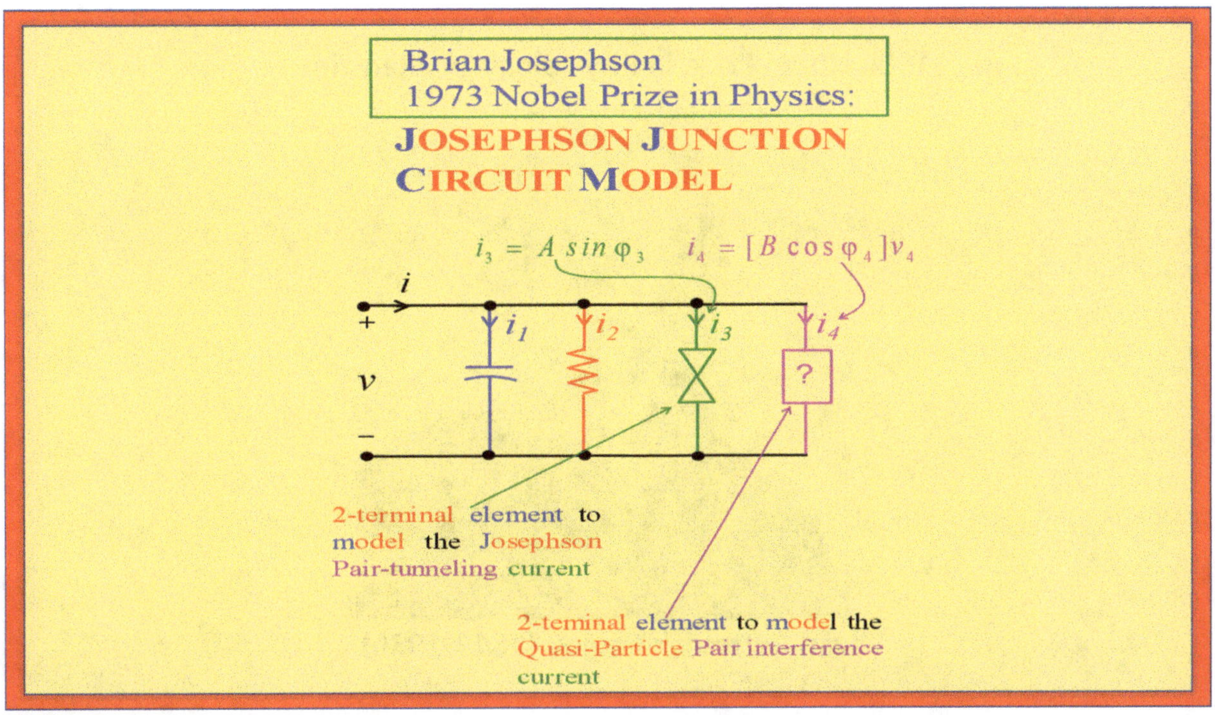

Exhibit 47

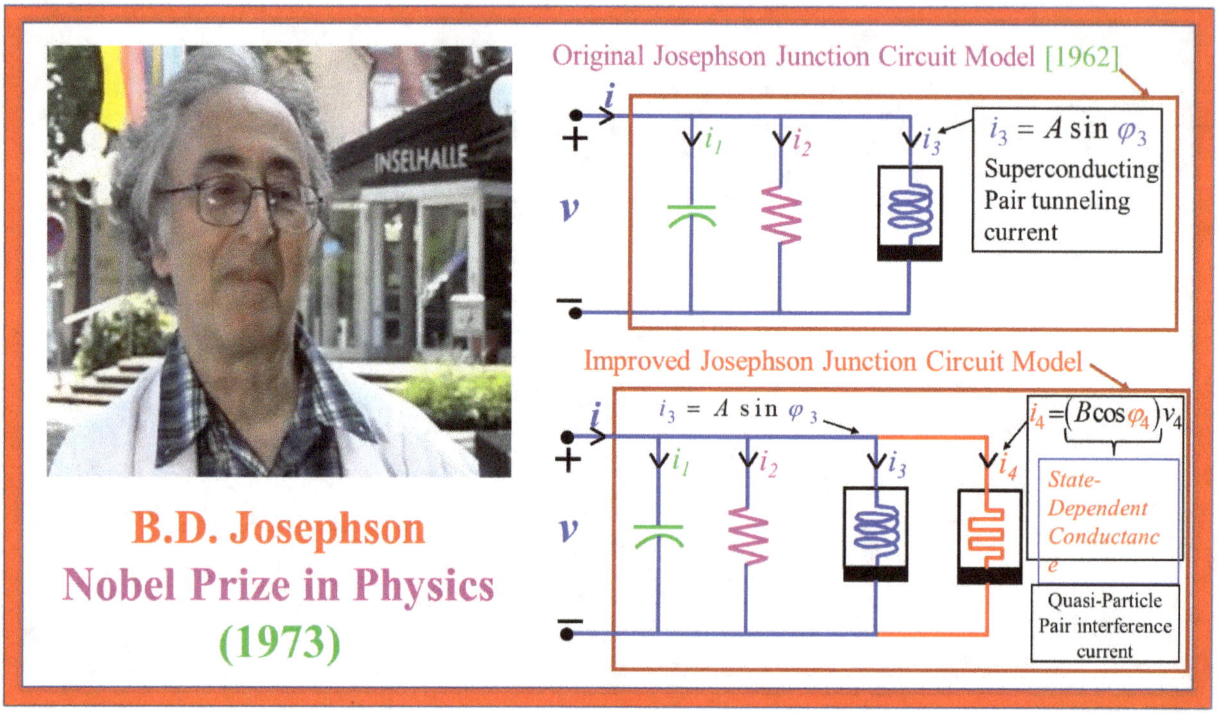

Exhibit 48

Exhibit 49

Pavlov, assistants, visitors, and dog. Pavlovian (classical) conditioning is the simplest associative learning paradigm.

Exhibit 50

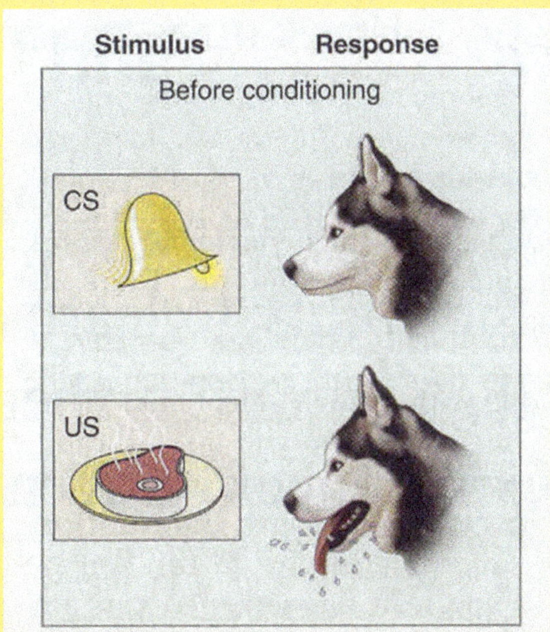

Exhibit 51

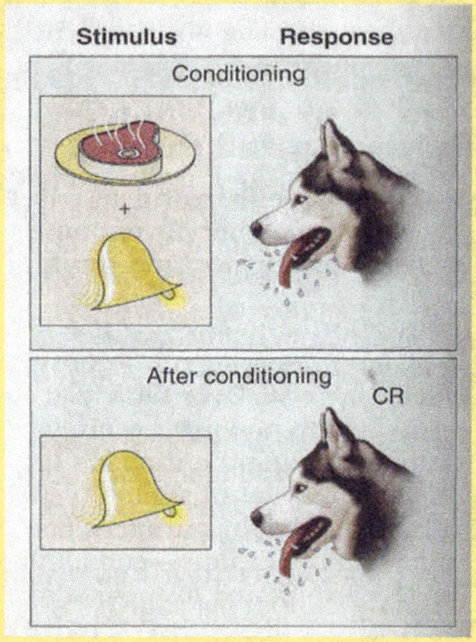

Exhibit 52

Associative Memory and Learning

Associative memory and learning is defined as the ability to learn and remember the relationship between unrelated experiences or events.

Exhibit 53

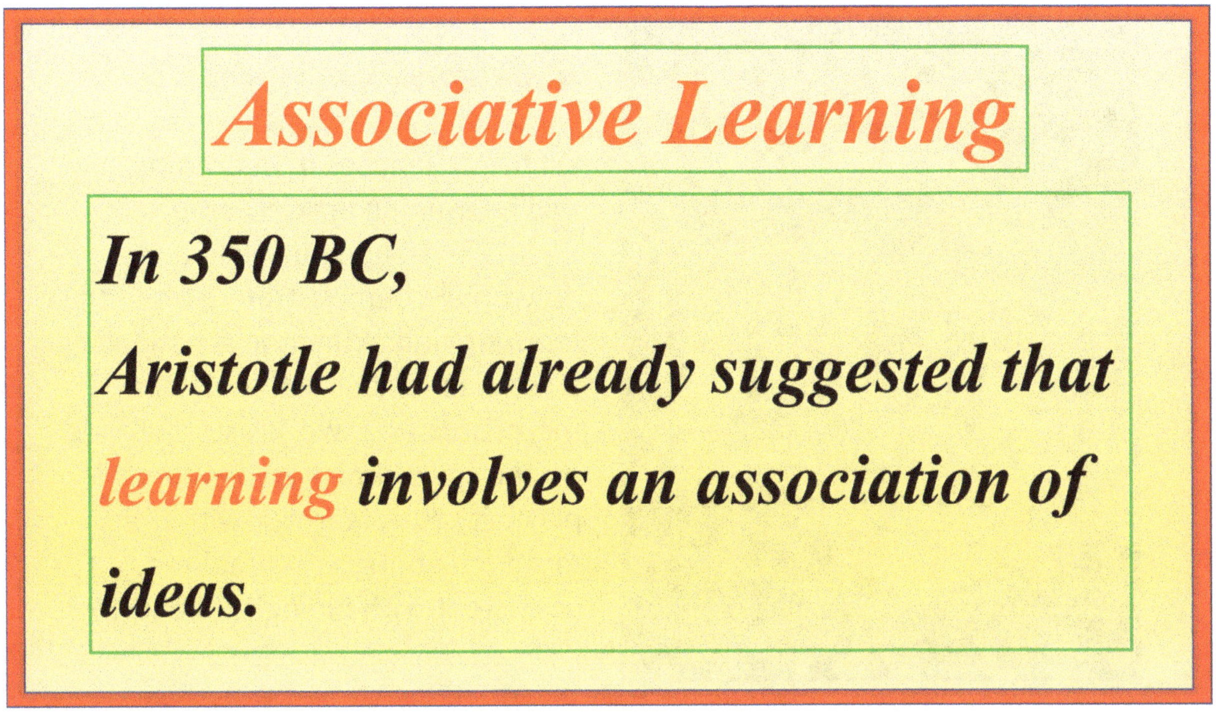

Exhibit 54

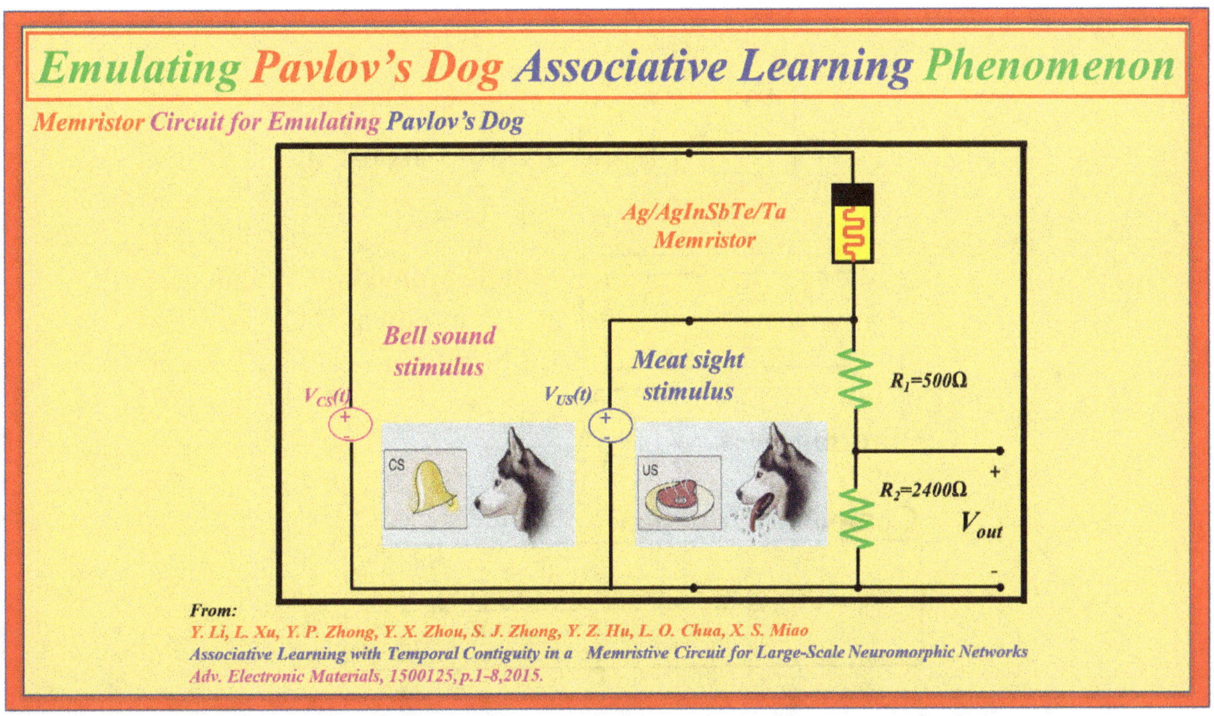

Exhibit 55

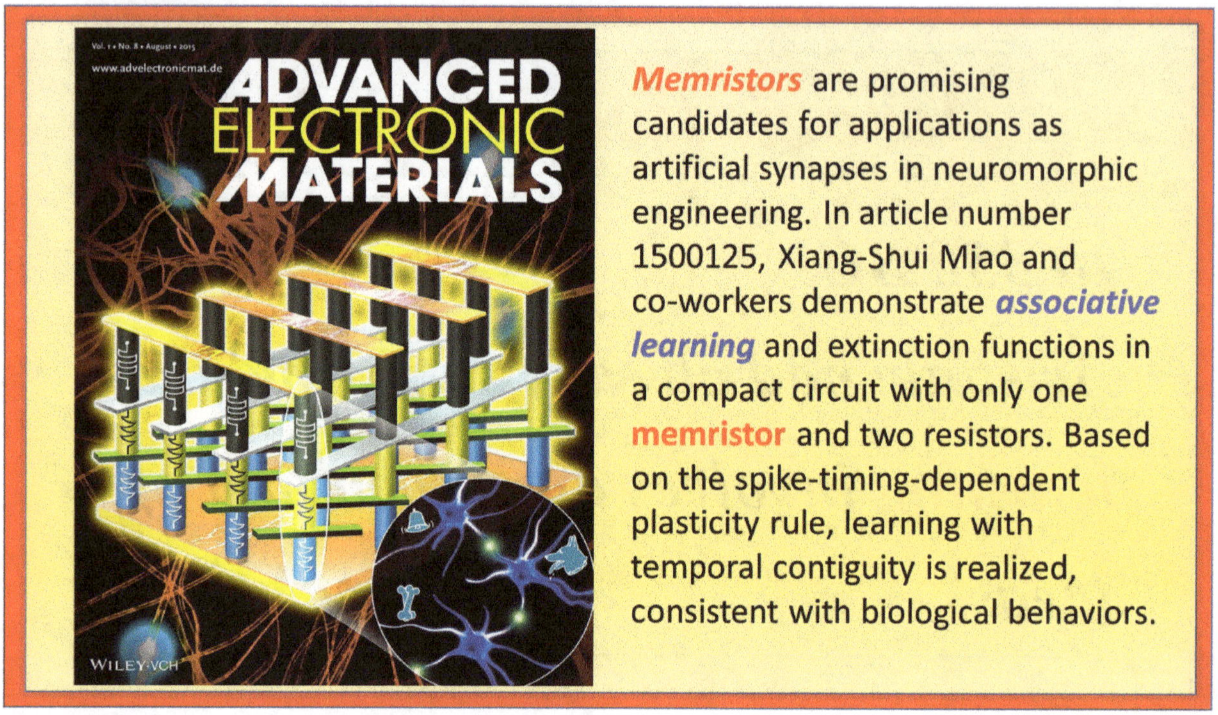

Exhibit 56

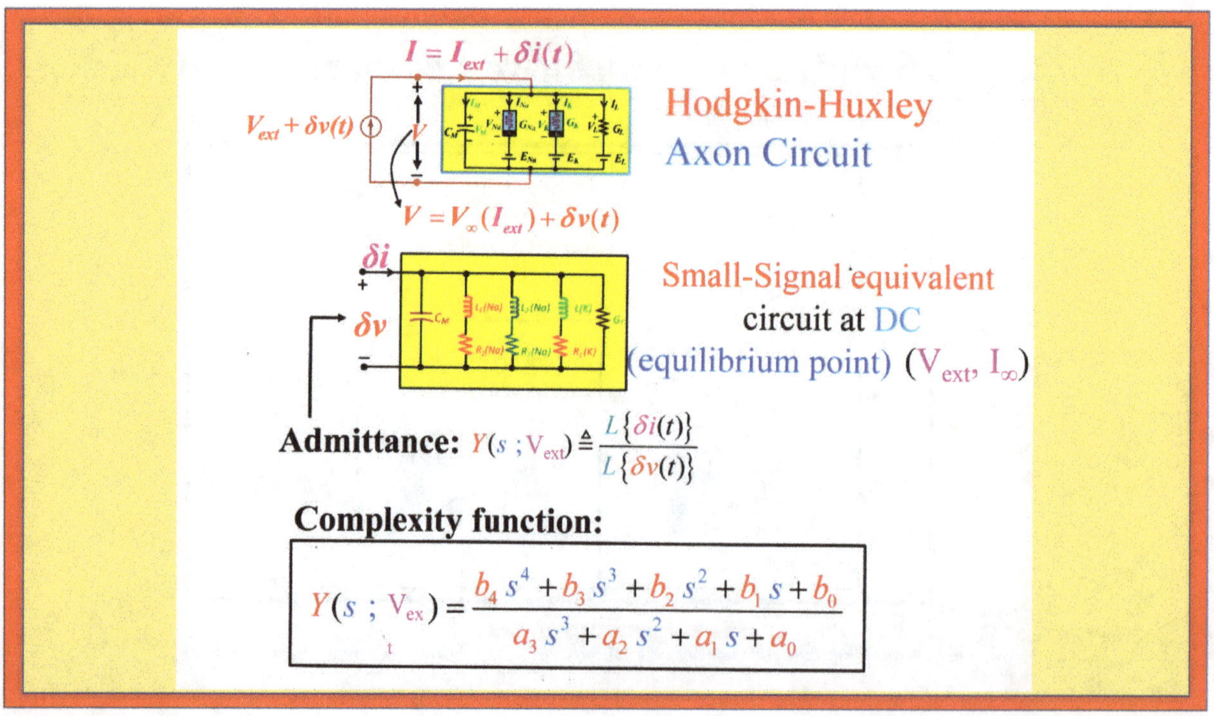

Exhibit 57

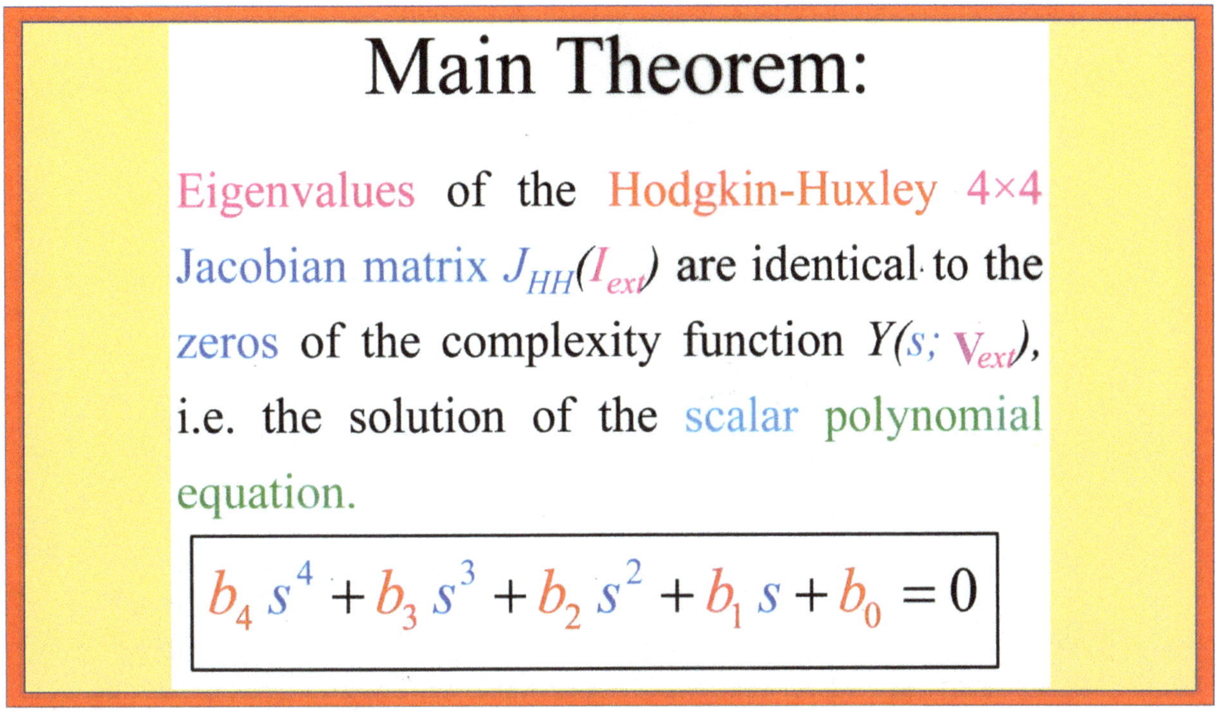

Exhibit 58

Exhibit 59

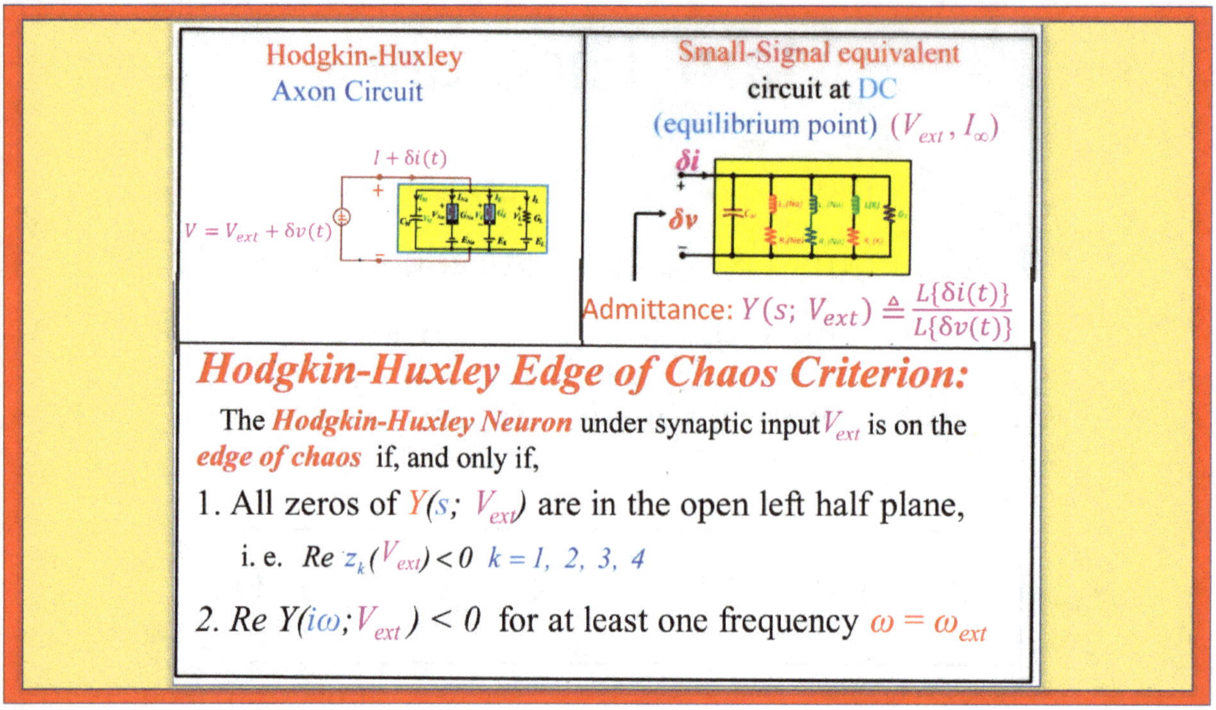

Exhibit 60

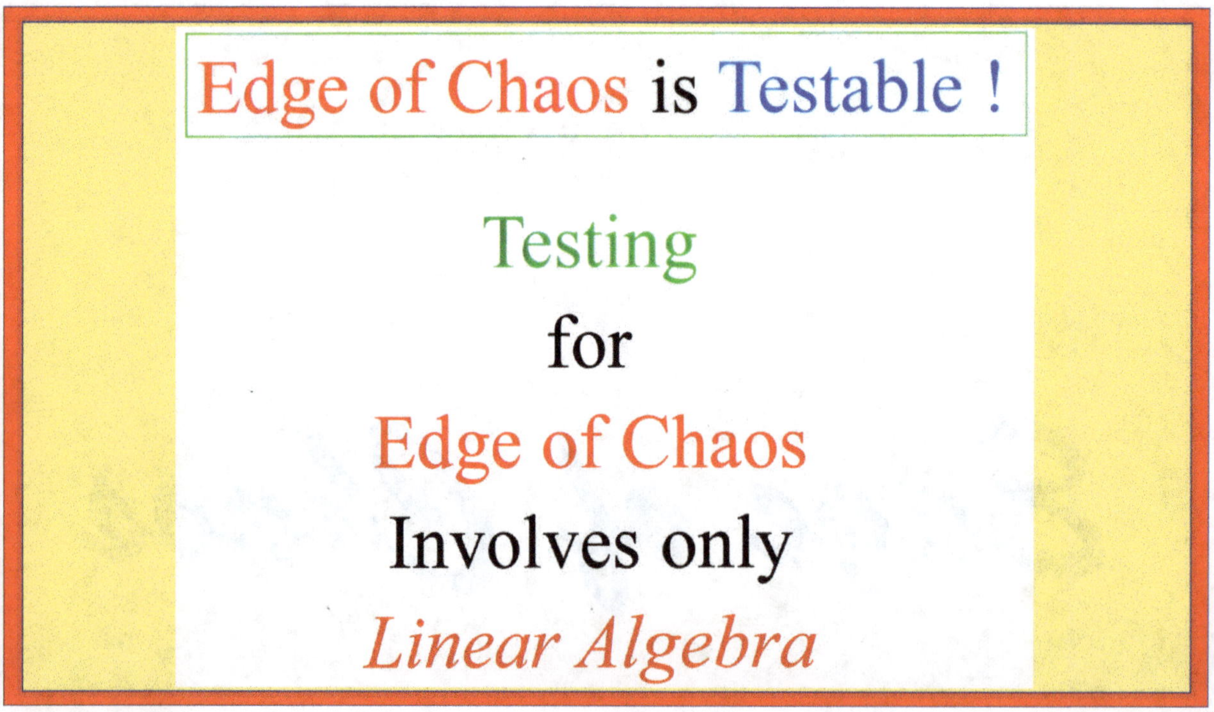

Exhibit 61

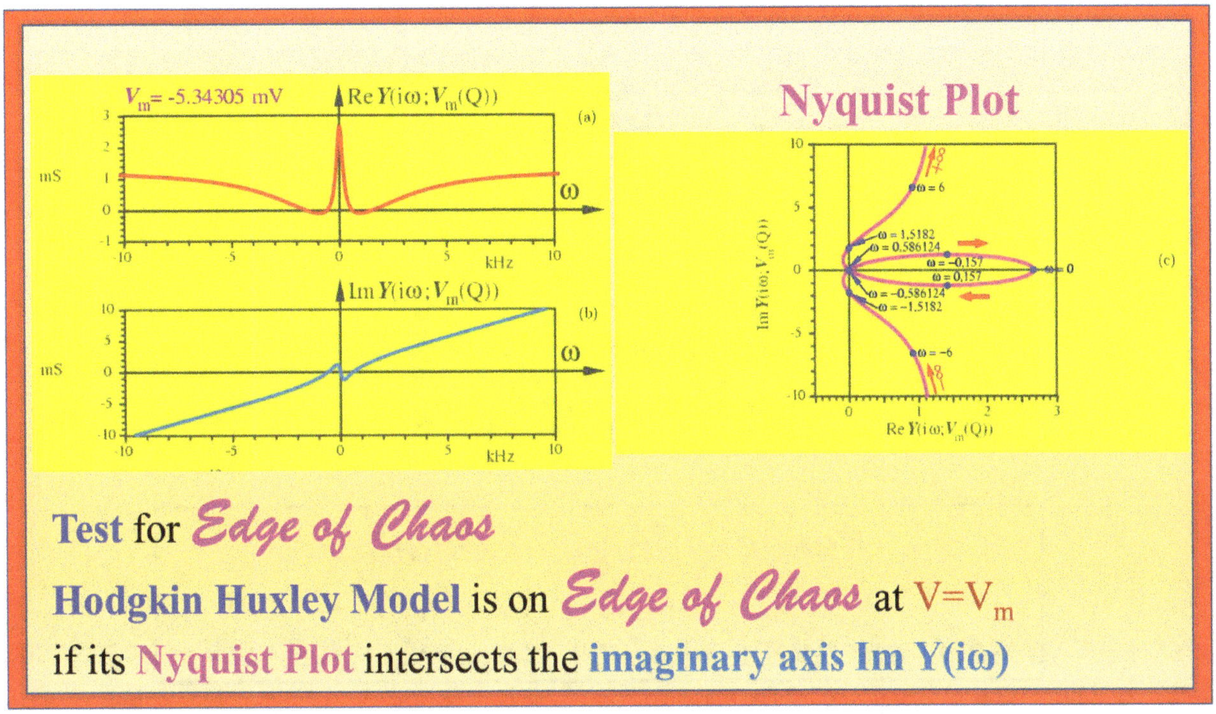

Exhibit 62

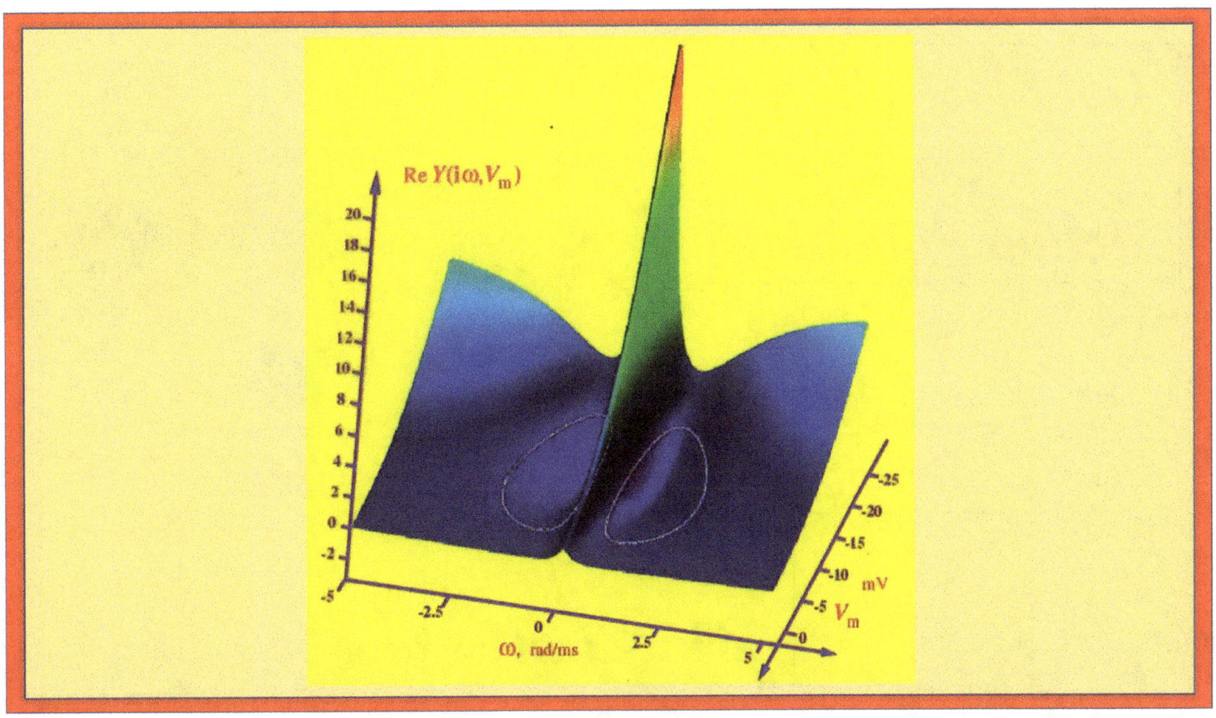

Exhibit 63

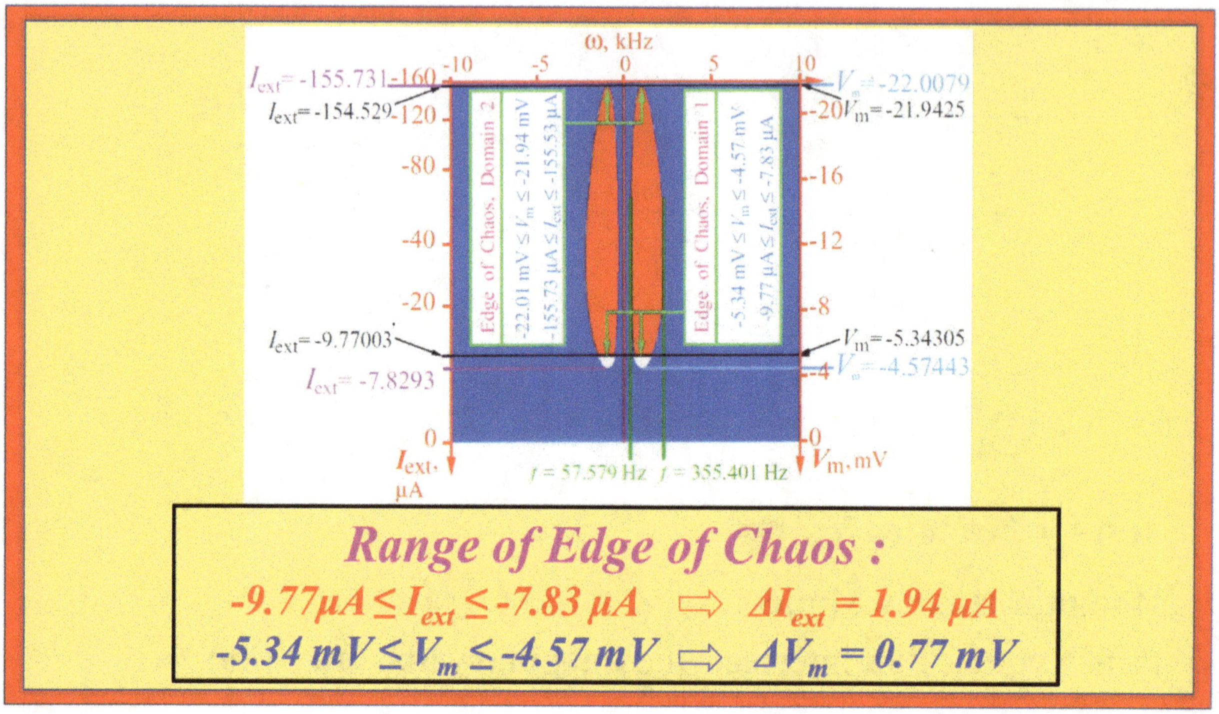

Exhibit 64

Exhibit 65

Galvani's Irritability *Doctrine* is a manifestation of the *Edge of Chaos*, the *crown jewel* of the *Principle of Local Activity*

Galvani was well aware of the ***doctrine of "irritability,"*** one of the most important conceptual elaborations of **18th century physiology.**

According to ***Galvani***, in the framework of ***doctrine of irritability***, external electricity was acting as an ***excitatory stimulus*** (as a ***trigger*** we would say now) for the **contraction**, and ***not as the direct "efficient" cause of the observed phenomena***.

How could it be that such a tiny electrical force would produce muscle contraction if it were not setting in motion some ***internal force***, and **triggering** some extremely ***mobile principle*** existing in nerves which then ***excites the action*** of the nerve-muscular force?

Exhibit 66

The Local Activity Principle

All 2-terminal electrical devices can be classified into either "*locally passive*", or "*locally active*". Occasionally, a tiny subset of "*locally active*" devices is blessed with an "*edge of chaos*" domain capable of exhibiting fascinating phenomena and attributes, such as "*learning*", "*artificial intelligence*", and even "*life*" itself.

Exhibit 67

Proof of the Principle of Local Activity requires non-trivial applications of the following mathematical results:

1. **Nonlinear Monotone Operator in Function Space**
2. **High-dimensional Analog of the Mean-Value Theorem**
3. **LaSalle's Complete Stability Theorem**
4. **Poisson's Integral on half-plane**
5. **Theory of Positive-Real Function of a complex Variable**
6. **Tellegen's Theorem**

Exhibit 68

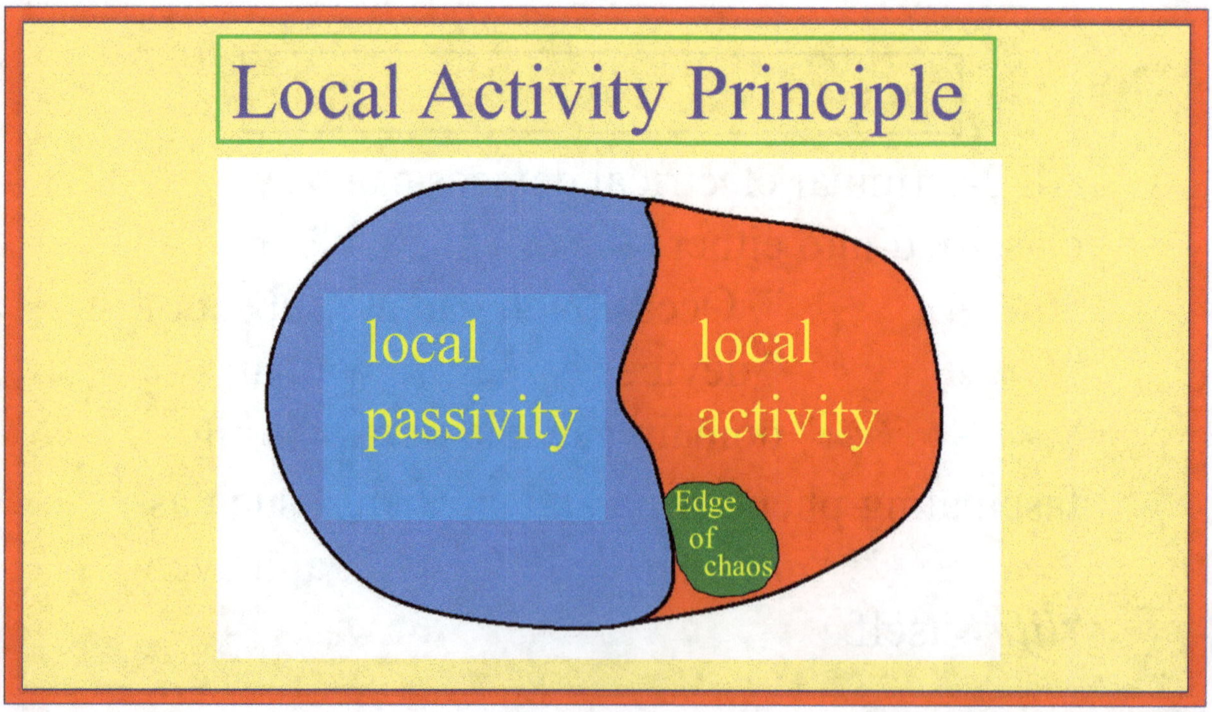

Exhibit 69

The "*Edge of Chaos*" is the
"**Pearl**" amidst the
"***locally active domain***"

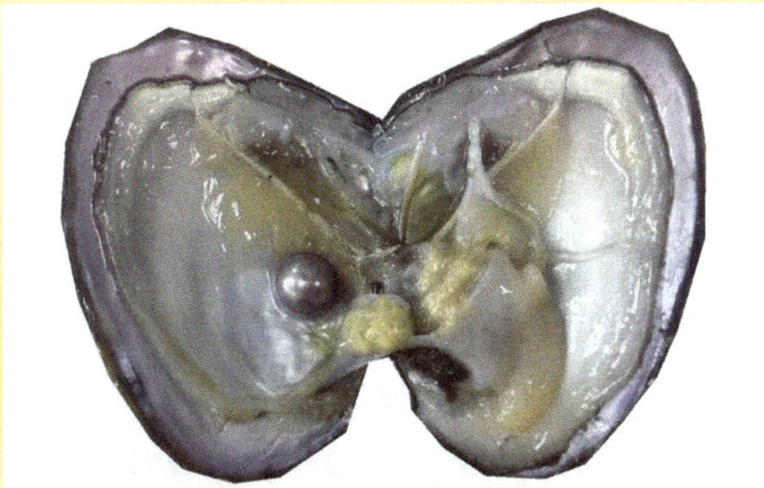

Exhibit 70

Local Activity Theorem

A ***reaction-diffusion equation*** with one diffusion coefficient is ***Locally active*** at $Q \Leftrightarrow$ any one of the following conditions holds :

1. ***Z_Q (s) has a pole in Re [s] > 0.***
2. ***Re Z_Q ($i\omega$) < 0 for at least one $\omega = \omega_0$.***
3. ***Z_Q (s) has a simple pole $s = \omega_P$ on the imaginary axis i. e.,***

$$Z_Q(s)=\frac{Z_1(s)}{(s-i\omega_p)^m Z_2(s)}, \quad m=1$$

and

$$\lim_{s \to i\omega_p} (s-i\omega_p)\, Z_Q(s) < 0.$$

4. ***Z_Q (s) has a multiple pole on the imaginary axis.***

Exhibit 71

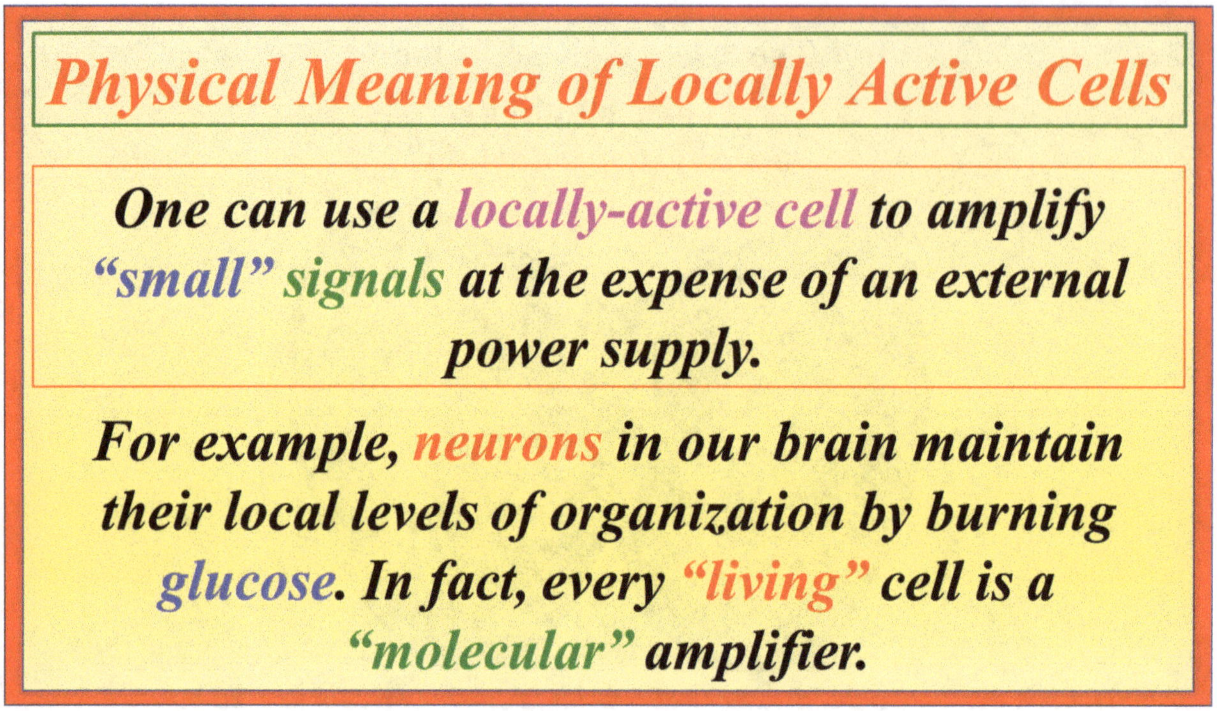

Exhibit 72

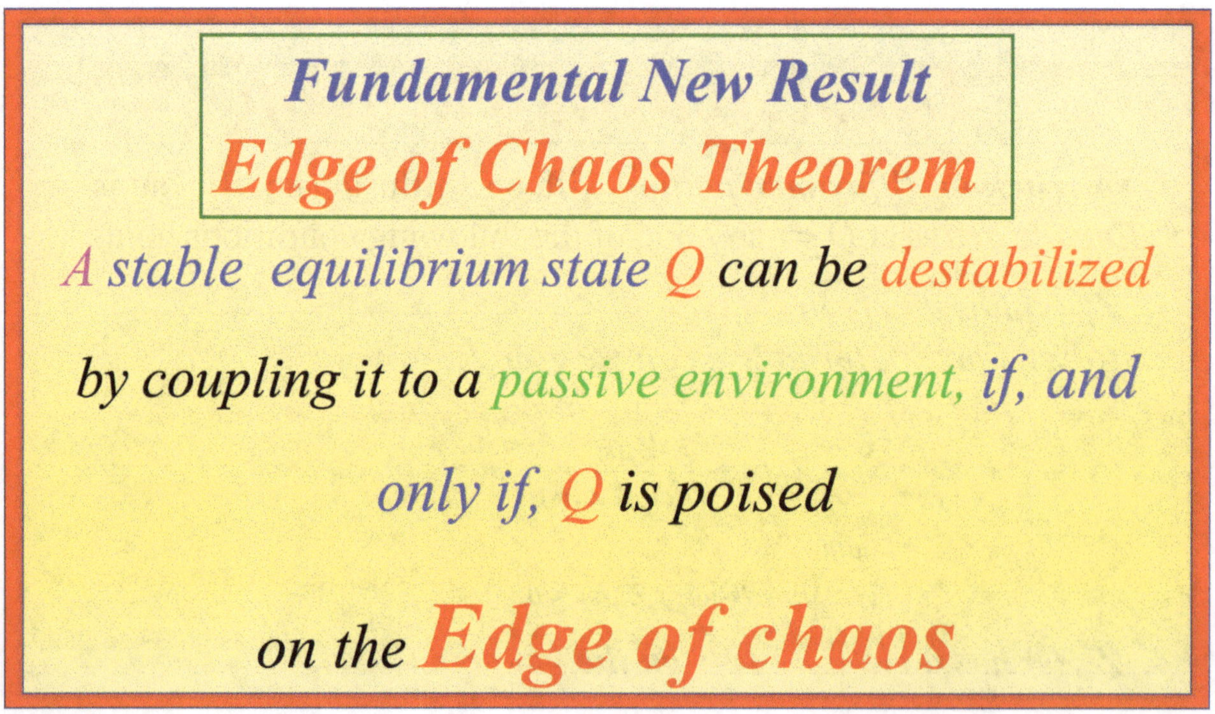

Exhibit 73

Action Potential
is
Impossible
without
Edge of Chaos

Exhibit 74

Neurons

are poised near the

Edge of Chaos

Exhibit 75

Local Activity

is the

Origin of Complexity

Exhibit 76

Exhibit 77

Exhibit 78

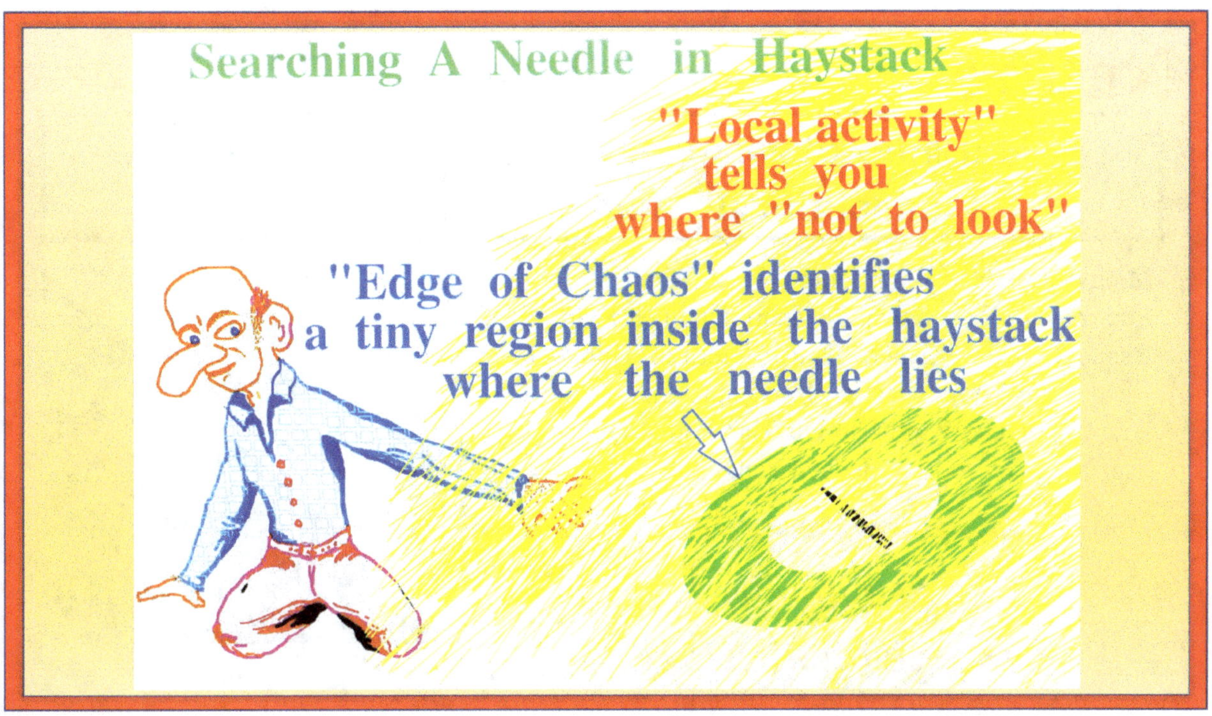

Exhibit 79

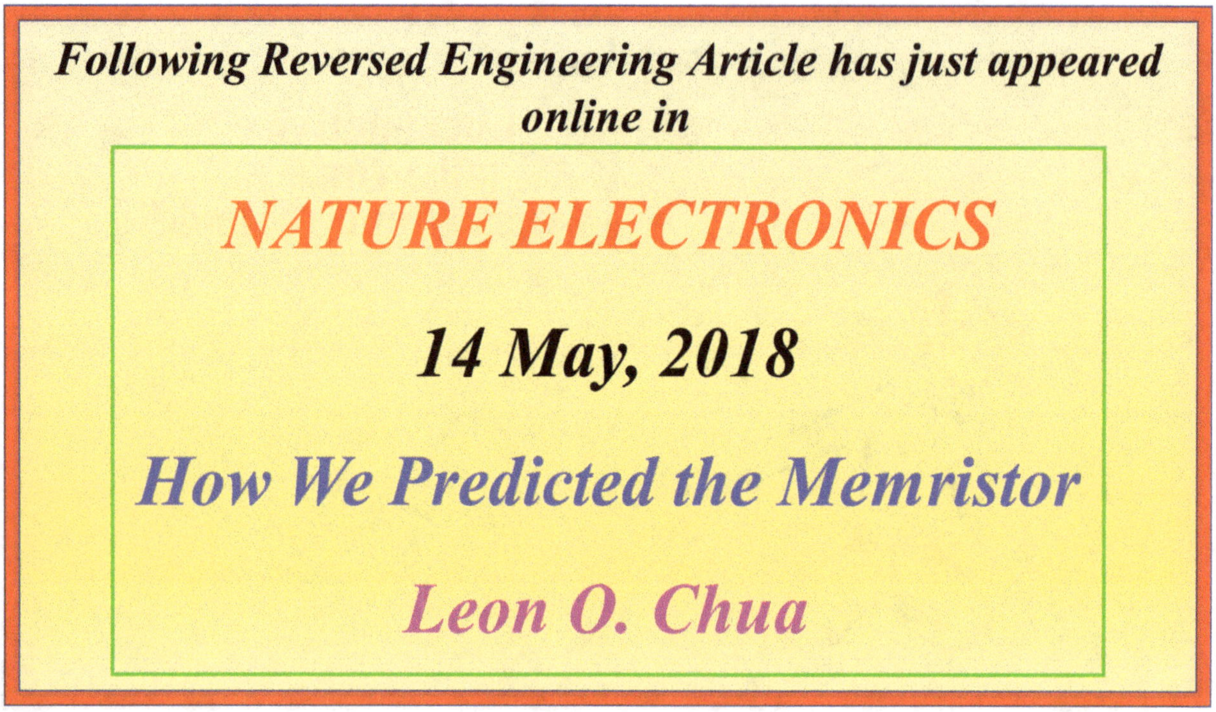

Exhibit 80

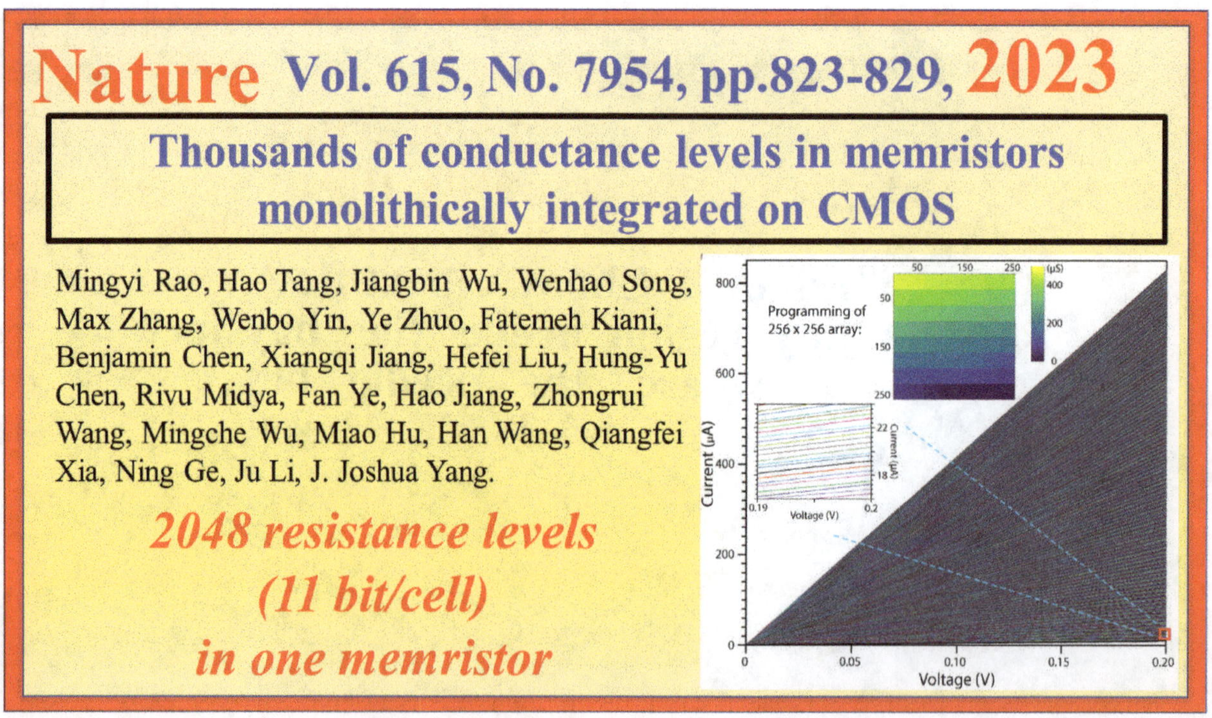

Exhibit 81

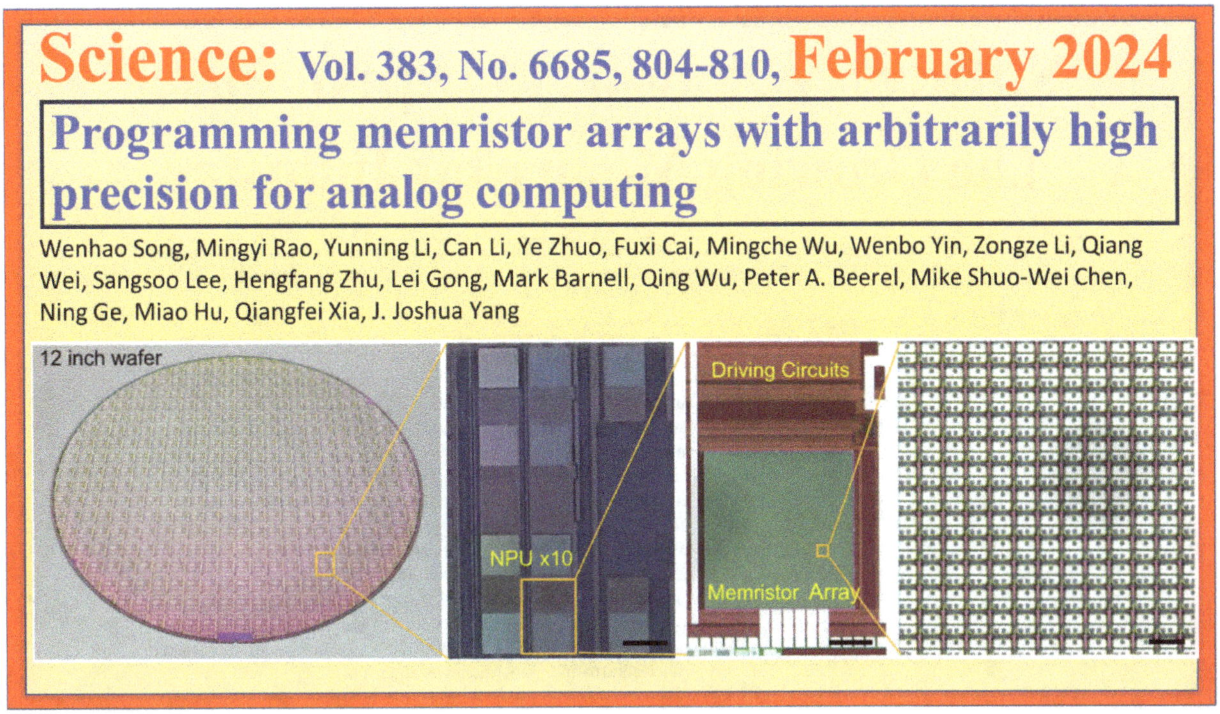

Exhibit 82

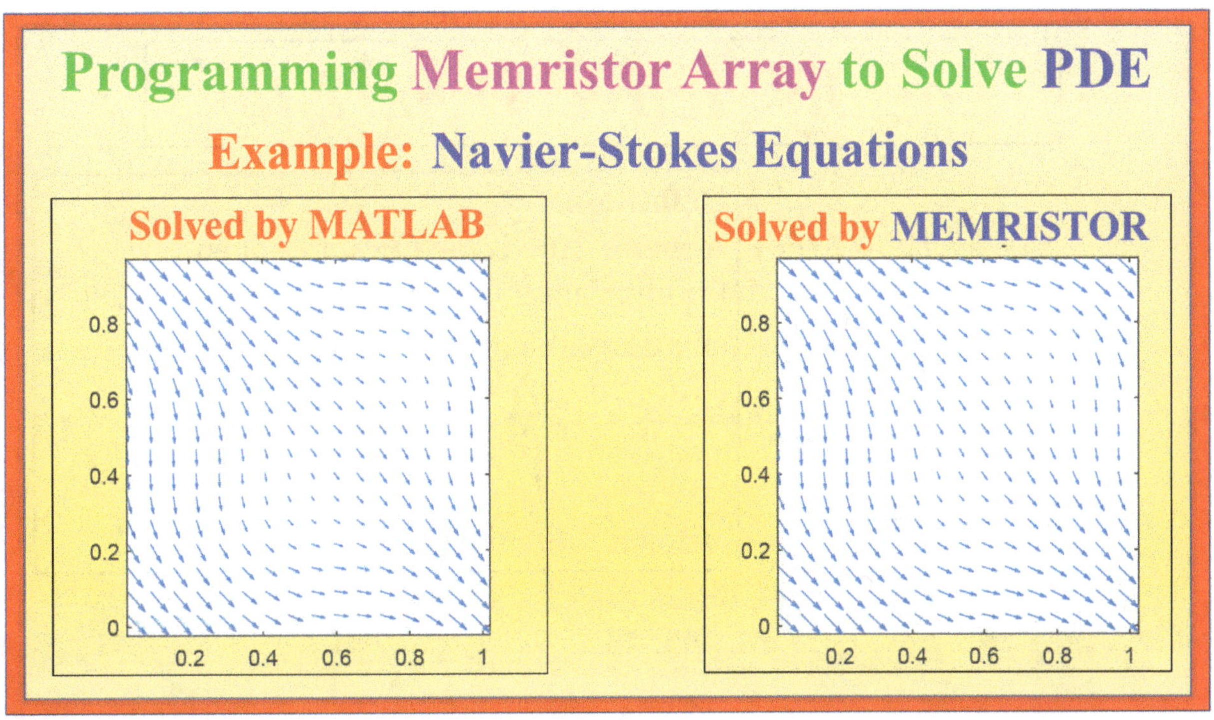

Exhibit 83

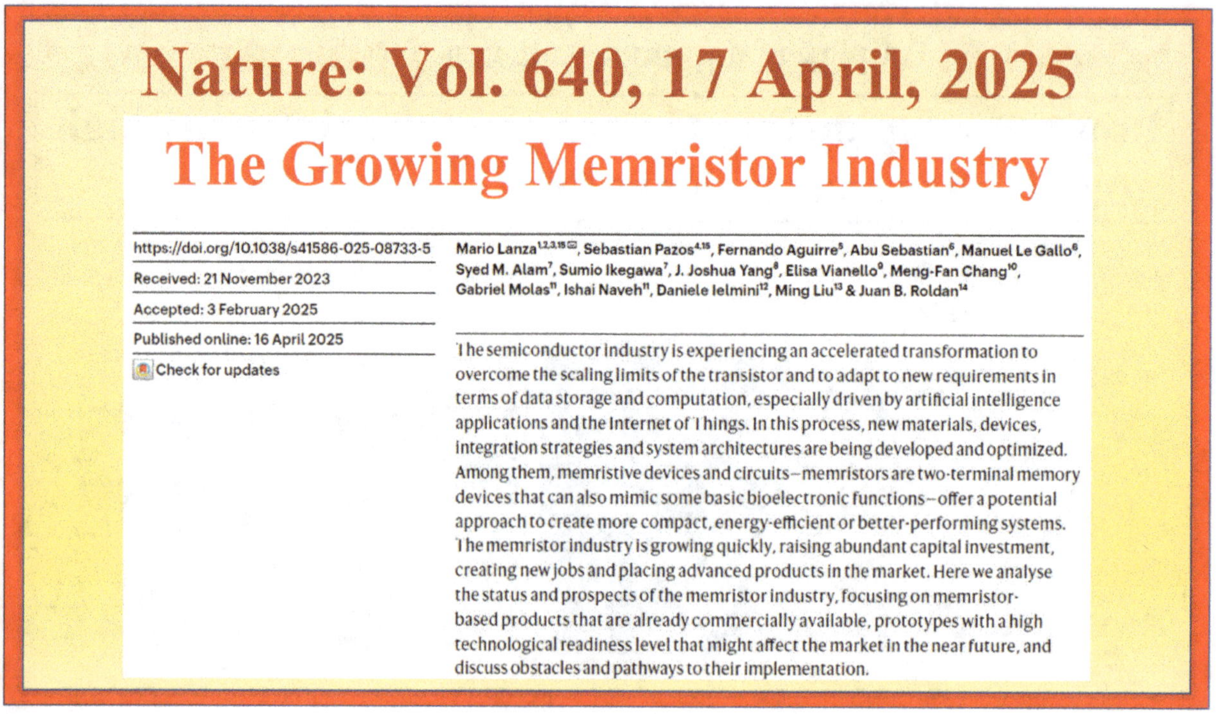

Exhibit 84

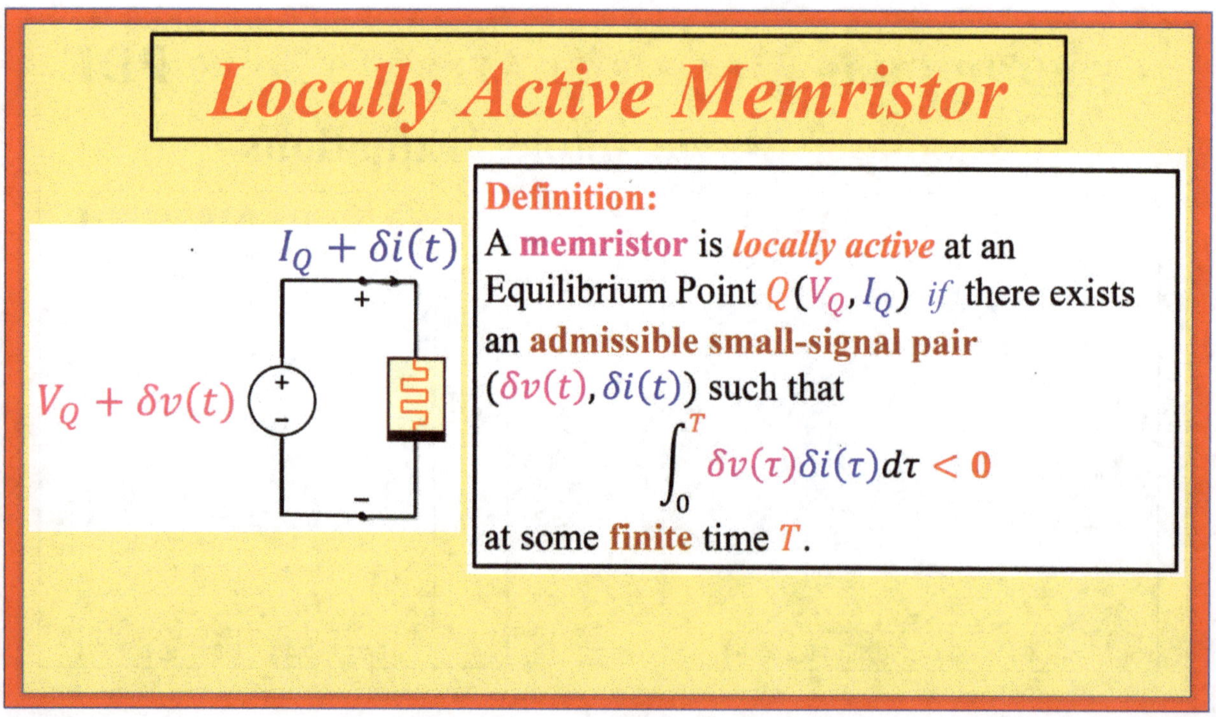

Exhibit 85

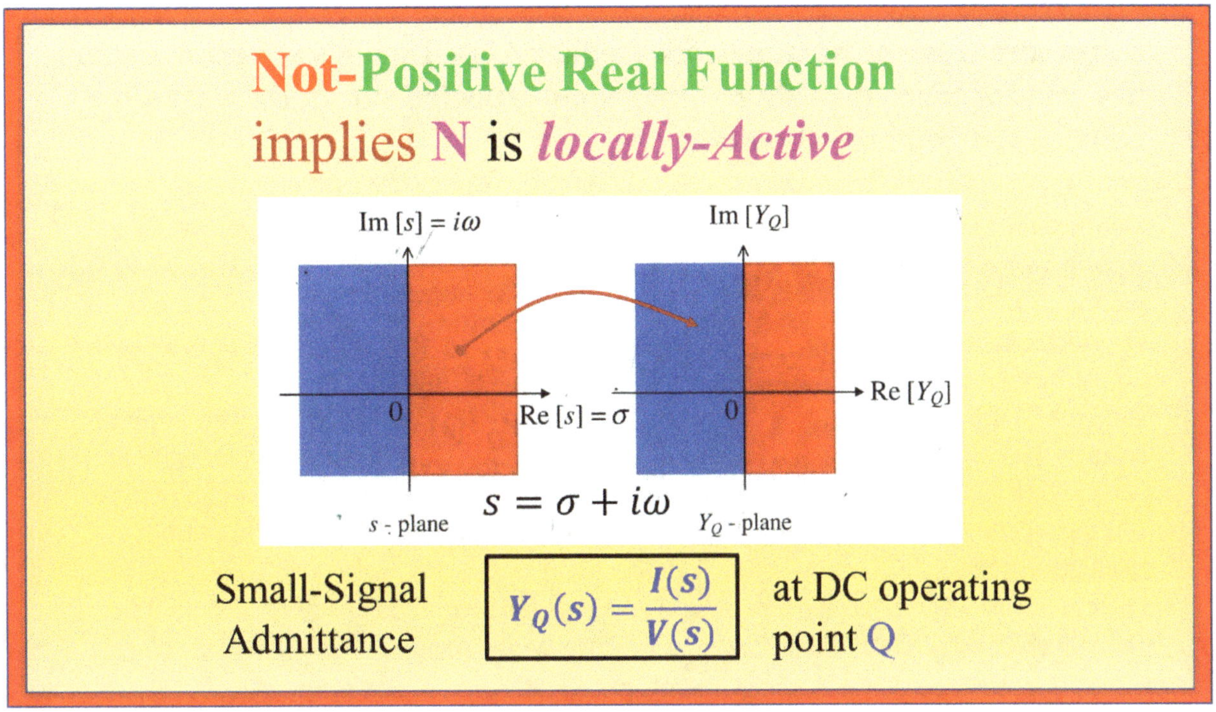

Exhibit 86

Exhibit 87

A treasure trove of wartime codebreaker Alan Turing's scientific papers has sold for more than three times their expected sale price... All after being saved from the shredder.

A collection of rare scientific papers written by mathematician, computer scientist and Second World War codebreaker Alan Turing has sold for a record £465,400 (€544,400) at auction in Lichfield, UK.
The documents were discovered in a loft at a property in Bermondsey, London, and were almost destroyed during a house clearance.
Some of the rare items that went under the hammer included a personal signed copy of Turing's 1938 PhD dissertation, "Systems of Logic Based on Ordinals", which sold for £110,500 (€129,200), as well as his paper "On Computable Numbers" - also known as "Turing's Proof" - which introduced the world to the idea of a universal computing machine in 1936.
The collection also included "The Chemical Basis Of Morphogenesis", which sold for £19,500 (€22,800). Dating from 1952, it is Turing's last major published work.

Exhibit 88

What is TURING INSTABILITY ?

2 Identical Uncoupled Cells

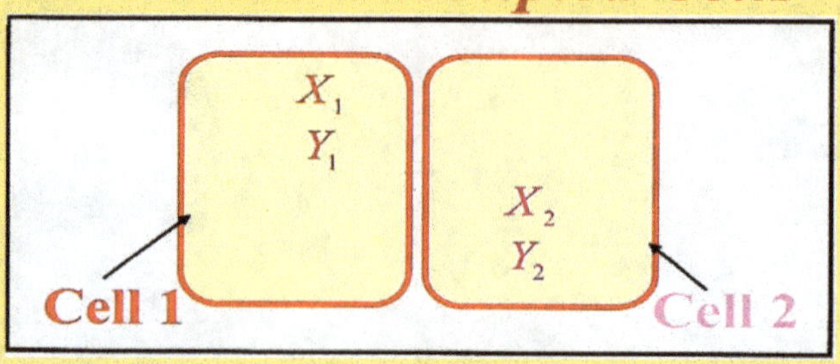

dynamics inside | *dynamics inside*

Cell 1

$$\dot{X}_1 = (5X_1 - 6Y_1 + 1)$$
$$\dot{Y}_1 = (6X_1 - 7Y_1 + 1)$$

Cell 2

$$\dot{X}_2 = (5X_2 - 6Y_2 + 1)$$
$$\dot{Y}_2 = (6X_2 - 7Y_2 + 1)$$

Equilibrium Solution of Cell 1:

$$X_1 = 1, \ Y_1 = 1$$

Equilibrium Solution of Cell 2:

$$X_2 = 1, \ Y_2 = 1$$

Exhibit 89

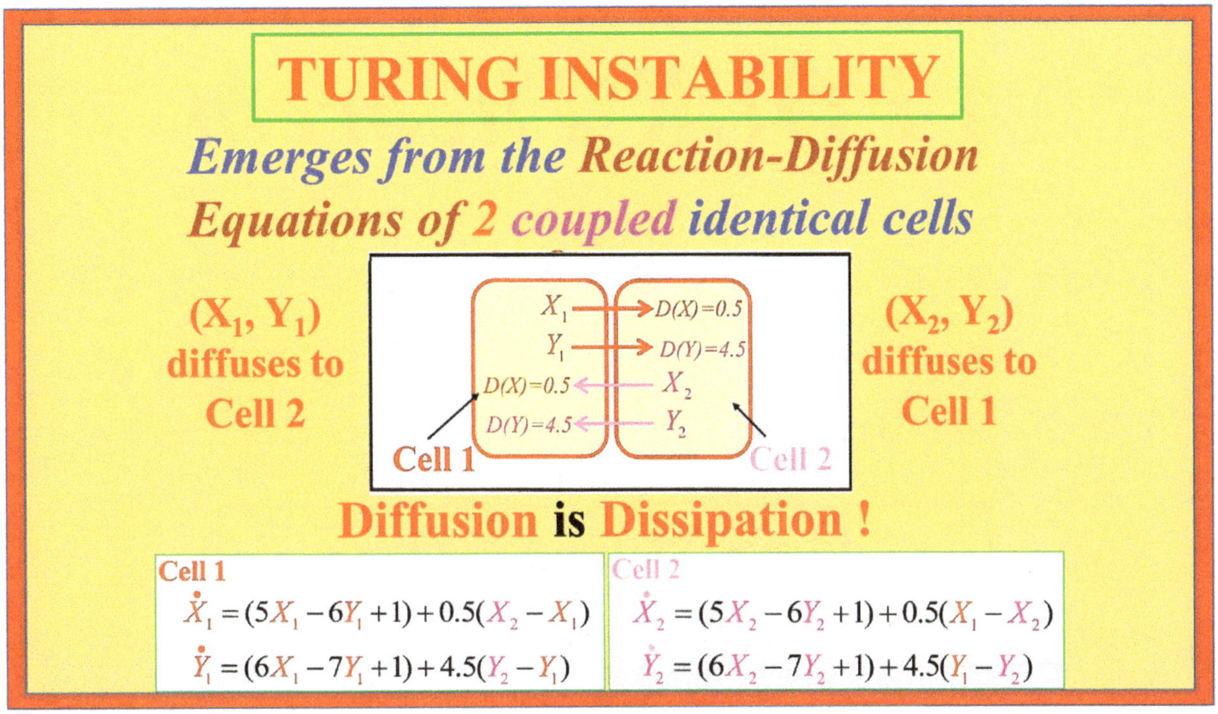

Exhibit 90

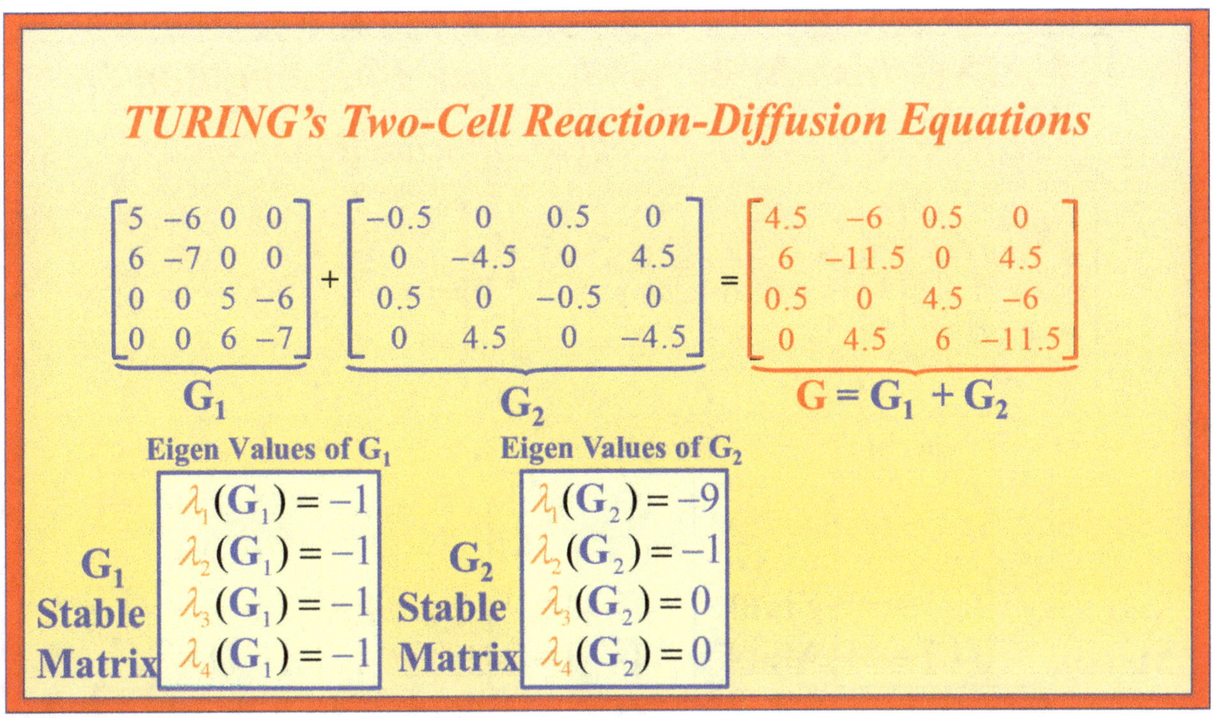

Exhibit 91

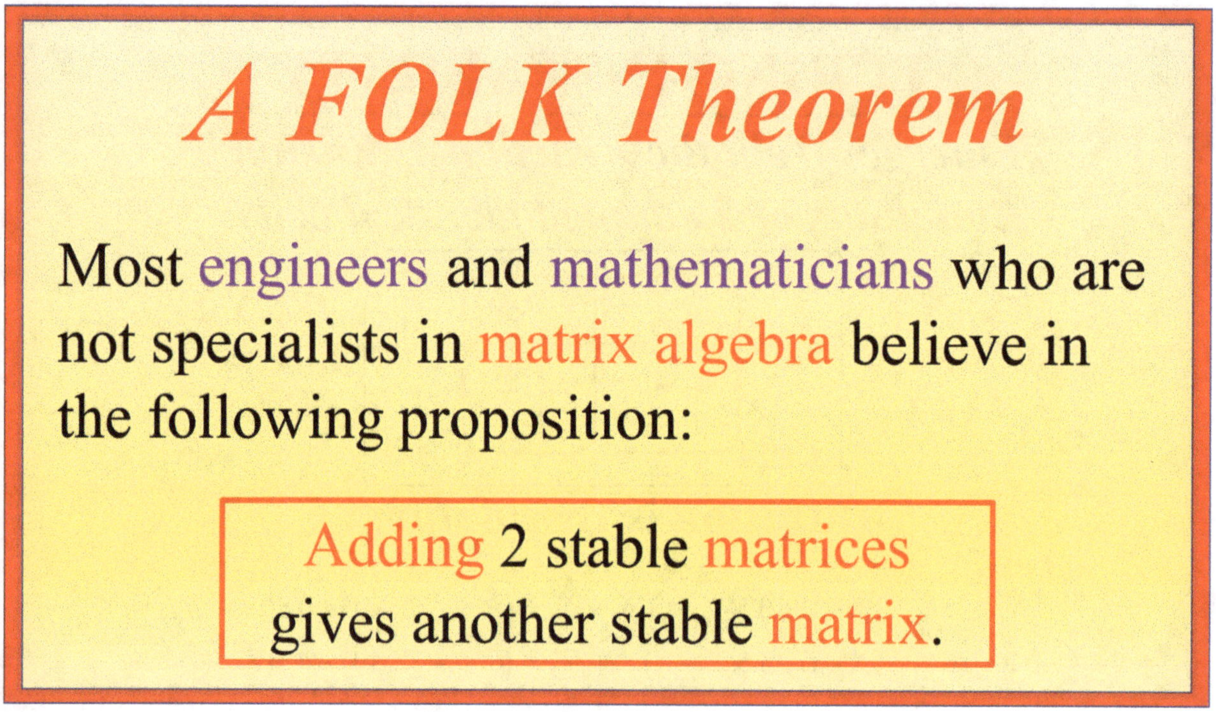

Exhibit 92

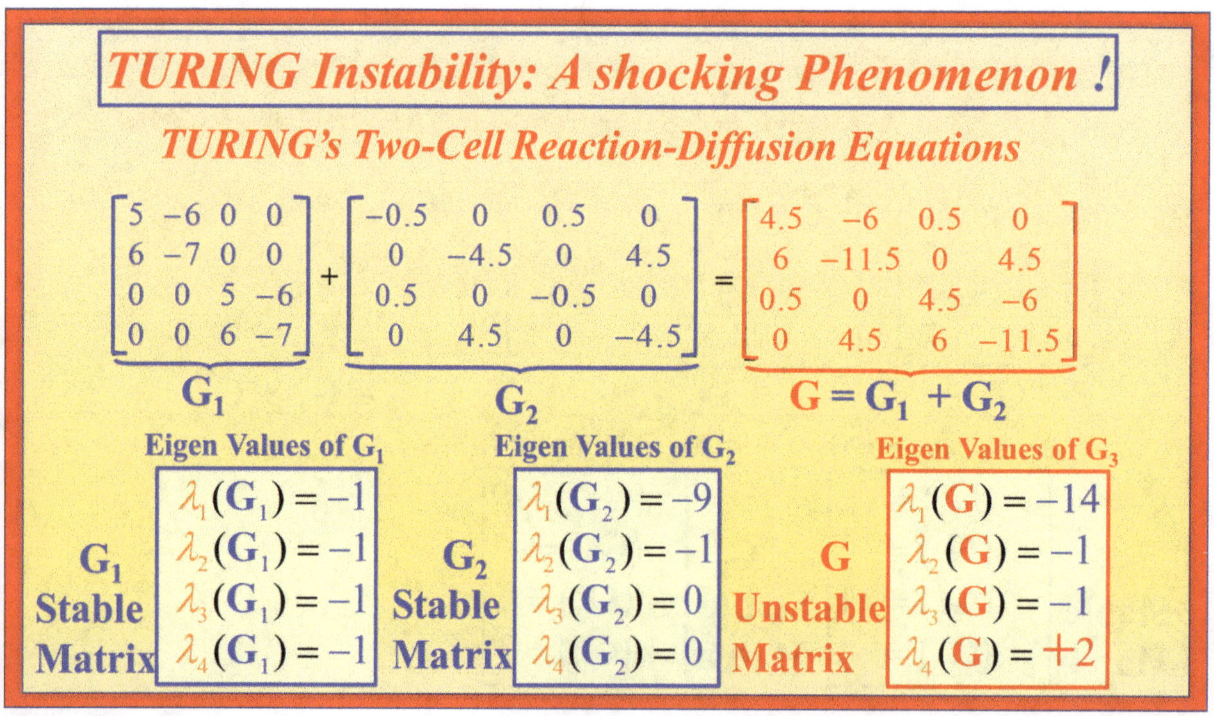

Exhibit 93

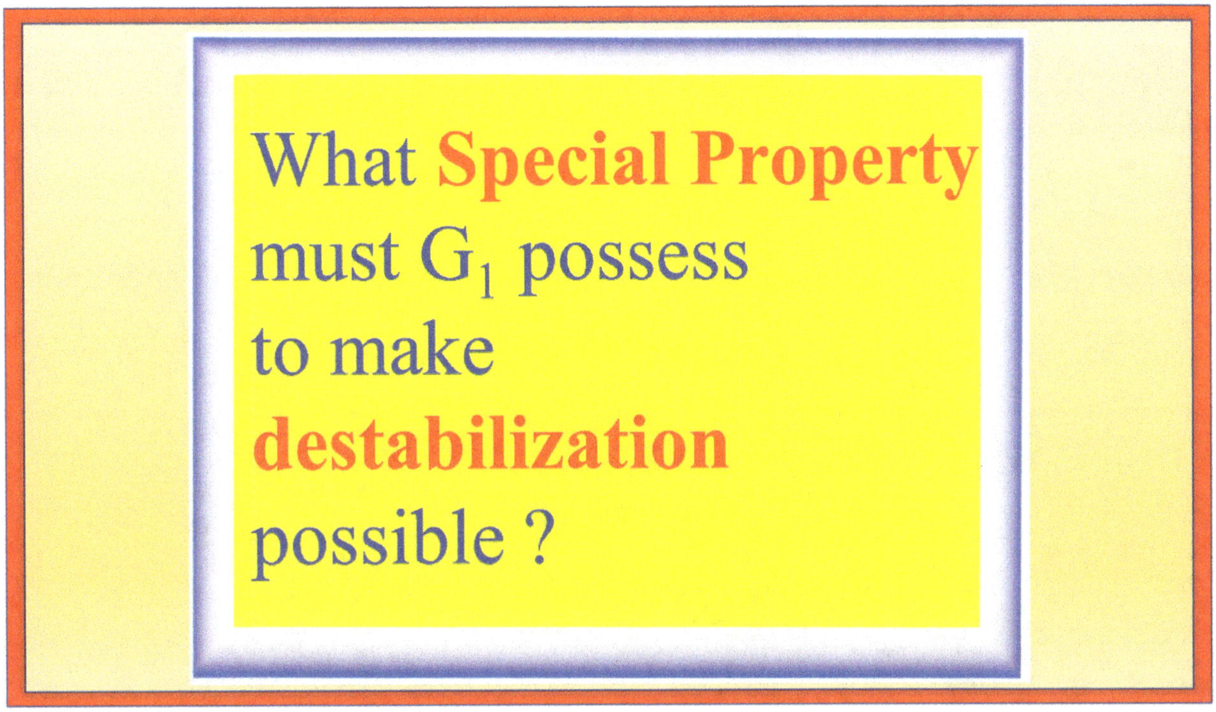

Exhibit 94

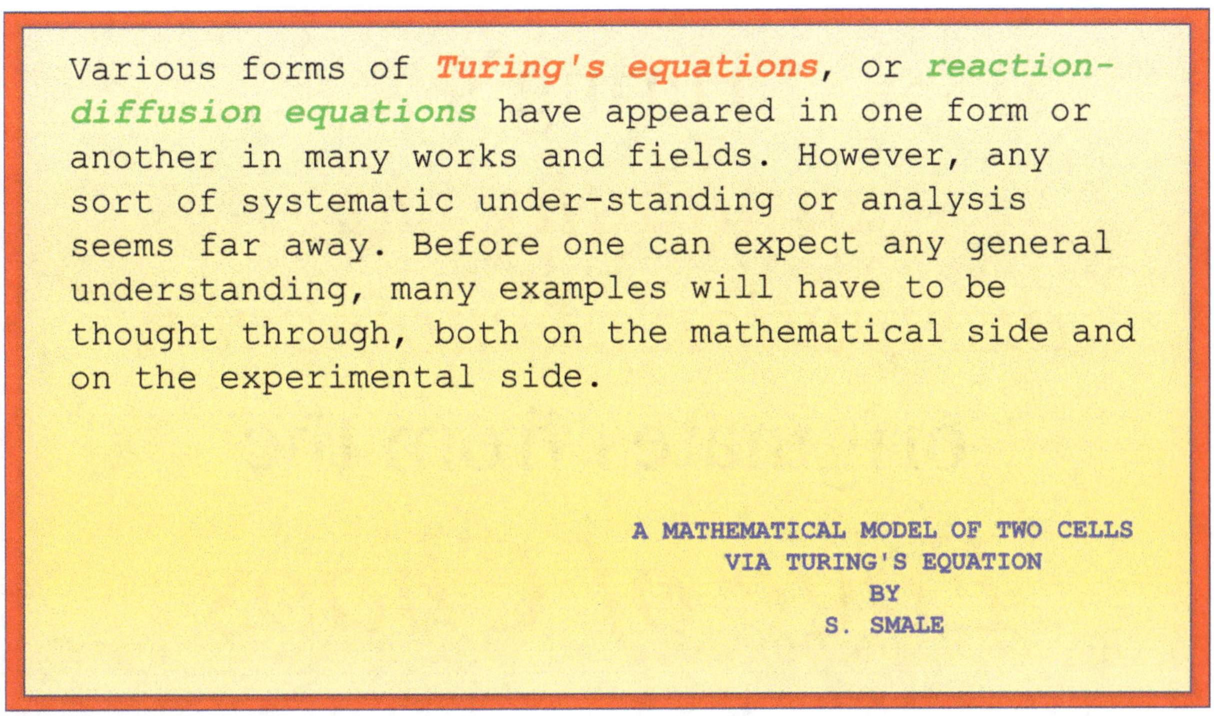

Exhibit 95

IEEE TRANSACTIONS ON CIRCUITS AND SYSTEMS–I: REGULAR PAPERS 1

Edge of Chaos is Sine Qua Non for Turing Instability

A. Ascoli *Senior Member, IEEE*, A.S. Demirkol, R. Tetzlaff *Senior Member, IEEE*, and L. Chua *Fellow, IEEE*

***Abstract*—Diffusion-driven instabilities with pattern formation may occur in a network of identical, regularly-spaced, and resistively-coupled cells if and only if the uncoupled cell is poised on a locally-active and stable operating point in the Edge of Chaos domain. This manuscript presents the simplest ever-reported two-cell neural network, combining together only 7 two-terminal components, namely 2 batteries, 3 resistors, and 2 volatile NbO_x memristive threshold switches from NaMLab, and subject to diffusion-driven instabilities with the concurrent emergence of Turing patterns. Very remarkably, this is the first time an homogeneous cellular medium, with no other dynamic element than 2 locally-active memristors, hence the attribute *all-memristor* coined to address it in this paper, is found to support complex phenomena. The destabilization of the homogeneous solution occurs in this second-order two-cell array if and only if the uncoupled cell circuit parameters are chosen from the Edge of Chaos domain. A deep circuit- and system-theoretic investigation, including linearization analysis and phase portrait investigation, provides a comprehensive picture for the local and global dynamics of the bio-inspired network, revealing how a theory-assisted approach may guide circuit design with inherently non-linear memristive devices.**

***Index Terms*—Smale Paradox, Turing Instability, Prigogine Symmetry-Breaking, Destabilization of the Homogeneous, Emergent Phenomena, Pattern Formation, Two-Cell Reaction-Diffusion System, Cellular Nonlinear/Neural/Nanoscale Network, Bio-Inspired Memristor Oscillator, NaMLab NbO_x Memristor, Threshold Switch, Theory of Local Activity, Edge of Chaos Principle**

implementation of computing paradigms, which could resolve the current limitations of purely-CMOS electronic systems. Moreover, some volatile devices, called threshold switches in the device physics community, are blessed with the extraordinary capability to act as local energy sources under suitable polarization, a property which earns them the attribute of *locally-active* [7]. Since, similarly as the sodium and potassium ion channels in neuronal axon membranes [8], devices of this kind feature a negative differential resistance (NDR) branch along their DC current-voltage characteristic, a signature for their small-signal amplification capability, their adoption in circuit design allows to design more reliable electronic neuron implementations [9], and to develop artificial cellular neural networks, which, processing information according to biological principles, enable the solution of challenging computational problems more efficiently than current software- or hardware-based solutions [10]. Recently [11], the adoption of locally-active NbO_x memristors [12] from NaMLab [13] has enabled to develop the simplest ever-reported reaction-diffusion network, employing only four dynamicas states to reproduce the counterintuitive phenomenon, which mesmerised the American luminary Stephen Smale in 1974 [14], as he reported how two biological cells, mathematically-dead on their own, were unexpectedly found to pulse indefinitely under diffusive coupling. As a follow-up study, this manuscript

Exhibit 96

Turing's

instability via

dissipation phenomenon

originates from the

Edge of Chaos

Exhibit 97

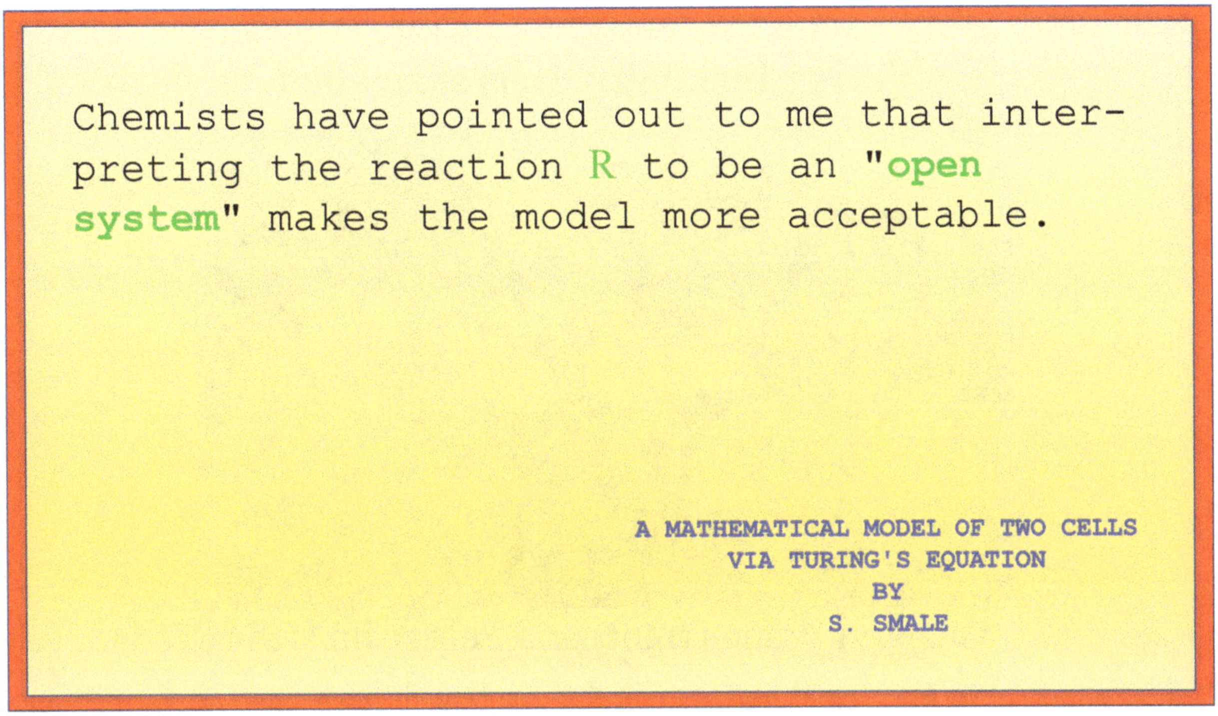

Exhibit 98

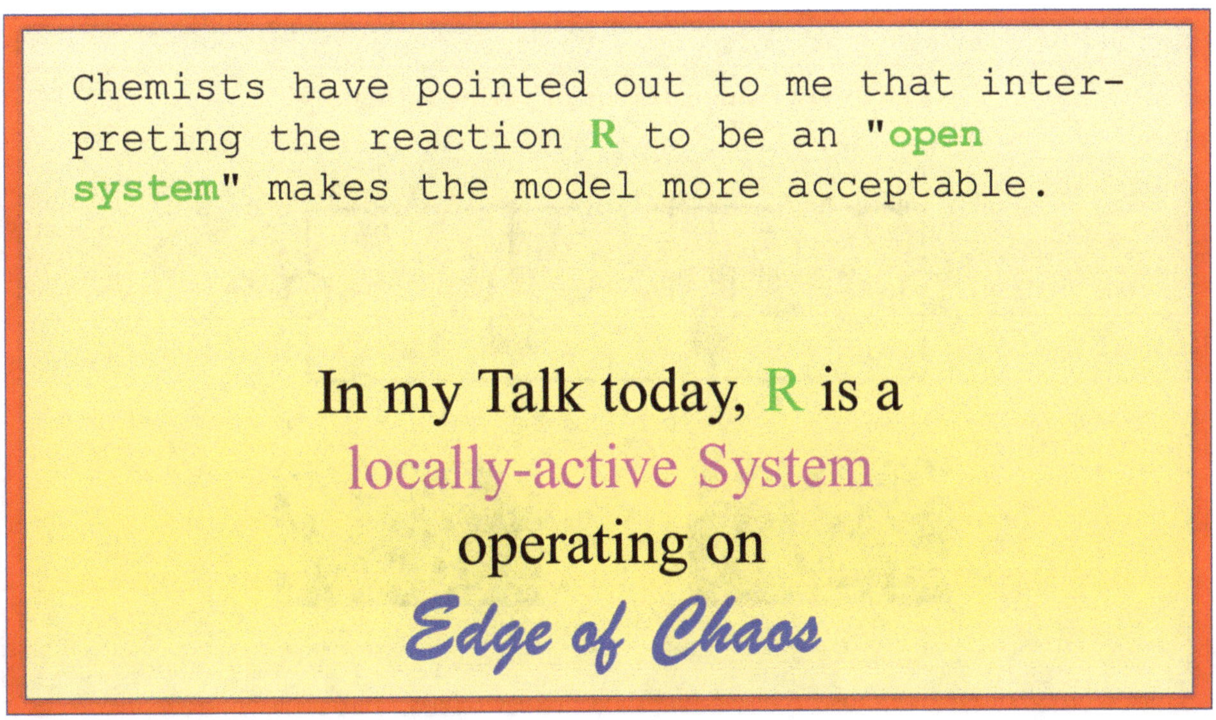

Exhibit 99

Exhibit 100

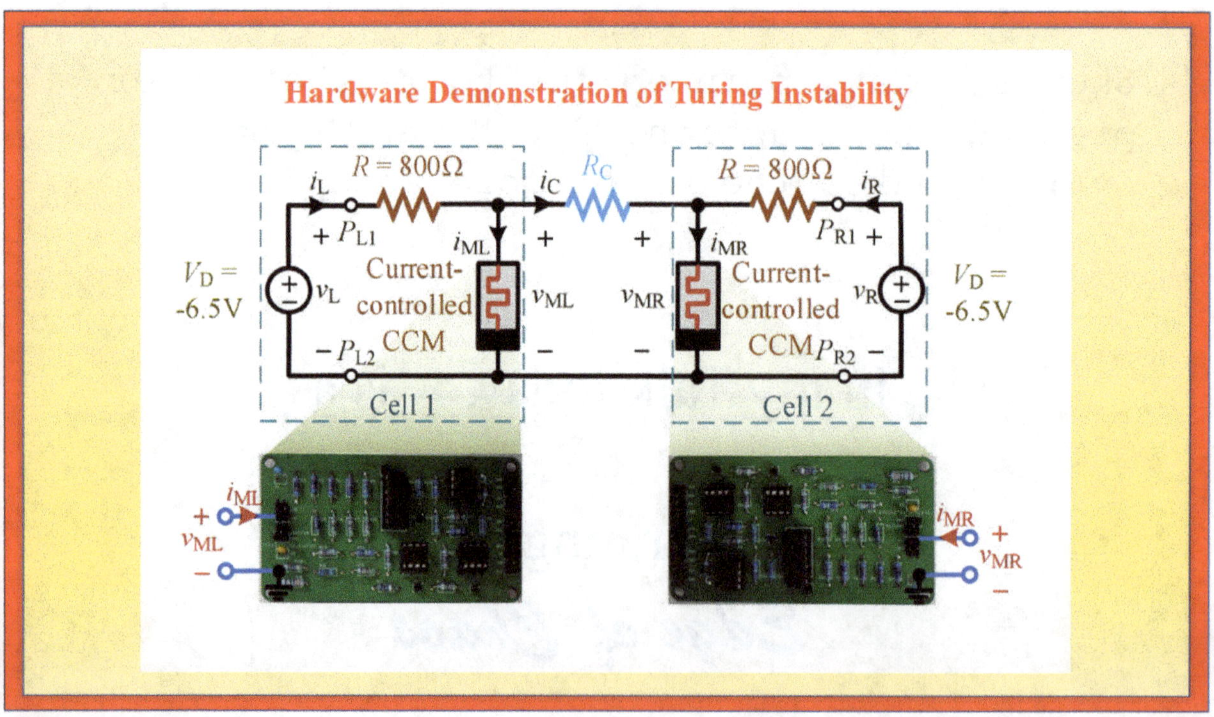

Exhibit 101

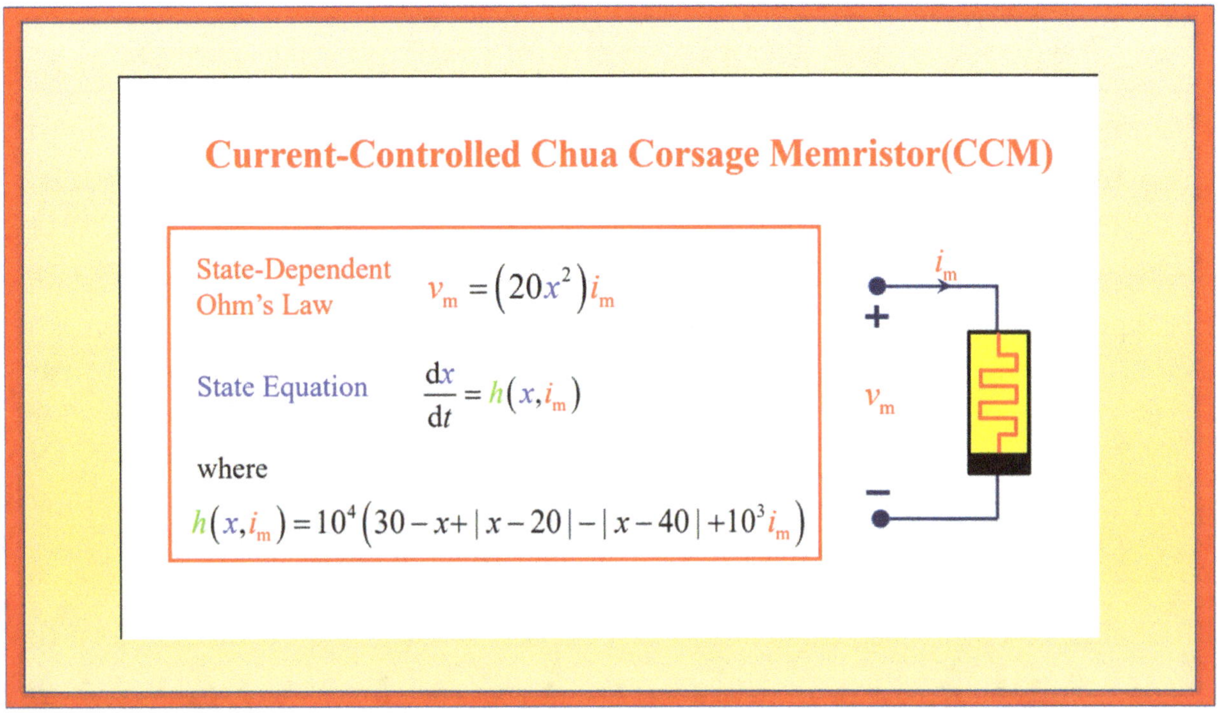

Exhibit 102

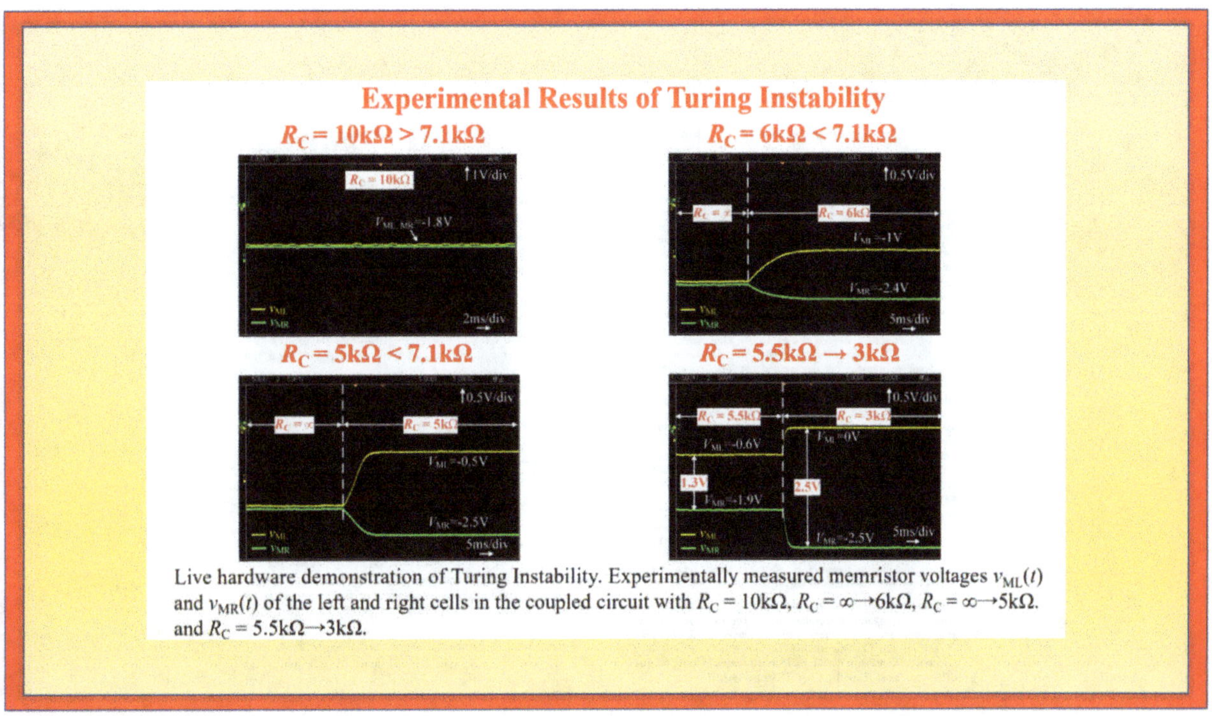

Live hardware demonstration of Turing Instability. Experimentally measured memristor voltages $v_{ML}(t)$ and $v_{MR}(t)$ of the left and right cells in the coupled circuit with R_C = 10kΩ, $R_C = \infty \to 6$kΩ, $R_C = \infty \to 5$kΩ. and R_C = 5.5kΩ→3kΩ.

Exhibit 103

There is a paradoxical aspect to the example. One has two dead (mathematically dead) cells interacting by a diffusion process which has a tendency in itself to equalize the concentrations. Yet in interaction, a state continues to pulse indefinitely.

A MATHEMATICAL MODEL OF TWO CELLS VIA TURING'S EQUATION BY S. SMALE

Exhibit 104

1252 IEEE TRANSACTIONS ON CIRCUITS AND SYSTEMS—I: REGULAR PAPERS, VOL. 69, NO. 3, MARCH 2022

Edge of Chaos Theory Resolves Smale Paradox

Alon Ascoli, *Member, IEEE*, Ahmet Samil Demirkol, Ronald Tetzlaff, *Senior Member, IEEE*, and Leon Chua, *Life Fellow, IEEE*

***Abstract*—No isolated system may ever support complexity. Emergent phenomena may however appear in an open system, if, as established by the Edge of Chaos theory, some of its constitutive elements feature the capability to amplify infinitesimal fluctuations in energy, provided an external source supplies them with a sufficient amount of DC power, which is known to be a signature for locally-active behaviour. In particular, complex behaviours, including static and dynamic pattern formation, may emerge in arrays of identical diffusively-coupled cells, if and only if the basic unit is poised on a particular sub-domain of the Local Activity regime, referred to as Edge of Chaos, within which a quiet state hides in fact a high degree of excitability. Here we show, for the first time, that these counterintuitive phenomena may emerge in a basic memristor cellular neural network, consisting of two identical diffusively-coupled second-order cells. The proposed bio-inspired array represents the simplest ever-reported open system, which reproduces the shocking phenomenon, reported by Smale in 1974, when, while studying a model from cellular biology, he observed two identical reaction cells, "mathematically dead" on their own, pulsating together upon diffusive coupling. Impressively, the bio-inspired two-cell reaction-diffusion network contains only nine circuit elements, specifically two DC voltage sources, three linear resistors, two linear capacitors, and two functional niobium oxide (NbO) memristors from NaMLab. Applying the theory of Local Activity to an accurate model of the memristor oscillator, a comprehensive picture for its local and global dynamics may be drawn, providing a systematic method to tune the design parameters of the two-cell array to enable diffusion-driven instabilities therein.**

***Index Terms*—Smale paradox, turing instability, Prigogine symmetry-breaking, destabilization of the homogeneous, emergent phenomena, pattern formation, two-cell reaction-diffusion system, cellular nonlinear/neural/nanoscale network, bio-inspired memristor oscillator, NaMLab niobium oxide (NbO) memristor, theory of Local Activity, Edge of Chaos principle.**

lies within a particular sub-domain of the Local Activity regime, referred to as *Edge of Chaos*. A neuronal axon in either the brain [3] or the heart [4], for example, may be modelled as a one-dimensional array of identical diffusively-coupled Hodgkin-Huxley cells. The propagation of an all-or-nothing spike, known as action potential, through this transmission line – an emergent phenomenon "par excellence" [5] – may occur when the single Hodgkin-Huxley cell, hosting biological memristors, namely the sodium and potassium ion channels [6], which are capable to amplify infinitesimal fluctuations in energy under suitable bias conditions, is poised near the Edge of Chaos.

The upcoming end of Moore's era calls for the development of new materials, processing paradigms, and computing architectures, which may allow to keep the integrated circuit performance growth in the years to follow, despite further CMOS transistor size reduction will no longer be feasible. The disruptive memristive nano-technologies ([7], [8]) offer the key to open up unprecedented opportunities in circuit design. Besides their extensive adoption in crossbar configuration for storing data [9] or for accelerating machine learning algorithms [10], which is of great appeal to the electronics industry, especially for artificial intelligence applications [11], non-volatile (volatile) memristor physical realisations may also be employed to design plausible electronic implementations of biological synapses [12] (neurons [13], [14]). Gaining a deep insight into the nonlinear dynamics of these memristors [15] is a fundamental preliminary step toward the ultimate goal to synthesise neuromorphic circuits capable to resemble as close as possible biological neural networks [16]. A compre-

Exhibit 105

Exhibit 106

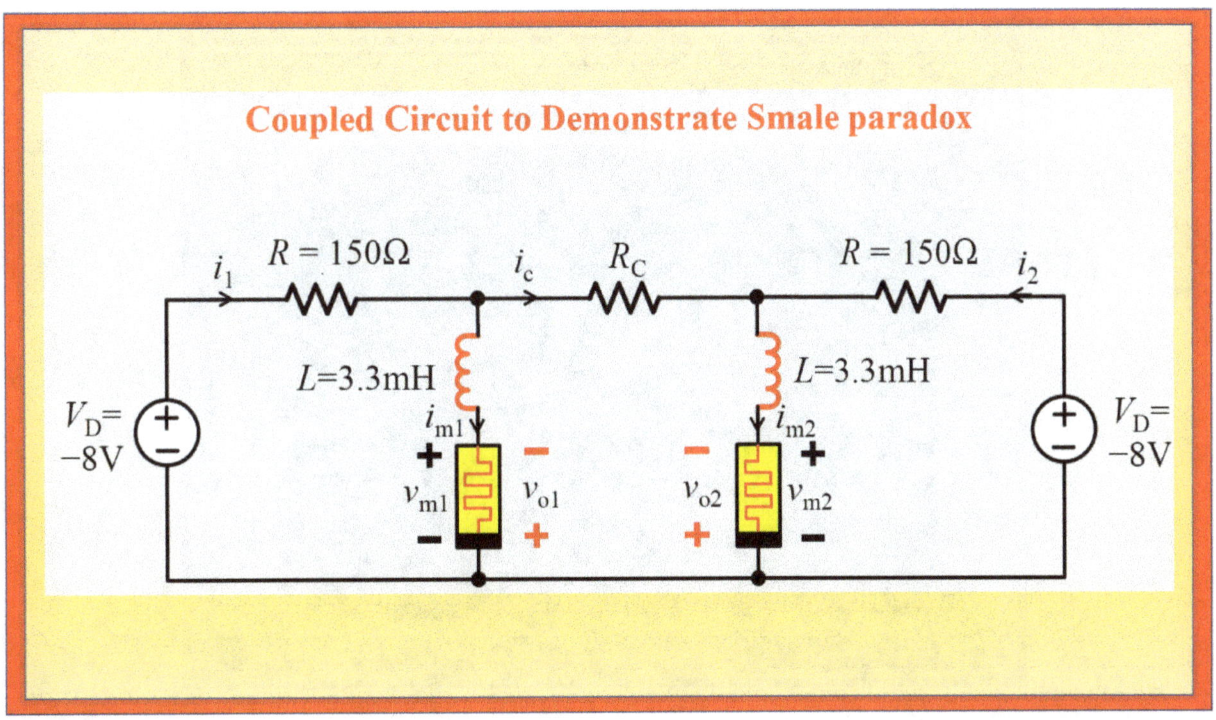

Exhibit 107

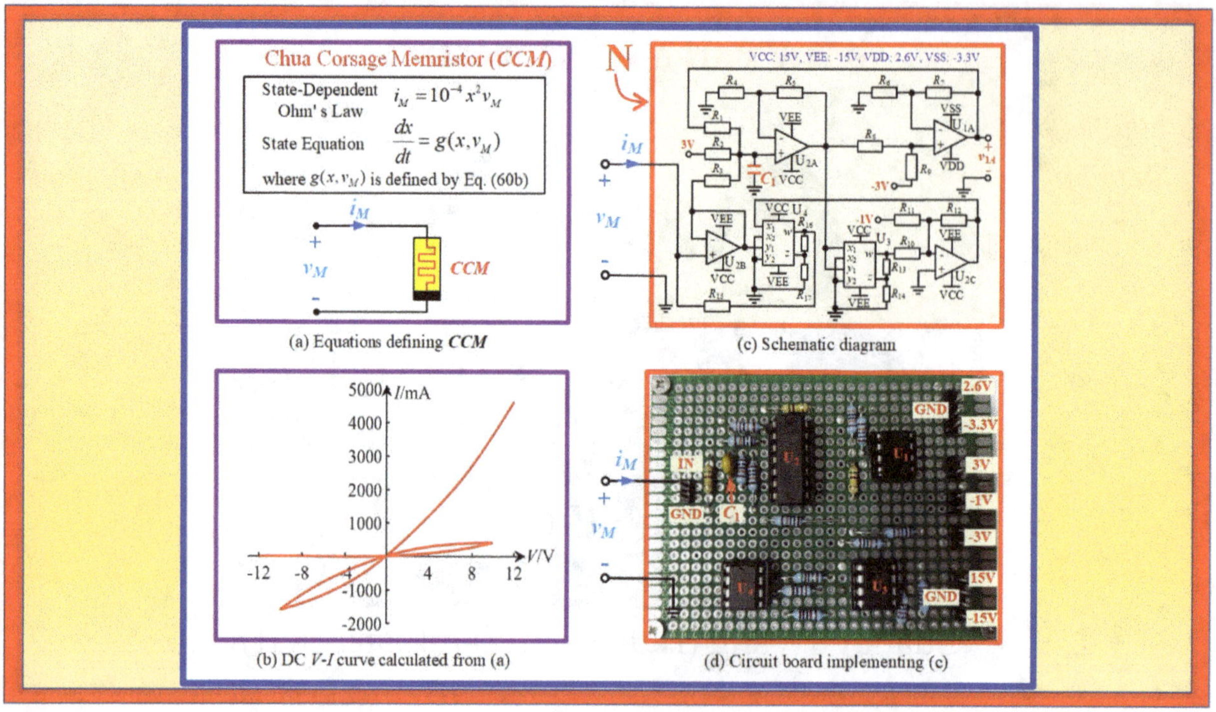

Exhibit 108

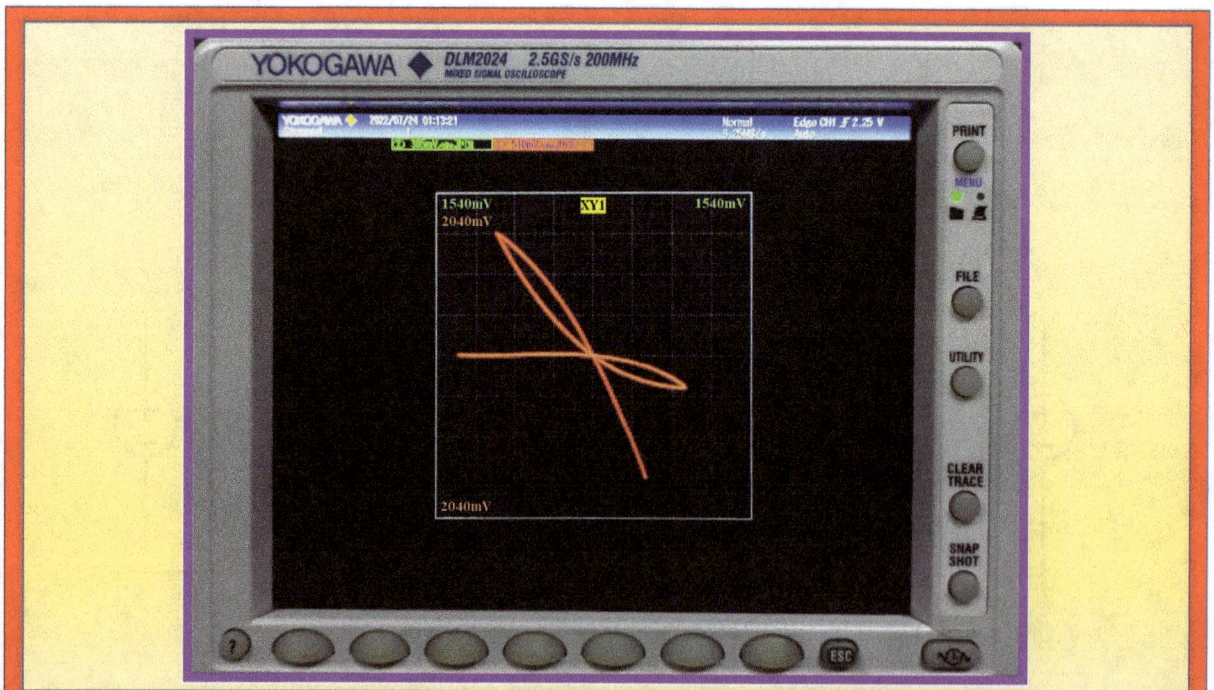

Exhibit 109

Exhibit 110

Exhibit 111

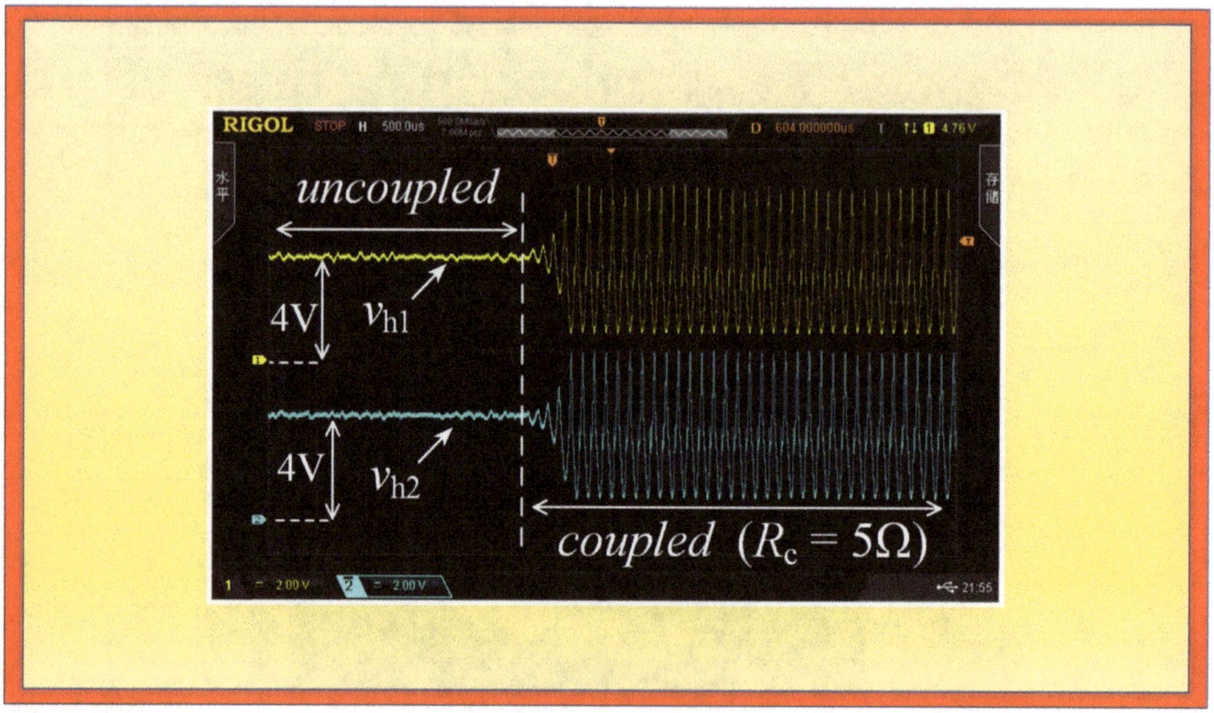

Exhibit 112

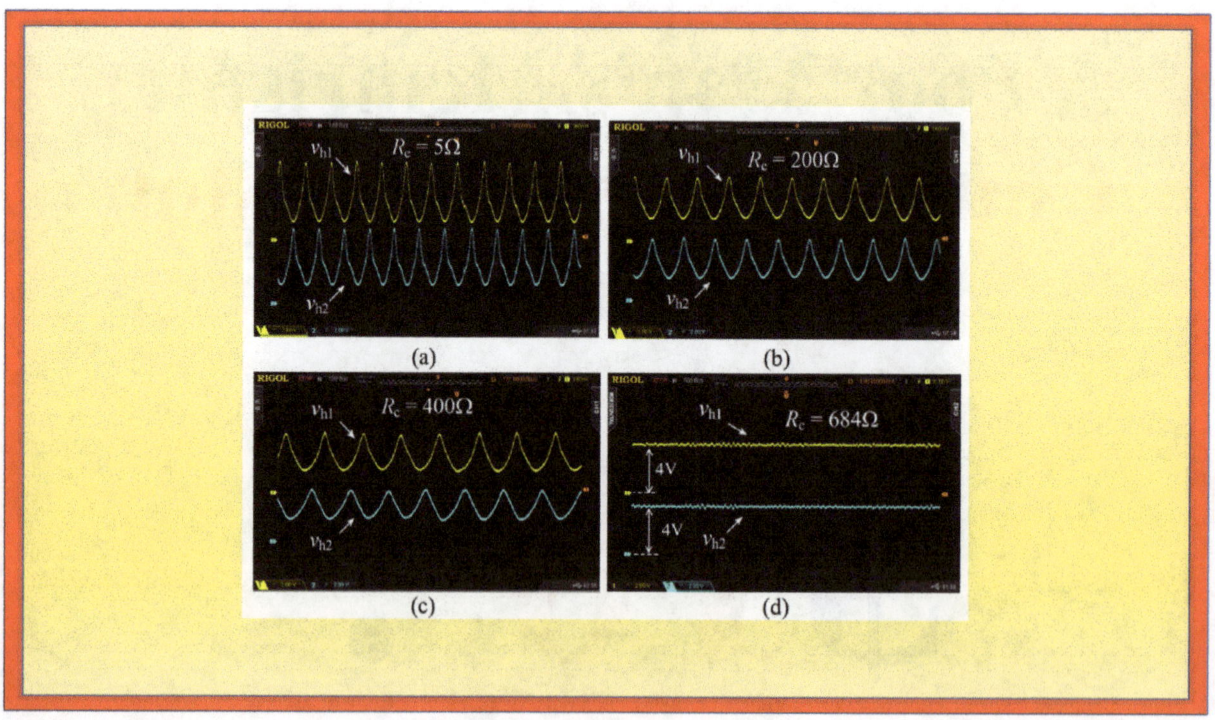

Exhibit 113

Exhibit 114

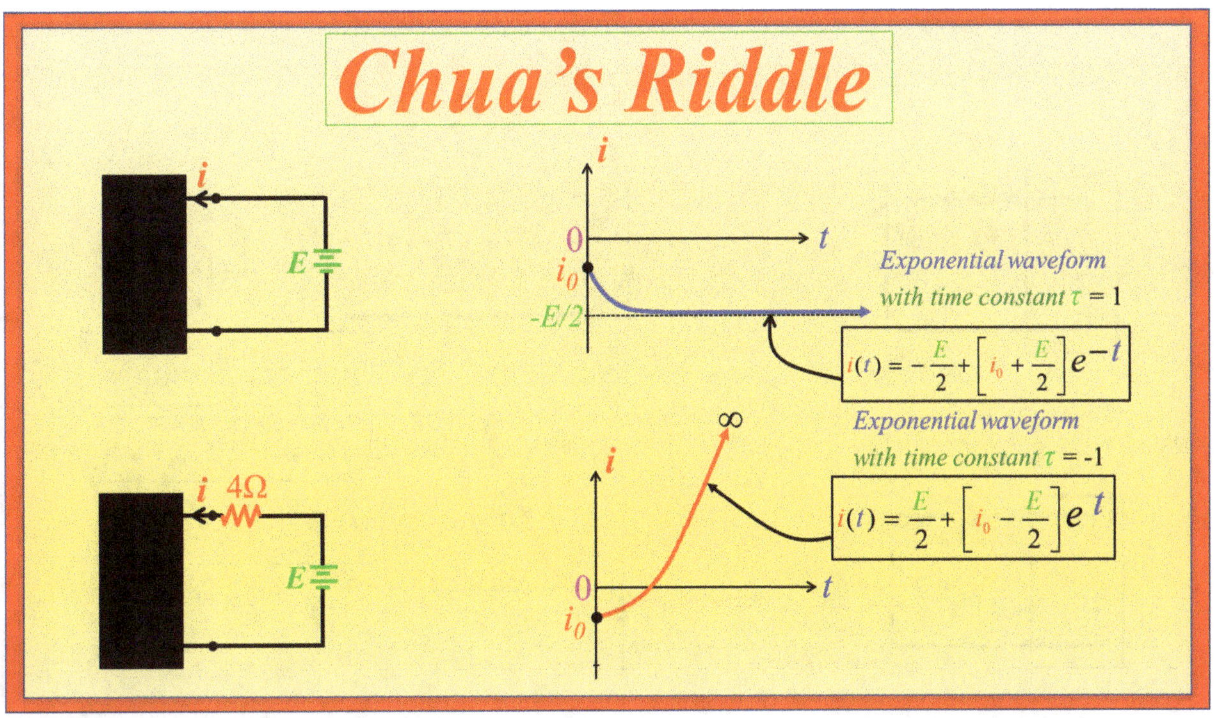

Exhibit 115

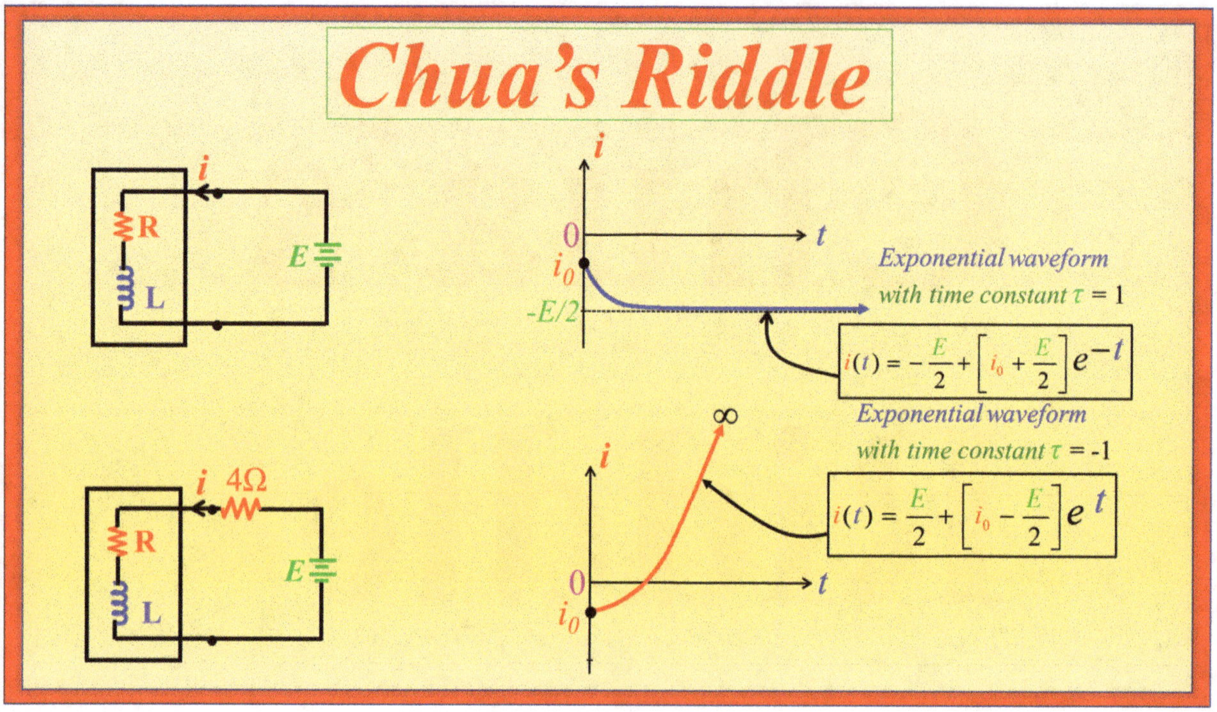

Exhibit 116

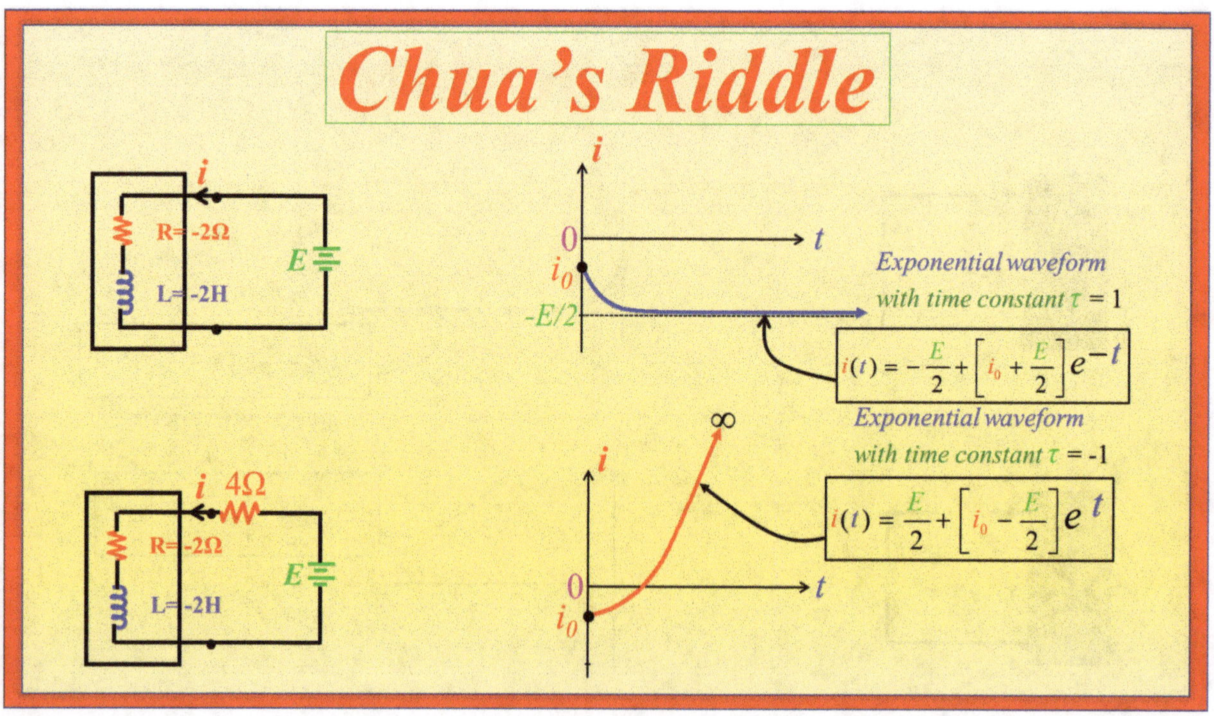

Exhibit 117

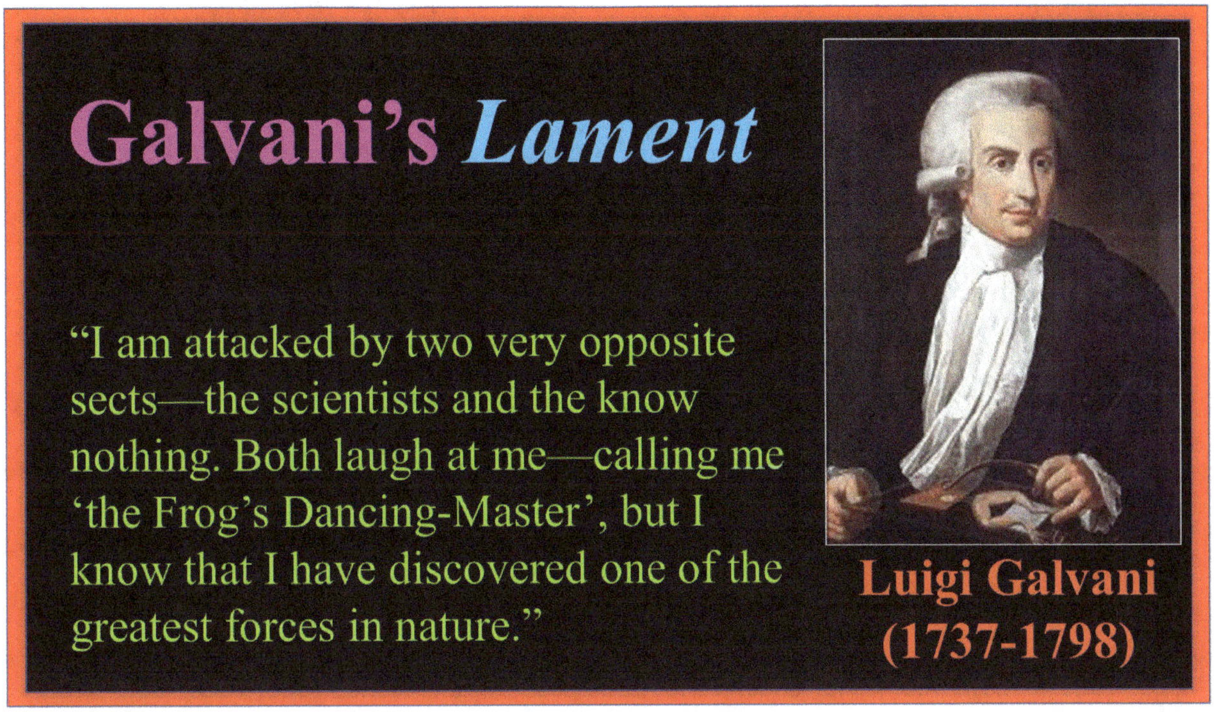

Exhibit 118

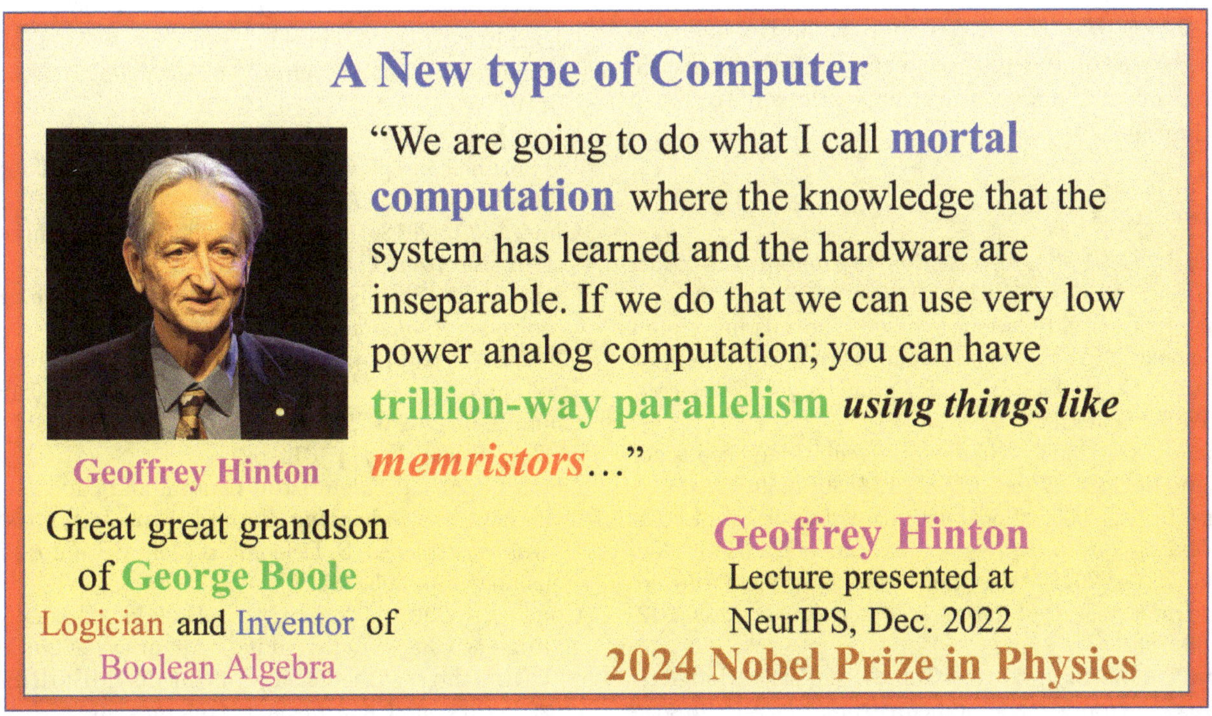

Exhibit 119

Exhibit 120

Acknowledgment

The author would like to thank Muhammad Umraiz for typing the entire paper, as well as retrieving and laying out the slides in Sec. 3. He has also made, revised, and improved many slides in Sec. 3. Moreover, his many other suggestions have been invaluable.

References

Chua, L. O. [1969] *Introduction to Nonlinear Network Theory* (McGraw-Hill).

Chua, L. O. [1971] "Memristor — The missing circuit element," *IEEE Trans. Circuit Theory* **18**, 507–519.

Chua, L. O. [1998] *CNN*: *A Paradigm for Complexity* (World Scientific).

Chua, L. O. [2003] "Nonlinear circuit foundations for nano devices," *Proc. IEEE* **91**, 1830–1859.

Chua, L. O. [2005] "Local activity is the origin of complexity," *Int. J. Bifurcation and Chaos* **15**, 3435–3456.

Chua, L. O. [2011] "Resistance switching memories are memristors," *Appl. Phys. A*: *Mater. Sci. Process.* **102**, 765–783.

Chua, L., Sbitnev, V. & Kim, H. [2012a] "Hodgkin–Huxley axon is made of memristors," *Int. J. Bifurcation and Chaos* **22**, 1230011.

Chua, L., Sbitnev, V. & Kim, H. [2012b] "Neurons are poised near the edge of chaos," *Int. J. Bifurcation and Chaos* **22**, 1250098.

Chua, L. O. [2013] "Memristor, Hodgkin–Huxley, and edge of chaos," *Nanotechnology* **24**, 383001.

Chua, L. O. [2014] "If it's pinched it's a memristor," *Semicond. Sci. Technol.* **29**, 104001–104042.

Chua, L. [2015a] "Everything you wish to know about memristors but are afraid to ask," *Radio Eng.* **24**, 119–368.

Chua, L. O. [2015b] "The Chua lectures," https://www.youtube.com/playlist?list=PLWIjtByUJvcvT3IgfDyej4yEpLgyAn2L9.

Chua, L. O. [2018a] "How we predicted the memristor," *Nat. Electron.* **1**, 522.

Chua, L. [2018b] "Five non-volatile memristor enigmas solved," *Appl. Phys. A* **124**, 563.

Chua, L. [2022] "Hodgkin–Huxley equations implies edge of chaos kernel," *Jpn. J. Appl. Phys.* **61**, SM0805.

Chua, L. O. [2024] "Memristors on 'edge of chaos'," *Nat. Rev. Electr. Eng.* **1**, 614–627.

Chua, L. [2025a] "Memristor, neuron, and edge of chaos kernel," in *Outlooking Beyond Nanoelectronics and Nanosystems*, ed. S. Deleonibus (Jenny Stanford Publishing), pp. 265–335.

Chua, L. [2025b] "Memristor, Hodgkin–Huxley, and Edge of Chaos, Quos Vadis?" Memrisys 2025. Video URL https://media.ed.ac.uk/media/MEMRISYS+2025_+Leon+Chua_plenary/1_ukym02a6.

Feynman, R. P., Leighton, R. B. & Sands, M. [1963] *Feynman Lectures on Physics* (Addison-Wesley).

Gell-Mann, M. [1994] "Complex adaptive systems," in *Complexity*: *Metaphors, Models, and Reality*, Santa

Fe Institute Studies in the Sciences of Complexity, Vol. 19 (Addison-Wesley), pp. 17–45.

Gell-Mann, M. [1995] *The Quark and the Jaguar: Adventures in the Simple and the Complex* (Macmillan).

Haken, H. [1988] *Synergetics. An Introduction* (Springer).

Huang, Y., He, C., Ling, Y., Ge, N., Yang, J. J., Hu, M., Katehi, L. & Xia, Q. [2025] "Radiofrequency signal processing with a memristive system-on-a-chip," *Nat. Electron.* **8**, 587–596.

Krestinskaya, O., James, A. P. & Chua, L. O. [2019] "Neuromemristive circuits for edge computing: A review," *IEEE Trans. Neural Netw. Learn. Syst.* **31**, 4–23.

Langton, C. G. [1990] "Computation at the edge of chaos: Phase transitions and emergent computation," *Physica D* **42**, 12–37.

Mainzer, K. [2007] *Thinking in Complexity: The Computational Dynamics of Matter, Mind and Mankind* (Springer).

Mainzer, K. & Chua, L. [2013] *Local Activity Principle* (Imperial College Press, London, UK).

Packard, N. H. [1988] "Adaptation toward the edge of chaos," in *Dynamic Patterns in Complex Systems*, Vol. 212 (World Scientific), pp. 293–301.

Premsankar, G., Di Francesco, M. & Taleb, T. [2018] "Edge computing for the Internet of Things: A case study," *IEEE Internet Things J.* **5**, 1275–1284.

Prigogine, I. [1980] *From Being to Becoming* (San Francisco).

Rao, M., Tang, H., Wu, J., Song, W., Zhang, M., Yin, W., Zhuo, Y., Kiani, F., Chen, B., Jiang, X., *et al.* [2023] "Thousands of conductance levels in memristors integrated on CMOS," *Nature* **615**, 823–829.

Schrödinger, E. [1948] *What is Life? Aspect of the Living Cell & Mind and Matter* (Cambridge University Press).

Smale, S. [1974] "A mathematical model of two cells via Turing's equation," Lectures in Applied Mathematics, Vol. 6 (American Mathematical Society), pp. 15–26.

Song, W., Rao, M., Li, Y., Li, C., Zhuo, Y., Cai, F., Wu, M., Yin, W., Li, Z., Wei, Q. *et al.* [2024] "Programming memristor arrays with arbitrarily high precision for analog computing," *Science* **383**, 903–910.

Strukov, D. B., Snider, G. S., Stewart, D. R. & Williams, R. S. [2008] "The missing memristor found," *Nature* **453**, 80–83.

Tour, M. M. & He, J. H. T. [2008] "The fourth element," *Nature* **453**, 12–43.

Turing, A. M. [1952] "The chemical basis of morphogenesis," *Philos. Trans. R. Soc. Lond. B* **237**, 37–72.

Volkov, A. G. [2018] "Memristors and electrical memory in plants," in *Memory and Learning in Plants* (Springer), pp. 139–161.

Williams, R. S. [2008] "How we found the missing memristor," *IEEE Spectr.* **45**, 28–35.

Williams, R. S. [2017] "What's new?" *Comput. Sci. Eng.* **19**, 7–13.

Zhang, X., Wu, Z. & Chua, L. [2020] "Hearts are poised near the edge of chaos," *Int. J. Bifurcation and Chaos* **30**, 2030023.

https://doi.org/10.1142/9789819830633_0002

Chapter 2

Standing on the Shoulders of Galileo, Newton, and Poincaré: A Brief Reasoned History of the Chaotic Dynamical Systems and the People Who Shaped Them[#]

René Lozi
Laboratory J. A. Dieudonné, CNRS,
Université Côte d'Azur, Nice 06108, France
Rene.Lozi@univ-cotedazur.fr; rlozi@unice.fr

This paper briefly describes the history of chaotic dynamical systems since Henri Poincaré's original work on celestial mechanics and more particularly on the three-body problem in the 1880s–1890s. Three periods are distinguished: the first period "the dawn of chaotic dynamical systems" from 1885 to 1970 which includes the Kolmogorov–Arnold–Moser theorem on the persistence of quasi-periodic motions under small perturbations extended by Moser in 1962, with its proof achieved in 1963 by Arnold, the discovery of the Lorenz attractor in 1963, the Sharkovsky order (1964), the Shilnikov bifurcation (1965), the horseshoe and the fractals in 1967. The second period "the intermediate period 1971–1989", begins with the definition of the strange attractors by Ruelle and Takens, and ends just before the introduction of the synchronization of chaotic systems in 1990 by Pecora and Carroll. The third period extends from 1990 to the present day, which is a flourishing time of applications in several sciences of this mathematical theory. This period corresponds to the publication of 35 volumes (415 issues) of the *International Journal of Bifurcation and Chaos in Applied Sciences and Engineering*, since the 35 years of existence of this journal. This history of chaotic dynamical systems is presented in a reasoned manner, which provides some insight into the driving forces behind their evolution. The people who shaped them are also featured. An epistemological analysis structures the analysis of these three periods. It takes into account the notions of causality, determinism, predictability, unpredictability and control.

Keywords: Chaos; dynamical systems; attractor; bifurcation; history; epistemology; causality; determinism; predictability; unpredictability; control.

1. Introduction

Thirty-five years have passed since the publication of the first issue of the first volume of the *International Journal of Bifurcation and Chaos in Applied Sciences and Engineering*. Since the creation of this journal, 35 Volumes (415 issues) have been published under the guidance of two editors-in-chief, Leon O. Chua and Guanrong (Ron) Chen. The number of issues published in each annual volume has grown steadily: four issues in the first two years, 6 in the following three years, 12 from 1996 to 2014, 14 from 2015 to 2019, and 16 since 2020. This underscores the importance of this journal, which covers chaos theory and nonlinear science across a wide range of applied science and engineering fields, in promoting this topic to a broad community of researchers. This journal encompasses experimental, computational, and theoretical aspects of bifurcations and chaos as well as complexity, memristors, Chaotic Neural Networks (CNNs), nonlinear

[#]To cite this article, please refer to its earlier version published in the *International Journal of Bifurcation and Chaos*, Vol. 35, No. 14 (2025) 2540002 (46 pages), DOI: 10.1142/S0218127425400024.

dynamics, complex network dynamics, chaotic time series, chaotic spreading, chaotic encryption, chaotic secure communications, chaotic signal detection, chaotic vibration, nonlinear mechanics, nonlinear spatiotemporal systems, pattern formation, nonlinear circuits, bio-system dynamics, system multistabilities, emergence, and AI techniques for any of the above.

With this in mind, I have attempted, in this paper, to briefly describe the history of chaotic dynamical systems[1] and, where possible, some elements of the lives of the people involved in them. Due to its limited length, this paper does not allowed me to expand as much as I would like on the contributions of all the outstanding scientists who have helped make chaos research so important today ("total history [of dynamical systems] looms as a tempting, but obviously unachievable goal" [Aubin & Dahan Dalmedico, 2002]). It does, however, recall some of the historical origins of this discipline and show the directions in which it is evolving, as well as the reasons behind it. It is why I added the word "reasoned" to the title.

The history of chaotic dynamical systems has already been described in several articles such as the very detailed article written by [Holmes, 2007] for the period 1885–2007 or the one he published in this journal [Holmes, 2005], or the less detailed one but intended for a wider and less specialized audience by [Oestreicher, 2007], or the partial points of view of [Sharkovsky, 2014] on the history of One-Dimensional (1D) dynamics, and of [Coullet & Pomeau, 2016] from a French (physicist) perspective, and the story of the discovery of the first chaotic attractor by [Ueda, 1978, 1992]. Aubin and Dahan Dalmedico [2002] after reviewing the historiography of dynamical systems theory highlighted the pioneering work of a few individuals (Steve Smale, Edward Lorenz, and David Ruelle) in their 67-page very comprehensive epistemological article. Alongside these historical articles, there are recollections from the mathematicians' personal lives about what led them to study dynamical systems [Smale, 1980; Martynyuk, 2014; Lozi *et al.*, 2025; Letellier *et al.*, 2021]. Hence, it seems difficult to write a better article on the subject. However, the detailed histories by Holmes, Aubin, and Dahan Dalmedico were published 20 years ago. Over the past two decades, both theory and applications have advanced considerably. Moreover, these histories focus on mathematics and partly on physics. It is necessary to consider the history of dynamical systems from a more general perspective, not only to describe the evolution of the theory, but also to provide some insight into the driving forces behind this evolution, which is what I will attempt to do.

It seemed to me that for the title of this paper I could parody the expression used by Isaac Newton in 1675, whose work is the first basis of dynamical systems, in his letter "if I have seen further [than others], it is by standing on the shoulders of giants" [Newton, 1675]. This expression seems based on the concept "*nos sumus quasi nani gigantum humeris insidentes — we are like dwarfs sitting on the shoulders of giants.*" (which has been dated to the 12th century and attributed to Bernard of Chartres according to John of Salisbury his disciple [Spina, 2004; Salisbury, 1991]). However, as it is customary today to use English, unlike the scientific usage of the Middle Ages using Latin, I have refrained from using this language for the title of this paper which could have been "*Stantes super Humeris Galilei, Newtoni, et Poincaré*: *Brevis et Rationata Historia Systematum Dynamicorum Chaoticorum et Virorum Qui Ea Formaverunt*"!

In Sec. 2, I emphasized in a few preliminary words the importance of the IJBC for me since its creation and my scientific links with the two editors-in-chief during the 35 years of existence of this journal. In the Sec. 3, I searched for an adequate definition of dynamical systems in accordance with the IJBC editorial line that would allow me to write a reasonably long article. In Sec. 4, I introduced a model which helps to understand the evolution of the theory and applications of dynamical systems and provides some insight into the driving forces behind this evolution. In Sec. 5, it is explained that dynamical systems follow not only their internal problematic but also the dual streams (theory and applications) which guide the development of mathematics. The history of chaotic dynamical systems is developed in the following three sections: in Sec. 6, for the period 1885–1970, in Sec. 7, for 1971–1989, and in Sec. 8, for the last period from 1990 to today characterized by a flourishing era of applications. The conclusions are given in Sec. 9.

[1] Aubin and Dahan Dalmedico [2002] remark that mathematicians prefer to call this scientific domain "dynamical systems theory", while others, including most physicists still talk about "deterministic chaos systems".

2. Preliminary Words

When I received the invitation to contribute to the special issue of the *International Journal of Bifurcation and Chaos in Applied Sciences and Engineering* (IJBC) from the Editor-in-Chief, Professor Guanrong Chen, an old memory of a series of discussions between 1988 and 1991 with Leon O. Chua, about the establishment of this journal suddenly came to mind (see Fig. 1).

Fig. 1. Leon O. Chua and Guanrong Chen in 2012.

Chua was the founder and first Editor-in-Chief of this journal. He was very proud of this journal allowing color picture with no charges, and glossy paper. I first met him at a conference in Issy-les-Moulineaux, near Paris, on December 3 and 4, 1986, about five years before the publication of the first issue of the first volume of IJBC, dated March 1991. During this conference, I discovered the Chua's circuit. One year after my first encounter, on February 24, 1988, as I was privately traveling in California, Chua invited me to give a talk in his Laboratory at University of California, Berkeley, and to spend one month there in November of the same year. At this time, I was working with Shigehiro Ushiki from Kyoto University on slow–fast Ordinary Differential Equations (ODEs) [Ushiki & Lozi, 1987]. With Ushiki, we first tried to use directly our method of *confinor* and *anti-confinor*[2] of slow–fast equations to Chua's system, and found that it was not adequate. We then adapted this method using half-Poincaré map, isochronic lines and selecting curves and published the result in the first issue of the first volume of IJBC [Lozi & Ushiki, 1991a] (see Fig. 2). It was a lengthy paper using these mathematical tools created especially for the Chua's equation (see next sections). The idea of isochronic lines

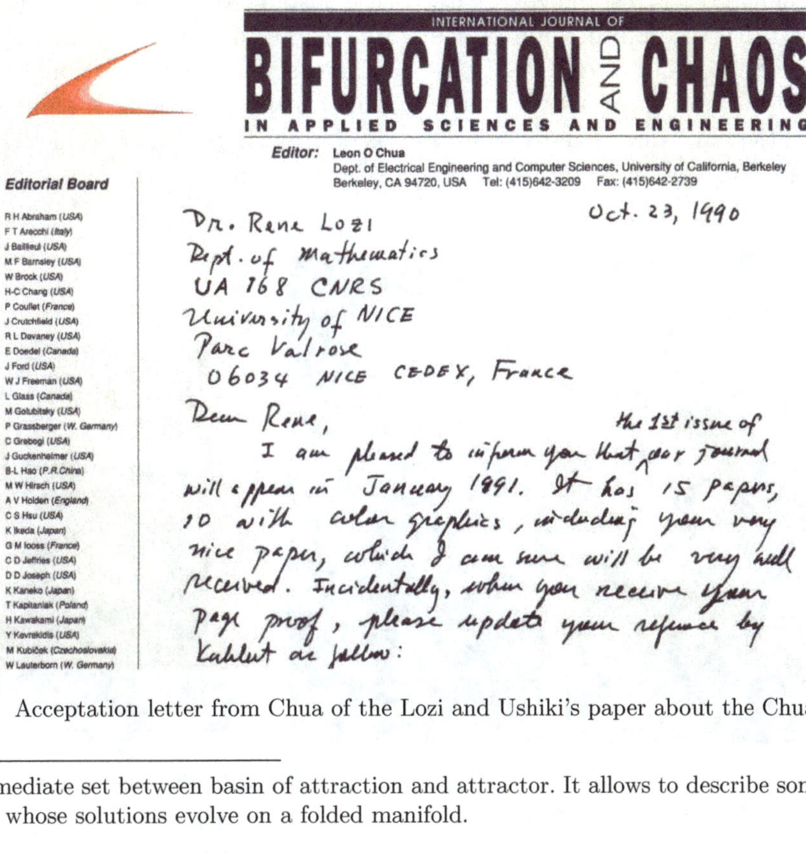

INTERNATIONAL JOURNAL OF

BIFURCATION AND CHAOS

IN APPLIED SCIENCES AND ENGINEERING

Editor: Leon O Chua
Dept. of Electrical Engineering and Computer Sciences, University of California, Berkeley
Berkeley, CA 94720, USA Tel: (415)642-3209 Fax: (415)642-2739

Oct. 23, 1990

Dr. Rene Lozi
Dept. of Mathematics
UA 168 CNRS
University of NICE
Parc Valrose
06034 NICE CEDEX, France

Dear Rene,

I am pleased to inform you that the 1st issue of our journal will appear in January 1991. It has 15 papers, 10 with color graphics, including your very nice paper, which I am sure will be very well received. Incidentally, when you receive your page proof, please update your reference by Kahlert as follow:

Fig. 2. Acceptation letter from Chua of the Lozi and Ushiki's paper about the Chua circuit.

[2]A confinor is an intermediate set between basin of attraction and attractor. It allows to describe some patterns of oscillation for slow–fast equations whose solutions evolve on a folded manifold.

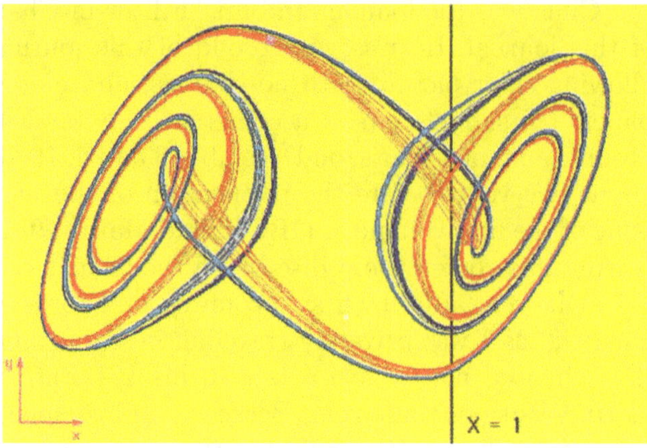

Fig. 3. Three coexisting Chua attractors.

was suggested to me by Claus Kahlert that I met in Chua's laboratory during the one-month stay in November 1988. He had previously introduced the curves of equal mapping duration [Kahlert & Chua, 1987]. Half-Poincaré map was introduced by [Uehleke & Rössler, 1983] and selecting curves by [Kahlert & Rössler, 1987]. Kahlert from Tübingen was a former student of Rössler at the Institute of Physical and Theoretical Chemistry. He was Humboldt Fellow for two years in Berkeley.

I discussed many times between 1988 and 1990 with Chua (who visited me in August 1989), and I had an extensive exchange of correspondence with him about this paper. After the publication of the first issues of IJBC, on September 5–7, 1991, Chua visited me one more time and gave a talk in the Institute of Mathematics in Nice. I particularly remember a discussion that took place on Saturday the 7th, when, while driving him to visit a tourist spot on the French Riviera, near Antibes (my home town), he told me how happy he was with the launch of this Journal.

The same year with Ushiki, we discovered a special value of parameters ($\alpha = 15.60$, $\beta = 28.58$, $m_0 = -1/7$, $m_1 = 2/7$) for which there is coexistence of three distinct Chua attractors (see Fig. 3). This new phenomenon may have escaped earlier detection because the three attractors are located very close to each other and require the "zooming" ability of our confinor approach to uncover them. The result was published in the Issue 4 of the first volume [Lozi & Ushiki, 1991b]. At this epoch, the computers were not as well-performing as they are today, and we were excited to create a picture using the *ray tracing* technique. For the Issue 2 of Volume 3, Chua chose one of our pictures showing a part of a "solid" Chua attractor composed of tubes and balls (see Fig. 4). In this issue, we published an accurate analysis of bifurcations and attractors of his circuit [Lozi & Ushiki, 1993], and in 2000, with a former PhD student Soraya Boughaba (now full professor at the University Frères Mentouri, Constantine, Algeria) — a method based on isochronic lines, for fitting trapping regions for Chua's attractors [Boughaba & Lozi, 2000]. During the last 35 years, I published many other articles in this journal with both Leon Chua and Guanrong Chen (see following sections). That is why writing this history for the 35th anniversary of this journal is of great importance to me. Of course, this history of chaotic dynamical systems that I write in this

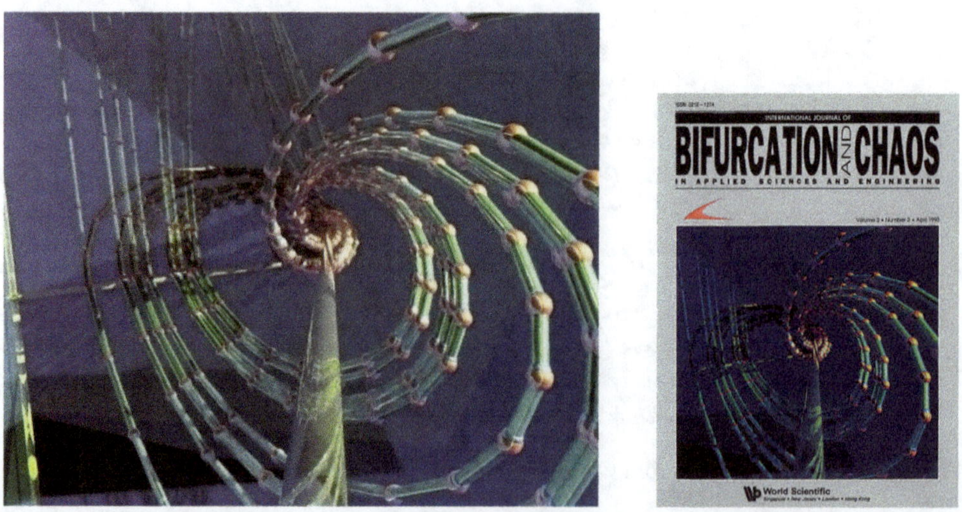

Fig. 4. Left: A part of a "solid" Chua attractor plotted composed of tubes and balls using the *ray tracing* technique. Right: Cover of Issue 2 of Volume 3 of IJBC displaying this image.

paper is not objective, but will be colored by my personal point of view and guided by the encounters I had with the researchers I met in my academic life.

3. In Search of an Adequate Definition of Dynamical Systems

The field of dynamical systems comprises several major disciplines, such as arithmetic and non-Archimedean dynamical systems, dynamical systems on Berkovich spaces, dynamics of holomorphic maps, Teichmüller flows and algebraic geometry, solitons, noncommutative dynamical systems, celestial mechanics and differential equations, etc. The terms "dynamical systems" are spread in more than 130 subsections of the Mathematics Subject Classification (MSC) 2020 of the American Mathematical Society. It is therefore an unreasonable challenge to claim to write a history of dynamical systems, even reduced to chaotic ones. Having studied (chaotic) dynamical systems for many years, sometimes, looking at the titles of articles by some academics working in this field, I am surprised that I have no idea what their research area is. As a preliminary remark on his short history of dynamical system, Holmes [2007] said that ""chaos theory" lacks the status of, say, the quantum or relativity theories, and that "nonlinear science" is not a science in the manner of physics, chemistry, or biology. Dynamical systems theory neither addresses specific phenomena nor proposes particular models of (parts of) reality; it is, rather, a loosely related set of methods for analyzing ODEs and iterated mappings, the canonical problems addressed having the forms

$$x_j(n+1) = F_j(x_1(n), \ldots, x_p(n); \mu_1, \ldots, \mu_q) \quad \text{or}$$
$$\dot{x}_j = f_j(x_1, x_2, \ldots, x_p; \mu_1, \ldots, \mu_q), \tag{1}$$

where $x_1, x_2, \ldots, x_p$ are *state variables* and $\mu_1, \ldots, \mu_q$ are *external control parameters*, usually regarded as fixed for the purpose of solving (1) to obtain *orbits* $\mathbf{x}(t) = (x_1(t), x_2(t), \ldots, x_n(t))$ or $\{\mathbf{x}(l)\}_{l=0}^{\infty}$".

In search of an adequate definition of what I would like to describe the history of, and one that is consistent with the themes addressed by the IJBC, I feel comfortable with the definition given to this field by Holmes in this paper: "Dynamical systems theory (also known as nonlinear dynamics or chaos theory) comprises a broad range of analytical, geometrical, topological, and numerical methods for analyzing differential equations and iterated mappings". The methods used in the qualitative theory of dynamical systems emphasize the study of *all* orbits or solution curves of (1), the dependence of the set of solutions or *phase portrait* on the parameters, and the description of *qualitative* properties such as the existence of periodic solutions, rather than derivation of explicit closed form expressions or approximations. Such theory is largely built on the pillars of analysis, geometry, and topology, the first and last of which, in turn, had their origins in Newtonian mechanics.

A less strictly mathematical definition I could use, might be the one given by [Oestreicher, 2007] "A dynamical system is a system that changes over time in both a causal and a deterministic manner, i.e., its future depends only on phenomena from its past and its present (causality) and each given initial condition will lead to only one given later state of the system (determinism). Systems that are noisy or stochastic, in the sense of showing randomness, are not dynamical systems, and the probability theory is the one to apply to their analysis".

Such a definition seems clear to any scientist and even to a nonscientist, but it is rooted in the fact that everyone accepts that we can predict, at least to some extent, the future of physical events using scientific laws. If we know which station a train is at, we can predict which station it will be at after a certain time provided we know its average speed and the distance between these stations. However, this conception of things that we call *determinism* is relatively recent, dating back only four centuries. Without determinism, no theory of dynamical systems can exist! Determinism is the root of predictability and therefore, a dynamical system. Moreover, determinism is based on causality.

4. Causality, Determinism, Predictability, Unpredictability, and Control

I introduce a model which will help to understand the evolution of the theory and applications of dynamical systems over their 150-year history, and also to provide some insight into the driving forces behind this evolution. This model is based on causality, determinism, predictability or unpredictability and control. It is described in Fig. 5 and will be explained in the following sections.

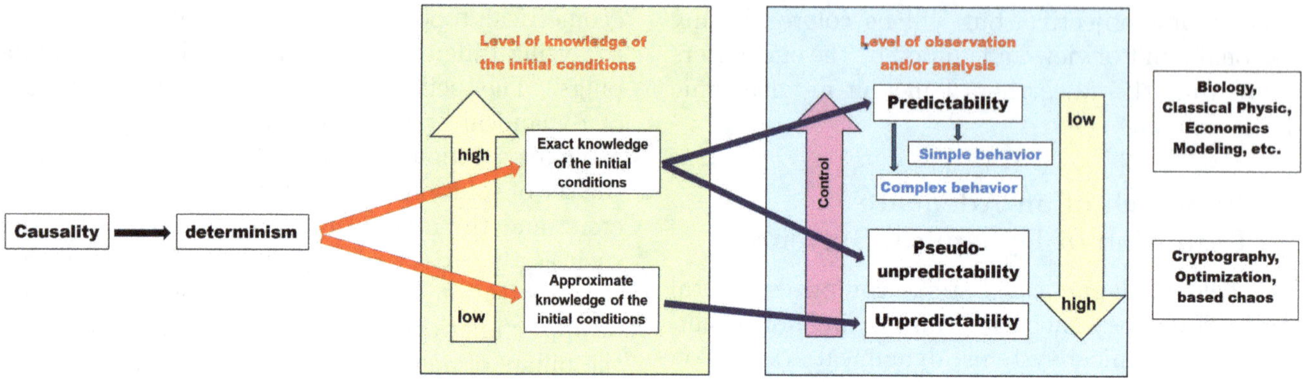

Fig. 5. Framework of the theory and applications of dynamic systems.

4.1. *Causal determinism*

It is important to go back to the 16th century with Galileo to question the notion of determinism (and beyond, the foundations of dynamical systems) that emerged with him and Newton. As [Ghys, 2023] points out in his book on chaos theory, these two geniuses introduced this notion what can be called one of the greatest scientific ideas. However, many philosophers still doubt nowadays this incredible idea: why would the present determine the future? Following [Bunge, 1959], certain thinkers, whom he calls romantic, have denied that the causal relation connects one event with another event; and they have done so on what he considers the correct grounds that every event is connected in innumerable ways with all sorts of other events and therefore that an event is not simply nicely and neatly connected with another as, say, its cause. Nevertheless, prediction is at the heart of science. Galileo was one of the first to understand this concept, climbing to the top of the Tower of Pisa, according to legend, to experiment with his ideas. The astronomer predicts, for example, the positions of the planets at any time in the future. Newton was the first to be able to make such a prediction and to understand, through calculation, why, how and how fast the planets revolve around the Sun, where Kepler had only observed this movement. Classical physics, that is to say physics which is not concerned by quantum mechanics, is based on the notion of predictability. If we know all the characteristics of a physical system with infinite precision, we can predict its evolution. But no matter how accurate the measurements are, there will always be small errors in the measurements. Before the development of Chaos theory, it was thought that these errors did not greatly influence predictions. The discovery of strange attractors (now called chaotic attractors) popularized by the "butterfly effect" changed the point of view. Small causes can have very big effects.

At this point, it is important to be more precise about determinism and its relationship with causality.[3] Oestreicher [2007] found a direct derivation of what he called "perhaps the most basic of all principles of physics" to the philosophy of René Descartes in his 1641 Third Meditation "The principle of causality is nonrefutable, i.e., not confirmed by experience, since it is an axiom that precedes experiences. For example, this principle is accepted a priori in physics. In a simple form, it reads: 'Every effect has a cause'" [Descartes, 1641, 1988]. Oestreicher highlighted that in 1687 Isaac Newton consolidated the causality principle by asserting that the two concepts of initial conditions and law of motion had to be considered separately [Newton, 1687, 1972].

If we refer to [Hoefer, 2024] in the Stanford Encyclopedia of Philosophy it might be more accurate to speak about causal determinism: "Causal determinism is, roughly speaking, the idea that every event is necessitated by antecedent events and conditions together with the laws of nature. The idea is ancient, but first became subject to clarification and mathematical analysis in the eighteenth

[3]It is sometimes difficult to distinguish between causality and causation. I will adopt the definition of [Sawesi *et al.*, 2022] who provide a general synopsis on how "causality" and "causation" are conceptualized and ontologically represented. Causality: "a general concept is that there is such a thing as a cause-effect relationship between entities, in which the cause makes a difference to the occurrence of the effect" and causation "a process that connects causes to their effects".

century". Hoefer adopts the following definition: "Determinism is true of the world if and only if, given a specified way things are at a time t, the way things go thereafter is fixed as a matter of natural law". The debate among philosophers over the relationship between causality, determinism, and predictability is not over. Pursuing this philosophical debate is well beyond the scope of this paper, but I will explore this relationship from the perspective of chaotic dynamical systems, following the assumptions of our framework. However, this perspective is not so far removed from philosophical concerns since in his article [Hoefer, 2024] explores the links between determinism and chaos: "The popularization of chaos theory in the relatively recent past perhaps made it seem self-evident that nature is full of genuinely chaotic systems. In fact, it is far from self-evident that such systems exist, other than in an approximate sense. Nevertheless, the mathematical exploration of chaos in dynamical systems helps us to understand some of the pitfalls that may attend our efforts to know whether our world is genuinely deterministic or not". I could summarize the difference between causality and determinism by saying that *causality* is the principle that an effect has a cause, and *determinism* is that the effect must follow from that cause, and refer the reader to the book of [Bunge, 1959] *causality* for more in-deep analysis.

4.2. *Predictability and unpredictability*

Predicting the future has been a fundamental human desire since the dawn of time. In ancient times, these predictions were based on the observation of natural phenomena, shape of clouds, flight of flocks of birds, etc. Concerned about their destiny, the Romans took divination very seriously. Endowed with great influence, augurs, haruspices inspecting the entrails of sacrificed animals, and soothsayers intervened in everything from everyday concerns to wars and public affairs. This desire to know the future was the basis of the observation of the planets and therefore of astronomy. With the birth of astronomy (independently in several civilizations) comes the beginning of the scientific method. Claudius Ptolemy (c. 100–160s/170s AD) introduced the mathematically sound geocentric model of the solar system in his book *Almagest* [Toomer, 1984]. More precisely, the accurate predictions allowed by this model were not based on physical laws. This situation continued until Newton's publications. Since then, predictions and predictability were based on causality and determinism. However, predictability is based on two mechanisms: on the one hand, the level of observation or prediction and on the other hand, the level of knowledge of the initial conditions of the system considered. As an example of the importance of both mechanisms, we can consider the movement of the Earth around the Sun for which the level of observation/prediction is the duration considered and the initial conditions are the orbital position and speed of the Earth and the other planets of the inner solar system. Since Galileo, who considered the Copernican system to be real, this movement has been known with great precision over periods of a few thousand years (previously, on the contrary, it was thought for millennia that the Sun revolved around the Earth [Abubakr, 1955] and it was even a dogma for certain religions). But is it known with the same precision over periods of several hundred million years? This was indirectly the question posed for the international prize competition sponsored by Oscar II, King of Sweden and Norway, to mark his 60th birthday on January 21, 1889. More specifically, the question posed in 1885, among three others, was "For a system of arbitrarily many mass points that attract each other according to Newton's law, assuming that no two points ever collide, find a series expansion of the coordinates of each point in known functions of time converging uniformly for any period of time". The prize jury consisted only of three members: Mittag-Leffler, acting as administrative and coordinative liaison with his mentors and friends Karl Weierstrass in Berlin and Charles Hermite in Paris. "Weierstrass nourished great hopes that there would be a method of solution based on a simple fundamental idea, and that such a solution could help conclude whether the system was stable. In applying this formulation of the n-body problem as a model of the solar system, could it be excluded that the planetary orbits might alter radically over a long time, or even that one planet be thrown out of the system?" [Rågstedt, 1976]. Finally, Poincaré [1890] discovered that due to the imprecision of the initial conditions of the planets known at that time (position and velocity) in the solar system, the solution was not predictable for a very long time. A century later Lachkar [1989] confirmed this result using accurate numerical computations.[4] Therefore, the Earth's orbit can be considered predictable or unpredictable depending on the length of the prediction considered and the level of accuracy of the initial conditions.

4.3. *Randomness and pseudo-unpredictability*

If predicting the future has been a fundamental human hope since the dawn of time, accepting chance or even provoking it (that is, being unable to predict the future) has always been an important element of human societies. These two opposing ways of seeing the future are not contradictory, because they depend on the situations in which people find themselves. On the one hand, men have often wanted to know their destiny, but they have also often been attracted to games of chance. Numerous Roman emperors, such as Augustus and Claudius were enthusiasts in gambling. Claudius wrote a now-lost book about playing dice [Suetonius, 2016]. We can consider that randomness is equivalent to unpredictability for a physical system. Indeed, traditionally, it is considered that tossing a coin produces a random result because it is impossible to predict whether the coin will land on heads or tails.[5] The growing need to produce long series of random numbers for online games or cryptography has led to the consideration of chaotic dynamical systems as a possible way to achieve this. However, a distinction should be made between true random numbers produced by physical means and Pseudo-Random Numbers (PRN) obtained through simple mathematical formulas. This is why I introduced in Fig. 5 the neologism pseudo-unpredictability, which means that chaotic dynamical systems can produce PRNs mimicking true random numbers [Lozi, 2011]. The chaos-based PRN Generators (PRNG) use exact initial conditions. It is also the case for [Daza *et al.*, 2024] who consider multistability as a possible mechanism of unpredictability. They give as an example the decision-making pendulum toy, a magnetic pendulum, which has three possible outcomes: yes, no, and maybe. "The complex interplay between the initial conditions and the multiple ending positions makes the pendulum a multistable system [...] The final answer given by the pendulum may seem random, but the system can be modeled by three basins of attraction [...] Knowledge of the initial release point will inform where the resting point will be". In conclusion, unpredictability of pseudo-unpredictability can be the result of approximate or exact knowledge of initial conditions (in the case of PRN, the use of electronic devices such as a micromechanical resonator to produce them is done with approximate initial conditions [Defoort *et al.*, 2021]).

4.4. *Control*

So far, we have considered three paths in Fig. 5, represented by the three blue arrows. The last question is therefore: can we define a path linking approximate knowledge of the initial condition to predictability? In fact, this can be achieved using control theory. One of the first examples of control of a mechanism is the famous centrifugal regulators invented by Christiaan Huygens and used to regulate the distance and pressure between millstones in windmills in the 17th century, and adapted in 1788 by James Watt to control his steam engine where it regulates the admission of steam into the cylinder(s) [Hannavy, 2021]. Since then, control theory has become a separate field of physics and mathematics [Zabczyk, 2008]. This theory was first adapted to chaotic dynamical systems by [Ott *et al.*, 1990] using what is now called the Ott–Grebogi–Yorke (OGY) method.

Any chaotic attractor contains an infinite number of unstable, periodic orbits. Control of chaos is the stabilization, by means of small system perturbations, of one of these Unstable Periodic Orbits (UPOs). In the OGY method, small, wisely chosen kicks are applied to the system once per cycle to maintain it near the desired UPO. Control of chaos is now routinely used for many chaotic systems, not only in physics but also in biology, economy, industry, etc. [Boccaletti *et al.*, 2000].

4.5. *The statistical point of view*

Ghys wanted to demonstrate in his book that the probability of these significant "butterfly" effects appearing in chaotic systems is negligible from a

[4] "The motion of the solar system is thus shown to be chaotic, not quasi-periodic. In particular, predictability of the orbits of the inner planets, including the Earth, is lost within a few tens of millions of years" [Lachkar, 1989].

[5] This is a widely used example to introduce the study of statistics and probability, however, some recent results suggest a possible bias [Bartoš *et al.*, 2025].

statistical point of view. The flapping of a butterfly's wings can change the timing of a hurricane, but not their total number over the course of a year. The future, although sensitive to initial conditions, could perhaps be statistically insensitive to them. He thus fought the idea of "catastrophism" that a poor understanding of chaotic dynamical systems can generate among the public. Therefore, he recalled how a probabilistic point of view was introduced by [Laplace, 1814] in his "Philosophical essay on probability". It is from this period that determinism will be gradually replaced by probability theory [Kahane, 2008]. I will leave it to Ghys to develop this point of view, which is not to be rejected, but I will approach the following sections by returning to the history of dynamical systems as I circumscribed them in Sec. 3. It is nevertheless necessary to mention the use of probabilistic concepts mobilized on several occasions by Poincaré who, from the point of view of [Aubin & Dahan Dalmedico, 2002] "ought to be included among his basic innovations in the theory of dynamical systems, since they have become essential for the understanding of chaotic phenomena. In the course of his study of the stability of trajectories, he showed that there were an infinite number of particular solutions, which were unstable in the sense of Poisson, but *exceptional* in the sense of probability theory (i.e. of measure zero)".

5. Theory and Application: The Dual Currents that Drive the Development of Mathematics

In Fig. 5, I presented a model explaining the development of dynamical systems. I will detail it in the following sections. This model is specific to our topic. However, dynamical systems, from my point of view, belong to mathematics and their development follows its own evolution. For millennia, this development has been driven by two currents: theory and application. It is also necessary to take this evolution into account, which encompasses the proposed model. Metaphorically, the development of dynamical systems can be compared to the evolution of galaxies under gravity (evolution following the model), which combines with the expansion of the universe (evolution of mathematics).

Humanity has been using mathematics for at least 5000 years, if not longer, to record time and establish calendars, as well as for trade and taxation. The Akkadian Empire and the city-states of Sumer and Assyria in Mesopotamia began using arithmetic, algebra, and geometry around 3000 BC. The Levantine state of Ebla and ancient Egypt closely followed them [Friberg, 1986]. The oldest mathematical texts we have discovered come from these regions: in Mesopotamia, the ancient Babylonian clay tablet commonly known as Plimpton 322, circa 2000–1900 BC [Friberg, 1981]; and in Egypt, the Moscow Mathematical Papyrus, circa 1890 BC, and the Rhind Mathematical Papyrus, circa 1800 BC [Neugebauer, 1969]. Later, in the sixth century BC, the Greeks introduced mathematics as a "demonstrative discipline" [Heath, 1931] (see [Hoyrup, 2011)] for a comparison between the two approaches). Since its emergence in human history, the research and development of mathematics have been guided by two distinct streams: theory and applications, in other words, by beauty and utility. These dual currents of research have been in force since antiquity and still operate today in a competitive–cooperative mode.

For example, the great German mathematician Karl Gustav Jacobi (1804–1851), in a letter to the French mathematician Adrien-Marie Legendre, wrote these memorable words "[...] Fourier believed that the main goals of mathematics were to serve the common good and to explain natural phenomena. However, a philosopher like him should have known that science is only intended for the *honor of the human spirit.* Therefore, a question about numbers is as important as a question about the world" [Mashaal, 2017]. Jean Alexandre Dieudonné was my Professor at the university of Nice (France) and was one of the founding members of the Bourbaki group. He took Jacobi's phrase as his motto, changing the word "science" by "mathematics", and published a book with that title [Dieudonné, 1987].

It is widely accepted that the theory of dynamical systems originated in [Poincaré, 1890]'s work on celestial mechanics and specifically on the three-body problem first published in his (initially flawed) paper and expanded after in his book (see Sec. 4.2). The methods he has developed therein laid the basis for the local and global analysis of nonlinear differential equations, including many mathematical tools still used nowadays. Even if a tricky astronomical problem for the epoch (will the Earth continue to orbit around the Sun forever?) is at the origin of his reflection, no practical application of his "Méthodes nouvelles de la mécanique céleste" has guided his mind. On the contrary, fifty years after, A. A. Andronov (1901–1952), a student of L. Mandelstam (1879–1944), established a strong group in dynamics in the USSR at Gorky (now, Nizhny Novgorod), and he and L. S. Pontryagin (1908–1988) introduced the key idea of *structural stability* under the name "systèmes grossiers" (coarse systems) [Andronov & Pontryagin, 1937]. This notion, now a central theme of the theory, asks what properties are necessary and sufficient for the qualitative behavior of the flow comprising all solutions of a given ODE to survive a small perturbation to the vector field defining it [Holmes, 2007]. Andronov *et al.* [1937, 1966]'s work was motivated by radio and electronic applications, and it led to further detailed studies of specific nonlinear oscillators, summarized in a classic text.

The same remark can also be made about two other mathematical objects of the "nonlinear science"[6] like chaos introduced at the same time: solitons and fractals.

In his "Early History of Solitons (Solitary Waves)" [Allen, 1998], Allen recalled that the story of solitons began, in the science of hydrodynamics, when in August 1834, John Scott Russell, a Victorian engineer inspecting the Union Canal, outside Edinburgh, observed "a most beautiful and extraordinary phenomenon", a shape-preserving wave that traveled miles ahead of the boat Russell was on, just after it had suddenly stopped. The definitive theory was not published until 1895 by [Korteweg & de Vries, 1895]. The subject reappeared in plasma physics in 1958 with the discovery by [Adlam & Allen, 1958] of solitary waves in a collisionless plasma containing a magnetic field. Zabusky and Kruskal [1965] introduced the term "soliton" to replace the term "solitary wave" when they discovered through numerical study that such waves retained their identity after colliding, exhibiting particle-like behavior. Nowadays, "Solitons have many applications in pure and applied mathematics, particularly in areas such as differential equations, Lie groups, Lie algebras, differential and algebraic geometry. [...] Equations that admit soliton solutions have profound mathematical structures and properties" [Manukure & Booker, 2021].

It is generally accepted that Benoît Mandelbrot coined the word "fractal" while working at IBM labs in Armonk, New York, in 1975 [Mandelbrot, 1975], although in the early 20th century, mathematicians Pierre Fatou and Gaston Julia discovered fractal patterns while looking at complex mathematical systems (see next section). It should be noted that from 1945 to 1947 Mandelbrot attended the École Polytechnique, where he studied under the direction of Gaston Julia (see Sec. 6). In 1967, he had already published his essay "How Long Is the Coast of Britain? Statistical Self-Similarity and Fractional Dimension", in which he argued that coastlines were "statistically self-similar" [Mandelbrot, 1967]. Later, he was the first to point out that fractals could be an ideal tool in applied mathematics for modeling a variety of phenomena ranging from physical objects to stock market behavior [Mandelbrot, 1997]. An important mathematical tool introduced and developed from Mandelbrot's work is the notion of *fractal dimension*, pivotal in the study of chaotic dynamical systems. Several formal mathematical definitions of different types of fractal dimension are used today, some of them being introduced before the term fractal was coined by Mandelbrot: box-counting dimension also known as Minkowski–Bouligand dimension [Bouligand, 1925; Pašić, 2003], information dimension (first introduced by [Rényi, 1959]), correlation dimension [Grassberger & Procaccia, 1983], Hausdorff dimension, Higuchi dimension [Higuchi, 1988], etc. They can be measured theoretically and empirically.

Thus, the three "pillars" of current nonlinear science: chaos, solitons and fractals have benefited from the synergy between theory and application to progress.

6. The Dawn of Chaotic Dynamical Systems 1885–1970

Having defined the framework within which I consider the qualitative theory of dynamic systems to evolve, it is possible to describe their history. In their presentation, Aubin and Dahan Dalmedico [2002] considered three periods: the first, a relatively long one, up to the 1960s, for their historic foundation, the second, centered on the three pioneering figures (Steve Smale, Edward Lorenz, and David Ruelle) "who in their field introduced elements of reconfiguration" and the third, beginning around 1974 until 2002, "introducing a rupture which gave rise to chaos". Modifying their plan slightly, I will consider the first period "the dawn

[6] A general view of nonlinear science is developed in the forthcoming book by [Sanjuán, 2025].

of chaotic dynamical systems" from 1885 to 1970 which includes in the last decade the proof of the Kolmogorov–Arnold–Moser (KAM) theorem about the persistence of quasi-periodic motions under small perturbations extended by Moser in 1962, with its proof achieved by [Arnold, 1963], the discovery of the Lorenz attractor in 1963, which seems to be of no interest in this period, the Sharkovsky order (1964), the Shilnikov bifurcation (1965), the horseshoe [Smale, 1967] and the fractals in 1967. I define the second period "the intermediate period 1971–1989", which begins with the definition of the strange attractors by [Ruelle & Takens, 1971], and finishes just before the introduction of the synchronization of chaotic systems by [Pecora & Carroll, 1990]. Then, I consider the third period from 1990 up to now. Since the first two periods being well documented in the chaos histories already published [Holmes, 2007; Oestreicher, 2007; Aubin & Dahan Dalmedico, 2002; Sharkovsky, 2014], I will only give an overview of them and develop the third further.

As said in the previous section, Poincaré was at the origin of this theory.[7] The methods he has developed in his paper [Poincaré, 1890] laid the foundations for qualitative analysis of nonlinear differential equations, and began to develop a coherent set of mathematical tools for their study. His paper, published in 1890 for the King Oscar II prize (see Sec. 4.2), described the use of first return (Poincaré) maps for the study of periodic motions; defined stable and unstable manifolds; discussed stability issues at length, developed perturbation methods, and included the (Poincaré) recurrence theorem. A global view of his work can be found in the three volumes [Poincaré, 1892, 1893, 1899].[8] He published later the first explicit example of what is now called chaos, creating a perturbation theory to detect it [Barrow-Green, 1997; Holmes, 1990]. Like his younger cousin Raymond Poincaré (President of the French Republic from 1913 to 1920), Henri Poincaré was not only a scientist, but also a famous public figure. As Fields Medalist 2010 Cédric Villani wrote in his preface to [Ginoux & Gérini, 2012]'s book **Henri Poincaré, Une Biographie au(x) Quotidien(s)**(Henri Poincaré, A Biography in Everyday Life): "Henri Poincaré left us a hundred years ago. He is often presented, and perhaps rightly so, as the last universal expert in mathematics–arguably the last to master all branches of mathematics, before the skyrocket rise of this discipline made this feat impossible. [...] But Poincaré was much more than that. Through his clear writings, which place him among the most important philosophers of the 20th century, he is also a symbol of the unity of human thought. [...] Thanks to his writings, and simply thanks to his way of being, Poincaré achieved the feat of making himself known to the general public without sacrificing his research in the least. During his lifetime, his philosophical writings enjoyed genuine popular success, making him a symbol of the integration of the scientific world into society". This book (in French) presents an original biography of this scholar, through press articles in numerous newspapers during his lifetime. The title of this book is a play on words that cannot be translated into English: au(x) quotidien(s) meaning both: in "everyday life" and "in newspapers".

Following his initial work, several researchers continued the research in this novel domain. We can cite among many Jacques Hadamard's studies of geodesic flows [Hadamard, 1898][9] and the pioneer works of Pierre Fatou (1878–1929) [Fatou, 1917] and Gaston Julia (1893–1978) [Julia, 1918] related to 1D maps with a complex variable (see Fig. 6), nearly a century ago. The results of Julia on the iteration of rational fractions (obtained simultaneously by Fatou) were brought to light in the 1970s by Mandelbrot, the Julia and Mandelbrot sets, being closely related [Douady, 1986].[10] The study of the frighteningly complicated solutions discovered by

[7] "Whatever conflicts have arisen between historiographical viewpoints, the recognition of Poincaré as the true founder and major theoretician of the domain has been unanimous" [Aubin & Dahan Dalmedico, 2002].

[8] Surveys of the immense literature on Poincaré's contribution can be found in [Chabert & Dahan Dalmenico, 1992; Bottazzini, 2000].

[9] Martin Zerner, one of my professors at the University of Nice, who introduced me in the numerical study of differential equations in 1971, was chosen, when working as "attaché de recherches" at CNRS, to help Jacques Hadamard to finalize the last books of his complete works in 1958–1959.

[10] Andrien Douady was one of the great French mathematicians of the last century. In a 1966 article, he definitively resolved a complex question in analytical geometry posed by Alexandre Grothendieck, which he began to study in his thesis under the supervision of Henri Cartan. His work led him to study the dynamics of complex polynomials and to extend the work of Pierre Fatou and Gaston Julia on iteration in the complex domain [Douady & Hubbard, 1985]. I met him when I was a student and he was Professor at the university of Nice and during several years after [Lozi, 2026].

Fig. 6. Example of Julia set.

Poincaré continued quietly for almost 80 years in several directions including conservative and dissipative dynamical systems, differential and difference equations. Among these studies, we can mention those of Christian Mira and Igor Gumowski, who began their mathematical research in 1958; those of [Birkhoff, 1927] [infinite sequence of periodic orbits whose periods approach infinity, near any homoclinic point of a Two-Dimensional (2D) map],[11] the strong group in dynamics developed by Aleksandr Aleksandrovich Andronov at Gorky in the USSR (for a recent and vivid overview of some of the activities of this group, see the testimonies of V. Belykh and L. Efremova [Lozi *et al.*, 2024a]), etc. (see Fig. 7), the continuous attractors of [Lorenz, 1963] (2)

$$\begin{cases} \dot{x} = \sigma(y - x), \\ \dot{y} = rx - y - xz, \\ \dot{z} = xy - bz, \end{cases} \tag{2}$$

(Fig. 8), and the fractals introduced in 1967 by Benoît Mandelbrot (1924–2010) (see previous section), etc. Also worth mentioning during this period is the work of Andrey Nikolaevich Kolmogorov from Moscow, who played a central role in the creation

Fig. 7. Henri Poincaré, Jacques Hadamard, George David Birkhoff, and Aleksandr Aleksandrovich Andronov.

of modern probability theory, but is also related to dynamical systems, and the first example of chaotic attractor for the Duffing equation, found by Ueda prior to the Lorenz attractor in November 1961, but only published in 1978 (see Sec. 7.2).

Gumowski and Mira [1980] have produced considerable work on iteration theory (they have

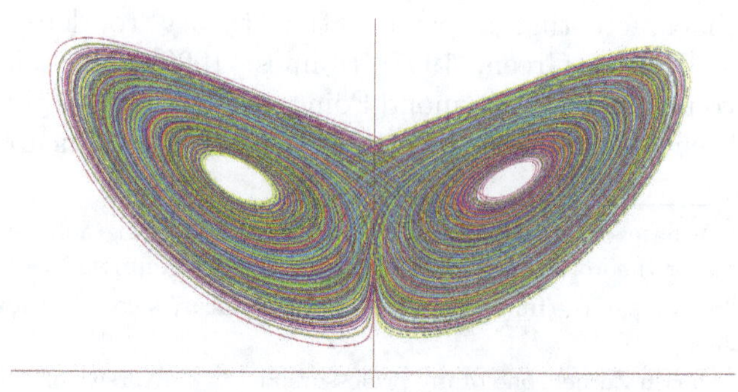

Fig. 8. Lorenz attractor for $\sigma = 10.2$, $b = 2.2$, and $r = 27$.

[11] At the beginning of his career [Birkhoff, 1913] had proved Poincaré's "last theorem" on fixed points of annulus maps. He subsequently proved that annulus maps having points of two distinct periods contain complex limit sets separating their domains of attraction Birkhoff [1932], thus providing a key clue for Cartwright and Littlewood in their studies of the van der Pol equation [Holmes, 2007].

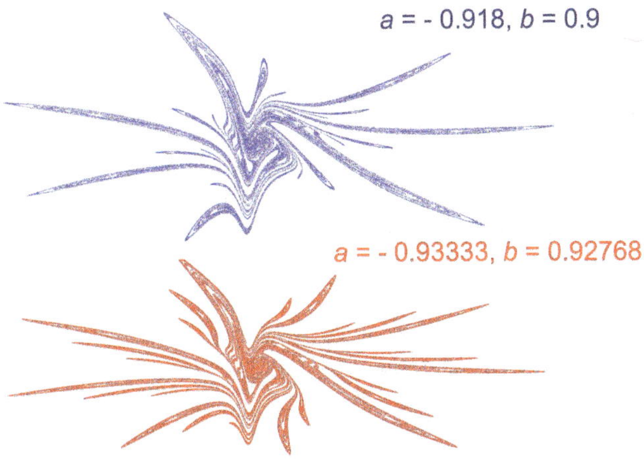

Fig. 9. Gumowski–Mira attractors for $a = -0.918$, $b = 0.9$ (blue) and $a = -0.93333$, $b = 0.92768$ (red).

invented the box-within-a-box theory, for example). Among their discoveries, one can emphasize on their family of attractors from an aesthetic point of view [the Gumowski–Mira map (3) in 1980]. The Gumowski–Mira attractors are sensitive to slight changes of parameters a and b (see Fig. 9), however, the idea of the butterfly effect was not in the minds of mathematicians at this time! (for more aesthetic images of attractors see [Mira, 1987])

$$\begin{cases} x_{n+1} = f(x_n) + by_n \\ \qquad \text{with } f(x_n) = ax + 2(1-a)\dfrac{x^2}{1+x^2}, \\ y_{n+1} = f(x_{n+1}) - x_n. \end{cases} \tag{3}$$

Mira had great admiration for Gaston Julia (see Fig. 10). He recounted his life in a journal dedicated to the French being forced to leave Algeria during the independence of this former French colony, Mira and Julia both having been born in North Africa (Sidi Bel Abbès, Algeria for Julia and Meknés, Morocco for Mira). The title of his article (in French) is "Gaston Julia, one of the greatest mathematicians of the 20th century" [Mira, 2003].

During this first period, special attention must be paid to the study of the logistic map, first motivated by the study of population dynamics. "In population dynamics, the study of the logistic equation, for instance, in the work of mathematicians like Vito Volterra, a zoologist like Umberto D'Ancona, and a statistician like Alfred J. Lotka is the starting point of a disciplinary lineage that can lead to population biologist Robert May, who figured prominently in the formation of a community of "chaologists" in the mid-1970 [...] From 1958 to 1963, Finnish mathematician P. J. Myrberg published a series of papers investigating the bifurcation properties of the quadratic map $x_{n+1} = x_n^2 - \lambda$ as λ went from $-1/4$ to 2. Myrberg was the first to study the limit cycles of this equation, thereby identifying the cascade of period-doubling bifurcations that have since made Feigenbaum's name famous (see Sec. 7.1). For the most part published in a Finnish journal, Myrberg's results received little notice. An article in the *Journal de Mathématiques Pures et Appliquées* [Myrberg, 1962] was, however, known to Mira" [Aubin & Dahan Dalmedico, 2002].

Concerning the school in dynamics developed by Andronov at Gorky (now Nizhny Novgorod,[12]

Pierre Fatou
1878-1929

Gaston Julia
1893-1978

Igor Gumowski
1928-2019

Christian Mira
1934-

Fig. 10. Pierre Fatou, Gaston Julia, Igor Gumowski, and Christian Mira.

[12]Between 1932 and 1990, the town was named Gorky, in honor of the famous Russian writer Maxim Gorky, who was born there in 1868.

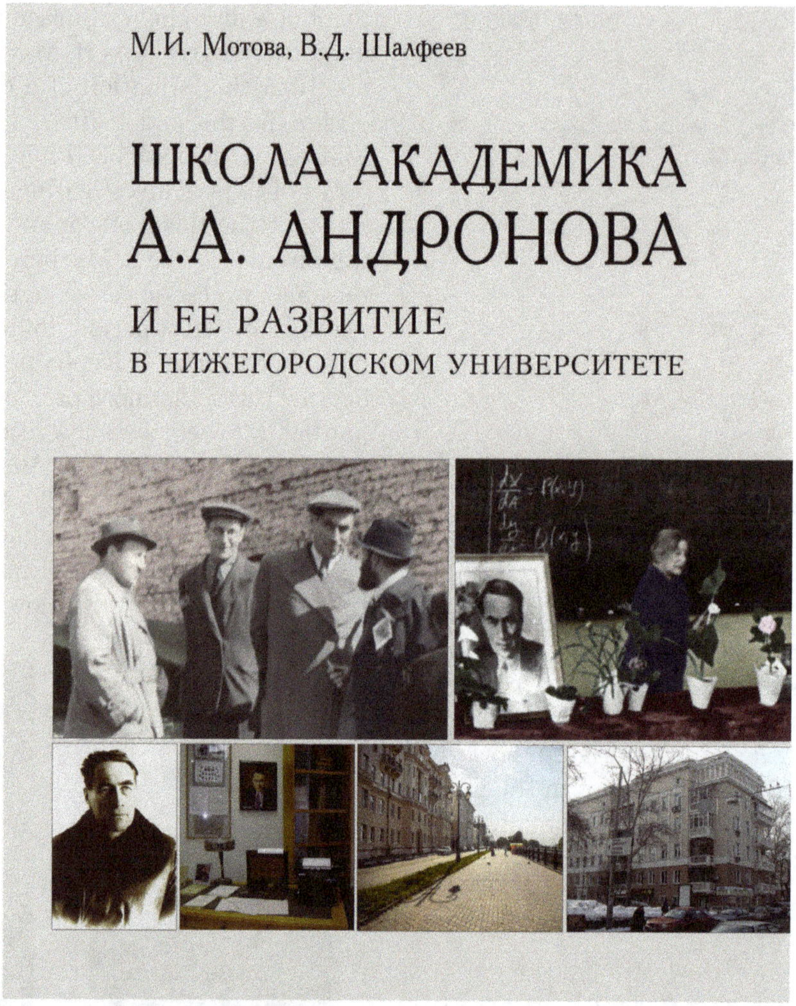

Fig. 11. Academic school Andronov. In the upper photo: left, Academician of the USSR Academy of Sciences Andronov (second from right) and Professor Mayer (first from right) with PhD students (Ya. Nikolaev and S. Bellustin), right, Professor Leontovich-Andronova presents a report on Andronov's mathematical works.

see Fig. 11) in the USSR, Vladimir Belykh in [Lozi *et al.*, 2024a] recalls that "I studied mathematics and physics in the Chair of Oscillation Theory and Automatic Control, named after its founding director, Alexander A. Andronov. I was fortunate to take classes from distinguished professors like Dmitry A. Gudkov, Yuri I. Neimark, Nikolai A. Fufaev, and Nikolai A. Zheleztsov, the brilliant team of Andronov's disciples. [...] in 1964, the University established a new large research unit, the Research Institute of Applied Mathematics & Cybernetics (NII PMK as abbreviated in Russian), directed by my future PhD adviser, Liudmila N. Belyustina. [...] I was accepted into the Division of Differential Equations headed by Evgeniya A. Andronova-Leontovich, Andronov's spouse and close collaborator and a sister of another celebrated Soviet physicist, Mikhail Leontovich of Moscow State University. [...] I was fortunate to meet Leonid Pavlovich Shilnikov (L. P., as friends called him), who worked in the same division. Over time, we became friends, and one day L. P. explained to me his famous theorem on the bifurcation of a homoclinic orbit of a saddle focus, which became a widely used criterion for the existence of chaotic dynamics in saddle-focus dynamical systems [Shil'nikov, 1965]".[13]

A remarkable result about the persistence of quasi-periodic motions under small perturbations the KAM theorem was initiated by [Kolmogorov,

[13]The Shilnikov bifurcation stands for the homoclinic bifurcation of a saddle-focus equilibrium state that elicits the onset of complex dynamics in a system.

1954] from Moscow, who made the first progress in solving this problem. It was then rigorously proved and extended by [Moser, 1962] (for smooth twist maps) and [Arnold, 1963] (for analytic Hamiltonian systems).

Aubin and Dahan Dalmedico [2002] said about this theorem: "the theorem presented by Kolmogorov at the Amsterdam International Congress of Mathematicians in 1954, and proved in the following decade by Arnold and Möser — the so-called KAM theorem — established that Birkhoff's conjecture was mistaken: ergodic conservative systems are not dense in the set of all Hamiltonian systems [...] This meant that the relationship between order and disorder, stability and instability in celestial mechanics had to be rethought. [...] When perturbations were sufficiently small, KAM theorem showed that majority of orbits were stable and quasi-periodic". It is not possible in this limited article to describe the immense influence of Kolmogorov in dynamical systems.[14] However, I would like to cite Michał Misiurewicz describing the origin of the study of dynamical systems in Warsaw. "At some moment two young Warsaw mathematicians, Wiesław Szlenk and Karol Krzyżewski, were sent by Stanisław Mazur to Moscow for a postdoctoral training. They were specialists in functional analysis and went to Andrey Kolmogorov for the advise of what they should do. Kolmogorov told them that the great days of functional analysis were already over, but there was a new interesting discipline, called dynamical systems (together with the ergodic theory). Then, he sent them to Yakov Sinai (who had been his student) to learn it. [...] Krzyżewski and Szlenk, after learning the basics of the dynamical systems, returned to Warsaw. They were followed by Sinai, who came and gave a cycle of lectures in ergodic theory" [Lozi *et al.*, 2024a]. In 1961, Steven Smale traveled to the Soviet Union where he met Kolmogorov and "in his word, an extraordinarily gifted group of mathematicians, Anosov, Novikov and Sinai" [Palis, 1993]. Aubin and Dahan Dalmedico added that "On this occasion, Smale stated a conjecture according to which geodesic flows on compact manifolds of negative curvature and algebraic automorphisms of the torus were coarse. The announcement made public by D.V. (Dmitri Viktorovitch) Anosov at the 1962 International Congress of Mathematicians in Stockholm that he had proved Smale's conjecture let to important advances in dynamical systems theory. Indeed, with so-called Anosov systems (a large class of hyperbolic stable systems), the widespread opinion holding that conservative systems could not be coarse was contradicted" [Aubin & Dahan Dalmedico, 2002].

Leaving the domain of differential equations to return to that of discrete dynamical systems, it is worth mentioning the astonishing result known as the "Sharkovsky's order" published by [Sharkovsky, 1964] (1936–2022). Sharkovsky [2014] explained the difficulty he had in communicating with James Yorke about this result in 1975 during the seventh International conference on nonlinear oscillations (ICNO-1975) in Berlin. I report how conflictual their relationship still was in 2009 in [Lozi, 2026]. In that article I recall also my relationships with Sharkovsky (which lasted many years [Lozi *et al.*, 2004]) and Shilnikov (see Fig. 12) starting in 1993, and my brief encounter in 2000 with two other important figures of the dynamical systems in USSR, Romen Plykin (1935–2010) and Evgenii Sataev (1950–), both from Obninsk State Technical University, in Katsiveli in 2000 (see Fig. 13). They worked in particular on the Lorenz attractor [Klinshpont *et al.*, 2005].

In 1967, we find again Smale discovering a fundamental mathematical tool for chaotic dynamical

Fig. 12. Alexander Sharkovsky, Leonid Shilnikov, and René Lozi upon their departure in Honolulu Airport to San Francisco on December 12, 1993.

[14]According to the *Mathematical Genealogy Project*, Andrei Kolmogorov has 82 students and 4562 descendants.

Fig. 13. From left to right, René Lozi, Romen Vasilievich Plykin, and Evgenii Sataaev on August 2000, during the Summer School on Dynamical System, Katsiveli, Crimea, Ukraine.

systems: the horseshoe map, while studying the behavior of the orbits of the van der Pol oscillator [Smale, 1967]. The action of the map is defined geometrically by squashing the square, then stretching the result into a long strip, and finally folding the strip into a horseshoe shape [Chen & Huang, 2011]. Smale's horseshoe map plays a crucial role in chaos theory and dynamical systems as it serves as an effective model for understanding and investigating these systems.

I have briefly described the first long period. Due to the limited length of this paper, I have undoubtedly omitted several important scientists and results, such as the work of Aleksandr Lyapunov on stability theory,[15] the study of Balthasar van der Pol and other radio engineers on relaxation oscillations, or sustained periodic solutions to nonlinear dissipative equations with a few degrees of freedom, and the monumental work of René Thom. From the late 1950s onward, Thom embarked on a research program in differential topology exhibiting many similarities with the program of Smale. He introduced the notion of genericity in global analysis, and shared with Smale the attractor concept. He developed the famous list of the seven "elementary catastrophes" in the following decade.[16] I should also mention the work of [Peixoto, 1955] on structural stability, and that of [Morse & Hedlund, 1938] on symbolic dynamics, etc. I refer the reader interested in a more complete description of this period to [Aubin & Dahan Dalmedico, 2002].

The driving forces behind the development of the theory of dynamical systems, chaotic or not, during this first period can be considered to be represented in Fig. 5 by the path: determinism → approximate knowledge of the initial conditions → unpredictability, for the Poincaré's work on n-body systems, and for the Lorenz attractor which was discovered thanks to the approximate initial condition of integration of the system when Lorenz relaunched his calculation. And by the path: determinism → exact knowledge of the initial conditions → predictability → complex behavior, for all the other works of Andronov, Anosov, Birkhoff, Fatou, Hadamard, Julia, Gumowski, Mira, Sharkovsky, Smale, etc. Most of these research studies were guided by theory, except for the research done by the Andronov school, who were motivated by radio and electronic applications, as discussed in Sec. 5. We can find in the book [Andronov *et al.*, 1937, 1966] many different cases of electric circuit like a valve generator (pp. 446, 626), but not only: the problem on gliding flight taking place in a vertical plane is also examined (p. 436) or a self-oscillating system using a very simple dynamic model of a steam engine (p. 559), etc.

7. The Intermediate Period 1971–1989

In the 1970's a change of paradigm in the study of nonlinear dynamical system and the model of turbulence occurred with the work of [Ruelle & Takens, 1971] who introduced the concept of strange attractor to overcome the classical theory of Landau promoting an infinite sequence of bifurcations in tori of increasing dimension corresponding to quasiperiodic motion with an increasing number of frequencies [Landau, 1944] (see Fig. 14). They shaded to light the "butterfly effect" revealed by [Lorenz, 1963] but forgotten for seven years. The "sexier"

[15]Jean Mawhin, who introduced the *coincidence index* wrote the interesting article "The early reception in France of the work of Poincaré and Lyapunov in the qualitative theory of differential equations" [Mawhin, 1996].

[16]Thom assisted me on several occasions for my "Thèse d'État" in 1983, my Habilitation in 1986, and later, to apply for higher position in the university of Nice. I was surprisingly chosen along seven prestigious scholars for a tribute in his honor in April 1990 in Strasbourg, on the occasion of his retirement from this university [Lozi, 1990].

Fig. 14. Lev Landau, Edward Lorenz, David Ruelle, and Floris Takens.

Fig. 15. Michel Hénon, Otto Rössler, James A. (Jim) Yorke, and Hiroshi Kawakami.

word "chaos" was coined by [Li & Yorke, 1975] (see Fig. 15). It has brought global awareness of these concepts, often not actually understood by the public.

Immediately thereafter, hundreds of examples of strange attractors, called chaotic attractors nowadays, appeared. These attractors, of which I will mention only the main ones here, still studied fifty years later, profoundly transformed research on chaotic dynamical systems [Lozi *et al.*, 2024b]. Among them, we find the continuous attractors of [Rössler, 1976, 2020] (4), (Fig. 16), in 1983 [Chua, 1992], and the discrete attractors of [Hénon, 1976; Belykh, 1976; Lozi, 1978; Ikeda, 1979]. In addition to the seminal articles by Ruelle and Takens and Li and Yorke, one of the reasons for this proliferation is the increasing use of computers that were more readily available for this research. This second period was a flourishing time for the study of very simple examples leading to profound theories.

$$\begin{cases} \dot{x}_1 = -x_2 - x_3, \\ \dot{x}_2 = x_1 + a x_2, \\ \dot{x}_3 = b + x_3(x_1 - c). \end{cases} \tag{4}$$

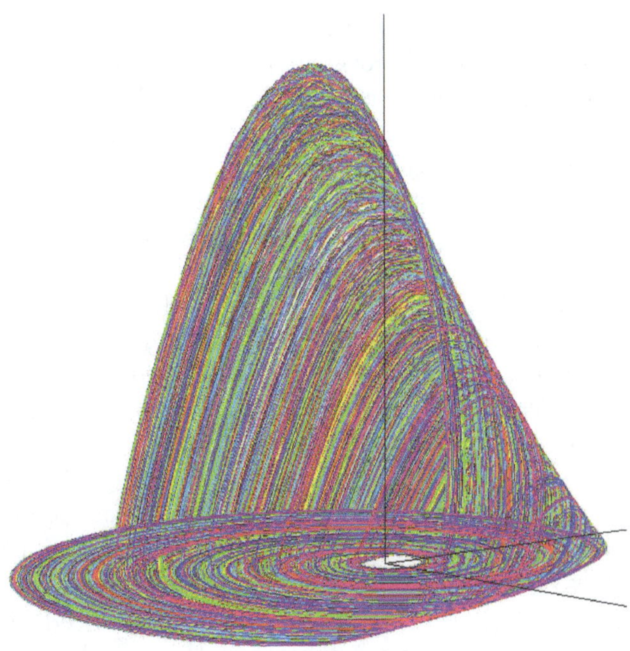

Fig. 16. Rössler attractor for $a = 0.35$, $b = 0.4$, and $c = 6.4$.

7.1. *The Hénon, Lozi, and Belykh attractors*

Alongside these chaotic attractors, during the first years of this second period, the logistic map was again explored by Mitchell Feigenbaum, who, using an HP-65 programmable pocket calculator (see Hénon's work below) investigated the bifurcation sequence of the logistic map

$$x_{n+1} = 4\lambda x_n(1 - x_n). \tag{5}$$

Like Myrberg, he noticed that the period of the stable attracting orbits doubled each time the parameter passed a bifurcation value λ_i, with x oscillating between 2^i values. He pointed out that this series converged towards a critical value λ_∞ at a rate defined as

$$\delta = \lim_{i\to\infty} \frac{\lambda_{i+1} - \lambda_i}{\lambda_{i+2} - \lambda_{i+1}} = 4.669201609\ldots. \tag{6}$$

This number was universal, that is, characteristic not only for the logistic map, but also of a large class of functions [Feigenbaum, 1978]. That phenomenon was discovered at the same time by Charles Tresser and Pierre Coullet from the University of Nice (the same university of Hénon) [Coullet & Tresser, 1978].

However, the chaotic behavior of maps of the real line (logistic and tent [Diamond & Pokrovskii, 1996]) are relatively simple compared to the behavior of maps on the plane which could be seen as a simple expansion in a phase space with increased dimension of both logistic and tent maps. It is not the case, Lozi, Hénon, and Belykh attractors have been discovered not in this ascending-dimensional way, instead they come in descending way as metaphor of Poincaré map of Three-Dimensional (3D)-continuous dynamical systems.

In order to study numerically the properties of the Lorenz attractor, Michel Hénon (see Fig. 15), an astronomer of the observatory of Nice in the south of France (see Fig. 17), introduced in 1976 a metaphoric model of the Poincaré map of this attractor. Hénon used a IBM 7040 computer to numerically simulate the Lorenz map (2) [Lorenz, 1963]. He found it difficult to highlight its inner nature due to the very high dissipativity of this map whose rate of volume contraction is determined by the Lie derivative of the Lorenz equations which gives with the parameters chosen by Lorenz, $V(t) = V(0)e^{\frac{-41}{3}}$. Hence, volumes are reduced by a factor 10^6, after one time unit. Hénon, who was a specialist of the motion of planets, had a good experience of Poincaré's map which inspired him to find a way to build the metaphoric model (7) [Hénon, 1976]

Fig. 17. The Nice Observatory in the South of France where Michel Hénon worked (buildings on the right).

$$\begin{cases} x_{n+1} = 1 - ax_n^2 + y_n, \\ y_{n+1} = bx_n. \end{cases} \tag{7}$$

This model can be represented by the iterates of any initial point $(x_0, y_0)^T$ belonging to the basin of attraction, by the map $\mathcal{H}_{a,b} : \mathbb{R}^2 \to \mathbb{R}^2$ (8). With the Lorenz attractor being in dimension-3, the corresponding Poincaré map is a map from the plane $\mathbb{R}^2$ into itself. Hence, the Hénon map is then also defined in dimension-2

$$\mathcal{H}_{a,b}\begin{pmatrix} x \\ y \end{pmatrix} = \begin{pmatrix} 1 - ax^2 + y \\ bx \end{pmatrix}. \tag{8}$$

The contracting properties of this map are only determined by the parameter b. When $b = 0.3$, the contraction in one iteration is mild enough that the sheaves of the attractors are distinguishable on a graphic. For the parameter value $a = 1.4$ and $b = 0.3$, Hénon pointed out numerically that there exists an attractor with fractal structure (see Fig. 18). Finally, Hénon observed on a series of images drawn by the computer, the astonishing fractal structure of the attractor which amazed the research community. This was the first example of a strange attractor previously introduced by Ruelle and Takens for a mapping defined by an analytic formula. The like-Cantor set structure in one direction orthogonal to the invariant manifold of one of the fixed points in this simple mapping was a dramatic surprise in the community of physicists and mathematicians.

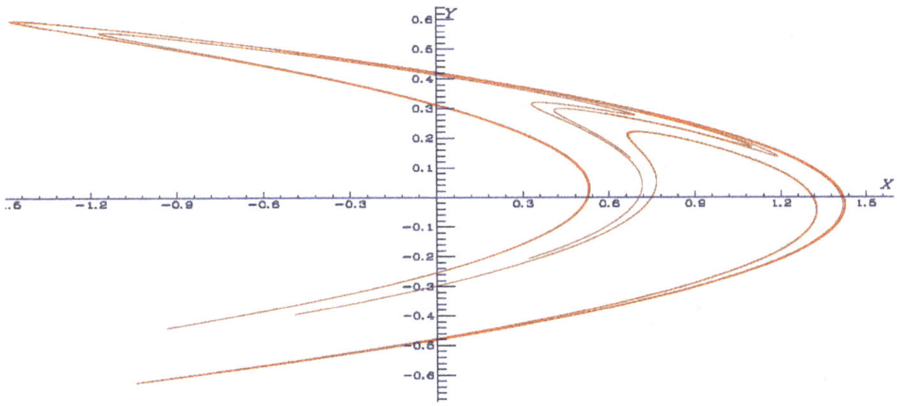

Fig. 18. Hénon attractor, $a = 1.2$, $b = 0.42$.

Nowadays, hundreds of research papers have been published on this prototypical map in order to fully understand its innermost structure. However, like for 1D-dynamical systems, there is a discrepancy between the mathematical properties of this map in the plane and the numerical computations done using (IEEE-754) double precision numbers. If we call *megaperiodic* orbits [Lozi, 2013] those whose length of the period belongs to the interval of natural numbers $[10^6, 10^9[$, and *gigaperiodic* orbits, those whose length of the period belongs to the interval $[10^9, 10^{12}[$, Hénon map possesses both mega- and gigaperiodic orbits. On a Dell computer with a Pentium IV microprocessor running at 1.5 GHz frequency, using a Borland C compiler and computing with ordinary (IEEE-754) double precision numbers, one can find for $a = 1.4$ and $b = 0.3$, one attracting period of length 3 800 716 788. This periodic orbit is numerically slowly attracting. Starting with the initial value $(x_0, y_0) = (0.35766, 0.14722356)$ one winds up

$$\begin{aligned}(x_{11\,574\,730\,767}, y_{11\,574\,730\,767}) &= (x_{15\,375\,447\,555}, y_{15\,375\,447\,555}) \\ &= (1.27297361350955662, \\ &\quad -0.0115735710153616837).\end{aligned}$$

It is not the only one [Lozi, 2013]. The discovery of the strange properties of this map can be considered as miraculous, considering that if Hénon tried to test rigorously his model nowadays with powerful computers he should only find long periodic orbits. In 1976, Hénon used the electronic pocket calculator built by Hewlett-Packard company HP-65 [see Fig. 19(left)] and one of the only two computers available at the university of Nice a IBM 7040 [see Fig. 19(right)]. The other one was a IBM 1130 (with 4 kB RAM), slower. The HP-65, first introduced in 1974, was worth 750€ (equivalent to 4200€ nowadays), the IBM 7040, was worth 1 100 000€ (equivalent to 9 400 000€ nowadays), it proved popular for use at universities, due to its comparatively low price [Wikipedia, 2025].

He concluded "The simple mapping (4) [i.e. Eq. (8) in this paper] appears to have the same basic properties as the Lorenz system. Its numerical exploration is much simpler: in fact, most of the exploratory work for the present paper was carried out with a programmable pocket computer (HP-65). For the more extensive computations, we used a IBM 7040 computer, with 16-digit accuracy. The solutions can be followed over a much longer time than in the case of a system of differential equations. The accuracy is also increased since there are no integration errors".

In June 1977, I have attended the *International Conference on Mathematical Problems in Theoretical Physics* in Roma. During the inaugural talk on June 6, Ruelle [1978] conjectured that, for the Hénon chaotic map, the theoretical entropy should be equal to the characteristic exponent. As explained in [Letellier *et al.*, 2021; Lozi, 2026], this is how I learned of the first example of a chaotic and strange attractor in the plane: the Hénon attractor (see Fig. 18). Surprisingly, I had never met before Michel Hénon, who worked a few kilometers apart from the math department of the university at the Nice Observatory (Fig. 17). At that time, I was a young researcher trying to understand bifurcation theory in the field of numerical analysis [Lozi, 1975]. This area was not very well known in France.

I was working on the discretization of nonlinear problems, using the finite element method

Fig. 19. (Left) Hewlett-Packard 65; (right) IBM 7040.

in which nonlinear functions are approximated by piecewise-linear functions. During the Roma conference, I tried without success to apply the spirit of the method of finite element to the Hénon attractor. Back to Nice after this conference, I eventually decided, using paper and pencil, to change the square function of the Hénon attractor, which is U-shaped, into the absolute value function, which has a V shape, implying folding property (folding property is important for horseshoe — a main ingredient of chaos, as highlighted by [Smale, 1967]). I was the owner of a small desktop computer HP 9820 linked to a HP 9862 plotter on which I tested this change of function and found a similar attractor of the Hénon attractor, with straight lines instead of curves (see Fig. 20)

$$\mathcal{L}_{a,b}\begin{pmatrix} x \\ y \end{pmatrix} = \begin{pmatrix} 1 - a|x| + y \\ bx \end{pmatrix}. \tag{9}$$

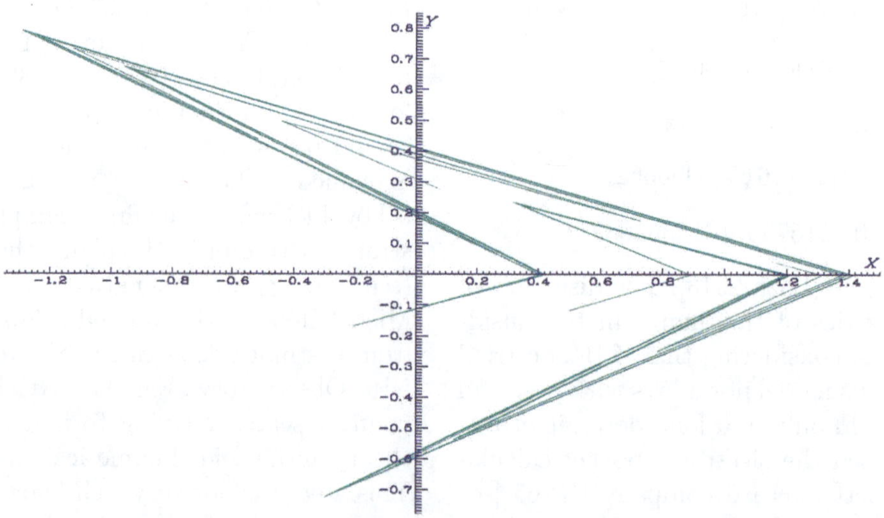

Fig. 20. Lozi attractor, for the parameter value $a = 1.62$, $b = 0.56$, initial value $x_0 = 0$, $y_0 = 0$.

In 1979, Michal Misiurewicz presented a rigorous proof that for a set of parameters $\mathcal{M}$, this map has a strange attractor, coining at this occasion the name "Lozi map" [Misiurewicz, 1980]. This set in the plane of parameters (a, b) is defined by

$$\mathcal{M} = \left\{ (a,b) \,|\, b > 0, a\sqrt{2} < b + 2, \right.$$
$$\left. b < \frac{a^2 - 1}{2a + 1}, 2a + b < 4 \right\}. \qquad (10)$$

Later, Misiurewicz and Stimac [2016] slightly enlarged this set of parameter $\mathcal{M}$ (defined only for $b > 0$) (as $\mathcal{M}^+$ in Fig. 21) in the upper edge of the triangle, by removing the condition $b < \frac{a^2-1}{2a+1}$

$$\mathcal{M}^+ = \{(a,b) \,|\, b > 0, a\sqrt{2} < b + 2, 2a + b < 4\}. \qquad (11)$$

The case $b < 0$ was studied by [Cao & Liu, 1998] and recently by [Kucharski, 2023], who found a quasi-symmetric set of parameters $\mathcal{M}^-$ (Fig. 21). However, the right-hand side boundary of $\mathcal{M}^-$ is not completely symmetric to the right-hand side boundary of $\mathcal{M}^+$

$$\mathcal{M}^- = \left\{ (a,b) \,|\, -1 < b < 0, a\sqrt{2} > b + 2, \right.$$
$$\left. b > \frac{-8a + 3a^2 + \sqrt{-16a^3 + 9a^4}}{8} \right\}. \qquad (12)$$

The symbolic dynamics of the Hénon and Lozi map were intensively studied by many authors including Wei-Mou Zheng (see Fig. 22) and Hong Zhao [Zhao & Zheng, 1993; Zheng, 1991, 1992].

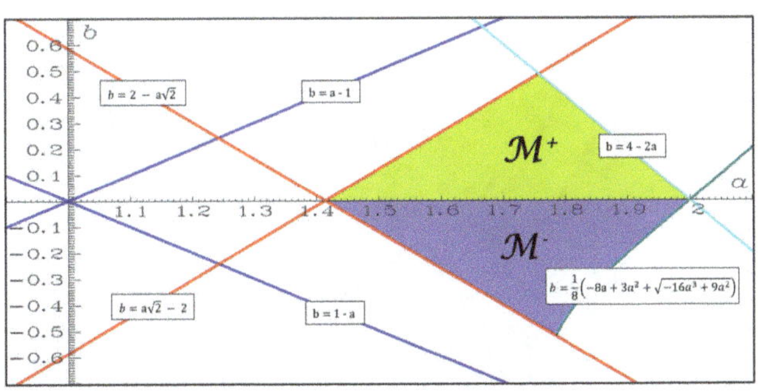

Fig. 21. The sets $\mathcal{M}^+$ and $\mathcal{M}^-$ of the plane of parameters where Lozi map has a strange attractor in the second proof of Misiurewicz and Stimac (11) and the proof of Kucharski (12). Blue lines $b = a - 1$ and $b = 1 - a$, red lines $b = a\sqrt{2} - 2$ and $b = 2 - a\sqrt{2}$, cyan line $b = 4 - 2a$, dark green line $b = \frac{-8a+3a^2+\sqrt{-16a^3+9a^4}}{8}$.

Fig. 22. Wei-Mou Zheng (left) and M. A. F. Sanjuán (right) at the Institute of Theoretical Physics of the Chinese Academy of Sciences in Beijing, China, in 2011.

The same year that Hénon found his attractor (1976), Vladimir Belykh (see Fig. 23) realized that his work on the Phase-Locked Loop (PLL) could lead to a new attractor. In [Lozi *et al.*, 2025] he said "My diploma research (an analog of an M.S. degree, in 1966) was on the analysis of homoclinic orbits bifurcations in a 2D-pendulum-type dynamical system on a cylinder, modeling a phase-locked loop (PLL), a control system with a nonlinear filter. [...] 1974 marked a significant year as I contributed to the theory of digital PLLs, yielding publications [Belykh & Maksakov, 1976a, 1976b] and, along with colleagues Vladimir D. Shalfeev and Valery P. Ponomarenko, the prestigious Lenin Komsomol Prize (a Soviet science award for young researchers under 33). [...] In September 1976, following my return from Denmark, Valentin (Valya) Afraimovich, a Shilnikov's pupil, and L. P. showed me a paper on the original and structure of the Lorenz attractor, which they, together with Vadim Bykov, submitted to a reputable Russian journal [Afraimovich *et al.*, 1977]. I soon realized that my model of a discrete-time PLL [Belykh, 1976] in the

Fig. 23. Vladimir Belykh (left) and Dmitry Anosov (right) at a conference in Moscow.

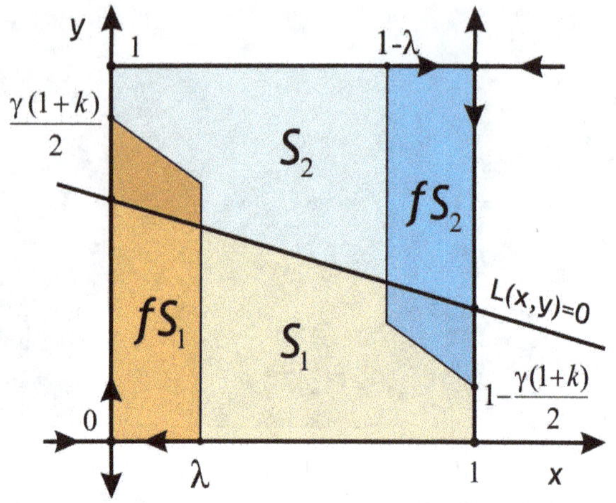

Fig. 24. First image of the unit square $S = S_1 \cup S_2$ under the map f (13) and (14), where S_1 and S_2 are separated by the discontinuity line $L(x, y) = 0$. One iterate of f transforms trapezoids S_1 (light brown) and S_2 (light blue) into fS_1 (dark brown) and fS_2 (dark blue), respectively.

case of a piecewise-linear nonlinearity becomes a discontinuous 2D map that acts like theirs. I showed this map with a chaotic attractor to Valya and L. P., and Valya said: "Here is the Belykh attractor", L. P. confirmed. It was during this exchange that the Belykh attractor was coined". While there are several variations of this map [Belykh, 1976, 1982, 1995], its standard form reads [(13), (14)] [Belykh & Belykh, 2011; Belykh *et al.*, 2023] (see Fig. 24).

$$\mathcal{B}_{1\gamma,\lambda}\begin{cases}\bar{x} = \lambda x \\ \bar{y} = \gamma y\end{cases} \quad \text{if } L(x,y) \le 0, \tag{13}$$

$$\mathcal{B}_{2\gamma,\lambda}\begin{cases}\bar{x} = \lambda(x-1)+1 \\ \bar{y} = \gamma(y-1)+1\end{cases} \quad \text{if } L(x,y) > 0, \tag{14}$$

where $L(x, y) = k(2x - 1) + 2y - 1$.[17] Pesin [1992] and Sataev [1992] proved the existence of the Sinai–Ruelle–Bowen measure for the chaotic Belykh map which can also be embedded into the phase space of a system of ODEs.

7.2. *Attractors linked to electric circuit realizations, Ikeda, Chua, and Alpazur circuits*

Before the applications of chaotic circuits were published, research was conducted to see if these strange objects were merely mathematical artifacts or if they could be present in physics experiments. In Japan, the Hayashi school (with disciples like Ikeda, Ueda, and Kawakami) at the same time, was motivated by the simulation of chaotic dynamics by electrical and electronic circuits. Chaotic maps have been used as models of electrical circuit behavior [the Ikeda map 1979 (15), see Fig. 25] [Ikeda, 1979] and ODE for the *Japanese attractor* found by Yoshisuke Ueda in the Duffing equation

$$\begin{cases} x_{n+1} = 1 + u(x_n\cos(t_n) - y_n\sin(t_n)) \\ \quad \text{with } t_n = 0.4 - \dfrac{6}{1 + x_n^2 + y_n^2}, \\ y_{n+1} = u(x_n\sin(t_n) + y_n\cos(t_n)). \end{cases} \tag{15}$$

In fact, the Japanese attractor (named after Ruelle) was discovered by Ueda in November 1961, before the Lorenz attractor. However, he struggled for many years to have the chaotic nature of this attractor recognized.[18] He said in [Ueda, 1978, 1992]

Fig. 25. Ikeda attractor for $u = 0.87$.

[17]The map is defined on the square $S = \{(x, y) : 0 \le x \le 1, 0 \le y \le 1\}$. The square S is cut into two parts, S_1 and S_2, by the line $L(x, y) = k(2x - 1) + 2y - 1 = 0$. The dynamics of the map f is governed by (13) on the lower part S_1, where $L(x, y) \le 0$, and by (14) on the upper part S_2.

[18]See the interesting meeting between Ueda and Saber Elaydi, with Ueda telling this story with great emotion in [Elaydi, 2023].

"At present, people say that the data I was collecting with my analog computer on the 27th of November 1961, is the oldest example of chaos discovered in a second-order nonautonomous periodic system. Around the same time, it was Lorenz who made the discovery of chaos in a third-order autonomous system. [...] At that time, I was a third-year graduate student a Kyoto University, working on the phenomenon of frequency entrainment under the guidance of Professor Chihiro Hayashi. [...] On that day, the 27th of November 1961, when I changed the parameter (frequency of the driving input), and the condition shifted from frequency entrainment to asynchronization, the oscillation phenomena portrayed by analog computer was chaotic indeed". However, due to the rigid guidance of Prof. Hayashi who did not recognize the chaotic nature of his discovery, it is only in 1978 that Ueda was able to publish his paper. He said "It was around that time that I first heard the term 'chaos'".

Regarding chaotic autonomous differential equations, Leon O. Chua, Professor at University of California, Berkeley (see Fig. 1), invented a very simple electric circuit producing chaos (16) in October 1983. He was Visiting Professor in the Matsumoto's laboratory, at Waseda University, Tokyo. In [Chua, 1992] he said "Prior to 1983, the conspicuous absence of a reproducible functioning chaotic circuit or system seems to suggest that chaos is a pathological phenomenon that can exist only in mathematical abstractions, and in computer simulations of contrived equations. Consequently, electrical engineers in general, and nonlinear circuit theorists in particular, have heretofore paid little attention to a phenomenon which many had regarded as an esoteric curiosity. Such was the state of mind among the nonlinear circuit theory community, around 1983". The only two systems that were considered during this period, were the Lorenz and the Rössler equations (2) and (4). The nonlinearity in both systems is the product function of two variables which is very difficult to build in electronic circuit. Chua added "I was to have witnessed a live demonstration of presumably the world's first successful electronic circuit realization of the Lorenz Equations, on which Professor Matsumoto's research group had toiled for over a year. It was indeed a remarkable piece of electronic circuitry. It was painstakingly breadboarded to near perfection, exposing neatly more than a dozen IC components, and embellished by almost as many potentiometers and trimmers for fine-tuning and tweaking their incredibly sensitive circuit board. There would have been no need for inventing a more robust chaotic circuit had Matsumoto's Lorenz Circuit worked. **It did not**. The fault lies on the dearth of a critical nonlinear IC component with a near-ideal characteristic and a sufficiently large dynamic range; namely, the analog multiplier. Unfortunately, this component was the key to building an autonomous chaotic circuit in 1983". He concluded by explaining how the idea for this now universally known and used circuit came to him "Matsumoto's Lorenz Circuit was to have turned the tide of indifference among nonlinear circuit theorists. Viewed from this historical perspective and motivation, the utter disappointments that descended upon all of us on that uneventful October afternoon was quite understandable. So profound was this failure that the wretched feeling persisted in my subconscious mind till about bedtime that evening. Suddenly it dawned upon me that since the main mechanism which gives rise to chaos, in both the Lorenz and the Rössler equations, is the presence of at least two unstable equilibrium points — 3 for the Lorenz Equations and 2 for the Rössler Equations — it seems only prudent to design a simpler and more robust circuit having these attributes. Having identified this alternative approach and strategy, it becomes a simple exercise in elementary nonlinear circuit theory to enumerate systematically all such circuit candidates, of which there were only 8 of them, and then to systematically eliminate those that, for one reason or another, cannot be chaotic".

The advantage of Chua's circuit [see Fig. 26(a)] is that the variables in the mathematical equations correspond to voltage and current and can be visualized on the screen of an oscilloscope [see Fig. 26(b)]

$$\begin{cases} \dot{x} = \alpha(y - \Phi(x)), \\ \dot{y}_2 = x - y + z, \\ \dot{x}_3 = -\beta y, \end{cases} \tag{16}$$

with $\Phi(x) = m_1 x + \frac{1}{2}(m_0 - m_1)[|x+1| - |x-1|]$.

The Chua's attractor is very interesting and nowadays it is thoroughly studied. However, it is not a slow–fast system, it is therefore not possible to apply all the mathematical tools developed for such category of systems (see Sec. 2). I first met Shigehiro Ushiki and Hiroshi Kawakami in 1982,

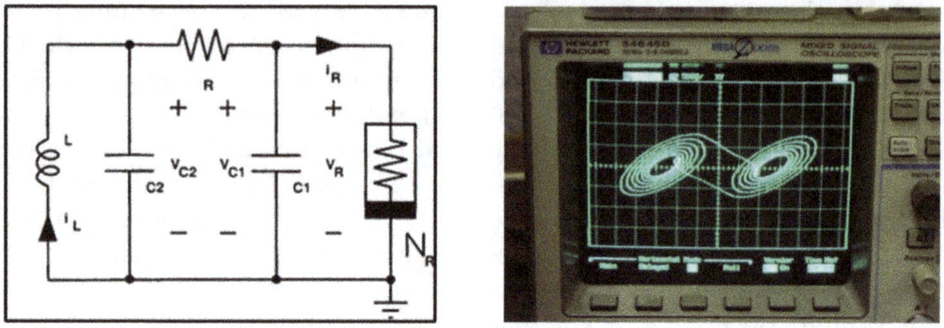

(a) (b)

Fig. 26. (a) Chua circuit and (b) Chua attractor on oscilloscope.

during a meeting in Toulouse organized by Mira [Lozi, 1982], and again in Kyoto during the summer 1983 invited by his advisor the professor Masaya Yamaguti[19] to a grant from the Japan Society for the Promotion of Science (see Fig. 27). Like Ushiki, Kawakami was fluent in French, mainly because he spent two years of internship at the CNRS Center of Physics researches in Marseilles in 1970–1972. Later in 1984, he was invited professor for one year at INSA of Toulouse, to work with Mira. In 1988, during my first long stay in Berkeley, Chua invited him, and we discussed a lot on such slow–fast electric circuits. Two years after, I attended the International Congress of Mathematicians at Kyoto on August 21–29, 1990 and spent three weeks in Japan, working with Ushiki. Both of us did a short visit to Kawakami in his university of Tokushima from August 31 to September 1. Eventually, I found a CNRS grant and invited him in Nice for one month in December 1991. During this month, we finalized the Alpazur oscillator[20] (see Fig. 28). Kawakami gave this weird name to this electric circuit because it was the name of the hotel where he stood in Nice. Referring to Rössler, we described this oscillator as follows [Kawakami & Lozi, 1992] "The Alpazur oscillator is a two-dimensional piecewise circuit with a switch controlled by state variables. Two circuit equations or vector fields which describe the dynamics of the circuit are abruptly interchanged each other by changing the position of the switch. If the characteristics of the switch have hysteresis, then the composite flow generated by the mixed vector fields is not necessarily continuous on initial conditions, which allows chaotic behavior of the state variables. Since an electronic switch acts

Fig. 27. Masaya Yamaguti, Shigehiro Ushiki, and René Lozi during the IUTAM symposium on "Turbulence and chaotic phenomena in fluids" (September 5–10, 1983) in Kyoto.

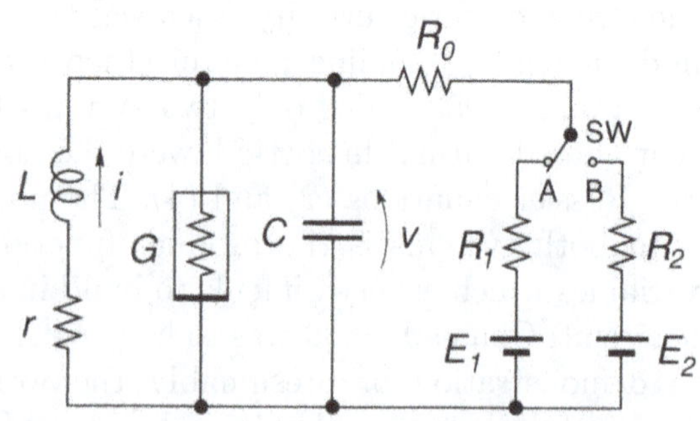

Fig. 28. The Alpazur circuit.

[19]Masaya Yamaguti's father Kayo Yamaguti (1899–1984) was a famous animal and flower bird painter. His son was upset that these paintings were too much expensive, he could not buy at least one https://www.momak.go.jp/Japanese/exhibitionArchive/2012/395.html.

[20]I included this attractor in the second period, even though the discovery was made two years after the end of this period, because it is closely related to the Chua circuit.

instantaneously, this model serves almost ideal realization of composite dynamical system proposed by [Rössler, 1979] as well as constrained systems analyzed by [Ushiki & Lozi, 1987; Lozi & Ushiki, 1988; Lozi & Saidi, 1989, 1990]".

The circuit dynamics of the Alpazur oscillator shown in Fig. 28 is given by

$$\begin{cases} L\dot{I} = -V - rI, \\ C\dot{V} = -G(V) + I + \dfrac{E_1 - V}{R_0 + R_1}, \end{cases} \tag{17}$$

when the switch SW is thrown into contact with position A and

$$\begin{cases} L\dot{I} = -V - rI, \\ C\dot{V} = -G(V) + I + \dfrac{E_2 - V}{R_0 + R_2}, \end{cases} \tag{18}$$

when SW is in position B, with

$$G(V) = -a_1 V + a_3 V^3, \quad a_1, a_3 > 0. \tag{19}$$

The function of the switch SW is controlled by the capacitor's voltage and has hysteresis. This functional characteristics can be realized by electronic switches controlled by comparator and operational amplifiers. Spiral, screw, and composite chaos stated by [Rössler, 1979] may be easily realized. The Alpazur oscillator did not have the expected success, however, some papers were published analyzing its properties. Miino and Ueta [2015], 23 years later, showed some results of laboratory experiment for bifurcation phenomena on the forced Alpazur oscillator. Chirita *et al.* [2015] and Stefanescu *et al.* [2013] studied the influence of computer simulation on a chaotic circuit. The Alpazur oscillator is being used in their study and the results are interpreted precisionwise and timewise (time needed to run one simulation).

Before 1990, computers were not as powerful as they are today. Therefore, many experimenters, like Rössler (see Fig. 15), still used analog electrical systems to study the behavior of chaotic dynamical systems (map or ODE) [Letellier & Messager, 2010]. Rodriguez-Vazquez *et al.* [1987] presented a special-purpose analog computer consisting of a switched capacitor circuit to analyze chaos and bifurcation phenomena in nonlinear discrete dynamical systems modeled by discrete maps. They published the results of four maps: a piecewise-linear map, the logistic map, the Hénon map (8), and the Lozi map (9). For each of these maps, they built a fairly complicated circuit design (see [Rodriguez-Vazquez *et al.*, 1987, Figs. 5, 8, 12, and 25]) and compared the attractor measured from those circuits with the corresponding numerical simulations and found good agreement between them. Although this work is not strictly speaking a true application of the chaotic map for electrical purposes, it constitutes one of the first examples of solid realization.

The driving forces behind the development of the theory of dynamical systems, during this second period can be considered to be represented in Fig. 5 by the path: determinism → exact knowledge of the initial conditions → predictability → complex behavior. As in the first period, most of this research was guided by theory, although the design of chaotic electric circuits can be seen as a first step towards future applications.

As with the first period, due to the limited length of this paper, I have certainly omitted several renowned scientists and their work, such as the work of [Pesin, 1989] who introduced a theory of smooth hyperbolic dynamical systems, the study by [Kloeden, 1976] of density of chaotic equations, etc.

8. The Last Period 1990 up to Today: The Flourishing Era of Applications

As stated in the introduction, with reference to [Aubin & Dahan Dalmedico, 2002], "total history [of dynamical systems] appears as a tempting, but manifestly unattainable, goal". This statement formulated 20 years ago is even more true today for this last period. Thus, instead of describing the complete history of dynamical systems over the last 35 years, which would require an incredible number of pages, beyond the scope of this short paper, I will simply indicate some lines of research in theory and applications of chaotic dynamical systems. This period corresponds to the creation and the 35 years of the journal IJBC, which has greatly contributed to the development of research on chaotic dynamical systems. It is marked by flourishing applications of chaotic attractors.

The driving forces driving these applications are represented in Fig. 5 by the path

- for secure communications (Sec. 8.1): determinism → exact knowledge of the initial conditions → pseudo-unpredictability;

- for Chen and multiscroll attractors (Sec. 8.2): → exact knowledge of the initial conditions → pseudo-unpredictability → control;
- for cryptography-based chaos and hash functions (Secs. 8.3.1 and 8.3.2): determinism → exact knowledge of the initial conditions → pseudo-unpredictability;
- for memristors and optimization (Secs. 8.3.3 and 8.3.4): determinism → exact knowledge of the initial conditions → predictability;
- for economy and human health (Secs. 8.4 and 8.5): determinism → approximate knowledge of the initial conditions → unpredictability.

However, the flourishing development of these applications has not prevented progress in pure mathematics. It is difficult to list all the improvements made to the theory of chaotic dynamical systems and bifurcation theory in recent decades. We can only mention a few, such as the concepts of border-collision bifurcation, general usage of fractional derivatives, fractional maps [Wu *et al.*, 2025], chimeras and solitary states, partial control theory, hidden attractors (discovered by [Leonov & Kuznetsov, 2013]), etc. They are driven by all the paths described in Fig. 5.

8.1. *Secure communications using chaotic devices*

The discovery of the synchronization of chaotic electrical circuits by [Pecora & Carroll, 1990] was the key event that triggered research into chaos-based secure communications. Two years later, Kocarev *et al.* [1992] built the first experimental secure communication system using chaotic synchronization with two Chua circuits (one used as master and the other as slave). However, the recovered signal from this system using the Chua circuit inevitably contained noise that degraded the fidelity of the original message. By cascading the receiver output of the original system to an identical copy of this receiver, the system was improved in 1993 [Lozi & Chua, 1993] (see Fig. 29). This cascading process was extended to multiple copies and analyzed using filtering theory [Lozi, 1995] in the case of a multi-tone signal.

Dmitriev *et al.* [2000] discussed a principle of multiple access in satellite communication systems or cellular telephony based on the fine structure of chaotic attractors, using control of special chaotic trajectories. They experimentally verified the proposed approach for asynchronous packet data transmission. In their approach, they considered that a chaotic attractor can be viewed as a countable set of particular trajectories: UPOs and transitions between these orbits. The irregular, chaotic behavior was generated by the instability of the periodic orbits and transient trajectories between them. They used the set of the unstable "skeleton" periodic trajectories, constituting the structure of the strange attractor (or a part of this set), as a "reservoir" of potential codes for multiuser communication systems and discovered that the number of users provided with individual code sets can be unlimited due to the multitude of the codes from a certain "reservoir" for communications which is practically infinite. To evaluate the realization of their method, they considered 20 period-16 (UPO) of the Lozi map (9) for $a = 1.7$ and

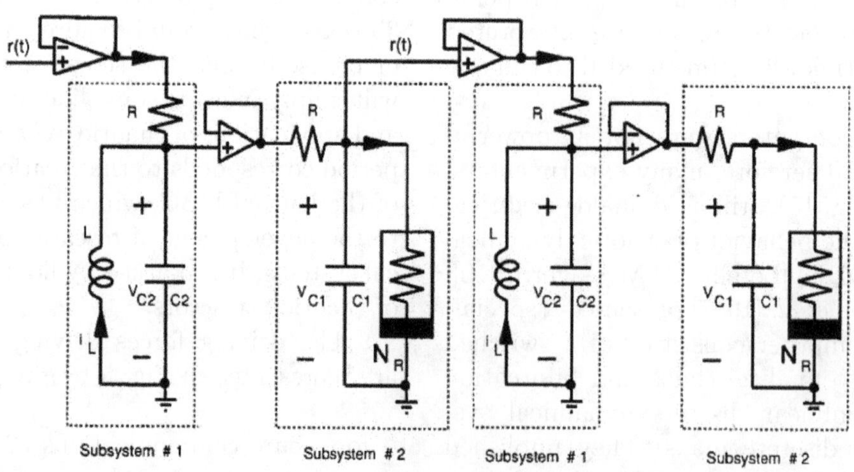

Fig. 29. Cascade of Chua circuits.

$b = 0.5$. They displayed the switching between them in Fig. 2 of [Dmitriev *et al.*, 2000] and showed from this diagram that the forming of all successive cycles (10-times repeating) is practically instantaneous. Moreover, they noticed that UPOs can be used not only to encode all transmitted information, but also to attribute it to a particular group of users, that is, they play the role of "chaotic markers". This observation allowed them to improve their initial method. The idea of using the system of UPOs as markers has been applied to the problem of asynchronous transmission of data packets from multiple users via a single common communication channel. They concluded that, thanks to modern digital methods, the generation and control of the UPO can be achieved in a rather high-frequency band.

8.2. *New attractors: Chen and multiscroll attractors*

In 1998, Guanrong Chen, second Editor-in-Chief of IJBC (see Fig. 1), studying the Lorenz attractor to benefit secure communication discovered a new chaotic attractor, now called the Chen Attractor. He recalled the discovery in [Lozi *et al.*, 2025, Chap. 2] "With control theory and systems engineering background, I first worked on chaos control by means of stabilizing chaotic dynamics when they are harmful, using various feedback control techniques. [...] In the early 1990s, I also worked on engineering applications of chaos, specifically using chaos to benefit secure communication, data encryption and liquid mixing. For those applications, we expect to generate or sustain chaos by means of control, which I called anti-control of chaos, or chaotification. One day in 1998, I looked at the classic Lorenz system again (2) and (20)

$$\begin{cases} \dot{x} = a(y - x), \\ \dot{y} = cx - xz - y, \\ \dot{z} = xy - bz, \end{cases} \tag{20}$$

where a, b, c are real parameters. As is well known, when $a = 10.2$, $b = 2.2$, $c = 27$, the system is chaotic, with an attractor as shown in Fig. 8. With the parameters a, b, c not in the range of chaos, I applied my anti-control technique to add a linear controller

$$u = \alpha x + \beta y + \gamma z, \tag{21}$$

to the right-hand side of the second equation of the Lorenz system (20), where α, β, γ are real constants to be determined. Then, under the Shilnikov condition for having chaos, I determined these three coefficients such that the resultant controlled system is chaotic. It turned out that one simple choice is $\alpha = -a$, $\beta = c + 1$, $\gamma = 0$, yielding the so-called "Chen system", named by others afterwards, described by

$$\begin{cases} \dot{x} = a(y - x), \\ \dot{y} = (c - a)x - xz + cy, \\ \dot{z} = xy - bz, \end{cases} \tag{22}$$

where a, b, c are the Lorenz parameters". When $a = 36.1$, $b = 3.1$, $c = 28.1$, the system is chaotic, with an attractor as shown in Fig. 30 [Chen & Ueta, 1999; Ueta & Chen, 2000]. The Chen system appears to be similar in formulation, yet nevertheless is not mathematically equivalent to the Lorenz system. In fact, in the sense of algebraic structures it is a dual system to the Lorenz system and it has much more complex dynamics [Yu *et al.*, 2007].

Rosalie [2016] developed the method of template for analyzing chaotic attractors using their periodic orbits to characterize their topological properties. He constructed the template of both Lorenz and Chen attractors [Rosalie, 2016] (see Fig. 31) and concluded that "The attractor of the Chen system is more foliated and consequently it exhibits more complicated dynamics. On the other hand, the attractor of the Lorenz system contains only two foliations that enable us to establish these templates", confirming what said Yu, Yao, and Chen.

Zhong and Tang [2002] proposed a realization with an electronic circuit of the Chen attractor (Fig. 32). It consists of three channels to conduct the integration of the three state variables, x, y, and z, respectively. Operational amplifiers, analog multipliers and linear resistors and capacitors are

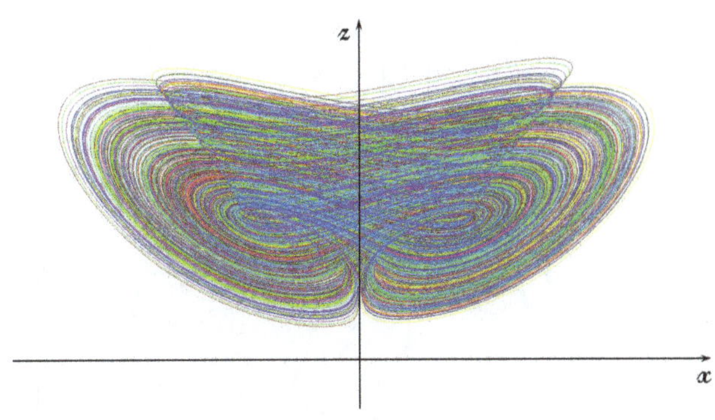

Fig. 30. Chen attractor for $a = 36.0$, $b = 3.1$, and $c = 28.1$.

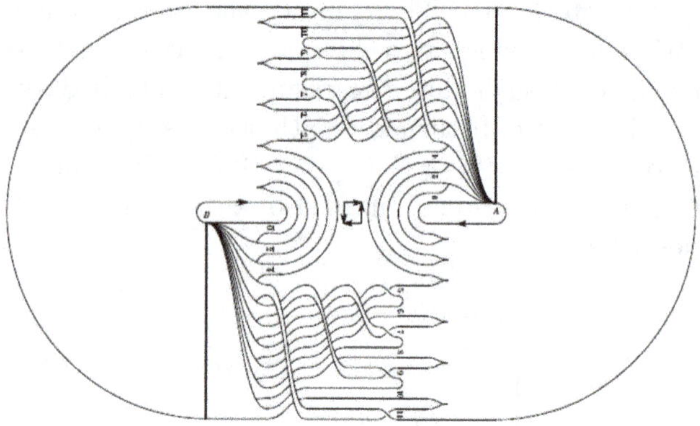

Fig. 31. Template of the Chen attractor.

employed to perform the required addition, subtraction and multiplication operations.

Later, Chen with his colleagues Simin Yu and also Jinhu Lu developed an effective methodology to generate all sorts of multiscroll and multiwing chaotic attractors from electronic circuits, using multiple saturated functions, as demonstrated by the example shown in Fig. 33, for which we can control the number, location and orientation of the scrolls and wings of the chaotic attractor [Lu & Chen, 2006]. In 2019, with him and Menacer I explored hidden bifurcation routes to multiscroll chaotic attractors generated by saturated function series [Zaamoune *et al.*, 2019].

8.3. *Applications of chaotic dynamical systems in technology*

The first "rudimentary" attempts to use chaotic attractors in telecommunications described in Sec. 8.1 opened the field for in-depth research into the use of chaotic dynamical systems for several technologies. The methods have been refined and today the use of chaos has become a very effective tool in several areas.

8.3.1. *Cryptography-based chaos*

Humanity today lives in a digital world where cryptography is an essential means of protecting communications. Modern technologies enable rapid communication between billions of devices via complex networks (satellite, mobile telephony, internet, etc.). The main concern with these complex and tangled networks is their protection against passive and active attacks that could compromise public

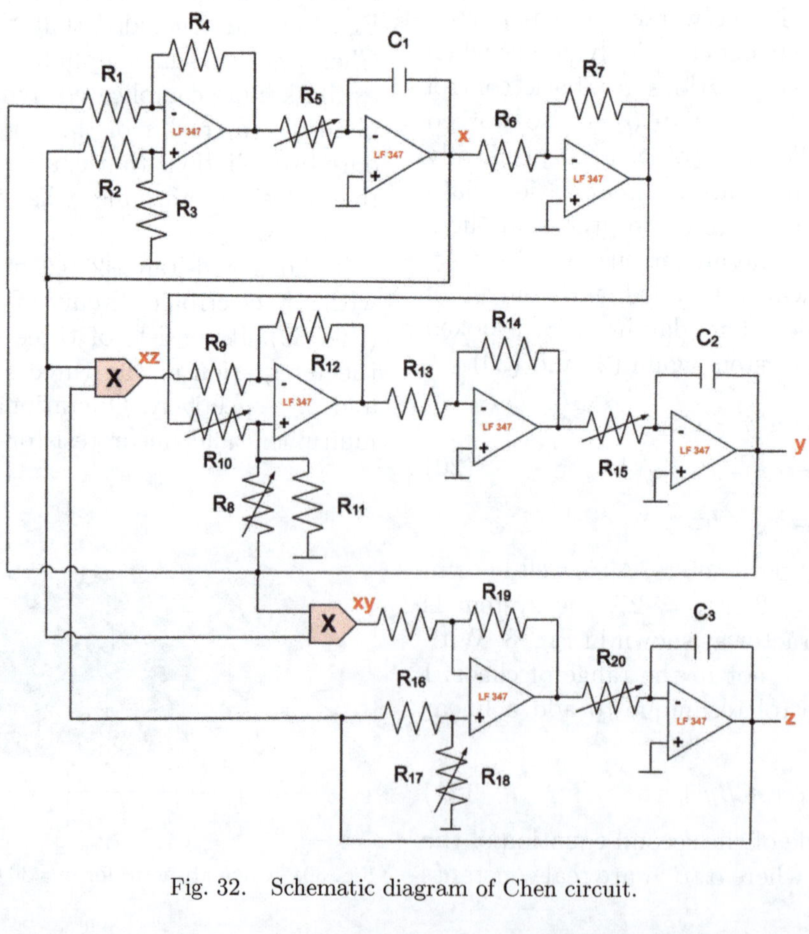

Fig. 32. Schematic diagram of Chen circuit.

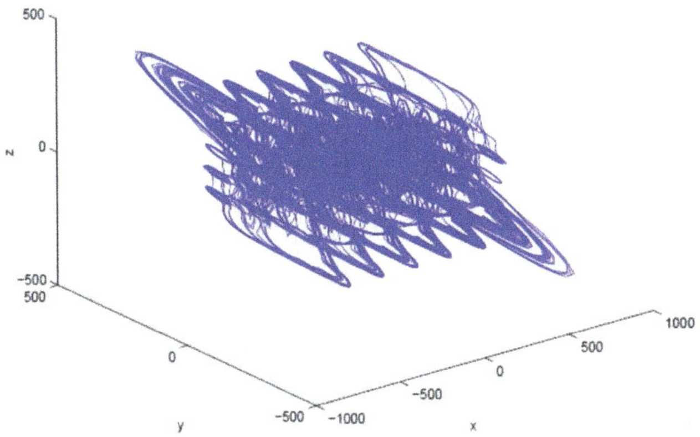

Fig. 33. A chaotic attractor with multiscrolls in three directions.

safety and privacy. Cryptography has been around for over 2000 years with the famous Caesar cipher employed by Emperor Julius Caesar and his army during the 1st century BC. It served as a means to encrypt military and official messages. Operating as a simple substitution cipher, it involved replacing each letter in the message with another letter positioned a fixed number of places away in the alphabet. Caesar's elementary cipher was followed by many other ciphers, such as that of Vigenère. In the early 20th century, electrical machines were used to simplify and automate coding, such as the *Enigma* machine invented in 1918 by Arthur Scherbius [Singh, 2002]. Today, modern ciphers rely on mathematical objects such as graphs, elliptic curves, etc. Nevertheless, the properties of chaotic attractors are recognized as the basis of certain cryptographic methods.

Among the many algorithms based on chaotic dynamical systems, we can cite, in relation to secure communications, the use of hyperchaotic systems [Khalifa & Benrejeb, 2017] which is an improvement of the methods presented in Sec. 8.1. In their method, the plain message encryption uses two discrete-time hyperchaotic systems on the transmitter side. Then, on the receiver side, two discrete-time observers are designed to fulfill the synchronization process. Thus, the decryption process is performed when the synchronization process is activated between the two-master and two-slave hyperchaotic systems, respectively. In the specific application of the method to ECG encryption, the hyperchaotic 3D-Hénon and Four-Dimensional (4D)-Hénon map (23) are chosen as the master systems. Hyperchaotic systems exhibit more complex behaviors than chaotic systems, with more positive Lyapunov exponents and more state variables

$$\begin{cases} x_{m1}^1(k+1) = a - (x_{m2}^1(k))^2 - bx_{m3}^1(k), \\ x_{m2}^1(k+1) = x_{m1}^1(k), \\ x_{m3}^1(k+1) = x_{m2}^1(k), \\ y_m^1(k) = Cx_m^1(k), \end{cases} \tag{23}$$

where C is a constant matrix.

Chaos is also used for image encryption algorithms, such as the optical color image encryption scheme based on fingerprint key and three-step phase-shifting digital holography, which was proposed by [Su *et al.*, 2021]. In this scheme, the fingerprint serves directly as the secret key. Random phase masks generated from the fingerprint using the Secure Hash Algorithm (SHA-256) and the chaotic Lozi map are simply used as intermediate variables. Using fingerprint-based random phase masks located in the linear canonical transform domain and three-step phase-shifting digital holography, the primary color image hidden in a grayscale carrier image can be encrypted into three noise-like holograms. Additionally, the parameters of Lozi chaotic map and linear canonical transform can also provide additional security to the proposed encryption scheme.

Recently, Dridi *et al.* [2022] have proposed a block cipher based on a secure chaotic generator. Their encryption/decryption system is based on a secure PRNG of Chaotic Sequences (PRNG-CS) and a strong key-dependent Substitution-Box (S-box).[21] The design of the proposed chaos-based cryptosystem takes into account the weaknesses and strengths of several current systems. The proposed PRNG-CS is able to prevent the divide-and-conquer and the side-channel attacks. This PRNG-CS is used to build the key-dependent S-box.

Many other examples of cryptography-based chaos can be found in [Kocarev & Lian, 2011; El Assad *et al.*, 2022].

[21] In cryptography, an S-box is a basic component of symmetric key algorithms which performs substitution. In block ciphers, they are typically used to obscure the relationship between the key and the ciphertext, thus ensuring Shannon's property of confusion. It also provides the encryption/decryption keys necessary for the substitution and permutation operation.

8.3.2. *Hash functions*

Cryptographic hash functions play a fundamental role in modern cryptography, and most secure communication systems rely on them. There are five main services needed for developers regarding information exchange over nonsecure channels (e.g. internet): confidentiality, authenticity, integrity, nonrepudiate, and availability. Hash functions are one of the most useful primitives in cryptography that play an important role in data security. This function is one-way and converts the data of an arbitrarily long message into a fixed-length hash value. It must guarantee certain security properties, such as message sensitivity, key sensitivity, confusion diffusion, pre-image, second pre-image, and collision resistance. Additionally, it should be immune to brute force and cryptanalytical attacks. Hash functions are classified into two classes: unkeyed and keyed.

Unkeyed cryptographic hash functions are a sub-type of the cryptographic hash functions. They take a variable-length input and convert it to a fixed-length output, and the length depends on the type of function used. The function's output is also known as Modification Detection Code (MDC). MDC represents the digest of the data sent as the input. The most popular standard secure unkeyed hash functions are Secure Hash Algorithms SHA-2 [Secure Hash Standard and Federal Information Processing Standard (FIPS) Publication 2002] and SHA-3, commonly used by many Secure Sockets Layer (SSL) certificate authorities.[22]

A keyed hash function (also known as a Hash Message Authentication Code, or HMAC) is an algorithm that uses a cryptographic key *and* a cryptographic hash function to produce a message authentication code that is encrypted and hashed. Algorithms for these functions include: Very fast Message Authentication Code — VMAC, Keyed-Hash MAC — HMAC, Galios/Counter Mode — GCM, Cipher-based MAC — CMAC, Destination MAC — DMAC, Cipher Block Chaining Message Authentication Code — CBC-MAC, and BLAKE 2 [Abdoun *et al.*, 2020]. In 2002, a new direction in the construction of chaos-based hash functions emerged. Due to the strong nonlinearity of chaotic systems and neural network structures, some designers usually combine these two systems to build robust hash functions (see [Xiao *et al.*, 2011] for a general presentation of the construction of chaos-based hash function).

Twenty years later, the domain is still fully explored with dozens of algorithms published. For example, Abdoun *et al.* [2020] designed, implemented, and analyzed the performance, in terms of security and speed, of two proposed keyed CNN hash functions based on Merkle–D̊amgard (MD) construction with three output schemes: CNN-Matyas–Meyer–Oseas, Modified CNN-Matyas–Meyer–Oseas, and CNN-Miyaguchi–Preneel. The first hash function's structure is composed of two-layer CNN while the structure of the second hash function is formed of one-layer CNN followed by nonlinear layer functions. The results obtained from several statistical tests and cryptanalytic analyses highlight the robustness of the proposed keyed CNN hash functions, which is fundamentally due to the strong nonlinearity of chaotic systems and neural networks. Comparison of performance analysis with some chaos-based hash functions from the literature and with standard hash functions makes the proposed hash functions suitable for data integrity, message authentication and digital signature applications.

8.3.3. *Memristors using fractal maps*

Before 1971, electrical circuit theory considered only three passive circuit elements: resistor, inductor and capacitor. In a paper published in 1971, L. O. Chua predicted the existence of a fourth missing element characterized by a nonlinear constitutive relationship between the charge q and the flux φ. He called this new element "memristor", meaning it is a resistor with memory [Chua, 1971]. This concept was forgotten until 2008, when a physical model of a two-terminal *hp* device behaving like a memristor was announced [Strukov *et al.*, 2008] sparkling intense research with thousands of papers published to date [Chua *et al.*, 2022]. Following this enthusiasm, the Chua Memristor Center was established at the University of Dresden, Germany, in 2016 [Chua Memristor Center, 2016]. It is a network of scientists in computer science, electrical engineering, physics and chemistry who conduct research in the field of memristors.

[22] In cryptography, a Certificate Authority or Certification Authority (CA) is an entity that stores, signs, and issues digital certificates. A digital certificate certifies the ownership of a public key by the named subject of the certificate.

A general theory, integrating the four devices memristor, resistor, inductor and capacitor, in a so-called first-order family of memfractance elements was published 11 years ago [Abdelouahab *et al.*, 2014]. In this theory, the existence of second- and higher-orders families is introduced together with a general Ohm's law using fractional derivatives (24).

Generalized Ohm's law [Abdelouahab *et al.*, 2014]: The voltage across a memfractance element can be expressed by the relation

$$v(t) = D_t^{1-\alpha_1}(F_M^{\alpha_1,\alpha_2}(t)D_t^{\alpha_2-1}i(t)), \qquad (24)$$

where $v(t)$ is the voltage, $i(t)$ the intensity, $D_t^{1-\alpha_1}$ is the fractional derivative at the order $1 - \alpha_1$, and $D_t^{\alpha_2-1}$ is the fractional derivative at the order α_2-1. $F_M^{\alpha_1,\alpha_2}(t)$ is called the memfractance: when $\alpha_1 = 1$, $\alpha_2 = 0$ memfractance concerns a memcapacitor, when $\alpha_1 = 1$, $\alpha_2 = 1$ a memristor, when $\alpha_1 = 0$, $\alpha_2 = 1$ a meminductor, and when $\alpha_1 = 0$, $\alpha_2 = 0$ a memristor.

Nowadays, fractal maps are used to model new families of memristors. Many researchers have devoted themselves to the analysis of chaotic phenomena in discrete memristors which is known as a research hotspot [Boudjerida *et al.*, 2023]. Recently, hidden attractors have also been discovered in some discrete memristor-based maps [Zhang *et al.*, 2004]. Wang *et al.* [2022] included a discrete-time memristor to create a memristive Lozi map. This new 3D-memristor-based Lozi map was established by coupling a discrete memristor to the original 2D-Lozi map (9)

$$\begin{cases} x_{n+1} = 1 - a|x_n| + y_n, \\ y_{n+1} = bx_n + ky_n \sin(z_n), \\ z_{n+1} = y_n + z_n, \end{cases} \qquad (25)$$

where k is a real-valued control parameter coupling gain between the discrete-time memristor and the Lozi map.

Since there are no fixed points but hyperchaos can emerge, the memristor-based Lozi map is a hidden hyperchaotic map. For some specific control parameters, the memristor-based 3D-Lozi map can exhibit heterogeneous and homogeneous hidden multistability. It should be noted that heterogeneous hidden multistability implies the coexisting behavior of multiple hidden attractors of different stability types, while homogeneous hidden multistability indicates the coexisting behavior of multiple hidden attractors of the same stability type but only in different dynamic intervals. In addition to the coexistence of these heterogeneous hidden attractors, the memristor-based Lozi map is very likely to produce the coexistence of homogeneous hidden hyperchaotic attractors, i.e. homogeneous hidden multistability. Therefore, the homogeneous hidden hyperchaotic attractors of the 3D-memristor-based Lozi map can be robustly controlled by the memristor initial conditions. Furthermore, Wang *et al.* [2022] implemented this memristor in a digital circuit based on a high-performance microcontroller. They physically obtained an image of the hyperchaotic hidden attractors using a digital oscilloscope. Finally, a digital platform was exploited and its experimental phase portraits were obtained to confirm the digital portraits.

8.3.4. *Optimization algorithms based on chaos theory*

Since the 17th century, mathematics has become an indispensable complement to the field of science and technology. The role of mathematics in industries is undoubtedly important and has a major role to play in their future success [Jain, 2023]. In particular, mathematics will help in solving engineering problems. Most of them can be defined as optimization problems, e.g. the finding of the optimal thickness of steel in pressure vessels, the optimal set of parameters for controllers, an optimal trajectory for a robot arm, optimal relations or fuzzy sets in fuzzy models, etc. Solutions to such problems are usually difficult to find, because their parameters usually include variables of different types, such as floating-point or integer variables.

It took longer for applications of chaotic maps to appear in the now flourishing field of optimization than for applications in monitoring electrical devices. This certainly comes from a paradigm shift in optimization algorithms. Long-used deterministic algorithms, such as the gradient method, steepest descent or simulated annealing, are not effective in optimizing high-dimensional problems involving up to thousands of variables. In contrast, heuristic algorithms introduced a few decades ago and based on an imitation of Darwin's theory of the evolution of species are more effective. It took three decades for this paradigm shift to occur in the study of chaotic maps (logistic, symmetric tent, Belykh, Hénon, Lozi, etc.). Instead of focusing on the theoretical study of their mathematical properties or on

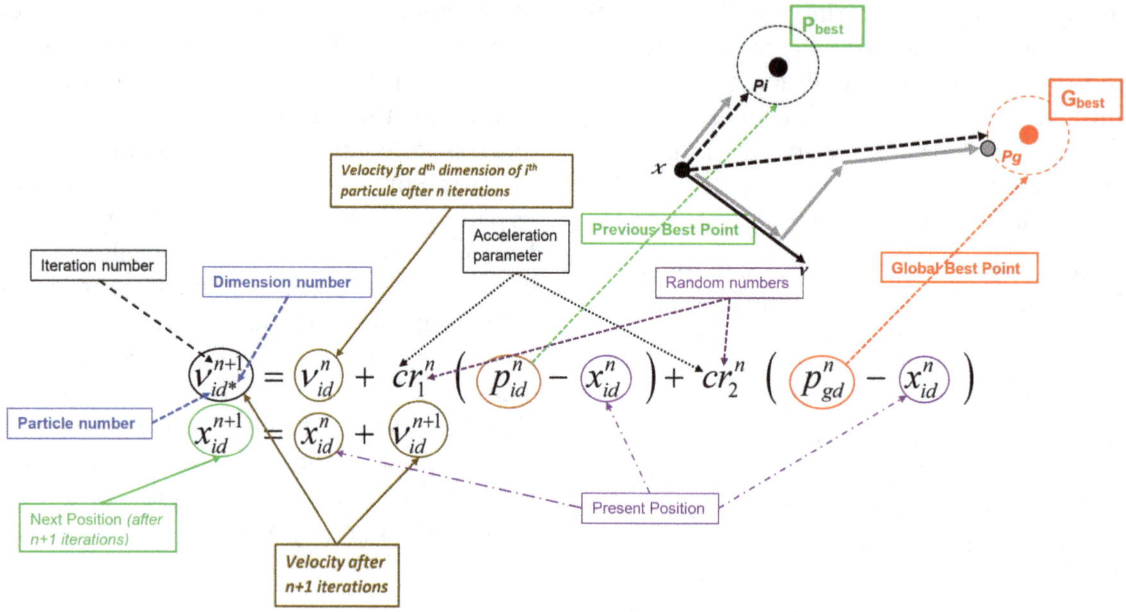

Fig. 34. Schematic representation of the PSO algorithm.

finding generalizations, Araujo and Coelho [2008] used them as a kernel for Particle Swarm Optimization (PSO) (see Fig. 34).

Chaos-based optimization algorithms are methodologies for finding optimal solutions that differ from all existing traditional stochastic optimization techniques. Due to the unpredictability of chaos, it can perform global searches in the solution space at higher speeds compared to stochastic ergodic searches, whose computation is based on probabilities. This numerical observation was made by [Caponetto *et al.*, 2003], who, five years before Araujo and Coelho, discovered that chaotic sequences improved the performance of evolutionary algorithms.

The PSO method has been used for many purposes. An example is the control of the thermal-vacuum system used at the Brazilian National Institute for Space Research (INPE), where an original controller was designed to control the temperature on the shroud (set of pipes) of a chamber where satellites are tested [Marinke *et al.*, 2005]. This method was also used by [Pluhacek *et al.*, 2012], who considered a Partial–Integral–Derivative (PID) controller for a Direct Current (DC) motor system to achieve optimal settings.[23] The optimization process involving the PSO algorithm was applied to minimize the errors of the output transfer function which can indicate the quality of regulation of such a controller.

Differential Evolution (DE) is another evolutionary optimization algorithm that has been used by [Davendra *et al.*, 2010] for the same purpose, and by [Senkerik *et al.*, 2013] in the task of optimizing the geometry of batch chemical reactors.

Self-Organizing Migrating Algorithm (SOMA) was proposed by [Zelinka, 2004]. It is a new class of stochastic optimization algorithms. Unlike evolutionary algorithms, which operate on populations of candidate solutions evolving over generations (two parents create a new individual — the offspring) and in which only the fittest individuals have a chance of survival, SOMA, which can also operate on a population of individuals, relies on the self-organized behavior of groups of individuals in a "social environment", such as a herd of animals foraging for food. A group of animals like wolves or other predators can be a good example. If they are searching for food, they usually cooperate and compete so that if one member of the group is successful (has found food or shelter), the other animals in the group change their trajectories towards the more successful member. If one member of this group is more successful than the previous best (found more food, etc.), then all members change their trajectory again towards the new successful member. This

[23] A DC motor is any of a class of rotary electrical motors that converts DC electrical energy into mechanical energy. PID control is the most common control algorithm used in industry and has been universally accepted in industrial control.

process is repeated until all members gather around a single food source. This principle, taken from the real world, is of course greatly simplified. However, it can be said that it is the competitive–cooperative behavior of intelligent agents that allows SOMA to conduct very fruitful research. Recently, Zelinka *et al.* [2023] used SOMA for the design of quantum computing circuits for future quantum computers.

It is not possible and is not the aim to present within the limited scope of this section all the dozens of algorithms using chaotic attractors (see [Lozi, 2023b] for an overview).

8.4. *Application in economy*

In order not to bore the reader with an encyclopedic presentation, we limit the panorama of the other applications of chaotic dynamic systems to only two additional domains: economics and human health. However, many applications exist in other fields, such as biology, disease transmission, ecology, population dynamics, etc. (see [Lozi, 2023a, 2023b] for a survey).

Applications of chaotic dynamical systems in economy have emerged two decades ago. For example, the case of Parrondo's game is interesting. The main idea of Parrondo's paradox, expounded in 1996, is that two individually losing games can be combined to win via a deterministic or nondeterministic mixture of games [Harmer *et al.*, 2001]. Tang *et al.* [2004] conducted an analysis of Parrondo games with different chaotic switching strategies. The performance of Parrondo's games was compared with random and periodic switching strategies. To this end, a fair method for comparing random and chaotic Parrondo's games was generalized. 1D-logistic, tent, sinusoidal, and Gaussian maps were considered, as well as Hénon and Lozi maps. To play Parrondo's chaotic games, one of these chaotic generators being chosen, a sequence that it generates from an initial value is considered. Then, every nth iteration of this sequence determines whether game A or B is played. The results of the Parrondo's game are affected by the different switching strategies applied and the initial value chosen. The proportion of games A and B played is kept equal for all switching strategies for fair comparison. In conclusion, the authors found that chaotic Parrondo games can generate a higher win rate than random switching strategies. This result is reminiscent of the observation made by [Caponetto *et al.*, 2003] that chaotic sequences can improve the performance of evolutionary algorithms compared to random sequences. It seems to be the same for Parrondo's games.

Another example can be found in [Commendatore *et al.*, 2015] in which Commendatore *et al.* proposed a New Economic Geography (*NEG*) model that describes the long-term spatial distribution of industrial activity in three identical regions based on the balance of agglomeration and dispersion forces. The *NEG* approach aims to explain how industrial economic activity tends to spread or agglomerate across space. It is defined by a 2D-piecewise smooth map $Z(\lambda_1, \lambda_2)$ depending on four parameters $\lambda_1, \lambda_2, M_r, M_s$ (26). It is shown that in the phase plane of this Z-map, there are several invariant lines on which the 2D Z-map is reduced to piecewise smooth 1D maps which can have quite complicated dynamics, including the coexistence of multiple attractors and chaos. In Fig. 35(left), an example of the basins of attraction is shown, where attracting (in Milnor sense) core–periphery fixed points $(0, 0)$, $(1, 0)$ and $(0, 1)$ coexist with three 2-cyclic and three 6-cyclic chaotic attractors

$$Z(\lambda_1, \lambda_2) = \begin{cases} Z_1(\lambda_1, \lambda_2), \\ Z_2(\lambda_1, \lambda_2), \end{cases}$$

were

$$Z_r(\lambda_1, \lambda_2) = \begin{cases} 0 & \text{for } M_r \leq 0, \\ M_r & \text{for } M_r > 0, \\ & \quad M_s > 0,\ M_r + M_s < 1, \\ \dfrac{M_r}{M_r + M_s} & \text{for } M_r > 0, \\ & \quad M_s > 0,\ M_r + M_s \geq 1, \\ \dfrac{M_r}{1 - M_s} & \text{for } M_r > 0, \\ & \quad M_s \leq 0,\ M_r + M_s < 1, \\ 1 & \text{for } M_r > 0, \\ & \quad M_s \leq 0,\ M_r + M_s \geq 1, \end{cases} \tag{26}$$

with

$$\begin{pmatrix} r = 1 \\ s = 2 \end{pmatrix} \quad \text{and} \quad \begin{pmatrix} r = 2 \\ s = 1 \end{pmatrix}.$$

$M_r = M_r(\lambda_1, \lambda_2)$, $M_s = M_s(\lambda_1, \lambda_2)$, indicate the shares of entrepreneurs.

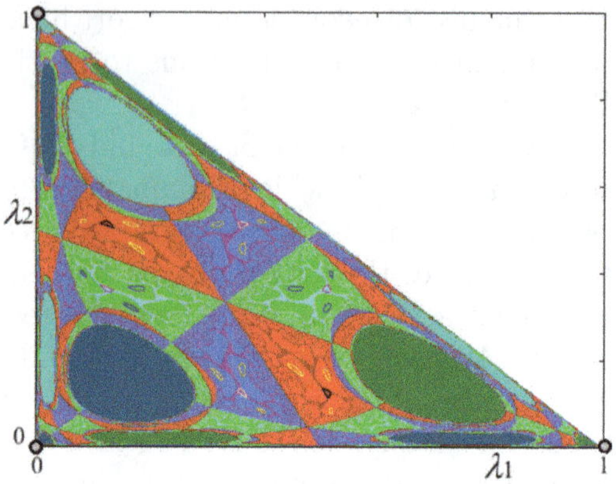

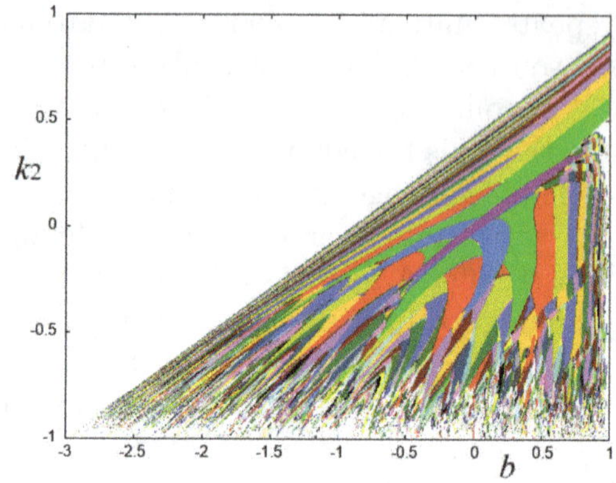

Fig. 35. Left: 2D-piecewise smooth map Z governing dynamics of a three-region *NEG* model. Basins of attraction of the fixed points $(0,0)$, $(1,0)$, $(0,1)$ (attracting in Milnor sense) coexisting with three 2-cyclic and three 6-cyclic chaotic attractors. Right: Periodicity regions (where different colors are related to attracting cycles of different periods) in the (b, k_2)-parameter plane for $k_1 = -1$, $m = 1.9$, $k = 0.1$.

A third example of the use of chaotic attractors in economics can be found in [Sushko & Tramontana, 2025] in which the dynamics of a financial market model with trend followers and contrarians is defined again by a 2D-piecewise-linear discontinuous mapping F_c given by

$$\begin{cases} x_{n+1} = (b - k_1)x_n + k_1 x_{n-1} \\ \qquad\qquad \text{if } |x_n - x_{n-1}| < k, \\ x_{n+1} = (b - k_2)x_n + k_2 x_{n-1} + d \\ \qquad\qquad \text{if } |x_n - x_{n-1}| > k. \end{cases} \tag{27}$$

Analysis of this model shows that despite the linearity of the map's components, its dynamics can be quite complex, with various attracting cycles and chaotic attractors possibly coexisting. In Fig. 35(right) periodicity regions related to attracting cycles of different periods are shown by different colors in the (b, k_2)-parameter plane for $k_1 = -1$, $k = 0.05$ and $d = 1$.

The authors concluded that "The presence of periodic and chaotic dynamics, together with the scenarios of multistability, are of extreme relevance for the possibility of replicating important stylized facts of financial markets. Chaotic motion causes the unpredictability typical of asset prices and the alternation of bubbles and crashes. Even more interestingly can be the scenarios where different attractors coexist. In particular, when an attractor characterized by low variability of the dynamics (periodic of low period, or even a stable equilibrium) coexists with another stable attractor causing price fluctuations of large amplitude (periodic of high period or chaotic attractor), this can be the perfect basis to replicate some other stylized facts of financial markets, such as volatility clustering, simply by introducing some stochastic element into the model".

8.5. *Chaotic dynamical systems for human health*

Another interesting type of application of chaotic dynamical systems is that they can be used as a reference for some dysfunctions of the human body or to model certain diseases.

8.5.1. *Chaos as an indicator of health or disease*

Drawing on recent advances in physiological chaos and disease dynamics, Boubaker [2024] has published a survey study that attempts to answer the crucial question: when is chaos a sign of health or disease? She wrote "In the context of disease, like for example growth of cancer cell populations, chaos often refers to irregular and unpredictable patterns. In such cases, chaos signatures can be used to prove the existence of some pathologies. However, for other physiological behaviors, chaos is a form of order disguised as disorder and can be a signature of healthy physiological functions. This is for example the case of human brain behavior. As the boundary between health and disease is not always clear-cut in chaotic systems in physiology, some conditions may involve transitions between ordered and chaotic states. Understanding these transitions and identifying critical points can be crucial for predicting healthy vs. pathological chaos". In this paper, chaos is considered as a positive or negative fashion. Healthy chaotic patterns are found in: healthy cardiovascular and respiratory systems, healthy neuronal activity of the brain, healthy cardiovascular and respiratory systems. Healthy chaos is found in the gait and locomotion system, and also in the immune system. However, chaos is also present in disease: cancer progression, metabolic disorders, neurological disorders, cardiovascular and respiratory disorders, viral diseases impacting the immune system.

Regarding cancer progression, it is worth noting that one of the most interesting models was proposed by [Itik & Banks, 2010]. In this nondimensional model a three-cell population is considered and its dynamics is described by

$$\begin{cases} \dot{x}_1 = x_1(1-x_1) - a_1x_1x_2 - a_2x_2x_3, \\ \dot{x}_2 = a_3x_2(1-x_2) - a_4x_1x_2, \\ \dot{x}_3 = \dfrac{a_5x_1x_3}{x_1+a_6} - a_7x_1x_3 - a_8x_3, \end{cases} \tag{28}$$

where x_1 is the number of tumor cells, x_2 stands for the number of host cells and x_3 the number of effectors cells in the single tumor compartment. The a_i $(i = 1, \ldots, 8)$ are the parameters of the system.

A patient is healthy when the effector cells are equal to zero, more precisely when the chaotic-cancer system converges to an equilibrium point. Concerning cancer again, Denis and Letellier [2012] clarified the main points of connection between chaos and carcinogenesis and highlighted some promising research perspectives, particularly in radiotherapy.

Several other models are examined in [Boubaker, 2024]. In particular, an integer-order model for human glucose–insulin regulatory system is described by [Shabestari *et al.*, 2018]

$$\begin{cases} \dfrac{dx_1}{dt} = a_1x_2(t-\tau_1)x_3(t-\tau 1) \\ \qquad - a_2x_1 + a_3x_3(t-\tau_1), \\ \dfrac{dx_2}{dt} = \dfrac{a_4}{x_3} - a_5x_1(t-\tau_2) + a_6, \\ \dfrac{dx_3}{dt} = a_7(x_2-\hat{x}_2)(T-x_3) \\ \qquad + a_8x_3(T-x_3) - a_9x_3, \end{cases} \tag{29}$$

where x_1, x_2, x_3, and $\hat{x}_2$ are the insulin concentration, glucose concentration, β-cells number and $\hat{x}_2$ the difference between glucose metabolism fasting level and its basal level, respectively. The delay in insulin secretion in response to increased blood glucose levels, based on clinical evidence, is τ_1, and τ_2 is the delay in glucose drop due to increased insulin based on clinical evidence. The parameters corresponding to several partial metabolisms are a_i $(i = 1, \ldots, 9)$.

8.5.2. *A chaotic model for migraine disease*

Migraine Disease (MD) is one of the most common primary headache disorders, which can disrupt patients' daily lives. Its global prevalence has increased from 721.9 million in 1990 to 1.1 billion in 2019 [Vos *et al.*, 2019]. Despite this high prevalence, its pathophysiology remains poorly understood. Garehdaghi *et al.* [2024] consider the brain as a dynamical system and propose using the Chua attractor (16) as a metaphorical explanation for the onset of this disease. When this attractor has a one-scroll attractor, it represents a healthy brain and when the attractor is a double-scroll, it represents a migraine sufferer brain. They believe that MD and Chua systems share some behavioral similarities. The attractor boundaries are sensitive brain areas where the slightest trigger can trigger the critical phase of migraine. The transition from the inter-ictal to the ictal phase in migraine patients occurs due to a decrease in serotonin levels when the brain is within the boundaries of the first attractor. Furthermore, the transition from the ictal phase to the inter-ictal phase in a migraine sufferer brain is due to a disruption in the coordination of brain structures, which persists for a certain period of time in each migraine patient. This explanation could offer new methods for preventing or curing MD, by studying the behavior of the Chua attractor.

8.5.3. *Hypertension*

In several articles, Moreno draw a chaotic analysis for pseudo- and Resistant Hypertension (RHT) [Moreno, 2022; Yugar-Toledo *et al.*, 2022; Dinamarco *et al.*, 2022]. He said "Systemic Blood Pressure (BP) may oscillate for homeostatic needs (equilibrium by constancy) or just as shifts in other intrinsic and extrinsic variables known as allostatic changes. This transitory pressure often alerts physicians about out-of-control hypertension or even hypertensive crisis. There is a very complex theory underlying these stochastic phenomena, which physicists and mathematicians translate into a single word: chaos. [...] Based on the chaos theory, small changes at the initial BP (baseline overtime) values could disturb the homeostasis[24] leading to extreme BP chaotic shifts. These almost insignificant oscillations may also affect other variables and systems, leading to the misdiagnosis of hypertension, "out-of-control" BP levels, and RHT. Thus, these unpredictable and transient increases in BP values may be improperly diagnosed as the white coat and masked or RHT. [...] When evaluating BP levels obtained by any method, physicians might incorporate this concept".

8.5.4. *Cardiac dysrhythmias*

Guevara and Glass [1982] presented a mathematical model of a periodically driven oscillator: a theory for the entrainment of biological oscillators and the genesis of cardiac dysrhythmias. This biological oscillator model is described in 2D-polar coordinates by the equations

$$\begin{cases} \dfrac{d\Phi}{dt} = 2\pi, \\ \\ \dfrac{dr}{dt} = ar(1-r), \end{cases} \tag{30}$$

where Φ is the angular coordinate ($-\infty < \Phi < \infty$), r is the radial coordinate and a is a positive real number.

This model has previously been discussed by [Winfree, 1975, 1980; Scott, 1979]. Guevara and Glass [1982] suggested that cardiac dysrhythmias might arise from the desynchronization of two or more spontaneously oscillating regions of the heart. In addition to phase-locked dynamics, chaotic dynamics were observed in certain cases and could correspond to phenomena observed in electrocardiography.

8.6. *New theoretical trends*

We have shown that chaotic attractors have been used in this last period lasting 35 years for applications in different fields. This does not mean that they did not advance pure mathematics. As said in the introduction of Sec. 8 we will consider only few theoretical trends among many.

8.6.1. *Improved attractors*

Chaotic maps such as logistic, Hénon, and Lozi maps have been generalized in many ways (see [Lozi, 2023b, 2023c]). New generalizations continue to generate interest and research, and are constantly being proposed. Gupta and Chandramouli [2021] studied the q-deformation of the logistic map using deformation on the right side as well as on the left side. They did not find any qualitatively different behavior, the period-doubling cascade being independent of the deformation parameter. However, the q-deformation of maps can be achieved in many ways. Jaganathan and Sinh [2005] using different q-deformations for the same map found many differences in the bifurcation diagram, including the absence of a complete period doubling cascade. Moreover, q-deformed logistic map showed multistability which is rare in 1D maps.

For 2D maps, Patidar *et al.* [2011] studied q-deformed Hénon map (31), which was also analyzed by [Gupta & Chandramouli, 2022]

$$\mathcal{H}_{a,b,\epsilon}\begin{pmatrix} x \\ y \end{pmatrix} = \begin{pmatrix} F_{a,\epsilon}(x) - by \\ \dfrac{x}{1+\epsilon-\epsilon x} \end{pmatrix}, \tag{31}$$

where

$$F_{a,\epsilon}(x) = a - \frac{x^2}{(1+\epsilon-\epsilon x)^2},$$

for $a > 0$ and $0 \leq b \leq 1$. They demonstrated that chaos is suppressed if the q-deformation values are chosen appropriately.

Gaiki *et al.* [2024] considered q-deformed Lozi map (32) under q-deformation of the x variable only, the y variable only and both x and y variables, for the control parameter values being fixed at $a = 1.4$ and $a = 1.7$ with different values of b. The underlying map is a rational function for such deformation and exhibits singularities

$$\begin{pmatrix} x_{n+1} \\ y_{n+1} \end{pmatrix} = \begin{pmatrix} f([x_n]_{q_x}, [y_n]_{q_y}) \\ g([y_n]_{q_y}, [y_n]_{q_y}) \end{pmatrix}, \tag{32}$$

where

$$[x]_{q_x} = \frac{x}{1+(1-q_x)(1-x)}$$

and

$$[y]_{q_y} = \frac{y}{1+(1-q_y)(1-y)}.$$

[24]Homeostasis is a self-regulation processes to maintain stability while adjusting to optimal conditions.

The bifurcation diagrams, Lyapunov exponents, phase portraits, and basins of the attractor were used to examine the underlying multistability in each case. A long list of result is presented in their article. In addition, they have studied chaotic synchronization in two doubly q-deformed Lozi maps.

8.6.2. *Chimeras and solitary states*

As said in Sec. 5 there is a dual stream of research in mathematics (in fact in every domains of it), which has been operating since Antiquity in a competition-cooperation mode. The best relevant example of a theory–application–theory interaction is that of chimeras and solitary states. Following the discovery of the synchronized chaotic attractors (theory), research focused on a network of attractors with several topologies for multiple purposes such as creating PRNG for cryptography [Garasym *et al.*, 2017] (application). By describing the dynamical synchronization properties of such networks, peculiar solutions called "chimeras" and "solitary states" have been discovered (theory).

A complex system consisting of three coupled rings of nonlocally coupled chaotic maps was considered by [Rybalova *et al.*, 2018]. This multilayer network is described by the following equations:

$$\begin{cases} x^i_{n+1} = f(x^i_n, y^i_n) \\ \qquad + \dfrac{\sigma_1}{2P} \displaystyle\sum_{j=i-P}^{j=i+P} [f(x^j_n, y^j_n) - f(x^i_n, y^i_n)] + \gamma_1 F^i_n, \\ y^i_{n+1} = bx^i_n, \\ u^i_{n+1} = f(u^i_n, v^i_n) \\ \qquad + \dfrac{\sigma_2}{2R} \displaystyle\sum_{j=i-R}^{j=i+R} [f(u^j_n, v^j_n) - f(u^i_n, v^i_n)] + \gamma_2 G^i_n, \\ v^i_{n+1} = bx^i_n, \\ z^i_{n+1} = f(z^i_n, s^i_n) \\ \qquad + \dfrac{\sigma_3}{2T} \displaystyle\sum_{j=i-T}^{j=i+T} [f(z^j_n, s^j_n) - f(z^i_n, s^i_n)] + \gamma_3 H^i_n, \\ s^i_{n+1} = bx^i_n. \end{cases} \tag{33}$$

The first system of equations in (33) describes a ring network of nonlocally coupled Hénon maps with f defined by (34)

$$f(x_n, y_n) = 1 - ax_n^2 + y_n, \tag{34}$$

with $a = 1.4$, $b = 0.3$, $\sigma_1 = 0.72$, and $P = 320$. The second pair of equations concerns the ring of nonlocally coupled Lozi maps with f defined by (35)

$$f(x_n, y_n) = 1 - a|x_n| + y_n \tag{35}$$

and is analyzed for $a = 1.4$, $b = 0.3$, $\sigma_2 = 0.206$, and $R = 180$. The third pair of equations determines again the ring of nonlocally coupled Hénon maps with $a = 1.4$, $b = 0.3$, $\sigma_1 = 0.295$, and $T = 320$.

The first two rings are inertially coupled via the coupling functions

$$F^i_n = -G^i_n = u^i_n - x^i_n,$$

using the coupling coefficients γ_1 and γ_2. The third ring node is connected unidirectionally with the first ring units by the coupling term $\gamma_3 H^i_n$ where

$$H^i_n = f(x^i_n, y^i_n) - f(z^i_n, s^i_n), \tag{36}$$

defines the diffusive coupling with the coupling coefficient γ_3. N is the number of elements in the ensemble of coupled equations in each ring. The coupling parameters σ_i, $i = 1, 2, 3$, characterize the coupling strength, and $2P, 2R, 2T$ are the number of neighbors on each ring (P (*resp.*, R, T) neighbors on the either side of the ith element). The initial conditions are chosen to be randomly distributed in the interval $[-0.5, 0]$ for all the variables of the network (33).

Using numerical simulation, Rybalova *et al.* [2018] have demonstrated that the network of two symmetrically coupled sets of Hénon and Lozi maps can exhibit a new type of chimera state, a *Solitary State Chimera* (*SSC*), when the coupling between them is weak. This particular structure appears when the Lozi ensemble presents a developed regime of solitary states. The *SSC* is quite stable and is observed over a finite range of parameter variation. If the two-layer network of nonlocally coupled Hénon and Lozi maps in the SSC is unidirectionally coupled to the third ring of nonlocally coupled Hénon maps, then the effect of external synchronization can be observed within a finite range of the coupling coefficient γ_3.

8.6.3. *Partial control theory*

Finally, we can point out that chaotic dynamical systems often exhibit transient chaos, where trajectories behave chaotically for a short period before escaping towards an external attractor. Maintaining transient chaotic dynamics under perturbations is challenging but desirable for many applications. A survey of the very recent methods developed by Capeáns and Sanjuán (see Fig. 22) that develop partial control theory for this purpose is presented in [Capeáns & Sanjuán, 2024]. The partial control approach exploits the inherent symmetry and geometric structure of chaotic saddles, the topological object responsible for transient chaos, to enable surprising control with only small perturbations. It relies on the existence of specific regions, called safe sets, embedded in the phase space containing the chaotic saddle. Informally, safe sets are areas near the saddle where trajectories can be sustained indefinitely through occasional small controls, despite disturbances (noise). To find safe sets, we assume that the dynamics in the region Q containing the chaotic saddle can be described by the map

$$q_{n+1} = f(q_n) + \xi_n + u_n.$$

Here, $q_n \in \mathbb{R}^n$ represents a certain phase state of the system, and they assume that the map f acts on a region Q like a horseshoe map. The disturbance ξ affecting the map is considered to be bounded such that $|\xi_n| \leq \xi_0$. To be realistic, the control term u is also limited to $|u_n| \leq u_0$. The safe set is defined as the subset $Q_\infty \subset Q$. Any safe point $q \in Q_\infty$ satisfies that the orbit $q_{n+1} = f(q_n) + \xi_n + u_n$ can be sustained in Q_∞ indefinitely. At each iteration, the control $|u_n| \leq u_0$ is chosen based on $f(q_n) + \xi_n$ to keep the orbit in Q_∞.

They applied their methods to Hénon (37) and Lozi maps (38)

$$\begin{cases} x_{n+1} = a - by_n - x_n^2 + \xi_n^x + u_n^x, \\ y_{n+1} = x_n + \xi_n^y + u_n^y, \end{cases} \tag{37}$$

with $a = 6$, $b = 0.4$, $Q = [-4, 4] \times [-4, 4]$, $\|\xi_n^x, \xi_n^y\| \leq \xi_0$, and $\|u_n^x, u_n^y\| \leq u_0$

$$\begin{cases} x_{n+1} = 1 - a|x_n| + by_n + \xi_n^x + u_n^x, \\ y_{n+1} = x_n + \xi_n^y + u_n^y, \end{cases} \tag{38}$$

with $a = 2$, $b = 0.5$, $Q = [-2, 2] \times [-2, 2]$, $\|\xi_n^x, \xi_n^y\| \leq \xi_0$, and $\|u_n^x, u_n^y\| \leq u_0$.

Additionally, they introduced escape functions, designed to intentionally force the escape of trajectories from the Q region in a predictable manner, using small controls. Escape sets based on escape functions allow escaping chaotic regions in precise steps. These new functions expand the potential applications of partial control.

9. Conclusion

In this paper, I have briefly described the history of chaotic dynamical systems since Henri Poincaré's original work on celestial mechanics and more particularly on the three-body problem in the 1880s–1890s. I have distinguished three periods: the first period "the dawn of chaotic dynamical systems" from 1885 to 1970 which includes the KAM theorem on the persistence of quasi-periodic motions under small perturbations extended by Moser in 1962, with its proof achieved in 1963 by Vladimir Arnold, the discovery of the Lorenz attractor in 1963, the Sharkovsky order (1964), the Shilnikov bifurcation (1965), the horseshoe and the fractals in 1967. The second period "the intermediate period 1971–1989", begins with the definition of the strange attractors by Ruelle and Takens, and ends just before the introduction of the synchronization of chaotic systems in 1990 by Pecora and Carroll. The third period extends from 1990 to the present day, which is a flourishing time of applications in several sciences of this mathematical theory. This period corresponds to the publication of 35 volumes (415 issues) of the *International Journal of Bifurcation and Chaos in Applied Sciences and Engineering*, since the 35 years of existence of this journal.

This history of chaotic dynamical systems is presented in a reasoned manner, which provides some insight into the driving forces behind their evolution. The people who shaped them are also featured. An epistemological analysis structures the analysis of these three periods. It takes into account the notions of causality, determinism, predictability, unpredictability and control through a model described in Fig. 5 which help to understand the evolution of the theory and applications of dynamical systems during their 150 years old history, and also to provide some insight into the driving forces behind this evolution.

Theory was the main driving force behind research progress in the field of chaotic dynamical systems during the first two periods of the century, from 1890 to 1990, with the exception of research carried out by the Andronov School in Gorky (now Nizhny Novgorod) in the USSR, which was motivated by radio and electronic applications, but not only. Since then, it is applications that have allowed an extraordinary development of mathematics in this field, even if theoretical research has continued. Given that several scientific fields are concerned with the applications of these dynamical systems, it is understandable that the number of researchers involved has increased dramatically as well as the results obtained. The creation of the *International Journal of Bifurcation and Chaos in Applied Sciences and Engineering* during this period contributed greatly to disseminating these results and encouraged many researchers around the world to take an interest in this research.

The first research on chaotic attractors using computers during this last period marked the mind of the public by the beauty of the images that these mathematical objects made it possible to draw. Nowadays, applications of chaotic attractors in several domains is a flourishing domain of research (see [Lozi, 2023a, 2023b] for a survey) and can nevertheless produce wonderful images (Fig. 35). In the meantime, theoretical research is still very much alive and offers new mathematical tools such as chimeras and solitary states, dedicated fractional differential equations and fractional mappings, partial control theory, hidden attractors, etc., which in turn will allow the development of new applications in several domains like electronics mentioned by [Minati, 2024] in chaotic electric circuits. One peculiarity of these circuits is the ability to generate, despite their structural simplicity, complex spatiotemporal patterns when several of them are brought towards synchronization via coupling mechanisms. Minati [2024] presented a selection of results exemplifying potential applications, including in robot control, distributed sensing, reservoir computing, and data augmentation. He highlighted that despite their interesting properties, the industrial applications of these circuits remain largely to be realized, seemingly due to a variety of technical and organizational factors including a paucity of design and optimization techniques. Therefore, chaotic attractors are definitely not a mathematical curiosity. It is possible that they will be used for several more decades.

In concluding this paper on the history of a scientific field that has occupied a large part of my academic life, I have only one regret: not having cited all the scientists involved in this field and all of their work. But a large book would not be enough to achieve this.

Acknowledgments

The author thanks Laura Gardini from University of Urbino (Italy) and Iryna Sushko from the National Academy of Sciences of Ukraine, Kyiv (Ukraine) for drawing two special figures (Fig. 35) for this paper.

ORCID

René Lozi
https://orcid.org/0000-0003-0451-4255

References

Abdelouahab, M.-S., Lozi, R. & Chua, L. O. [2014] "Memfractance: A mathematical paradigm for circuit elements with memory," *Int. J. Bifurcation and Chaos* **24**, 1430023.

Abdoun, N., El Assad, S., Deforges, O., Assaf, R. & Khalil, M. [2020] "Design and security analysis of two robust keyed hash functions based on chaotic neural networks," *J. Ambient Intell. Humaniz. Comput.* **11**, 2137–2161.

Abubakr, A. M. [1955] "Divine boats of ancient Egypt," *Archaeology* **8**, 96–101.

Adlam, J. H. & Allen, J. E. [1958] "The structure of strong collision-free hydromagnetic waves," *Philos. Mag.: J. Theor. Exp. Appl. Phys.* **39**, 448–455.

Afraimovich, V. S., Bykov, V. V. & Shilnikov, L. P. [1977] "On the origin and structure of the Lorenz attractor," *Akad. Nauk SSSR Dokl.* **234**, 336–339.

Allen, J. E. [1998] "The early history of solitons (solitary waves)," *Phys. Scr.* **57**, 436–441.

Andronov, A. A. & Pontryagin, L. [1937] "Systèmes grossiers," *Dokl. Akad. Nauk SSSR* **14**, 247–251.

Andronov, A. A., Vitt, A. A. & Khaikin, S. E. [1937, 1966] *Theory of Oscillators*, 1st edition (Pergamon, London, UK), Translated from the Russian by F. Immerzi, Moscow.

Araujo, E. & Coelho, L. S. [2008] "Particle swarm approaches using Lozi map chaotic sequences to fuzzy modeling of an experimental thermal-vacuum system," *Appl. Soft Comput.* **8**, 1354–1364.

Arnold, V. I. [1963] "Proof of a theorem of A. N. Kolmogorov on the preservation of conditionally periodic motions under a small perturbation of the Hamiltonian," *Usp. Mat. Nauk* **18**, 9–36.

Aubin, D. & Dahan Dalmedico, A. [2002] "Writing the history of dynamical systems and chaos: Longue durée and revolution, disciplines and cultures," *Hist. Math.* **29**, 1–67.

Barrow-Green, J. [1997] *Poincaré and the Three Body Problem* (American Mathematical Society, Providence, RI).

Bartoš, F. *et al.* [2025] "Fair coins tend to land on the same side they started: Evidence from 350,757 flips," *J. Am. Stat. Assoc.*, 1–10, https://doi.org/10.1080/01621459.2025.2516210.

Belykh, V. N. [1976] "On models of phase synchronization systems and their study," *Din. Syst.* **11**, 23–32.

Belykh, V. N. & Maksakov, V. P. [1976a] "Dynamics of the simplest digital system of phase synchronization," *J. Commun. Technol. Electron.* **21**, 2155–2163.

Belykh, V. N. & Maksakov, V. P. [1976b] "Dynamics of discrete-time systems of digital phase synchronization," *Din. Syst.* **11**, 130–143 (in Russian).

Belykh, V. N. [1982] "Models of discrete systems of phase synchronization," in *Systems of Phase Synchronization*, eds. Shakhildyan, V. V. & Belyustina, L. N. (Radio i Svyaz, Moscow), pp. 161–216.

Belykh, V. N. [1995] "Chaotic and strange attractors of a two-dimensional map," *Sb. Math.* **186**, 311–326.

Belykh, V. N. & Belykh, I. V. [2011] "Belykh map," *Scholarpedia* **6**, 5545.

Belykh, V. N., Barabash, N. V. & Grechko, D. A. [2023] "Existence proofs for strange attractors in piecewise-smooth nonlinear Lozi-Hénon and Belykh maps," *J. Diff. Eqs. Appl.* **29**, 994–1014.

Birkhoff, G. D. [1913] "Proof of Poincaré's geometric theorem," *Trans. Am. Math. Soc.* **14**, 14–22.

Birkhoff, G. D. [1927] *Dynamical Systems* (American Mathematical Society, Providence, RI), Reprinted with an introduction by J. Moser and a preface by M. Morse, 1966.

Birkhoff, G. D. [1932] "Sur quelques courbes fermées remarquables," *Bull. Soc. Math. France* **60**, 1–26.

Boccaletti, S., Grebogi, C., Lai, Y.-C., Mancini, H. & Maza, D. [2000] "The control of chaos: Theory and applications," *Phys. Rep.* **329**, 103–197.

Bottazzini, U. [2000] *Poincaré, Philosophe et Mathématicien*, Les Génies de la Science, Vol. 4 (Pour la Science, Paris).

Boubaker, O. [2024] "Chaos in physiological control systems: Health or disease?" *Chaos Theory Appl.* **6**, 1–12.

Boudjerida, N., Abdelouahab, M.-S. & Lozi, R. [2023] "Nonlinear dynamics and hyperchaos in a modified memristor-based Chua's circuit and its generalized discrete system," *J. Diff. Eqs. Appl.* **29**, 1369–1390.

Boughaba, S. & Lozi, R. [2000] "Fitting trapping regions for Chua's attractors — A novel method based on isochronic lines," *Int. J. Bifurcation and Chaos* **10**, 205–225.

Bouligand, G. [1925] "Dimension, étendue, densité," *C. R. Acad. Sci. Paris* **180**, 246–248.

Bunge, M. [1959] *The Place of the Causal Principle in Modern Science* (Harvard University Press, Cambridge), 380 pp.

Cao, Y. & Liu, Z. [1998] "Strange attractors in the orientation-preserving Lozi map," *Chaos Solit. Fract.* **9**, 1857–1863.

Capeáns, R. & Sanjuán, M. A. F. [2024] "Partial control and beyond: Controlling chaotic transients with the safety function," *Symmetry* **16**, 338.

Caponetto, R., Fortuna, L., Fazzino, S. & Xibilia, M. G. [2003] "Chaotic sequences to improve the performance of evolutionary algorithms," *IEEE Trans. Evol. Comput.* **7**, 289–304.

Chabert, J.-L. & Dahan Dalmenico, A. [1992] "Les idées nouvelles de Poincaré," in *Chaos et Déterminisme*, eds. Dahan, A., Chanert, J.-L. & Chemla, K. (Éditions du Seuil, Paris), pp. 274–305.

Chen, G. & Ueta, T. [1999] "Yet another chaotic attractor," *Int. J. Bifurcation and Chaos* **9**, 1465–1466.

Chen, G. & Huang, Y. [2011] "The Smale horseshoe," in *Chaotic Maps*, Synthesis Lectures on Mathematics & Statistics (Springer).

Chirita, D., Stefanescu, V., Florea, B. C. & Stoichescu, D. A. [2015] "An analogical control method of the Alpazur oscillator," in *2015 9th Int. Symp. Advanced Topics in Electrical Engineering* (*ATEE*) (IEEE), pp. 854–857.

Chua, L. O. [1971] "Memristor — The missing circuit element," *IEEE Trans. Circuit Theory* **18**, 507–519.

Chua, L. O. [1992] "The genesis of Chua's circuit," *AEÜ* **46**, 250–257.

Chua Memristor Center [2016] *Chua Memristor Center*, https://cmc-dresden.org/.

Chua, L. O., Tetzlaff, R. & Slavova, A. [2022] *Memristor Computing Systems* (Springer, Berlin), 298 pp.

Commendatore, P., Kubin, I. & Sushko, I. [2015] "I.: Typical bifurcation scenario in a three region symmetric new economic geography model," *Math. Comput. Simul.* **108**, 63–80.

Coullet, P. & Tresser, C. [1978] "Iterations d'endomorphismes et groupe de renormalisation," *C. R. Acad. Sci. Paris, Ser. A-B* **287**, A577–A580.

Coullet, P. & Pomeau, Y. [2016] "History of chaos from a French perspective," in *The Foundations of Chaos Revisited*: *From Poincaré to Recent Advancements*, Understanding Complex Systems, ed. Skiadas, C. (Springer), pp. 91–101.

Davendra, D., Zelinka, I. & Senkerik, R. [2010] "Chaos driven evolutionary algorithms for the task of PID control," *Comput. Math. Appl.* **60**, 1088–1104.

Daza, Á., Wagemakers, A. & Sanjuán, M. A. F. [2024] "Multistability and unpredictability," *Phys. Today* **77**, 44–50.

Defoort, M., Rufer, L., Fesquet, L. & Basrour, S. [2021] "A dynamical approach to generate chaos in a micromechanical resonator," *Microsyst. Nanoeng.* **7**, 17.

Denis, F. & Letellier, C. [2012] "Chaos theory: A fascinating concept for oncologists," *Cancer Radiothér.* **16**, 230–236.

Descartes, R. [1641, 1988] *Meditationes de Prima Philosophia*, eds. Cottingham, J., Stoothoff, R. & Murdoch, D., Vols 1–3. (Cambridge University Press, Cambridge, UK), Trans. *The Philosophical Writings of Descartes.*

Diamond, P. & Pokrovskii, A. [1996] "Statistical laws for computational collapse of discretized chaotic mappings," *Int. J. Bifurcation and Chaos* **6**, 2389–2399.

Dieudonné, J. A., [1987] *Pour l'Honneur de l'Esprit Humain: Les Mathématiques Aujourd'hui*, Coll. Pluriel (Hachette, Paris), 298 pp.

Dinamarco, N., Rodrigues, B., Yugar-Toledo, J. C. & Moreno, H. [2022] "Chaos a complex noise on blood pressure orchestration," *Cardiol. Res. Cardiovasc. Med.* **7**, 174–179.

Dmitriev, A. S., Panas, A. I. & Starkov, S. O. [2000] "Multiple access communication based on control of special chaotic trajectories," *Proc. 2nd Int. Conf. Control of Oscillations and Chaos* (*COC-2000*) (IEEE), pp. 518–522.

Douady, A. & Hubbard, J. [1985] "On the dynamics of polynomial-like mappings," *Ann. Sci. École Norm. Sup.* (*4*) **18**, 287–343.

Douady, A. [1986] "Julia sets and the Mandelbrot set," in *The Beauty of Fractals* (Springer, Berlin), pp. 161–174.

Dridi, F., El Assad, S., Youssef, W. E., Machhout, M. & Lozi, R. [2022] "Design, implementation, and analysis of a block cipher based on a secure chaotic generator," *Appl. Sci.* **12**, 9952.

El Assad, S., Lozi, R. & Puech, W. [2022] "Special issue on cryptography and its applications in information security," *Appl. Sci.* **12**, 2588.

Elaydi, S. [2023] "Chapter 7: Encounter with the Japanese attractor," in *Chaos Theory, Complexity Theory, Catastrophe Theory, and the Enigmatic Concept of Chaoplexity* (Independent), 166 pp.

Fatou, P. [1917] "Sur les substitutions rationnelles," *C. R. Acad. Sci. Paris* **164**, 806–808; **165**, 992–995.

Feigenbaum, M. [1978] "Quantitative universality for a class of nonlinear transformations," *J. Stat. Phys.* **19**, 25–52.

Friberg, J. [1981] "Methods and traditions of Babylonian mathematics. Plimpton 322, Pythagorean triples, and the Babylonian triangle parameter equations," *Hist. Math.* **8**, 277–318.

Friberg, J. [1986] "The early roots of Babylonian mathematics 3: Three remarkable texts from ancient Ebla," *Vicino Oriente* **6**, 3–25.

Gaiki, P. M., Bhoyar, P. D., Joshi, D. D. & Gade, P. M. [2024] "Existence of multistability in the dynamical behavior of q-deformed Lozi map," *Indian J. Phys.* **98**, 3641–3659.

Garasym, O., Lozi, J. P. & Lozi, R. [2017] "How useful randomness for cryptography can emerge from multicore-implemented complex networks of chaotic maps," *J. Diff. Eqs. Appl.* **23**, 821–859.

Garehdaghi, F., Sarbaz, Y. & Baradari, E. [2024] "A novel hypothesis for migraine disease mechanism: The creation of a new attractor responsible for migraine disease symptoms," *Chaos Theory Appl.* **6**, 19–25.

Ghys, E. [2023] *La Théorie du Chaos*, Les Grandes Voix de la Recherche (Editions du CNRS, Paris), 64 pp.

Ginoux, J.-M. & Gérini, C. [2012] *Henri Poincaré: Une Biographie Au(x) Quotidien(s)* (Editions Ellipses, Paris), 298 pp.

Grassberger, P. & Procaccia, I. [1983] "Measuring the strangeness of strange attractors," *Physica D* **9**, 189–208.

Guevara, M. R. & Glass, L. [1982] "Phase locking, period doubling bifurcations and chaos in a mathematical model of a periodically driven oscillator: A theory for the entrainment of biological oscillators and the genesis of cardiac dysrhythmias," *J. Math. Biol.* **14**, 1–23.

Gumowski, I. & Mira, C. [1980] *Recurrences and Discrete Dynamic Systems: An Introduction*, Lecture Notes in Mathematics, Vol. 809 (Springer, Berlin), 250 pp.

Gupta, D. & Chandramouli, V. V. M. S. [2021] "An improved q-deformed logistic map and its implications," *Pramana J. Phys.* **95**, 175.

Gupta, D. & Chandramouli, V. V. M. S. [2022] "Dynamics of deformed Hénon-like map," *Chaos Solit. Fract.* **155**, 111760.

Hadamard, J. [1898] "Les surfaces à courbures opposées et leurs lignes géodésiques," *J. Math. Pures Appl.* **4**, 27–73.

Hannavy, J. [2021] *The Governor: Controlling the Power of Steam Machines* (Pen & Sword Transport, Barnsley, SYK), pp. 60–78.

Harmer, G. P., Abbott, D., Taylor, P. G. & Parrondo, J. M. R. [2001] "Brownian ratchets and Parrondo's games," *Chaos* **11**, 705A–714A.

Heath, J. [1931] *A Manual of Greek Mathematics* (Oxford University Press, Oxford).

Hénon, M. [1976] "A two-dimensional mapping with a strange attractor," *Commun. Math. Phys.* **50**, 69–77.

Higuchi, T. [1988] "Approach to an irregular time-series on the basis of the fractal theory," *Physica D* **31**, 277–283.

Hoefer, C. [2024] "Causal determinism," *The Stanford Encyclopedia of Philosophy*, eds. Zalta, E. N. & Nodelman, U., Summer 2024 edition (Stanford University Press).

Holmes, P. [1990] "Poincaré, celestial mechanics, dynamical systems theory and 'chaos'," *Phys. Rep.* **193**, 137–163.

Holmes, P. [2005] "Ninety plus thirty years of nonlinear dynamics: Less is more and more is different," *Int. J. Bifurcation and Chaos* **15**, 2703–2716.

Holmes, P. [2007] "A short history of dynamical systems theory: 1885–2007," in *History of Mathematics*, Vol 1 (UNESCO–EOLSS).

Hoyrup, J. [2011] "Mesopotamian calculation: Background and contrast to Greek mathematics," *Contribution to IX Congresso della Società Italiana di Storia della Matematica*, Genova, pp. 1–36.

Ikeda, K. [1979] "Multiple-valued stationary state and its instability of the transmitted light by a ring cavity system," *Opt. Commun.* **30**, 257–261.

Itik, M. & Banks, S. P. [2010] "Chaos in a three-dimensional cancer model," *Int. J. Bifurcation and Chaos* **20**, 71–79.

Jaganathan, R. & Sinh, S. [2005] "A q-deformed nonlinear map," *Phys. Lett. A* **338**, 277–287.

Jain, K. [2023] "Role of mathematics in industries," *Int. J. Creat. Res. Thoughts* **11**, a909–a911.

Julia, G. [1918] "Mémoire sur l'itération des fonctions rationnelles," *J. Math. Pures Appl.* **8**, 47–245.

Kahane, J.-P. [2008] "Hasard et déterminisme chez Laplace," *Cahiers Rationalistes* **593**, 237–238.

Kahlert, C. & Chua, L. O. [1987] "Transfer maps and return maps for piecewise-linear three-region dynamical systems," *Int. J. Circuit Theory Appl.* **15**, 23–49.

Kahlert, C. & Rössler, O. E. [1987] "Analytical properties of Poincaré halfmaps in a class of piecewise-linear dynamical systems," *Z. Naturforsch. A* **40**, 1011–1025.

Kawakami, H. & Lozi, R. [1992] *Switched Dynamical Systems: Dynamics of a Class of Circuits with Switch*, ed. Ushiki, S., Advanced Series in Dynamical Systems, Vol. 11 (World Scientific Publishing, Singapore), pp. 39–58.

Khalifa, N. & Benrejeb, M. [2017] "On secure communication scheme-based mixed discrete-time hyperchaotic systems," *J. Diff. Eqs. Appl.* **29**, 1429–1448.

Klinshpont, E., Sataev, E. A. & Plykin, R. V. [2005] "Geometrical and dynamical properties of Lorenz type system," *J. Phys.: Conf. Ser.* **23**, 96–104.

Kloeden, P. E. [1976] "Chaotic difference equations are dense," *Bull. Aust. Math. Soc.* **15**, 371–379.

Kocarev, L., Halle, K. S., Eckert, K. & Chua, L. O. [1992] "Experimental demonstration of secure communication via chaotic synchronization," *Int. J. Bifurcation and Chaos* **2**, 709–713.

Kocarev, L. & Lian, S. [2011] *Chaos-Based Cryptography: Theory, Algorithms and Applications*, Studies in Computational Intelligence, Vol. 354 (Springer, Berlin), 396 pp.

Kolmogorov, A. N. [1954] "On the conservation of conditionally periodic motions under small perturbation of the Hamiltonian," *Dokl. Akad. Nauk SSSR* **98**, 527–530.

Korteweg, D. J. & de Vries, G. [1895] "On the change of form of long waves advancing in a rectangular canal and a new type of long stationary waves," *Philos. Mag. Ser.* **39**, 422–443.

Kucharski, P. [2023] "Strange attractors for the family of orientation preserving Lozi maps," *Chaos* **33**, 113121.

Lachkar, J. [1989] "A numerical experiment on the chaotic behaviour of the solar system," *Nature* **338**, 237–238.

Landau, L. D. [1944] "On the problem of turbulence," *Dokl. Akad. Nauk SSSR* **44**, 339–342 (in Russian).

Laplace, P.-S. [1814] *Essai Philosophique sur les Probabilités* (Courcier, Paris), 97 pp.

Leonov, G. A. & Kuznetsov, N. V. [2013] "Hidden attractors in dynamical systems. From hidden oscillations in Hilbert–Kolmogorov, Aizerman, and Kalman problems to hidden chaotic attractor in Chua circuits," *Int. J. Bifurcation and Chaos* **3**, 1330002.

Letellier, C. & Messager, V. [2010] "The deep Otto E. Rössler behind his pictorial way of writing," *Int. J. Bifurcation and Chaos* **20**, 3585–3616.

Letellier, C., Abraham, R., Shepelyansky, D. L. *et al.* [2021] "Some elements for a history of the dynamical systems theory," *Chaos* **31**, 053110.

Li, T.-Y. & Yorke, J. A. [1975] "Period three implies chaos," *Am. Math. Mon.* **82**, 985–992.

Lorenz, E. N. [1963] "Deterministic nonperiodic flow," *J. Atmos. Sci.* **20**, 130–141.

Lozi, R. [1975] "Analyse numérique de certains problèmes de bifurcation," PhD thesis, University of Nice, France.

Lozi, R. [1978] "Un attracteur étrange (?) du type attracteur de Hénon," *J. Phys.* **39**, C5-9–C5-10.

Lozi, R. [1982] "Dimensional bifurcations between thread and sheet strange attractors," in *Théorie de l'Itération et ses Applications: Colloques Internationaux du CNRS*, ed. Thibault, R. (Editions du CNRS), pp. 145–152.

Lozi, R. & Ushiki, S. [1988] "Organized confinors and anticonfinors and their bifurcations in constrained "Lorenz" system," *Ann. Télécom.* **43**, 187–208.

Lozi, R. & Saidi, A. [1989] "Étude des motifs dans les oscillations de la réaction de Belousov-Zhabotinsky I. Existence d'un confineur principal pour un modèle contraint particulier," *Ann. ENIT* **3**, 69–98.

Lozi, R. [1990] "Motifs et confineurs organisés dans la réaction de Bélousov-Zhabotinskii et le 'double scroll'," *Les Rencontres Physiciens–Mathématiciens de Strasbourg — RCP25* (Université Louis Pasteur, Strasbourg), 62 pp.

Lozi, R. & Saidi, A. [1990] "Étude des motifs dans les oscillations de la réaction de Belousov–Zhabotinsky II. Un algorithme de partition de l'espace des paramètres pour une famille d'équations différentielles contraintes," *Ann. ENIT* **4**, 91–124.

Lozi, R. & Ushiki, S. [1991a] "Confinors and bounded-time patterns in Chua's circuit and the double scroll family," *Int. J. Bifurcation and Chaos* **1**, 119–138.

Lozi, R. & Ushiki, S. [1991b] "Coexisting attractors in Chua's circuit," *Int. J. Bifurcation and Chaos* **1**, 923–926.

Lozi, R. & Chua, L. O. [1993] "Secure communications via chaotic synchronization II: Noise reduction by cascading two identical receivers," *Int. J. Bifurcation and Chaos* **13**(5), 1319–1325.

Lozi, R. & Ushiki, S. [1993] "The theory of confinors in Chua's circuit: Accurate analysis of bifurcations and attractors," *Int. J. Bifurcation and Chaos* **3**, 333–361.

Lozi, R. [1995] "Secure communications via chaotic synchronization in Chua's circuit and Bonhoeffer-Van der Pol equation: Numerical analysis of the errors of the recovered signal," *IEEE Int. Symp. Circuits and Systems* (IEEE), pp. 684–687.

Lozi, R., Sousa Ramos, J. & Sharkovsky, A. [2004] "One-dimensional dynamics generated by boundary value problems for the wave equation," *Grazer Math. Ber.* **346**, 255–270.

Lozi, R. [2011] "Complexity leads to randomness in chaotic systems," in *Mathematics in Science and Technology: Mathematical Methods, Models and Algorithms in Science and Technology*, eds. Siddiqi, A. H., Singh, R. C. & Manchanda, P. (World Scientific, Singapore), pp. 93–125.

Lozi, R. [2013] "Can we trust in numerical computations of chaotic solutions of dynamical systems?" in *Topology and Dynamics of Chaos*, ed. Letellier, C. (World Scientific Publishing, Singapore), pp. 63–98.

Lozi, R. [2023a] "Research on actual or industrial applications of difference equations: A chimerical task?" *J. Diff. Eqs. Appl.* **30**, 1438–1492.

Lozi, R. [2023b] "Survey of recent applications of the chaotic Lozi map," *Algorithms* **16**, 491.

Lozi, R. [2023c] "Coexistence of thread and sheet chaotic attractors for three-dimensional Lozi map," *Dynamics* **3**, 315–344.

Lozi, R., Belykh, V., Cushing, J. M., Efremova, L., Elaydi, S., Gardini, L., Misiurewicz, M., Schöll, E. & Strelkova, G. [2024a] "The paths of nine mathematicians to the realm of dynamical systems," *J. Diff. Eqs. Appl.* **30**, 1–89.

Lozi, R., Efremova, L., Abdelouahab, M.-S. & El Assad, S. [2024b] *Lozi, Hénon and Belykh Chaotic Attractors: New Results Fifty Years On* (CRC Press, Boca Raton), 270 pp.

Lozi, R. [2026] "How chaos shaped my academic life," in *Recollections from the Early Days of Chaos Theory*, ed. Letellier, C. (World Scientific Publishing, Singapore), p. 480, https://doi.org/10.1142/14581.

Lozi, R., El Assad, S. & Abdelouahab, M.-S. [2025] *Dynamical Systems: Paths of Twelve Mathematicians and Physicists to Chaos and Its Applications* (CRC Press, Boca Raton), 340 pp.

Lu, J. & Chen, G. [2006] "Generating multiscroll chaotic attractors: Theories, method and applications," *Int. J. Bifurcation and Chaos* **16**, 775–858.

Mandelbrot, B. [1967] "How long is the coast of Britain? Statistical self-similarity and fractional dimension," *Science* **156**, 636–638.

Mandelbrot, B. [1975] *Les Objets Fractals: Forme, Hasard et Dimension*, Coll. Nouvelle Bibliothèque Scientifique (Flammarion, Paris).

Mandelbrot, B. [1997] *Fractals and Scaling in Finance* (Springer, NY), 551 pp.

Manukure, S. & Booker, T. [2021] "A short overview of solitons and applications," *Partial Diff. Eqs. Appl. Math.* **4**, 100140.

Marinke, R., Araujo, J. E., Coelho, L. S. & Matko, I. [2005] "Particle swarm optimization (PSO) applied to fuzzy modeling in a thermal-vacuum system," *Proc. 5th Int. Conf. Hybrid Intelligent Systems (HIS'05)* (IEEE), pp. 67–72.

Martynyuk, A. A. [2014] *Revisiting the Past with Neither Indifference Nor Resentment* (Phoenicx, Kiev), 139 pp.

Mashaal, M. [2017] *Bourbaki. Une Société Secrète de Mathématiciens*, Coll. Alpha (Belin, Paris), 116 pp.

Mawhin, J. [1996] "The early reception in France of the work of Poincaré and Lyapunov in the qualitative theory of differential equations," *Philos. Sci.* **1**, 119–133.

Miino, Y. & Ueta, T. [2015] "Laboratory experiment for bifurcation phenomena of the forced Alpazur oscillator," *Int. Symp. Nonlinear Theory and Its Applications NOLTA2015* (Kowloon, Hong Kong, China), pp. 169–175.

Minati, L. [2024] "Chaos and synchronization — Potential ingredients of innovation in analog circuit design?" *IEICE Trans. Electron.* **107**, 376–391.

Mira, C. [1987] *Chaotic Dynamics. From the One-Dimensional Endomorphism to the Two-Dimensional Diffeomorphism* (World Scientific), 472 pp.

Mira, C. [2003] "Gaston Julia, l'un des plus grands mathématiciens du 20eme siècle," *L'Algérianiste* **104**, 61–67.

Misiurewicz, M. [1980] "Strange attractors for the Lozi mappings," *Ann. N. Y. Acad. Sci.* **357**, 348–358.

Misiurewicz, M. & Stimac, S. [2016] "Symbolic dynamics for Lozi maps," *Nonlinearity* **29**, 3031–3046.

Moreno, H. [2022] "Pseudo and resistant hypertension: A chaotic perspective," *J. Clin. Hypertens.* **24**, 698–703.

Morse, M. & Hedlund, G. A. [1938] "Symbolic dynamics," *Am. J. Math.* **60**, 815–866.

Moser, J. [1962] "On invariant curves of area-preserving mappings of an annulus," *Nachr. Akad. Wiss. Göttingen, Math. Phys., Kl. IIa* **1**, 1–20.

Myrberg, P. J. [1962] "sur l'itération des polynômes réels quadratiques," *J. Math. Pures Appl. (9)* **41**, 339–351.

Neugebauer, O. [1969] "Chapter IV. Egyptian mathematics and astronomy," in *The Exact Sciences in Antiquity* (Dover Publications), pp. 71–96.

Newton, I. [1675] "Letter from Sir Isaac Newton to Robert Hooke," Historical Society of Pennsylvania, Retrieved June 7, 2018.

Newton, I. [1687, 1972] *Philosophiae Naturalis Principia Mathematica*, eds. Koyré, A. & Cohen, I. B., facsimile edition (Cambridge University Press, Cambridge, UK).

Oestreicher, C. [2007] "A history of chaos theory," *Dialogues Clin. Neurosci.* **9**, 279–289.

Ott, E., Grebogi, C. & Yorke, J. A. [1990] "Controlling chaos," *Phys. Rev. Lett.* **64**, 1196; Erratum **64**, 2837.

Palis, J. [1993] "On the contribution of Smale to dynamical systems," in *From Topology to Computation*, eds. Hirsch, M. W., Marsden, J. E. & Shub, M. (Springer-Verlag, NY), pp. 165–178.

Pašić, M. [2003] "Minkowski–Bouligand dimension of solutions of the one-dimensional p-Laplacian," *J. Diff. Eqs.* **190**, 268–305.

Patidar, V., Purohit, G. & Sud, K. K. [2011] "Dynamical behavior of q-deformed Hénon map," *Int. J. Bifurcation and Chaos* **21**, 1349–1356.

Pecora, L. M. & Carroll, T. L. [1990] "Synchronization in chaotic systems," *Phys. Rev. Lett.* **64**, 821–824.

Peixoto, M. M. [1955] "Note on structurally stable systems," *An. Acad. Bras. Cienc.* **27**, 35.

Pesin, Y. [1989] "General theory of smooth hyperbolic dynamical systems," in *Dynamical Systems II*, Encyclopaedia of Mathematical Sciences, Vol. 2 (Springer), pp. 108–151 (in Russian).

Pesin, Ya. B. [1992] "Dynamical systems with generalized hyperbolic attractors: Hyperbolic, ergodic and topological properties," *Ergodic Th. Dynam. Sys.* **12**, 123–151.

Pluhacek, M., Senkerik, R., Davendra, D. & Zelinka, I. [2012] "Designing PID controller for DC motor system by means of enhanced PSO algorithm with discrete Lozi map," *Proc. 26th European Conf. Modelling and Simulation, ECMS 2012*, eds. Troitzsch, K. G., Möhring, M. & Lotzmann, U. (European Council for Modelling and Simulation, Koblenz), pp. 405–409.

Poincaré, H. J. [1890] "Sur le problème des trois corps et les équations de la dynamique," *Acta Math.* **13**, 1–270.

Poincaré, H. J. [1892, 1893, 1899] *Les Méthodes Nouvelles de la Mécanique Cèleste*, Vols. 1–3 (Gauthiers-Villars, Paris), English translation edited by D. Goro, published by the American Institute of Physics, NY, 1993.

Rågstedt, M. [1976] "From order to chaos: The prize competition in honour of king Oscar II," https://www.mittag-leffler.se/about-us/history/prize-competition/.

Rényi, A. [1959] "On the dimension and entropy of probability distributions," *Acta Math. Acad. Sci. Hung.* **10**, 193–215.

Rodriguez-Vazquez, A., Huertas, J. L., Perez-Verdu, B. & Chua, L. O. [1987] "Chaos from switched-capacitor circuits: Discrete maps," *Proc. IEEE*, **75**, 1090–1106.

Rosalie, M. [2016] "Templates of two foliated attractors — Lorenz and Chen systems," *Int. J. Bifurcation and Chaos* **26**, 1650037.

Rössler, O. E. [1976] "Chaotic behavior in simple reaction system," *Z. Naturforsch. A* **31**, 259–264.

Rössler, O. E. [1979] "Continuous chaos: Four prototype equations," *Ann. New York Acad. Sci.* **316**, 376–392.

Rössler, O. E. [2020] "On the Rössler attractor," *Chaos Theory Appl.* **2**, 1–2.

Ruelle, D. & Takens, F. [1971] "On the nature of turbulence," *Commun. Math. Phys.* **20**, 167–192.

Ruelle, D. [1978] "Dynamical systems with turbulent behavior," in *Mathematical Problems in Theoretical Physics*, Dell'Antonio, G., Doplicher, S. & Jona-Lasinio, G., Lecture Notes in Physics, Vol. 80 (Springer, Berlin, Germany), pp. 341–360.

Rybalova, E., Semenova, N. & Anishchenko, V. [2018] "Solitary state chimera: Appearance, structure, and synchronization," *Int. Symp. Nonlinear Theory and Its Applications NOLTA2018* (Tarragona, Spain), pp. 601–604.

Salisbury, J. [1991] *Ioannis Saresberiensis Metalogicon*, eds. Hall, J. B. & Keats-Rohan, K. S. B., Turnhout: Brepols, CCCM 98, https://www.brepols.net/products/IS-9782503039817-1.

Sanjuán, M. A. F. [2025] *Interdisciplinarity in the Sciences. Nonlinear Dynamics, Chaos and Complexity* (World Scientific Publishing, Singapore), 208 p.

Sataev, E. A. [1992] "Invariant measures for hyperbolic maps with singularities," *Russian Math. Surveys* **47**, 191–251.

Sawesi, S., Rashrash, M. & Dammann, O. [2022] "The representation of causality and causation with ontologies: A systematic literature review," *Online J. Public Health Inf.* **14**, e4.

Scott, S. W. [1979] "Stimulation simulations of young yet cultured beating hearts," PhD thesis, State University of New York, Buffalo.

Senkerik, R., Davendra, D., Zelinka, I., Pluhacek, M. & Kominkova Oplatkova, Z. [2013] "Chaos driven differential evolution with Lozi map in the task of chemical reactor optimization," *Int. Conf. Artificial Intelligence and Soft Computing* (Springer), pp. 56–66.

Shabestari, P. S., Rajagopal, K., Safarbali, B., Jafari, S. & Duraisamy, P. [2018] "A novel approach to numerical modeling of metabolic system: Investigation of chaotic behavior in diabetes mellitus," *Complexity* **2018**, 6815190.

Sharkovsky, A. N. [1964] "Co-existence of cycles of a continuous mapping of the line into itself," *Ukr. Math. J.* **16**, 61–71.

Sharkovsky, A. N. [2014] "On the history of one-dimensional dynamics," *ESAIM*: *Proc. Surv.* **46**, 83–85.

Shil'nikov, L. P. [1965] "A case of the existence of a denumerable set of periodic motions," *Sov. Math. Dokl.* **6**, 163–166 (in Russian).

Singh, S. [2002] *The Code Book*: *The Science of Secrecy from Ancient Egypt to Quantum Cryptography* (Delacorte Press, NY), 263 pp.

Smale, S. [1967] "Differentiable dynamical systems," *Bull. Am. Math. Soc.* **73**, 747–817.

Smale, S. [1980] "On how I got started in dynamical systems, 1959–1962," *The Mathematics of Time*: *Essays on Dynamical Systems, Economic Processes and Related Topics* (Springer-Verlag, NY).

Spina, L. [2004] "Nains et géants: Une dialectique antique," *Inf. Litt. Ét. Crit.* **1**, 28–33.

Stefanescu, V., Stoichescu, D., Frunzete, M. & Florea, B. [2013] "Influence of computer computation precision in chaos analysis," *Univ. Politehn. Buchar. Sci. Bull. A* **75**, 151–162.

Strukov, D. B., Snider, G. S., Stewart, D. R. & Williams, R. S. [2008] "The missing memristor found," *Nature* **453**, 80–83.

Su, Y., Xu, T., Li, T., Zhao, J. & Liu, S. [2021] "Optical color image encryption based on fingerprint key and phase-shifting digital holography," *Opt. Lasers Eng.* **140**, 106550.

Suetonius, T. [2016] "The lives of the twelve Caesars," The Project Gutenberg, Produced by Tapio Riikonen and David Widger, https://www.gutenberg.org/files/6400/6400-h/6400-h.htm.

Sushko, I. & Tramontana, F. [2025] "Regular and chaotic dynamics in a 2D discontinuous financial market model with heterogeneous traders," *Math. Comput. Simul.* **230**, 386–399.

Tang, T. W., Allison, A. & Abbott, D. [2004] "Investigation of chaotic switching strategies in Parrondo's games," *Fluct. Noise Lett.* **4**, L585–L596.

Toomer, G. J. [1984] *Ptolemy's Almagest* (Duckworth, London), Translated and annotated by G. J. Toomer.

Ueda, Y. [1978] "Random phenomena resulting from nonlinearity — In the system described by Duffing's equation," *Trans. Inst. Electr. Eng. Jpn. A* **98**, 167–173 (in Japanese).

Ueda, Y. [1992] "Strange attractor and the origin of chaos," in *The Road to Chaos*, eds. Ueda, Y., Abraham, R. H. & Stewart, H. B. (Aerial Press), pp. 185–215.

Uehleke, B. & Rössler, O. E. [1983] "Complicated Poincaré half-maps in a linear system," *Z. Naturforsch. A* **38**, 1107–1113.

Ueta, T. & Chen, G. [2000] "Bifurcation analysis of Chen's attractor," *Int. J. Bifurcation and Chaos* **10**, 1917–1931.

Ushiki, S. & Lozi, R., [1987] "Confinor and anti-confinor in constrained "Lorenz" system," *Jpn. J. Ind. Appl. Math* **4**, 433–454.

Vos, T. *et al.* [2019] "Global burden of 369 diseases and injuries in 204 countries and territories, 1990–2019: A systematic analysis for the global burden of disease study," *Lancet* **396**, 1204–1222.

Wang, J., Gu, Y., Rong, K., Xu, Q. & Zhang, X. [2022] "Memristor-based Lozi map with hidden hyperchaos," *Mathematics* **10**, 3426.

Wikipedia [2025] https://en.wikipedia.org/wiki/IBM_7040.

Winfree, A. T. [1975] "Resetting biological clocks," *Phys. Today* **28**, 34–39.

Winfree, A. T. [1980] *The Geometry of Biological Time* (Springer, NY).

Wu, G.-C., Hou, H.-S. & Lozi, R. [2025] "Neimark–Sacker bifurcation of discrete fractional chaotic systems," *Int. J. Bifurcation and Chaos* **35**, 2550049.

Xiao, D., Liao, X. & Deng, S. [2011] "Chaos based hash function," in *Chaos-Based Cryptography: Theory, Algorithms and Applications*, Studies in Computational Intelligence, Vol. 354 (Springer, Berlin), pp. 137–203.

Yu, P., Yao, W. & Chen, G. [2007] "Analysis on topological properties of the Lorenz and the Chen attractors using GCM," *Int. J. Bifurcation and Chaos* **17**, 2791–2797.

Yugar-Toledo, J. C., Dinamarco, N., Rodrigues, B. & Moreno, H. [2022] "Chaos, resistant and pseudoresistant hypertension "thousands of butterflies in the BP control system"," *J. Cardiovasc. Med. Cardiol.* **9**, 006–010.

Zaamoune, F., Menacer, T., Lozi, R. & Chen, G. [2019] "Symmetries in hidden bifurcation routes to multiscroll chaotic attractors generated by saturated function series," *J. Adv. Eng. Comput.* **3**, 511–522.

Zabusky, N. J. & Kruskal, M. D. [1965] "Interaction of "solitons" in a collisionless plasma and the recurrence of initial states," *Phys. Rev. Lett.* **15**, 240–243.

Zabczyk, J. [2008] *Mathematical Control Theory: An Introduction* (Birkhäuser, Boston).

Zelinka, I. [2004] "Self-organizing migrating algorithm," *Stud. Fuzziness Soft Comput.* **141**, 167–217.

Zelinka, I., Kojecky, L., Lampart, M., Nowakova, M. J. & Plucar, J. [2023] "iSOMA swarm intelligence algorithm in synthesis of quantum computing circuits," *Appl. Soft Comput.* **142**, 110350.

Zhang, L. P., Wei, Z. C., Jiang, H. B., Lyu, W. P. & Bi, Q. S. [2004] "Extremely hidden multistability in a class of a two dimensional maps with a cosine memristor," *Chin. Phys. B.* **31**, 100503.

Zhao, H. & Zheng, W.-M. [1993] "Symbolic analysis of the Hénon map at $a = 1.4$ and $b = 0.3$," *Commun. Theor. Phys.* **19**, 21–26.

Zheng, W.-M. [1991] "Symbolic dynamics for the Lozi map," *Chaos Solit. Fract.* **1**, 243–248.

Zheng, W.-M. [1992] "Admissibility conditions for symbolic sequences of the Lozi map," *Chaos Solit. Fract.* **2**, 461–470.

Zhong, C. Q. & Tang, W. K. S. [2002] "Circuitry implementation and synchronization of Chen's attractor," *Int. J. Bifurcation and Chaos* **12**, 1423–1427.

https://doi.org/10.1142/9789819830633_0003

Chapter 3

Synchronous Propagation of Periodic Signals in Feedforward Networks of Standard Model Neurons#

Ian Stewart* and David Wood†
Mathematics Institute, University of Warwick, Coventry CV4 7AL, UK
**i.n.stewart@warwick.ac.uk*
†*David.Wood@warwick.ac.uk*

Periodic signals propagating along chains are common in biology, for example in locomotion and peristalsis, and are also of interest for continuum robots. In a previous work, we constructed such networks as "feedforward lifts" of a Central Pattern Generator (CPG). When the CPG undergoes periodic oscillations, created by Hopf bifurcation or other mechanisms, it can then transmit periodic signals along one or more feedforward chains in a synchronous or phase-synchronous manner. We proved necessary and sufficient conditions for the stability of these lifted periodic orbits, in several senses. Here we examine the implications of the resulting theory for chains of neurons, using several standard neuron models: FitzHugh–Nagumo, Morris–Lecar, Hindmarsh–Rose, and Hodgkin–Huxley. We compare different notions of transverse stability, and summarize some numerical simulations showing that for all these neuron models, the propagating signal can be transversely Floquet stable. Finally, we discuss implications for less idealized models.

Keywords: Network; synchrony; symmetry; balanced coloring; quotient network; bifurcation.

1. Introduction

This paper is a continuation of [Stewart & Wood, 2024], which discussed theoretical aspects of signal propagation in a *feedforward lift.* This is a network in which a small subnetwork [Central Pattern Generator (CPG)] sends signals to a longer chain of repeating modules, constructed so that nodes along the chain can be synchronous with nodes of the CPG. (For simplicity we often use the term "chain" without implying strict linear topology. The results apply to any feedforward lift, for example, a tree. Thus the propagating signal can branch.) The CPG dynamic can be steady, oscillatory, or even chaotic. We concentrate on the periodic case, where nodes can also be phase-synchronous; that is, synchronous except for a phase shift. One natural way to generate periodic oscillations is via Hopf bifurcation but other mechanisms are also possible; our results apply to any periodic state, however it is created.

The analysis in [Stewart & Wood, 2024] is based on the formal theory of network dynamics of [Stewart *et al.*, 2003; Golubitsky *et al.*, 2005; Golubitsky & Stewart, 2023], and the CPG is both a quotient network and a subnetwork of the full network. The formal theory is "model-independent" in the sense that the results do not depend on specific model equations. However, they cannot be *completely* independent of the model; certain generic hypotheses must be satisfied, in the same way that Hopf bifurcation to a periodic orbit is generic if

#To cite this article, please refer to its earlier version published in the *International Journal of Bifurcation and Chaos*, Vol. 35, No. 14 (2025) 2540003 (28 pages), DOI: 10.1142/S0218127425400036.
*Author for correspondence

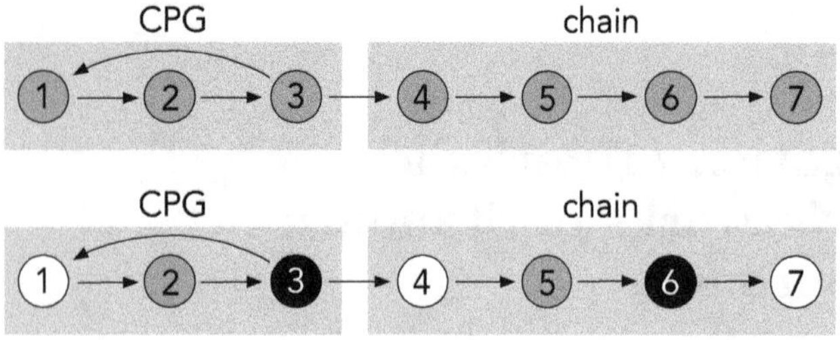

Fig. 1. A 7-node network with one node-type and one arrow-type. *Top*: A balanced coloring corresponding to dynamics in which all nodes are synchronous. When nodes oscillate this gives a standing wave along the chain. *Bottom*: A balanced coloring corresponding to dynamics in which the nodes fall into three synchrony clusters $\{1,4,7\}$, $\{2,5\}$, and $\{3,6\}$, with colors white, gray, and black. The CPG $\{1,2,3\}$ has cyclic group symmetry $\mathbb{Z}_3$, so, this network naturally propagates signals such that nodes of the same color are in synchrony and successive nodes are phase-shifted by $\frac{1}{3}$ of the period. When nodes oscillate, this appears to create a traveling wave along the chain.

critical eigenvalues are simple, purely imaginary, nonresonant, and satisfy the eigenvalue-crossing condition [Guckenheimer & Holmes, 1983; Hassard *et al.*, 1981].

Figure 1 is a simple "toy" example used in [Stewart & Wood, 2024] to illustrate the main ideas, and we continue to use it here. The CPG is the subnetwork with nodes $\{1,2,3\}$ and the arrows connecting them; the chain comprises the remainder of the network. We emphasize that the stability computations reported below for this model remain valid if the chain is replaced by *any* feedforward lift. However, the results depend on the chosen model for CPG dynamics.

Intuitively, nodes 1 and 4 receive the same signal from node 3 and have the same internal dynamics, so they can synchronize. Then, nodes 2 and 5 receive synchronous signals, so they can synchronize. Similarly node 6 can synchronize with node 3, node 7 with node 4, and so on for longer chains. More formally, each coloring shown is "balanced": nodes of the same color have inputs that are color-isomorphic [Golubitsky & Stewart, 2023, Sec. 10.3]. The top figure corresponds to complete synchrony, with all nodes having identical waveforms. The bottom figure corresponds to cluster synchrony with three clusters.

Moreover, the CPG $\{1,2,3\}$ has cyclic group symmetry $\mathbb{Z}_3$. This implies that a synchrony-breaking Hopf bifurcation can create a phase pattern in which successive nodes are phase-shifted by $\frac{1}{3}$ of the period [Golubitsky & Stewart, 2002, 2023]. This synchrony pattern corresponds to the 3-coloring in the bottom figure. More precisely, the phase shift in a fixed time direction is $\pm\frac{1}{3}$ of the period, depending on parameters: for simplicity, we ignore the $\pm$ sign throughout.

1.1. *Motivation for feedforward lifts*

The structure of a feedforward lift is motivated by common aspects of animal physiology, in which oscillations propagate along a linear chain of neurons (or populations of neurons) in synchrony with each other, or with specific phase relations. This results in the longitudinal propagation of rhythmic time-periodic patterns. Complete synchrony gives the appearance of a standing wave, and phase synchrony gives the appearance of a traveling wave. Some biological examples are discussed in Sec. 1.3. Similar waves of motion are used to propel continuum robots for exploration [Mu *et al.*, 2017] and medicine [Jha & Chauhan, 2019; Seok *et al.*, 2010; Zhong *et al.*, 2020]. Synchrony patterns also occur in gene regulatory networks, where chains of essentially identical regulatory links can act as delay lines [Leifer *et al.*, 2020]. The states concerned are usually equilibria, but periodic oscillations can also occur [Purcell *et al.*, 2010] and phase-synchronous patterns are found, for example, in the repressilator, a synthetic genetic circuit that has $\mathbb{Z}_3$ symmetry in an idealized model [Elowitz & Leibler, 2000].

As in [Golubitsky & Stewart, 2023; Golubitsky *et al.*, 2005; Stewart *et al.*, 2003; Stewart & Wood, 2024], we adopt a strong definition of synchrony. Two nodes are *synchronous* if their waveforms (time series) are identical, and they are *phase-synchronous* if their waveforms are identical except for a phase shift (time translation). These definitions are idealizations, but they open up a powerful mathematical approach with useful implications. Real systems can be considered as perturbations of idealized ones, and much of the interesting structure persists in an appropriately approximate form; see Sec. 8. For example, Elowitz and Leibler [2000] demonstrate approximate $\frac{1}{3}$-period phase shifts in a more realistic model in which the equations lack $\mathbb{Z}_3$ symmetry. Many other concepts of synchrony have been studied [Uhlhaas *et al.*, 2009].

Figures 4 and 5 of Sec. 4.1 are examples of such signals for the network of Fig. 1, and, respectively, show complete synchrony and a $\frac{1}{3}$-period phase pattern. The corresponding balanced colorings are those in Fig. 1. More generally, if the symmetry group of the CPG is a cyclic group $\mathbb{Z}_k$ of order k, then there can exist phase patterns with phase shifts that are integer multiples of T/k, where T is the period [Golubitsky & Stewart, 2002; Golubitsky *et al.*, 1988]. These patterns are *rigid*, which is a form of structural stability. Specifically, rigidity means that the phase shifts are preserved, as fractions of the period, after any sufficiently small admissible perturbation of the Ordinary Differential Equation (ODE). Stability of these synchronous and phase-synchronous states depends on nonlinear terms on the ODE, usually in a complicated manner. It also depends on the notion of stability concerned.

As remarked in [Stewart & Wood, 2024], the use of chains to propagate signals is scarcely novel; it is simple, natural, and obvious. However, the formal setting for network dynamics leads to rigorous proofs of some general model-independent results. Paramount among them are conditions for the propagating signals to be stable. For feedforward lifts, stability decomposes into stability against synchrony-preserving perturbations and stability against synchrony-breaking perturbations. The first requires the orbit of the CPG to be stable in the state space for the CPG nodes; the second requires stability transverse to the synchrony subspace concerned.

1.2. *Stability*

There are several distinct notions of stability, even for equilibria, and further distinctions arise for periodic orbits because of their global nature. The most important stability concept for a periodic orbit comes from Floquet theory [Floquet, 1883; Hassard *et al.*, 1981], and we use the term "Floquet stable" for clarity. It is the stability notion adopted in this paper.

More generally, four notions of transverse stability are considered in [Stewart & Wood, 2024]: transverse Lyapunov, asymptotic, and Floquet stability of the periodic orbit, and transverse stability of the synchrony subspace. Floquet stability of signals propagating along arbitrarily long chains reduces to certain stability conditions related to the CPG alone. Thus, if the propagating signal is stable one step along the chain, it remains stable for a chain of any length. A similar property holds for Lyapunov stability if this is established using a Lyapunov function, subject to reasonable technical conditions; see [Stewart & Wood, 2024, Sec. 5.7].

Transverse Floquet stability and transverse stability of the synchrony subspace are properties of the periodic orbit on the CPG. Transverse Lyapunov and asymptotic stability are properties of orbits near the periodic orbit on the CPG. These concepts characterize stability (in the corresponding sense) of the periodic orbit to synchrony-breaking perturbations. For example, the lifted periodic orbit is Floquet stable if and only if the CPG orbit is Floquet stable on the CPG state space and transversely Floquet stable for each node of the chain [Stewart & Wood, 2024]. In general, transverse Floquet multipliers must be computed numerically. Simulations, described below, show that the transverse stability condition is often satisfied in standard models of neuron dynamics, which also tend to be Floquet stable. A weaker condition, *transverse stability on average*, is more common, and is also associated with Floquet stability.

A different stability measure, transverse stability of the synchrony subspace — either globally or on (hence also near) the periodic orbit — occurs when the transverse eigenvalues of the Jacobian have negative real parts at every point of the synchrony subspace (or just of the periodic orbit). This is a tempting stability criterion, because the transverse eigenvalues of the Jacobian, or the signs of their real parts, can often be computed analytically, in contrast to Floquet multipliers. However, it must be treated with care. Transverse stability in this sense implies transverse Floquet stability for equilibria with node spaces of any dimension, and for periodic orbits when node spaces are one-dimensional. However, the Markus–Yamabe example [Markus & Yamabe, 1960] shows that for node spaces of higher dimension, this condition need not imply Floquet stability of a periodic orbit: in this example, a nearby trajectory diverges to infinity. This counterintuitive result disproves a conjecture that was widely believed at the time.

1.3. *Biological motivation*

To set the scene, we begin with some typical examples of propagating phase-synchronous signals in biological systems. We discussed two such examples in [Stewart & Wood, 2024]: peristalsis in the gut and peristalsis in the locomotion of *Drosophila* larvae. Here we add three more: the heartbeat of the medicinal leech, legged locomotion, and locomotion of the nematode *Caenorhabditis elegans*.

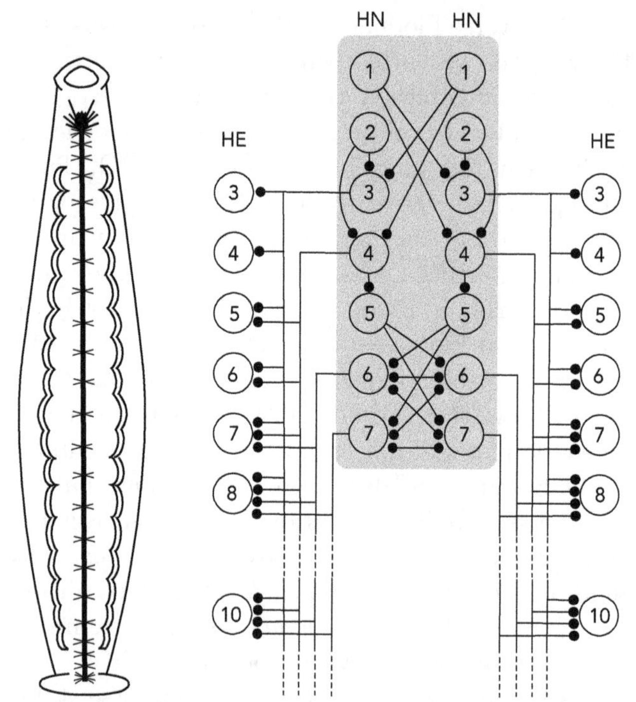

Fig. 2. *Left*: Medicinal leech showing two linear series of heart chambers, one on each side, redrawn from [Calabrese & Peterson, 1983]. *Right*: Timing signal CPG for the leech heartbeat (shaded box) and propagating chains of neurons, redrawn from [Kristan Jr. *et al.*, 2005, Figs. 4A and 4B], HE = heart excitatory motor neuron. HN = heart interneuron. All connections shown are inhibitory.

Example 1.1. The leech *Hirudo medicinalis* has two series of heart chambers, one on each side, Fig. 2(left). The chambers on one side of the animal contract in synchrony, while those on the other side contract in a traveling wave. After 30–40 heartbeat cycles, the two sides swap patterns [Calabrese & Peterson, 1983; Thompson & Stent, 1976; Wenning *et al.*, 2004]. Figure 2(right) shows an experimentally observed CPG [Kristan Jr. *et al.*, 2005] and a chain (nodes 8 onwards) along which the timing signals propagate. This chain forms a feedforward lift of the CPG together with nodes HE 8 (left and right), since the inputs to nodes 9 onwards are copied from node 8.

The CPG shown is a refinement of that described in [Calabrese & Peterson, 1983; Calabrese *et al.*, 1995]. The architecture is essentially identical in different animals, except for connection strengths [Roffman *et al.*, 2012]. A mathematical model that treats the dynamics as a mode interaction is analyzed in [Buono & Palacios, 2004], and an alternative model for such behavior is proposed in [Parker & Stewart, 2008].

Example 1.2. A second class of examples is animal locomotion, especially in linear creatures such as arthropods, snakes, worms, eels, and lampreys. Generally speaking, forward locomotion arises from a wave of muscle (hence limb, where appropriate) movements traveling from the rear of the animal to the front. This compresses the animal near the rear, and the compression zone moves forwards, propelling successive segments of the body forwards. Backward locomotion involves a similar wave traveling in the opposite direction.

Even in animals with few limbs, the muscle and limb movements can be modeled as traveling waves along short chains, closed up into cycles. These networks often exhibit synchrony or phase-synchrony, in characteristic patterns known as *gaits* [Gambaryan, 1974; Muybridge, 1899]. It is widely believed that these patterns are created by a CPG, and that in vertebrates, this is located in the spinal column [Grillner, 2003; Grillner & El Manira, 2020; Grillner & Jessell, 2009]. For some organisms, such as the lamprey [Grillner, 2003], this has been verified directly; see also [Goulding, 2009; Kiehn, 2006]. A recent survey is [Berkowitz, 2019], and the prevailing paradigm of alternating activity in excitatory and inhibitory neurons is laid out in [Arber & Costa, 2018; McCrea & Rybak, 2008]. New observations challenging this paradigm are described in [Lindén *et al.*, 2022].

Gaits in bipeds, quadrupeds, and hexapods have characteristic phase patterns, many of which are typical in certain symmetric networks [Buono, 2001; Collins & Stewart, 1993a, 1993b; Golubitsky *et al.*, 1998; Golubitsky *et al.*, 1999; Pinto & Golubitsky, 2006; Stewart, 2014]. Experiments verify that circuits of this kind can generate basic gait rhythms [Collins & Richmond, 1994], with potential applications to legged robots [In *et al.*, 2022]. Analogous networks generate gaits typical of myriapods such as centipedes and millipedes. Laterally related limbs move half a gait period out of phase (anti-synchrony) in centipedes, and in phase in millipedes [Gray, 1968, Chap. 11]. The CPG architecture suggested in [Golubitsky *et al.*, 1998; Golubitsky *et al.*, 1999] predicts that centipede locomotion should occur with wavelengths of half an odd integer relative to the length of the organism, consistent with observations. These phase relations between segments are distinct from the phase relations between extensor and flexor motions of a single limb. The early "half-center" architecture assumed strict alternation of extensor and flexor actions, but they can also partially overlap or synchronize [McCrea & Rybak, 2008, Sec. 1].

The symmetric CPG architectures proposed in the above papers need not be realized *directly* as neural circuits. Instead, a distributed network can have a synchrony pattern such that the corresponding "quotient network" of synchronized clusters has symmetry, even though the original network has no symmetry. A similar point is made in [Nicosia *et al.*, 2013]. Indeed, general theorems in network dynamics imply that (subject to certain technical conditions) any structurally stable (or "rigid") phase pattern — one that persists after small perturbations to the model equations — should be generated by a network that has a quotient with cyclic group symmetry; see Sec. 3. The "feedforward lift" construction of [Stewart & Wood, 2024] at the core of this paper leads to networks of this kind, composed of feedforward chains of simpler modules. This suggests an alternative to the small CPG networks in the above papers, more suited to realization within the spinal column [Stewart & Wood, 2025].

Example 1.3. A widely studied example is the nematode worm *C. elegans*, whose cellular structure is known in great detail (see the WormAtlas [Hall *et al.*, 2025]). In particular, the connectome (neural wiring diagram) is known. It includes two distinct neuronal circuits for forward and backward movement [Morone & Makse, 2019, Figs. 2 and 3], which drive muscle groups distributed along the body of the nematode. Its motion has been modeled using linear chains of repeating modules [Boyle *et al.*, 2012; Izquierdo & Beer, 2018; Olivares *et al.*, 2021; Sakamoto *et al.*, 2021].

Many networks in the literature have a similar repetitive feedforward structure; see for example [Mehring *et al.*, 2003; Setareh *et al.*, 2018]. Similar but less constrained network topology occurs in synfire chains [Abeles *et al.*, 2004; Aertsen *et al.*, 1996; Hanuschkin *et al.*, 2011; Zheng & Triesch, 2014].

Remark 1.1. The ability of feedforward lifts to propagate synchronous or phase-synchronous signals in a natural, often stable, manner suggests a plausible, though speculative, evolutionary route from small CPG networks to networks with long chains that automatically propagate CPG signals stably. Namely: begin by evolving the CPG network and add a short chain. If the signal is stable, copy the modules repeatedly to form a longer chain. Since stability is automatically preserved, this process is guaranteed to propagate the signals stably, low-hanging fruit for evolution, which can readily create repetitive structures once the basic module has evolved, as found, for example, in arthropods.

1.4. *Summary of paper*

Section 2 is included for convenience. It begins with a short summary of the network dynamics formalism of [Stewart *et al.*, 2003; Golubitsky *et al.*, 2005; Golubitsky & Stewart, 2023], condensed from [Stewart & Wood, 2024, Sec. 2]. It then considers feedforward lifts, the basic construction of that paper and this one, and transverse Floquet multipliers, which determine the stability of the lifted periodic orbit to perturbations that break synchrony or phase-synchrony. Section 3.1 discusses a method for computing Floquet multipliers, which we use in subsequent numerical calculations.

Sections 4–7 analyze two transverse stability conditions (transverse Floquet stability and transverse stability of the synchrony subspace) for four common model neurons: FitzHugh–Nagumo, Morris–Lecar, Hodgkin–Huxley, and Hindmarsh–Rose. For each type of model neuron, we consider stability in the Floquet sense and transverse stability of the synchrony subspace, either globally or on the periodic orbit. In some cases, the periodic state is Floquet stable even when the synchrony subspace is not transversely stable, an interesting counterpoint to the Markus–Yamabe example [Markus & Yamabe, 1960], which shows that transverse stability need not imply Floquet stability. For all of these models, except Hodgkin–Huxley, we briefly mention a form of "transverse stability on average" [Stewart & Wood, 2024, Sec. 5.2]. This concept has not been made rigorous. We analyze transverse stability analytically, when possible, and otherwise use numerics. For simplicity, we study the "toy" network of Fig. 1, which has a very simple CPG that robustly generates oscillations with a regular phase pattern, and a long enough chain to illustrate the theory.

Section 8 addresses the issue whether the propagating signals persist after synchrony-breaking modifications or perturbations of the model. This is important for applications, since the conditions assumed in an idealized feedforward chain, and the strict definition of synchrony, apply only approximately in real systems. We give qualitative theoretical reasons to suppose that the results are reasonably robust, and support them by quantitative numerical simulations.

Finally, we summarize the main conclusions in Sec. 9.

The numerical calculations in Secs. 4–7 were carried out using Mathematica. Those in Sec. 8 were performed in MatLab.

2. Summary of Network Dynamics Formalism

We work in a formal framework for analyzing networks of coupled dynamical systems (ODEs) [Golubitsky & Stewart, 2023; Golubitsky *et al.*, 2005; Stewart *et al.*, 2003]. Such systems are represented by directed graphs in which nodes and directed edges ("arrows") are labeled with "types". (In graph theory, these are called *colored graphs* [Wright, 1961] or *multilayered graphs* [Hammoud & Kramer, 2020], but we reserve the term "color" to define sets of synchronous nodes.) Nodes of the same type have the same state space, and arrows of the same type represent identical types of coupling. Nodes with isomorphic sets of input arrows obey identical ODEs when corresponding couplings are identified. We outlined this formalism in [Stewart & Wood, 2024, Sec. 2]; it is a natural generalization of the formalism for equivariant dynamical systems [Golubitsky *et al.*, 1988] when applied to symmetric networks. We restate a few basic concepts here for convenience.

2.1. *Admissible ODEs*

Each network diagram is associated with a class of *admissible ODEs*, which are the most general ODEs that respect the network structure. Number the nodes from 1 to n and let x_c be the state variable for node c. We assume that $x_c \in \mathbb{R}^{n_c}$, a finite-dimensional real vector space. (Other choices are possible; in particular, the circle $\mathbb{S}^1$ applies to phase oscillator networks [Kuramoto, 1984].) The overall state of the system is then $x = (x_1, \ldots, x_n)$. The dynamics takes the form

$$\dot{x}_c = f_c(x_c, x_{i_1}, \ldots, x_{i_k}), \quad 1 \leq c \leq n, \tag{1}$$

where $i_1, \ldots, i_k$ are the tail nodes of arrows that input to node c.

Remark 2.1. In this approach to network dynamics, an output arrow comes into play only as an input to its head node, a convention sometimes referred to as the *influence network* of the model. In some areas of application, such as biochemical networks and epidemiology, different conventions are standard [Golubitsky & Stewart, 2023, Examples 2.2–2.4]. ODEs constructed using those conventions remain admissible for the influence network and therefore fall into the scope of that formalism. However, unlike many of these conventions, arrows in the network diagram do not correspond to specific *terms* in the associated model ODEs, and the formalism does not (as sometimes assumed) include only two-body interactions, as Eq. (1) clearly demonstrates. In fact, the network diagram in this formalism can best be viewed as a very general type of hypergraph [Berge, 1973], in which bunches of input arrows of the same type correspond to hyperarcs satisfying certain symmetry constraints. These hyperarcs determine the functional form of the ODE, not specific terms in its component functions [Makse *et al.*, 2025, Sec. 9.3.3].

For convenience, we assume that f_c is of class C^∞ (infinitely differentiable), but less stringent smoothness conditions are possible with no change to proofs. Nodes of the same "input-type" have the same function f_c, applied to their respective sets of inputs. (The same node may send multiple inputs, so, these concepts are formalized in terms of the input arrows.) For example, the admissible ODEs for Fig. 1 have the following form:

$$\begin{aligned} &\dot{x}_1 = f(x_1, x_3), \quad \dot{x}_2 = f(x_2, x_1), \\ &\dot{x}_3 = f(x_3, x_2), \quad \dot{x}_4 = f(x_4, x_3), \\ &\dot{x}_5 = f(x_5, x_4), \quad \dot{x}_6 = f(x_6, x_5), \\ &\dot{x}_7 = f(x_7, x_6). \end{aligned} \tag{2}$$

So, for example, $f_1(x_1, x_3) = f(x_1, x_3)$, and so on. These variables need not be, and often are not, one-dimensional. The same function f is used to specify all components, because all nodes have the same node-type and all arrows have the same arrow-type. The domains of these copies f_i of f are technically different, being Cartesian products of state spaces of the respective nodes, but they are canonically identified with the domain of f. By convention, the first variable in f is the node variable; the other variables correspond to the input nodes. In this case, the input node is unique so there is only one other variable. See [Stewart & Wood, 2024, Example 2.2] for full details and caveats.

2.2. *Colorings, synchrony, and quotient networks*

Synchrony relations are determined by a *coloring* κ of the nodes, associating to each node c a *color*, which is an element $\kappa(c)$ of some specified set $\mathcal{K}$. Synchronous nodes are assigned the same color, so c, d being synchronous means that $\kappa(c) = \kappa(d)$. The key concept here is that of a *balanced* coloring. Balance requires nodes of the same color to have input sets with matching colors, in a manner that preserves node- and arrow-types. The coloring in Fig. 1 is balanced, because all nodes have the same input-type, every white node receives one black input, every gray node receives one white input, and every black node receives one gray input.

The corresponding *synchrony subspace*

$$\Delta_\kappa = \{x : \kappa(c) = \kappa(d) \Rightarrow x_c = x_d\},$$

corresponds to states with the given synchrony pattern. Balance implies that Δ_κ is invariant under any admissible ODE. Conversely, any subspace invariant under every admissible ODE is equal to Δ_κ for some balanced coloring κ. The restriction of the ODE to this subspace determines the dynamics of the synchronized *clusters*, where a cluster is the set of all nodes with a given color. Identifying the nodes in each cluster, and copying arrows according to the colors of their heads and tails, lead to the *quotient network* for the coloring. For example, the quotient for the 3-coloring in Fig. 1(bottom) is a 3-node directed ring. In this case, it is isomorphic to the CPG $\{1, 2, 3\}$, but in general, a quotient network need not be a subnetwork.

Any admissible ODE on the full network restricts to one on the quotient network; conversely any admissible ODE on the quotient network *lifts* to one on the full network. Synchronous or phase-synchronous solutions restrict and lift in the same manner, preserving synchrony and phase synchrony.

2.3. *Feedforward lift*

Let $\mathcal{G}$ be any network. A *feedforward lift* of $\mathcal{G}$ is a network $\tilde{\mathcal{G}}$ such that $\mathcal{G}$ is both a subnetwork and a quotient network of $\tilde{\mathcal{G}}$ by some balanced coloring, and all arrows of $\tilde{\mathcal{G}}$ other than those in $\mathcal{G}$ are feedforward. That is, we can order the nodes not in $\mathcal{G}$ so that the tail $\mathcal{T}(e)$ of every arrow e is either in $\mathcal{G}$, or is less than the head $\mathcal{H}(e)$ in that ordering [Golubitsky & Stewart, 2023, Theorem 4.11]. Moreover, every node in $\tilde{\mathcal{G}}$ has the same color as one in $\mathcal{G}$. Again Fig. 1 is an example, with a suitable ordering indicated by the numbers. Here, $\mathcal{G}$ is the initial CPG with nodes $\{1, 2, 3\}$ and the three arrows between them, which form a cycle.

As in [Stewart & Wood, 2024], the central topic is a periodic orbit $A = \{a(t)\}$ for some $\mathcal{G}$-admissible ODE, lifted to a periodic orbit $\tilde{A} = \{\tilde{a}(t)\}$ for $\tilde{\mathcal{G}}$. Any synchrony or phase pattern on A lifts to one on $\tilde{A}$.

2.4. *Stability and transverse Floquet multipliers*

If A is stable on the state space of the CPG nodes, then the lift $\tilde{A}$ is stable to *synchrony-preserving* perturbations within the synchrony subspace. However, $\tilde{A}$ might be unstable to *synchrony-breaking* perturbations, transverse to the synchrony subspace. We therefore investigate its stability, considering two main senses of that word. The feedforward structure of the network implies that the Jacobian (linearization) of the ODE is block-triangular; the first block corresponds to the CPG and subsequent blocks correspond to nodes running through the chain in feedforward order [Stewart & Wood, 2024, Corollary 3.8]. Thus, its eigenvalues are those of the diagonal blocks. For a periodic orbit, an analogous result applies to Floquet stability: the Floquet operator is (conjugate to) a block-triangular matrix, so the Floquet multipliers are those of the individual blocks [Stewart & Wood, 2024, Sec. 5.1]. The *transverse Floquet multipliers* for A come from all blocks except that for the CPG. The main result states that a necessary and sufficient condition for $\tilde{A}$ to be Floquet stable is that A is Floquet stable and all transverse Floquet multipliers have absolute value less than 1 [Stewart & Wood, 2024, Theorem 4.4].

In more detail, we need:

Definition 2.1. Let f_c be the cth component of an admissible ODE, $[c]$ a node in the CPG with the same color as node c, and $\{a(t)\}$ a periodic orbit of the CPG that is lifted to a periodic orbit $\{\tilde{a}(t)\}$ for the full network. Denote the partial derivative with respect to $x_{[c]}$ by $\mathrm{D}_{[c]}$. Then:

(a) With the above notation, the *transverse Floquet equation* of A, for node $c \in \mathcal{C}^*$ is

$$\dot{y}_c = \mathrm{D}_{[c]} f_{[c]}|_{a(t)} y_c. \tag{3}$$

Here, Eq. (3) depends only on the ODE for the CPG and the periodic orbit $\{a(t)\}$. This is why we refer to the transverse Floquet equation of A, rather than of $\tilde{A}$.

(b) By Floquet theory [Hassard *et al.*, 1981], every solution of Eq. (3) has the form

$$y_c(t) = P_c(t)e^{B_c t}v,$$

for a constant vector v. Here, $P_c(t)$ is T-periodic and B_c is a constant matrix. The matrix $M_c = e^{B_c T}$ is the *transverse Floquet matrix* for node c.

(c) The matrix B_c is the *transverse Floquet exponent matrix* for node c.

(d) The periodic orbit $A = \{a(t)\}$ is *transversely Floquet stable* at node c if all eigenvalues of M_c have an absolute value < 1. Equivalently, all eigenvalues of B_c have negative real part.

(e) The eigenvalues of M_c are the *transverse Floquet multipliers* of A for node c.

(f) The eigenvalues of B_c are the *transverse Floquet exponents* of A for node c.

One main result of [Stewart & Wood, 2024] is:

Theorem 1. *Let $\{\tilde{a}(t)\}$ be a feedforward lift of the periodic orbit $\{a(t)\}$ on $P_{\mathcal{C}}$. Then*:

(a) *The Floquet multipliers for $\{\tilde{a}(t)\}$ are the Floquet multipliers for $a(t)$, together with the transverse Floquet multipliers of $a(t)$ for all $c \in \mathcal{C}^*$.*

(b) *The transverse Floquet multipliers for $c \in \mathcal{C}^*$ are the same as those for $[c] \in \mathcal{C}$.*

(c) *$\{\tilde{a}(t)\}$ is stable on P if and only if $\{a(t)\}$ is stable on $P_{\mathcal{C}}$ and, for all nodes in $\mathcal{C}$, all transverse Floquet multipliers of $a(t)$ have absolute value < 1.*

The Floquet multipliers for $a(t)$ always include an eigenvalue 1 corresponding to the periodic orbit [Hassard *et al.*, 1981, Chap. 1, Sec. 4]. This is already present for the lifted orbit $\{\tilde{a}(t)\}$, so the transverse Floquet multipliers generically do not include 1.

Remark 2.2. It is well known that symmetry can create multiple eigenvalues generically [Golubitsky *et al.*, 1988, Chap. XII, Sec. 2], and this applies in particular to symmetric networks. In network dynamics, other aspects of network topology can also create multiple eigenvalues, and feedforward lifts provide examples of this phenomenon. Conditions (a) and (b) imply that long chains give rise to Floquet multipliers of high multiplicity, even when the overall network $\tilde{\mathcal{G}}$ has trivial symmetry. Analogous effects for critical eigenvalues at equilibria are well known; see [Golubitsky & Stewart, 2006, Example 11.1; Golubitsky & Stewart, 2023, Sec. 21.1]. High multiplicity of transverse critical eigenvalues implies that synchrony-*breaking* bifurcations for these networks can be very complex.

3. Phase Synchrony

If the CPG has cyclic group symmetry, suitable admissible ODEs naturally support discrete rotating waves, with phase shifts related to a generator of the cyclic group [Golubitsky & Stewart, 2002; Golubitsky *et al.*, 1988; Stewart & Parker, 2008]. Such phase patterns are *rigid*. A hyperbolic periodic orbit persists after any C^1-small perturbation of the ODE. Rigidity means that the phase pattern also persists, with phase shifts equal to the same fractions of the period. Conversely, subject to certain technical conditions, rigid phase patterns conjecturally imply cyclic symmetry on the quotient network by the synchrony coloring. This conjecture has been proved for a broad class of networks [Golubitsky *et al.*, 2010, 2012] (the results are stated there for all networks, but a technical condition was overlooked, see [Stewart, 2022, Sec. 1.4] or [Golubitsky & Stewart, 2023, Sec. 15.1]). It has also been proved in full generality under hypotheses slightly stronger than hyperbolicity [Stewart, 2022].

For example, the CPG of Fig. 1, a cycle of length 3, has $\mathbb{Z}_3$ symmetry. It supports periodic states where phases of successive nodes are shifted by $\frac{1}{3}$ of the period. Figure 3 shows superposed numerically computed examples of time series for nodes 1, 2, 3, for FitzHugh–Nagumo, Morris–Lecar, Hodgkin–Huxley, and Hindmarsh–Rose models. For details, see the corresponding sections below, where the same figures appear as part of a more general investigation.

When lifted to $\tilde{\mathcal{G}}$, such orbits resemble traveling waves. The theory generalizes to these phase patterns. Indeed, phase relations not caused by symmetries are also copied along the chain, and similar remarks apply, although we do not pursue this possibility here. Subject to some minor technical conditions, the setting is as follows; see [Stewart & Wood, 2024, Sec. 6] for details.

Let $\mathcal{G}$ be a CPG with nodes $\{1, \ldots, k\}$ and $\mathbb{Z}_k$ symmetry generated by a permutation α, which cycles the nodes in the CPG so that $\alpha(1) = 2, \alpha(2) = 3, \ldots, \alpha(k) = 1$. That is, $\alpha = (1\,2\,3\,\ldots\,k)$ in cycle notation. Choose an admissible map f so that the ODE $\dot{x} = f(x)$ has a discrete rotating wave state $u(t)$ satisfying the

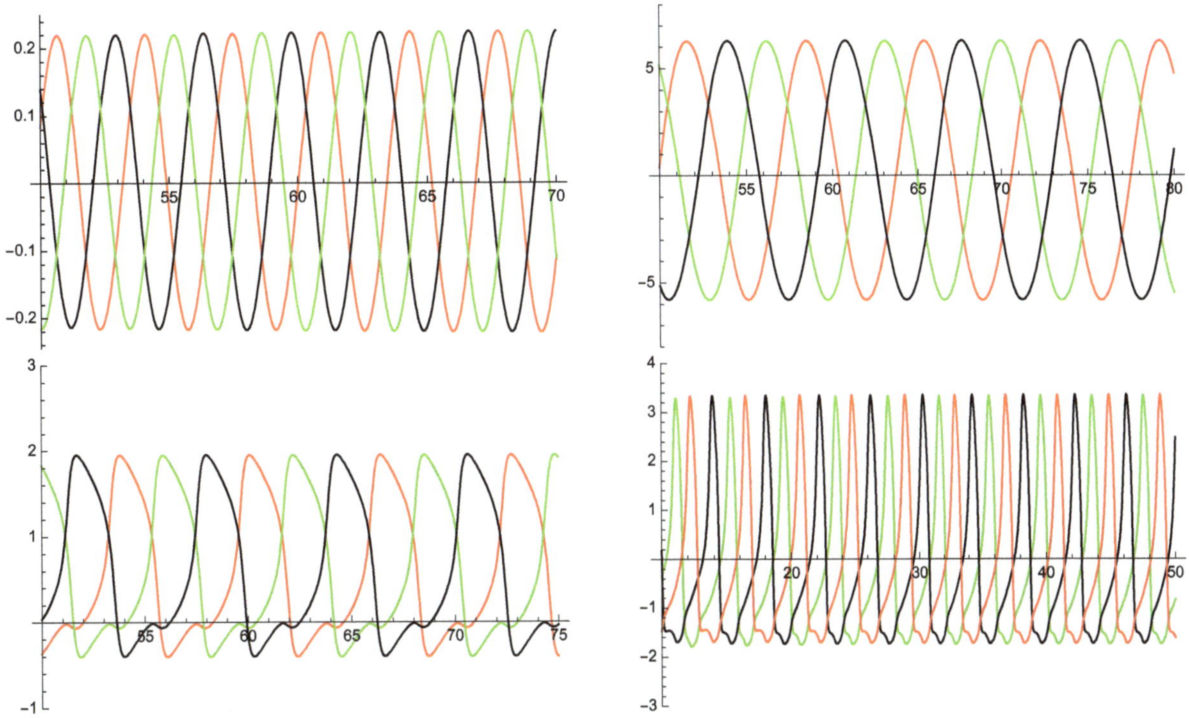

Fig. 3. $\mathbb{Z}_3$ traveling wave periodic states in a 3-node ring, for four standard model neurons. Each subfigure shows three superposed time series of corresponding node variables. *Top left*: FitzHugh–Nagumo. *Top right*: Morris–Lecar. *Bottom left*: Hodgkin–Huxley. *Bottom right*: Hindmarsh–Rose.

phase relation $\alpha u(t) = u(t + rT/k)$, where r is an integer and $0 \leq r \leq k-1$. Choose a *module* $\mathcal{M}$ whose nodes are a set of representatives of the $\mathbb{Z}_k$-orbits. Assign phase 0 to these nodes, so that the other $\mathbb{Z}_k$-orbits correspond to phase shifts $rT/k, 2rT/k, \ldots, (k-1)rT/k$. Copy the module (along with any arrows whose heads and tails lie in the module) to obtain $\mathcal{M}_{k+1}, \mathcal{M}_{k+2}, \ldots, \mathcal{M}_l$. Assign phases $0, rT/k, \ldots, (l-k)rT/k$ to nodes in $\mathcal{M}_{k+1}, \mathcal{M}_{k+2}, \ldots, \mathcal{M}_l$. Assign input arrows to these nodes, preserving the arrow-type and the phase relations in $\mathcal{G}$. Do so in a manner that makes all new arrows feedforward. Finally, rewire internal arrows in $\mathcal{M}_{k+1}, \mathcal{M}_{k+2}, \ldots, \mathcal{M}_l$, preserving the arrow-type and the phase relations in $\mathcal{G}$. Do so in a manner that makes all rewired arrows feedforward.

In this context, assuming the above notation, the following result is proved in [Stewart & Wood, 2024, Theorem 6.1].

Theorem 2. *Assume that* $\mathcal{G}$ *has nodes* $\mathcal{C} = \{1, \ldots, m\}$ *with a cyclic automorphism group* $\mathbb{Z}_k = \langle \alpha \rangle$, *such that* $n = mk$ *and* α *acts like the cycle* $(1\,2\,\ldots\,k)$ *on all of its orbits on* $\mathcal{G}$. *Let* $u(t)$ *be a* T*-periodic solution of an admissible ODE with a discrete rotating wave phase pattern induced by the symmetry, and let the module* $\mathcal{M}$ *be a set of orbit representatives. Let* $\tilde{\mathcal{G}}$ *be obtained by lifting appropriate copies of translates of this module by* $\mathbb{Z}_k$. *Then,*

(a) *The periodic state* $u(t)$ *on* $\mathcal{G}$ *lifts to a* T*-periodic traveling wave state* $\tilde{u}(t)$ *for* $\tilde{\mathcal{G}}$ *with phases corresponding to the extra copies* $\mathcal{M}_{k+1}, \mathcal{M}_{k+2}, \ldots, \mathcal{M}_l$ *of* $\mathcal{M}$.
(b) *The Floquet exponents (evaluated at any point) are those on the module* $\mathcal{M}$, *together with the transverse Floquet exponents for* $\mathcal{M}$.
(c) *If the Floquet exponents and the transverse Floquet exponents on* $\mathcal{M}$ *have negative real part, then* $\tilde{u}(t)$ *is Floquet stable.*

In particular, the transverse Floquet multipliers need be computed only for nodes in the module $\mathcal{M}$, however large the feedforward lift $\tilde{\mathcal{G}}$ may be. This remark applies to any periodic state of the CPG, and does not require phase relations to be caused by symmetry.

3.0.1. *Comparison of stability concepts*

Floquet multipliers generally cannot be computed analytically, except when node spaces are one-dimensional. A more tractable, but less rigorous, stability condition is *transverse stability* of $\tilde{A}$ relative to the synchrony subspace Δ_κ. This requires all transverse eigenvalues of the Jacobian — those not coming from the CPG — to have negative real part when evaluated at any point of A. When node spaces are one-dimensional, this is equivalent to the transverse Floquet multipliers having absolute value less than 1, but this result can be false for higher-dimensional node spaces [Stewart & Wood, 2024, Examples 5.3 and 5.4].

The main focus of this paper is to compare the behavior of transverse Floquet multipliers and transverse eigenvalues of the Jacobian, for a variety of standard neuron equations — namely, FitzHugh–Nagumo, Morris–Lecar, Hodgkin–Huxley, and Hindmarsh–Rose neurons. To standardize the method, we mainly do this for the network of Fig. 1. We also discuss Floquet stability of the CPG orbit A for FitzHugh–Nagumo neurons. Pragmatically, if numerical simulations produce traveling waves with the $\frac{1}{3}$-period phase pattern for randomly chosen initial conditions, the periodic orbit concerned should be stable. Otherwise, this solution would not be observed. Thus, the Floquet stability results below should not come as a surprise.

3.1. *Computation of Floquet multipliers*

We recall the classical stability criterion of Floquet theory [Hassard *et al.*, 1981, Chap. 1, Sec. 4] and describe how we compute Floquet multipliers. Linearizing the ODE about the periodic orbit A gives a nonautonomous ODE

$$\dot{y} = M(t)y, \tag{4}$$

where $M(t) = \mathrm{D}_y f|_{a(t)}$ is T-periodic. A *fundamental matrix* for (4) is a time-dependent matrix $Y(t)$ such that any solution $y(t)$ has the form $y(t) = Y(t)v$ for some constant vector v. Floquet's theorem [Floquet, 1883] implies the existence of a fundamental matrix. Moreover, there is a T-periodic matrix function $P(t)$ and a matrix B such that every fundamental matrix can be written as

$$Y(t) = P(t)\mathrm{e}^{Bt}. \tag{5}$$

The eigenvalues β_i of B are *Floquet exponents* and the eigenvalues ρ_i of e^{BT} are *Floquet multipliers.* The condition for Floquet stability is that all Floquet exponents β_i have negative real part, except for a simple eigenvalue 0 corresponding to the orbit. Equivalently, all ρ_i are in the interior of the unit circle, except for a single eigenvalue 1. In modern dynamical systems theory, the matrix e^{BT} is replaced by the Jacobian of a Poincaré map at the fixed point determined by the periodic orbit [Guckenheimer & Holmes, 1983, Sec. 1.5]. This reduces the dimension by one, and excludes the eigenvalue 1 whose eigenvector is tangent to the periodic orbit.

It is well known that Floquet exponents/multipliers can seldom be found analytically. However, numerical calculations are straightforward for specific examples. In Secs. 4–7, we present numerical results on some Floquet multipliers. The corresponding Floquet exponents can be found by taking logarithms.

Our method is based on (5). This implies that, for any solution $Y(t)$

$$Y(t_0) = P(t_0)\mathrm{e}^{Bt_0},$$

$$Y(t_0+T) = P(t_0+T)\mathrm{e}^{B(t_0+T)} = P(t_0)\mathrm{e}^{BT}\mathrm{e}^{Bt_0},$$

since Bt_0 and BT commute. Therefore,

$$\begin{aligned} Y(t_0+T) &= P(t_0)\mathrm{e}^{BT}\mathrm{e}^{Bt_0} \\ &= (P(t_0)\mathrm{e}^{BT})((t_0)^{-1}Y(t_0)) \\ &= [P(t_0)\mathrm{e}^{BT}P(t_0)^{-1}]Y(t_0). \end{aligned}$$

The matrix $E = P(t_0)\mathrm{e}^{BT}P(t_0)^{-1}$ is conjugate to e^{BT}, so it has the same eigenvalues. These are the required Floquet multipliers.

The period T can be computed numerically. Having done so, one way to compute E is to take n linearly independent initial conditions $Y_i(0)$, such as the column vectors $\mathbf{e}_i$ with 1 in the ith place and 0 everywhere else. Compute $Y(t_0)$ and $Y(t_0+T)$ numerically. In each case, these sets of vectors are also linearly independent, so we can solve the system

$$EY_i(t_0) = Y_i(t_0+T), \quad 1 \leq i \leq n,$$

for E, whose eigenvalues are the Floquet multipliers. Explicitly, consider the Y_i as column vectors and let

$$U = [Y_1(t_0)|\cdots|Y_n(t_0)],$$

$$V = [Y_1(t_0+T)|\cdots|Y_n(t_0+T)],$$

be the matrices obtained by concatenating the indicated columns. Then, $EU = V$, so $E = VU^{-1}$.

This approach lets us start the integration from time 0 and then choose t_0 large enough for the initial state to have settled down close to the periodic cycle, so the computed $a(t)$ effectively lies on the cycle. These values can then be fed into (4). This method yields both the Floquet multipliers for the periodic orbit A and the transverse Floquet multipliers for $\tilde{A}$.

Some care must be taken to avoid numerical instabilities, and to ensure that the computed oscillation has settled down sufficiently close to the actual periodic orbit.

Remark 3.1. For a specific small network such as the 7-node chain, both the Floquet multipliers for the cycle A and the transverse Floquet multipliers can be found directly, by computing the Floquet multipliers for the lifted periodic orbit. Theorems 1 and 2 imply that the same computation gives the stability of $\tilde{a}(t)$ for *any* feedforward lift $\tilde{\mathcal{G}}$, however large it may be.

4. FitzHugh–Nagumo Model

We now examine two main notions of transverse stability for feedforward lifts whose nodes are standard model neurons, beginning with FitzHugh–Nagumo neurons. These stability notions are transverse Floquet stability and transverse stability of the synchrony subspace. We also consider Floquet stability of the CPG periodic orbit A for FitzHugh–Nagumo neurons. All time series and Floquet multipliers in Secs. 4–7 are computed using Mathematica.

For all of the standard models, we use the same running example as [Stewart & Wood, 2024], namely Fig. 1. This 7-node network is a feedforward chain with a single feedback connection from node 3 to node 1. Nodes $\{1,2,3\}$ and connecting arrows are interpreted as a CPG with $\mathbb{Z}_3$ symmetry, and the rest of the network is the chain in a feedforward lift. The "colors" of the nodes (white, gray, black) indicate a synchrony pattern that is preserved by the lift. They also correspond to synchronous nodes in a phase pattern with $\frac{1}{3}$-period relative phase shifts.

The FitzHugh–Nagumo model of a single neuron [FitzHugh, 1961; Nagumo *et al.*, 1962] takes the form

$$\dot{V} = V(a-V)(V-1) - W + I, \tag{6}$$

$$\dot{W} = bV - \gamma W, \tag{7}$$

where V is the membrane potential, W is a recovery variable, I is the magnitude of stimulus current, $0 < a < 1$ is a parameter and $b, \gamma > 0$ are parameters.

In principle, the voltage V and input I can have either sign.

The transverse Floquet equations for FitzHugh–Nagumo neurons are derived explicitly in [Stewart & Wood, 2024, Example 5.3]. We discuss the results for the special case of Fig. 1, but since the transverse Floquet equations depend only on the CPG and its periodic orbit $\{a(t)\}$, the same results apply to any feedforward lift, such as a linear chain of n FitzHugh–Nagumo neurons for any n, or a branching tree of such chains.

4.1. *Simulations*

We model Fig. 1 when the nodes are FitzHugh–Nagumo neurons with voltage coupling. By [Stewart & Wood, 2024, Theorem 2], it suffices to calculate the transverse eigenvalues and transverse Floquet multipliers at a single node, which plays the role of the module $\mathcal{M}$.

Let

$$G(V,W,I) = V(a-V)(V-1) - W + I,$$
$$H(V,W) = bV - \gamma W.$$

Model equations are

$$\begin{aligned}
\dot{V}_1 &= G(V_1,W_1,I) + cV_3, & \dot{W}_1 &= H(V_1,W_1),\\
\dot{V}_2 &= G(V_2,W_2,I) + cV_1, & \dot{W}_2 &= H(V_2,W_2),\\
\dot{V}_3 &= G(V_3,W_3,I) + cV_2, & \dot{W}_3 &= H(V_3,W_3),\\
\dot{V}_4 &= G(V_4,W_4,I) + cV_3, & \dot{W}_4 &= H(V_4,W_4),\\
\dot{V}_5 &= G(V_5,W_5,I) + cV_4, & \dot{W}_5 &= H(V_5,W_5),\\
\dot{V}_6 &= G(V_6,W_6,I) + cV_5, & \dot{W}_6 &= H(V_6,W_6),\\
\dot{V}_7 &= G(V_7,W_7,I) + cV_6, & \dot{W}_7 &= H(V_7,W_7).
\end{aligned} \tag{8}$$

Here, c is the strength of the voltage coupling, V_c is the voltage at node c, and W_c is the recovery variable at node c.

Figure 4 shows a synchronous periodic state. Parameters are

$$I = 0, \quad a = 0.05, \quad b = 2.5, \quad \gamma = 0.3, \quad c = 0.4. \tag{9}$$

Figure 5 shows a $\mathbb{Z}_3$ traveling wave periodic state. Parameters are

$$I = 0, \quad a = 0.05, \quad b = 2.5, \quad \gamma = 0.3, \quad c = -0.6. \tag{10}$$

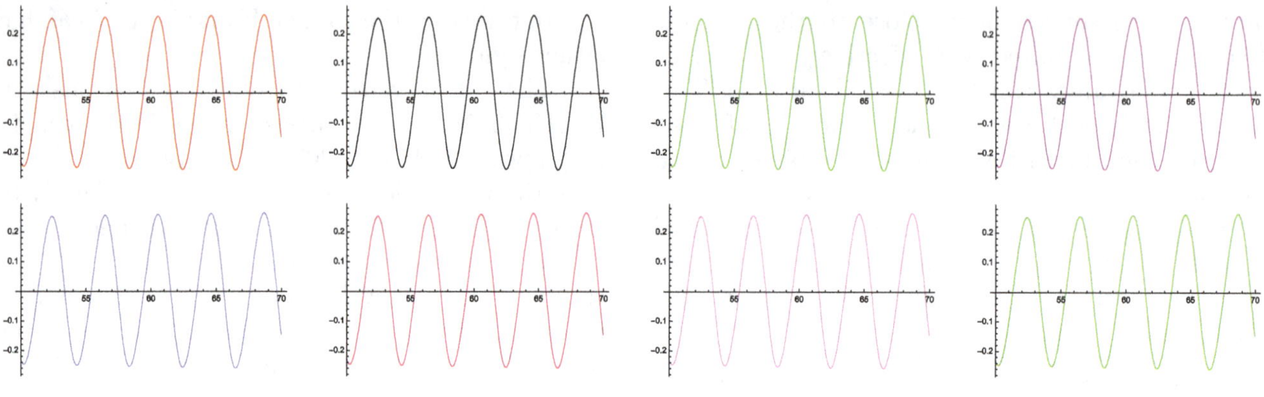

Fig. 4. Synchronous periodic state. Time range = [50, 70] to eliminate transients. *Top left to bottom right*: Nodes 1–7; all nodes superposed.

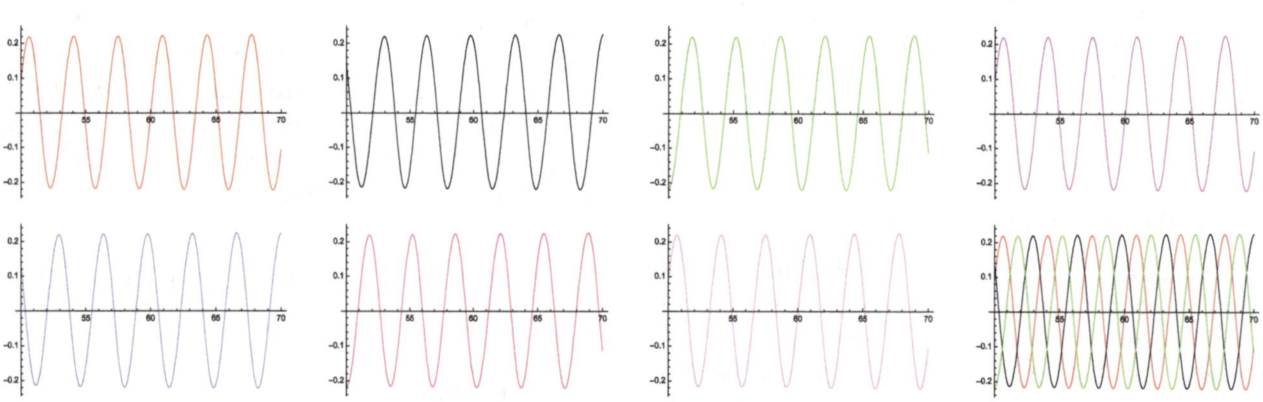

Fig. 5. $\mathbb{Z}_3$ traveling wave periodic state. Time range = [50, 70] to eliminate transients. *Top left to bottom right*: Nodes 1–7; all nodes superposed.

4.2. *Transverse eigenvalues*

We computed the transverse eigenvalues along the periodic orbit for the above parameter values. The results are shown graphically in Figs. 6 and 7, and confirm that the eigenvalues of the Jacobian J on the periodic cycle all have negative real part, indicating transverse stability. In Sec. 4.3, we show (numerically) that these periodic orbits are transversely Floquet stable.

Remark 4.1. If instead we use parameters

$$I = 0, \quad a = 0.05, \quad b = 2.5, \quad \gamma = 0.3, \quad c = -0.8, \tag{11}$$

then numerical simulations show a traveling wave very similar to Fig. 5, which therefore appears to be stable. However, numerical calculations, Fig. 7(left), show that the transverse eigenvalues have (very small) *positive* real part for a short time interval.

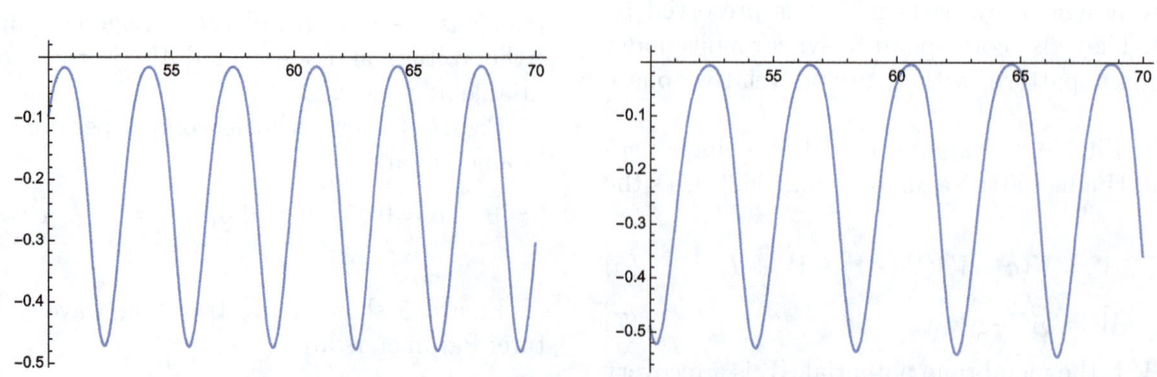

Fig. 6. Cases where real parts of eigenvalues of J at times $t \in [50, 70]$ are always negative. Both eigenvalues have the same real parts because the eigenvalues are complex conjugates. *Left*: Parameters (9). *Right*: Parameters (10).

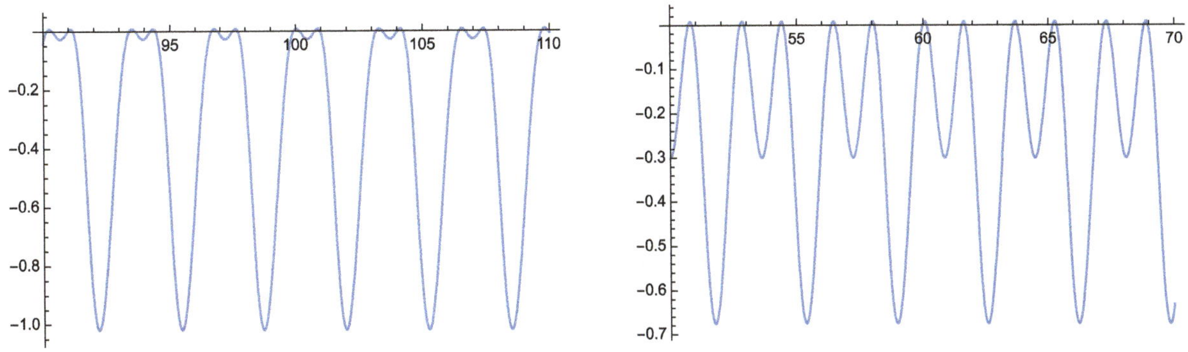

Fig. 7. For Floquet stable periodic states, real parts of eigenvalues of J at times $t \in [50, 70]$ can sometimes be positive; see tips of double peaks. Both eigenvalues have the same real parts because the eigenvalues are complex conjugates. *Left*: Parameters (11). *Right*: Parameters (12).

This suggests that in practice, the periodic cycle can be transversely Floquet stable if the region of the cycle in which the transverse dynamic fails to be attractive is sufficiently small. Numerical calculations summarized below in Table 1 confirm that the CPG periodic orbit is Floquet stable and the lifted periodic orbit is transversely Floquet stable.

Similar results apply if the input I is nonzero. Negative I seems to damp down the oscillations. Positive (not too large) I leads to $\mathbb{Z}_3$ traveling waves, which are apparently stable, but again there can be a short interval of times for which the transverse eigenvalues have (very small) positive real part. Figure 7(right) shows this for parameters

$$I = 2, \quad a = 0.05, \quad b = 2.5, \quad \gamma = 0.3, \quad c = -0.4. \tag{12}$$

4.3. *CPG and transverse Floquet multipliers*

We computed the Floquet multipliers for the periodic state $a(t)$ on the CPG $\mathcal{G}$. We consider the four parameter sets (9)–(12). The results are listed in Table 1. The CPG periodic orbit is Floquet stable and the lifted periodic orbit is transversely Floquet stable, for all four parameter sets.

Comparing with the transverse stability condition, as in Sec. 4.2, we see that for parameters (11)

Table 1. CPG and transverse Floquet multipliers for FitzHugh–Nagumo examples. The CPG periodic orbit and the lifted periodic orbit are stable for all four parameter sets. All entries stated to three significant figures after the decimal point. abs. = absolute value of multiplier in previous column.

Param.	T	CPG Multipliers	Abs.	Transverse Multipliers	Abs.
(9)	4.070	1	1	0.435	0.435
		0.812	0.812	0.366	0.366
Stable		$0.222 \pm 0.168\,\mathrm{i}$	0.279		
		$0.183 \pm 0.174\,\mathrm{i}$	0.252		
(10)	3.420	1	1	$0.290 \pm 0.388\,\mathrm{i}$	0.485
		0.868	0.868		
Stable		$-0.167 \pm 0.696\,\mathrm{i}$	0.715		
		$0.0957 \pm +0.141\,\mathrm{i}$	0.171		
(11)	3.261	1	1	$0.0850 + 0.315\,\mathrm{i}$	0.327
		0.507	0.507		
Stable		$-0.444 \pm 0.401\,\mathrm{i}$	0.598		
		$-0.00298 \pm 0.0820\,\mathrm{i}$	0.0820		
(12)	3.620	1	1	$0.371 + 0.163\,\mathrm{i}$	0.405
		0.385	0.385		
Stable		$0.233 \pm 0.532\,\mathrm{i}$	0.581		
		$0.156 \pm 0.098\,\mathrm{i}$	0.184		

and (12), transverse eigenvalues can have positive real parts on some time intervals, without loss of transverse Floquet stability of the lifted periodic orbit. This suggests that some form of "transverse stability on average", as discussed in [Stewart & Wood, 2024, Sec. 5.2], might be operating, whereby short periods of repulsion away from the synchrony subspace are compensated for by longer periods of attraction. However, it is unclear how to make this suggestion rigorous when node spaces have dimension greater than 1.

5. Morris–Lecar Model

The Morris–Lecar model of a neuron [Morris & Lecar, 1981] was originally devised to model voltage oscillations in the muscles of barnacles. It is a simplified variant of the Hodgkin–Huxley equations, and takes the form

$$\begin{aligned} C\dot{V} &= -g_{\mathrm{Ca}}M_{\mathrm{ss}}(V)(V - V_{\mathrm{Ca}}) - g_{\mathrm{K}}W(V - V_{\mathrm{K}}) \\ &\quad - g_L(V - V_L) + I_{\mathrm{app}}, \\ \dot{W} &= \frac{W_{\mathrm{ss}}(V) - W}{T_W(V)}, \end{aligned} \tag{13}$$

where V is the membrane potential, W is a recovery variable (almost always the normalized K^+-ion conductance), I_{app} is the magnitude of stimulus current and $C > 0$ is a time constant for the voltage response.

The conductance functions are

$$\begin{aligned} M_{\mathrm{ss}}(V) &= \frac{\left(1 + \tanh\left[\frac{(V - v_1)}{v_2}\right]\right)}{2}, \\ W_{\mathrm{ss}}(V) &= \frac{\left(1 + \tanh\left[\frac{(V - v_3)}{v_4}\right]\right)}{2}, \end{aligned} \tag{14}$$

and the time constant for the K^+-channel relaxation is

$$T_W(V) = T_0 \operatorname{sech}\left[\frac{(V - v_3)}{2v_4}\right], \tag{15}$$

where v_1, v_2, v_3, v_4 are constants and T_0 determines the time scale for the recovery process. (The v_i are usually denoted by V_1, V_2, V_3, V_4, but this conflicts with our notation for admissible ODEs.)

Again, we simulate the 7-node network of Fig. 1, with a 3-node $\mathbb{Z}_3$-symmetric CPG $\{1, 2, 3\}$ feeding forward to a chain of four nodes $\{4, 5, 6, 7\}$. Analytic results are not tractable for Eqs. (13)–(15), so we exhibit two representative numerical plots of oscillatory states with $\frac{1}{3}$-period phase patterns.

For each node $c \in \{1, \ldots, 7\}$, let $P_c = \mathbb{R}^2$ with node coordinates (V_c, W_c). Assume that each node obeys the Morris–Lecar equations with identical parameters (to make the coloring balanced), and couple them via the voltage variables V_c. Then, the lifted equations have the form

$$\begin{aligned} \dot{V}_c &= \frac{[-g_{\mathrm{Ca}}M_{\mathrm{ss}}(V_c)(V_c - V_{\mathrm{Ca}}) - g_{\mathrm{K}}W_c(V_c - V_{\mathrm{K}}) - g_L(V_c - V_L) + I_{\mathrm{app}}]}{C} + aV_{I(c)}, \\ \dot{W}_c &= \frac{W_{\mathrm{ss}}(V_c) - W_c}{T_W(V_c)}. \end{aligned}$$

Here, V_c is the voltage for node c and W_c is the recovery variable for node c. The input nodes $I(c)$ are defined by $I(1) = 3$ and $I(c) = c - 1$ for $2 \le c \le 7$. (The notation $I(c)$ should not to be confused with I_{app}.) The parameter a is a coupling constant, which we make negative to break synchrony and obtain the pattern of $\frac{1}{3}$-period phase shifts. Positive a tends to create synchronous states, which we do not illustrate.

5.1. *Simulations*

Figure 8 shows two time series for the Morris–Lecar model, namely V_c and W_c where $1 \le c \le 7$. These plots confirm the required phase pattern on both CPG and chain. Only three waves appear, phase-shifted by multiples of $\frac{1}{3}$ of the period, because of the synchrony pattern induced by the balanced 3-coloring. They have the same waveform but differing phases because of spontaneous $\mathbb{Z}_3$ symmetry breaking. The parameters used are similar to those appearing in some examples in [Morris & Lecar, 1981], and are listed in (16) and (17). The time interval is $t \in [50, 80]$. In this case, the waveforms are approximately sinusoidal. Figure 8 is obtained using the parameter set

$$\begin{aligned} &g_{\mathrm{Ca}} = 5, \quad g_{\mathrm{K}} = 8, \quad g_L = 3, \quad V_{\mathrm{Ca}} = 7, \\ &V_{\mathrm{K}} = -70, \quad V_L = 50, \quad v_1 = 1, \quad v_2 = 1, \end{aligned}$$

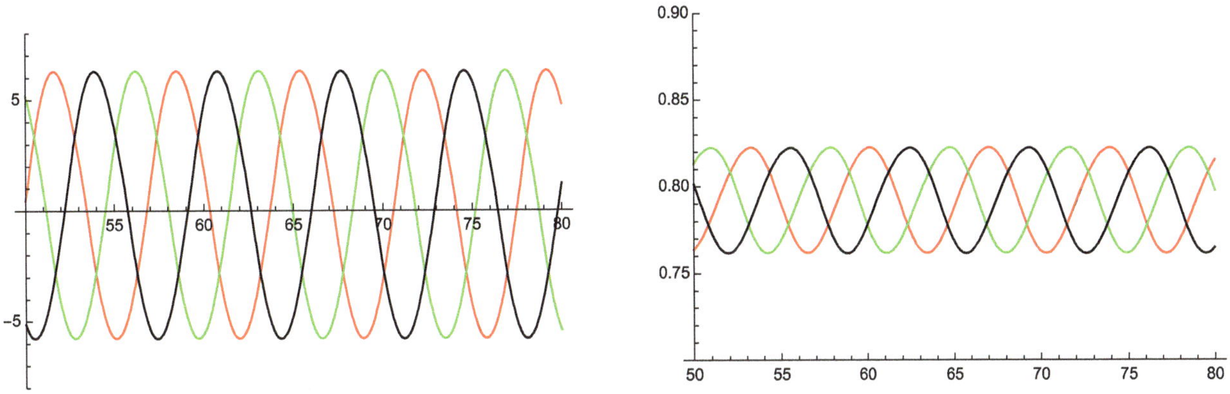

Fig. 8. Morris–Lecar model. *Left*: Superposed time series for V_c where $1 \leq c \leq 7$. *Right*: Superposed time series for W_c where $1 \leq c \leq 7$. Parameter values are (16).

$$\begin{aligned} v_3 &= -10, \quad v_4 = 14.5, \quad C = 20, \\ I_{\text{app}} &= 295, \quad a = -0.9, \quad T_0 = 5. \end{aligned} \tag{16}$$

In contrast, Fig. 9 is obtained using

$$\begin{aligned} g_{\text{Ca}} &= 5, \quad g_{\text{K}} = 8, \quad g_L = 3, \quad V_{\text{Ca}} = 7, \\ V_{\text{K}} &= -70, \quad V_L = 50, \quad v_1 = 1, \quad v_2 = 1, \\ v_3 &= -10, \quad v_4 = 14.5, \quad C = 1, \\ I_{\text{app}} &= 300, \quad a = -0.9, \quad T_0 = 5. \end{aligned} \tag{17}$$

The time interval is now $t \in [50, 75]$ and vertical scales are different for clarity. The waveforms are no longer approximately sinusoidal, but the same phase pattern appears.

5.2. *Transverse eigenvalues*

Figure 10 shows two time series for this model, namely the real and imaginary parts of the eigenvalues of $\mathrm{D}_{x_1} f$ on the first node-component of the CPG periodic orbit $x(t)$. All other components are either in synchrony with this component or are phase-shifted versions of it. The parameters concerned are (16). The time interval is $t \in [50, 80]$. In this case, the waveforms are approximately sinusoidal. Both eigenvalues of $\mathrm{D}_{x_1} f$ have negative real parts on the entire periodic orbit, ensuring transverse stability of the synchrony subspace.

Figure 11 shows the same time series for parameters (17). Now, the real parts of some eigenvalues are positive over very short portions of the periodic orbit, but negative on most of the orbit. This corresponds to transverse stability on average, relative to the unique invariant probability measure on the periodic orbit.

It is also possible to find parameter values giving approximately sinusoidal oscillations that are only transversely stable on average, so the less regular shape of Fig. 9(left) is not directly related to the occurrence of intervals of eigenvalues with positive real part.

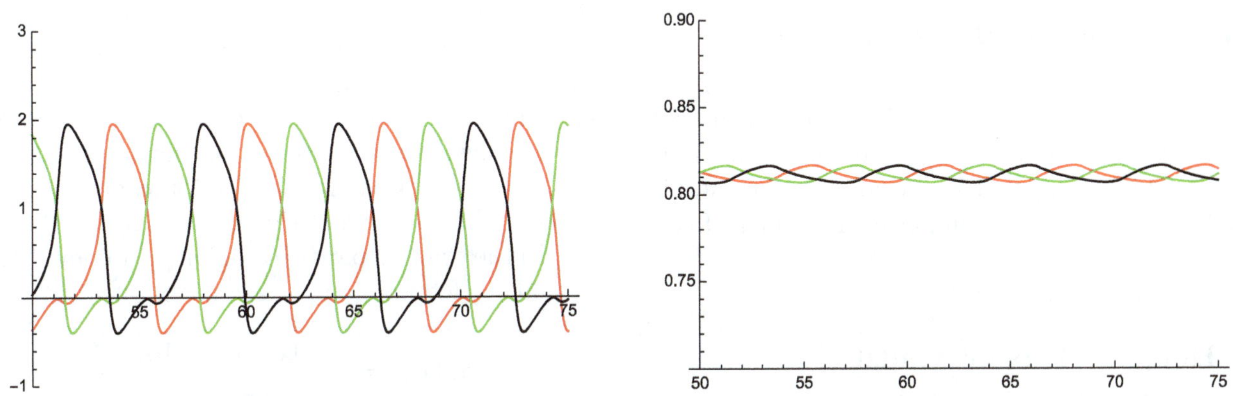

Fig. 9. Morris–Lecar model. *Left*: Superposed time series for V_c where $1 \leq c \leq 7$. *Right*: Superposed time series for W_c where $1 \leq c \leq 7$. Parameter values are (17).

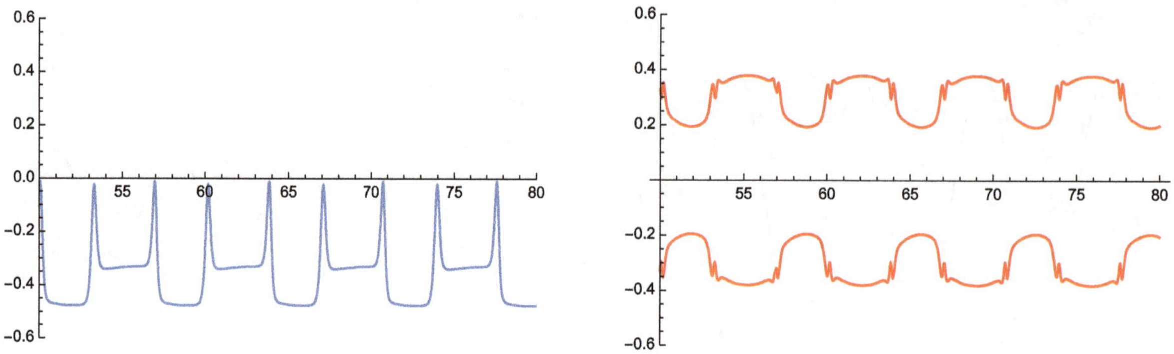

Fig. 10. Morris–Lecar model. *Left*: Real parts of eigenvalues of $D_{x_1}f$ on the periodic orbit $x(t)$. *Right*: Imaginary parts of eigenvalues of $D_{x_1}f$ on the periodic orbit $x(t)$. Parameter values are (16).

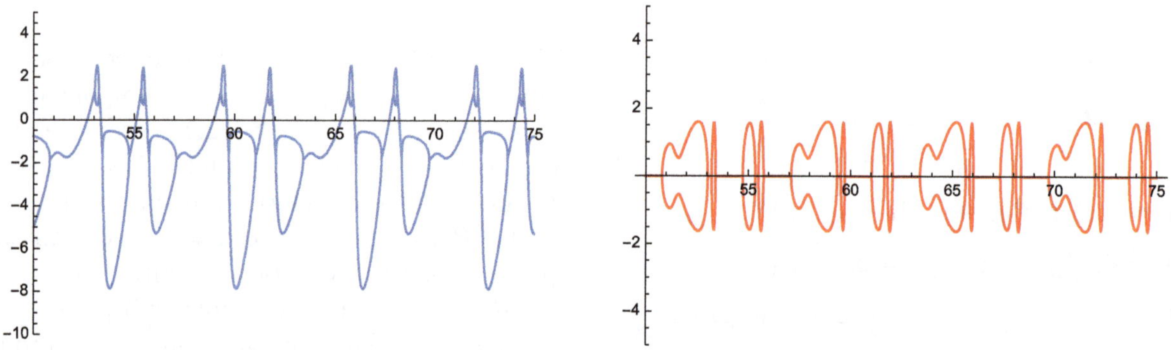

Fig. 11. Morris–Lecar model. *Left*: Real parts of eigenvalues of $D_{x_1}f$ on the periodic orbit $x(t)$. *Right*: Imaginary parts of eigenvalues of $D_{x_1}f$ on the periodic orbit $x(t)$. Parameter values are (17).

Table 2. Periods and transverse Floquet multipliers for Morris–Lecar model. All entries stated to three significant figures after the decimal point.

Parameters	Period	Transverse Floquet Multipliers	Absolute Value
(16)	6.892	$-0.609 \pm 0.0575i$	0.0838
(17)	6.309	0.0986	0.0986
		2.71×10^{-8}	2.71×10^{-8}

5.3. *Transverse Floquet multipliers*

The periods and transverse Floquet multipliers for the parameters above are listed in Table 2. In both cases, the eigenvalues lie close to the origin, indicating a high degree of transverse Floquet stability. This seems to be common in the Morris–Lecar model.

6. Hodgkin–Huxley Model

The Hodgkin–Huxley model of a neuron [Hodgkin & Huxley, 1952] takes the form

$$\begin{aligned} C_m\dot{V}_m &= I - [\bar{g}_K n^4(V_m - V_K) + \bar{g}_{Na}m^3h \\ &\quad \times (V_m - V_{Na}) + \bar{g}_l(V_m - V_l)], \\ \dot{n} &= \alpha_n(V_m)(1-n) - \beta_n(V_m)n, \\ \dot{m} &= \alpha_m(V_m)(1-m) - \beta_m(V_m)m, \\ \dot{h} &= \alpha_h(V_m)(1-h) - \beta_h(V_m)h. \end{aligned}$$

Here, the variables are V_m is the membrane potential (voltage), n, m, and h are dimensionless quantities between 0 and 1 associated with potassium channel activation, sodium channel activation, and sodium channel inactivation, respectively. There are also numerous parameters C_m is the membrane capacitance, I is the current per unit area, α_i and β_i are rate functions for the ith ion channel and $\bar{g}_n$ is the maximal value of the conductance.

In the original paper [Hodgkin & Huxley, 1952], α_i and β_i (for $i = n, m, h$) are given by

$$\alpha_n(V_m) = \frac{0.01(V_m + 10)}{\exp\left(\dfrac{V_m + 10}{10}\right) - 1},$$

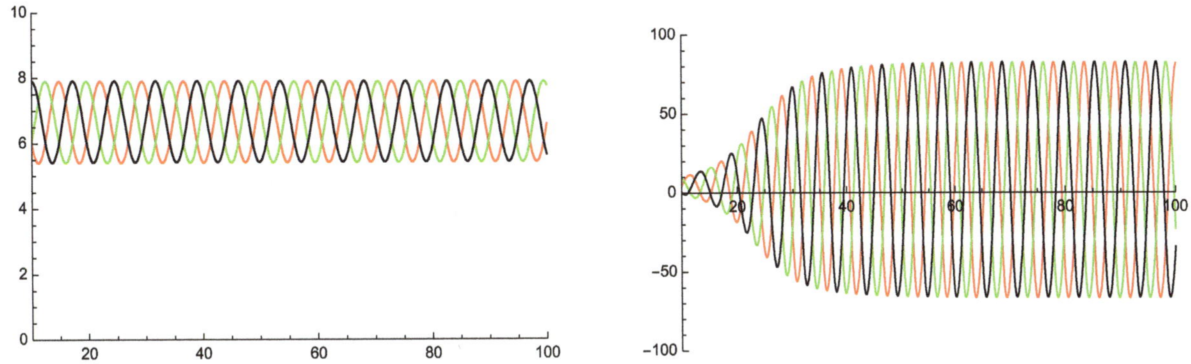

Fig. 12. Hodgkin–Huxley. Superposed time series for V_c where $1 \leq c \leq 7$. *Left*: Parameter values as in (18). *Right*: Parameter values as in (19); initial transients are included to illustrate convergence onto the periodic state.

$$\alpha_m(V_m) = \frac{0.1(V_m + 25)}{\exp\left(\dfrac{V_m + 25}{10}\right) - 1},$$

$$\alpha_h(V_m) = 0.07 \exp\left(\frac{V_m}{20}\right),$$

$$\beta_n(V_m) = 0.125 \exp\left(\frac{V_m}{80}\right),$$

$$\beta_m(V_m) = 4 \exp\left(\frac{V_m}{18}\right),$$

$$\beta_h(V_m) = \frac{1}{\exp\left(\dfrac{V_m + 30}{10}\right) + 1}.$$

6.1. *Simulations*

We simulated the 7-node network for the Hodgkin–Huxley model, where each node is a Hodgkin–Huxley neuron with identical parameters, voltage-coupled with coupling constant a. Again, we make a negative to break synchrony and obtain the pattern of $\frac{1}{3}$-period phase shifts. The parameters we used are

$$\begin{aligned} &\bar{g}_{\mathrm{K}} = 10, \quad \bar{g}_{\mathrm{Na}} = 20, \quad \bar{g}_l = 20, \\ &V_{\mathrm{K}} = -150, \quad V_{\mathrm{Na}} = 100, \quad V_l = 20, \\ &C_m = 40, \quad I = 0, \quad a = -1, \end{aligned} \tag{18}$$

$$\begin{aligned} &\bar{g}_{\mathrm{K}} = 100, \quad \bar{g}_{\mathrm{Na}} = 50, \quad \bar{g}_l = 20, \\ &V_{\mathrm{K}} = 50, \quad V_{\mathrm{Na}} = 10, \quad V_l = 20, \\ &C_m = 40, \quad I = -50, \quad a = -1.3. \end{aligned} \tag{19}$$

Figure 12 shows superposed time series for this model, for the voltages V_c, where $1 \leq c \leq 7$, all superposed. As usual we see only three traces, with $\frac{1}{3}$-period phase shifts, confirming the required phase pattern.

6.2. *Transverse eigenvalues*

Figure 13 shows the corresponding time series for the real parts of eigenvalues of $\mathrm{D}_{x_1} f$ on the periodic orbit $x(t)$. (Here, $x(t)$ can be either the CPG

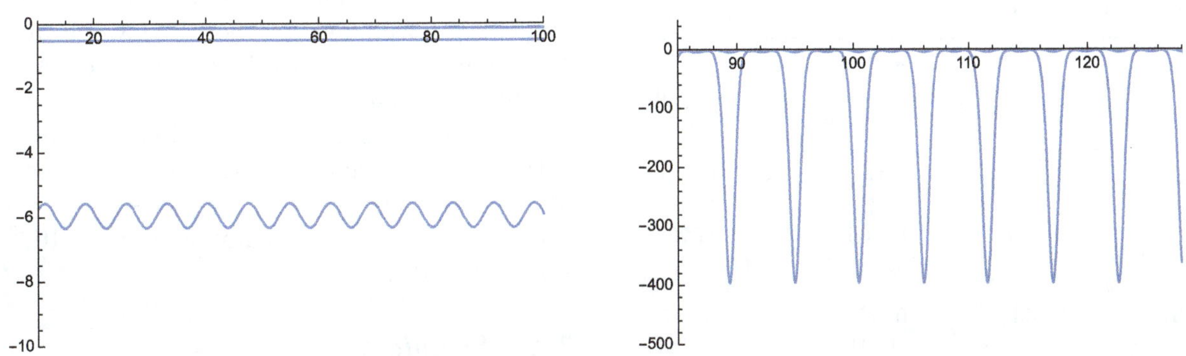

Fig. 13. Hodgkin–Huxley. Real parts of eigenvalues of $\mathrm{D}_{x_1} f$ on the periodic orbit $x(t)$. *Left*: Parameter values as in (18). *Right*: Parameter values as in (19).

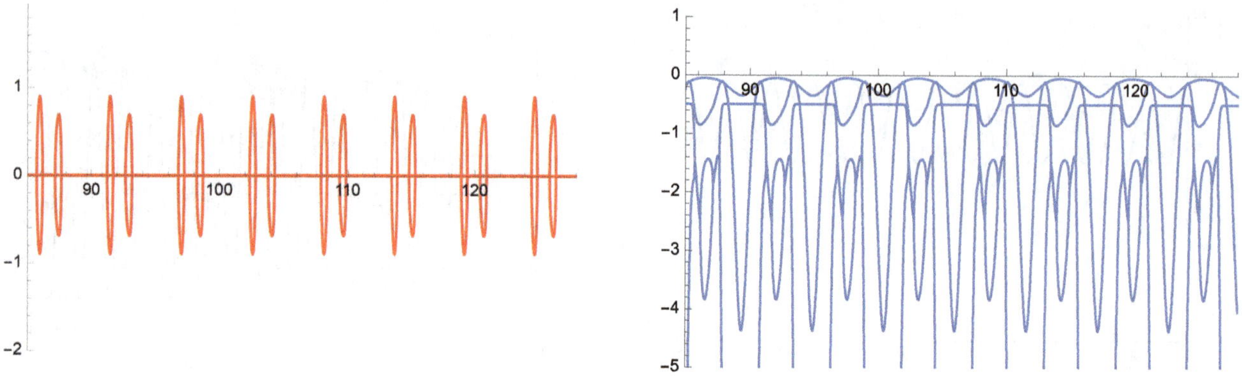

Fig. 14. Hodgkin–Huxley. Parameter values as in (19). *Left*: Imaginary parts of eigenvalues of $\mathrm{D}_{x_1}f$ on the periodic orbit $x(t)$. *Right*: Larger scale time series for real parts of eigenvalues of $\mathrm{D}_{x_1}f$ on the periodic orbit $x(t)$.

orbit or the lift, since all seven nodes are phase-synchronous. For the same reason, we do not need to compute eigenvalues of $\mathrm{D}_{x_c}f$ for $c > 1$.) The figure appears to show only three eigenvalues in each case, but inspection of the numerical values shows that the top line actually consists of two lines very close together representing two eigenvalues that are almost the same, but not exactly so. All four eigenvalues have negative real part on the entire orbit, implying transverse stability of the synchrony subspace.

For parameters (18), the imaginary parts of these eigenvalues are all zero, so are not plotted. For parameters (19), the imaginary parts of these eigenvalues are not all zero, and are plotted in Fig. 14(left). The real parts, Fig. 14(right), are all negative.

We have not found parameters giving a periodic orbit of the Hodgkin–Huxley model with only transverse stability on average. The transverse eigenvalues have negative real parts in all cases we have computed.

Table 3. Periods and transverse Floquet multipliers for Hodgkin–Huxley model. All entries stated to three significant figures after the decimal point.

Parameters	Period	Transverse Floquet Multipliers	Absolute Value
(18)	7.253	0.410	0.410
		0.373	0.373
		0.0265	0.0265
		-8.03×10^{-11}	8.03×10^{-11}
(19)	5.546	0.387	0.387
		0.0941	0.0941
		1.88×10^{-6}	1.88×10^{-6}
		2.47×10^{-8}	2.47×10^{-8}

6.3. *Transverse Floquet multipliers*

Table 3 shows computed transverse Floquet multipliers for the Hodgkin–Huxley model with parameters (18) and (19). Both cases are transversely Floquet stable.

7. Hindmarsh–Rose Model

The Hindmarsh–Rose neuron model was originally introduced in [Hindmarsh & Rose, 1984] as a simplification of the Hodgkin–Huxley equations, with the main focus on spiking activity. The equations have the form

$$\begin{aligned}\dot{x} &= y - ax^3 + bx^2 - z + I,\\ \dot{y} &= c - dx^2 - y, \qquad (20)\\ \dot{z} &= r[s(x - x_R) - z].\end{aligned}$$

Here, x is the membrane potential and y, z are variables modeling ion channel dynamics.

All other quantities are parameters a, b, c, d model fast ion channels, r models slow ion channels, I is the input current to the neuron and x_R sets an equilibrium point.

The Hindmarsh–Rose equations can generate a variety of dynamic behavior, including chaos. Standard parameters are $s = 4, x_R = -1.6, a = 1, b = 3, c = 1, d = 5$. The parameter $r \sim 10^{-3}$ and $-10 \leq I \leq 10$.

7.1. *Simulations*

Figure 15 shows three simulations of the 7-node chain. We use Hindmarsh–Rose dynamics (20) on

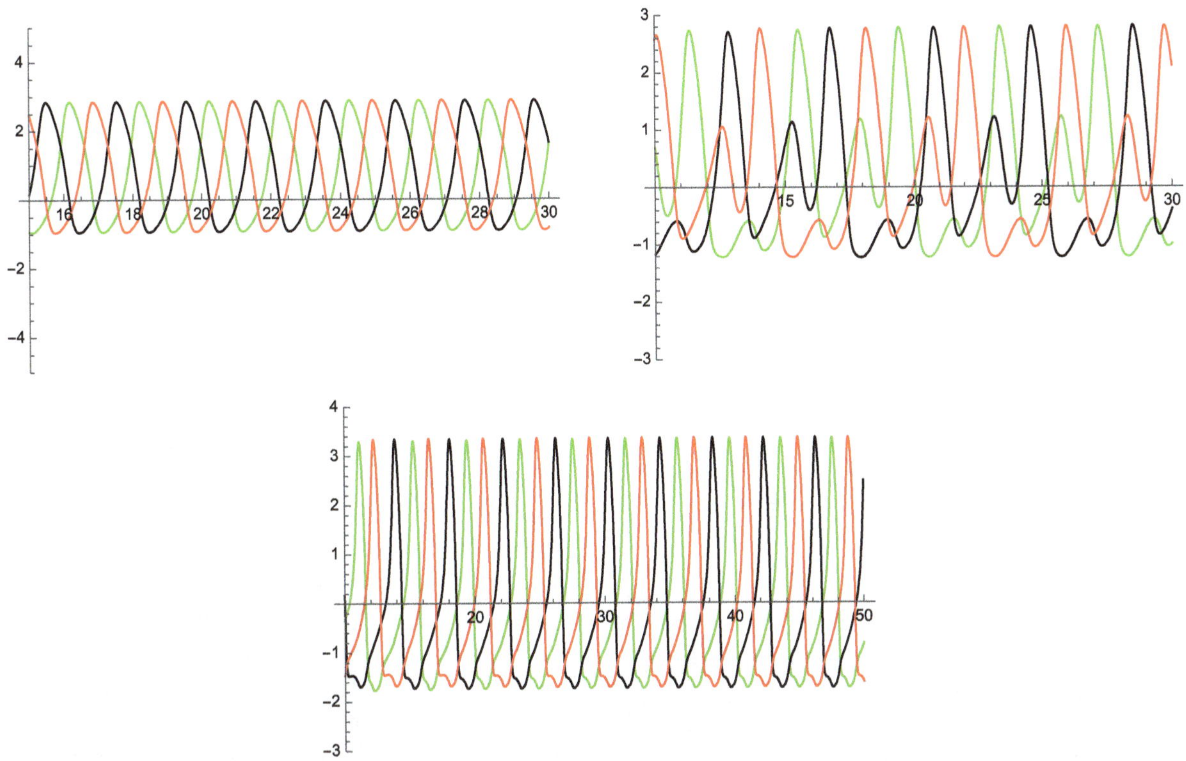

Fig. 15. Three simulations of the 7-node feedforward chain of Fig. 1 with Hindmarsh–Rose dynamics (20) on each node. Parameters are: *Top left*: (21). *Top right*: (22). *Bottom*: (23).

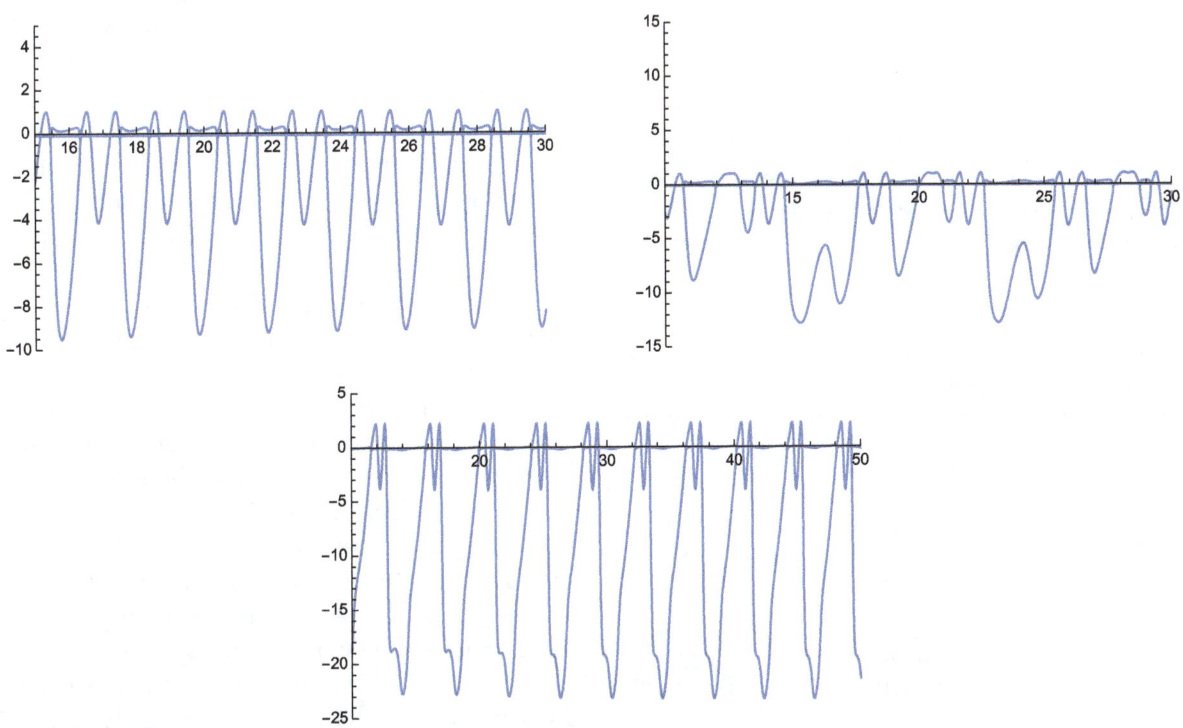

Fig. 16. Transverse eigenvalues for the 7-node feedforward chain of Fig. 1 with Hindmarsh–Rose dynamics (20) on each node. Parameters are: *Top left*: (21). *Top right*: (22). *Bottom*: (23).

each node c with variables (x_c, y_c, z_c). Coupling is by adding a term $gx_{i(c)}$ to the x_c-component of each node, where $i(c)$ is the (unique) node inputting to c.

The figure is a superposition of the x-coordinates of nodes 1–7, verifying the lifted property of the $\frac{1}{3}$-period phase pattern. The respective parameter values are

$$\begin{aligned} &a = 1, \quad b = 3, \quad c = 1, \quad d = 5, \quad r = 0.1, \\ &s = 0, \quad x_R = -1.6, \quad I = 10, \quad g = -1, \end{aligned} \tag{21}$$

$$\begin{aligned} &a = 1, \quad b = 3, \quad c = 1, \quad d = 5, \quad r = 0.1, \\ &s = 0, \quad x_R = -1.6, \quad I = 5, \quad g = -2, \end{aligned} \tag{22}$$

$$\begin{aligned} &a = 1, \quad b = 4, \quad c = 2, \quad d = 5, \quad r = 0.05, \\ &s = 2, \quad x_R = -1.6, \quad I = 5, \quad g = -2. \end{aligned} \tag{23}$$

The top figure shows roughly sinusoidal waveforms. The middle figure shows more complex waveforms. Between each peak of the time series for a single node, two different waveforms alternate — possibly some kind of period-doubling effect. The bottom figure shows spiking waveforms, which was the original aim of the Hindmarsh–Rose model.

7.2. *Transverse eigenvalues*

Figure 16 plots the real parts of the three transverse eigenvalues. In all three cases, there are short intervals over which the real part of at least one eigenvalue is positive, and shorter intervals over which all real parts of eigenvalues are negative. However, the real parts of the negative eigenvalues are much larger on those intervals. Presumably this leads to transverse stability on average, in some suitable sense.

Table 4. Periods and transverse Floquet multipliers for Hindmarsh–Rose model. All entries stated to three significant figures after the decimal point.

Parameters	Period	Transverse Floquet Multipliers	Absolute Value
(21)	1.973	0.820	0.820
		0.187	0.187
		0.00465	0.00465
(22)	3.554	0.700	0.700
		0.270	0.270
		2.78×10^{-6}	2.78×10^{-6}
(23)	3.919	0.887	0.0887
		0.534	0.534
		4.40×10^{-9}	4.40×10^{-9}

7.3. *Transverse Floquet multipliers*

Table 4 shows computed transverse Floquet multipliers for the Hindmarsh–Rose model with parameters (21)–(23). All three cases are transversely Floquet stable.

8. Breaking Synchrony

The assumption that some nodes and arrows in the network have the same types, leading to a balanced coloring in which nodes are *exactly* synchronized, is an idealization. In real networks, especially biological ones, nodes and arrows may be very similar, but not identical, and nodes are considered synchronous if they exhibit very similar — but not identical — time series. Formalizing suitable notions of "approximate symmetry" and "approximate synchrony" is possible (such as the "pseudosymmetries" of [Morone & Makse, 2019]), but lacks the elegant mathematical structure of the exact versions.

The term "symmetry breaking" is used in two distinct ways [Golubitsky & Schaeffer, 1983]. In *spontaneous* symmetry-breaking, the symmetry of the model equations is exact but solutions have less symmetry because the fully symmetric state becomes unstable. In *forced* (or *induced*) symmetry-breaking, the symmetry of the model equations is only approximate. A similar distinction applies to "synchrony-breaking". Transverse (Floquet) stability prevents spontaneous synchrony-breaking in an idealized feedforward chain. We now study forced synchrony-breaking.

We prefer to consider the idealized models as useful approximations, whose solutions, suitably interpreted, offer insights into the more messy reality. If the ideal model has exactly synchronous nodes, small changes to the model that break synchrony generally lead to approximately synchronous nodes. By the Kupka–Smale theorem [Kupka, 1963, 1964; Smale, 1963; Peixoto, 1966], steady or periodic states of a dynamical system are generically hyperbolic: see [Guckenheimer & Holmes, 1983] for equilibria and [Katok & Hasselblatt, 1995] for equilibria and periodic states. All Floquet stable periodic orbits are hyperbolic. Hyperbolic states persist under sufficiently small C^1 perturbations [Hirsch *et al.*, 1977]. An equivariant version of the Kupka–Smale theorem is proved in [Field, 1977, 1980], which applies to symmetric dynamical systems. It is plausible that a similar genericity theorem applies to networks, but this has been proved only for a few small networks [Stewart, 2022].

These theoretical considerations are supported by numerical experiments, in which idealized equations are modified to break symmetry and synchrony. These experiments suggest that, in practice, the traveling wave states obtained from balanced colorings are quite robust, and similar states persist if the nodes and couplings differ — though not by too large an amount. Exact synchrony or phase relations are replaced by approximate ones. There are many variations with similar properties. In the following subsections, we expand on these theoretical considerations and report results for some typical examples of such experiments.

8.1. *Theoretical considerations*

To keep the discussion within bounds, we consider only Floquet stability. One important feature of a feedforward lift is that it propagates periodic CPG signals exactly, and that transverse stability one step along the chain implies transverse stability along the entire chain. When combined with stability of the CPG signal, this guarantees that the entire propagating signal is stable. Another important feature is that regular phase patterns in the CPG generate the same phase patterns in the chain, so that discrete rotating waves in the CPG generate apparent traveling waves along the chain.

These results are proved above only for the idealized feedforward chain, where all nodes and couplings in the chain are identical to appropriate ones in the CPG. In a real biological system, such conditions at best hold approximately. So we can ask: what effect do such approximations have on the existence, stability, phase patterns, and amplitudes of the propagating signal?

These questions can be formulated in many ways. In this subsection, we offer some theoretical reasons why we expect the results to be relatively robust, provided exact equalities of waveform are replaced by suitable approximate equalities. These considerations address only "sufficiently small" changes to dynamics and couplings. *How* small is sufficient depends on the model ODE. We discuss some typical quantitative answers, obtained by simulation, in the following subsection. These simulations, and many others not mentioned here, indicate that, in practice, the feedforward chain propagation mechanism is remarkably robust, provided that the CPG oscillation that drives the chain continues to exist and has a reasonably large amplitude.

When the oscillation is Floquet stable, the corresponding periodic orbit is hyperbolic. Therefore, this orbit persists, remains stable, and perturbs continuously, for any continuous family of C^1-small perturbations of the ODE [Hirsch *et al.*, 1977]. Such perturbations include

(a) Changes to CPG node dynamics.
(b) Changes to CPG couplings.
(c) Changes to node dynamics in the chain.
(d) Changes to couplings in the chain.

Moreover, such changes can be made independently for each node and arrow.

A small deformation of the periodic orbit projects onto node coordinates to give small deformations of the waveforms observed at each node. Therefore, phases and amplitudes change by small amounts. In other words, the main qualitative features of the propagating signal persist, with exact synchrony replaced by approximate synchrony, and changes in quantitative features are small.

Indeed, we expect these features to persist until the perturbation becomes large enough to cause a bifurcation. If it is a local bifurcation, which is a reasonable assumption in many cases, it is caused by some Floquet multiplier becoming unstable. The numerical calculations reported in Secs. 4–7 show that the transverse Floquet multipliers are often *very* stable; that is, well inside the unit circle. For FitzHugh–Nagumo neurons, the largest (absolute value of a) transverse eigenvalue in Table 1 is 0.435. For Morris–Lecar neurons, the largest transverse eigenvalue in Table 2 is 0.0986. For Hodgkin–Huxley neurons, the largest transverse eigenvalue in Table 3 is 0.410. For Hindmarsh–Rose neurons, the largest transverse eigenvalue in Table 4 is 0.820. Typically, the transverse eigenvalues with smaller absolute value are much smaller.

These figures suggest that a signal propagating along a chain of Hindmarsh–Rose neurons will be more fragile (to perturbations of the model) than one propagating along a chain of FitzHugh–Nagumo or Hodgkin–Huxley neurons, while one propagating along a chain of Morris–Lecar neurons is likely to be more robust. Possibly, the largest absolute value of a transverse eigenvalue could be used as a rule-of-thumb "stability index", with smaller indices suggesting more robust propagation. This would be analogous to the use of the maximal Lyapunov exponent in the "master stability function" for synchronous equilibria [Arenas *et al.*, 2008; Pecora & Carroll, 1998].

The key question in this paper is: given a stable CPG oscillation, how can it be propagated stably along a chain, in a sufficiently robust manner for biological and other applications? Perturbations (a) and (b) above are changes to the CPG dynamics, and can be dealt with by recalculating the transverse Floquet multipliers. That is, we can work with a specified CPG, and assume that it is fixed. Attention therefore focuses on (c) and (d), changes along the chain.

The most important case is a change that affects the *first* module along the chain. If the first change occurs at a later step, this is equivalent to replacing the CPG oscillation by the signal transmitted by the previous module. Initially, it therefore makes sense to consider (c) and (d) for the first module in the chain. That is, node 4 in Fig. 1 and the arrow into it from node 3. We therefore focus on this case, which illustrates the main issue of robustness to perturbations by terms that are not admissible and thus break synchrony. However, we also examine some features for longer chains (up to 30 nodes). Numerical experiments on other modifications could be continued indefinitely.

8.2. *Forced symmetry breaking: FitzHugh–Nagumo*

As an exemplar, we take the system of equations for the 7-node model using FitzHugh–Nagumo equations for each node as outlined in Sec. 4, Eqs. (8), with the completely balanced case having parameters (9) for each node. We reproduce these equations here for convenience

$$G(V,W,I) = V(a-V)(V-1) - W + I,$$
$$H(V,W) = bV - \gamma W,$$

with

$$\dot{V}_1 = G(V_1,W_1,I) + cV_3, \quad \dot{W}_1 = H(V_1,W_1),$$
$$\dot{V}_2 = G(V_2,W_2,I) + cV_1, \quad \dot{W}_2 = H(V_2,W_2),$$
$$\dot{V}_3 = G(V_3,W_3,I) + cV_2, \quad \dot{W}_3 = H(V_3,W_3),$$
$$\dot{V}_4 = G(V_4,W_4,I) + cV_3, \quad \dot{W}_4 = H(V_4,W_4),$$
$$\dot{V}_5 = G(V_5,W_5,I) + cV_4, \quad \dot{W}_5 = H(V_5,W_5),$$
$$\dot{V}_6 = G(V_6,W_6,I) + cV_5, \quad \dot{W}_6 = H(V_6,W_6),$$
$$\dot{V}_7 = G(V_7,W_7,I) + cV_6, \quad \dot{W}_7 = H(V_7,W_7).$$

Initially we assumed parameter values

$$I = 0, \quad a = 0.05, \quad b = 2.5, \quad \gamma = 0.3, \quad c = -0.6.$$

These equations were then run through MatLab using its `ode89` solver, an implementation of Verner's "most robust" Runge–Kutta 9(8) pair with an eighth-order continuous extension [Verner, 2006]. We used random initial conditions, whilst for each run step, changing one of the parameters in turn, b, γ, c for the fourth node in the network by factors from 0.5 up to 2. Thus, we are looking at the effect of changing both the internal dynamics of, and the coupling strength to, this node. As remarked in the previous subsection, there is nothing special about this choice of node; doing the same to any node further down the chain will lead to an identical response both for that node and for nodes downstream from that one. Additional runs were made for a selection of parameter choices for much longer chains, extending the 7-node chain to up to 30 nodes to confirm how these later nodes behave with respect to the changes to node 4.

Before perturbing any of the parameters, the time series for the seven nodes were superposed, again demonstrating a traveling wave pattern with phase differences of one third of a period, see Fig. 17(top left).

In all cases, the periods of all nodes remained almost constant to a remarkable extent, so that phase difference remained a meaningful concept even in the case of different waveforms (chiefly different maximum amplitudes). We used the times of maximum amplitude to calculate phase differences between nodes.

8.2.1. *Changing the coupling term c*

The maximum amplitude for all nodes when there are no forced symmetry breaking perturbations is 0.233. Changing the coupling term from node 3 to node 4 by the above factors alters the maximum amplitude of node 4 from 0.119 to 0.433, with an approximately linear increase as the value of c is increased from the lower to upper values [see Fig. 17(top right)]. This change in amplitude

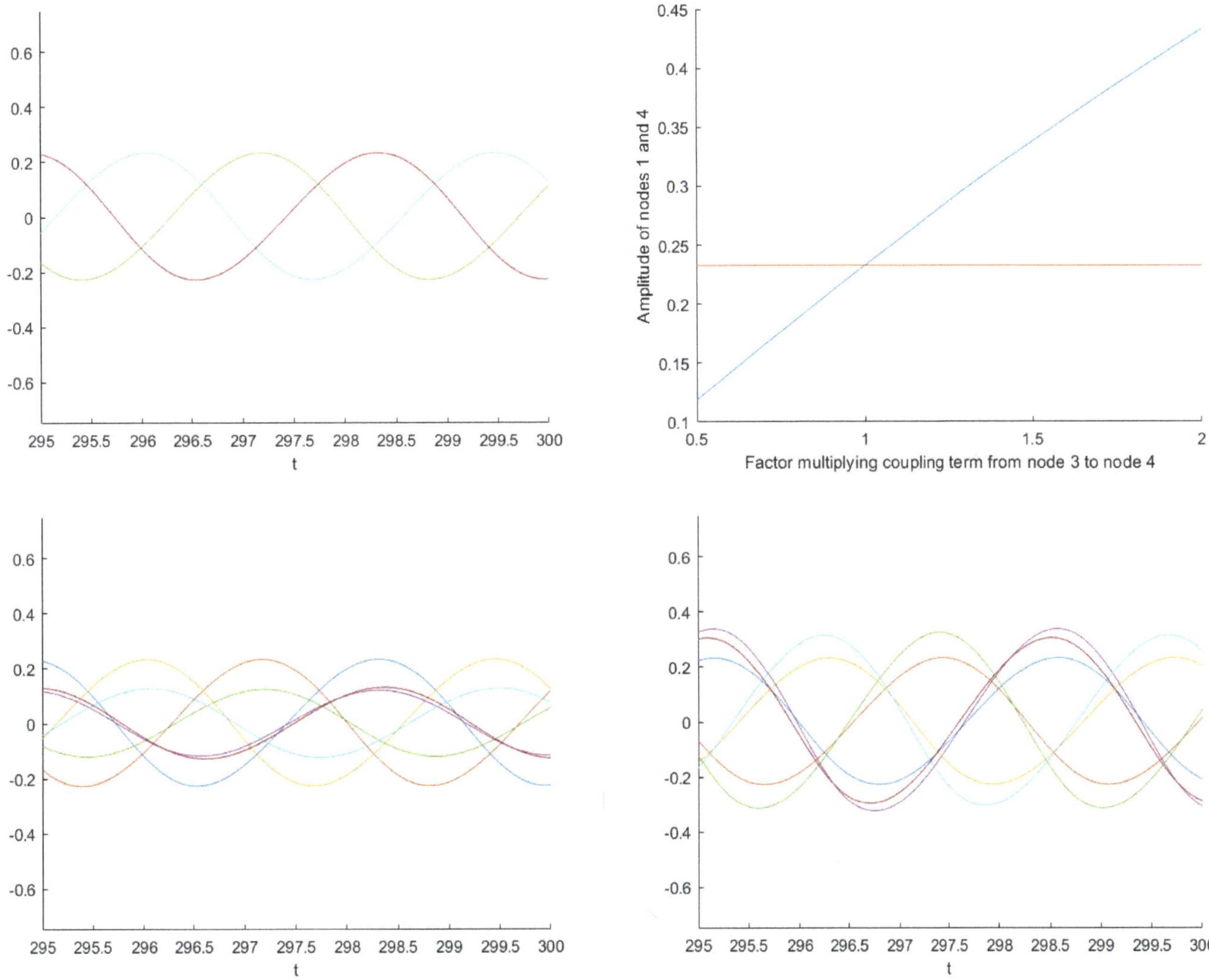

Fig. 17. Traveling wave periodic states in the 7-node FitzHugh–Nagumo feedforward network. The top subfigures show (*left*) all seven nodes superposed before perturbing parameters and (*right*) the effect on the amplitude of the fourth node when changing the coupling from the third node (blue) compared to the amplitude of nodes 1–3 (red). The bottom subfigures show all seven nodes superposed when the same coupling term is multiplied by a factor of 0.5 (*left*) and 2 (*right*).

propagates along the chain, gradually converging towards the original values; for example, the amplitude of node 7 varies from 0.13 to 0.338. As more nodes are added, the amplitudes further along the chain appear to converge towards their original values. Phase differences are largely unchanged along the chain, with differences of only around 5% at the two extremes of perturbations to c. This can be seen in the bottom two subfigures of Fig. 17, along with the differences in amplitude of the nodes.

8.2.2. *Changing parameter b in only node 4*

As with the case of changing the coupling, the most noticeable change to patterns of oscillation when changing parameter b in node 4 is to the amplitude of that node. The difference is that at the two extremes the amplitude is decreased (to 0.116 and 0.145, *resp.*) but as b is increased this rises to a value larger than the unperturbed case before falling again, reaching a maximum of 0.337 approximately around a multiplying factor of 1.4 [Fig. 18(top left)].

Between a factor of 0.5 and 1.1, the phase difference between nodes 1 and 4 remains largely unchanged, but as b is increased further, the phase difference starts to drift. By the time b is multiplied by a factor of 2, the phase difference is 0.676, and it seems to be converging to become in phase with node 2 [see Fig. 18(top right)]. However, this

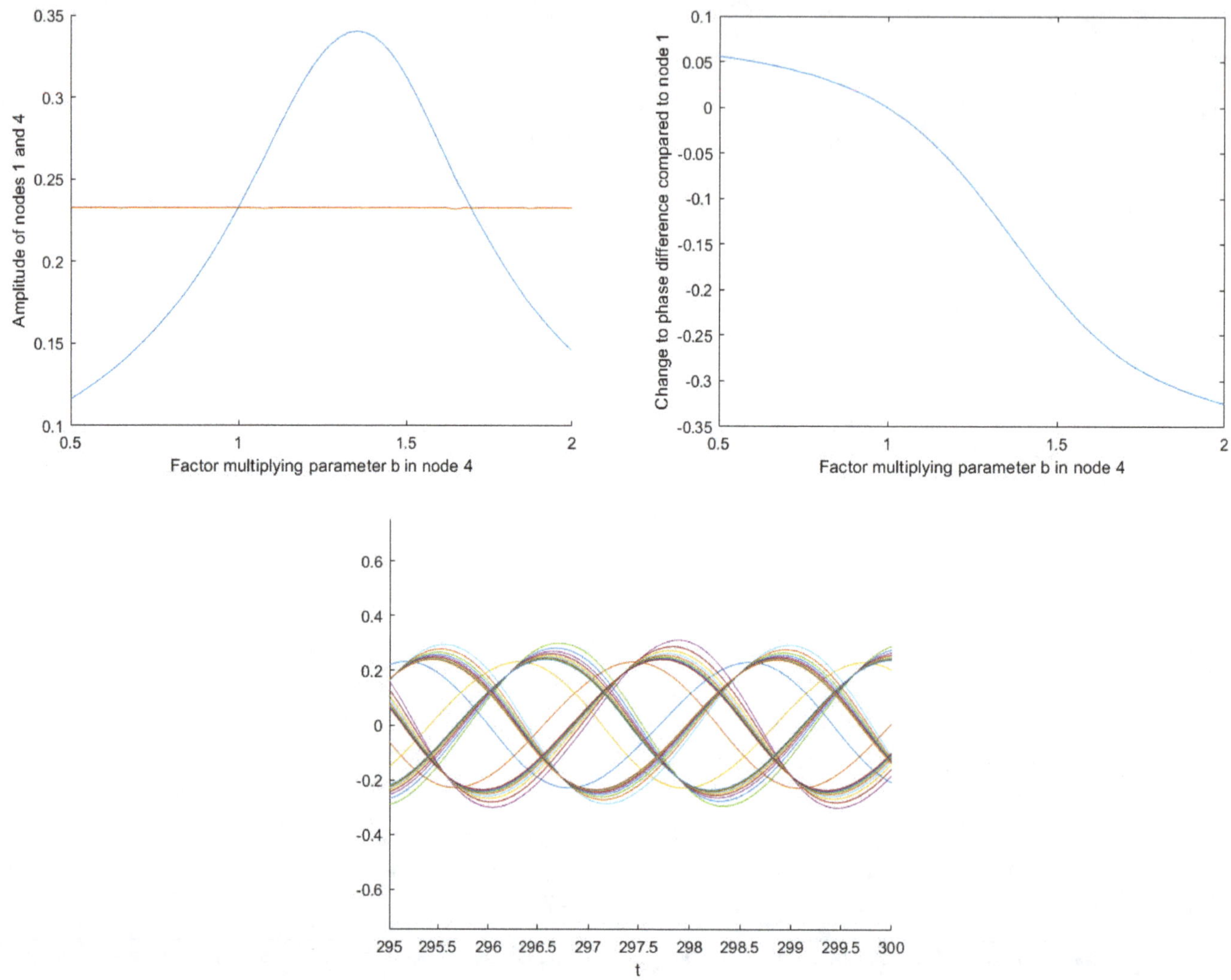

Fig. 18. Graphs showing (*top left*) the effect of altering the parameter b in node 4 on the amplitude of node 4, and (*top right*) phase difference between nodes 4 and 1. *Bottom*: the feedforward network extended to 30 nodes, whose time series are superposed, when parameter b is multiplied by a factor of 1.5.

apparent convergence is an artifact of stopping at a factor of 2. When b is increased further, the phase difference continues to increase, albeit more slowly. A more important feature, when thinking of feedforward chains as a mechanism to propagate patterns, is that although the phase difference between node 1 and 4 is changing, the phase difference between successive nodes in the chain beyond node 4 remains fairly consistently within a few per cent of the unperturbed value 0.667.

Figure 18(bottom) shows an extended chain of 30 nodes whose time series are superposed. Three clumps of approximately in-phase oscillations are evident, as well as changes in amplitude further downstream. These clumps are separated by phase differences close to one third of a period.

8.2.3. *Changing parameter γ in only node 4*

Finally, changing the value of parameter γ in node 4, again by a factor of 0.5–2, also changes the maximum amplitude of node 4 (from 0.258 down to 0.183) monotonically as γ increases. In contrast, changes to phase differences are minimal, within 3% of the normal value at the two extremes (see Fig. 19). Again, the amplitudes of the upstream nodes converge back towards their unperturbed values.

8.2.4. *Other modifications*

We chose the 3-node ring as one of the simplest GPGs that generates periodic discrete rotating

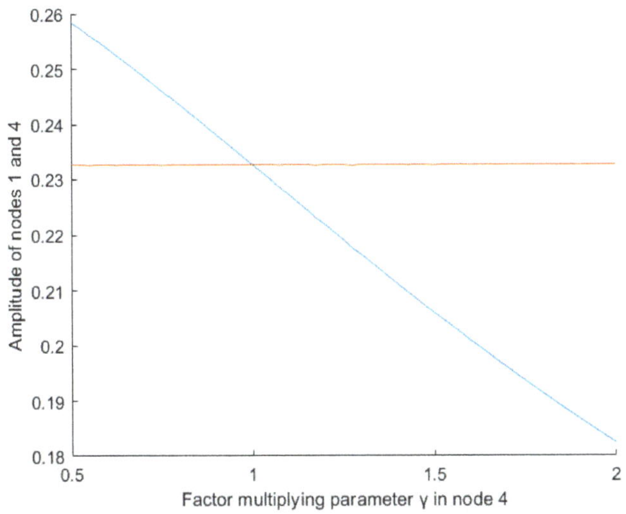

Fig. 19. Graph showing the effect on the amplitude of node 4 (*blue*) compared to that of nodes 1–3 (*red*) when changing the parameter γ in node 4.

waves, leading to traveling waves in the chain. The length of the chain, here 4, is an arbitrary choice that illustrates the main points and is unrelated to the number of nodes in the CPG. These choices were made as "proof of concept" rather than for their biological significance. In further work [Stewart & Wood, 2025], we apply feedforward lifts to animal locomotion, using more complex and more realistic topologies for both the CPG and the chain.

Experience suggests that similar results can be expected for any CPG with cyclic group symmetry that has periodic states with corresponding phase relations. Experiments with rings with more than three nodes (not reported here) are consistent with this expectation for a broad range of model ODEs. The topology of the "chain" can be more complex; all that is needed is the feedforward lift structure.

9. Conclusions

Propagation of synchronous or phase-synchronous states along linear chains is important in biology, medicine, and robotics. Such signals can be generated by a CPG and propagated along a feedforward chain.

Given the CPG, many different architectures for propagation are possible, but those with short-range connections are particularly useful and natural, both in engineering and evolutionary terms.

An important issue is the stability of the propagating signals. This was discussed in [Stewart & Wood, 2024] in terms of four stability notions: transverse stability of the synchrony subspace, and transverse Lyapunov, asymptotic, or Floquet stability of the periodic orbit. Analytic calculations and simulations show that four standard models of neurons — namely FitzHugh–Nagumo, Morris–Lecar, Hodgkin–Huxley, and Hindmarsh–Rose neurons — satisfy transverse stability conditions, either universally or over broad parameter ranges. They can satisfy the transverse Floquet condition even when the synchrony subspace is transversely unstable at some points of the periodic orbit.

Simulations and some theoretical considerations suggest that a more general condition, transverse stability on average, might also be a useful indicative test for stability, perhaps in a weaker sense such as the "Milnor attractor" condition that most nearby initial conditions converge to the orbit [Milnor, 1985].

Real-world systems break the idealized conditions assumed in the theoretical model. Simulations suggest that breaking synchrony through a variety of different modifications leads to states that approximate the idealized ones in many important respects, provided the modifications are not too great.

Phase relations in periodic signals appear to be more robust to such modifications than amplitudes.

Feedforward lifts provide a simple, effective, and robust way to propagate signals with specific synchrony and phase patterns in a stable manner.

ORCID

Ian Stewart https://orcid.org/0000-0002-3759-0142
David Wood https://orcid.org/0009-0005-8537-0397

References

Abeles, M., Hayon, G. & Lehmann, D. [2004] "Modeling compositionality by dynamic binding of synfire chains," *J. Comput. Neurosci.* **17**, 179–201.

Aertsen, A., Diesmann, M. & Gewaltig, M. O. [1996] "Propagation of synchronous spiking activity in feedforward neural networks," *J. Physiol.* **90**, 243–247.

Arber, S. & Costa, R. M. [2018] "Connecting neuronal circuits for movement," *Science* **360**, 1403–1404.

Arenas, A., Díaz-Guilera, A., Kurths, J., Moreno, Y. & Zhou, C. [2008] "Synchronization in complex networks," *Phys. Rep.* **469**, 93–153.

Berge, C. [1973] *Graphs and Hypergraphs* (North Holland, Amsterdam).

Berkowitz, A. [2019] "Expanding our horizons: Central pattern generation in the context of complex activity sequences," *J. Exp. Biol.* **222**, 192054.

Boyle, J. H., Berri, S. & Cohen, N. [2012] "Gait modulation in *C. elegans*: An integrated neuromechanical model," *Front. Comput. Neurosci.* **6**, 10.

Buono, P.-L. [2001] "Models of central pattern generators for quadruped locomotion: II. Secondary gaits," *J. Math. Biol.* **42**, 327–346.

Buono, P.-L. & Palacios, A. [2004] "A mathematical model of motorneuron dynamics in the heartbeat of the leech," *Physica D* **188**, 292–313.

Calabrese, R. L. & Peterson, E. [1983] "Neural control of heartbeat in the leech *Hirudo medicinalis*," in *Neural Origin of Rhythmic Movements*, eds. Roberts, A. & Roberts, B. (Society for Experimental Biology, Bailrigg), pp. 195–221.

Calabrese, R. L., Nadim, F. & Olsen, O. H. [1995] "Heartbeat control in the medicinal leech: A model system for understanding the origin, coordination, and modulation of rhythmic motor patterns," *J. Neurobiol.* **27**, 390–402.

Collins, J. J. & Stewart, I. [1993a] "Hexapodal gaits and coupled nonlinear oscillator models," *Biol. Cybern.* **68**, 287–298.

Collins, J. J. & Stewart, I. [1993b] "Coupled nonlinear oscillators and the symmetries of animal gaits," *J. Nonlinear Sci.* **3**, 349–392.

Collins, J. J. & Richmond, S. A. [1994] "Hard-wired central pattern generators for quadrupedal locomotion," *Biol. Cybern.* **71**, 375–385.

Elowitz, M. B. & Leibler, S. [2000] "A synthetic oscillatory network of transcriptional regulators," *Nature* **403**, 335–338.

Field, M. J. [1977] "Transversality in G-manifolds," *Trans. Am. Math. Soc.* **231**, 429–450.

Field, M. J. [1980] "Equivariant dynamical systems," *Trans. Am. Math. Soc.* **259**, 185–205.

FitzHugh, R. [1961] "Impulses and physiological states in theoretical models of nerve membrane," *Biophys. J.* **1**, 445–466.

Floquet, G. [1883] "Sur les équations différentielles linéaires à coefficients périodiques," *Ann. Sci. Éc. Norm. Sup.* **12**, 47–89.

Gambaryan, P. [1974] *How Mammals Run: Anatomical Adaptations* (Wiley, NY).

Golubitsky, M. & Schaeffer, D. [1983] "A discussion of symmetry and symmetry breaking," *Proc. Symp. Pure Math.* **40**, 499–515.

Golubitsky, M., Stewart, I. & Schaeffer, D. G. [1988] *Singularities and Groups in Bifurcation Theory*, Vol. II, Applied Mathematics Series, Vol. 69 (Springer, NY).

Golubitsky, M., Stewart, I., Buono, P.-L. & Collins, J. J. [1998] "A modular network for legged locomotion," *Physica D* **115**, 56–72.

Golubitsky, M., Stewart, I., Collins, J. J. & Buono, P.-L. [1999] "Symmetry in locomotor central pattern generators and animal gaits," *Nature* **401**, 693–695.

Golubitsky, M. & Stewart, I. [2002] *The Symmetry Perspective*, Progress in Mathematics, Vol. 200 (Birkhäuser, Basel).

Golubitsky, M., Stewart, I. & Török, A. [2005] "Patterns of synchrony in coupled cell networks with multiple arrows," *SIAM J. Appl. Dyn. Syst.* **4**, 78–100.

Golubitsky, M. & Stewart, I. [2006] "Nonlinear dynamics of networks: The groupoid formalism," *Bull. Am. Math. Soc.* **43**, 305–364.

Golubitsky, M., Romano, D. & Wang, Y. [2010] "Network periodic solutions: Full oscillation and rigid synchrony," *Nonlinearity* **23**, 3227–3243.

Golubitsky, M., Romano, D. & Wang, Y. [2012] "Network periodic solutions: Patterns of phase-shift synchrony," *Nonlinearity* **25**, 1045–1074.

Golubitsky, M. & Stewart, I. [2023] *Dynamics and Bifurcation in Networks* (SIAM, Philadelphia).

Goulding, M. [2009] "Circuits controlling vertebrate locomotion: Moving in a new direction," *Nat. Rev. Neurosci.* **10**, 507–518.

Gray, J. [1968] *Animal Locomotion* (Weidenfeld & Nicolson, London).

Grillner, S. [2003] "The motor infrastructure: From ion channels to neuronal networks," *Nat. Rev. Neurosci.* **4**, 573–586.

Grillner, S. & Jessell, T. M. [2009] "Measured motion: Searching for simplicity in spinal locomotor networks," *Curr. Opin. Neurobiol.* **19**, 572–586.

Grillner, S. & El Manira, A. [2020] "Current principles of motor control, with special reference to vertebrate locomotion," *Physiol. Rev.* **100**, 271–320.

Guckenheimer, J. & Holmes, P. [1983] *Nonlinear Oscillations, Dynamical Systems, and Bifurcations of Vector Fields* (Springer, NY).

Hall, D. H. *et al.* (eds.) [2025] "WormAtlas," www.wormatlas.org/.

Hammoud, Z. & Kramer, F. [2020] "Multilayer networks: Aspects, implementations, and application in biomedicine," *Big Data Anal.* **5**, 2.

Hanuschkin, A., Herrmann, J. M., Morrison, A. & Diesmann, M. [2011] "Compositionality of arm movements can be realized by propagating synchrony," *J. Comput. Neurosci.* **30**, 675–697.

Hassard, B. D., Kazarinoff, N. D. & Wan, Y.-H. [1981] *Theory and Applications of Hopf Bifurcation*, London Mathematical Society Lecture Note Series, Vol. 41 (Cambridge University Press, Cambridge).

Hindmarsh, J. L. & Rose, R. M. [1984] "A model of neuronal bursting using three coupled first order differential equations," *Proc. R. Soc. Lond. B* **221**, 87–102.

Hirsch, M. W., Pugh, C. C. & Shub, M. [1977] *Invariant Manifolds*, Lecture Notes in Mathematics, Vol. 583 (Springer, NY).

Hodgkin, A. L. & Huxley, A. F. [1952] "A quantitative description of membrane current and its application to conduction and excitation in nerve," *J. Physiol.* **117**, 500–544.

In, V., Kho, A., Longhini, P., Neff, J. D., Palacios, A. & Buono, P.-L. [2022] "Meet ANIBOT: The first biologically-inspired animal robot," *Int. J. Bifurcation and Chaos* **32**, 2230001.

Izquierdo, E. J. & Beer, R. D. [2018] "From head to tail: A neuromechanical model of forward locomotion in *Caenorhabditis elegans*," *Philos. Trans. R. Soc. Lond. B* **373**, 20170374.

Jha, M. & Chauhan, N. R. [2019] "A review on snake-like continuum robots for medical surgeries," *IOP Conf. Ser.: Mater. Sci. Eng.* **691**, 012093.

Katok, A. & Hasselblatt, B. [1995] *Introduction to the Modern Theory of Dynamical Systems* (Cambridge University Press, Cambridge).

Kiehn, O. [2006] "Locomotor circuits in the mammalian spinal cord," *Annu. Rev. Neurosci.* **29**, 279–306.

Kristan Jr., W. B., Calabrese, R. L. & Friesen, W. O. [2005] "Neuronal control of leech behavior," *Prog. Neurobiol.* **76**, 279–327.

Kupka, I. [1963] "Contribution à la théorie des champs génériques," *Contrib. Differ. Equ.* **2**, 457–484.

Kupka, I. [1964] "Contribution à la théorie des champs génériques: Addendum and corrections," *Contrib. Differ. Equ.* **3**, 411–420.

Kuramoto, Y. [1984] *Chemical Oscillations, Waves, and Turbulence* (Springer, Berlin).

Leifer, I., Morone, F., Reis, S. D. S., Andrade Jr., J. S., Sigman, M. & Makse, H. A. [2020] "Circuits with broken fibration symmetries perform core logic computations in biological networks," *PLoS Comput. Biol.* **16**, e1007776.

Lindén, H., Petersen, P. C., Vestergaard, M. & Berg, R. W. [2022] "Movement is governed by rotational neural dynamics in spinal motor networks," *Nature* **610**, 526–531.

Makse, H. A., Boldi, P., Sorrentino, F. & Stewart, I. [2025] *Symmetries of Living Systems* (Cambridge University Press, Cambridge), to appear.

Markus, L. & Yamabe, H. [1960] "Global stability criteria for differential systems," *Osaka J. Math.* **12**, 305–317.

McCrea, D. A. & Rybak, I. [2008] "Organization of mammalian locomotor rhythm and pattern generation," *Brain Res. Rev.* **57**, 134–146.

Mehring, C., Hehl, U., Kubo, M., Diesmann, M. & Aertsen, A. [2003] "Activity dynamics and propagation of synchronous spiking in locally connected random networks," *Biol. Cybern.* **88**, 395–408.

Milnor, J. [1985] "On the concept of attractor," *Commun. Math. Phys.* **99**, 177–195.

Morone, F. & Makse, H. A. [2019] "Symmetry group factorization reveals the structure-function relation in the neural connectome of *Caenorhabditis elegans*," *Nat. Commun.* **10**, 4961.

Morris, C. & Lecar, H. [1981] "Voltage oscillations in the barnacle giant muscle fiber," *Biophys. J.* **35**, 193–213.

Mu, Z., Wang, H., Xu, W., Liu, T. & Wang, H. [2017] "Two types of snake-like robots for complex environment exploration: Design, development, and experiment," *Adv. Mech. Eng.* **9**, 1687814017721854.

Muybridge, E. [1899] *Animals in Motion* (Chapman & Hall, London), Reprinted in Dover, NY, 1957.

Nagumo, J., Arimoto, S. & Yoshizawa, S. [1962] "An active pulse transmission line simulating nerve axon," *Proc. IRE* **50**, 2061–2070.

Nicosia, V., Valencia, M., Chavez, M., Díaz-Guilera, A. & Latora, V. [2013] "Remote synchronization reveals network symmetries and functional modules," *Phys. Rev. Lett.* **110**, 174102.

Olivares, E., Izquierdo, E. J. & Beer, R. D. [2021] "A neuromechanical model of multiple network rhythmic pattern generators for forward locomotion in *C. elegans*," *Front. Comput. Neurosci.* **15**, 572339.

Parker, M. & Stewart, I. [2008] "A new mechanism for intermittency in rings of cells," *Int. J. Bifurcation and Chaos* **18**, 675–687.

Pecora, L. M. & Carroll, T. L. [1998] "Master stability functions for synchronized coupled systems," *Phys. Rev. Lett.* **80**, 2109–2112.

Peixoto, M. M. [1966] "On an approximation theorem of Kupka and Smale," *J. Differ. Equ.* **3**, 214–227.

Pinto, C. A. & Golubitsky, M. [2006] "Central pattern generators for bipedal locomotion," *J. Math. Biol.* **53**, 474–489.

Purcell, O., Savery, N. J., Grierson, C. S. & di Bernardo, M. [2010] "A comparative analysis of synthetic genetic oscillators," *J. R. Soc. Interface* **7**, 1503–1524.

Roffman, R. C., Norris, B. J. & Calabrese, R. L. [2012] "Animal-to-animal variability of connection strength in the leech heartbeat central pattern generator," *J. Neurophysiol.* **107**, 1681–1693.

Sakamoto, K., Soh, Z., Suzuki, M., Iino, Y. & Tsuji, T. [2021] "Forward and backward locomotion patterns in *C. elegans* generated by a connectome-based model simulation," *Sci. Rep.* **11**, 13737.

Seok, S., Onal, C. D., Wood, R., Rus, D. & Kim, S. [2010] "Peristaltic locomotion with antagonistic actuators in soft robotics," *2010 IEEE Int. Conf. Robotics and Automation* (IEEE), pp. 1228–1233.

Setareh, H., Deger, M. & Gerstner, W. [2018] "Excitable neuronal assemblies with adaptation as a building block of brain circuits for velocity-controlled signal propagation," *PLoS Comput. Biol.* **14**, e1006216.

Smale, S. [1963] "Stable manifolds for differential equations and diffeomorphisms," *Ann. Sc. Norm. Sup. Pisa Cl. Sci.* **17**, 97–116.

Stewart, I., Golubitsky, M. & Pivato, M. [2003] "Symmetry groupoids and patterns of synchrony in coupled cell networks," *SIAM J. Appl. Dyn. Syst.* **2**, 609–646.

Stewart, I. & Parker, M. [2008] "Periodic dynamics of coupled cell networks II: Cyclic symmetry," *Dyn. Syst.* **23**, 17–41.

Stewart, I. [2014] "Symmetry-breaking in a rate model for a biped locomotion central pattern generator," *Symmetry* **6**, 23–66.

Stewart, I. [2022] "Overdetermined ODEs and rigid periodic states in network dynamics," *Port. Math.* **79**, 85–161.

Stewart, I. & Wood, D. [2024] "Stable synchronous propagation of periodic signals by feedforward networks," *SIAM J. Appl. Dyn. Syst.* **23**, 167–204.

Stewart, I. & Wood, D. [2025] "Stable synchronous propagation of signals by feedforward networks: Biped locomotion," in preparation.

Thompson, W. J. & Stent, G. S. [1976] "Neuronal control of heartbeat in the medicinal leech II: Intersegmental coordination of heart motor neuron activity by heart interneurons," *J. Comp. Physiol.* **111**, 281–307.

Uhlhaas, P. J., Pipa, G., Lima, B., Melloni, L., Neuenschwander, S., Nikolić, D. & Singer, W. [2009] "Neural synchrony in cortical networks: History, concept and current status," *Front. Integr. Neurosci.* **3**, 17.

Verner, J. [2006] "Jim Verner's refuge for Runge–Kutta pairs," uk.mathworks.com/help/matlab/ref/ode89.html.

Wenning, A., Hill, A. A. & Calabrese, R. L. [2004] "Heartbeat control in leeches. I: Fictive motor pattern," *J. Neurophysiol.* **91**, 397–409.

Wright, E. M. [1961] "Counting coloured graphs," *Can. J. Math.* **13**, 683–693.

Zheng, P. & Triesch, J. [2014] "Robust development of synfire chains from multiple plasticity mechanisms," *Front. Comput. Neurosci.* **8**, 66.

Zhong, Y., Hu, L. & Xu, Y. [2020] "Recent advances in design and actuation of continuum robots for medical applications," *Actuators* **9**, 142.

https://doi.org/10.1142/9789819830633_0004

Chapter 4

Multistability in a 3D Circulant Chaotic Flow: The Circulant Symmetry of Coexisting Dynamics#

Karthikeyan Rajagopal
Center for Research, SRM Easwari Engineering College, Chennai 600089, India
Center for Cognitive Science, Trichy SRM Medical College Hospital and Research Center, Trichy 621105, India
rkarthiekeyan@gmail.com

Fahimeh Nazarimehr[*,‡] and Sajad Jafari[*,†,§]
[*]Department of Biomedical Engineering, Amirkabir University of Technology (Tehran Polytechnic), Tehran 1591634311, Iran
[†]Health Technology Research Institute, Amirkabir University of Technology (Tehran Polytechnic), Tehran 1591634311, Iran
[‡]f_nazarimehr@aut.ac.ir; fahimenazarimehr@yahoo.com
[§]sajadjafari@aut.ac.ir

Julien C. Sprott
Department of Physics, University of Wisconsin–Madison, Madison, WI 53706, USA
csprott@wisc.edu

In this paper, a novel chaotic flow exhibiting circulant symmetry is modeled and analyzed. The chaotic flow's dynamical behavior is thoroughly investigated, revealing that it is dissipative over the studied parameter values. A basic equilibrium analysis shows that four of the system's equilibrium points are stable. Notably, the system exhibits two distinct sets of three coexisting chaotic attractors, each set displaying circulant symmetry. To the best of our knowledge, such a system has not been previously reported in the literature. Bifurcation analysis with respect to all three system parameters confirms the rich dynamical behavior, supported by corresponding Lyapunov exponent spectra. The basins of attraction reveal rare V-shaped boundaries, adding to the system's uniqueness. Furthermore, the circulant symmetry is clearly reflected in the basins of attraction of the coexisting attractors.

Keywords: Circulant symmetry; chaos; dissipative; multistability.

[#]To cite this article, please refer to its earlier version published in the *International Journal of Bifurcation and Chaos*, Vol. 35, No. 14 (2025) 2540004 (13 pages), DOI: 10.1142/S0218127425400048.

[‡]Author for correspondence

1. Introduction

Chaos theory is a fascinating field of research that is a cornerstone of nonlinear dynamics; it provides an insight into the understanding of deterministic yet unpredictable behaviors of real-world systems [Bob, 2007; Sprott, 2010; Danca, 2024; Pan *et al.*, 2025]. The main feature of chaotic systems is sensitive dependence on initial conditions, which makes them a proper choice for applications in diverse fields such as secure communication [Banerjee, 2010; Almatroud *et al.*, 2024], cryptography [Lawnik *et al.*, 2022; Tutueva *et al.*, 2022], and biological modeling [Panahi *et al.*, 2019; Wang *et al.*, 2024; Zhang *et al.*, 2025]. For many years, there was an idea that chaotic attractors are related to saddle equilibria [Rössler, 1976], until some counterexamples were proposed showing chaos without any equilibrium points [Wei, 2011] or with one stable equilibrium [Wang & Chen, 2012]. After that, the idea was broken, and many chaotic systems with various equilibrium points were proposed [Danca, 2017]. Various studies have been done to untangle the mystery of the generation of chaos in nonlinear dynamical systems [Xu *et al.*, 2020; Shukur *et al.*, 2025]. Chaotic dynamics of memristive neurons have attracted a lot of attention [Xu *et al.*, 2024a; Xu *et al.*, 2024b].

Despite many trials, there are many unknown mysteries in studying chaotic flows. Some systems have features like multistability or adjustable dynamics, while many have complex topologies that make them challenging to work with [Nazarimehr & Sprott, 2020; Khan *et al.*, 2025; Raza *et al.*, 2025]. Furthermore, organized systems, such as those with symmetry, are yet largely unexplored. One interesting type of these systems is circulant systems, which have cyclic permutation characteristics and rotational symmetry [Sprott, 2010]. They are especially attractive for new technologies because of their intrinsic symmetry, which helps streamline theoretical analysis and real-world applications.

Symmetry plays a crucial role in understanding the behavior of dynamical systems. Many chaotic systems exhibit reflectional symmetry [Kyurkchiev *et al.*, 2024]. More recent works have explored chaotic systems with permutation symmetry, where variables can be cyclically interchanged without altering the system's dynamics [Rajagopal *et al.*, 2019; Panahi *et al.*, 2021]. Notably, several algebraically simple chaotic flows with symmetry properties have been introduced [Li & Sprott, 2021; Li *et al.*, 2022]. However, few studies have focused on systems with circulant symmetry, where each variable is a cyclic permutation of the others in the governing equations [Sprott, 2010].

In this study, we present a new circulant chaotic system with cubic nonlinearities intended to show coexisting attractors and rich dynamical behavior. The system's nonlinear terms allow for intricate stretching and folding dynamics that are necessary for chaos, while its circulant structure guarantees rotational symmetry. We illustrate a variety of phenomena, such as multistability, period-doubling pathways to chaos, and crisis-induced transitions, by altering important constants. Our research advances our theoretical knowledge of circulant systems while simultaneously demonstrating its potential for real-world uses in secure communication and random number generation.

This is how the rest of the paper is structured. The mathematical description of the suggested system is shown in Sec. 2, focusing on its nonlinear structure and circulant symmetry. The dynamical behavior of the system, including phase portraits, bifurcation diagrams, and Lyapunov exponents, is examined in Sec. 3. In order to provide more light on the chaotic characteristics of the system, Sec. 3.4 examines the basin of attraction. A review of the main conclusions and possible future directions is provided in Sec. 4, which brings the paper to a close.

2. Mathematical Formulation

Here, we propose a circulant chaotic flow with cubic nonlinearities, designed to exhibit coexisting attractors. Circulant systems are defined by rotational symmetry in their governing equations, ensuring cyclic permutation of variables [Sprott, 2010]. Our system adheres to the general form

$$\begin{aligned} \dot{x} &= g(x, y, z), \\ \dot{y} &= g(y, z, x), \\ \dot{z} &= g(z, x, y), \end{aligned} \tag{1}$$

where $g(\cdot)$ is a nonlinear function preserving circulant symmetry. Through an exhaustive computational search for chaotic dynamics, we derive the following system:

$$\begin{aligned} \dot{x} &= ax + b(y + z)yz + cyz, \\ \dot{y} &= ay + b(z + x)zx + czx, \\ \dot{z} &= az + b(x + y)xy + cxy. \end{aligned} \tag{2}$$

Table 1. Equilibrium points and their stabilities of system (2).

#	Equilibrium Points			Eigen Values			Stability
	x	y	z	λ_1	λ_2	λ_3	
1	0.0000	0.0000	0.0000	−0.0556	−0.0556	−0.0556	Stable
2	−0.1516	−0.4819	−0.4819	−0.0643 + 0.2108i	−0.0643 − 0.2108i	−0.0381	Stable
3	−0.4819	−0.1516	−0.4819	−0.0643 + 0.2108i	−0.0643 − 0.2108i	−0.0381	Stable
4	−0.4819	−0.4819	−0.1516	−0.0643 + 0.2108i	−0.0643 − 0.2108i	−0.0381	Stable
5	0.2131	0.2131	−1.1654	−0.3594	0.0964 + 0.3785i	0.0964 − 0.3785i	Unstable
6	−0.2841	−0.2841	−0.6273	0.0671	−0.1169 + 0.0953i	−0.1169 − 0.0953i	Unstable
7	0.0529	0.0529	−0.0557	0.0571	−0.1141	−0.1097	Unstable
8	−0.4363	−0.4363	−0.4363	−0.3252	0.0793	0.0793	Unstable
9	−0.0637	−0.0637	−0.0637	−0.1071	−0.1071	0.0475	Unstable
10	−0.2841	−0.6273	−0.2841	0.0671	−0.1169 + 0.0953i	−0.1169 − 0.0953i	Unstable
11	−0.6273	−0.2841	−0.2841	−0.1169 + 0.0953i	−0.1169 − 0.0953i	0.0671	Unstable
12	0.05297	−0.0557	0.0529	0.0571	−0.1141	−0.1097	Unstable
13	−0.0557	0.0529	0.0529	−0.1097	0.0571	−0.1141	Unstable
14	0.2131	−1.1654	0.2131	−0.3594	0.0964 + 0.3785i	0.0964 − 0.3785i	Unstable
15	−1.1654	0.2131	0.2131	0.0964 + 0.3785i	0.0964 − 0.3785i	−0.3594	Unstable

Each equation is a cyclic permutation of variables. The cubic nonlinearities, such as $b(y+z)yz$ and cyz, introduce strong nonlinear interactions critical for chaos. The system can show chaotic dynamics where the parameters are $a = -\frac{1}{18}$, $b = -1$, $c = -1$. The divergence of the vector field is $3a = -\frac{1}{6}$ (for $a = -\frac{1}{18}$), ensuring volume contraction in phase space — a prerequisite for chaos. Parameter a governs linear damping, while b and c control nonlinear coupling, with b amplifying cross-terms like $(y+z)yz$. Cubic terms enable complex stretching/folding dynamics, while the linear term ax stabilizes trajectories since the a's sign is negative.

As the most fundamental feature of a chaotic flow, the equilibrium points of the system are investigated. These points are obtained by solving the following system of equations:

$$\begin{aligned} \dot{x} &= ax + b(y+z)yz + cyz = 0, \\ \dot{y} &= ay + b(z+x)zx + czx = 0, \\ \dot{z} &= az + b(x+y)xy + cxy = 0. \end{aligned} \tag{3}$$

Numerical methods implemented in MATLAB were utilized to calculate equilibrium points as presented in Table 1. To investigate the stability of each fixed point, the system's Jacobian matrix is calculated as follows:

$$\mathbf{J} = \begin{bmatrix} a & 2byz + bz^2 + cz & by^2 + 2byz + cy \\ bz^2 + 2bxz + cz & a & 2bzx + bx^2 + cx \\ 2bxy + by^2 + cy & bx^2 + 2bxy + cx & a \end{bmatrix}. \tag{4}$$

Then, for each equilibrium point, the characteristic equation is constructed as

$$\det(\lambda\mathbf{I} - \mathbf{J}|_{\text{FP}}) = 0,$$

where FP denotes the fixed point. The eigenvalues corresponding to each equilibrium point are presented in Table 1. The eigenvalue analysis reveals that the first four equilibrium points are stable, as the real components of all eigenvalues are negative. The remaining equilibrium points are unstable, with at least one eigenvalue containing a positive real component.

3. Dynamical Analysis

This section looks at the chaotic dynamics of the suggested circulant system (Eq. (2)). We analyze Lyapunov exponents, bifurcations, and coexisting attractors.

3.1. *Phase portrait*

The proposed circulant chaotic system has a diverse dynamical behavior, with six coexisting chaotic attractors and four stable equilibrium points. These

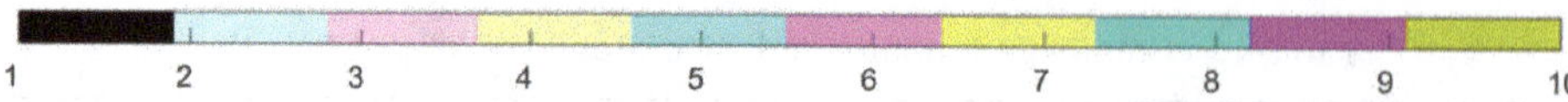

Fig. 1. The consistent color-coding scheme employed throughout the paper to separate the dynamics of each attractor and equilibrium point.

chaotic attractors are arranged in two groups of three, each with circulant symmetry. To aid visualization and analysis, a consistent color-coding scheme (shown in Fig. 1) is employed throughout the paper to separate the dynamics of each attractor and equilibrium point.

The four stable equilibrium points, denoted as FP_1 through FP_4, are assigned colors 1–4 in the color-coding scheme. Their coordinates are

$$\mathrm{FP}_1 = (0, 0, 0),$$

$$\mathrm{FP}_2 = (-0.1516, -0.4819, -0.4819),$$

$$\mathrm{FP}_3 = (-0.4819, -0.1516, -0.4819),$$

$$\mathrm{FP}_4 = (-0.4819, -0.4819, -0.1516).$$

These equilibrium points correspond to stable regions in the phase space, where trajectories converge for certain initial conditions.

The system also features six chaotic attractors, grouped into two distinct sets based on their initial conditions and symmetry properties. Each set consists of three attractors that are cyclic permutations of one another, reflecting the circulant symmetry of the system. The first group of chaotic attractors

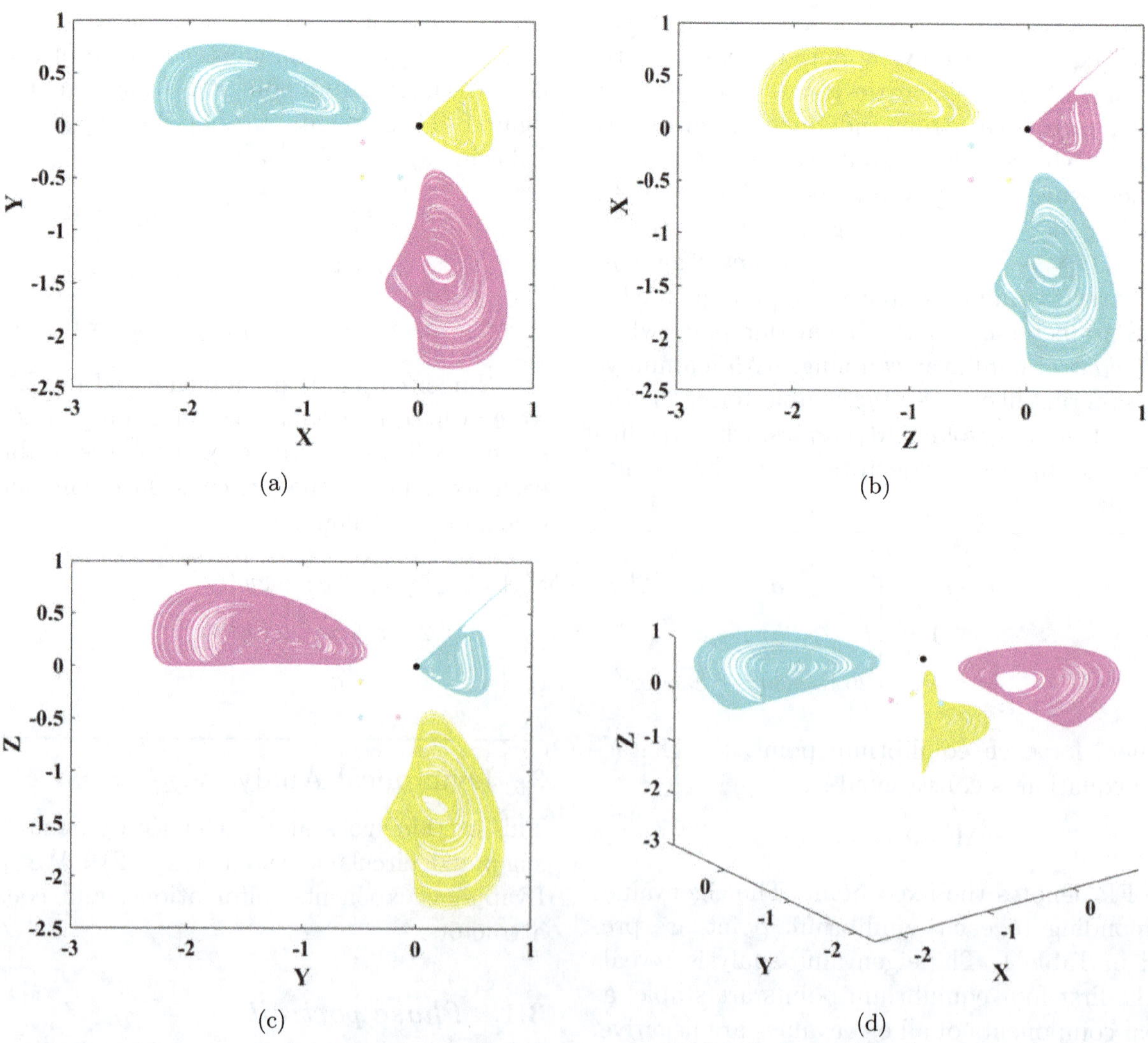

(a) (b) (c) (d)

Fig. 2. Four stable equilibria and chaotic dynamics of Eq. (2) correspond to the initial conditions $\mathrm{IC}_{05} = (-0.67, 0.92, 0.4)$, $\mathrm{IC}_{06} = (0.4, -0.67, 0.92)$, $\mathrm{IC}_{07} = (0.92, 0.4, -0.67)$, where colors 5–7 are used to represent the chaotic respective dynamics; the three attractors in (a) X–Y plane, (b) Z–X plane, (c) Y–Z plane and (d) X–Y–Z space.

corresponds to the initial conditions $IC_{05} = (-0.67, 0.92, 0.4)$, $IC_{06} = (0.4, -0.67, 0.92)$, $IC_{07} = (0.92, 0.4, -0.67)$. These attractors are plotted in Fig. 2, where colors 5–7 are used to represent their respective dynamics. The cyclic permutation of initial conditions ensures that the attractors exhibit symmetric behavior in the phase space. The second group of chaotic attractors corresponds to the initial conditions $IC_{08} = (-1.51, -0.275, 0)$, $IC_{09} = (0, -1.51, -0.275)$, $IC_{10} = (-0.275, 0, -1.51)$. These attractors are plotted in Fig. 3, with colors 8–10 distinguishing their dynamics. Similar to the first group, the symmetry of the system ensures that these attractors are cyclic permutations of one another.

The phase portraits in Figs. 2 and 3 highlight the intricate structure of the chaotic attractors, showcasing their circulant symmetry. The coexistence of multiple attractors underscores the multistability of the system, which has potential applications in secure communication, random number generation, and other fields requiring complex dynamical behavior.

3.2. *Bifurcation analysis*

Bifurcation diagrams are computed by adjusting three important parameters (a, b, and c) utilizing various forward and backward continuation techniques in order to investigate the dynamical transitions of the suggested circulant chaotic system. To investigate the dynamics of each attractor by changing each parameter, we plot bifurcation diagrams, starting from $a = -\frac{1}{18}$, $b = -1$, $c = -1$,

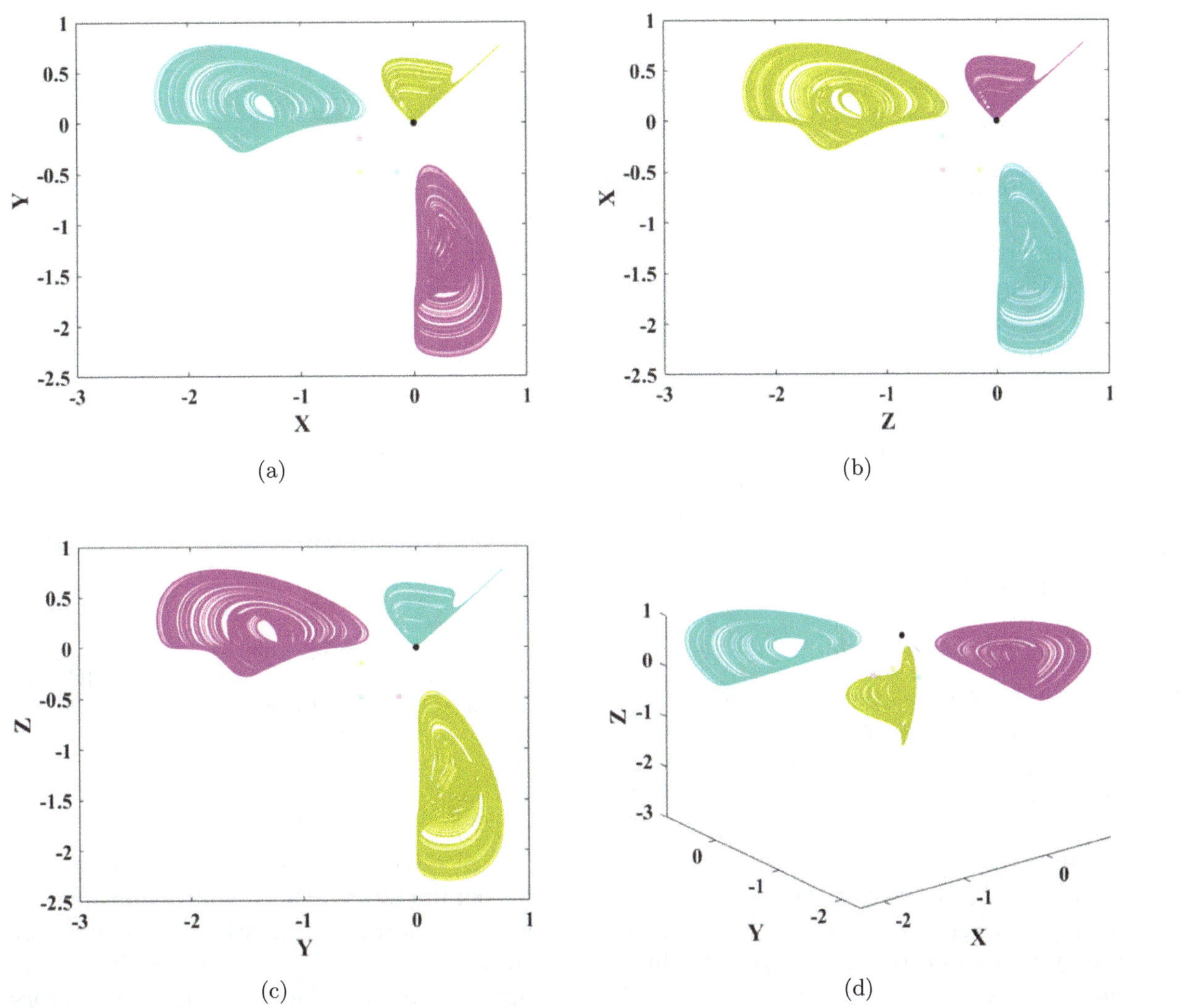

Fig. 3. Four stable equilibria and chaotic dynamics of Eq. (2) correspond to the initial conditions $IC_{08} = (-1.51, -0.275, 0)$, $IC_{09} = (0, -1.51, -0.275)$, $IC_{10} = (-0.275, 0, -1.51)$, with colors 8–10 distinguishing the chaotic dynamics; the three attractors in (a) X–Y plane, (b) Z–X plane, (c) Y–Z plane and (d) X–Y–Z space.

by increasing the bifurcation parameter in a forward method for larger values of the parameter or decreasing the bifurcation parameter in a backward method for smaller values. Then, we chose to plot the bifurcations that have more diverse dynamics. These graphs demonstrate the circulant symmetry of the attractors and show how sensitive the system is to changes in parameters.

Case 1. Varying Parameter a

In the first scenario, the bifurcation diagram by changing the parameter a is investigated. Bifurcation diagrams for the two sets of three chaotic attractors are shown in Figs. 4 and 5 as the parameter a is changed. In each figure, the bifurcations of three attractors of each group are plotted by their corresponding colors. Bifurcation diagrams with the first set of initial conditions IC_{05}, IC_{06}, IC_{07}, and forward continuation are shown in Fig. 4, which presents the circulant symmetry of attractors by varying a. Each panel of the figure shows the bifurcations from the viewpoint of one of the system's variables. Comparing the bifurcations from the viewpoint of various variables also discloses the circulant symmetry of the three attractors of the system. The circulant symmetry can be seen as the bifurcation diagram of the attractor, with color 7 in the $X_{\max}$ plot which is the same as for the attractor with color 5 in the $Y_{\max}$ plot and for the attractor with color 6 in the $Z_{\max}$ plot. Figure 5 shows the bifurcation diagrams of chaotic attractors with IC_{08}, IC_{09}, IC_{10}, and forward continuation by changing a. Since the dynamics for each group are cyclic permutations of each other under the same parameter change, both pictures show the circulant symmetry of the attractors.

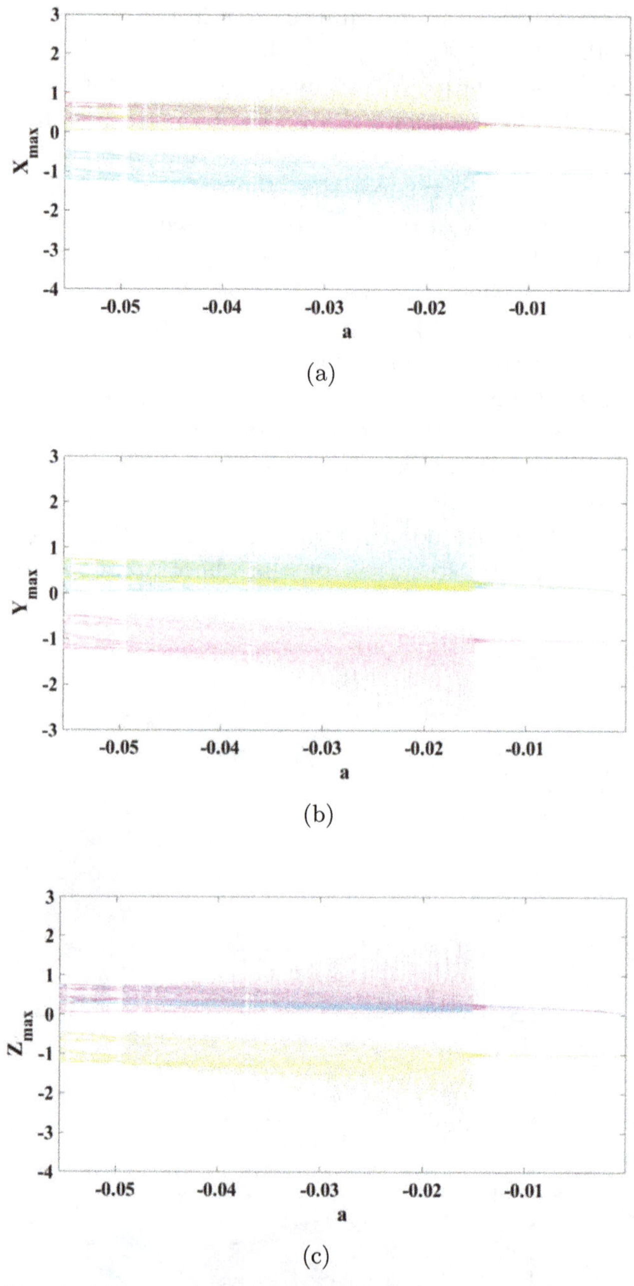

Fig. 4. Bifurcation diagram by changing the parameter a with the first set of initial conditions IC_{05}, IC_{06}, IC_{07}, and forward continuation; the maximum value of (a) X variable, (b) Y variable and (c) Z variable.

Case 2. Varying Parameter b

In the second scenario, the bifurcation diagram by changing the parameter b is investigated. Figures 6 and 7 present bifurcation diagrams obtained using the backward continuation method, where the parameter b is varied. Figure 6 represents the first group of attractors' bifurcations (IC_{05}, IC_{06}, IC_{07}), while Fig. 7 represents the second group of attractors (IC_{08}, IC_{09}, IC_{10}). These diagrams highlight the system's multistability and the influence of parameter b on the coexisting attractors. The backward continuation method reveals discontinuous transitions, providing insights into the system's complex behavior.

Case 3. Varying Parameter c

The third scenario analyzes bifurcations using varying parameters c. Bifurcation diagrams produced by the forward continuation approach, where the parameter c is changed, are shown in Figs. 8 and 9. Figure 8 identifies the first group of attractors (IC_{05}, IC_{06}, IC_{07}), and Fig. 9 identifies the second group of attractors (IC_{08}, IC_{09}, IC_{10}). By displaying routes to chaos (such as period-doubling cascades) and

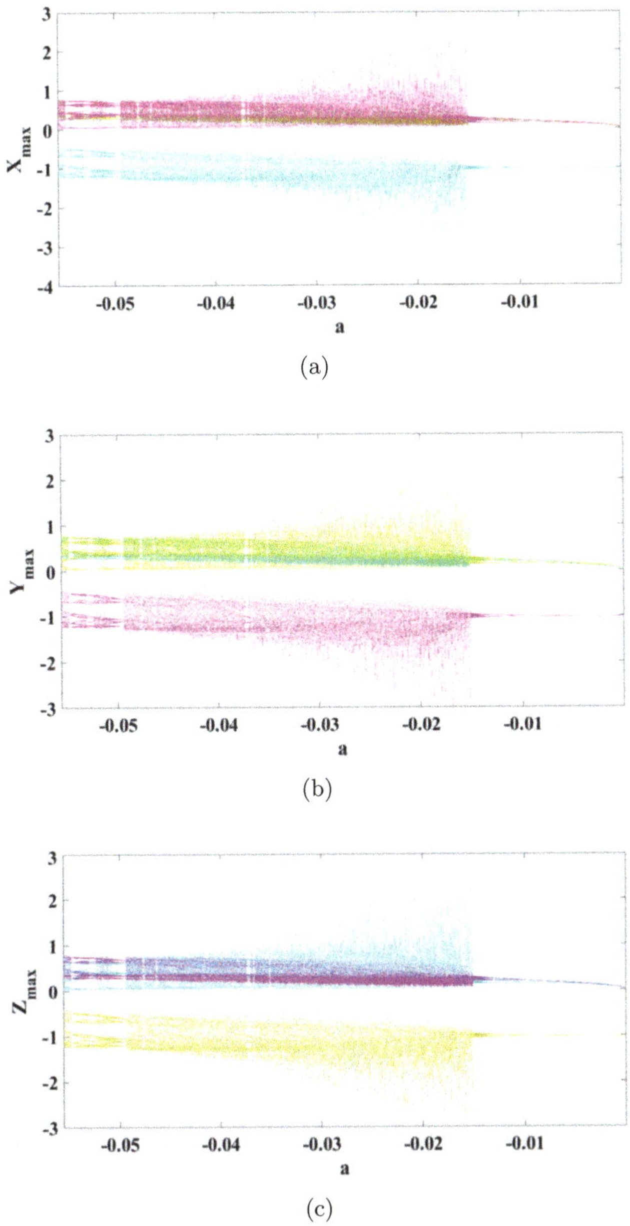

Fig. 5. Bifurcation diagram by changing the parameter a with the second set of initial conditions IC_{08}, IC_{09}, IC_{10} and forward continuation; the maximum value of (a) X variable, (b) Y variable and (c) Z variable.

the continuity of circulant symmetry across parameter changes, these diagrams point out the system's dependence on parameter c.

Bifurcation diagrams draw attention to the system's fundamental characteristics. The system's circulant symmetry is visible in every situation. Because the governing equations are inherently symmetric, the dynamics for each group of attractors are cyclic permutations of each other. Different dynamical behaviors emerge from varied initial conditions, confirming the coexistence of attractors. Random number generation and secure communication are two possible uses for this characteristic. Rich dynamical transitions are present in the system, such as periodic windows, chaotic regions, and abrupt changes. These changes highlight the intricacy of the system and its potential for real-world uses.

3.3. *Lyapunov spectrum*

We examine the circulant system's Lyapunov spectrum, a crucial property for comprehending the

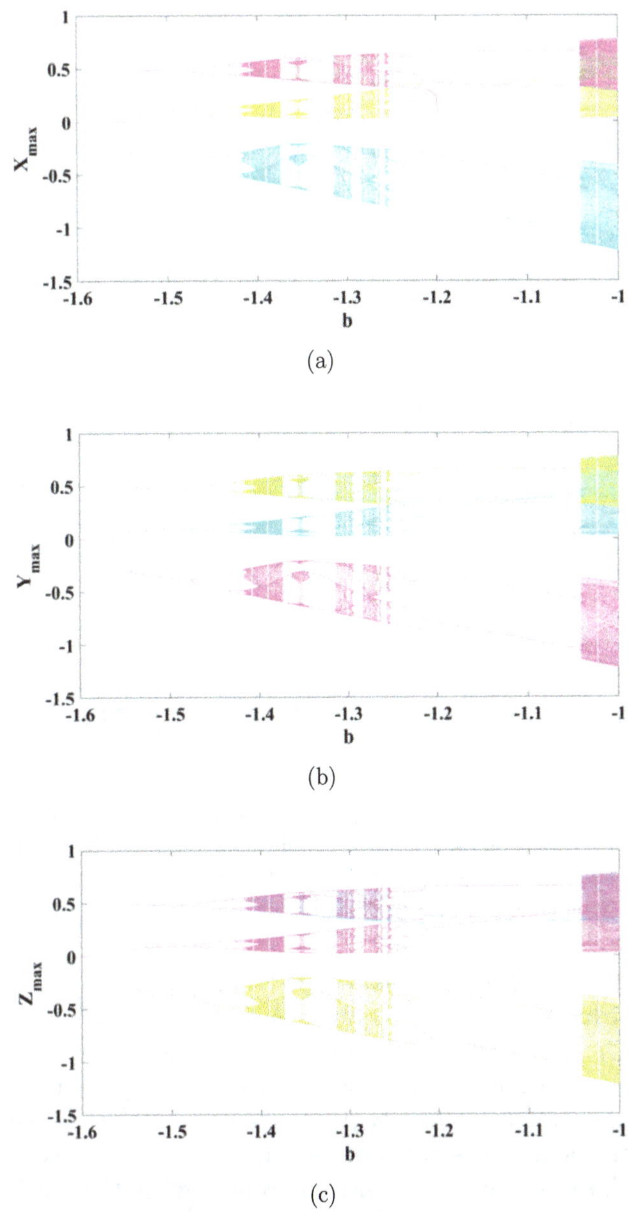

Fig. 6. Bifurcation diagram by changing the parameter b with the first set of initial conditions IC_{05}, IC_{06}, IC_{07} and backward continuation; the maximum value of (a) X variable, (b) Y variable and (c) Z variable.

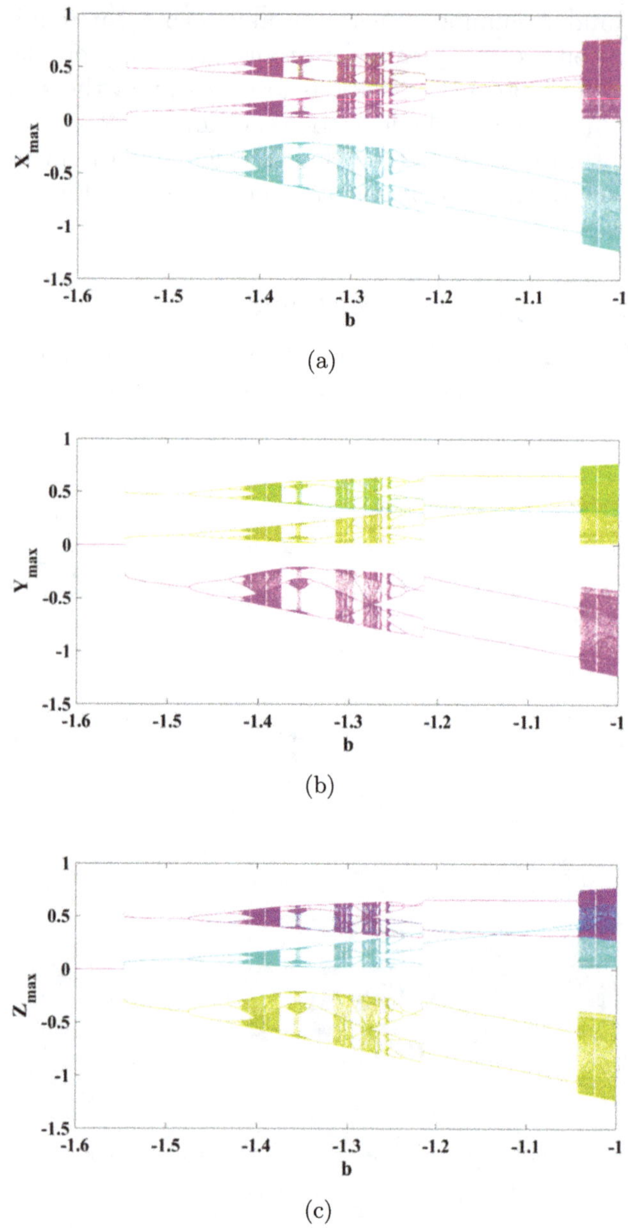

Fig. 7. Bifurcation diagram by changing the parameter b with the second set of initial conditions IC_{08}, IC_{09}, IC_{10} and backward continuation; the maximum value of (a) X variable, (b) Y variable and (c) Z variable.

stability and complexity of dynamical systems, in order to learn more about its chaotic dynamics. The Lyapunov exponents, which provide important information about periodicity, chaos, and bifurcation points, quantify the exponential divergence or convergence rates of neighboring trajectories in phase space. They are computed with the Wolf method [Wolf *et al.*, 1985] and runtime 20 000. Here, parameters a, b, and c are systematically changed to concentrate on the dynamic variations. Each set of three attractors has the same dynamics because of the system's circulant symmetry. Therefore, we examine the Lyapunov spectra of one typical attractor from each group (for example, the attractors that correspond to colors 5 and 8).

Case 1. Varying Parameter a

The Lyapunov spectra of the attractors corresponding to colors 5 and 8 are shown in Figs. 10(a) and 10(b) as parameter a is changed. Notably, the two attractors exhibit identical dynamics and Lyapunov exponents, consistent with the circulant

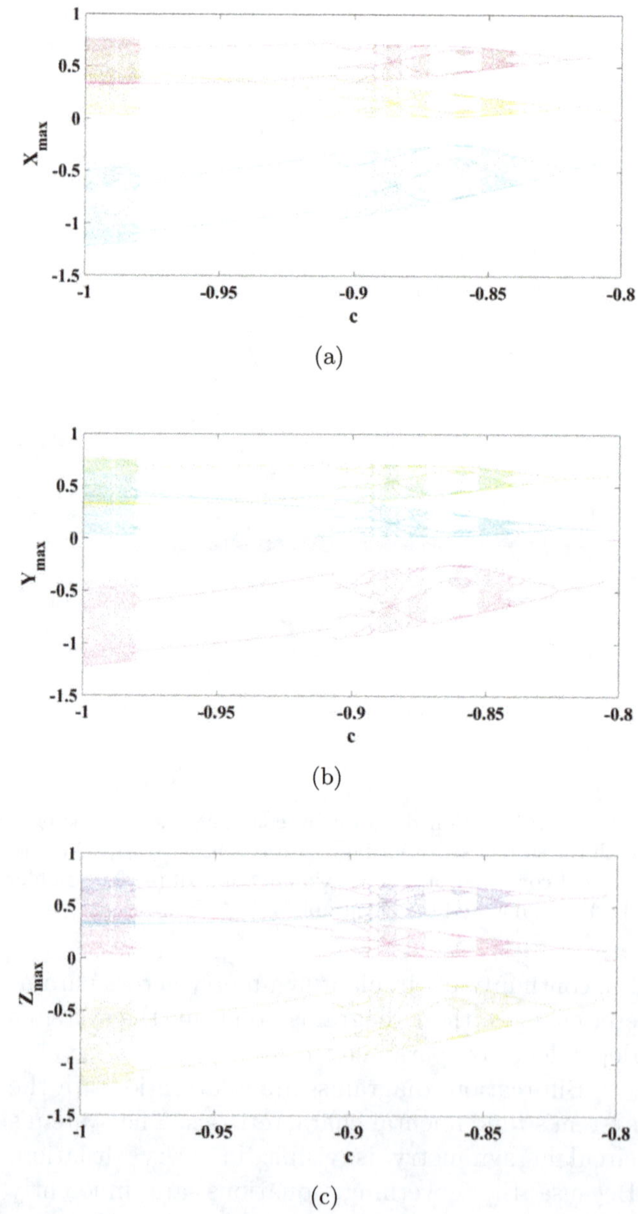

Fig. 8. Bifurcation diagram by changing the parameter c with the first set of initial conditions IC_{05}, IC_{06}, IC_{07}, and forward continuation; the maximum value of (a) X variable, (b) Y variable and (c) Z variable.

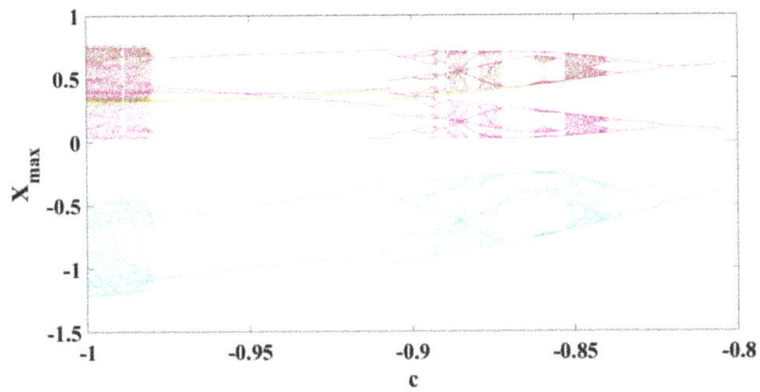

(a)

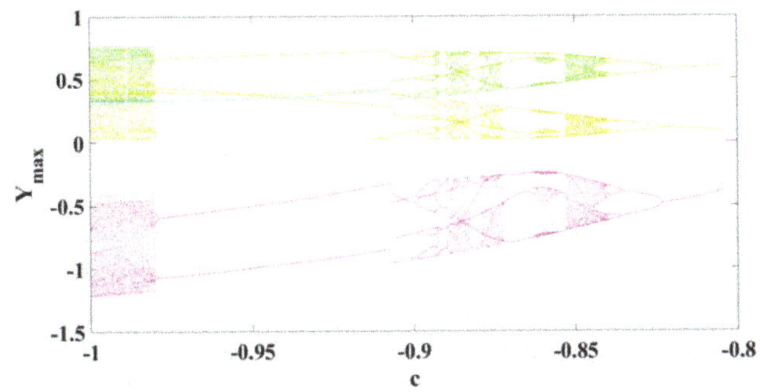

(b)

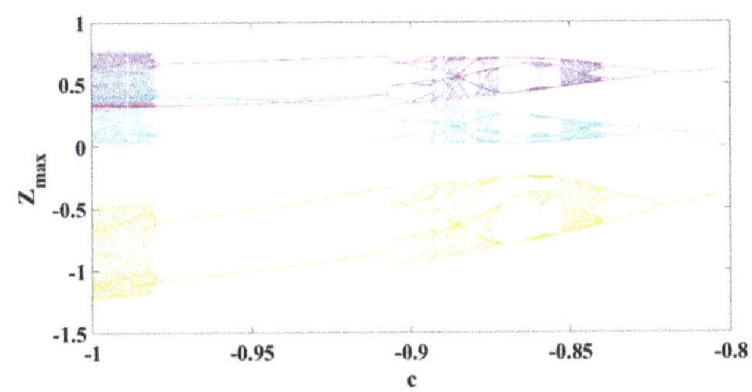

(c)

Fig. 9. Bifurcation diagram by changing the parameter c with the second set of initial conditions IC_{08}, IC_{09}, IC_{10}, and forward continuation; the maximum value of (a) X variable, (b) Y variable and (c) Z variable.

symmetry of the system. For small a, the system exhibits periodic windows (among the chaotic dynamics), characterized by all Lyapunov exponents being nonpositive. By intermediate a, a large chaotic attractor emerges as a result of a crisis. This transition is marked by the appearance of a positive Lyapunov exponent, indicating sensitive dependence on initial conditions. In large a, the chaotic attractor contracts after another crisis, and then there is an inverse period-doubling route to chaos. Also, the Lyapunov spectrum reflects the bifurcation points, with one exponent going toward zero and then returning. The dynamical variation is confirmed by the Lyapunov spectrum. Chaos is shown by one positive Lyapunov exponent, periodicity by one zero Lyapunov exponent and the rest negative, and equilibrium as all Lyapunov exponents are negative.

Case 2. Varying Parameter b

The Lyapunov spectra of the colors 5 and 8 attractors as parameter b is changed are shown in Fig. 11. Throughout the parameter range, the dynamics experience notable changes. In small b, all Lyapunov exponents are negative, indicating that the attractor corresponds to a stable equilibrium point. By intermediating b, a jump to a limit cycle occurs, marked by one zero Lyapunov exponent and the rest negative. Then, chaos, which is defined by one positive Lyapunov exponent, is followed by a period-doubling cascade. In large b, complex interactions between attractors are reflected in the system's alternating chaotic and periodic windows. A crisis eventually leads to a larger chaotic attractor.

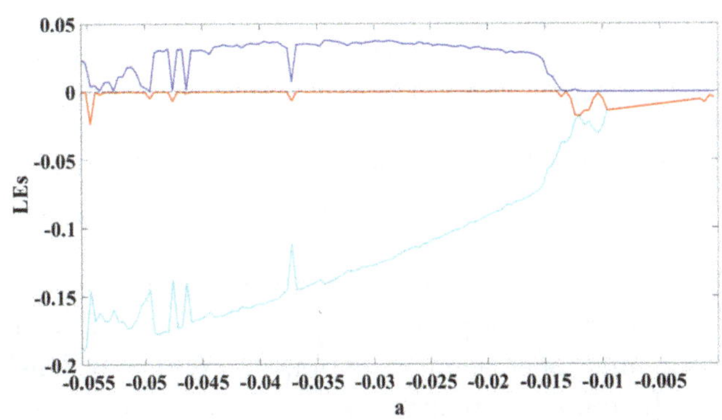

(a)

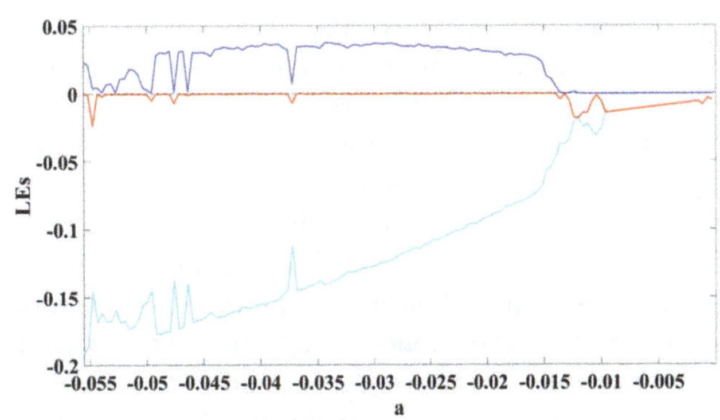

(b)

Fig. 10. The Lyapunov spectra of the attractors corresponding to colors (a) 5 and (b) 8 as parameter a is changed.

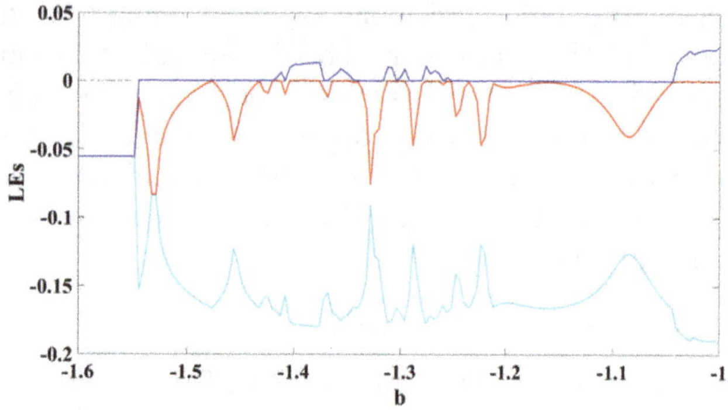

(a)

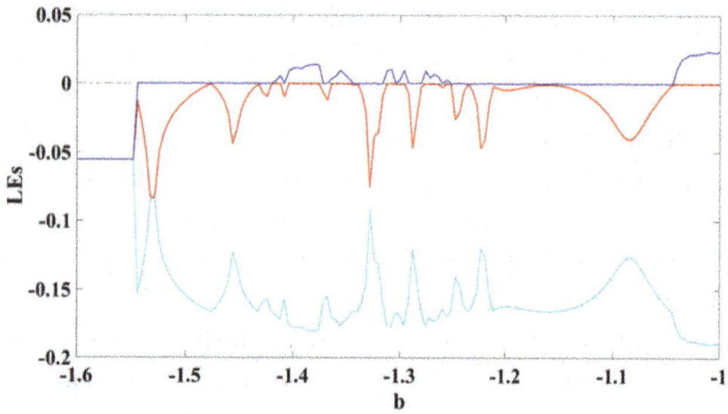

(b)

Fig. 11. The Lyapunov spectra of the attractors corresponding to colors (a) 5 and (b) 8 as parameter b is changed.

These transitions are highlighted by the Lyapunov exponents in Fig. 11. Stable equilibrium is observed when there are three negative exponents, limit cycle by one zero exponent and two negative exponents, and chaos by one positive exponent, one zero exponent, and one negative exponent. When at least one Lyapunov exponent goes to zero, it indicates a regime change.

Case 3. Varying Parameter c

As parameter c changes, the Lyapunov spectra for the identical attractors (colors 5 and 8) are displayed in Fig. 12. There are rich transitions in the dynamics. For small c, there is only one positive Lyapunov exponent, indicating that the system is chaotic. In intermediate c, the dynamics shift to a limit cycle, where the highest Lyapunov exponent drops to zero. For large c, periodic and chaotic windows alternate, followed by an inverse period-doubling route to chaos. Eventually, the dynamics collapse to a stable equilibrium point via a sudden transition from a limit cycle. The transitions which are captured by the Lyapunov spectrum in Fig. 12 are chaos when one exponent is positive, limit cycle when there are two negative exponents and one zero exponent, and equilibrium when all exponents are negative.

3.4. *Basin of attraction*

The basins of attraction provide significant insight into the system's multistability by illustrating how initial conditions influence which attractor a trajectory converges to. To depict these basins, we examine the circulant chaotic system in three orthogonal planes: x_0–y_0 (when $z_0 = 0$), y_0–z_0 (when $x_0 = 0$), and z_0–x_0 (when $y_0 = 0$). The results are shown in Fig. 13. Each color corresponds to one of the ten coexisting attractors identified earlier in the investigation. We use a consistent color scheme to make interpretation easier.

Several intriguing features emerge from the basins of attraction, highlighting the unique dynamical properties of the circulant chaotic system. One particularly noteworthy characteristic is the V-shaped boundaries observed in the basins. Such geometric patterns are extremely rare in the basins of conventional dynamical systems and reflect a

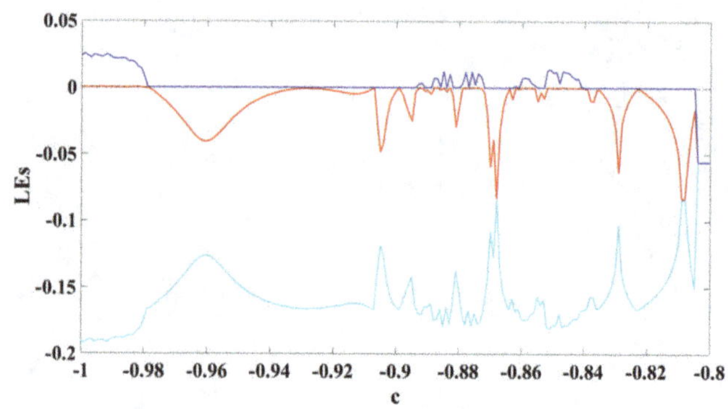

(a)

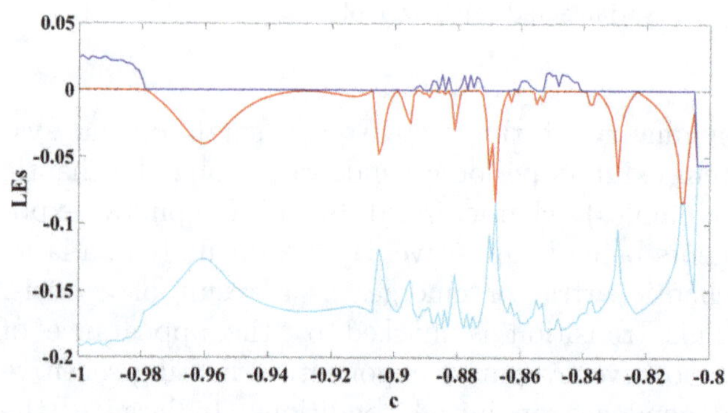

(b)

Fig. 12. The Lyapunov spectra of the attractors corresponding to colors (a) 5 and (b) 8 as parameter c is changed.

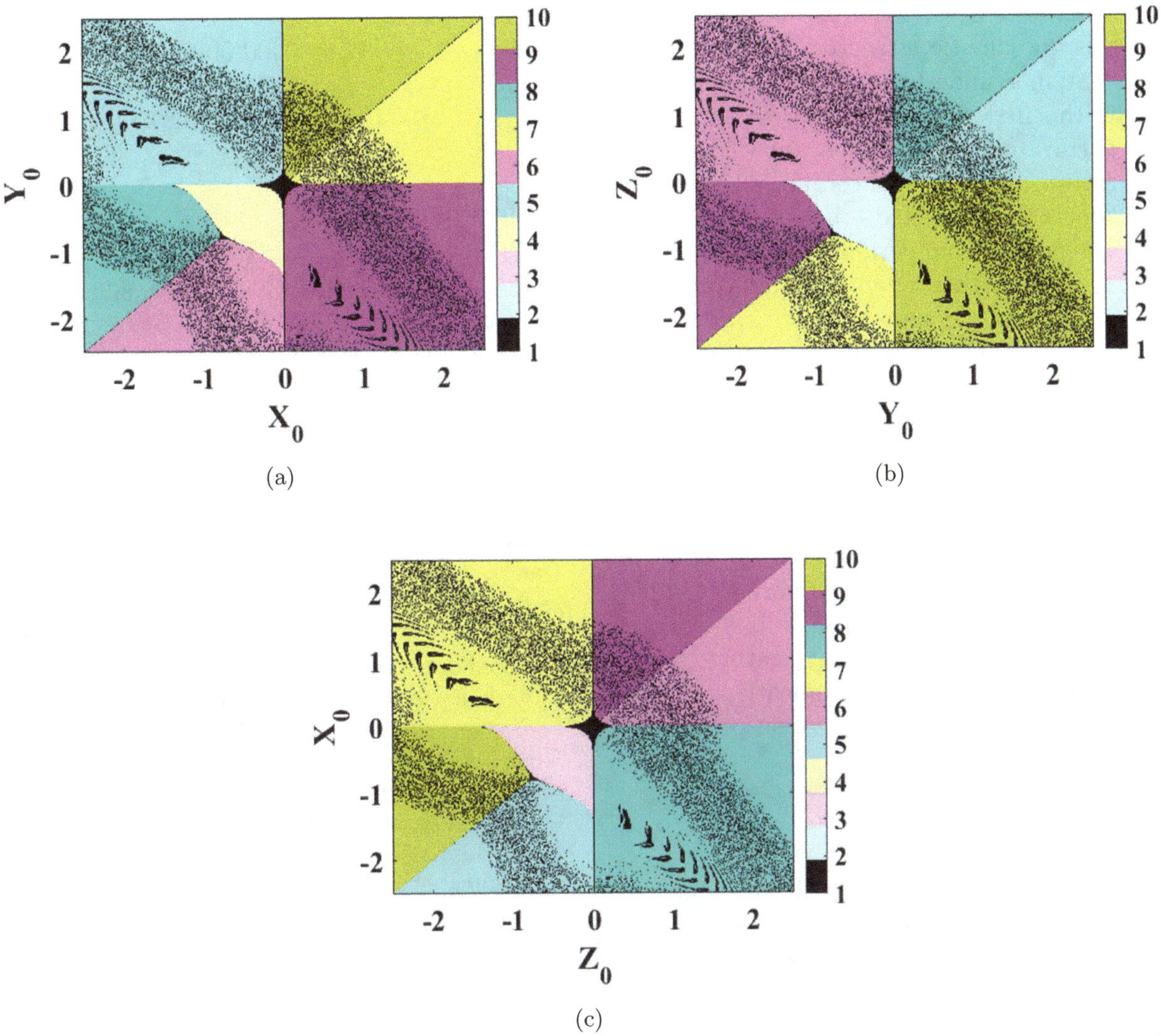

Fig. 13. The basins of attraction for the circulant chaotic system in three orthogonal planes; (a) in x_0–y_0 (when $z_0 = 0$), (b) in y_0–z_0 (when $x_0 = 0$) and (c) in z_0–x_0 (when $y_0 = 0$).

particular interplay between attractors. These V-shaped borders presumably develop due to the nonlinear interactions in the system's governing equations, which create sharp transitions between regions dominated by various attractors.

Another striking feature is the basin associated with the stable equilibrium at the origin. Unlike the other attractors, whose basins occupy contiguous regions in phase space, the basin of the origin appears as a "powder-like" distribution scattered throughout the basins of the other attractors. This fragmented structure implies that trajectories are particularly susceptible to perturbations, as they can be dragged into the origin depending on minor alterations in initial conditions. This phenomenon represents the delicate balance between stability and chaos in the system.

When comparing the three basin plots in Fig. 13 (corresponding to the x_0–y_0, y_0–z_0, and z_0–x_0 planes), the same patterns are obvious, underlining the circulant symmetry of the system. Consistent geometric shapes across all planes demonstrate that the system's dynamics are invariant by cyclic permutations of variables.

4. Conclusion

In this study, we introduced a novel Three-Dimensional (3D)-chaotic flow characterized by circulant symmetry. The system exhibits rich and complex dynamical behavior, including dissipative dynamics over the studied parameter range. A detailed equilibrium analysis revealed that four of the fifteen fixed points within the system are stable.

Notably, the system supports two distinct sets of three coexisting chaotic attractors, each maintaining the system's underlying circulant symmetry, making it the first reported chaotic system with such properties to the best of our knowledge.

Bifurcation diagrams concerning all three system parameters illustrated the wide range of dynamical regimes, from periodic to chaotic behavior, which was further confirmed by computing the corresponding Lyapunov exponent spectra. The basins of attraction exhibited unusual V-shaped boundaries, highlighting the complexity of the system's multistability. Moreover, the circulant symmetry was visually evident in the basin structures when projected onto the three orthogonal planes.

These findings not only expand the current understanding of symmetric chaotic systems but also open new directions for exploring complex dynamics in systems with structured symmetry. Future work may include investigating potential applications in secure communication, random number generation, or synchronization phenomena based on this unique chaotic flow.

Author's Contributions

Karthikeyan Rajagopal: Conceptualization, Investigation, Writing — original draft,

Fahimeh Nazarimehr: Conceptualization, Methodology, Writing — original draft,

Sajad Jafari: Methodology, Validation, Writing — review and editing,

Julien C. Sprott: Software, Supervision, Writing — review and editing.

Conflict of Interest

The authors declare that they have no conflict of interest.

Data Availability

Data generated during this study will be made available on reasonable request.

ORCID

Karthikeyan Rajagopal
https://orcid.org/0000-0003-2993-7182
Fahimeh Nazarimehr
https://orcid.org/0000-0002-2664-9006
Sajad Jafari
https://orcid.org/0000-0002-6845-7539
Julien C. Sprott
https://orcid.org/0000-0001-7014-3283

References

Almatroud, O. A., Shukur, A. A., Pham, V.-T. & Grassi, G. [2024] "Oscillator with line of equilibiria and nonlinear function terms: Stability analysis, chaos, and application for secure communications," *Mathematics* **12**, 1874.

Banerjee, S. [2010] *Chaos Synchronization and Cryptography for Secure Communications: Applications for Encryption* (IGI Global).

Bob, P. [2007] "Chaos, brain and divided consciousness," *Acta Univ. Carol. Med. Monogr.* **153**, 9–80.

Danca, M.-F. [2017] "Hidden chaotic attractors in fractional-order systems," *Nonlin. Dyn.* **89**, 577–586.

Danca, M.-F. [2024] "Chaotic hidden attractor in a fractional order system modeling the interaction between dark matter and dark energy," *Commun. Nonlin. Sci. Numer. Simul.* **131**, 107838.

Khan, A., Li, C., Zhang, X. & Cen, X. [2025] "A two-memristor-based chaotic system with symmetric bifurcation and multistability," *Chaos Fract.* **2**, 1–7.

Kyurkchiev, N., Zaevski, T., Iliev, A., Kyurkchiev, V. & Rahnev, A. [2024] "Generating chaos in dynamical systems: Applications, symmetry results, and stimulating examples," *Symmetry* **16**, 938.

Lawnik, M., Moysis, L. & Volos, C. [2022] "Chaos-based cryptography: Text encryption using image algorithms," *Electronics* **11**, 3156.

Li, C. & Sprott, J. C. [2021] "Multi-stability in symmetric systems," in *Chaotic Systems with Multistability and Hidden Attractors* (Springer International Publishing), pp. 311–329.

Li, C., Sprott, J. C., Zhang, X., Chai, L. & Liu, Z. [2022] "Constructing conditional symmetry in symmetric chaotic systems," *Chaos Solit. Fract.* **155**, 111723.

Nazarimehr, F. & Sprott, J. C. [2020] "Investigating chaotic attractor of the simplest chaotic system with a line of equilibria," *Eur. Phys. J. Spec. Top.* **229**, 1289–1297.

Pan, G., Li, C., Liu, W., Xue, Y. & Qi, X. [2025] "Countless coexisting chaotic attractors: From system construction to FPGA-based observation," *Chaos Solit. Fract.* **198**, 116610.

Panahi, S., Shirzadian, T., Jalili, M. & Jafari, S. [2019] "A new chaotic network model for epilepsy," *Appl. Math. Comput.* **346**, 395–407.

Panahi, S., Nazarimehr, F., Jafari, S., Sprott, J. C., Perc, M. & Repnik, R. [2021] "Optimal synchronization of circulant and non-circulant oscillators," *Appl. Math. Comput.* **394**, 125830.

Rajagopal, K., Akgul, A., Pham, V.-T., Alsaadi, F. E., Nazarimehr, F., Alsaadi, F. E. & Jafari, S. [2019] "Multistability and coexisting attractors in a new circulant chaotic system," *Int. J. Bifurcation and Chaos* **29**, 1950174.

Raza, N., Amjad, Z., Rani, B., Almalki, Y. & Bayram, M. [2025] "Exploring chaos, multistability, and interaction patterns in (3 + 1)-dimensional KdV-BBM model," *Nonlin. Dyn.* **113**, 11969–11986.

Rössler, O. [1976] "An equation for continuous chaos," *Phys. Lett. A* **57**, 397–398.

Shukur, A. A., Neamah, A. A., Pham, V.-T. & Grassi, G. [2025] "A novel chaotic system with one absolute term: Stability, ultimate boundedness, and image encryption," *Heliyon* **11**, e37239.

Sprott, J. C. [2010] *Elegant Chaos: Algebraically Simple Chaotic Flows* (World Scientific).

Tutueva, A., Nepomuceno, E. G., Moysis, L., Volos, C. & Butusov, D. [2022] "Adaptive chaotic maps in cryptography applications," in *Cybersecurity* (Springer International Publishing), pp. 193–205.

Wang, X. & Chen, G. [2012] "A chaotic system with only one stable equilibrium," *Commun. Nonlin. Sci. Numer. Simul.* **17**, 1264–1272.

Wang, N., Cui, M., Yu, X., Shan, Y. & Xu, Q. [2024] "Generation of no-equilibrium multi-fold chaotic attractor for image processing and security," *Appl. Math. Model.* **133**, 271–285.

Wei, Z. [2011] "Dynamical behaviors of a chaotic system with no equilibria," *Phys. Lett. A* **376**, 102–108.

Wolf, A., Swift, J. B., Swinney, H. L. & Vastano, J. A. [1985] "Determining Lyapunov exponents from a time series," *Physica D* **16**, 285–317.

Xu, Q., Tan, X., Zhu, D., Bao, H., Hu, Y. & Bao, B. [2020] "Bifurcations to bursting and spiking in the Chay neuron and their validation in a digital circuit," *Chaos Solit. Fract.* **141**, 110353.

Xu, Q., Wang, K., Chen, M., Parastesh, F. & Wang, N. [2024a] "Bursting and spiking activities in a Wilson neuron circuit with memristive sodium and potassium ion channels," *Chaos Solit. Fract.* **181**, 114654.

Xu, Q., Wang, K., Shan, Y., Wu, H., Chen, M. & Wang, N. [2024b] "Dynamical effects of memristive electromagnetic induction on a 2D Wilson neuron model," *Cogn. Neurodyn.* **18**, 645–657.

Zhang, X., Li, C., Lei, T., Iu, H. H.-C. & Kapitaniak, T. [2025] "Coexisting hyperchaos in a memristive neuromorphic oscillator," *IEEE Trans. Comput.-Aided Des. Integr. Circuits Syst.* **44**, 3179–3188.

https://doi.org/10.1142/9789819830633_0005

Chapter 5

Period Constants and Hopf Bifurcation#

Cinthya A. Bares* and Franco S. Gentile
Departamento de Matemática,
Universidad Nacional del Sur,
Instituto de Investigaciones en Ingeniería
Eléctrica – IIIE (UNS – CONICET),
B8000CPB Bahía Blanca, Argentina
cinthya.bares@uns.edu.ar

Guillermo L. Calandrini and Jorge L. Moiola
Departamento de Ingeniería Eléctrica y de Computadoras,
Universidad Nacional del Sur,
Instituto de Investigaciones en Ingeniería
Eléctrica – IIIE (UNS – CONICET),
B8000CPB Bahía Blanca, Argentina

The goal of this paper is to highlight the importance of certain coefficients that give a correction in the frequency of periodic solutions arising from Hopf bifurcation in dynamical systems. Those frequency coefficients are closely related to the so-called period constants found previously in the literature. The vanishing of the frequency coefficients aims at the detection and analysis of isochronous periodic solutions. The first frequency coefficients are computed for several typical examples that are well-known in the context of dynamical systems having an isochronous center. An example of a polycyclic configuration with a critical point in the period and in the amplitude is presented for illustration.

Keywords: Frequency coefficient; frequency-domain approach; isochronous periodic solution; Hopf bifurcation.

1. Introduction

The study of oscillations via methods related to feedback systems and harmonic balance has a very large tradition (see [Mees, 1981; Tesi *et al.*, 1996] and the references therein). The amplitude of oscillations under the variation of a distinguished parameter can also be obtained through the application of the so-called Graphical Hopf Bifurcation Theorem (GHBT, for brevity) in the frequency domain [Mees & Chua, 1979]. That method is capable of approximating the periodic solution branch, close to the Hopf bifurcation, using different higher-order harmonic balance formulas [Mees, 1981; Moiola & Chen, 1996].

Although the approximation is local as well as others provided by equivalent methods [Kuznetsov, 2004], it has a useful graphical interpretation for engineers and scientists who are familiar with Nyquist stability criterion. Furthermore, the local generation of periodic solutions under the variation of a single parameter has been extended to cover

#To cite this article, please refer to its earlier version published in the *International Journal of Bifurcation and Chaos*, Vol. 35, No. 14 (2025) 2540005 (15 pages), DOI: 10.1142/S021812742540005X.
*Author for correspondence

other singularities or degeneracies by considering the normal forms of multiple (unfoldings) parameters [Golubitsky & Langford, 1981].

One kind of degenerate Hopf bifurcation is when the first Lyapunov or focal value vanishes, and then enables the appearance of up to two cycles in the surroundings by appropriate perturbations of two parameters. This phenomenon is typical of a "weak" focus type concerning the variation of the amplitude of the limit cycle close to the singularity. A strong technical literature starting from [Bautin, 1954; Sibirskii, 1965] gives the expressions of the first focal values (see [Farr *et al.*, 1989] for explicit formulas) in terms of the system parameters.

In a similar context as the amplitude of the oscillation can be studied, the period of the limit cycle is also a very interesting issue. Again, it has a long history, especially when the center is isochronous, i.e. when all the periodic solutions have the same period. As in the dual context with the multiplicity of limit cycles, the period constants can be defined in order to evaluate the complexity of the center. In this regard, the period constants also have expressions in terms of system parameters. A significant contribution of the period constants has been given in [Chicone & Jacobs, 1989] related to Bautin system, and a special simplification on it, called dehomogenized Loud system. Following this landmark study for determining the number of critical points of the period functions in terms of system parameters for quadratic systems, several contributions have focused on extending it to third-degree polynomial as in [Yu & Han, 2009] and other higher-order degrees (see [Sánchez-Sánchez & Torregrosa, 2021], and the references therein). Similarly, other articles cover the period constants in order to study isochronous systems as in [Algaba *et al.*, 2000; Algaba & Reyes, 2009; Cima *et al.*, 1997; Gasull *et al.*, 1997; Mardešić *et al.*, 2006; Rousseau & Toni, 1993, 1997], to mention only a few.

Starting from the contribution of [Mees & Chua, 1979], this paper shows how the first period constant can be computed, improving the features of the method regarding the variation of the frequency. Another interesting feature is presented: an example of a polycyclic configuration with a critical point in the period is shown, enhancing the engineering view of the oscillatory systems.

The paper is organized as follows. Section 2 presents a brief about the computation of the second-order harmonic balance solution giving the first curvature coefficient index and the first period constant. In Sec. 3, several classical examples are expounded, remarking the fact of the eigenloci in the frequency domain. Finally, Sec. 4 gives some conclusions.

2. The Frequency-Domain Approach

To introduce briefly the frequency-domain approach, some background materials are provided first. Consider a general parametric n-dimensional nonlinear system:

$$\dot{\mathbf{x}}(t) = \mathbf{f}(\mathbf{x}(t); \mu), \quad \mathbf{x}(0) = \mathbf{0}, \tag{1}$$

where $\mathbf{x} \in \mathbb{R}^n$, μ is a real parameter and $\mathbf{f}\colon \mathbb{R}^n \times \mathbb{R} \to \mathbb{R}^n$ is a (generally nonlinear) function. System (1) can be reformulated in a state-variable form:

$$\begin{cases} \dot{\mathbf{x}}(t) = A(\mu)\mathbf{x}(t) + B(\mu)\mathbf{g}(\mathbf{e}(t); \mu), \\ \mathbf{e}(t) = -C(\mu)\mathbf{x}(t), \end{cases} \tag{2}$$

where $A(\mu)$ is an $n \times n$ constant matrix which can be arbitrarily chosen for convenience (such as nonsingular and stable), $B(\mu)$ and $C(\mu)$ are $n \times p$ and $m \times n$ constant matrices and $\mathbf{g}(\cdot)$ is a $p \times 1$ nonlinear function belonging to C^4. It is convenient to choose $A(\mu)$ in such a way that it is invertible for all values of μ. Formulation (2) is standard in the framework of feedback systems, where the function $\mathbf{g}$ represents a controller, so, it depends on the error signal $\mathbf{e}$, which is the difference between the desired behavior or reference input $\mathbf{r}$ and the output $\mathbf{y} = C(\mu)\mathbf{x}(t)$. For an autonomous system, the external reference is zero ($\mathbf{r} = \mathbf{0}$), the controller depends only on $\mathbf{e} = -\mathbf{y}$. System (2) represents a typical control setting with a linear plant described by the transfer matrix

$$G(s) = C(sI_n - A)^{-1}B, \tag{3}$$

with a nonlinear feedback given by $\mathbf{g}(\mathbf{e}(t); \mu)$. The $n \times n$ identity matrix is denoted as I_n. Here and from now on, the dependence of A, B, C and G on μ is omitted for short. If $\hat{\mathbf{x}}$ is an equilibrium solution of system (1), then $G(0)g(\hat{\mathbf{e}}; \mu) = -\hat{\mathbf{e}}$ is the corresponding equilibrium solution of the feedback system.

2.1. *Stability analysis: The Nyquist criterion*

The stability of the equilibrium solution(s) can be studied through linearizing the feedback system about $\hat{\mathbf{e}}$. Then, one obtains a linear loop, where the associated Jacobian matrix is given by

$$J = \left. \frac{\partial \mathbf{g}}{\partial \mathbf{e}} \right|_{\mathbf{e}=\hat{\mathbf{e}}}. \tag{4}$$

MacFarlane and Postlethwaite [1977] introduced the *generalized* Nyquist stability criterion, which is fundamental to the development of the frequency-domain approach. Recall that $G(s)$ and J are $m \times p$ and $p \times m$ matrices, respectively. The eigenvalues of $G(s)J$ satisfy the characteristic equation

$$|\lambda I_m - G(s)J| = \lambda^{m-k}[\lambda^k + a_{k-1}(s;\mu)\lambda^{k-1} + \cdots + a_0(s;\mu)] = 0, \tag{5}$$

where $k = \min(m,p)$ and $a_0, \ldots, a_{k-1}$ are rational functions in s (note that $\text{rank}(GJ) = k$). Equation (5) has k nonzero solutions $\lambda_1(s), \ldots, \lambda_k(s)$, and the stability of the equilibrium point can change according to the behavior of those solutions when the parameter μ varies. Fix $\mu = \tilde{\mu}$ and let $s = i\omega, -\infty < \omega < \infty$, as stated by the Nyquist stability criterion. Then, one obtains k curves on the complex plane determined by the functions $\lambda_1(\omega), \ldots, \lambda_k(\omega)$, known as eigenloci. A critical stability condition is reached if, for a given value of $\mu = \mu_0$, one of those eigenloci passes exactly through the point -1. For the moment, assume that there is only one solution of (5), denoted by $\hat{\lambda}(i\omega)$ which satisfies $\hat{\lambda}(i\omega_0) = -1$ for a given value of the parameter $\mu = \mu_0$. Here and in the rest of the paper, the direct analysis of the nonlinear Eq. (1) is referred to as *time domain* setting, and the corresponding study of the problem via the GHBT is considered as the *frequency domain* counterpart. Then, the following lemma establishes the correspondence between the critical stability conditions in the time domain and in the frequency domain:

Lemma 1. *If an eigenvalue of the corresponding Jacobian of the nonlinear system (1), in the time domain, assumes a purely imaginary value $i\omega_0$ at a particular value $\mu = \mu_0$, then the corresponding eigenvalue of the constant matrix $G(i\omega_0)J$ in the frequency domain assumes the value -1 at $\mu = \mu_0$.*

Remark 2.1. If a given system can be described in the form (2) with matrices B and C of dimensions $n \times 1$ and $1 \times n$, respectively, then the feedback system is called SISO (Single Input–Single Output), and the transfer function (3) is scalar. If B is $n \times 1$ or C is $1 \times n$, still $k = \min(m,p) = 1$ in (5). Thus, a unique characteristic function describes the stability and bifurcations of the equilibrium point, in contrast with the n eigenvalues of the classical linearization of (1). Systems studied in Sec. 3.1 below have this important feature, greatly simplifying the calculations.

2.2. *Approximating the periodic solution*

The GHBT establishes sufficient conditions for the appearance of a periodic solution branch related to this stability change of $\hat{\mathbf{e}}$. The basic scheme allows an approximation of the periodic solution with two harmonics; then, one needs to obtain both amplitude and frequency estimates. They can be found from the intersection between the loci of the characteristic function ($\hat{\lambda}$) and the one associated with $-1 + \xi_1\theta^2$ in the complex plane; i.e. the procedure has a graphical interpretation evocative of the describing function method in nonlinear system design [Gelb & Vander Velde, 1968]. Here, $\xi_1 \in \mathbb{C}$ is an auxiliary quantity that will be defined in short and $\theta \in \mathbb{R}^+$ is a measure of the amplitude of the sought periodic solution. In other words, one needs to solve

$$\hat{\lambda}(i\omega) = -1 + \xi_1(\omega)\theta^2, \tag{6}$$

for θ and ω, where the quantity ξ_1 is given by

$$\xi_1(\omega) = -\frac{\mathbf{u}^T G(i\omega)\mathbf{p}_1(i\omega)}{\mathbf{u}^T\mathbf{v}}, \tag{7}$$

where $\mathbf{u}$ and $\mathbf{v}$ are the left and right eigenvectors, respectively, of matrix $G(i\omega)J$ associated with the eigenvalue $\hat{\lambda}(i\omega)$, $(\cdot)^T$ is the conjugate transpose, and

$$\mathbf{p}_1(i\omega) = D_2\,\mathbf{v} \otimes \mathbf{V}_{02} + \frac{1}{2}D_2\,\bar{\mathbf{v}} \otimes \mathbf{V}_{22} + \frac{1}{8}D_3\,\mathbf{v} \otimes \mathbf{v} \otimes \bar{\mathbf{v}}, \tag{8}$$

where $\overline{(\cdot)}$ represents the complex conjugate and $\otimes$ is the tensor product operator. Also

$$D_2 = \left. \frac{\partial^2 \mathbf{g}}{\partial \mathbf{e}^2} \right|_{\mathbf{e}=\hat{\mathbf{e}}}, \quad D_3 = \left. \frac{\partial^3 \mathbf{g}}{\partial \mathbf{e}^3} \right|_{\mathbf{e}=\hat{\mathbf{e}}},$$

are the (matrix) high-order derivatives of $\mathbf{g}$ and

$$\begin{aligned} \mathbf{V}_{02} &= -\frac{1}{4}H(0)D_2\mathbf{v} \otimes \bar{\mathbf{v}}, \\ \mathbf{V}_{22} &= -\frac{1}{4}H(2i\omega)D_2\mathbf{v} \otimes \mathbf{v}, \end{aligned} \tag{9}$$

are the zeroth- and second-order harmonics of the Fourier representation, with $H(s) = [I_m + G(s)J]^{-1}G(s)$ being the closed-loop transfer matrix of the linearized system. Note that vectors $\mathbf{V}_{02}$ and $\mathbf{V}_{22}$ can be calculated without knowing θ. Moreover, their computation can be further simplified by introducing the $p \times m$ auxiliary matrices

$$Q = \{Q_{jk}\} = \sum_{\ell=1}^{m} g^j_{\ell k} v_\ell, \quad 1 \le j \le p, \; 1 \le k \le m, \tag{10a}$$

$$L = \{L_{jk}\} = \sum_{\ell=1}^{m}\sum_{q=1}^{m} g^j_{\ell qk} v_\ell v_q, \quad 1 \le j \le p, \; 1 \le k \le m, \tag{10b}$$

where

$$g^j_{\ell k} = \left.\frac{\partial^2 g_j}{\partial e_\ell \, \partial e_k}\right|_{\hat{\mathbf{e}}}, \quad g^j_{\ell qk} = \left.\frac{\partial^3 g_j}{\partial e_\ell \, \partial e_q \, \partial e_k}\right|_{\hat{\mathbf{e}}},$$

and v_i is the ith component of $\mathbf{v}$, and the same applies for $\mathbf{g}$ and $\mathbf{e}$. By using (10), one obtains the alternative expressions

$$\begin{aligned} \mathbf{V}_{02} &= -\frac{1}{4}H(0)Q\bar{\mathbf{v}}, \\ \mathbf{V}_{22} &= -\frac{1}{4}H(i2\omega)Q\mathbf{v}. \end{aligned} \tag{11}$$

Also, $\mathbf{p}_1$ can be written as

$$\mathbf{p}_1(\mathbf{v}) = Q\mathbf{V}_{02} + \frac{1}{2}\bar{Q}\mathbf{V}_{22} + \frac{1}{8}L\bar{\mathbf{v}}. \tag{12}$$

In short, once $\hat{\lambda}(s)$ is known, one can compute $\mathbf{v}$, then obtain the expressions (11) and (12), and finally calculate $\mathbf{p}_1(\omega)$ (see [Mees & Chua, 1979] for more details).

If one needs to build up the periodic solution, its second-order approximation reads

$$\mathbf{e}(t) \simeq \Re\left\{\sum_{k=0}^{2} \mathbf{E}^k \exp(ik\omega t)\right\},$$

where $\mathbf{E}^0 = \theta^2\mathbf{V}_{02}$, $\mathbf{E}^1 = \theta\mathbf{v}$, $\mathbf{E}^2 = \theta^2\mathbf{V}_{22}$, where θ and ω form the solution pair of (6), and $\Re(\cdot)$ stands for the real part.

In addition, more accurate approximations, that include more harmonics, can be built up [Mees, 1981; Moiola & Chen, 1996]. It is important to point out that the GHBT has local validity, i.e. the obtained results become less accurate as the parameter value departs from the critical bifurcation value, worsening the predicted results.

2.3. *Curvature and frequency coefficients*

When only a second-order harmonic balance is used, the GHBT provides explicit formulas for the first curvature coefficient, used to predict the stability of the emerging periodic solution, and also for the first *frequency coefficient*, giving a correction of the oscillation frequency. Suppose that for a given value of the bifurcation parameter $\mu = \mu_0$, matrix $G(s)J$ has an eigenvalue assuming the value -1 with $s = i\omega_0$, and that when μ increases from μ_0, s moves to the right-half plane, taking the value $s = \alpha + i\omega$. If $\hat{\omega}$ is the frequency of the periodic solution, then [Mees & Chua, 1979] have shown that the following equation holds:

$$\alpha - i\delta\omega = \gamma_1\theta^2 + \mathcal{O}(\theta^3), \tag{13}$$

where $\delta\omega = \hat{\omega} - \omega_0$ and γ_1 is now defined as

$$\gamma_1 = \frac{\mathbf{u}^T G(i\omega_0)\mathbf{p}_1(i\omega_0)}{\mathbf{u}^T G'(i\omega_0)J\mathbf{v}}, \tag{14}$$

where $G'(i\omega_0) = \frac{dG}{ds}|_{s=i\omega_0}$. In (13), the real part is related to the stability of the emerging limit cycle, and then the first curvature coefficient yields

$$\sigma_1 = -\Re(\gamma_1).$$

Supposing that the equilibrium changes from stable to unstable as μ increases from μ_0, if σ_1 is negative (positive), then the emerging orbit is stable (unstable). On the other hand, by taking the imaginary part (denoted as $\Im(\cdot)$) of the right-hand side in (13), it follows that the frequency of the periodic orbit is $\hat{\omega} = \omega_0 - \Im(\gamma_1)\theta^2$, or equivalently $\hat{\omega} = \omega_0 + \phi_2\theta^2 + \mathcal{O}(\theta^4)$, where

$$\phi_2 = -\Im(\gamma_1), \tag{15}$$

and it is defined as the *first frequency coefficient.*

2.4. *Frequency coefficients and period constants*

Let p be the period of the oscillating solution. The period constants $p_0, p_2, p_4, \ldots$ are the coefficients in the power series

$$p = p_0 + p_2\theta^2 + p_4\theta^4 + p_6\theta^6 + \cdots,$$

and analogously, the frequency coefficients $\phi_2, \phi_4, \phi_6, \ldots$ are used in the power series for the frequency as

$$\omega = \omega_0 + \phi_2\theta^2 + \phi_4\theta^4 + \phi_6\theta^6 + \cdots,$$

where the relationship between p and ω is $p = \frac{2\pi}{\omega}$, or equivalently $2\pi = p\omega$. Thus, by replacing their power series and equating powers in θ gives

$$\begin{aligned} 2\pi &= (p_0 + p_2\theta^2 + p_4\theta^4 + p_6\theta^6 + \cdots) \\ &\quad \times (\omega_0 + \phi_2\theta^2 + \phi_4\theta^4 + \phi_6\theta^6 + \cdots) \\ &= p_0\omega_0 + (p_0\phi_2 + p_2\omega_0)\theta^2 \\ &\quad + (p_0\phi_4 + p_2\phi_2 + p_4\omega_0)\theta^4 \\ &\quad + (p_0\phi_6 + p_2\phi_4 + p_4\phi_2 + p_6\omega_0)\theta^6 + \cdots \end{aligned}$$

resulting in the equations

$$\begin{aligned} 2\pi &= p_0\omega_0, \\ 0 &= p_0\phi_2 + p_2\omega_0, \\ 0 &= p_0\phi_4 + p_2\phi_2 + p_4\omega_0, \\ 0 &= p_0\phi_6 + p_2\phi_4 + p_4\phi_2 + p_6\omega_0, \end{aligned}$$

thus one obtains

$$\begin{aligned} p_0 &= \frac{2\pi}{\omega_0}, \\ p_2 &= -2\pi\frac{\phi_2}{\omega_0^2}, \\ p_4 &= 2\pi\frac{\phi_2^2 - \phi_4\omega_0}{\omega_0^3}, \\ p_6 &= -2\pi\frac{\phi_2^3 - 2\phi_2\phi_4\omega_0 + \phi_6\omega_0^2}{\omega_0^4}, \\ &\vdots \end{aligned}$$

In the search for an isochronous periodic solution, the conditions $p_2 = p_4 = p_6 = \cdots = 0$ are imposed, so the frequency remains constant and independent of the amplitude of the limit cycle. Note that the vanishing of the first frequency coefficient ϕ_2 is equivalent to the vanishing of the first period constant p_2. Also, under the constraint $\phi_2 = 0$, the vanishing of ϕ_4 also gives $p_4 = 0$, and so on. However, in order to obtain $\phi_4, \phi_6, \ldots$, higher-order expansions should be used in (13).

3. Examples

3.1. *Systems with a single characteristic function*

3.1.1. *Example 1: One nonlinear function and one observable variable*

Consider the system introduced in [Cima *et al.*, 1997] given by

$$\begin{cases} \dot{x} = -y + p(x), \\ \dot{y} = x, \end{cases} \tag{16}$$

where $p(x) = a_2x^2 + a_3x^3 + a_4x^4 + a_5x^5 + a_6x^6 + \cdots$ and a_i, $i = 2, 3, \ldots, 6$ are system parameters. Note that this example has a very simple feedback representation, since there is a single nonlinearity and it depends only on the x state variable. It follows that $G(s)J = -\frac{s}{s^2+s+1}$ is scalar and the Hopf bifurcation condition reads $G(i\omega_0)J = -1$, giving the solution $\omega_0 = 1$. After some calculations given in Appendix A, it follows that $\gamma_1 = \frac{1}{2}p_1$, so $\sigma_1 = \frac{3}{8}a_3$ is the first curvature coefficient. On the other hand, the first frequency coefficient is $\phi_2 = -\frac{1}{6}a_2^2$, and then $p_2 = \frac{\pi}{3}a_2^2$.

The structure of this example makes it relatively simple to produce situations with multiple oscillations in the phase space, for proper combinations of parameters a_i, $i = 2, 3, \ldots$. Thus, this example will be explored with some detail. For simplicity, let $a_6 = a_7 = a_8 = \cdots = 0$, and consider a slight modification of (16) by introducing a new bifurcation parameter μ, thus the resulting system is

$$\begin{cases} \dot{x} = \mu x - y + p(x), \\ \dot{y} = x + \mu y. \end{cases}$$

Note that μ represents the real part of the eigenvalues, now producing a Hopf bifurcation when μ goes from negative to positive values. In the following, two different sets of values will be considered for a_2 to a_5 and the period and amplitude of the emerging limit cycles will be computed as long as μ varies. It is easy to determine a configuration of two nested limit cycles (the inner cycle is stable while the outer is unstable) by setting conveniently the values of a_3 and a_5 in a type of *weak focus* singularity. More importantly, Fig. 1 shows the increasing value of the period function for *case b* (red curve) in contrast to the increasing and then decreasing period for *case a* (blue curve).

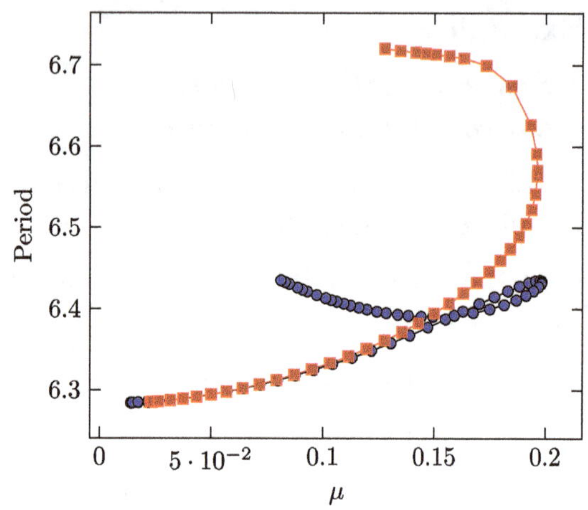

Fig. 1. Period versus parameter μ, for $a_3 = -0.30$, $a_5 = 0.05$. Case (a) (blue): $a_2 = 0$, $a_4 = 0$, and case (b) (red): $a_2 = -0.01$, $a_4 = 0.05$.

On the other hand, Fig. 2 shows the typical structure of nested cycles for a certain bifurcation parameter μ and the appearance of a fold point, which is characteristic of this type of singularity.

Returning to the geometrical locus of $\lambda = G(i\omega)J = -\frac{s}{s^2+s+1}$, a SISO system, this eigenvalue can be seen in Fig. 3(a) in order to compare it with the following examples.

3.1.2. *Example 2: One nonlinear function and one observable variable*

Consider the system introduced in [Algaba & Reyes, 2009] given by

$$\begin{cases} \dot{x} = -y, \\ \dot{y} = x + a_2 y^2 + a_4 y^4 + a_6 y^6, \end{cases} \tag{17}$$

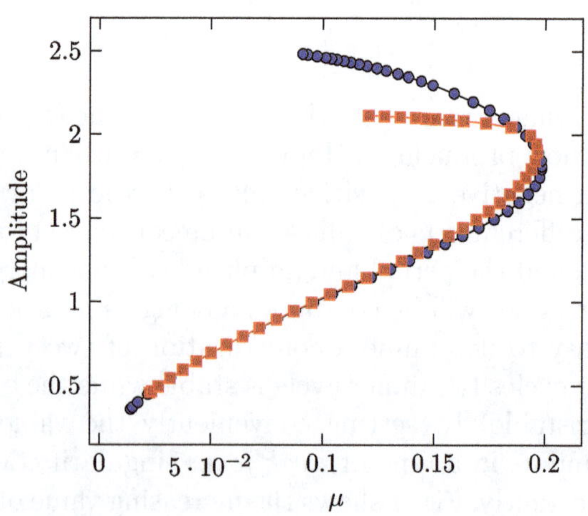

Fig. 2. Amplitude versus parameter μ, for $a_3 = -0.30$, $a_5 = 0.05$. Case (a) (blue): $a_2 = 0$, $a_4 = 0$, and case (b) (red): $a_2 = -0.01$, $a_4 = 0.05$.

where a_{2k}, $k = 1, 2$ and 3 are system parameters. The linear loop gain is $G(s)J = -\frac{s}{s^2+s+1}$ (scalar) and the Hopf bifurcation condition $G(i\omega_0)J = -1$, gives again $\omega_0 = 1$. The geometrical locus of $\lambda = G(i\omega)J$ is the one shown in Fig. 3(a). One can obtain $\gamma_1 = \frac{1}{6}a_2^2 i$. As $\Re(\gamma_1) = 0$, i.e. the first stability coefficient vanishes at the origin, while the first frequency coefficient is $\phi_2 = -\frac{1}{6}a_2^2$, and $p_2 = \frac{\pi}{3}a_2^2$. Note that these results are completely equivalent to those found in Example 1, setting $a_3 = a_5 = 0$ in (16). It occurs since (17) is a rotated version of (16), in other words, (17) can be obtained from (16) by the transformation $(x', y') = (-y, x)$.

3.1.3. *Example 3: Two nonlinear functions and one observable variable*

Consider the system introduced in [Gasull *et al.*, 1997] given by

$$\begin{cases} \dot{x} = -y + ax^2 + bx^3, \\ \dot{y} = x + dx^2 + gx^3 + hx^4, \end{cases} \tag{18}$$

where a, b, d, g and h are system parameters, and the trivial equilibrium point is considered. As shown in Appendix A, the characteristic function in the

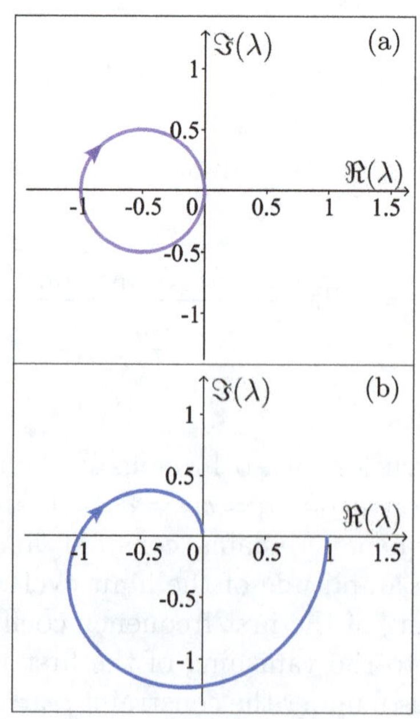

Fig. 3. Geometrical loci of the characteristic function(s) for all examples in Sec. 3.1. (a) Examples 1 and 2 and (b) Examples 3 and 4.

frequency domain is given by

$$\lambda(s) = G(s)J = -\frac{\left(s - \frac{1}{2}\right)}{s^2 + s + \frac{1}{2}}, \tag{19}$$

and the corresponding Nyquist diagram can be seen in Fig. 3(b). The Hopf bifurcation condition $\lambda(i\omega_0) = -1$ gives the solution $\omega_0 = 1$. Following the procedure described in Sec. 2 (see Appendix A for details), the expression γ_1 is found to be

$$\gamma_1 = -\frac{1}{24}[-6ad + 9b - i(4a^2 + 10d^2 - 9g)].$$

Thus, the first curvature coefficient $\sigma_1 = -\Re(\gamma_1)$ vanishes under the condition $2ad = 3b$. On the other hand, the first coefficient in the frequency expansion is $\phi_2 = -\Im(\gamma_1) = (-4a^2 - 10d^2 + 9g)/24$, then the first period constant is $p_2 = \pi(4a^2 + 10d^2 - 9g)/(12\omega_0^2)$ and it vanishes if $9g = 4a^2 + 10d^2$. These results are equivalent to those found in [Gasull *et al.*, 1997; Algaba *et al.*, 2000]. Very recently, a rather similar example has been fully analyzed in [Oliveira *et al.*, 2022] considering bi-weakness of the stability focal indexes and period constants.

3.1.4. *Example 4: One nonlinear function and two observable variables, a reduced Kukles system*

Consider the system introduced in [Rousseau & Toni, 1997] given by

$$\begin{cases} \dot{x} = -y, \\ \dot{y} = x + a_1x^2 + a_2xy + a_3y^2 + a_4x^3 \\ \qquad + a_5x^2y + a_6xy^2, \end{cases} \tag{20}$$

where a_i, $i = 1, \ldots, 6$ are system parameters. Again, following the procedure in Sec. 2, it is found that the characteristic equation in the frequency domain reads

$$h(s;\lambda) = \det(\lambda I_2 - G(s)J) = \lambda^2 + \frac{s - \frac{1}{2}}{\Delta(s)}\lambda = 0,$$

where $\Delta(s) = s^2 + s + 1/2$. The characteristic functions are $\lambda_1(s) \equiv 0$ and $\lambda_2(s) = (1/2 - s)/\Delta(s)$. Thus, a Hopf bifurcation can only occur if $\lambda_2(i\omega_0) = -1$, leading to the solution $\omega_0 = 1$. The corresponding Nyquist diagram can be seen in Fig. 3(b). As can be found in Appendix A, one then obtains the coefficient $\gamma_1 = \frac{i\sqrt{2}p_1}{2}$, thus, the first curvature coefficient is

$$\sigma_1 = -\frac{1}{16}[(a_1 + a_3)a_2 - a_5],$$

and the first coefficient in the frequency expansion is $\phi_2 = -\Im(\gamma_1)$, which in turn leads to

$$p_2 = \frac{\pi}{24}[10a_1^2 + a_2^2 + 4a_3^2 + 10a_1a_3 - 9a_4 - 3a_6].$$

3.2. *General cases*

3.2.1. *Example 5: Two nonlinear functions plus two observable variables*

Consider the system introduced in [Cima *et al.*, 1997] given by

$$\begin{cases} \dot{x} = -y + \alpha x^2 + \beta x^3, \\ \dot{y} = x + \gamma y^2 + \delta y^3, \end{cases} \tag{21}$$

where α, β, γ and δ are system parameters. The stability of the equilibrium point at the origin is analyzed via the eigenvalues of $G(s)J$ given by

$$\lambda_{1,2}(s) = \frac{-s - 1 \pm i}{s^2 + 2s + 2} = \frac{-1}{s + 1 \pm i}.$$

The Hopf bifurcation condition holds with $\lambda_2(i) = -1$, i.e. the solution frequency is $\omega_0 = 1$. The Nyquist diagrams for $\lambda_1(i\omega)$ and $\lambda_2(i\omega)$ are shown in Fig. 4(a). Note that a portion of the curve $\lambda_2(i\omega)$ completely overlaps the curve of $\lambda_1(i\omega)$ (it is clear that $\lim_{\omega\to\infty} \lambda_2(i\omega) = 0$). It can be obtained that

$$\gamma_1 = \frac{1}{48}[-9(\beta + \delta) + 4(\alpha^2 + \gamma^2)i].$$

Thus, the first curvature and frequency coefficients are $\sigma_1 = \frac{3}{16}(\beta + \delta)$ and $\phi_2 = -\frac{1}{12}(\alpha^2 + \gamma^2)$, respectively. Then, $p_2 = \frac{\pi}{6}(\alpha^2 + \gamma^2)$.

3.2.2. *Example 6: Bautin system*

Consider the original Bautin system [Bautin, 1954; Chicone & Jacobs, 1989] given by

$$\begin{cases} \dot{x} = \lambda_1 x - y - \lambda_3 x^2 + (2\lambda_2 + \lambda_5)xy + \lambda_6 y^2, \\ \dot{y} = x + \lambda_1 y + \lambda_2 x^2 + (2\lambda_3 + \lambda_4)xy - 2\lambda_2 y^2, \end{cases} \tag{22}$$

where λ_i, $i = 1, \ldots, 6$ are system parameters. The system comprises two nonlinear functions and two

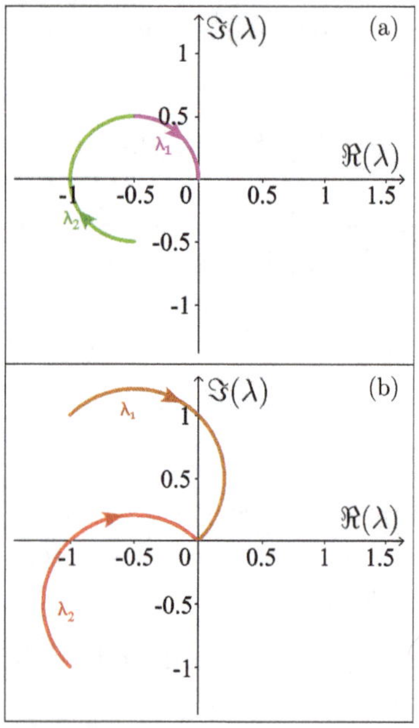

Fig. 4. Geometrical loci of the characteristic function(s) for all examples in Sec. 3.2. (a) Example 5 and (b) Example 6 (with parameter $\lambda_1 = 0$), Examples 7–9 (with parameter $\alpha = 0$).

observable variables, so it is very general. For this example, the eigenvalues of $G(s)J$ are

$$\lambda_{a,b}(s) = \frac{-1-\lambda_1 \pm i}{s+1}. \tag{23}$$

The Hopf bifurcation condition $\lambda_b(i\omega_0) = -1$ is satisfied with $\lambda_1 = 0$ and $\omega_0 = 1$. The characteristic loci of $\lambda_a(i\omega)$ and $\lambda_b(i\omega)$ are shown in Fig. 4(b) for the value of the parameter $\lambda_1 = 0$ (they are indicated as $\lambda_1(i\omega)$ and $\lambda_2(i\omega)$ in the diagram). As can be seen in Appendix A, the expression for γ_1 is quite complicated, and still the first curvature coefficient is found to be

$$\sigma_1 = -\Re(\gamma_1) = \frac{1}{16}\lambda_5(\lambda_6 - \lambda_3),$$

and the first frequency coefficient is

$$\phi_2 = -\frac{1}{48}(9\lambda_3\lambda_4 - 12\lambda_3\lambda_6 - \lambda_4\lambda_6 + 16\lambda_2^2 + \lambda_4^2 + 18\lambda_3^2 + \lambda_5^2 + 10\lambda_6^2 + 8\lambda_2\lambda_5),$$

thus $p_2 = -2\pi\phi_2$, since $\omega_0 = 1$.

3.2.3. *Example 7: Loud system*

Consider the Loud system [Chicone & Jacobs, 1989] given by

$$\begin{cases} \dot{x} = -y + B_a xy, \\ \dot{y} = x + Dx^2 + Fy^2, \end{cases} \tag{24}$$

where B_a, D and F are system parameters. For this example (see details in Appendix A), the stability of the trivial equilibrium point is determined by the eigenvalues of $G(s)J$, given by

$$\lambda_{1,2}(s) = \frac{-1 \pm i}{s+1},$$

and the Hopf bifurcation condition can be obtained via $\lambda_2(i\omega_0) = -1$, giving the solution $\omega_0 = 1$. Note that the characteristic functions coincide with those of the previous example with the parameter value $\lambda_1 = 0$ in (23), thus the Nyquist diagrams also correspond to Fig. 4(b). Following the procedure as shown in Appendix A, one obtains

$$\gamma_1 = \frac{i}{48}(1 - D - 5F + 10D^2 + 10DF + 4F^2),$$

and note that, since γ_1 is purely imaginary, the first curvature coefficient σ_1 is identically zero for all parameter values. The first coefficient in the frequency expansion is $\phi_2 = -\Im(\gamma_1)$ and then, p_2 yields

$$p_2 = \frac{\pi}{24}(1 - D - 5F + 10D^2 + 10DF + 4F^2).$$

3.2.4. *Example 8: Sibirskii system*

Consider the system introduced in [Sibirskii, 1965; Rousseau & Toni, 1993] given by

$$\begin{cases} \dot{x} = -y + (a - \omega_s - \theta_s)x^3 + (3\mu - \eta)x^2y \\ \qquad + (3\theta_s + \xi_s - 3\omega_s - 2a)xy^2 + (\nu - \mu)y^3, \\ \dot{y} = x + (\nu + \mu)x^3 + (3\omega_s + 3\theta_s + 2a)x^2y \\ \qquad + (\eta - 3\mu)xy^2 + (\omega_s - \theta_s - a)y^3, \end{cases} \tag{25}$$

where a, ν, μ, η, θ_s, ω_s and ξ_s are system parameters. It follows that the eigenvalues of $G(s)J$, considering the trivial equilibrium point, are

$$\lambda_{1,2}(s) = \frac{-1 \pm i}{s+1},$$

whose Nyquist diagrams are also shown in Fig. 4(b), and the Hopf bifurcation condition can be obtained via $\lambda_2(i\omega_0) = -1$, giving the solution $\omega_0 = 1$. After some calculations given in Appendix A, it can be obtained as $\gamma_1 = -\frac{1}{16}\xi_s - \frac{1}{8}\eta i$. Thus, the first frequency coefficient is given by $\phi_2 = -\Im(\gamma_1) = \frac{1}{8}\eta$, and then, p_2 yields $p_2 = -\frac{\pi}{4}\eta$.

3.2.5. *Example 9*: *Normal form of the Hopf bifurcation*

Consider the following system, corresponding to the normal form of the Hopf bifurcation

$$\begin{cases} \dot{x} = \alpha x - y + \beta x(x^2 + y^2) - \nu y(x^2 + y^2), \\ \dot{y} = x + \alpha y + \nu x(x^2 + y^2) + \beta y(x^2 + y^2), \end{cases} \tag{26}$$

where α, β and ν are system parameters. Particularly, α represents the real part of the bifurcating eigenvalues, thus $\alpha = 0$ produces a center at the origin. Letting $\alpha = 0$, it follows that the eigenvalues of $G(s)J$ at the trivial equilibrium point, are the same as in Examples 7 and 8. The corresponding Nyquist diagrams are also shown in Fig. 4(b), and the Hopf bifurcation condition can be obtained via $\lambda_2(i\omega_0) = -1$, giving the solution $\omega_0 = 1$. After some calculations given in Appendix A, it can be obtained as $\gamma_1 = \frac{1}{2}(-\beta - \nu i)$. Thus, the first frequency coefficient is given by $\phi_2 = \nu/2$, and then, p_2 yields $p_2 = -\pi\nu$. Note that the normal form in polar coordinates is given by

$$\dot{\rho} = \rho(\alpha + \beta\rho^2), \quad \dot{\varphi} = 1 + \nu\rho^2,$$

where $x = \rho\cos\varphi$, $y = \rho\sin\varphi$, so it is easy to appreciate the meaning of β and ν.

4. Conclusions

In this work, the computation of the first period constant is shown in several classical examples using a frequency domain method. The graphs of the eigenloci are remarked in order to appreciate some simplifications that occurred in the GHBT for SISO systems.

The results confirm that the GHBT framework can be effectively extended to extract not only amplitude information but also periodic features of limit cycles near bifurcation points. Ongoing work aims to extend this methodology to compute higher-order period constants, such as p_4, and to explore its applicability in more complex or multiparameter scenarios.

Acknowledgments

The authors are thankful for the support of Universidad Nacional del Sur (PGI 24/K097) and CONICET (PIP 11220200102076).

ORCID

Cinthya A. Bares
https://orcid.org/0009-0000-1662-051X
Franco S. Gentile
https://orcid.org/0000-0001-7989-8295
Guillermo L. Calandrini
https://orcid.org/0000-0002-4557-4556
Jorge L. Moiola
https://orcid.org/0000-0002-3657-8969

References

Agamennoni, G., Calandrini, G. L. & Moiola, J. L. [2008] "Some realizations in the study of oscillations with a frequency method," *Dyn. Contin. Discrete Impuls. Syst. B: Appl. Algorithms* **15**, 99–109.

Algaba, A., Freire, E. & Gamero, E. [2000] "Isochronicity via normal form," *Qual. Theory Dyn. Syst.* **1**, 133–156.

Algaba, A. & Reyes, M. [2009] "Characterizing isochronous points and computing isochronous sections," *J. Math. Anal. Appl.* **355**, 564–576.

Bautin, N. N. [1954] *On the Number of Limit Cycles Which Appear with the Variation of the Coefficients from an Equilibrium Position of Focus or Center Type*, American Mathematical Society Translations, Vol. 100 (American Mathematical Society, Providence, RI). Reprinted in [1962] *Stability and Dynamic Systems*, American Mathematical Society Translations Series 1, Vol. 5 (American Mathematical Society), pp. 396–413. Original in [1952] *Mat. Sb. (N.S.)* **30**, 181–196 (in Russian).

Chicone, C. & Jacobs, M. [1989] "Bifurcation of critical periods for plane vector fields," *Trans. Am. Math. Soc.* **312**, 433–486.

Cima, A., Gasull, A., Mañosa, V. & Mañosas, F. [1997] "Algebraic properties of the Liapunov and period constants," *Rocky Mt. J. Math.* **27**, 471–501.

Farr, W. W., Li, C. Z., Labouriau, I. S. & Langford, W. F. [1989] "Degenerate Hopf bifurcation formulas and Hilbert's 16th problem," *SIAM J. Math. Anal.* **20**, 13–30.

Gasull, A., Guillamon, A. & Mañosa, V. [1997] "An explicit expression of the first Liapunov and period constants with applications," *J. Math. Anal. Appl.* **211**, 190–212.

Gelb, A. & Vander Velde, W. E. [1968] *Multiple-Input Describing Functions and Nonlinear System Design* (McGraw-Hill Book Co., NY).

Golubitsky, M. & Langford, W. F. [1981] "Classification and unfoldings of degenerate Hopf bifurcations," *J. Differ. Equ.* **41**, 375–415.

Kuznetsov, Y. A. [2004] *Elements of Applied Bifurcation Theory*, Applied Mathematical Sciences, Vol. 112 (Springer).

MacFarlane, A. G. J. & Postlethwaite, I. [1977] "The generalized Nyquist stability criterion and multivariable root loci," *Int. J. Control* **25**, 81–127.

Mardešić, P., Marín, D. & Villadelprat, J. [2006] "The period function of reversible quadratic centers," *J. Differ. Equ.* **224**, 120–171.

Mees, A. I. & Chua, L. O. [1979] "The Hopf bifurcation theorem and its applications to nonlinear oscillations in circuits and systems," *IEEE Trans. Circuits Syst.* **26**, 235–254.

Mees, A. I. [1981] *Dynamics of Feedback Systems* (John Wiley & Sons, Chichester, UK).

Moiola, J. L. & Chen, G. [1996] *Hopf Bifurcation Analysis: A Frequency-Domain Approach* (World Scientific, Singapore).

Oliveira, R. D. S., Sánchez-Sánchez, I. & Torregrosa, J. [2022] "Simultaneous bifurcations of limit cycles and critical periods," *Qual. Theory Dyn. Syst.* **21**, 20.

Rousseau, C. & Toni, B. [1993] "Local bifurcation of critical periods in vector fields with homogeneous nonlinearities of the third degree," *Can. Math. Bull.* **36**, 473–484.

Rousseau, C. & Toni, B. [1997] "Local bifurcations of critical periods in the reduced Kukles system," *Can. J. Math.* **49**, 338–358.

Sánchez-Sánchez, I. & Torregrosa, J. [2021] "Criticality via first order development of the period constants," *Nonlin. Anal.* **203**, 112164.

Sibirskii, K. S. [1965] "On the number of limit cycles in a neighborhood of singular points," *J. Differ. Equ.* **1**, 36–47 (translated from *Differ. Uvavn.* **1**, 53–66).

Tesi, A., Abed, E. H., Genesio, R. & Wang, H. O. [1996] "Harmonic balance analysis of period-doubling bifurcations with implications for control of nonlinear dynamics," *Automatica* **32**, 1255–1271.

Yu, P. & Han, M. [2009] "Critical periods of planar revertible vector field with third-degree polynomial functions," *Int. J. Bifurcation and Chaos* **19**, 419–433.

Appendix A

Here, the calculations needed to obtain the results given in Sec. 3, following the procedure described in Sec. 2, are provided.

Example 1. Consider the system given by

$$\begin{cases} \dot{x} = -y + p(x), \\ \dot{y} = x, \end{cases} \tag{A.1}$$

where

$$p(x) = a_2x^2 + a_3x^3 + a_4x^4 + a_5x^5 + a_6x^6 + \cdots.$$

In order to apply the GHBT described in Sec. 2, one can choose matrices

$$A = \begin{bmatrix} -1 & -1 \\ 1 & 0 \end{bmatrix}, \quad B = \begin{bmatrix} 1 \\ 0 \end{bmatrix} = C^T,$$

thus the nonlinear feedback becomes $g(e) = -e + p(e)$, where $e = -x$. Note that this is a SISO case, since the nonlinearity only appears in the first equation in (A.1), and this nonlinearity depends only on x. Thus, the transfer function is scalar and it is given by $G(s) = \frac{s}{s^2+s+1}$. The dynamics about the equilibrium point $\hat{e} = 0$ is considered, thus one has $J = -1$ and it follows that $G(s)J$ is also scalar. Note that a linear term has been considered into the nonlinear feedback function g. This choice is helpful in two aspects: It avoids the vanishing of the linear loop gain $G(s)J$ at the trivial equilibrium point, and it also prevents $G(s)$ having poles on the imaginary axis, which would make the stability analysis more involved. As the system is SISO, its dynamics can be studied through a single characteristic function $\lambda(s) = G(s)J$. The Hopf bifurcation condition can be obtained via $\lambda(i\omega_0) = G(i\omega_0)J = -1$, giving the solution $\omega_0 = 1$. The closed-loop transfer function is given by $H(s) = \frac{s}{s^2+1}$ thus, after some calculations, one can obtain $V_{02} = 0$, $V_{22} = \frac{1}{3}a_2i$, and also

$$\gamma_1 = -\frac{3}{8}a_3 + \frac{1}{6}a_2^2i.$$

It follows that $\sigma_1 = \frac{3}{8}a_3$ is the first curvature coefficient, and the first frequency coefficient is $\phi_2 = -\frac{1}{6}a_2^2$. It also gives the first period constant as $p_2 = -\frac{2\pi}{\omega_0^2}\phi_2 = \frac{\pi}{3}a_2^2$.

Example 2. Consider the system introduced in [Algaba & Reyes, 2009] given by

$$\begin{cases} \dot{x} = -y, \\ \dot{y} = x + a_2y^2 + a_4y^4 + a_6y^6. \end{cases} \tag{A.2}$$

Since the nonlinearity only appears in the second equation, and only y needs to be observed to construct g, one can choose the following matrices:

$$A = \begin{bmatrix} 0 & -1 \\ 1 & -1 \end{bmatrix}, \quad B = \begin{bmatrix} 0 \\ 1 \end{bmatrix} = C^T,$$

and the nonlinear feedback is then given by

$$g(e) = -e + a_2e^2 + a_4e^4 + a_6e^6,$$

where $e = -y$. The transfer function is given by $G(s) = \frac{s}{s^2+s+1}$ and considering the equilibrium point $e = 0$, one has $J = -1$. The Hopf bifurcation condition $G(i\omega_0)J = -1$ is verified with $\omega_0 = 1$. It is computed $H(s) = \frac{s}{s^2+1}$ and then one has $V_{02} = 0$, $V_{22} = \frac{1}{3}a_2 i$. Then, $p_1 = \frac{1}{3}a_2^2 i$ and finally $\gamma_1 = \frac{1}{6}a_2^2 i$. As $\Re(\gamma_1) = 0$, the origin is a center, and the first frequency coefficient is $\phi_2 = -\frac{1}{6}a_2^2$. As mentioned in the main text, system (A.2) can be thought of as a particular case of system (A.1) via the linear transformation $(x', y') = (-y, x)$ and under the constraint $a_3 = a_5 = 0$. Thus, the obtained value of ϕ_2 could have been predicted from the previous example.

Example 3. For the system described in Sec. 3.1.3, given by

$$\begin{cases} \dot{x} = -y + ax^2 + bx^3, \\ \dot{y} = x + dx^2 + gx^3 + hx^4, \end{cases} \tag{A.3}$$

there are two nonlinear functions, but they depend only on one state variable. The following realization is adequate in order to apply the GHBT:

$$A = \begin{bmatrix} -1 & -1 \\ \dfrac{1}{2} & 0 \end{bmatrix}, \quad B = I_2, \quad C = [1 \quad 0],$$

so the error is $e = -x$. With this choice of matrices, the nonlinear feedback becomes

$$\mathbf{g}(e) = \begin{bmatrix} -e + ae^2 - be^3 \\ -\dfrac{1}{2}e + de^2 - ge^3 + he^4 \end{bmatrix}.$$

Note that, again, the choice of A is not obvious from (A.3), and some linear terms are left in $\mathbf{g}$. This choice, once more, prevents the vanishing of the linearization about the trivial equilibrium, and the appearance of poles of $G(s)$ in the imaginary axis. It also follows that the linear sub-system is fully observable and controllable [Agamennoni *et al.*, 2008], and its transfer function is given by

$$G(s) = C(sI_2 - A)^{-1}B = \frac{1}{s^2 + s + \dfrac{1}{2}}[s \quad -1].$$

The equilibrium points verify that $G(0)\mathbf{g}(e) = -e$, which leads to

$$e - 2de^2 + 2ge^3 - 2he^4 = -e,$$

thus $\hat{e} = 0$ or $1 - de + ge^2 - he^3 = 0$, but here only the dynamics related to the trivial equilibrium point is considered. For $\hat{e} = 0$, one has $J = [-1 \quad -1/2]^T$, so the loop gain is a scalar, and it is given by

$$\lambda = G(s)J = -\frac{\left(s - \dfrac{1}{2}\right)}{s^2 + s + \dfrac{1}{2}}.$$

Thus, even if the system is not SISO, its stability is determined by a single characteristic function. It is easy to see that the open-loop poles lie in the left-half plane and also that the bifurcation condition $\lambda(i\omega_0) = -1$ gives the solution $\omega_0 = 1$. Since $G(s)J$ is scalar, its eigenvectors are simply $v = 1$ and $u^T = 1$, and the closed-loop transfer function is given by

$$H(s) = (1 + G(s)J)^{-1}G(s) = \frac{1}{s^2+1}[s \quad -1],$$

thus one obtains $V_{02} = \frac{d}{2}$ and $V_{22} = \frac{1}{6}(-d + 2ai)$. Then, the expression γ_1 yields

$$\gamma_1 = -\frac{1}{24}[-6ad + 9b - i(4a^2 + 10d^2 - 9g)].$$

Thus, the first curvature coefficient $\sigma_1 = -\Re(\gamma_1)$ vanishes under the condition $2ad = 3b$. On the other hand, the first coefficient in the frequency expansion is $\phi_2 = -\Im(\gamma_1) = (-4a^2 - 10d^2 + 9g)/24$, then the first period constant is $p_2 = \pi(4a^2 + 10d^2 - 9g)/(12\omega_0^2)$ and it vanishes if $9g = 4a^2 + 10d^2$.

Example 4. For the system described in Sec. 3.1.4, given by

$$\begin{cases} \dot{x} = -y, \\ \dot{y} = x + a_1x^2 + a_2xy + a_3y^2 + a_4x^3 \\ \qquad + a_5x^2y + a_6xy^2, \end{cases} \tag{A.4}$$

the following realization is chosen in order to apply the GHBT:

$$A = \begin{bmatrix} 0 & -1 \\ \dfrac{1}{2} & -1 \end{bmatrix}, \quad B = \begin{bmatrix} 0 \\ 1 \end{bmatrix}, \quad C = I_2,$$

so the output is $\mathbf{y} = [y_1 \quad y_2]^T = [x \quad y]^T$ and the error is $\mathbf{e} = [e_1 \quad e_2]^T = -\mathbf{y}$. With this choice of matrices, the nonlinear feedback becomes

$$g(\mathbf{e}) = -\frac{1}{2}e_1 - e_2 + a_1e_1^2 + a_2e_1e_2 + a_3e_2^2 - a_4e_1^3 - a_5e_1^2e_2 - a_6e_1e_2^2.$$

It also follows that the linear sub-system is fully observable and controllable, and its transfer function is given by

$$G(s) = C(sI_2 - A)^{-1}B = \frac{1}{s^2 + s + \frac{1}{2}}\begin{bmatrix}-1\\ s\end{bmatrix}.$$

It is again considered that the equilibrium point $\hat{\mathbf{e}} = \mathbf{0}$. Thus, one has $J = [-1/2 \quad -1]$, so the loop gain is given by

$$G(s)J = \frac{1}{s^2 + s + \frac{1}{2}}\begin{bmatrix}\frac{1}{2} & 1\\ \frac{-s}{2} & -s\end{bmatrix}.$$

The characteristic function in the frequency domain reads

$$\begin{aligned}h(s;\lambda) &= \det(\lambda I_2 - G(s)J)\\ &= \lambda^2 + \frac{\left(s - \frac{1}{2}\right)}{\Delta(s)}\lambda = 0,\end{aligned}$$

where $\Delta(s) = s^2 + s + 1/2$. The characteristic functions are $\lambda_1(s) \equiv 0$ and $\lambda_2(s) = (1/2 - s)/\Delta(s)$. For this example, even if the system is not SISO, the nonlinearity affects only the second equation in (A.4), and it produces that there is only one nonzero characteristic function. Thus, a Hopf bifurcation can only occur if $\lambda_2(i\omega_0) = -1$, leading to the solution $\omega_0 = 1$.

The right and left eigenvectors of $G(s)J$ associated with $\lambda_2(s)$ are

$$\mathbf{v} = \frac{1}{\sqrt{2}}\begin{bmatrix}1\\ -i\end{bmatrix}, \quad \mathbf{u}^T = \frac{\sqrt{2}(1+2i)}{5}[1 \quad 2],$$

correspondingly. Note that they are normalized in such a way that $\mathbf{u}^T\mathbf{v} = 1$. The closed-loop transfer function is given by

$$H(s) = (I_2 + G(s)J)^{-1}G(s) = \frac{1}{s^2+1}\begin{bmatrix}-1\\ s\end{bmatrix},$$

and one obtains

$$\mathbf{V}_{02} = \frac{a_1 + a_3}{4}\begin{bmatrix}1\\ 0\end{bmatrix}, \quad \mathbf{V}_{22} = \frac{a_1 - a_3 - a_2 i}{12}\begin{bmatrix}-1\\ 2i\end{bmatrix}.$$

Then, one can obtain the expression

$$\begin{aligned}\mathbf{p}_1 = \frac{1}{48\sqrt{2}}[&20a_1^2 + 2a_2^2 + 8a_3^2 + 20a_1a_3 - 18a_4\\ &- 6a_6 + i6(-a_1a_2 - a_2a_3 + a_5)],\end{aligned}$$

and finally $\gamma_1 = \frac{i\sqrt{2}\mathbf{p}_1}{2}$. Thus, the first curvature coefficient $\sigma_1 = -\Re(\gamma_1)$ can be obtained as

$$\sigma_1 = -\frac{1}{16}[(a_1 + a_3)a_2 - a_5].$$

On the other hand, the first frequency coefficient is $\phi_2 = -\Im(\gamma_1)$, resulting in

$$\phi_2 = -\frac{1}{48}[10a_1^2 + a_2^2 + 4a_3^2 + 10a_1a_3 - 9a_4 - 3a_6],$$

and $p_2 = -2\pi\phi_2$.

Example 5. Consider the system introduced in [Cima *et al.*, 1997] given by

$$\begin{cases}\dot{x} = -y + \alpha x^2 + \beta x^3,\\ \dot{y} = x + \gamma y^2 + \delta y^3.\end{cases}$$

One can choose the following matrices:

$$A = \begin{bmatrix}-1 & -1\\ 1 & -1\end{bmatrix}, \quad B = C = I_2,$$

and thus the nonlinear feedback is

$$\mathbf{g}(\mathbf{e}) = \begin{bmatrix}-e_1 + \alpha e_1^2 - \beta e_1^3,\\ -e_2 + \gamma e_2^2 - \delta e_2^3\end{bmatrix}.$$

The transfer function is given by

$$G(s) = \frac{1}{s^2 + 2s + 2}\begin{bmatrix}s+1 & -1\\ 1 & s+1\end{bmatrix}.$$

It is again considered that the equilibrium point $\hat{\mathbf{e}} = \mathbf{0}$. Thus, one has

$$J = \begin{bmatrix}-1 & 0\\ 0 & -1\end{bmatrix},$$

and it follows that the eigenvalues of $G(s)J$ are

$$\lambda_{1,2}(s) = \frac{-s - 1 \pm i}{s^2 + 2s + 2} = \frac{-1}{s + 1 \pm i}.$$

It follows that the Hopf bifurcation condition holds with $\lambda_2(i) = -1$, i.e. the solution frequency is $\omega_0 = 1$. The (normalized) right and left eigenvectors of

$G(s)J$ associated with $\lambda_2(s)$ are

$$\mathbf{v} = \frac{1}{\sqrt{2}}\begin{bmatrix} 1 \\ -i \end{bmatrix}, \quad \mathbf{u}^T = \frac{1}{\sqrt{2}}[1 \quad i],$$

correspondingly. The closed-loop transfer function is given by

$$H(s) = \frac{1}{s^2+1}\begin{bmatrix} s & -1 \\ 1 & s \end{bmatrix},$$

and thus the zeroth- and second-order harmonics are computed as

$$\mathbf{V}_{02} = -\frac{1}{4}\begin{bmatrix} -\gamma \\ \alpha \end{bmatrix}, \quad \mathbf{V}_{22} = \frac{1}{12}\begin{bmatrix} \gamma + 2\alpha i \\ \alpha - 2\gamma i \end{bmatrix}.$$

It is then computed $\mathbf{p}_1$ as

$$\mathbf{p}_1 = \frac{1}{24\sqrt{2}}\begin{bmatrix} 14\alpha\gamma - 9\beta + 4\alpha^2 i \\ 4\gamma^2 + (14\alpha\gamma + 9\delta)i \end{bmatrix}.$$

Now, after some calculations, it can be obtained that

$$\gamma_1 = \frac{\mathbf{u}^T G(i)\mathbf{p}_1}{\mathbf{u}^T G'(i) J\mathbf{v}} = \frac{1}{48}[-9(\beta+\delta) + 4(\alpha^2+\gamma^2)i].$$

Thus, the first curvature and frequency coefficients are $\sigma_1 = \frac{3}{16}(\beta+\delta)$ and $\phi_2 = -\frac{1}{12}(\alpha^2+\gamma^2)$.

Example 6. For the system described in Sec. 3.2.2, given by

$$\begin{cases} \dot{x} = \lambda_1 x - y - \lambda_3 x^2 + (2\lambda_2 + \lambda_5)xy + \lambda_6 y^2, \\ \dot{y} = x + \lambda_1 y + \lambda_2 x^2 + (2\lambda_3 + \lambda_4)xy - 2\lambda_2 y^2, \end{cases}$$

the following matrices are chosen: $A = -I_2$, $B = C = I_2$, so the output is $\mathbf{y} = [y_1 \quad y_2]^T = [x \quad y]^T$ and the error is $\mathbf{e} = [e_1 \quad e_2]^T = -\mathbf{y}$. The nonlinear feedback becomes

$$\mathbf{g}(\mathbf{e}) = \begin{bmatrix} -(\lambda_1+1)e_1 + e_2 - \lambda_3 e_1^2 + (2\lambda_2+\lambda_5)e_1e_2 + \lambda_6 e_2^2 \\ -e_1 - (\lambda_1+1)e_2 + \lambda_2 e_1^2 + (2\lambda_3+\lambda_4)e_1e_2 - \lambda_2 e_2^2 \end{bmatrix}.$$

The transfer function is given by

$$G(s) = \frac{1}{s+1}I_2.$$

It is again considered that the equilibrium point $\hat{\mathbf{e}} = \mathbf{0}$. Thus, one has

$$J = \begin{bmatrix} -(\lambda_1+1) & 1 \\ -1 & -(\lambda_1+1) \end{bmatrix},$$

and it follows that the eigenvalues of $G(s)J$ are

$$\lambda_{a,b}(s) = \frac{-1-\lambda_1 \pm i}{s+1},$$

where now the characteristic functions are noted as $\lambda_a(s)$ and $\lambda_b(s)$ to avoid misunderstanding with the system parameters. The Hopf bifurcation condition is verified with $\lambda_1 = 0$ and $\lambda_b(i\omega_0) = -1$, leading to the solution $\omega_0 = 1$.

The (normalized) right and left eigenvectors of $G(s)J$ associated with $\lambda_2(s)$ are

$$\mathbf{v} = \frac{1}{\sqrt{2}}\begin{bmatrix} 1 \\ -i \end{bmatrix}, \quad \mathbf{u}^T = \frac{1}{\sqrt{2}}[1 \quad i],$$

correspondingly. The closed-loop transfer function is given by

$$H(s) = (I_2 + G(s)J)^{-1}G(s) = \frac{1}{s^2 - 2\lambda_1 s - \lambda_1^2 + 1}\begin{bmatrix} s-\lambda_1 & -1 \\ 1 & s-\lambda_1 \end{bmatrix},$$

thus, after some straightforward but time-consuming calculations, one can obtain

$$\mathbf{V}_{02} = -\frac{1}{4}\begin{bmatrix} 0 \\ \lambda_6 - \lambda_3 \end{bmatrix},$$

$$\mathbf{V}_{22} = \frac{1}{12}\begin{bmatrix} 2(\lambda_2+\lambda_5) + i(\lambda_4 - 2\lambda_6) \\ 3\lambda_3 + 2\lambda_4 - \lambda_6 + i(2\lambda_2 - \lambda_5) \end{bmatrix}.$$

Then, one can obtain some expressions needed in order to compute p_1. For example,

$$\frac{1}{2}\overline{Q}\mathbf{V}_{22} = \begin{bmatrix} p_{11}^a \\ p_{11}^b \end{bmatrix},$$

where

$$\Re(p_{11}^a) = \frac{1}{24\sqrt{2}}(2\lambda_2\lambda_3 - \lambda_3\lambda_5 - 2\lambda_2\lambda_6 + 2\lambda_2\lambda_4 + 3\lambda_5\lambda_6 + \lambda_4\lambda_5),$$

$$\Im(p_{11}^a) = \frac{1}{24\sqrt{2}}(-2\lambda_3\lambda_4 + 10\lambda_3\lambda_6 + 4\lambda_4\lambda_6 - 2\lambda_6^2 + 8\lambda_2^2 + \lambda_5^2 + 6\lambda_2\lambda_5),$$

$$\Re(p_{11}^b) = \frac{1}{24\sqrt{2}}(5\lambda_3\lambda_4 + 2\lambda_3\lambda_6 + \lambda_4\lambda_6 + 8\lambda_2^2 + \lambda_4^2 + 6\lambda_3^2 + 2\lambda_2\lambda_5),$$

$$\Im(p_{11}^b) = \frac{1}{24\sqrt{2}}(2\lambda_2\lambda_3 + 2\lambda_3\lambda_5 + 2\lambda_2\lambda_4 - 2\lambda_2\lambda_6 + \lambda_4\lambda_5),$$

also

$$Q\mathbf{V}_{02} = -\frac{(\lambda_6 - \lambda_3)}{4\sqrt{2}}\begin{bmatrix} 2\lambda_2 + \lambda_5 - 2\lambda_6 i \\ 2\lambda_3 + \lambda_4 + 2\lambda_2 i \end{bmatrix},$$

and finally the term $\frac{1}{8}Lv$ is zero since the third-order derivatives of $g(\cdot)$ are null. Thus,

$$\mathbf{p}_1 = \begin{bmatrix} p_{11}^a \\ p_{11}^b \end{bmatrix} - \frac{(\lambda_6 - \lambda_3)}{4\sqrt{2}}\begin{bmatrix} 2\lambda_2 + \lambda_5 - 2\lambda_6 i \\ 2\lambda_3 + \lambda_4 + 2\lambda_2 i \end{bmatrix}.$$

Now, since

$$\gamma_1 = \frac{\mathbf{u}^T G(i)\mathbf{p}_1}{\mathbf{u}^T G'(i) J\mathbf{v}},$$

the first curvature coefficient becomes

$$\sigma_1 = -\Re(\gamma_1) = \frac{1}{16}\lambda_5(\lambda_6 - \lambda_3),$$

and the first frequency coefficient is $\phi_2 = -\Im(\gamma_1)$, resulting in

$$\phi_2 = -\frac{1}{48}(9\lambda_3\lambda_4 - 12\lambda_3\lambda_6 - \lambda_4\lambda_6 + 16\lambda_2^2 + \lambda_4^2 + 18\lambda_3^2 + \lambda_5^2 + 10\lambda_6^2 + 8\lambda_2\lambda_5),$$

which gives:

$$p_2 = -2\pi\frac{\phi_2}{\omega_0^2} = \frac{\pi}{24}(9\lambda_3\lambda_4 - 12\lambda_3\lambda_6 - \lambda_4\lambda_6 + 16\lambda_2^2 + \lambda_4^2 + 18\lambda_3^2 + \lambda_5^2 + 10\lambda_6^2 + 8\lambda_2\lambda_5).$$

Example 7. The Loud system is given by

$$\begin{cases} \dot{x} = -y + B_a xy, \\ \dot{y} = x + Dx^2 + Fy^2, \end{cases}$$

when B_a, D and F are parameters, and assigning $B_a = 1$: this particular case is called *dehomogenized* Loud system [Chicone & Jacobs, 1989]. One can choose matrices $A = -I_2$, $B = I_2$, $C = I_2$ and the nonlinear feedback

$$\mathbf{g}(\mathbf{e}) = \begin{bmatrix} -e_1 + e_2 + e_1 e_2 \\ -e_1 - e_2 + De_1^2 + Fe_2^2 \end{bmatrix}.$$

The transfer function is given by $G(s) = \frac{1}{s+1}I_2$. It is again considered that the equilibrium point $\hat{\mathbf{e}} = \mathbf{0}$. Thus, one has

$$J = \begin{bmatrix} -1 & 1 \\ -1 & -1 \end{bmatrix},$$

and it follows that the eigenvalues of $G(s)J$ are

$$\lambda_{1,2}(s) = \frac{-1 \pm i}{s+1},$$

and the Hopf bifurcation condition can be obtained via $\lambda_2(i\omega_0) = -1$, giving the solution $\omega_0 = 1$.

The (normalized) right and left eigenvectors of $G(s)J$ associated with $\lambda_2(s)$ are

$$\mathbf{v} = \frac{1}{\sqrt{2}}\begin{bmatrix} 1 \\ -i \end{bmatrix}, \quad \mathbf{u}^T = \frac{1}{\sqrt{2}}[1 \quad i],$$

correspondingly. The closed-loop transfer function is given by

$$H(s) = \frac{1}{s^2+1}\begin{bmatrix} s & -1 \\ 1 & s \end{bmatrix},$$

thus, after some calculations, one can obtain

$$\mathbf{V}_{02} = \frac{1}{4}\begin{bmatrix} D+F \\ 0 \end{bmatrix}, \quad \mathbf{V}_{22} = \frac{1}{12}\begin{bmatrix} 2 - D + F \\ (-1 + 2D - 2F)i \end{bmatrix}.$$

Following the procedure, it yields

$$\mathbf{p}_1 = \frac{1}{24\sqrt{2}}\begin{bmatrix} (1 - 5D - 7F)i \\ 4D + 10D^2 + 10DF + 2F + 4F^2 \end{bmatrix}$$

and

$$\gamma_1 = \frac{i}{48}(1 - D - 5F + 10D^2 + 10DF + 4F^2).$$

Note that, since γ_1 is purely imaginary, the first curvature coefficient σ_1 is identically zero for all parameter values. The first frequency coefficient ϕ_2 becomes

$$\phi_2 = -\Im(\gamma_1) = -\frac{1}{48}(1 - D - 5F + 10D^2 + 10DF + 4F^2),$$

and then, $p_2 = -2\pi\phi_2$.

Example 8. Consider the system given by

$$\begin{cases} \dot{x} = -y + (a - \omega_s - \theta_s)x^3 + (3\mu - \eta)x^2y + (3\theta_s + \xi_s - 3\omega_s - 2a)xy^2 + (\nu - \mu)y^3, \\ \dot{y} = x + (\nu + \mu)x^3 + (3\omega_s + 3\theta_s + 2a)x^2y + (\eta - 3\mu)xy^2 + (\omega_s - \theta_s - a)y^3, \end{cases}$$

one can choose matrices $A = -I_2$, $B = I_2$, $C = I_2$ and the nonlinear feedback

$$\mathbf{g}(\mathbf{e}) = \begin{bmatrix} -e_1 + e_2 - (a - \omega_s - \theta_s)e_1^3 - (3\mu - \eta)e_1^2e_2 - (3\theta_s + \xi_s - 3\omega_s - 2a)e_1e_2^2 - (\nu - \mu)e_2^3, \\ -e_1 - e_2 - (\nu + \mu)e_1^3 - (3\omega_s + 3\theta_s + 2a)e_1^2e_2 - (\eta - 3\mu)e_1e_2^2 - (\omega_s - \theta_s - a)e_2^3 \end{bmatrix}.$$

The transfer function $G(s) = \frac{1}{s+1}I_2$ is the same as Example 7, as well as the equilibrium point at the origin, the Jacobian J, the eigenvalues, the eigenvectors, and the Hopf bifurcation condition. Since the second-order derivatives are zero at the equilibrium point, one has $V_{02} = V_{22} = 0$. Then, p_1 reduces to $\mathbf{p}_1 = \frac{1}{8}L\bar{\mathbf{v}}$, resulting in

$$\mathbf{p}_1 = \frac{1}{16\sqrt{2}} \begin{bmatrix} -2a - 2\xi_s + 12\omega_s + i(6\nu - 2\eta) \\ -6\nu - 2\eta + i(-2a + 12\omega_s) \end{bmatrix}.$$

Now, after some calculations, it can be obtained that $\gamma_1 = -\frac{1}{16}\xi_s - \frac{1}{8}\eta i$.

Thus, the first curvature and frequency coefficients are given by $\sigma_1 = \frac{1}{16}\xi_s$ and $\phi_2 = -\frac{1}{6}\eta$.

Example 9. Consider the system given by

$$\begin{cases} \dot{x} = \alpha x - y + \beta x(x^2 + y^2) - \nu y(x^2 + y^2), \\ \dot{y} = x + \alpha y + \nu x(x^2 + y^2) + \beta y(x^2 + y^2), \end{cases} \tag{A.5}$$

which can be represented in the feedback form with

$$G(s) = \frac{1}{s+1}I_2,$$

and the nonlinear feedback

$$\mathbf{g}(\mathbf{e}) = \begin{bmatrix} -(\alpha+1)e_1 + e_2 - \beta e_1(e_1^2 + e_2^2) + \nu e_2(e_1^2 + e_2^2) \\ -e_1 - (\alpha+1)e_2 - \nu e_1(e_1^2 + e_2^2) - \beta e_2(e_1^2 + e_2^2) \end{bmatrix}.$$

In the following, it is considered that $\alpha = 0$ is the Hopf bifurcation condition. Thus, the Jacobian J, the eigenvalues, the eigenvectors are the same as in Example 7. Since the second-order derivatives are zero at the equilibrium point, one has $V_{02} = V_{22} = 0$. Then, p_1 reduces to $\mathbf{p}_1 = \frac{1}{8}L\bar{\mathbf{v}}$, resulting in

$$\mathbf{p}_1 = \frac{1}{2\sqrt{2}} \begin{bmatrix} -\beta - \nu i \\ -\nu - \beta i \end{bmatrix}.$$

Now, after some calculations, it can be obtained that

$$\gamma_1 = \frac{1}{2}(-\beta - \nu i).$$

Thus, the first curvature coefficient is $\sigma_1 = \beta/2$ and the first frequency coefficient is $\phi_2 = \nu/2$, which in turn gives $p_2 = -\pi\nu$.

https://doi.org/10.1142/9789819830633_0006

Chapter 6

Complexity of Grid-Connected Power Electronics Converters and Its Impacts on Modern Power Grids#

Jingxi Yang* and Chi K. Tse†
Department of Electrical Engineering,
City University of Hong Kong, Kowloon,
Hong Kong SAR, P. R. China
*jyang264@cityu.edu.hk
†chitse@cityu.edu.hk

Meng Huang
School of Electrical Engineering and Automation,
Wuhan University, Wuhan 430072, P. R. China
meng.huang@whu.edu.cn

Grid-connected power electronic converters play a vital role in integrating renewable energy sources into the power grid and enhancing the efficiency and flexibility of grid operations. However, due to their inherent nonlinear characteristics, these converters can introduce unexpected, diverse, and complex dynamic behaviors into the power system. Such behaviors may compromise normal grid operation and increase the risk of failures and blackouts, including voltage collapse, sub-synchronous oscillations, loss of synchronization, and frequency instability. The physical origins of these phenomena cannot be adequately captured using conventional linearization techniques, which typically neglect nonlinear terms and only provide insight into the stability of the nominal operating point. To address this limitation, this paper presents a comprehensive review of recent research on the complex dynamics of converter-based power systems. Topics include various routes to voltage instability through bifurcation sequences, transient stability enhancement via significant expansion of the stable basin of attraction, novel mechanisms for sub-synchronous oscillations involving hidden attractors, and frequency, voltage, and power fluctuations under the influence of distributed secondary control. Adopting this nonlinear perspective enables a deeper understanding of the fundamental mechanisms behind power system instabilities. It also supports the development of effective mitigation strategies to prevent blackouts and facilitates the integration of increasing levels of renewable energy and power electronic devices. As the penetration of power electronics continues to rise — alongside the adoption of inter-coupled control and communication technologies — the dynamic behavior of modern power grids is becoming increasingly complex. This complexity has profound implications for the reliable and efficient delivery of electrical power, underscoring the need for continued research in this evolving field.

Keywords: Grid-connected converter; power system stability; sustainable power distribution; net-zero initiatives; voltage stability; transient synchronization; sub-synchronous oscillation; frequency stability; bifurcation; grid-forming; grid-following; distributed control.

#To cite this article, please refer to its earlier version published in the *International Journal of Bifurcation and Chaos*, Vol. 35, No. 14 (2025) 2540006 (26 pages), DOI: 10.1142/S0218127425400061.

†Author for correspondence

1. Introduction

Power electronics are concerned with the processing and control of electrical energy across a wide range of applications, from small-scale energy harvesting and implantable systems to large-scale power generation and transmission [Erickson & Maksimović, 2000; Blaabjerg, 2018; Bose, 2022; Blaabjerg *et al.*, 2023; Farivar *et al.*, 2023; Rivera *et al.*, 2023; Liserre *et al.*, 2023]. In particular, grid-connected power electronic converters play a pivotal role in advancing net-zero initiatives by serving as key interfaces for integrating renewable energy sources and energy storage systems into the power grid [Strunz *et al.*, 2023; Singh *et al.*, 2022; Buticchi *et al.*, 2021; Kroposki *et al.*, 2017]. However, the nonlinear interactions among grid-connected converters can significantly alter the dynamic behavior of the power grid, potentially disrupting normal operations and increasing the risk of system failures and blackouts. Notable examples include sub-synchronous oscillation events in Hami, China (2015–2016) [Sun *et al.*, 2017; Cheng *et al.*, 2023]; frequency and voltage instability in South Australia following the sudden shutdown of approximately 52% of wind generation on September 28, 2016 [Yan *et al.*, 2018; AEMC, 2019]; severe undamped oscillations in the Hornsea offshore wind farm leading to a major blackout in Britain on August 9, 2019 [ESO, 2019]; and multiple solar power loss incidents in California, USA in 2021 [NERC, 2021]. These events underscore the urgency of investigating the complex nonlinear behavior of grid-connected converters and their impact on the stability of modern power systems. Despite this urgency, current research efforts have not sufficiently addressed the complexity inherent in converter-based systems. Most studies rely on linear tools such as eigenvalue analysis or impedance-based methods to assess system stability and identify instability mechanisms [Yang *et al.*, 2021a; Zou *et al.*, 2022; Zhu *et al.*, 2023; Wang *et al.*, 2018a, 2018b; Rosso *et al.*, 2020; Wan *et al.*, 2015; Jia *et al.*, 2018; Shuai *et al.*, 2019a; Zhang *et al.*, 2023b; Zhou *et al.*, 2023; Chen *et al.*, 2022b; Liu *et al.*, 2015; Liu *et al.*, 2018]. While these approaches are valuable, they inherently neglect the nonlinear characteristics of the system and only provide insights near the assumed operating point. As a result, they fail to capture the full spectrum of complex and diverse nonlinear behaviors that are increasingly relevant in modern power grids [Tse *et al.*, 2020; Gu & Green, 2023; Yang & Tse, 2024; Tse, 2003; Reza & Lu, 2020; Tse & Li, 2011; Tse & Bernardo, 2002].

The study of complex behavior is an interdisciplinary field rooted in the analysis of nonlinear dynamical systems. Its origins trace back to classical physics, beginning with Newton's formulation of differential equations to describe universal gravitation. The field advanced significantly through the pioneering work of Henri Poincaré, whose investigation of the three-body problem is widely recognized as the earliest formal exploration of chaotic dynamics. Despite these early developments, chaos theory remained largely underappreciated until Edward Lorenz's discovery of deterministic aperiodic flows in 1963. This breakthrough sparked widespread interest in chaotic dynamics, oscillations, bifurcations, and their applications throughout the 1970s and beyond [Strogatz, 1994; Robinson, 2012; Peitgen *et al.*, 2004; Chua & Lin, 1990; Chua *et al.*, 1986; Chen & Ueta, 1999, 2002; Laskar, 1989; Chen, 2020; Chen & Dong, 1998; Ott, 1993; Small & Tse, 2012; Zhang *et al.*, 2020; Shen, 2023; Li *et al.*, 2023; Cristiano *et al.*, 2024; Chua, 2024].

Power electronic circuits, by their very nature, are nonlinear and often exhibit a wide range of complex behaviors. These include sudden regime shifts (and even cascading transitions), altered robustness to transient disturbances, sustained oscillations under small or transient perturbations, intermittent or persistent chaotic behavior (with various routes to chaos), and occasional instability triggered by specific parameter ranges, initial conditions, or final states. Such phenomena cannot be adequately described using linear analytical techniques [Tse *et al.*, 2020; Yang & Tse, 2024; Tse, 2003; Tse & Bernardo, 2002; Dranga *et al.*, 2003; Ding *et al.*, 2020; Peng *et al.*, 2019; Huang *et al.*, 2013a; Huang *et al.*, 2017a; Xiong *et al.*, 2015]. Although these complex behaviors are common in power electronic systems, they did not receive formal academic attention until the late 1980s [Hamill & Jefferies, 1988; Deane & Hamill, 1990, 1991; Tse, 1994a, 1994b]. Fundamentally, power electronic converters can be modeled as piecewise-switched dynamical systems. Depending on the level of detail in the switching dynamics, their complex behavior can be broadly categorized into two types: fast-scale behavior, which includes detailed switching actions, and slow-scale behavior, which abstracts away the switching details. Fast-scale behavior is typically analyzed using nonlinear difference equations (or iterated maps), which preserve the time-varying nature of switching events. In contrast, slow-scale behavior is studied through averaged nonlinear differential equations, which smooth out the switching effects to focus on the system's broader dynamic trends [Tse, 2003; Tse & Bernardo, 2002; Chen *et al.*, 2008; Iu *et al.*, 2003; Ji *et al.*, 2022; Yang *et al.*, 2020b; Cao *et al.*, 2022].

Over the past two decades, extensive research has focused on the normal and abnormal operation of individual converters or small groups of converters. However, as the number of grid-connected converters integrated into power systems has grown into the hundreds, recent attention has shifted toward understanding the impact of their collective complex behavior on overall system stability. In this context, slow-scale complex dynamics of converters have become increasingly relevant, as many power system instability phenomena are driven by slow-scale interactions. Consequently, the study of nonlinear dynamics at the individual converter level has evolved into the investigation of networked system dynamics — a rapidly developing area that warrants significantly more attention [Tse *et al.*, 2020; Zhang *et al.*, 2023a].

In this paper, we attempt to provide a review on the recent research progress of the complex behavior and instability mechanisms of converter-based power systems. We first present the basic structures of grid-forming control, grid-following control and distributed secondary control. Then, the stability framework of converter-based power systems, and the essential pathways to instability are summarized. On this basis, the practical impact of converters' slow-scale complex behavior on the voltage stability, transient synchronization stability, sustained oscillations, and frequency stability is systemically exposed.

2. Basic Control Strategies

For the system (e.g. microgrid) containing a number of grid-connected power electronic converters, a hierarchical control structure is usually adopted, which consists of three different levels, namely, primary control level, secondary control level, and tertiary control level [Guerrero *et al.*, 2011].

In the primary control level, all control algorithms are executed in the local controllers, and there is no communication between different controllers in this level. From the synchronization's point of view, the primary control can be classified into two categories, namely, grid-forming control and grid-following control. The grid-forming control includes droop control, power synchronization control, virtual synchronous control, and virtual oscillator control [Guerrero *et al.*, 2011; Zhang *et al.*, 2010; Zhong & Weiss, 2011; Pan *et al.*, 2020; Yu *et al.*, 2021; Lu, 2022]. The grid-following control includes a Phase-Locked-Loop-based (PLL-based) vector-current control and a direct-power control without using PLL [Zou *et al.*, 2021; Gui *et al.*, 2019; Liu *et al.*, 2024]. By using the grid-forming control, the converter is controlled as a voltage source, and synchronizes with the grid or other converters indirectly, whereas through the grid-following control, the converter is controlled as a current source and synchronizes with the grid directly. In practice, grid-connected converters serve as the intermediaries to connect renewable energy sources or batteries to the power grid, which can be controlled as Grid-Forming Converters (GFMCs), or Grid-Following Converters (GFLCs), as shown in Fig. 1. Moreover, the interface power electronic converters, which are widely used in transmitting power to loads, are usually controlled as Grid-Following Rectifiers (GFLRs), as shown in Fig. 2.

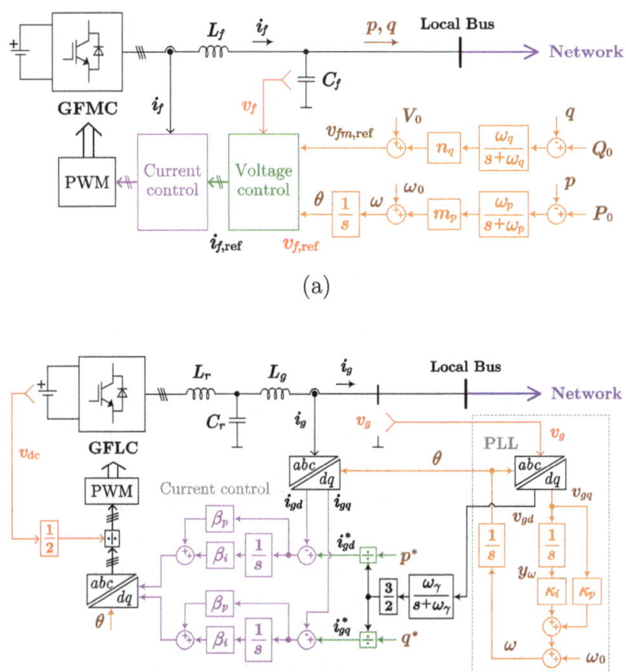

(b)

Fig. 1. Typical control strategy for GFMCs and GFLCs: (a) droop control for GFMCs and (b) PLL-based vector-current control for GFLCs.

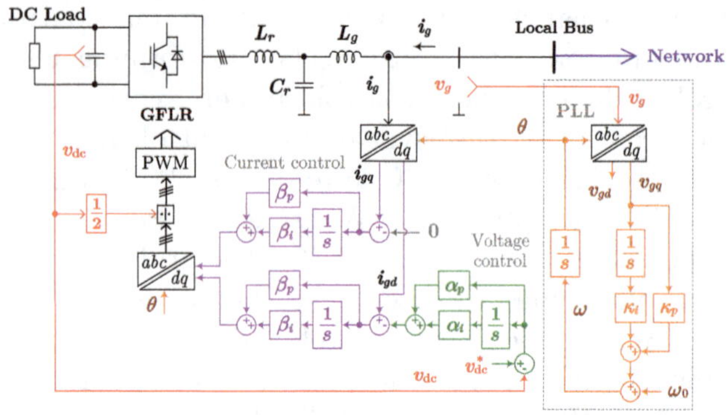

Fig. 2. Typical control strategy for GFLRs.

From Fig. 1(a), we have

$$\frac{\mathrm{d}^2\theta}{\mathrm{d}t^2} = \omega_p\left(-\frac{\mathrm{d}\theta}{\mathrm{d}t} + \omega_0 + m_p(P_0 - p)\right), \quad (1)$$

$$\frac{\mathrm{d}v_{fm,\mathrm{ref}}}{\mathrm{d}t} = \omega_q(-v_{fm,\mathrm{ref}} + V_0 + n_q(Q_0 - q)), \quad (2)$$

where p and q are the output active and reactive powers of GFMC, respectively; ω, θ and $v_{fm,\mathrm{ref}}$ are the frequency, phase angle and magnitude of GFMC's reference output voltage, respectively; P_0, Q_0, ω_0 and V_0 are the nominal values of p, q, ω and $v_{fm,\mathrm{ref}}$, respectively; ω_p and ω_q are the cutoff frequencies of the low-pass filters, and m_p and n_q are the droop coefficients [Pan *et al.*, 2020; Yang *et al.*, 2021b].

From Fig. 1(b), we have

$$\frac{\mathrm{d}\theta}{\mathrm{d}t} = \omega_0 + \kappa_p v_{gq} + \kappa_i y_\omega, \quad (3)$$

$$\frac{\mathrm{d}y_\omega}{\mathrm{d}t} = v_{gq}, \quad (4)$$

where ω, θ and v_{gq} are the frequency, phase angle and q-axis component of GFLC's grid-connected-point voltage, respectively; y_ω, κ_p and κ_i are the internal variable, proportional coefficient and integral coefficient of PLL, respectively. Notably, v_{gq} is determined by the GFLC's injection current and the network [Zou *et al.*, 2021; Wu & Wang, 2020].

In the secondary control level, restoration of the microgrid frequency, compensation of the voltage deviation, and reactive power sharing are the three key objectives [Yang *et al.*, 2024a, 2024b; Han *et al.*, 2017]. Depending on the topologies of the communication networks, the secondary control structures can be classified under decentralized control structure, centralized control structure, and distributed control structure [D'Silva *et al.*, 2020]. The distributed control structure is shown in Fig. 3. In this structure, each local controller communicates with its neighboring local controllers, and only a small number of local controllers are selected to communicate with the microgrid controller. From Fig. 3, we have

$$\frac{\mathrm{d}\Delta\omega_{0\ell}}{\mathrm{d}t} = g_\ell\chi_{mc}(\omega^*_{\mathrm{MG}} - \omega_{\mathrm{MG}}) + \chi_{lc}\sum_{k=1}^{N} a_{\ell k}(\Delta\omega_{0k} - \Delta\omega_{0\ell}), \quad (5)$$

$$\frac{\mathrm{d}\Delta v_{0\ell}}{\mathrm{d}t} = g_\ell\gamma_{mc}(v^*_{\mathrm{MG}} - v_{\mathrm{MG}}) + \gamma_{lc}\sum_{k=1}^{N} a_{\ell k}\left(\frac{n_{qk}}{n_{q1}}q_k - \frac{n_{q\ell}}{n_{q1}}q_\ell\right), \quad (6)$$

where ω^*_{MG} and v^*_{MG} are the reference frequency and reference voltage, respectively; ω_{MG} and v_{MG} are the frequency and voltage, respectively; $\Delta\omega_{0\ell}$, $\Delta v_{0\ell}$ and q_ℓ are the frequency compensation term, voltage compensation term and output reactive power of GFMC-ℓ, respectively, where $\ell = 1, 2, \ldots, N$ (assuming that there are N GFMCs in the system); χ_{mc}, γ_{mc}, χ_{lc} and γ_{lc} are the control parameters in secondary control level; g_ℓ is the element of the pinning vector, and $a_{\ell k}$ is the element of the adjacency matrix of the communication network. Notably, $g_\ell = 1$ if there is a communication link between the ℓth local controller and the microgrid controller, and $g_\ell = 0$ otherwise. Also, $a_{\ell k} = 1$ if there is a communication link between the ℓth local controller and the kth local controller, and $a_{\ell k} = 0$ otherwise.

According to (5) and (6), the distributed secondary control contains two algorithms, namely, a pinning-based algorithm and a consensus-based algorithm. For the pinning-based algorithm, only a fraction of the local controllers will receive the control signals from the microgrid controller (i.e. $\chi_{mc}(\omega^*_{\mathrm{MG}} - \omega_{\mathrm{MG}})$ and $\gamma_{mc}(v^*_{\mathrm{MG}} - v_{\mathrm{MG}})$) to compensate for the frequency and voltage deviations. For the consensus-based algorithm, the information of $(n_{qk}q_k)/n_{q1}$ is shared among all the local controllers to reach an agreement that $n_{q1}q_1 = n_{q2}q_2 = \cdots = n_{qN}q_N$. Basically, the distributed secondary control shown in Fig. 3 is a typical real application of the network pinning control [Yang *et al.*, 2024b; Wu *et al.*, 2020; Chen, 2022; Wang & Chen, 2002; Olfati-Saber *et al.*, 2007; Chen *et al.*, 2015].

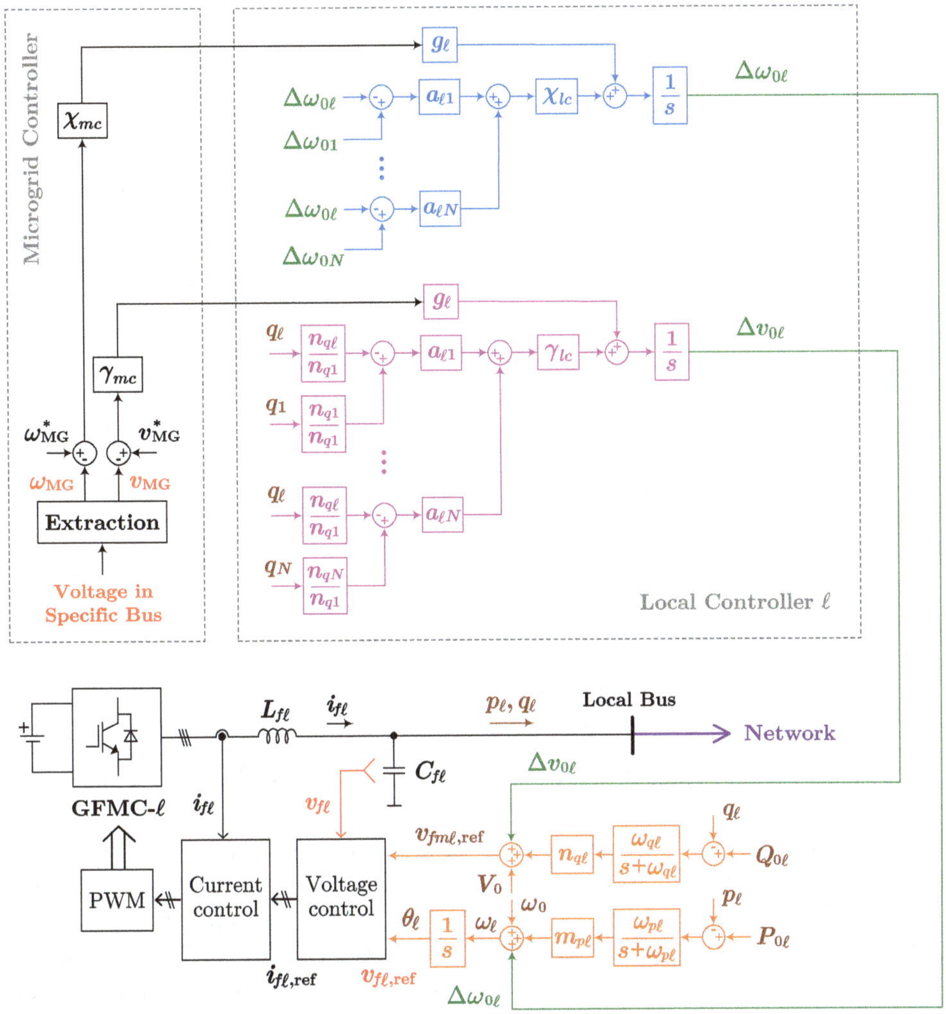

Fig. 3. Distributed control. As an example, only the information exchange between the microgrid controller and the ℓth local controller is exhibited. Here, $\ell = 1, 2, \ldots, N$.

3. Stability and Complex Behavior

3.1. *Stability framework*

The stability framework for converter-dominated power systems is shown in Fig. 4. According to the physical nature of the resulting modes, the instability problems are generally classified into four categories, i.e. synchronization stability, voltage stability, frequency stability, and broadband oscillations [Yang & Tse, 2024; Kundur *et al.*, 2004; Hatziargyriou *et al.*, 2021; Farrokhabadi *et al.*, 2020; Kundur, 1993].

Synchronization stability is concerned with the ability of the grid-connected converters and synchronous generators in a power system to remain in synchronism after being disturbed. In terms of timescale, synchronization stability is concerned with the synchronism of line-frequency waveforms. Thus, it is considered as short-term stability. Depending on the relative size of the disturbance, synchronization stability can be classified into two categories, namely, small-disturbance synchronization stability and large-disturbance synchronization stability (i.e. transient synchronization stability) [Yang & Tse, 2024; Wang *et al.*, 2020].

Voltage stability refers to the ability of a power system to maintain a steady voltage at all buses in the system after being disturbed [Kundur *et al.*, 2004; Hatziargyriou *et al.*, 2021]. From the viewpoint of the relative timescale, voltage stability can be classified as short-term type and long-term type. For the short-term type, it is concerned with the ability of a single bus to maintain its voltage in a normal range [Yang & Tse, 2024]. For the long-term type, it is also called voltage collapse, where cascaded events accompanying short-term voltage

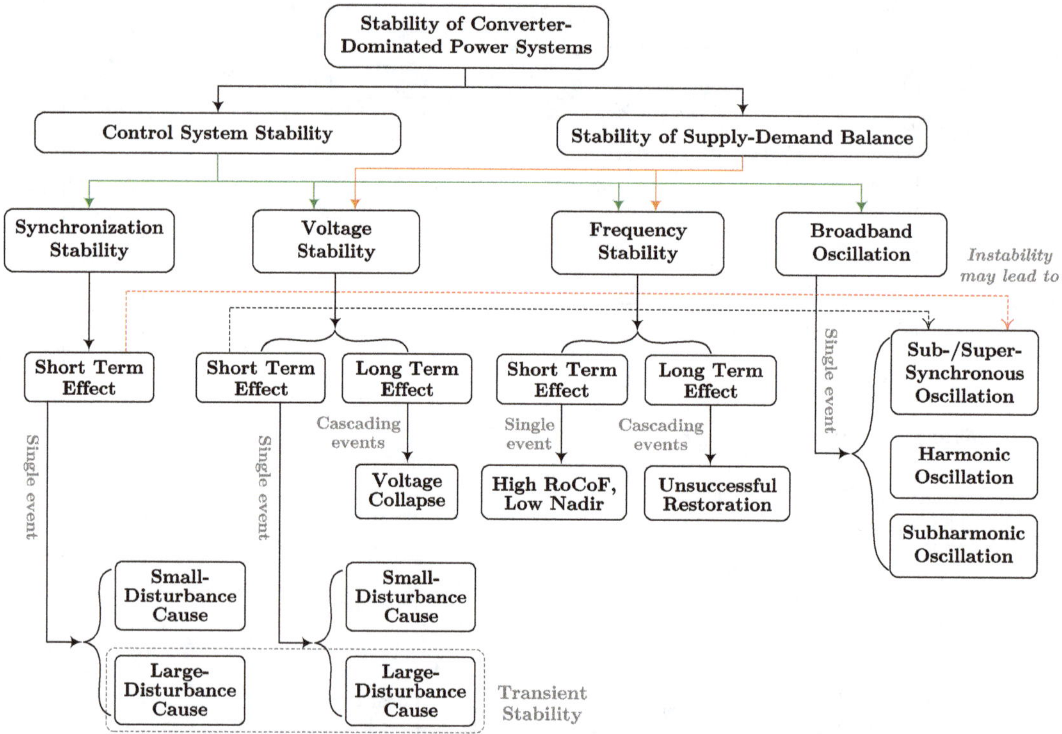

Fig. 4. Classification of the stability of converter-dominated power systems.

instability lead to a blackout or abnormally low voltage for a significant part of the power system [Li *et al.*, 2024a]. Depending on the relative size of the disturbance, short-term voltage stability can also be categorized as small-disturbance voltage stability and large-disturbance voltage stability (i.e. transient voltage stability) [Kundur *et al.*, 2004; Hatziargyriou *et al.*, 2021].

Frequency stability refers to the ability of a power system to maintain a steady frequency following an incident where the supply-demand active power balance is significantly destroyed [Yang & Tse, 2024; Kundur *et al.*, 2004; Hatziargyriou *et al.*, 2021]. From the viewpoint of the relative timescale, frequency stability can be short-term type or long-term type. Short-term frequency stability is represented by the Rate of Change of Frequency (RoCoF) and the frequency nadir, whereas long-term frequency stability is concerned with the ability of successful frequency restoration [Yang *et al.*, 2024b; Fang *et al.*, 2019; Liu *et al.*, 2022].

When a converter-dominated system becomes unstable, sustained broadband oscillations may be generated [Yang & Tse, 2024; Wang *et al.*, 2023]. Depending on the value of the oscillation frequency, broadband oscillation can be classified into three categories, i.e. sub-/super-synchronous oscillation [Sun *et al.*, 2017; Cheng *et al.*, 2023; Li *et al.*, 2020; NERC, 2017; Fan *et al.*, 2022; Liu *et al.*, 2017; Xu *et al.*, 2020], harmonic oscillation [Wan *et al.*, 2015; Ruan *et al.*, 2018; Wang & Blaabjerg, 2019; Yoon *et al.*, 2016], and subharmonic oscillation [Tse, 2003; Iu *et al.*, 2003; Ji *et al.*, 2022; Yang *et al.*, 2020b; Cao *et al.*, 2022]. For the first type, sub-synchronous and super-synchronous oscillations coexist in the current or voltage, but only sub-synchronous oscillation exists in the active/reactive power. For the second type, the oscillation frequency is typically in the range from hundreds to a few thousands hertz, which is intermediate between the double-line frequency and the switching frequency. For the third type, the oscillation frequency is usually slightly lower than the switching frequency.

According to the characteristics of the aforementioned types of instability, synchronization instability, voltage instability, frequency instability, sub-/super-synchronous oscillation, and some harmonic oscillation are associated with slow-scale interaction, where the converters' switching behavior is not involved. Moreover, some harmonic oscillation and subharmonic oscillation are associated with fast-scale interaction, where the converters' switching behavior can have a significant impact on the system's stability [Yang & Tse, 2024].

3.2. *Physical origin behind instability events*

Bifurcation is a fundamental concept in nonlinear dynamics, referring to a qualitative change in a system's behavior due to variations in one or more system parameters, specifically when the phase portrait becomes topologically nonequivalent [Yang & Tse, 2024; Tse, 2003; Kuznetsov, 1998]. Bifurcations are closely tied to changes in system stability and can be broadly classified into two categories: smooth bifurcations and nonsmooth bifurcations [Tse, 2003]. The term "smooth" has a mathematical origin, which relates to the differentiability of the function that describes the system [Tse, 2003]. In other words, smooth bifurcation is an intrinsic behavior of the differential equations, where no nonsmooth component (i.e. boundary condition) is involved, and the system is essentially structurally unchanged. Smooth bifurcation can be further divided into local bifurcation and global bifurcation [Yang & Tse, 2024; Kuznetsov, 1998]. Local bifurcations, which occur near the usual operating point, are detected by analyzing a small neighborhood around it. Common examples include saddle-node bifurcations and supercritical/subcritical Hopf bifurcations. Global bifurcations involve changes in the structure of multiple attractors and their basins of attraction. These cannot be identified by local analysis alone. Typical examples include homoclinic and heteroclinic bifurcations [Yang & Tse, 2024; Kuznetsov, 1998]. In contrast, nonsmooth bifurcations arise in systems that undergo structural changes, often due to discontinuities or switching mechanisms [Tse, 2003]. Typical examples include catastrophic bifurcations and border collision. Catastrophic bifurcations occur when the Pulse-Width Modulation (PWM) saturates, pushing the system into an overmodulation regime [Huang *et al.*, 2017b]. Moreover, border collision occurs in dynamical systems where two or more structurally different systems operate for different parameter ranges [Tse, 2003]. This type of bifurcation is common in converters switching between Continuous Conduction Mode (CCM) and Discontinuous Conduction Mode (DCM) [Tse, 2003]. Border collisions are particularly relevant in grid-connected converter systems, where frequent mode transitions occur. Examples include switching from grid-forming to grid-following control due to current limiting [Gao *et al.*, 2023; Paquette & Divan, 2015; He & Li, 2011; Qoria *et al.*, 2023], and transitioning between islanded and grid-connected operation modes in microgrids [D'Silva *et al.*, 2020]. These bifurcation phenomena are critical for understanding the stability and dynamic performance of modern power electronic systems, especially as they become increasingly complex and interconnected.

Based on the stability framework illustrated in Fig. 4, two fundamental research questions can be identified [Yang & Tse, 2024]:

(1) How does the system become unstable?
(2) What are the post-disturbance phenomena?

Both synchronization stability and short-term voltage stability are closely related to the first question, as they represent two general pathways to instability. To investigate these pathways, visualization tools such as phase portraits are commonly employed to analyze the complex behavior of converter-based systems. As shown in the phase portrait in Fig. 5, the system typically operates around a Stable Equilibrium Point (SEP), referred to as the usual or nominal operating point. This SEP is surrounded by a basin of attraction, meaning that any trajectory originating within this region will eventually converge to the SEP. Outside this basin, other dynamic entities may exist, such as Unstable Equilibrium Points (UEPs), attractors, or repellers. In practical applications, it is crucial to maintain the stability of the usual operating point and ensure that, under disturbances, the system trajectory remains within the SEP's basin of attraction. Therefore, the SEP and its basin of attraction are of primary interest in stability analysis. From this viewpoint, two general pathways to instability are indicated as follows [Yang & Tse, 2024]:

(1) *Pathway to instability via small disturbance*: Under a small disturbance (e.g. the change of the grid impedance), the trajectory is perturbed from the SEP but does not escape the region that initially belongs to the SEP's basin of attraction. However, the SEP becomes a UEP, accompanied by the emergence of a Stable Periodic Orbit (SPO), as shown in Fig. 5(a). Notably, the SPO takes the role of the original SEP and dominates the original SEP's basin of attraction. In this case, the system becomes unstable due to the change of the stability status of the usual operating point. This type of instability is usually associated with a local bifurcation, e.g. Hopf bifurcation [Wan *et al.*, 2015; Huang *et al.*, 2013a; Yang *et al.*, 2022b;

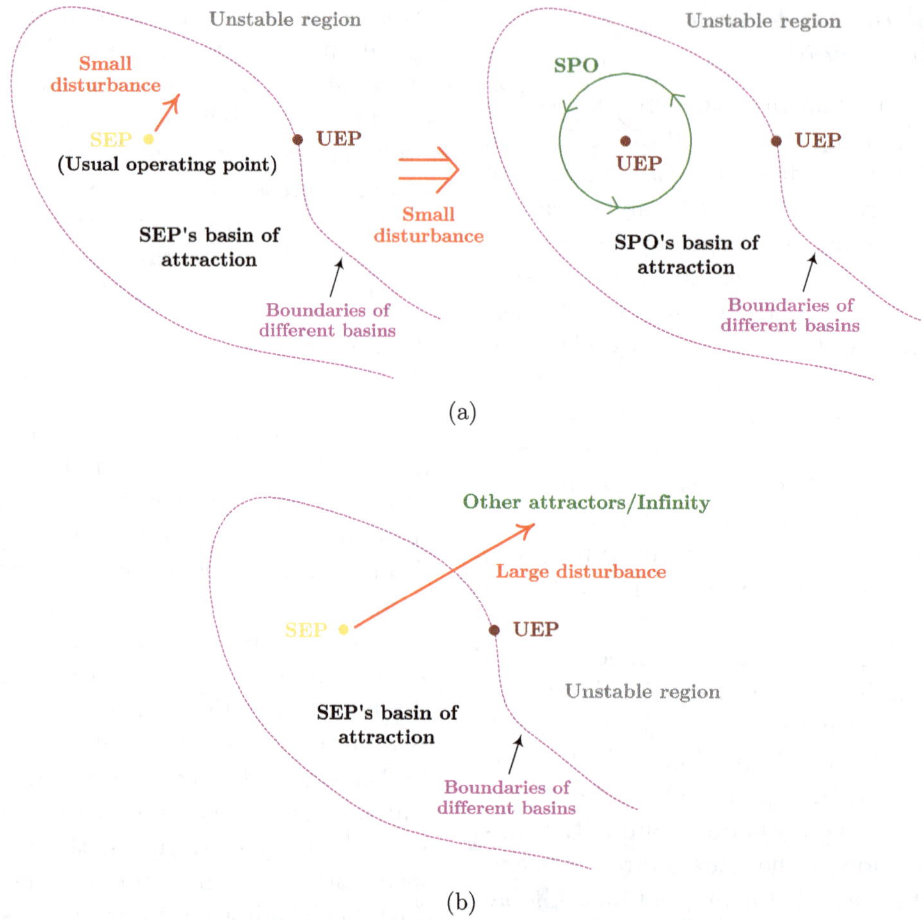

Fig. 5. Phase portraits. (a) Under a small disturbance, the usual operating point becomes a UEP and an SPO. (b) Under a large disturbance, the stability of the usual operating point remains unchanged, but the trajectory moves outside the SEP's basin of attraction.

Moutevelis *et al.*, 2022; Lenz *et al.*, 2018; Shuai *et al.*, 2017; Shuai *et al.*, 2019b].

(2) *Pathway to instability via large disturbance*: Under a large disturbance (e.g. transient short-circuit fault), the usual operating point remains an SEP. However, upon removal of the disturbance, the trajectory escapes the SEP's basin of attraction, and may converge to other attractors or diverge to infinity, as shown in Fig. 5(b). In this case, the system becomes unstable due to the escape of the trajectory out of the SEP's basin of attraction. In this type of instability, the SEP's basin of attraction is of great interest. A large SEP's basin of attraction means high robustness against large disturbances. The sudden change of the appearance of the SEP's basin of attraction is usually associated with a global bifurcation, e.g. homoclinic bifurcation [Yang *et al.*, 2021b; Yang *et al.*, 2024c; Hu *et al.*, 2019; Zhao *et al.*, 2020; Zhao *et al.*, 2021; Ma *et al.*, 2022; Chen *et al.*, 2022a; Shuai *et al.*, 2019c; Shen *et al.*, 2021; Taul *et al.*, 2019].

Moreover, as illustrated in Fig. 4, the emergence of sub-synchronous oscillations corresponds to the second research question and reveals two general pathways through which sub-synchronous oscillations can be generated [Yang & Tse, 2024]. We again use Fig. 5 to illustrate these pathways to instability.

(1) *Pathway to sub-synchronous oscillation through Hopf bifurcation*: As described above, the original SEP becomes a UEP and an SPO upon the onset of a Hopf bifurcation. The trajectory finally converges to an SPO, which manifests as sub-synchronous oscillation, as shown in Fig. 5(a) [Yang *et al.*, 2022b].

(2) *Pathway to sub-synchronous oscillation through existence of multiple attractors and an escape from the SEP's basin of attraction*: In

converter-dominated power systems, multiple attractors, including the SEP, may coexist simultaneously. Each attractor possesses its own basin of attraction, defining the region in which system trajectories will converge to that attractor. If the system is initially operating at the SEP, a specific disturbance may cause the trajectory to exit the SEP's basin and enter the basin of another attractor. In such cases, even after the disturbance is removed, the trajectory cannot return to the SEP; instead, it converges to the alternative attractor, which may manifest as sub-synchronous oscillations, as illustrated in Fig. 5(b) [Yang *et al.*, 2023b]. Notably, this pathway to instability, via basin switching and attractor migration, has received limited attention within the power systems community, despite early reports of sub-synchronous oscillation events triggered by network-switching operations dating back to the 1970s [Kundur, 1993; NERC, 2017; IEEE, 1982].

4. Voltage Stability

Maintaining bus voltages within a prescribed range is essential for ensuring reliable power supply and distribution [Kundur *et al.*, 2004; Hatziargyriou *et al.*, 2021]. Voltage stability is typically illustrated using a simple radial system, as shown in Fig. 6, where the load is modeled as a dynamic active/reactive power load with voltage-dependent characteristics. In this setup, the load maintains constant active and/or reactive power absorption in the steady state. Power-voltage curves derived from the radial system are commonly used to analyze voltage stability. These curves reveal that voltage instability is fundamentally caused by a saddle-node bifurcation, as depicted in Fig. 6, where the system loses its stable operating point due to the merging and disappearance of equilibrium solutions [Kundur, 1993; Cañizares, 1995; Cañizares *et al.*, 1996; Simpson-Porco *et al.*, 2016; Chen *et al.*, 2019; Aik & Andersson, 2019]. However, it is important to note that converter-based loads differ from the aforementioned types of loads, as they interact nonlinearly with the grid, resulting in various routes to voltage instability in converter-dominated systems [Huang *et al.*, 2013a, 2013b; Huang *et al.*, 2017b; Yang *et al.*, 2020a; Yang *et al.*, 2023a; Wan *et al.*, 2013].

Saturation is a common kind of nonlinearity inherent in many engineering systems, e.g. the limiting control embedded in the PWM of GFLR, as shown in Fig. 2. Due to such saturating nonlinearity, the DC-link output current of GFLR will saturate when the GFLR enters the overmodulation mode, giving rise to a special nonlinear relationship between the control command and the output current. In other words, the system exhibits a structural change that triggers the onset of a catastrophic bifurcation [Huang *et al.*, 2013b]. This bifurcation is catastrophic in the sense that the GFLR exhibits an irreversible collapse of DC-link output voltage or/and AC-side input voltage, which is accompanied by a sudden expansion of the magnitude of the reactive current that may damage the system [Tse *et al.*, 2020]. The catastrophic bifurcation was first discovered in a GFLR connected to an ideal grid [Huang *et al.*, 2013b], and later, was commonly found in power grids with significant grid impedances [Huang *et al.*, 2017b]. It has been confirmed that the catastrophic bifurcation plays a crucial role in power system voltage instability, as shown in Fig. 8 [Yang *et al.*, 2023a]. Such catastrophic bifurcation is unique for grid-connected power electronic converters, and can be examined by using the state-space averaging models that discard the switching details [Huang *et al.*, 2013b].

Moreover, it was found that an SPO and an Unstable Periodic Orbit (UPO) might coexist in a GFLR. Therefore, four types of bifurcation

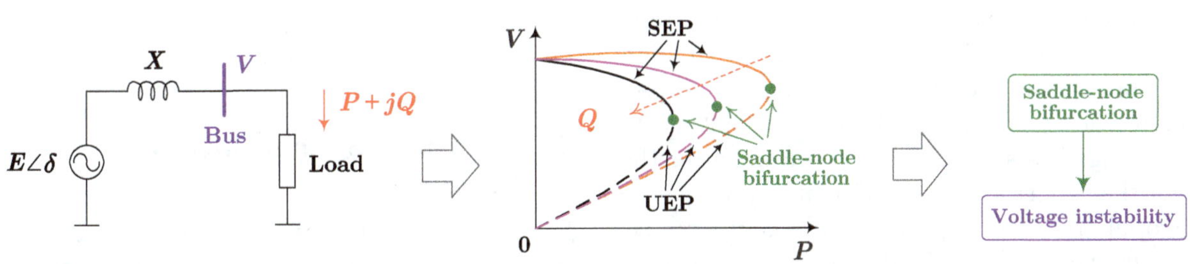

Fig. 6. Conventional study of power system voltage stability using a simple radial system and associated power-voltage curves.

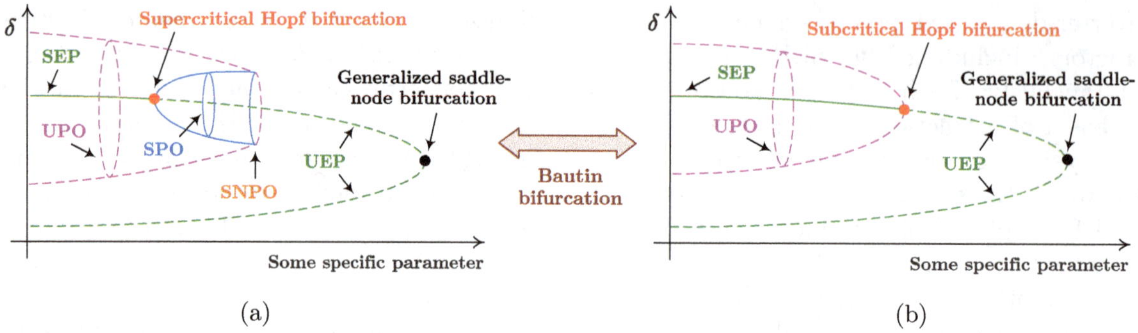

Fig. 7. Codimension-1 bifurcation diagrams corresponding to (a) supercritical Hopf bifurcation and SNPO and (b) subcritical Hopf bifurcation in the GFLR-based system. Between (a) and (b), a Bautin bifurcation occurs.

may occur, namely, supercritical Hopf bifurcation, Saddle-Node bifurcation of Periodic Orbits (SNPO), subcritical Hopf bifurcation, and Bautin bifurcation, as shown in Fig. 7 [Yang *et al.*, 2023a]. From Fig. 7(a), the SPO is generated from a supercritical Hopf bifurcation. Then, its magnitude increases continuously until it merges with the UPO, where an SNPO occurs. After the onset of the SNPO, there is no periodic orbit, or more precisely, there is no attractor. From Fig. 7(b), the SEP absorbs the UPO and becomes a UEP after the onset of the subcritical Hopf bifurcation. Between these two cases shown in Fig. 7, a Bautin bifurcation occurs, which changes the Hopf bifurcation from a supercritical type to a subcritical type, or vice versa. As the UPO is located outside the SPO, this Bautin bifurcation occurs in a subcritical mode. Furthermore, due to the presence of the UPO, the phase space is separated into two regions, one being the SEP/SPO's basin of attraction. Thus, if the initial state is in this region, the trajectory will be attracted to the SEP/SPO. On the contrary, if the initial state is outside this region, the trajectory will diverge to infinity.

For the GFLR-based power system, the trajectory cannot diverge to infinity in practice but be bounded, e.g. limited by the modulation of GFLRs. Hence, as shown in Fig. 7, after the onset of an SNPO, the grid current will increase dramatically and eventually reaches a critical value that triggers a catastrophic bifurcation. Then, the DC-link output voltage and AC-side input voltage collapse, leading to voltage instability [Yang *et al.*, 2023a]. As shown in Fig. 7, a similar situation occurs after the onset of the subcritical Hopf bifurcation, where the catastrophic bifurcation and the voltage instability will occur eventually. It is also worth noting that for some specific ranges of parameters, supercritical

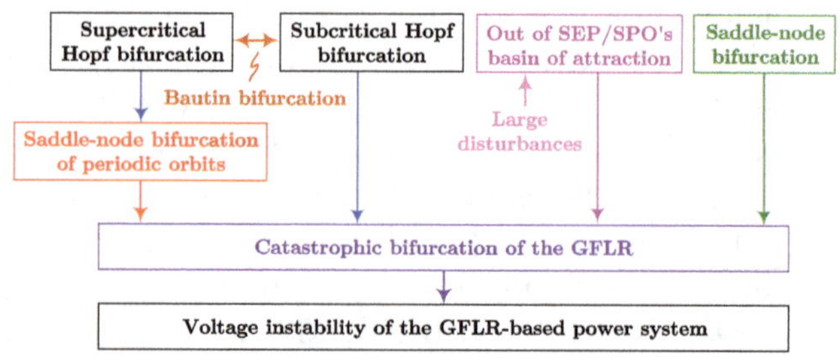

Fig. 8. Routes to the voltage instability in the GFLR-based system.

or subcritical Hopf bifurcation may not occur. The system will operate normally until the onset of a saddle-node bifurcation. As no attractor exists after the onset of the saddle-node bifurcation, the system collapses eventually, also through the onset of a catastrophic bifurcation. Furthermore, under a transient disturbance (e.g. grid voltage dip), the trajectory may be pushed out of the SEP/SPO's basin of attraction, resulting in a catastrophic bifurcation and voltage instability. The routes to the voltage instability are summarized in Fig. 8 [Yang *et al.*, 2023a].

From Figs. 7 and 8, the physical mechanism of voltage instability is fully illustrated through a range of complex behavior of the GFLR-based power system. The diversified routes to voltage instability cannot be exposed using the traditional analytical tools such as impedance-based approach, as such tools can only show the stability status near an assumed operating point.

5. Transient Synchronization Stability

In the study of the transient synchronization stability, the SEP's basin of attraction is of great interest. Large SEP's basin of attraction implies high robustness against large disturbances [Yang & Tse, 2024], as shown in Fig. 5(b).

5.1. *Homoclinic bifurcation and cross-cycle resynchronization*

Suppose the phase angle of the AC-side output voltage of grid-connected converters is θ, as shown in Fig. 1. When a converter is connected to the infinite bus, the phase difference is defined as $\delta = \theta - \theta_s$, where θ_s is the phase angle of the voltage of the infinite bus. Although the SEP is unique, $\delta = \delta_0 + k \times 2\pi$, where δ_0 is constant and $k = 0, \pm 1, \pm 2, \ldots$. Usually, we use $\delta = \delta_0$ to represent the SEP. However, during transient, δ may increase beyond 2π, and finally become $\delta_0 + 2\pi$, which corresponds to the same SEP. In order to distinguish these two different scenarios, we refer to *intra-cycle resynchronization* as the phenomenon that δ starts from δ_0, and eventually returns to δ_0 under transient disturbances, and also refer to *cross-cycle resynchronization* as the phenomenon that δ starts from δ_0, and eventually returns to $\delta_0 + k \times 2\pi$ $(k \neq 0)$ under transient disturbances. The synchronization behavior for these two phenomena is shown in Fig. 9. In practice, intra-cycle resynchronization is commonly known as *first-swing stability*, and it has been known that cross-cycle behavior implies desynchronization. However, cross-cycle resynchronization has remained largely unexplored until the recent discovery of homoclinic bifurcation behavior in grid-connected converter systems.

To study the essential synchronization behavior, both GFLC and GFMC are modeled as second-order systems [Yang *et al.*, 2021b; Yang *et al.*, 2024c]. On this basis, it has been found that there are three UPOs and two homoclinic bifurcations in the GFLC, while there are only one SPO and one homoclinic bifurcation in the GFMC. For the first homoclinic bifurcation, different behaviors of GFLC and GFMC can be explained using the saddle quantity and stable/unstable manifolds, as shown in Fig. 10. The sum of the two eigenvalues of the saddle (i.e. UEP) is defined as the saddle quantity σ. If $\sigma > 0$ on the homoclinic bifurcation point, a UPO will emerge before the onset of homoclinic bifurcation. If $\sigma < 0$ on the homoclinic bifurcation point, an SPO will emerge after the onset of homoclinic bifurcation. Moreover, the UEP has two branches of stable manifolds $\boldsymbol{W}^s$ and two branches of unstable manifolds $\boldsymbol{W}^u$. The manifold is a trajectory with a particular initial condition or final state. For the stable manifold $\boldsymbol{W}^s$, the final state is a UEP. For the unstable manifold $\boldsymbol{W}^u$, the initial condition is a UEP.

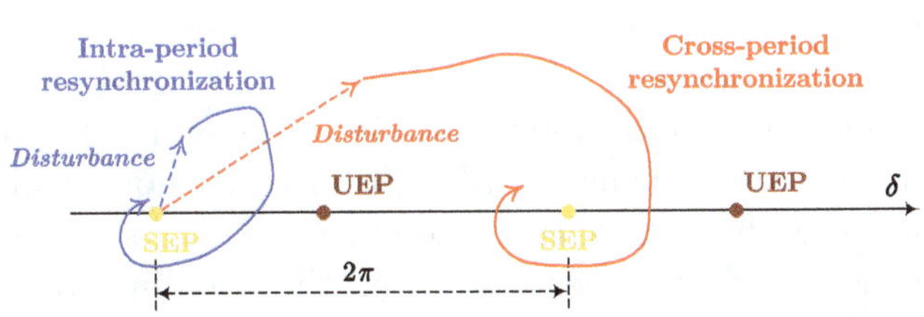

Fig. 9. Intra-cycle resynchronization and cross-cycle resynchronization.

For Fig. 10, the onset direction of the first homoclinic bifurcation is defined as: $\boldsymbol{W}^s$ is on the top of $\boldsymbol{W}^u$ before the homoclinic bifurcation, while $\boldsymbol{W}^s$ is on the bottom of $\boldsymbol{W}^u$ after the homoclinic bifurcation. It is calculated that $\sigma > 0$ for the GFLC, and $\sigma < 0$ for the GFMC, all the time, implying the duality property between the GFLC and GFMC. In other words, for the GFLC, a UPO emerges before the onset of homoclinic bifurcation, and no UPO exists after the onset of homoclinic bifurcation; while for the GFMC, no SPO exists before the onset of homoclinic bifurcation, and an SPO emerges after the onset of homoclinic bifurcation, as shown in Fig. 10.

From Fig. 10, before the onset of the homoclinic bifurcation, the SEP's basin of attraction is continuous in the power-angle direction. Thus, GFLCs and GFMCs can resynchronize with the grid within several cycles of the power angle. In this case, both GFLCs and GFMCs can achieve the cross-cycle resynchronization. On the other hand, after the onset of the homoclinic bifurcation, the SEP's basin of attraction is discontinuous in the power-angle direction. Thus, GFLCs and GFMCs may desynchronize with the grid beyond one cycle of the power angle. In this case, both GFLCs and GFMCs can only achieve the intra-cycle resynchronization, but not the cross-cycle resynchronization.

Therefore, the converter-based system shows much higher robustness against transient disturbance if appropriate parameters can be selected to avoid the onset of the first homoclinic bifurcation.

5.2. *Impacts of extra control loops*

For GFLCs, an ideal current loop is assumed, where the AC-side current perfectly follows its control reference (i.e. constant current source with infinite output admittance). For GFMCs, an ideal

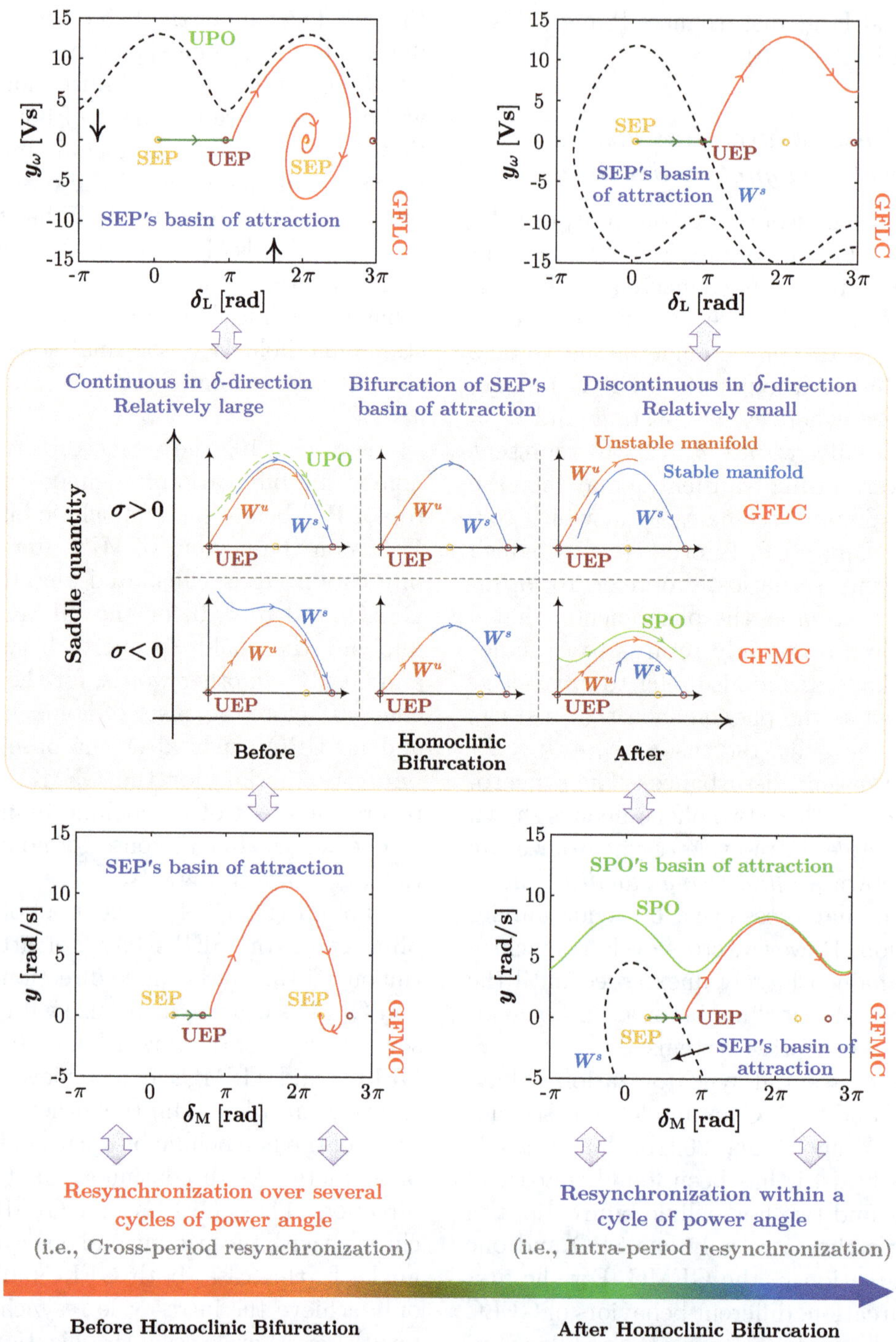

Fig. 10. Homoclinic bifurcation behaviors of GFLCs and GFMCs.

voltage loop is assumed, where the AC-side voltage perfectly follows its control reference (i.e. constant voltage source with zero output impedance). Although such simplifying assumptions allow the essential synchronization behavior to be retained in the second-order models, the transient stability assessment may not be accurate enough for practical purposes because: (1) the inner loops of GFLCs and GFMCs (i.e. current inner loops for GFLCs [Liu *et al.*, 2023; Chen *et al.*, 2020; Hu *et al.*, 2021], and predominately voltage inner loops for GFMCs [Yang *et al.*, 2022b]) are not ideal; (2) the current control reference of GFLCs is not constant, but generated from the DC-link voltage control [Yang & Tse, 2024], while the voltage control reference of GFMCs is regulated by the reactive power droop control and the virtual impedance control [Yang *et al.*, 2021b; Yang *et al.*, 2022b]; (3) there are variants of grid-forming/grid-following controls (e.g. virtual oscillator control for GFMCs [Yu *et al.*, 2021; Lu, 2022], and direct power control for GFLCs [Liu *et al.*, 2024]), which may show significantly different dynamical behaviors, compared with typical droop/PLL-based control schemes; and (4) limitations inherent in the control loops (e.g. current limit) may lead to more complicated dynamics [Gao *et al.*, 2023; Paquette & Divan, 2015; He & Li, 2011; Qoria *et al.*, 2023], which cannot be captured by the original second-order models.

Basically, these practical factors (such as nonideal current loops) may enhance or worsen the transient synchronization stability by reshaping the SEP's basin of attraction [Liu *et al.*, 2023]. An example is presented as follows. From Sec. 5.1, there are two homoclinic bifurcations in the GFLC. Here, we focus on the impact of the current loop and voltage feedforward control of GFLCs on the second homoclinic bifurcation, as shown in Fig. 11. From Fig. 11, some observations are summarized as follows:

(1) If the bandwidth of the current loop is relatively wide, and/or the coefficient of the voltage feedforward control is relatively large, the complex behavior of the full-order system (considering the nonideal current loop) is consistent with that of the PLL-based second-order model. In this case, there is only one UPO surrounding the SEP. This UPO is generated upon a subcritical Hopf bifurcation and subsides via a homoclinic bifurcation. After the onset of the homoclinic bifurcation, the SEP's basin of attraction enlarges significantly. Therefore, in this scenario, the impact of the current loop and voltage feedforward control on the variation of SEP's basin of attraction is relatively small. Thus, neglecting the dynamics of the current loop and voltage feedforward control does not reduce the accuracy of the transient stability assessment.

(2) If the bandwidth of the current loop is relatively narrow, and/or the coefficient of the voltage feedforward control is relatively small, the complex behavior of the full-order system (considering the nonideal current loop) is distinct from that of the PLL-based second-order model. For the full-order system, there are two UPOs that surround the SEP. The blue UPO is generated from the onset of a homoclinic bifurcation, while the red UPO is generated from the onset of a subcritical Hopf bifurcation. These two UPOs merge and disappear through the onset of an SNPO, leading to a significant enlargement of the SEP's basin of attraction. Thus, in this scenario, the impact of the current loop and voltage feedforward control on the variation of SEP's basin of attraction is considerably more significant. In other words, the dynamics of the current loop and voltage feedforward control should be retained to guarantee the accuracy of transient stability prediction.

5.3. *Analytical methods*

Usually, when we attempt to identify the complexity of a multiconverter system and assess the system's stability, the following analytical procedure

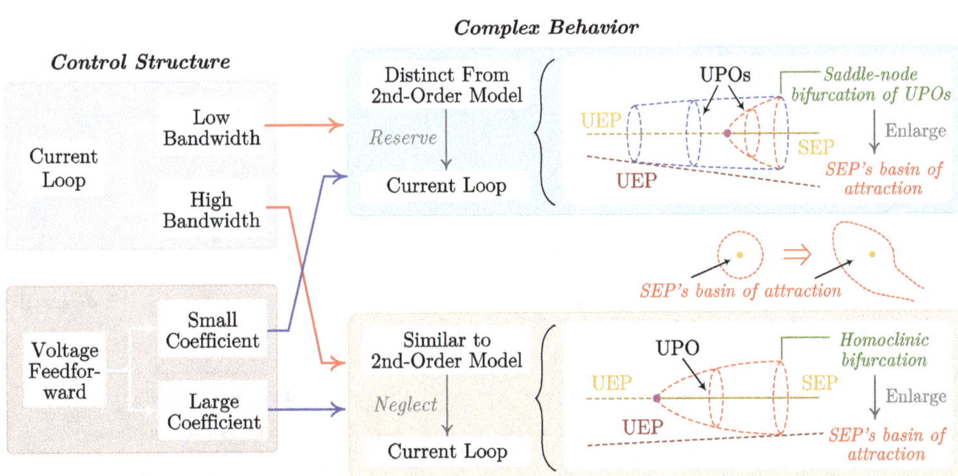

Fig. 11. Impact of the nonideal current loops of GFLCs on the complex behavior and transient synchronization stability.

is adopted, i.e. (1) modeling the multiconverter system as a set of ordinary differential equations [Yang & Tse, 2024; Pogaku *et al.*, 2007]; (2) applying an appropriate order reduction technique to consider the dynamical behavior of interest; (3) applying a suitable analytical method to solve the differential equations established in the previous step to study the characteristics of the system.

For a multiconverter system, a reduced-order model offers the advantage of capturing essential dynamical information while maintaining computational efficiency. Typically, two approaches are employed to achieve order reduction. In the first approach, the dynamics of control loops that have minimal impacts are ignored. This approach involves ignoring the dynamics of control loops that have negligible influence on the behavior of interest. For instance, the current loops of GFMCs do not affect the Hopf bifurcation associated with the onset of sub-synchronous oscillations. As a result, these current loops can be excluded when studying Hopf bifurcation behavior [Yang *et al.*, 2022b; Yang *et al.*, 2023b; Nikolakakos *et al.*, 2018; Vorobev *et al.*, 2018; Luo & Dhople, 2014; Rasheduzzaman *et al.*, 2015; Gu *et al.*, 2018]. The second approach involves model aggregation, truncation, and residualization. By aggregating subsystems, truncating unimportant dynamics, or residualizing fast-mode state variables associated with eigenvalues far from the imaginary axis, the system model can be greatly simplified while preserving the salient characteristics [Shuai *et al.*, 2019a; Purba *et al.*, 2019; Taul *et al.*, 2021; Liao *et al.*, 2019; Skogestad & Postlethwaite, 2001]. With the reduced-order model established, the next step is to analyze it. Generally, two distinct analytical methods are available for this purpose.

The first approach involves directly studying the differential equations of the system. This includes identifying some or all attractors and repellers, defining the basins of attraction for the attractors, and analyzing bifurcation behaviors as one or more system parameters are varied. Once this information is gathered, the next step is to uncover the relationships between these behaviors and the overall stability of the system. For example, this analysis can reveal pathways to voltage instability through a sequence of bifurcations [Yang *et al.*, 2023a], as discussed in Sec. 4, or demonstrate how transient synchronization can be enhanced via the onset of a homoclinic bifurcation [Yang *et al.*, 2021b; Yang *et al.*, 2024c], as elaborated in Sec. 5.1.

The second approach involves employing appropriate scalar functions (i.e. real-valued energy functions) to assess the stability of the system's usual operating point. If the operating point is determined to be stable (i.e. a stable equilibrium point, or SEP), the basin of attraction of the SEP can then be identified, which represents the system's domain of attraction [Shuai *et al.*, 2019c; Liu *et al.*, 2021; Liu *et al.*, 2025; Kabalan *et al.*, 2017; Fu *et al.*, 2021; Fu *et al.*, 2023; Zhang *et al.*, 2022b; Cheng *et al.*, 2020; Shuai, 2021; Tian *et al.*, 2022; Tian *et al.*, 2023; Li *et al.*, 2024b; Li *et al.*, 2025]. Commonly used energy functions include the equal-area criterion function, mixed-potential function, and Lyapunov function. These functions are typically constructed by combining various energy terms. However, due to their inherent conservativeness, additional numerical iterations and optimization

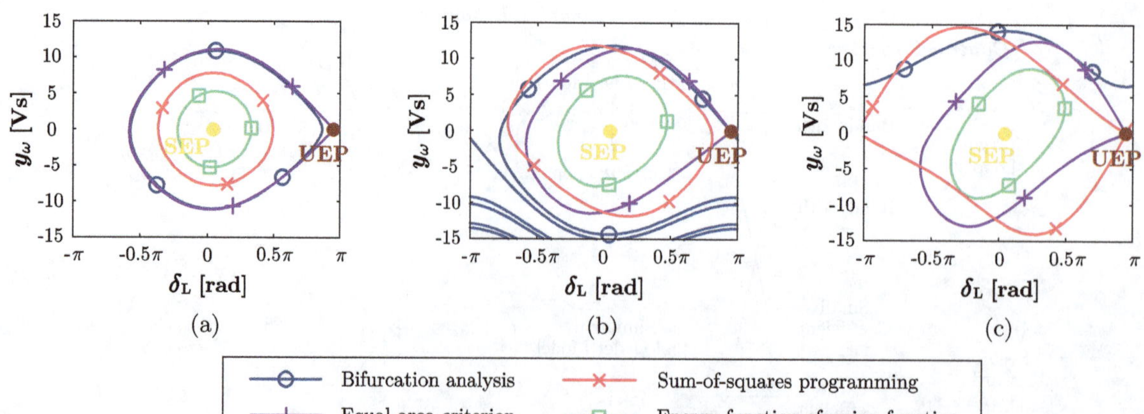

Fig. 12. Estimation of the SEP's basin of attraction in GFLC-based system with varying values of the PLL proportional coefficient, and proportional coefficient being (a) small, (b) medium, and (c) large.

techniques are often required to enhance their accuracy. Examples of such techniques include the Takagi–Sugeno multimodeling method and the sum-of-squares programming method [Liu *et al.*, 2025; Cheng *et al.*, 2020; Marx *et al.*, 2012; Zhang *et al.*, 2022a].

The two categories of analytical methods described above are complementary in uncovering essential physical mechanisms and enabling practical applications in power systems. For instance, energy-based methods (the second category) often face limitations such as conservativeness and an inability to capture sudden changes in the shape of the Stability Equilibrium Point (SEP) basin of attraction, as illustrated in Fig. 12 [Liu *et al.*, 2025]. These shortcomings can be addressed by investigating global bifurcation behavior. However, power grid operators may find it challenging to directly apply the theoretical insights derived from bifurcation analysis to tasks such as online stability margin prediction, critical clearing time/angle calculation, and related applications. To overcome this limitation, constructing an energy function to estimate the domain of attraction offers a practical solution.

6. Double Effects of Hidden Attractors

In the study of power system stability, significant research efforts have focused on assessing the stability of a single assumed operating point. Analytical techniques such as small-signal modeling, impedance-based methods, and eigenvalue-based methods are commonly employed for this purpose. However, real-world power systems often exhibit the coexistence of multiple attractors, with the usual operating point being just one of them, as illustrated in Fig. 5(b). Despite the widespread occurrence of multiple attractors, limited research has been conducted to investigate their characteristics and impacts in power systems [Tse, 2003; NERC, 2017; Banerjee, 1997; Geng *et al.*, 2023]. For instance, sub-synchronous oscillations caused by network-switching operations have been observed since the early 1970s [Kundur, 1993; NERC, 2017; IEEE, 1982]. However, the fundamental mechanisms behind these oscillations remained unclear until recent years, when they were found to be strongly related to the coexistence of multiple attractors [Yang *et al.*, 2023b].

An important concept related to "multiple attractors" is the notion of "hidden attractor", which stems from efforts to uncover the fundamental mechanisms responsible for generating attractors, particularly chaotic ones. In the nonlinear science community, the emergence of conventional chaotic attractors, such as those in the Lorenz system, Rössler system, and Chen system, is typically linked to one or more UEPs in the phase space [Wang *et al.*, 2021]. However, in the early 2010s, researchers reported the existence of a new type of chaotic attractor in Chua's circuit, distinct from the well-known Chua's attractor. This newly discovered chaotic attractor could not be associated with the neighborhood of any UEPs. In other words, the UEPs had no role in the generation of this attractor, rendering it the classification of a hidden attractor [Leonov *et al.*, 2011; Leonov & Kuznetsov, 2013]. Around the same time, it was discovered that in a simple three-dimensional autonomous quadratic system, no UEPs existed, yet a SEP, a SPO, and a chaotic attractor coexisted [Wang & Chen, 2012]. Both the SPO and the chaotic attractor were classified as hidden attractors due to the absence of UEPs [Leonov & Kuznetsov, 2013; Sprott *et al.*, 2013; Yang *et al.*, 2022a; Bhindwar *et al.*, 2024; Foroni, 2023]. This discovery expanded the understanding of hidden attractors and highlighted their distinctive nature in dynamical systems.

In recent years, numerous hidden attractors have been identified in multiconverter systems [Yang *et al.*, 2023b; Yang *et al.*, 2025]. These hidden attractors can have a dual impact on the transient stability of such systems.

6.1. *Harmful effect*

In an islanded microgrid comprising several GFMCs, various types of hidden attractors have been identified, coexisting with the SEP and manifesting as sub-synchronous oscillations in the time domain [Yang *et al.*, 2023b]. These hidden attractors neither emerge nor vanish via a homoclinic bifurcation or a Hopf bifurcation. An example is shown in Fig. 13, where the two quasiperiodic orbits and the chaotic attractor are hidden attractors. Under transient disturbances, such as Short-Circuit Faults (SCFs), the system trajectory can be perturbed away from the SEP and may eventually converge to one of these hidden attractors, as shown in Fig. 14. Specifically, Fig. 14(a) depicts convergence to a hidden quasiperiodic orbit, while Fig. 14(b) shows convergence to a hidden chaotic attractor.

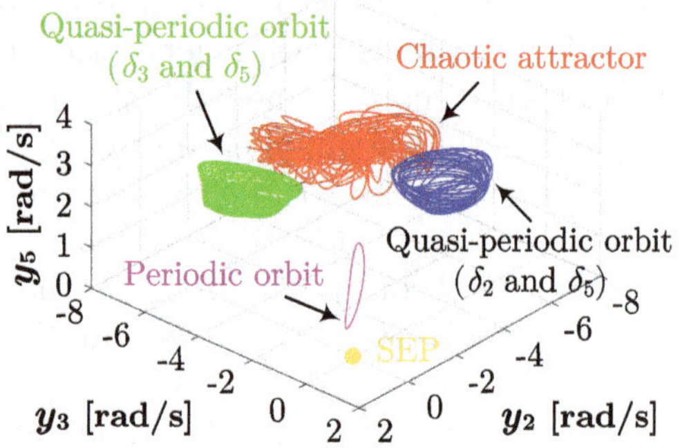

Fig. 13. Several hidden attractors in the phase space including two quasiperiodic orbits and a chaotic attractor.

Moreover, it was found that the hidden periodic orbit with several phase-locked angles may become a hidden quasiperiodic orbit through an SNPO [Yang *et al.*, 2023b]. The SNPO here is different from that described in Secs. 4 and 5.2. The SNPO here focuses on the synchronization status of power angles, whereas the SNPO in the previous sections is restricted to a specific power angle.

6.2. *Beneficial effect*

For an islanded microgrid comprising multiple GFMCs and GFLCs, its transient synchronization stability is investigated by analyzing the characteristics of hidden attractors and their associated bifurcation behaviors. It has been found that the microgrid's transient stability can be enhanced by sustaining a special type of hidden attractor and/or enlarging the region of the SEP's basin of attraction [Yang *et al.*, 2025]. In the presence of this special type of hidden attractor, the PLL functions properly, and the GFLCs remain synchronized with the microgrid. To provide some intuition, a simple example is shown in Fig. 15. In this example, the SEP's basin of attraction is significantly enlarged, as illustrated in Figs. 15(a) and 15(c). Initially, the system operates at the SEP, as shown in Fig. 15(a). Then, a SCF is applied to the microgrid. During the SCF, no equilibrium point exists — including the SEP — but a hidden attractor emerges, where the PLL continues to function properly, as shown in Fig. 15(b). In this case, the system trajectory transitions from the pre-fault SEP to the hidden attractor. Notably, this hidden attractor is located near the pre-fault SEP and lies within its basin of attraction. After the SCF is removed, the hidden attractor subsides, the SEP reappears, and its basin of attraction is restored, as shown in Fig. 15(c). Since the trajectory after the SCF is removed remains within the post-fault SEP's basin

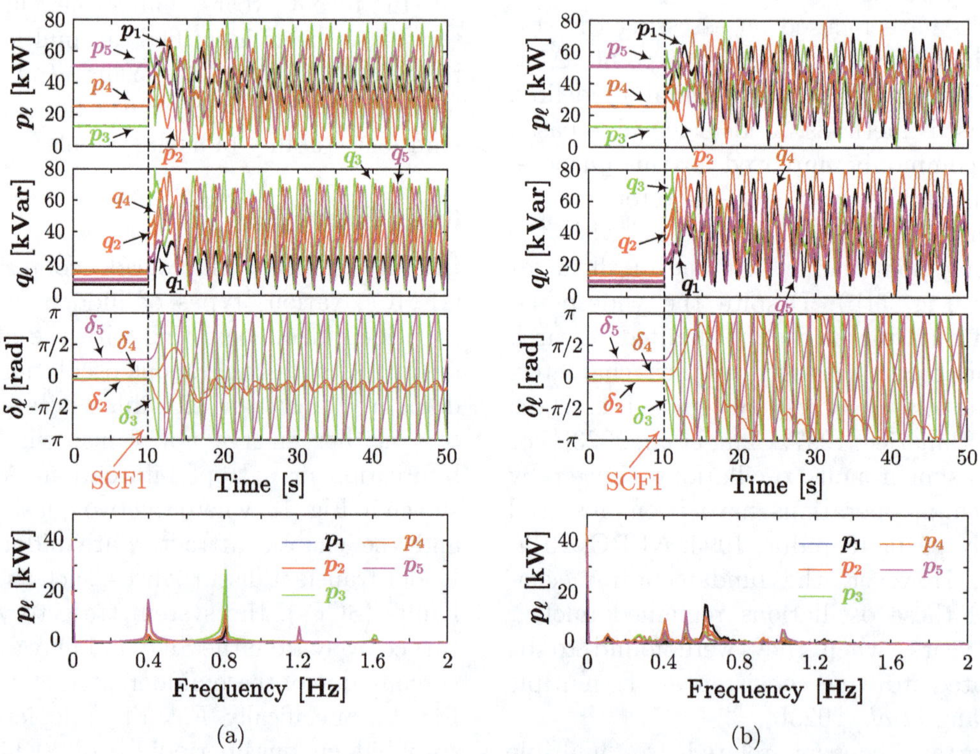

Fig. 14. Sub-synchronous oscillations due to transient SCFs, with fault duration being (a) 0.8 sec, and (b) 1 sec.

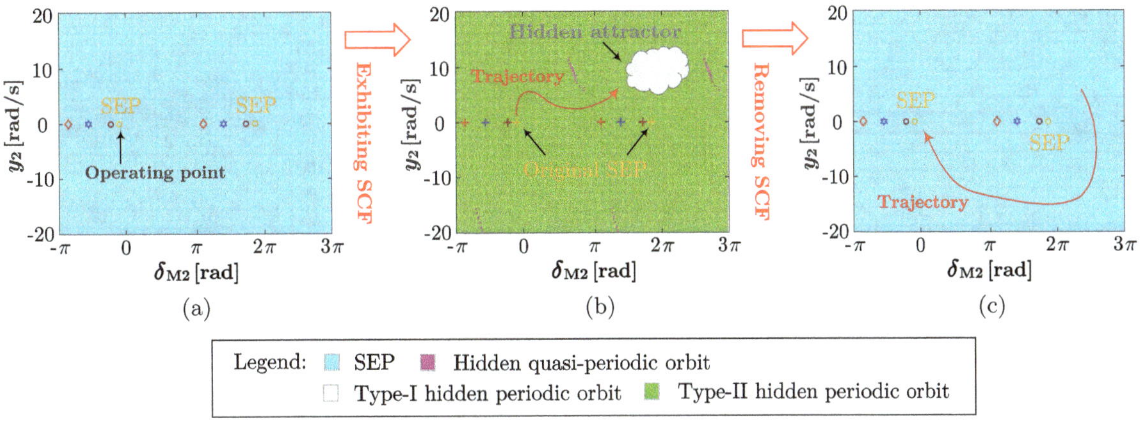

Fig. 15. Example of positive impact of the hidden attractors. Basin of attractions (a) before applying the SCF to the system; (b) during the SCF; and (c) after removing the SCF.

of attraction, it eventually returns to the SEP. As a result, the microgrid's transient synchronization stability is preserved. This example demonstrates the beneficial effect of hidden attractors in maintaining transient synchronization stability.

Moreover, the aforementioned hidden attractor has been found to originate from an SNPO [Yang *et al.*, 2025]. The onset point of the SNPO is closely related to the injected active and reactive powers/currents of the GFLCs. A moderate injection of active and reactive power by the GFLCs can trigger the onset of the SNPO, thereby sustaining the hidden attractor and enhancing the microgrid's transient synchronization stability.

6.3. *Fractals and multiattractors*

For a single grid-following or GFMC connected to an infinite bus, it has been reported that the structure of the SEP's basin of attraction may repeat itself infinitely in certain regions, resembling a fractal pattern [Yang *et al.*, 2024c]. This phenomenon becomes more pronounced in multiconverter systems, where multiple attractors may coexist simultaneously. The basins of attraction for different attractors can have fractal boundaries [Yang *et al.*, 2023b]. Any trajectory near these boundaries may exhibit transient chaotic behavior and high final-state sensitivity [Yang *et al.*, 2023b; Peitgen *et al.*, 2004]. In other words, if the initial condition lies near a fractal basin boundary, the trajectory may wander around the boundary for some time before eventually converging to a specific attractor. In the time domain, this results in transient chaos prior to settling into an attractor. Moreover, a slight adjustment to the initial condition can lead to convergence to a different final attractor — this phenomenon is known as final-state sensitivity. Furthermore, a hidden chaotic attractor may emerge when the aforementioned fractal basin boundary becomes dominant, typically due to the adjustment of specific system parameters. In such cases, the trajectory may remain trapped near the fractal boundary for an extended duration, resulting in the formation of a hidden chaotic attractor [Yang *et al.*, 2023b]. This represents a novel route to chaos, distinct from the four classical routes: period doubling, quasi-periodicity, intermittency, and crisis [Yang & Tse, 2024; Tse, 2003].

7. Impact of Distributed Secondary Control

From (5) and (6), the distributed secondary control consists of two components: a pinning-based algorithm and a consensus-based algorithm. In the former, only a small subset of local controllers receives control signals directly from the microgrid controller. These controllers are referred to as driver controllers, a term adopted from network science [Chen, 2022]. The selection of driver controllers, along with the topology of the communication network, plays a crucial role in determining the microgrid's stability — particularly its transient response and frequency stability [Yang *et al.*, 2024b].

7.1. *Transient response*

In the established microgrid model, the Jacobian matrix is typically nondiagonal, indicating that the state variables are interdependent. This

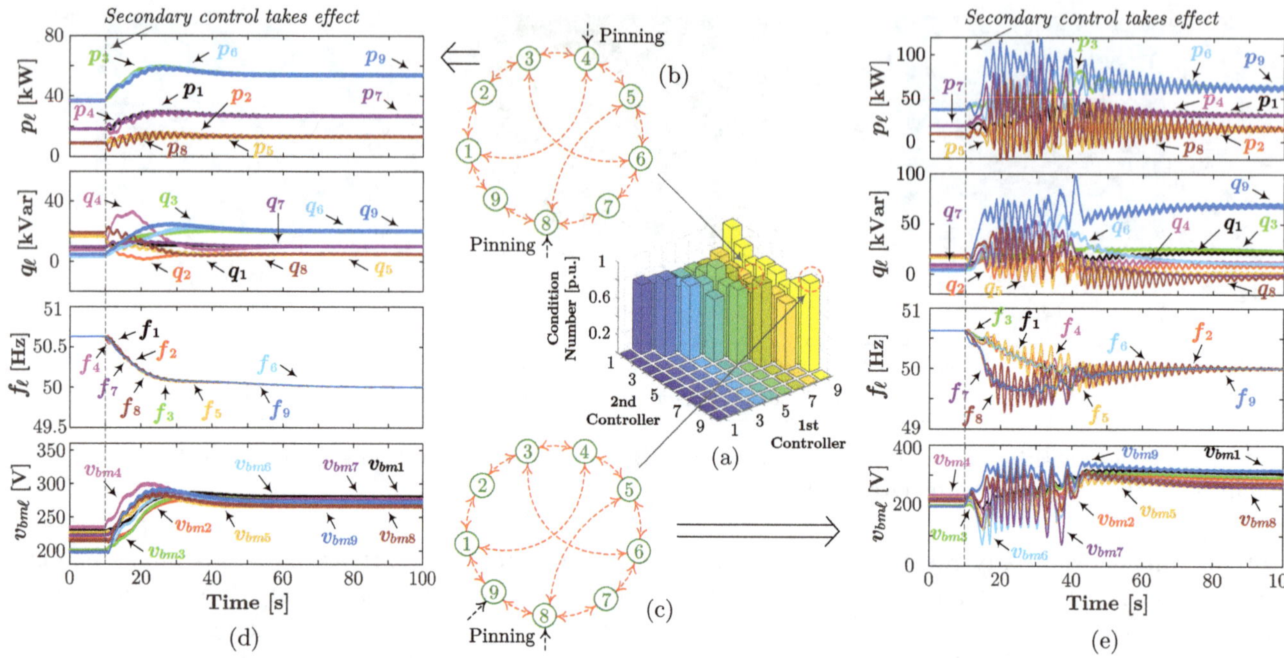

Fig. 16. For an islanded microgrid containing nine GFMCs, (a) values of the condition number for different selections of driver controllers; (b) selected communication network and driver controllers for the distributed secondary control, where the driver controllers are local controllers 4 and 8; (c) selected communication network and driver controllers for the distributed secondary control, where the driver controllers are local controllers 8 and 9; (d) time-domain waveforms of active powers p_ℓ, reactive powers q_ℓ, frequencies f_ℓ, and voltages $v_{bm\ell}$ for the distributed secondary control system shown in (b); (e) time-domain waveforms of active powers p_ℓ, reactive powers q_ℓ, frequencies f_ℓ, and voltages $v_{bm\ell}$ for the distributed secondary control system shown in (c).

interdependence can lead to pronounced transient oscillations, especially when conflicting control tasks interact. Although these oscillations may eventually subside, allowing the system trajectory to converge to the standard SEP, their intensity and duration can significantly affect system performance. To quantify the degree of interaction among state variables, the condition number of the Jacobian matrix is employed [Yang *et al.*, 2024b]. A higher condition number reflects stronger coupling between state variables, which in turn results in more intense and potentially destabilizing transient oscillations.

An islanded microgrid comprising nine GFMCs is used as an illustrative example. In this system, each GFMC is paired with a local controller, resulting in a total of nine local controllers. Among them, two are selected to act as driver controllers. The selection of these driver controllers significantly affects the system's condition number, as shown in Fig. 16(a). For example, when local controllers 4 and 8 are chosen as driver controllers, the condition number is 0.6775 p.u. In contrast, selecting controllers 8 and 9 results in a higher condition number of 0.9738 p.u., as illustrated in Figs. 16(a)–16(c). In the former case [Fig. 16(b)], the time-domain waveforms of active powers p_ℓ, reactive powers q_ℓ, frequencies f_ℓ, and voltages $v_{bm\ell}$ before and after the activation of the distributed secondary control are shown in Fig. 16(d). These waveforms indicate that the system quickly and smoothly converges to a stable operating state after the control takes effect. In the latter case [Fig. 16(c)], the corresponding waveforms are presented in Fig. 16(e). Here, after the secondary control is activated, p_ℓ, q_ℓ, f_ℓ and $v_{bm\ell}$ exhibit severe oscillations, posing a risk to the practical operation of the microgrid. This example clearly demonstrates that the selection of driver controllers has a substantial impact on the transient dynamics and overall stability of the microgrid.

7.2. *Frequency stability*

As discussed in Sec. 3, short-term frequency stability is primarily associated with the RoCoF and the frequency nadir, while long-term frequency stability relates to frequency restoration — specifically, the time it takes for the grid frequency to return to its nominal value [Yang *et al.*, 2024b; Fang *et al.*, 2019; Liu *et al.*, 2022]. For short-term frequency stability, the virtual inertia provided by the primary control plays a dominant role in reducing RoCoF and increasing the frequency nadir. Additionally, factors such as the communication network topology, the selection of driver controllers, and certain parameters of the secondary control (e.g. the cutoff frequencies of low-pass filters) also influence short-term frequency behavior [Yang *et al.*, 2024b]. For example, the structure of the communication network significantly affects the convergence speed of state variables in consensus-based algorithms. A slower convergence rate can lead to pronounced oscillations in grid frequency, resulting in a lower frequency nadir. Moreover, slow convergence typically prolongs the time required for the frequency to return to its nominal value, thereby adversely affecting long-term frequency stability. The impact of the communication network structure on convergence speed can be quantitatively assessed using the algebraic connectivity of the network [Yang *et al.*, 2024b]. In most multiconverter microgrids, the communication network is modeled as a connected, undirected graph. Consequently, the Laplacian matrix of the network is symmetric and positive semi-definite, with real, non-negative eigenvalues. The smallest eigenvalue is zero, and the second smallest, known as the algebraic connectivity, serves as a key indicator of the network's robustness [Yang *et al.*, 2024b; Chen, 2022; Chen *et al.*, 2015]. In practice, designing the communication network to maximize algebraic connectivity can significantly enhance frequency stability — resulting in a higher frequency nadir and faster frequency restoration.

8. Conclusion

The penetration of power electronics into power generation and distribution systems is expected to continue increasing. In this context, the inherently nonlinear nature of power electronic devices will significantly alter power system dynamics, potentially leading to a higher incidence of instability events and power blackouts. Given this trend, it is imperative to shift away from conventional linear perspectives and methodologies in order to gain a more accurate and comprehensive understanding of emerging instability phenomena. As demonstrated in this paper, the nonlinearity of grid-connected converters enriches the dynamic behavior of power systems — manifesting in unexpected, diverse, complex, and often intriguing phenomena that play a critical role in many instability events. This new perspective enables a deeper understanding of the fundamental mechanisms behind system instability and supports the development of optimal mitigation strategies to prevent blackouts and facilitate the integration of renewable energy sources and advanced power electronic technologies. The analysis of complex behaviors and instability mechanisms in grid-connected power electronic systems remains a vital and evolving area of research. It deserves greater attention and continued investigation to ensure the resilience and reliability of future power systems.

Acknowledgment

This work was supported by Hong Kong Research Grant Council GRF Project 11201524.

ORCID

Jingxi Yang
https://orcid.org/0000-0003-3137-7444
Chi K. Tse
https://orcid.org/0000-0002-0462-3999
Meng Huang
https://orcid.org/0000-0002-0139-9178

References

AEMC [2019] "Review of the South Australian black system event," Final report, https://www.aemc.gov.au/sites/default/files/documents/.

Aik, D. L. H. & Andersson, G. [2019] "Fundamental analysis of voltage and power stability of single-infeed voltage-source converter HVDC systems," *IEEE Trans. Power Delivery* **34**, 365–375.

Banerjee, S. [1997] "Coexisting attractors, chaotic saddles, and fractal basins in a power electronic circuit," *IEEE Trans. Circuits Syst.* **44**, 847–849.

Bhindwar, M., Varshney, V., Kumarasamy, S., Shrimali, M. D. & Prasad, A. [2024] "Role of UPOs in characterizing the hidden attractors: A comparison with self-excited attractors," *Int. J. Bifurcation and Chaos* **34**, 2430016.

Blaabjerg, F. [2018] *Control of Power Electronic Converters and Systems* (Academic Press, London, UK).

Blaabjerg, F., Yang, Y., Kim, K. A. & Rodriguez, J. [2023] "Power electronics technology for large-scale renewable energy generation," *Proc. IEEE* **111**, 335–355.

Bose, B. K. [2022] "Power electronics: My life and vision for the future [my view]," *IEEE Ind. Electron. Mag.* **16**, 65–72.

Buticchi, G., Lam, C.-S., Ruan, X., Liserre, M., Barater, D., Benbouzid, M., Gomis-Bellmunt, O., Ramos-Paja, C. A., Kumar, C. & Zhu, R. [2021] "The role of renewable energy system in reshaping the electrical grid scenario," *IEEE Open J. Ind. Electron. Soc.* **2**, 451–468.

Cañizares, C. A. [1995] "On bifurcations, voltage collapse and load modeling," *IEEE Trans. Power Syst.* **10**, 512–522.

Cañizares, C. A., Souza, A. C. Z. D. & Quintana, V. H. [1996] "Comparison of performance indices for detection of proximity to voltage collapse," *IEEE Trans. Power Syst.* **11**, 1441–1450.

Cao, H., Wang, F. & Liu, J. [2022] "Analysis of fast-scale instability in three-level T-type single-phase inverter feeding diode-bridge rectifier with inductive load," *IEEE Trans. Power Electron.* **37**, 15066–15083.

Chen, G. & Dong, X. [1998] *From Chaos to Order: Methodologies, Perspectives and Applications* (World Scientific, Singapore).

Chen, G. & Ueta, T. [1999] "Yet another chaotic attractor," *Int. J. Bifurcation and Chaos* **9**, 1465–1466.

Chen, G. & Ueta, T. [2002] *Chaos in Circuits and Systems* (World Scientific, Singapore).

Chen, Y., Tse, C. K., Qiu, S. S., Lindenmuller, L. & Schwarz, W. [2008] "Coexisting fast-scale and slow-scale instability in current-mode controlled DC/DC converters: Analysis, simulation and experimental results," *IEEE Trans. Circuits Syst.-I: Regul. Pap.* **55**, 3335–3348.

Chen, G., Wang, X. & Li, X. [2015] *Fundamentals of Complex Networks: Models, Structures and Dynamics* (Wiley, Singapore).

Chen, J., Milano, F. & O'Donnell, T. [2019] "Assessment of grid-feeding converter voltage stability," *IEEE Trans. Power Syst.* **34**, 3980–3982.

Chen, G. [2020] "Generalized Lorenz systems family," arXiv preprint arXiv:2006.04066.

Chen, J., Liu, M., O'Donnell, T. & Milano, F. [2020] "Impact of current transients on the synchronization stability assessment of grid-feeding converters," *IEEE Trans. Power Syst.* **35**, 4131–4134.

Chen, G. [2022] "Pinning control of complex dynamical networks," *IEEE Trans. Consum. Electron.* **68**, 336–343.

Chen, J., Liu, M., Geng, H., Donnell, T. O. & Milano, F. [2022a] "Impact of PLL frequency limiter on synchronization stability of grid feeding converter," *IEEE Trans. Power Syst.* **37**, 2487–2490.

Chen, S., Ai, Y., Shi, Y., Zhao, S. & Liu, J. [2022b] "A novel online inductance identification of traction network and adaptive control method for cascaded H-bridge converter," *IEEE Trans. Power Electron.* **37**, 6475–6487.

Cheng, H., Shuai, Z., Shen, C., Liu, X., Li, Z. & Shen, Z. J. [2020] "Transient angle stability of paralleled synchronous and virtual synchronous generators in islanded microgrids," *IEEE Trans. Power Electron.* **35**, 8751–8765.

Cheng, Y., Fan, L., Rose, J., Huang, F., Schmall, J., Wang, X., Xie, X., Shair, J., Ramamurthy, J., Modi, N., Li, C., Wang, C., Shah, S., Pal, B. C., Miao, Z., Isaacs, A., Mahseredjian, J. & Zhou, Z. J. [2023] "Real-world subsynchronous oscillation events in power grids with high penetrations of inverter-based resources," *IEEE Trans. Power Syst.* **38**, 316–330.

Chua, L. O., Komuro, M. & Matsumoto, T. [1986] "The double scroll family," *IEEE Trans. Circuits Syst.* **33**, 1072–1118.

Chua, L. O. & Lin, G.-N. [1990] "Canonical realization of Chua's circuit family," *IEEE Trans. Circuits Syst.* **37**, 885–902.

Chua, L. O. [2024] "Memristors on 'edge of chaos'," *Nat. Rev. Electr. Eng.* **1**, 614–627.

Cristiano, R., Henao, M. M., Pagano, D. J. & Ponce, E. [2024] "Chaos through sliding bifurcations in a DC–DC boost power converter," *Int. J. Bifurcation and Chaos* **34**, 2430033.

D'Silva, S., Shadmand, M., Bayhan, S. & Abu-Rub, H. [2020] "Towards grid of microgrids: Seamless transition between grid-connected and islanded modes of operation," *IEEE Open J. Ind. Electron. Soc.* **1**, 66–81.

Deane, J. H. B. & Hamill, D. C. [1990] "Instability, subharmonics and chaos in power electronics systems," *IEEE Trans. Power Electron.* **5**, 260–268.

Deane, J. H. B. & Hamill, D. C. [1991] "Chaotic behavior in a current-mode controlled DC/DC converter," *Electron. Lett.* **27**, 1172–1173.

Ding, L., Wong, S. C. & Tse, C. K. [2020] "Bifurcation analysis of a current-mode-controlled DC cascaded system and applications to design," *IEEE J. Emerging Sel. Top. Power Electron.* **8**, 3214–3224.

Dranga, O., Tse, C. K., Iu, H. H. C. & Nagy, I. [2003] "Bifurcation behavior of a power-factor-correction boost converter," *Int. J. Bifurcation and Chaos* **13**, 3107–3114.

Erickson, R. W. & Maksimović, D. [2000] *Fundamentals of Power Electronics*, 2nd edition (Kluwer Academic/Plenum Publishers, NY, USA).

ESO [2019] "Appendices to the technical report on the events of 9 August 2019," https://www.nationalgrideso.com/document/152351/download.

Fan, L., Miao, Z., Shah, S., Cheng, Y., Rose, J., Huang, S.-H., Pal, B., Xie, X., Modi, N., Wang, S. & Zhu, S. [2022] "Real-world 20-Hz IBR subsynchronous oscillations: Signatures and mechanism analysis," *IEEE Trans. Energy Convers.* **37**, 2863–2873.

Fang, J., Li, H., Tang, Y. & Blaabjerg, F. [2019] "On the inertia of future more-electronics power systems,"

IEEE J. Emerging Sel. Top. Power Electron. **7**, 2130–2146.

Farivar, G. G., Manalastas, W., Tafti, H. D., Ceballos, S., Sanchez-Ruiz, A. & Lovell, E. C. [2023] "Grid-connected energy storage systems: State-of-the-art and emerging technologies," *Proc. IEEE* **111**, 397–420.

Farrokhabadi, M., Cañizares, C. A., Simpson-Porco, J. W., Nasr, E., Fan, L., Mendoza-Araya, P. A., Tonkoski, R., Tamraka, U., Hatziargyriou, N., Lagos, D., Wies, R. W., Paolone, M., Lissarre, M., Meegahapola, L., Kabalan, M., Hajimiragha, A. H., Peralta, D., Elizondo, M. A., Schneider, K. P., Tuffner, F. K. & Reilly, J. [2020] "Microgrid stability definitions, analysis, and examples," *IEEE Trans. Power Syst.* **35**, 13–29.

Foroni, I. [2023] "Border collision bifurcations and coexisting attractors in an economic bimodal map," *Int. J. Bifurcation and Chaos* **33**, 2330003.

Fu, X., Sun, J., Huang, M., Tian, Z., Yan, H., Iu, H. H. C., Hu, P. & Zha, X. [2021] "Large-signal stability of grid-forming and grid-following controls in voltage source converter: A comparative study," *IEEE Trans. Power Electron.* **36**, 7832–7840.

Fu, X., Huang, M., Tse, C. K., Yang, J., Ling, Y. & Zha, X. [2023] "Synchronization stability of grid-following VSC considering interactions of inner current loop and parallel-connected converters," *IEEE Trans. Smart Grid* **14**, 4230–4241.

Gao, X., Zhou, D., Anvari-Moghaddam, A. & Blaabjerg, F. [2023] "Seamless transitions between grid-following and grid-forming control: A novel switching method," *2023 Int. Conf. Power Electronics-ECCE Asia, Jeju, Korea* (IEEE), pp. 1154–1160.

Geng, X., Xie, F., Zhang, B., Qiu, D., Chen, Y. & Wang, R. [2023] "Identifying coexisting attractors and quantifying dynamics characteristics of DC–DC converter based on improved variational mode decomposition and wavelet transform," *IEEE Trans. Power Electron.* **38**, 3928–3938.

Gu, Y., Bottrell, N. & Green, T. C. [2018] "Reduced-order models for representing converters in power system studies," *IEEE Trans. Power Electron.* **33**, 3644–3654.

Gu, Y. & Green, T. C. [2023] "Power system stability with a high penetration of inverter-based resources," *Proc. IEEE* **111**, 832–853.

Guerrero, J. M., Vasquez, J. C., Matas, J., de Vicuna, L. G. & Castilla, M. [2011] "Hierarchical control of droop-controlled AC and DC microgrids — A general approach toward standardization," *IEEE Trans. Ind. Electron.* **58**, 158–172.

Gui, Y., Wang, X., Blaabjerg, F. & Pan, D. [2019] "Control of grid-connected voltage-source converters: The relationship between direct-power control and vector-current control," *IEEE Ind. Electron. Mag.* **13**, 31–40.

Hamill, D. C. & Jefferies, D. J. [1988] "Subharmonics and chaos in a controlled switched-mode power converter," *IEEE Trans. Circuits Syst.* **35**, 1059–1061.

Han, Y., Li, H., Shen, P., Coelho, E. A. A. & Guerrero, J. M. [2017] "Review of active and reactive power sharing strategies in hierarchical controlled microgrids," *IEEE Trans. Power Electron.* **32**, 2427–2451.

Hatziargyriou, N., Milanovic, J., Rahmann, C., Ajjarapu, V., Cañizares, C. A., Erlich, I., Hill, D., Hiskens, I., Kamwa, I., Pal, B., Pourbeik, P., Sanchez-Gasca, J., Stankovic, A., Cutsem, T. V., Vittal, V. & Vournas, C. [2021] "Definition and classification of power system stability — Revisited and extended," *IEEE Trans. Power Syst.* **36**, 3271–3281.

He, J. & Li, Y. W. [2011] "Analysis, design, and implementation of virtual impedance for power electronics interfaced distributed generation," *IEEE Trans. Ind. Appl.* **47**, 2525–2538.

Hu, Q., Fu, L., Ma, F. & Ji, F. [2019] "Large signal synchronizing instability of PLL-based VSC connected to weak AC grid," *IEEE Trans. Power Syst.* **34**, 3220–3229.

Hu, Q., Fu, L., Ma, F., Ji, F. & Zhang, Y. [2021] "Analogized synchronous generator model of PLL-based VSC and transient synchronizing stability of converter dominated power system," *IEEE Trans. Sustainable Energy* **12**, 1174–1185.

Huang, M., Tse, C. K., Wong, S. C., Wan, C. & Ruan, X. [2013a] "Low-frequency Hopf bifurcation and its effects on stability margin in three-phase PFC power supplies connected to non-ideal power grid," *IEEE Trans. Circuits Syst.-I: Regul. Pap.* **60**, 3328–3340.

Huang, M., Wong, S., Tse, C. K. & Ruan, X. [2013b] "Catastrophic bifurcation in three-phase voltage-source converters," *IEEE Trans. Circuits Syst.-I: Regul. Pap.* **60**, 1062–1071.

Huang, M., Ji, H., Sun, J., Wei, L. & Zha, X. [2017a] "Bifurcation-based stability analysis of photovoltaic-battery hybrid power system," *IEEE J. Emerging Sel. Top. Power Electron.* **5**, 1055–1067.

Huang, M., Peng, Y., Tse, C. K., Liu, Y., Sun, J. & Zha, X. [2017b] "Bifurcation and large-signal stability analysis of three-phase voltage source converter under grid voltage dips," *IEEE Trans. Power Electron.* **32**, 8868–8879.

IEEE [1982] "Effects of switching network disturbances on turbine-generator shaft systems," *IEEE Trans. Power Apparatus Syst.* **PAS-101**, 3151–3157.

Iu, H. H. C., Zhou, Y. F. & Tse, C. K. [2003] "Fast-scale instability in a PFC boost converter under average current-mode control," *Int. J. Circuit Theory Appl.* **31**, 611–624.

Ji, H., Xie, F., Shen, L., Yang, R. & Zhang, B. [2022] "Unstable behavior analysis and stabilization of double-loop proportional-integral control H-bridge inverter with inductive impedance load," *Int. J. Circuit Theory Appl.* **50**, 904–925.

Jia, L., Ruan, X., Zhao, W., Lin, Z. & Wang, X. [2018] "An adaptive active damper for improving the stability of grid-connected inverters under weak grid," *IEEE Trans. Power Electron.* **33**, 9561–9574.

Kabalan, M., Singh, P. & Niebur, D. [2017] "Large signal Lyapunov-based stability studies in microgrids: A review," *IEEE Trans. Smart Grid* **8**, 2287–2295.

Kroposki, B., Johnson, B., Zhang, Y., Gevorgian, V., Denholm, P., Hodge, B.-M. & Hannegan, B. [2017] "Achieving a 100% renewable grid: Operating electric power systems with extremely high levels of variable renewable energy," *IEEE Power Energy Mag.* **15**, 61–73.

Kundur, P. [1993] *Power System Stability and Control* (McGraw-Hill, NY, USA).

Kundur, P., Paserba, J., Ajjarapu, V., Andersson, G., Bose, A., Cañizares, C. A., Hatziargyriou, N., Hill, D., Stankovic, A., Taylor, C., Cutsem, T. V. & Vittal, V. [2004] "Definition and classification of power system stability," *IEEE Trans. Power Syst.* **19**, 1387–1401.

Kuznetsov, Y. A. [1998] *Elements of Applied Bifurcation Theory*, 2nd edition (Springer, NY, USA).

Laskar, J. [1989] "A numerical experiment on the chaotic behaviour of the Solar System," *Nature* **338**, 237–238.

Lenz, E., Pagano, D. J. & Pou, J. [2018] "Bifurcation analysis of parallel-connected voltage-source inverters with constant power loads," *IEEE Trans. Smart Grid* **9**, 5482–5493.

Leonov, G. A., Kuznetsov, N. V. & Vagaitsev, V. I. [2011] "Localization of hidden Chua's attractors," *Phys. Lett. A* **375**, 2230–2233.

Leonov, G. A. & Kuznetsov, N. V. [2013] "Hidden attractors in dynamical systems. From hidden oscillations in Hilbert-Kolmogorov Aizerman, and Kalman problems to hidden chaotic attractor in Chua circuits," *Int. J. Bifurcation and Chaos* **23**, 1330002.

Li, Y., Fan, L. & Miao, Z. [2020] "Wind in weak grids: Low-frequency oscillations, subsynchronous oscillations, and torsional interactions," *IEEE Trans. Power Syst.* **35**, 109–118.

Li, B., Sang, B., Liu, M., Hu, X., Zhang, X. & Wang, N. [2023] "Some jerk systems with hidden chaotic dynamics," *Int. J. Bifurcation and Chaos* **33**, 2350069.

Li, M. J., Tse, C. K. & Yi, M. [2024a] "Steady-state cascading failure model with voltage instability event detection," *IEEE Trans. Circuits Syst.-I: Regul. Pap.* **71**, 463–472.

Li, X., Feng, P., Tian, Z., Huang, M., Shu, S., Zha, X. & Hu, P. [2024b] "Transient stability analysis for grid-tied VSG considering high-order nonlinear interactions between active and reactive power control loops," *IEEE Trans. Power Electron.* **39**, 6974–6988.

Li, X., Feng, P., Zhang, R., Tian, Z., Huang, M., Zha, X., Xiong, X. & Hu, P. [2025] "Transient stability evaluation and decoupled control for grid-following converters considering nonideal alternating current control," *IEEE Trans. Power Electron.* **40**, 8990–9003.

Liao, S., Zha, X., Li, X., Huang, M., Sun, J., Pan, J. & Guerrero, J. M. [2019] "A novel dynamic aggregation modeling method of grid-connected inverters: Application in small-signal analysis," *IEEE Trans. Sustainable Energy* **10**, 1554–1564.

Liserre, M., Perez, M. A., Langwasser, M., Rojas, C. A. & Zhou, Z. [2023] "Unlocking the hidden capacity of the electrical grid through smart transformer and smart transmission," *Proc. IEEE* **111**, 421–437.

Liu, Z., Liu, J., Bao, W. & Zhao, Y. [2015] "Infinity-norm of impedance-based stability criterion for three-phase AC distributed power systems with constant power loads," *IEEE Trans. Power Electron.* **30**, 3030–3043.

Liu, H., Xie, X., He, J., Xu, T., Yu, Z., Wang, C. & Zhang, C. [2017] "Subsynchronous interaction between direct-drive PMSG based wind farms and weak AC networks," *IEEE Trans. Power Syst.* **32**, 4708–4720.

Liu, J., Yang, J., Zhang, J., Zhao, N. & Zheng, T. Q. [2018] "Voltage balance control based on dual active bridge DC/DC converters in a power electronic traction transformer," *IEEE Trans. Power Electron.* **33**, 1696–1714.

Liu, Y., Huang, M., Tse, C. K., Iu, H. H. C., Yan, Z. & Zha, X. [2021] "Stability and multiconstraint operating region of grid-connected modular multilevel converter under grid phase disturbance," *IEEE Trans. Power Electron.* **36**, 12551–12564.

Liu, D., Zhang, X. & Tse, C. K. [2022] "Effects of high level of penetration of renewable energy sources on cascading failure of modern power systems," *IEEE J. Emerging Sel. Top. Circuits Syst.* **12**, 98–106.

Liu, C. C., Yang, J., Tse, C. K. & Huang, M. [2023] "Transient synchronization stability of grid-following converters considering nonideal current loop," *IEEE Trans. Power Electron.* **38**, 13757–13769.

Liu, C. C., Tse, C. K. & Yang, J. [2024] "Nonlinear behavior and transient stability of grid-following converters using direct power control under weak grid," *IEEE Trans. Sustainable Energy* **15**, 2287–2298.

Liu, C. C., Tse, C. K., Huang, M., Wu, Z., Han, H. & Yang, J. [2025] "Design-oriented transient stability analysis of grid-connected converters: A comparative study of analysis methods," *IEEE Trans. Power Electron.* **40**, 749–763.

Lu, M. [2022] "Virtual oscillator grid-forming inverters: State of the art, modeling, and stability," *IEEE Trans. Power Electron.* **37**, 11579–11591.

Luo, L. & Dhople, S. V. [2014] "Spatiotemporal model reduction of inverter-based islanded microgrids," *IEEE Trans. Energy Convers.* **29**, 823–832.

Ma, R., Li, J., Kurths, J., Cheng, S. & Zhan, M. [2022] "Generalized swing equation and transient synchronous stability with PLL-based VSC," *IEEE Trans. Energy Convers.* **37**, 1428–1441.

Marx, D., Magne, P., Nahid-Mobarakeh, B., Pierfederici, S. & Davat, B. [2012] "Large signal stability analysis tools in DC power systems with constant power loads and variable power loads — A review," *IEEE Trans. Power Electron.* **27**, 1773–1787.

Moutevelis, D., Roldán-Pérez, J., Prodanovic, M. & Sanchez-Acevedo, S. [2022] "Bifurcation analysis of active electrical distribution networks considering load tap changers and power converter capacity limits," *IEEE Trans. Power Electron.* **37**, 7230–7246.

NERC [2017] "Reliability guideline — Forced oscillation monitoring and mitigation," https://www.nerc.com/comm/.

NERC [2021] "Multiple solar PV disturbances in CAISO," https://www.nerc.com/pa/rrm/ea/Documents/.

Nikolakakos, I. P., Zeineldin, H. H., El-Moursi, M. S. & Kirtley, J. L. [2018] "Reduced-order model for inter-inverter oscillations in islanded droop-controlled microgrids," *IEEE Trans. Smart Grid* **9**, 4953–4963.

Olfati-Saber, R., Fax, J. A. & Murray, R. M. [2007] "Consensus and cooperation in networked multi-agent systems," *Proc. IEEE* **95**, 215–233.

Ott, E. [1993] *Chaos in Dynamical Systems* (Cambridge University Press, UK).

Pan, D., Wang, X., Liu, F. & Shi, R. [2020] "Transient stability of voltage-source converters with grid-forming control: A design-oriented study," *IEEE J. Emerging Sel. Top. Power Electron.* **8**, 1019–1033.

Paquette, A. D. & Divan, D. M. [2015] "Virtual impedance current limiting for inverters in microgrids with synchronous generators," *IEEE Trans. Ind. Appl.* **51**, 1630–1638.

Peitgen, H.-O., Jürgens, H. & Saupe, D. [2004] *Chaos and Fractals: New Frontiers of Science*, 2nd edition (Springer, NY, USA).

Peng, D., Huang, M., Li, J., Sun, J., Zha, X. & Wang, C. [2019] "Large-signal stability criterion for parallel-connected DC–DC converters with current source equivalence," *IEEE Trans. Circuits Syst.-II: Exp. Briefs* **66**, 2037–2041.

Pogaku, N., Prodanovic, M. & Green, T. C. [2007] "Modeling, analysis and testing of autonomous operation of an inverter-based microgrid," *IEEE Trans. Power Electron.* **22**, 613–625.

Purba, V., Johnson, B. B., Rodriguez, M., Jafarpour, S., Bullo, F. & Dhople, S. V. [2019] "Reduced-order aggregate model for parallel-connected single-phase inverters," *IEEE Trans. Energy Convers.* **34**, 824–837.

Qoria, T., Wu, H., Wang, X. & Colak, I. [2023] "Variable virtual impedance-based overcurrent protection for grid-forming inverters: Small-signal, large-signal analysis and improvement," *IEEE Trans. Smart Grid* **14**, 3324–3336.

Rasheduzzaman, M., Mueller, J. A. & Kimball, J. W. [2015] "Reduced-order small-signal model of microgrid systems," *IEEE Trans. Sustainable Energy* **6**, 1292–1305.

Reza, C. M. F. S. & Lu, D. D. C. [2020] "Recent progress and future research direction of nonlinear dynamics and bifurcation analysis of grid-connected power converter circuits and systems," *IEEE J. Emerging Sel. Top. Power Electron.* **8**, 3193–3203.

Rivera, S., Goetz, S. M., Kouro, S., Lehn, P. W., Pathmanathan, M. & Bauer, P. [2023] "Charging infrastructure and grid integration for electromobility," *Proc. IEEE* **111**, 371–396.

Robinson, R. C. [2012] *An Introduction to Dynamical Systems: Continuous and Discrete*, 2nd edition (American Mathematical Society, Providence, RI, USA).

Rosso, R., Engelken, S. & Liserre, M. [2020] "Robust stability investigation of the interactions among grid-forming and grid-following converters," *IEEE J. Emerging Sel. Top. Power Electron.* **8**, 991–1003.

Ruan, X., Wang, X., Pan, D., Yang, D., Li, W. & Bao, C. [2018] *Control Techniques for LCL-Type Grid-Connected Inverters* (Springer, Singapore).

Shen, C., Shuai, Z., Shen, Y., Peng, Y., Liu, X., Li, Z. & Shen, Z. J. [2021] "Transient stability and current injection design of paralleled current-controlled VSCs and virtual synchronous generators," *IEEE Trans. Smart Grid* **12**, 1118–1134.

Shen, B.-W. [2023] "A review of Lorenz's models from 1960 to 2008," *Int. J. Bifurcation and Chaos* **33**, 2330024.

Shuai, Z., Hu, Y., Peng, Y., Tu, C. & Shen, Z. J. [2017] "Dynamic stability analysis of synchronverter-dominated microgrid based on bifurcation theory," *IEEE Trans. Ind. Electron.* **64**, 7467–7477.

Shuai, Z., Peng, Y., Liu, X., Li, Z., Guerrero, J. M. & Shen, Z. J. [2019a] "Dynamic equivalent modeling for multi-microgrid based on structure preservation method," *IEEE Trans. Smart Grid* **10**, 3929–3942.

Shuai, Z., Peng, Y., Liu, X., Li, Z., Guerrero, J. M. & Shen, Z. J. [2019b] "Parameter stability region analysis of islanded microgrid based on bifurcation theory," *IEEE Trans. Smart Grid* **10**, 6580–6591.

Shuai, Z., Shen, C., Liu, X., Li, Z. & Shen, Z. J. [2019c] "Transient angle stability of virtual synchronous generators using Lyapunov's direct method," *IEEE Trans. Smart Grid* **10**, 4648–4661.

Shuai, Z. [2021] *Transient Characteristics, Modelling and Stability Analysis of Microgrid* (Springer Nature, Singapore).

Simpson-Porco, J. W., Dörfler, F. & Bullo, F. [2016] "Voltage collapse in complex power grids," *Nat. Commun.* **7**, 10790.

Singh, R., Paniyil, P. & Zhang, Z. [2022] "Transformative role of power electronics: In solving climate emergency," *IEEE Power Electron. Mag.* **9**, 39–47.

Skogestad, S. & Postlethwaite, I. [2001] *Multivariable Feedback Control: Analysis and Design*, 2nd edition (Wiley, Singapore).

Small, M. & Tse, C. K. [2012] "Predicting the outcome of roulette," *Chaos* **22**, 033150.

Sprott, J., Wang, X. & Chen, G. [2013] "Coexistence of point, periodic and strange attractors," *Int. J. Bifurcation and Chaos* **23**, 1350093.

Strogatz, S. H. [1994] *Nonlinear Dynamics and Chaos: With Applications to Physics, Biology, Chemistry and Engineering* (Perseus Books, Massachusetts, USA).

Strunz, K., Almunem, K., Wulkow, C., Kuschke, M., Valescudero, M. & Guillaud, X. [2023] "Enabling 100% renewable power systems through power electronic grid-forming converter and control: System integration for security, stability, and application to europe," *Proc. IEEE* **111**, 891–915.

Sun, J., Li, M., Zhang, Z., Xu, T., He, J., Wang, H. & Li, G. [2017] "Renewable energy transmission by HVDC across the continent: System challenges and opportunities," *CSEE J. Power Energy Syst.* **3**, 353–364.

Taul, M. G., Wang, X., Davari, P. & Blaabjerg, F. [2019] "An overview of assessment methods for synchronization stability of grid-connected converters under severe symmetrical grid faults," *IEEE Trans. Power Electron.* **34**, 9655–9670.

Taul, M. G., Wang, X., Davari, P. & Blaabjerg, F. [2021] "Reduced-order and aggregated modeling of large-signal synchronization stability for multiconverter systems," *IEEE J. Emerging Sel. Top. Power Electron.* **9**, 3150–3165.

Tian, Z., Tang, Y., Zha, X., Sun, J., Huang, M., Fu, X. & Liu, F. [2022] "Hamilton-based stability criterion and attraction region estimation for grid-tied inverters under large-signal disturbances," *IEEE J. Emerging Sel. Top. Power Electron.* **10**, 413–423.

Tian, Z., Li, X., Zha, X., Tang, Y., Sun, P., Huang, M. & Yu, P. [2023] "Transient synchronization stability of an islanded AC microgrid considering interactions between grid-forming and grid-following converters," *IEEE J. Emerging Sel. Top. Power Electron.* **11**, 4463–4476.

Tse, C. K. [1994a] "Chaos from a buck switching regulator operating in discontinuous mode," *Int. J. Circuit Theory Appl.* **22**, 263–278.

Tse, C. K. [1994b] "Flip bifurcation and chaos in a three-state boost switching regulator," *IEEE Trans. Circuits Syst.* **42**, 16–23.

Tse, C. K. & Bernardo, M. D. [2002] "Complex behavior in switching power converters," *Proc. IEEE* **90**, 768–781.

Tse, C. K. [2003] *Complex Behavior of Switching Power Converters* (CPC Press, Boca Raton, FL, USA).

Tse, C. K. & Li, M. [2011] "Design-oriented bifurcation analysis of power electronics systems," *Int. J. Bifurcation and Chaos* **21**, 1523–1537.

Tse, C. K., Huang, M., Zhang, X., Liu, D. & Li, X. L. [2020] "Circuits and systems issues in power electronics penetrated power grid," *IEEE Open J. Circuits Syst.* **1**, 140–156.

Vorobev, P., Huang, P., Hosani, M. A., Kirtley, J. L. & Turitsyn, K. [2018] "High-fidelity model order reduction for microgrids stability assessment," *IEEE Trans. Power Syst.* **33**, 874–887.

Wan, C., Huang, M., Tse, C. K., Wong, S. C. & Ruan, X. [2013] "Nonlinear behavior and instability in a three-phase boost rectifier connected to a nonideal power grid with an interacting load," *IEEE Trans. Power Electron.* **28**, 3255–3265.

Wan, C., Huang, M., Tse, C. K. & Ruan, X. [2015] "Effects of interaction of power converters coupled via power grid: A design-oriented study," *IEEE Trans. Power Electron.* **30**, 3589–3600.

Wang, X. F. & Chen, G. [2002] "Pinning control of scale-free dynamical networks," *Physica A* **310**, 521–531.

Wang, X. & Chen, G. [2012] "A chaotic system with only one stable equilibrium," *Commun. Nonlin. Sci. Numer. Simul.* **17**, 1264–1272.

Wang, X., Harnefors, L. & Blaabjerg, F. [2018a] "Unified impedance model of grid-connected voltage-source converters," *IEEE Trans. Power Electron.* **33**, 1775–1787.

Wang, Y., Wang, X., Chen, Z. & Blaabjerg, F. [2018b] "Small-signal stability analysis of inverter-fed power systems using component connection method," *IEEE Trans. Smart Grid* **9**, 5301–5310.

Wang, X. & Blaabjerg, F. [2019] "Harmonic stability in power electronic-based power systems: Concept, modeling, and analysis," *IEEE Trans. Smart Grid* **10**, 2858–2870.

Wang, X., Taul, M. G., Wu, H., Liao, Y., Blaabjerg, F. & Harnefors, L. [2020] "Grid-synchronization stability of converter-based resources — An overview," *IEEE Open J. Ind. Appl.* **1**, 115–134.

Wang, X., Kuznetsov, N. V. & Chen, G. [2021] *Chaotic Systems with Multistability and Hidden Attractors* (Springer, Switzerland).

Wang, W., Li, G. & He, G. [2023] *Analysis and Mitigation of Broadband Oscillation in Renewable Energy Generation and AC/DC Transmission Systems* (Springer Nature, Singapore).

Wu, H. & Wang, X. [2020] "Design-oriented transient stability analysis of PLL-synchronized voltage-source converters," *IEEE Trans. Power Electron.* **35**, 3573–3589.

Wu, X., Xu, Y., He, J., Wang, X., Vasquez, J. C. & Guerrero, J. M. [2020] "Pinning-based hierarchical and distributed cooperative control for AC microgrid clusters," *IEEE Trans. Power Electron.* **35**, 9865–9885.

Xiong, X., Tse, C. K. & Ruan, X. [2015] "Bifurcation analysis and experimental study of a multi-operating-mode photovoltaic-battery hybrid power system," *IEEE J. Emerging Sel. Top. Circuits Syst.* **5**, 316–325.

Xu, Y., Gu, Z. & Sun, K. [2020] "Characterization of subsynchronous oscillation with wind farms using describing function and generalized nyquist criterion," *IEEE Trans. Power Syst.* **35**, 2783–2793.

Yan, R., Masood, N. A., Saha, T. K., Bai, F. & Gu, H. [2018] "The anatomy of the 2016 South Australia blackout: A catastrophic event in a high renewable network," *IEEE Trans. Power Syst.* **33**, 5374–5388.

Yang, Z., Ma, R., Cheng, S. & Zhan, M. [2020a] "Nonlinear modeling and analysis of grid-connected voltage-source converters under voltage dips," *IEEE J. Emerging Sel. Top. Power Electron.* **8**, 3281–3292.

Yang, L., Yang, L., Yang, F. & Ma, X. [2020b] "Slow-scale and fast-scale instabilities in parallel-connected single-phase H-bridge inverters: A design-oriented study," *Int. J. Bifurcation and Chaos* **30**, 2050005.

Yang, C., Huang, L., Xin, H. & Ju, P. [2021a] "Placing grid-forming converters to enhance small signal stability of PLL-integrated power systems," *IEEE Trans. Power Syst.* **36**, 3563–3573.

Yang, J., Tse, C. K., Huang, M. & Fu, X. [2021b] "Homoclinic bifurcation of a grid-forming voltage source converter," *IEEE Trans. Power Electron.* **36**, 13176–13187.

Yang, H., Jiang, J. & Hong, L. [2022a] "The birth of a hidden attractor through boundary crisis," *Int. J. Bifurcation and Chaos* **32**, 2230005.

Yang, J., Tse, C. K. & Liu, D. [2022b] "Nonlinear behavior and reduced-order models of islanded microgrid," *IEEE Trans. Power Electron.* **37**, 9212–9225.

Yang, J., Tse, C. K., Huang, M. & Liu, D. [2023a] "Bifurcations of grid-following rectifiers and routes to voltage instability in weak AC grids," *IEEE Trans. Power Syst.* **38**, 1702–1713.

Yang, J., Tse, C. K. & Liu, D. [2023b] "Sub-synchronous oscillations and transient stability of islanded microgrid," *IEEE Trans. Power Syst.* **38**, 3760–3774.

Yang, J. & Tse, C. K. [2024] *Complex Behavior of Grid-Connected Power Electronics Systems* (Springer Nature, Switzerland).

Yang, J., Liu, C. C. & Tse, C. K. [2024a] "Reactive power sharing and stability of islanded microgrid using centralized secondary control," *2024 IEEE 10th Int. Power Electronics and Motion Control Conference (IPEMC2024-ECCE Asia), Chengdu, China* (IEEE), pp. 2536–2540.

Yang, J., Liu, C. C., Tse, C. K. & Huang, M. [2024b] "Stability analysis and design of islanded microgrids integrating pinning-based and consensus-based distributed secondary control," *IEEE Trans. Power Electron.* **39**, 9490–9504.

Yang, J., Tse, C. K., Huang, M., Liu, D., Liu, C. C. & Fu, X. [2024c] "Comparison of homoclinic bifurcations between grid-following and grid-forming converters," *IEEE Trans. Ind. Electron.* **71**, 4731–4741.

Yang, J., Tse, C. K., Huang, M., Liu, D., Fu, X. & Liu, C. C. [2025] "Transient stability enhancement of multi-inverter microgrids," *IEEE Trans. Sustainable Energy* **16**, 2458–2472.

Yoon, C., Bai, H., Beres, R. N., Wang, X., Bak, C. L. & Blaabjerg, F. [2016] "Harmonic stability assessment for multiparalleled, grid-connected inverters," *IEEE Trans. Sustainable Energy* **7**, 1388–1397.

Yu, H., Awal, M. A., Tu, H., Husain, I. & Lukic, S. [2021] "Comparative transient stability assessment of droop and dispatchable virtual oscillator controlled grid-connected inverters," *IEEE Trans. Power Electron.* **36**, 2119–2130.

Zhang, L., Harnefors, L. & Nee, H.-P. [2010] "Power-synchronization control of grid-connected voltage-source converters," *IEEE Trans. Power Syst.* **25**, 809–820.

Zhang, X., Wu, Z. & Chua, L. O. [2020] "Hearts are poised near the edge of chaos," *Int. J. Bifurcation and Chaos* **30**, 2030023.

Zhang, Z., Schuerhuber, R., Fickert, L., Friedl, K., Chen, G. & Zhang, Y. [2022a] "Domain of attraction's estimation for grid connected converters with phase-locked loop," *IEEE Trans. Power Syst.* **37**, 1351–1362.

Zhang, Y., Zhang, C. & Cai, X. [2022b] "Large-signal grid-synchronization stability analysis of PLL-based VSCs using Lyapunov's direct method," *IEEE Trans. Power Syst.* **37**, 788–791.

Zhang, X., Wu, J., Fapojuwo, A. O., Galias, Z. & Tse, C. K. [2023a] "Guest editorial complex cyber-multitudinal-physical systems: Analysis, decision-making, and AI applications," *IEEE J. Emerging Sel. Top. Circuits Syst.* **13**, 641–645.

Zhang, C., Zong, H., Cai, X. & Molinas, M. [2023b] "On the relation of nodal admittance- and loop gain-model based frequency-domain modal methods for converters-dominated systems," *IEEE Trans. Power Syst.* **38**, 1779–1782.

Zhao, J., Huang, M. & Zha, X. [2020] "Nonlinear analysis of PLL damping characteristics in weak-grid-tied inverters," *IEEE Trans. Circuits Syst.-II: Exp. Briefs* **67**, 2752–2756.

Zhao, J., Huang, M., Yan, H., Tse, C. K. & Zha, X. [2021] "Nonlinear and transient stability analysis of phase-locked loops in grid-connected converters," *IEEE Trans. Power Electron.* **36**, 1018–1029.

Zhong, Q.-C. & Weiss, G. [2011] "Synchronverters: Inverters that mimic synchronous generators," *IEEE Trans. Ind. Electron.* **58**, 1259–1267.

Zhou, Y., Xin, H., Wu, D., Liu, F., Li, Z., Wang, G., Yuan, H. & Ju, P. [2023] "Small-signal stability assessment of heterogeneous grid-following converter power systems based on grid strength analysis," *IEEE Trans. Power Syst.* **38**, 2566–2579.

Zhu, Y., Gu, Y., Li, Y. & Green, T. C. [2023] "Impedance-based root-cause analysis: Comparative study of impedance models and calculation of eigenvalue sensitivity," *IEEE Trans. Power Syst.* **38**, 1642–1654.

Zou, Z., Besheli, B. D., Rosso, R., Liserre, M. & Wang, X. [2021] "Interactions between two phase-locked loop synchronized grid converters," *IEEE Trans. Ind. Appl.* **57**, 3935–3947.

Zou, Z., Tang, J., Wang, X., Wang, Z., Chen, W., Buticchi, G. & Liserre, M. [2022] "Modeling and control of a two-bus system with grid-forming and grid-following converters," *IEEE J. Emerging Sel. Top. Power Electron.* **10**, 7133–7149.

https://doi.org/10.1142/9789819830633_0007

Chapter 7

Chaos in Real Neurons#

Kazuyuki Aihara
International Research Center for Neurointelligence,
The University of Tokyo Institutes for Advanced Study,
The University of Tokyo, Tokyo 113-0033, Japan
kaihara@g.ecc.u-tokyo.ac.jp

In this review paper, I survey attractors including chaos and bifurcations to chaos experimentally observed in periodically forced oscillations of squid giant axons. I also discuss their possible applications briefly.

Keywords: Chaos; bifurcation; squid giant axon; forced oscillation; chaotic neural network.

1. Introduction

Squid giant axons are excellent materials for electrophysiological experiments to study nonlinear dynamics of neurons. They are literally giant with diameter between 400 μm and 900 μm and length between 4 cm and 8 cm [Matsumoto, 1981; Aihara, 1995]. These properties peculiar to squid giant axons made easier experimental procedures such as intracellular perfusion and intracellular recording of membrane potential by inserting a glass pipette electrode. Moreover, squid giant axons are structurally simple because they don't have myelin sheaths. Therefore, squid giant axons were widely used in electrophysiological experiments on nerve membranes in the 20th century. In particular, Hodgkin and Huxley [1952] showed experimentally and theoretically with squid giant axons that nonlinear dynamical characteristics of nerve membranes can be understood with a model of an equivalent electrical circuit with nonlinear and dynamical conductances g_{Na} and g_{K} for sodium and potassium ions, respectively.

Under normal physiological conditions, the squid giant axons are in a resting state where they stably maintain the constant membrane potential of about -60 mV that is called the resting potential. There is threshold potential for firing, namely generation of an action potential, which is an electrical spike with amplitude of about 100 mV and width of about 1 msec. The threshold potential is higher or depolarized by about 15 mV from the resting potential. Action potentials are thought to be the carrier of information in the brain, although how the information is coded with action potentials, e.g. either a firing rate or timing of spikes, remains an unsolved essential problem of neuroscience. Figure 1 schematically illustrates the waveform of an action potential in squid giant axons. The membrane potential returns quickly to the resting potential without any action potential after a perturbation that is below the threshold potential. However, if the strength of the depolarizing perturbation is above the threshold potential, a nerve membrane produces a spike of an action potential before returning to the resting state.

#To cite this article, please refer to its earlier version published in the *International Journal of Bifurcation and Chaos*, Vol. 35, No. 14 (2025) 2540007 (27 pages), DOI: 10.1142/S0218127425400073.

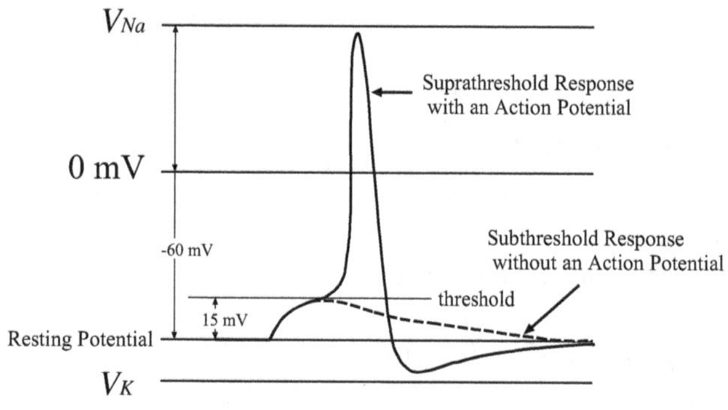

Fig. 1. The resting potential and the action potential in a squid giant axon. V_{Na} and V_{K} are the reversal potentials for sodium and potassium ions, respectively.

Neurons in the brain interact each other by means of chemical and electrical synapses, which are junctions between neurons. Each cortical neuron usually receives about ten thousand chemical synapses from other neurons. The chemical synapses are classified to excitatory and inhibitory synapses. While the former synapses increase or depolarize the membrane potential of a postsynaptic neuron, the latter ones decrease or hyperpolarize it. Then, each neuron takes a spatio-temporal summation of inputs from many chemical synapses. Simply speaking, if the sum of the inputs is more than the threshold potential, an action potential is triggered around the axon hillock of a neuron and propagated along the axon to stimulate connected postsynaptic neurons. It should be noted that rich electrical synapses also exist even among cortical neurons, especially interneurons in the brain [Benardo, 1997].

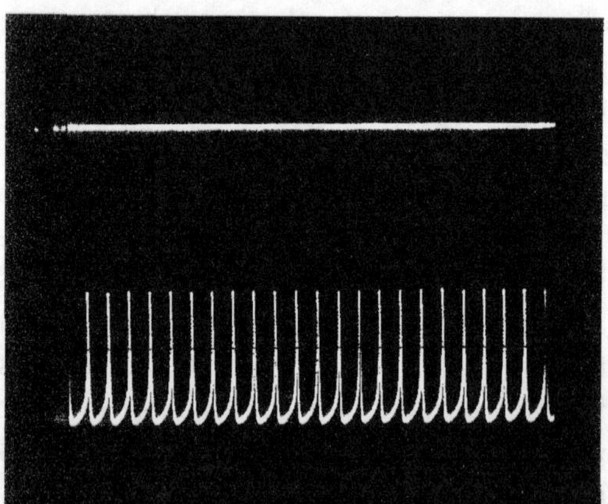

Fig. 2. A self-sustained oscillation with the natural frequency $f_N = 180\,\mathrm{Hz}$ in a squid giant axon. A 1:8 mixture of natural sea water and 550 mM NaCl was used as the external medium surrounding the axon to induced the self-sustained oscillation. The lower trace shows the time series of the membrane potential; each spike is an action potential. The upper trace of the horizontal line shows that there is no stimulation current, in other words, this repetitive firing emerges spontaneously.

By changing the external ionic concentrations surrounding squid giant axons in an *in vitro* experiment, it is possible to induce self-sustained oscillations with spontaneous and periodic firing of action potentials [Huxley, 1959; Matsumoto, 1981; Aihara & Matsumoto, 1982]. In such a situation, a nerve membrane behaves as a neural oscillator. Figure 2 shows an example of waveform of the membrane potential in a self-sustained oscillation experimentally observed with a squid giant axon. From the viewpoint of dynamical systems theory, nerve membranes are nonlinear dissipative systems as shown by ionic conductances in the equivalent circuit model [Hodgkin & Huxley, 1952]. Generally speaking, steady states of a nonlinear dissipative system can be characterized by a concept of attractors such as equilibrium points, limit cycles, tori, strange attractors, and others. The resting state and the self-sustained oscillatory state of squid giant axons correspond to a stable equilibrium point and a stable limit cycle, respectively. Moreover, experiments with squid giant axons [Matsumoto *et al.*, 1984; Matsumoto *et al.*, 1987; Aihara & Matsumoto, 1986, 1987; Aihara *et al.*, 1985, 1986; Takahashi *et al.*, 1990; Aihara, 1994] and other biological cells [Guevara *et al.*, 1981; Hayashi *et al.*, 1982; Glass & Mackey, 1988] have shown that deterministic chaos described by strange attractors can be observed in both resting and oscillatory cells when they are forced by periodic currents such as sinusoidal waves and periodic pulses.

I review sinusoidally forced neural oscillations experimentally observed with squid giant axons in this paper. Although these are my old studies, I believe that it is still worth leaving them as historical records.

2. Entrained Oscillations

Let us consider a general situation that a neural oscillator with the natural frequency f_N is stimulated by an externally applied periodic stimulation with the frequency f_F. The natural frequency f_N represents the number of action potentials generated during one second in the state of a spontaneously self-sustained oscillation in a squid giant axon without any external stimulation.

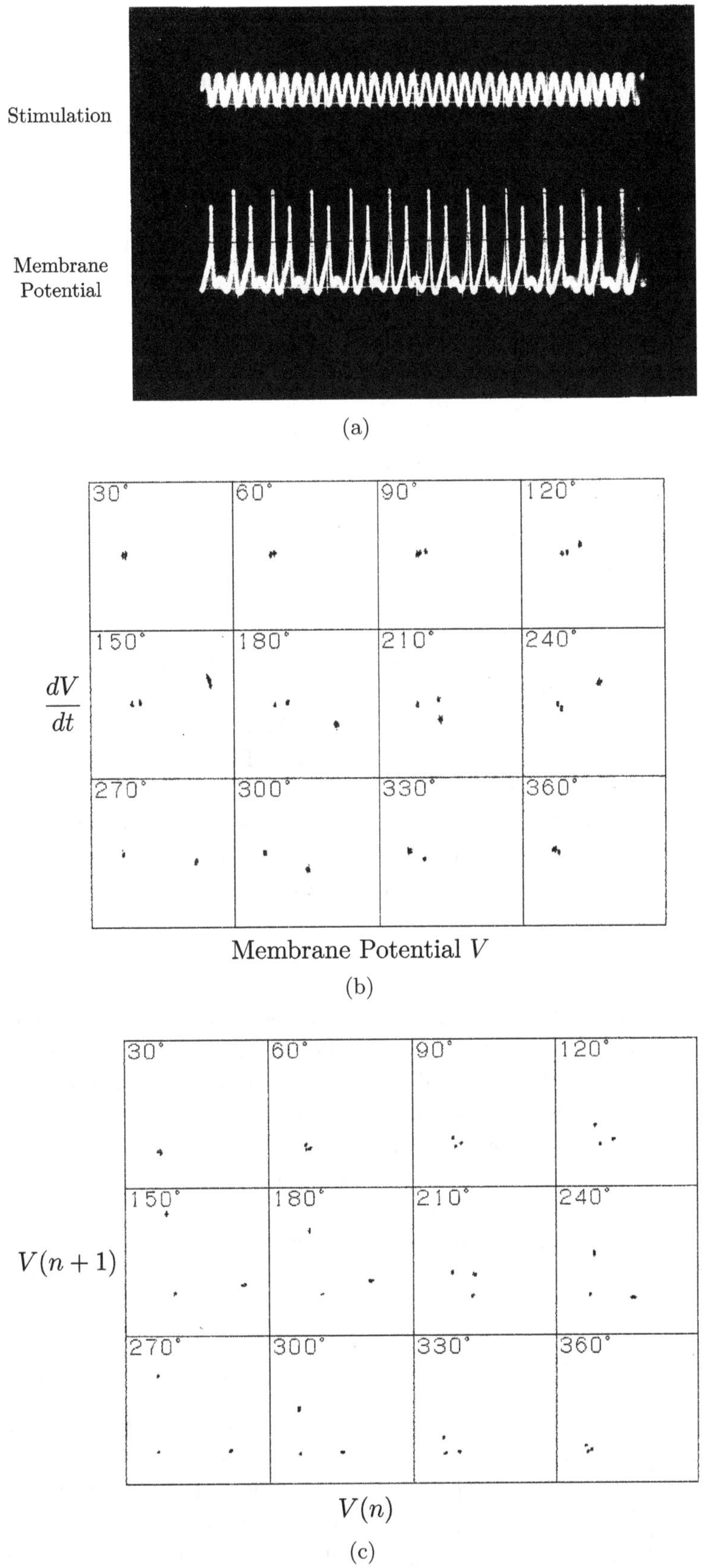

Fig. 3. A 2/3-entrained oscillation with $f_N = 197\,\text{Hz}$, $f_F = 332\,\text{Hz}$, and $A = 4\,\mu\text{A}$. (a) The waveforms (above: the stimulating current, below: the response of the membrane potential). (b) The stroboscopic plots where the number in each frame shows the phase at which the response is stroboscopically observed. (c) The return plots of the ordinate $V(n+1)$ versus the abscissa $V(n)$.

As a periodic force, we applied a sinusoidal current $A\sin(2\pi f_F t)$ for the sake of simplicity. Generally speaking, when the forcing frequency f_F is close enough to a simple rational number m/n times f_N with small natural numbers m and n, the neural oscillator responds to the periodic stimulation with an n/m-entrained (or phase-locked) oscillation. Here, the n/m-entrained (or phase-locked) oscillation is defined as a periodic oscillation in which n action potentials are generated with locked phases during the fundamental period of the forced oscillation that is equal to m times the forcing period $1/f_F$; the reciprocal number of the fundamental period is the fundamental frequency f_R of the forced periodic response. Thus, the excitation number of an n/m-entrained oscillation is exactly n/m, where the excitation number is defined as the ratio of the number of action potentials to the number of the forcing periods. It should be noted that the natural numbers m and n in an n/m-entrained oscillation are not always relatively prime. For example, a q/p-entrained oscillation changes to a $2q/2p$-entrained oscillation through a period-doubling bifurcation as shown in the following.

Figure 3 shows an example of an entrained oscillation experimentally recorded with a squid axon. In this case, the axon in the state of a self-sustained oscillation with the natural frequency $f_N = 197\,\mathrm{Hz}$ was forced by a sinusoidal current with frequency $f_F = 332\,\mathrm{Hz}$ and amplitude $A = 4\,\mu\mathrm{A}$. The periodic oscillation in Fig. 3 is a 2/3-entrained oscillation, where the fundamental period of the forced oscillation is three times the period of the sinusoidal force, and two action potentials are generated during the fundamental period. Figure 4 shows the method of stroboscopic observation analysis in which the membrane potential of the forced oscillation is stroboscopically observed at a fixed phase of the stimulating sinusoidal force,

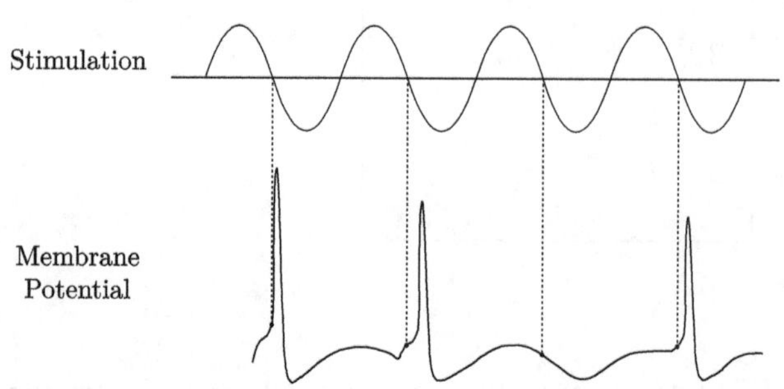

Fig. 4. Stroboscopic observation of a forced neural oscillation at the phase 180° of a stimulating sinusoidal current.

for example, at 180° in this figure. This analysis is equivalent to the observation at a Poincaré section and is also called the stroboscopic plot. By this stroboscopic analysis, we obtain the time series data of the membrane potential V and those of the time differential dV/dt of the membrane potential on a Poincaré section, which are denoted as $\{V(1), V(2), V(3), \ldots\}$ and $\{dV(1)/dt, dV(2)/dt, dV(3)/dt, \ldots\}$, respectively. Each frame in Fig. 3(b) gives plots $(V(n), dV(n)/dt)$ for $n = 1, 2, 3, \ldots$, showing the Poincaré section of the attractor at each corresponding phase. For example, the frame denoted as 180° in Fig. 3(b) is the Poincaré section at the phase of 180° observed as Fig. 4.

In other words, the stroboscopic analysis is understood as follows: the orbit of the forced oscillation representing a possible attractor is first reconstructed in a space of $\mathbf{R}^2$ (the membrane potential V and its temporal differential dV/dt) $\times S^1$ (the circumference representing the sinusoidal force) only from the experimentally obtained time series data of the membrane potential, and is then observed on its Poincaré section at a fixed phase of the periodic force. The Poincaré sections on space $V \times dV/dt$ observed by the stroboscopic analysis at each 30° in the 2/3-entrained oscillation are shown in Fig. 3(b). Each frame of Fig. 3(c), on the other hand, gives plots of $(V(n), V(n+1))$ for $n = 1, 2, 3, \ldots$, at each phase, which is called the return plot. In general, the orbit of an n/m-entrained oscillation reconstructed in space $V \times dV/dt \times S^1$ is a limit cycle whose Poincaré section and return plot are composed of distinct m points. Various n/m-entrained oscillations with different values of n and m have been actually observed in squid giant axons when the forcing frequency f_F is close to m times the natural frequency f_N of the self-sustained oscillation. When we change the amplitude A and the frequency f_F of the sinusoidal force as the bifurcation parameters, we can further observe more complicated forced oscillations that are not entrained.

3. Quasi-Periodic Oscillations

Even if a self-sustained neural oscillator is forced by a periodic stimulation, its response is not necessarily periodic. Some of these nonperiodic responses are understood as quasi-periodic oscillations with multiple independent frequencies, usually two frequencies, that are incommensurate with each other. Generally speaking, the membrane potential $V(t)$ in a quasi-periodic oscillation with n independent frequencies $f_1, f_2, \ldots,$ and f_n can be represented as follows [Glendinning, 1994; Kapitaniak & Bishop, 1999]:

$$V(t) = g(2\pi f_1 t, 2\pi f_2 t, \ldots, 2\pi f_n t), \tag{1}$$

where the function g is periodic with period 2π in each argument and the frequencies $f_1, f_2, \ldots,$ and f_n are not rationally related. In other words, none of the frequencies can be expressed as a linear combination of the others with rational coefficients [Ott *et al.*, 1994].

When the amplitude A of the sinusoidal force is sufficiently weak, a quasi-periodic oscillation is generally observed because the external force is not strong enough for entraining a neural oscillator to a phase-locked oscillation with very little influence on a neural oscillator, and two co-existing independent frequencies of the natural frequency f_N and the forcing one f_F are usually incommensurable with each other or irrationally related [Moser, 1969; Sell, 1981; Ueda & Akamatsu, 1981; Zeeman, 1982]. The reason is that two real numbers, if randomly chosen, are incommensurate almost always. Figure 5 shows an example of such a quasi-periodic oscillation with a small forcing amplitude. Figures 6–8 also show other kinds of quasi-periodic oscillations.

Figures 5(b), 6(b), 7(b), and 8(b) are the stroboscopic plots or the Poincaré sections of the

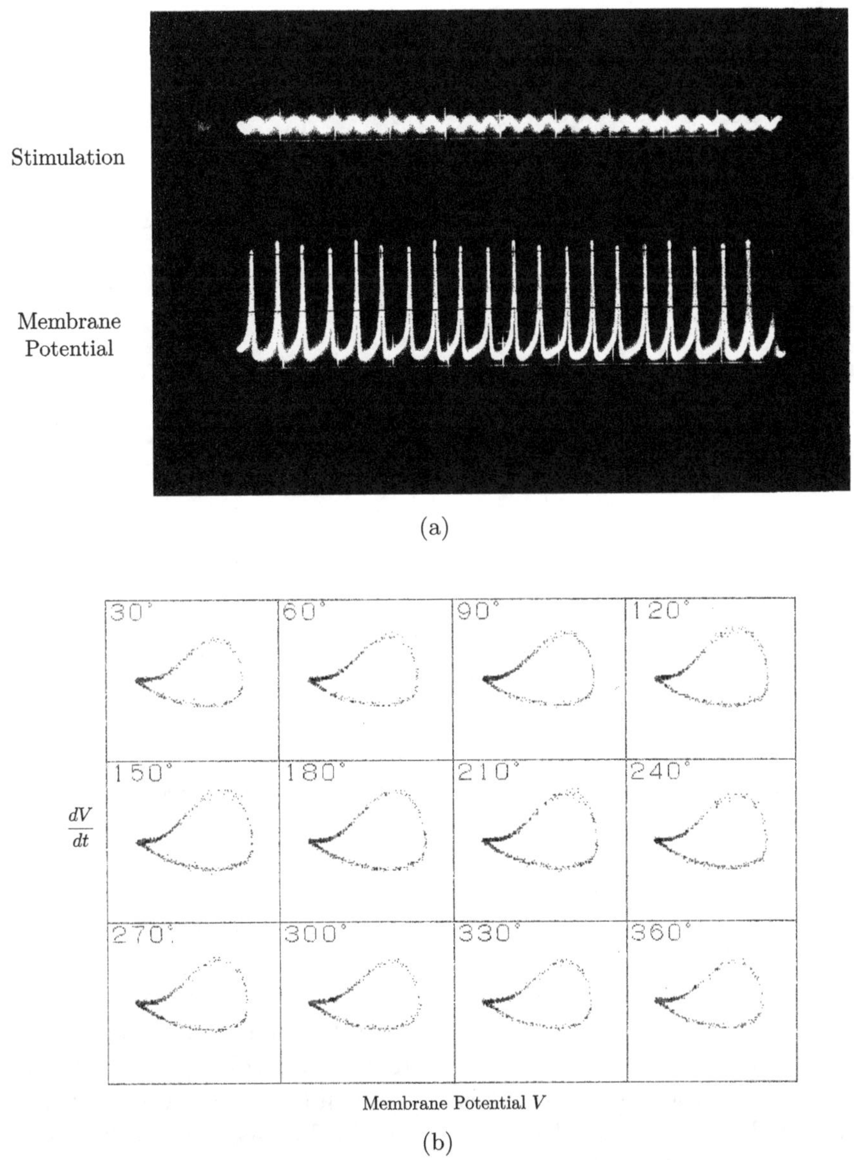

Fig. 5. A quasi-periodic oscillation with $f_N = 196\,\text{Hz}$, $f_F = 271\,\text{Hz}$, and a small forcing amplitude of $A = 0.5\,\mu\text{A}$. (a) The waveforms (above: the stimulating current, below: the response of the membrane potential). (b) The stroboscopic plots where the number in each frame shows the phase at which the response is stroboscopically observed.

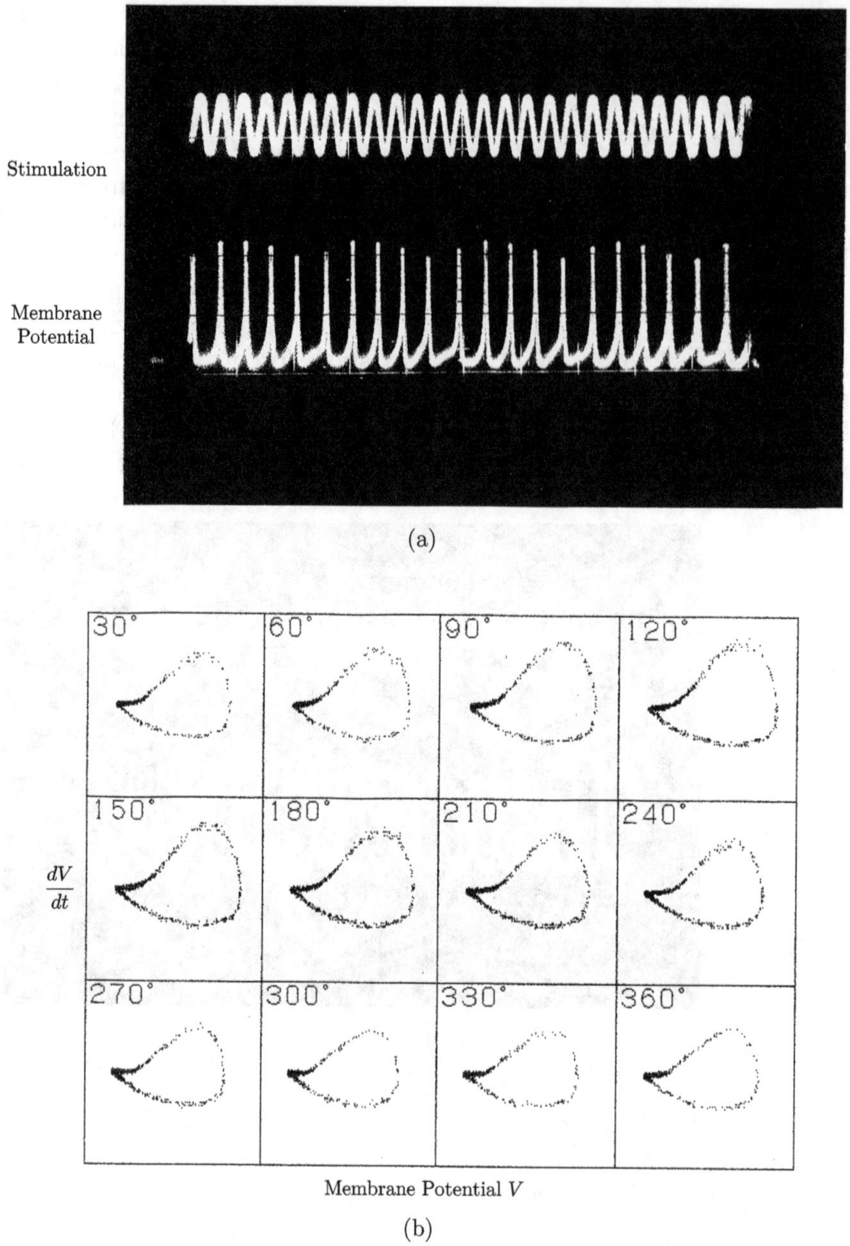

Fig. 6. A quasi-periodic oscillation with $f_N = 203\,\text{Hz}$, $f_F = 250\,\text{Hz}$, and $A = 1\,\mu\text{A}$. (a) The waveforms (above: the stimulating current, below: the response of the membrane potential). (b) The stroboscopic plots where the number in each frame shows the phase at which the response is stroboscopically observed.

quasi-periodic oscillations. Each stroboscopic plot asymptotically forms a closed curve [Matsumoto *et al.*, 1984; Aihara *et al.*, 1986]. The closed curve emerging in every Poincaré section means that the trajectory of the quasi-periodic oscillation in space $V \times dV/dt \times S^1$ is a two-dimensional torus.

The quasi-periodic oscillation in Fig. 6 is observed in a vicinity of the 1/1-entrained region. When an entrained forced oscillation bifurcates to a desynchronized oscillation with changing the forcing frequency f_F or reducing the forcing amplitude A, the forced oscillation is usually quasi-periodic with modulation of amplitudes and phases of action potentials as shown in Fig. 6(a).

Figures 7 and 8 show different kinds of quasi-periodic oscillations. The quasi-periodic oscillation in Fig. 7 is observed when the forcing frequency f_F is sufficiently lower than the natural frequency f_N. In this case, the modulation frequency is almost equal to f_F [see the waveforms of the membrane potential in Fig. 7(a)]. The quasi-periodic oscillation in Fig. 8, on the other hand, is observed when

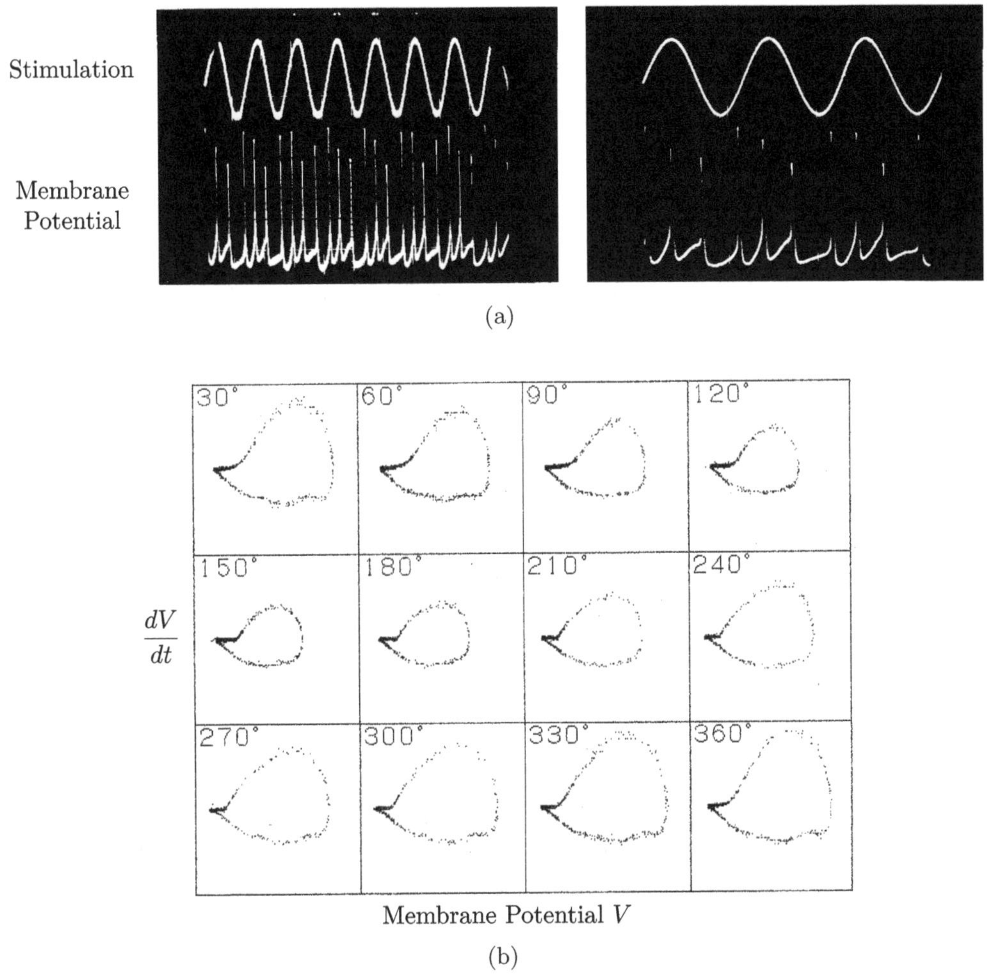

Fig. 7. A quasi-periodic oscillation with $f_N = 187\,\text{Hz}$, a low forcing frequency of $f_F = 60\,\text{Hz}$, and $A = 2\,\mu\text{A}$. (a) The waveforms (above: the stimulating current, below: the response of the membrane potential); the time axis of the right blows up 2.5 times that of the left. (b) The stroboscopic plots where the number in each frame shows the phase at which the response is stroboscopically observed.

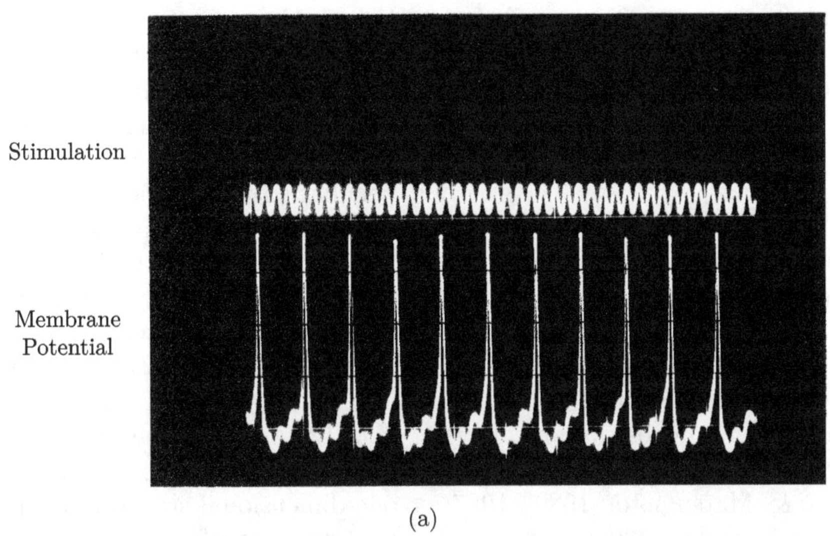

Fig. 8. A quasi-periodic oscillation with $f_N = 187\,\text{Hz}$, a high forcing frequency of $f_F = 800\,\text{Hz}$, and $A = 2\,\mu\text{A}$. (a) The waveforms (above: the stimulating current, below: the response of the membrane potential). (b) The stroboscopic plots where the number in each frame shows the phase at which the response is stroboscopically observed.

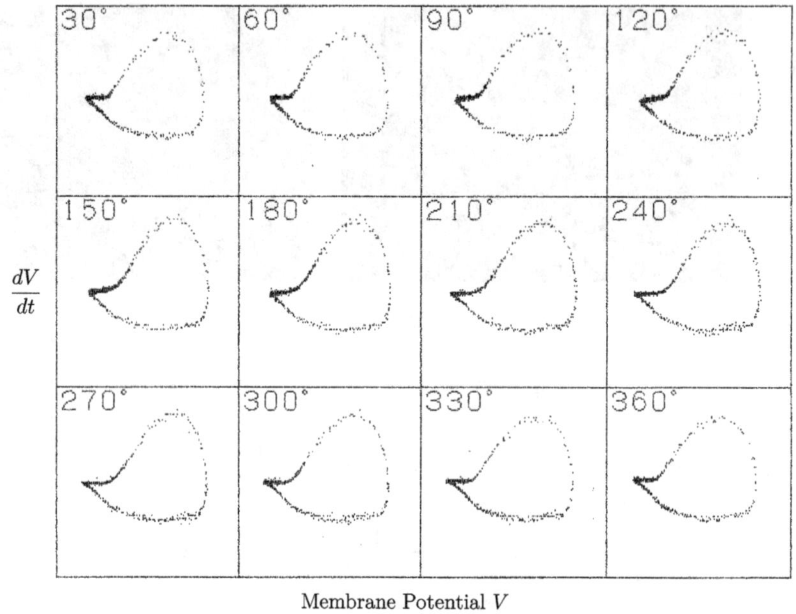

(b)

Fig. 8. (*Continued*)

f_F is sufficiently higher than f_N. While the closed curves asymptotically emerging in the stroboscopic plots in Fig. 8(b) are almost the same, the sizes of those in Fig. 7(b) are different, depending upon the phase of the sinusoidal force. The latter reflects the fact that the waveform of the membrane potential, especially the heights of the action potentials are modulated directly by the forcing frequency f_F in the quasi-periodic oscillation in Fig. 7. In the case of the quasi-periodic oscillation with very high f_F in Fig. 8(a), there are small peaks of the membrane potential between two adjacent spikes of action potentials; each peak is a response to each peak of the stimulating sinusoidal current [Aihara *et al.*, 1984].

4. Chaotic Oscillations

Nonperiodic oscillations that are more complex than quasi-periodic oscillations can be widely observed when the amplitude A and the frequency f_F of the sinusoidal current forcing squid giant axons in the state of the self-sustained oscillation are adjusted. Such nonperiodic oscillations have been classified as chaotic oscillations [Matsumoto *et al.*, 1984; Aihara & Matsumoto, 1986, 1987; Aihara *et al.*, 1985, 1986; Aihara, 1994].

Figures 9–12 show examples of chaotic oscillations observed in squid giant axons. The Poincaré sections in Fig. 9(b) show that the reconstructed attractor keeps almost cylindrical structure but a part of the surface is pinched like a beak, stretched, folded like a wing, and compressed during one forcing period [Abraham & Shaw, 1983; Thompson & Stewart, 1984, 1986]. Such transformation with stretching, folding, and compressing is typically observed in low-dimensional chaotic systems. This kind of a chaotic attractor with a structure of a two-dimensional torus with beaks and wings is also observed in forced oscillators such as the forced Van der Pol oscillator [Shaw, 1981] and called the Birkhoff–Shaw chaotic attractor [Thompson & Stewart, 1984]. Although the tubular structure is similar to that of a quasi-periodic oscillation, deformation of a usual two-dimensional torus peculiar to a quasi-periodic oscillation is necessary for emergence of a chaotic oscillation through quasi-periodicity. Figure 10 shows another example of a chaotic oscillation that is represented by a strange attractor with stretching, folding, and compressing while keeping an almost tubular structure. The strange attractors of chaotic oscillations in Figs. 11 and 12, on the other hand, don't keep cylindrical structures anymore and are crushed into nearly one-dimensional structure during phases from 360° to 120°. Transformation with stretching, folding, and compressing can be also seen in these strange attractors.

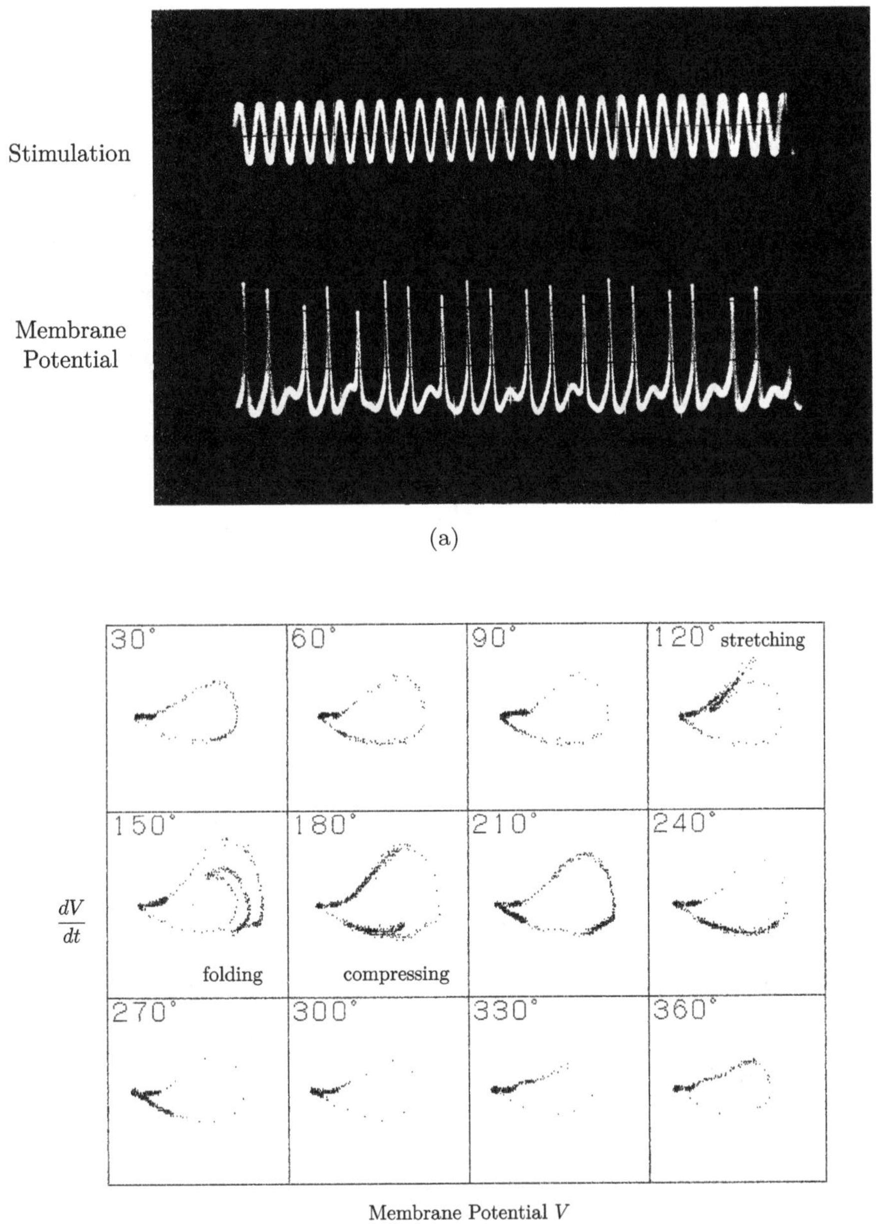

(b)

Fig. 9. A chaotic oscillation with $f_N = 183\,\text{Hz}$, $f_F = 270\,\text{Hz}$, and $A = 2\,\mu\text{A}$. (a) The waveforms (above: the stimulating current, below: the response of the membrane potential). (b) The stroboscopic plots where the number in each frame shows the phase at which the response is stroboscopically observed. (*Source*: Modified from Fig. 2 in [Aihara & Matsumoto, 1987] with permission.)

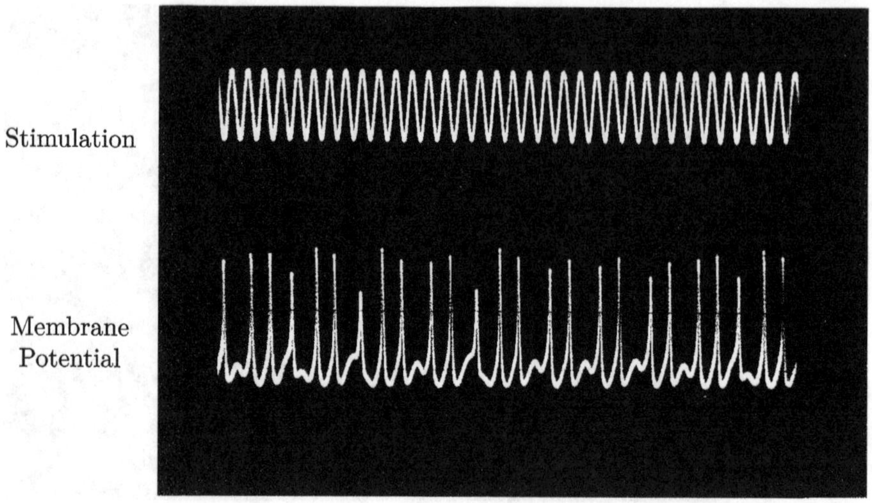

(a)

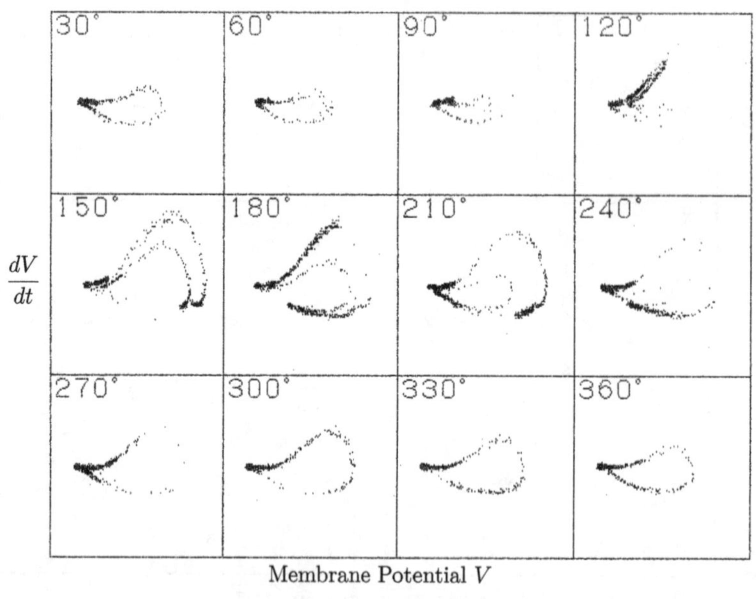

(b)

Fig. 10. A chaotic oscillation with $f_N = 183\,\text{Hz}$, $f_F = 270\,\text{Hz}$, and $A = 2.41\,\mu\text{A}$. (a) The waveforms (above: the stimulating current, below: the response of the membrane potential). (b) The stroboscopic plots where the number in each frame shows the phase at which the response is stroboscopically observed.

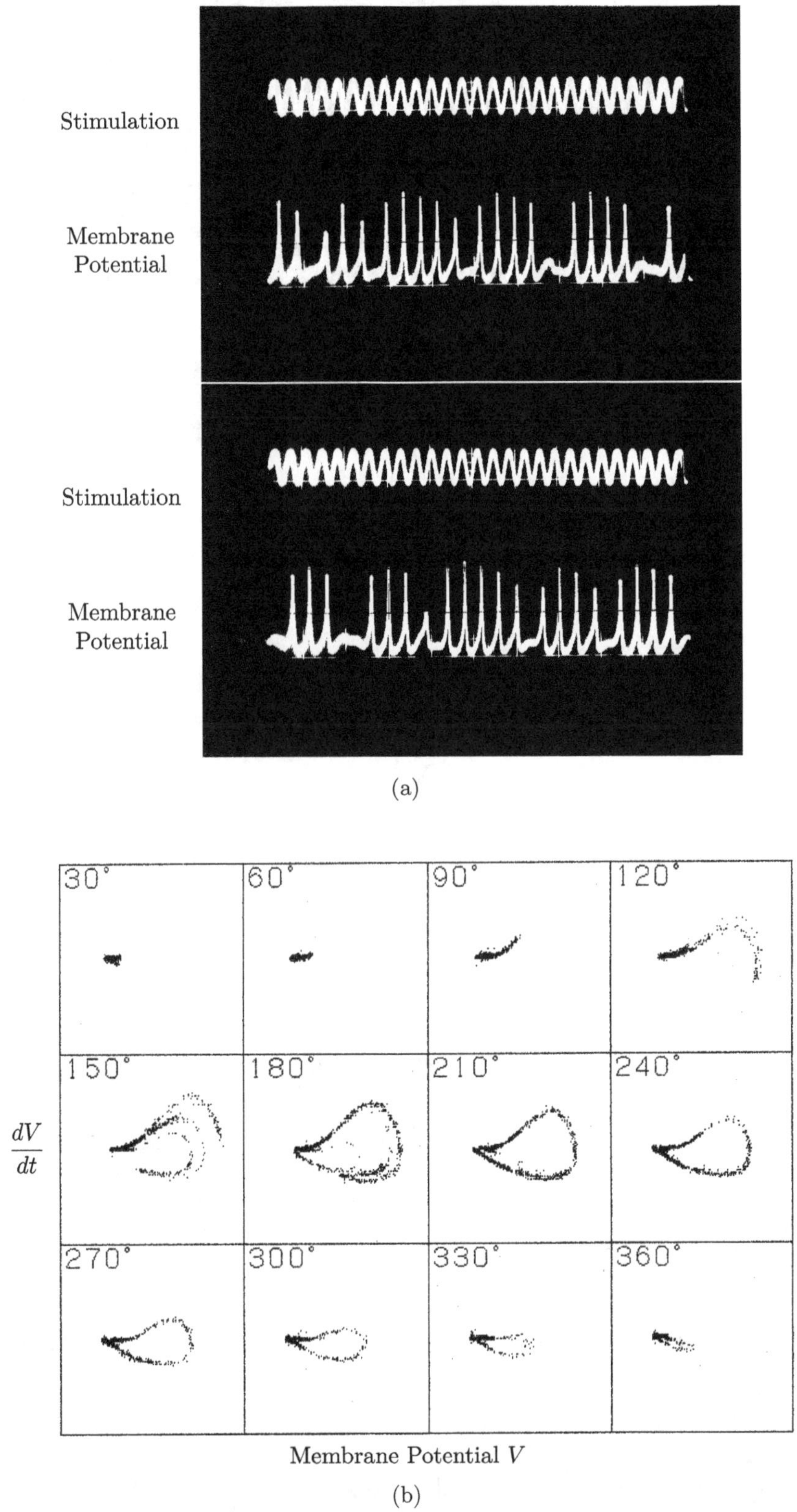

Fig. 11. A chaotic oscillation with $f_N = 184\,\text{Hz}$, $f_F = 262\,\text{Hz}$, and $A = 1.5\,\mu\text{A}$. (a) The waveforms (above: the stimulating current, below: the response of the membrane potential). (b) The stroboscopic plots where the number in each frame shows the phase at which the response is stroboscopically observed.

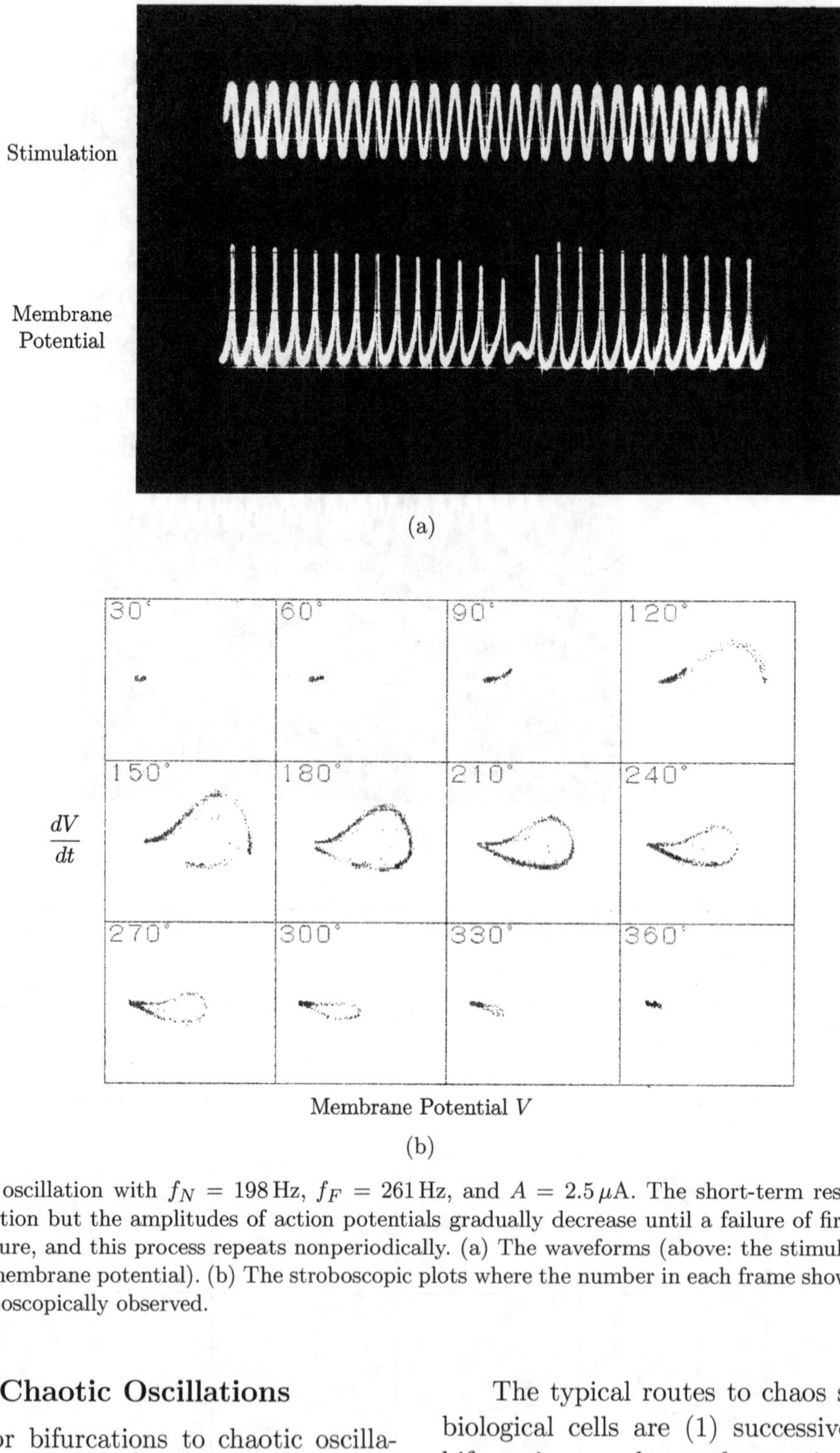

Fig. 12. A chaotic oscillation with $f_N = 198\,\text{Hz}$, $f_F = 261\,\text{Hz}$, and $A = 2.5\,\mu\text{A}$. The short-term response is similar to a 1/1-entrained oscillation but the amplitudes of action potentials gradually decrease until a failure of firing, recover to larger heights after the failure, and this process repeats nonperiodically. (a) The waveforms (above: the stimulating current, below: the response of the membrane potential). (b) The stroboscopic plots where the number in each frame shows the phase at which the response is stroboscopically observed.

5. Routes to Chaotic Oscillations

Various routes, or bifurcations to chaotic oscillations have been studied in detail (e.g. [Thompson & Stewart, 1986; Hilborn, 2000; Ott, 2002]). Although the whole story of routes to chaos is still far from fully understood, some typical routes are well known especially in low-dimensional systems. Forced neural oscillations analyzed in this paper can be also described as low-dimensional systems.

The typical routes to chaos so far observed in biological cells are (1) successive period-doubling bifurcations, where the period of oscillations increases in the form of 2^n times with $n \to \infty$, (2) intermittency, where chaotic behavior emerges intermittently and repetitively among apparently periodic oscillations, and (3) collapse of quasi-periodicity, where chaotic behavior is generated by deformation and collapse of a two-dimensional torus representing a quasi-periodic oscillation (e.g. [Kaneko, 1986; Aihara & Matsumoto, 1986; Glass & Mackey, 1988; Aihara, 1994, 1995]).

Figures 13–15 show an example of a route to chaos through successive period-doubling bifurcations where Figs. 13–15 are waveforms, Poincaré sections, and return plots, respectively [Aihara, 1994]. In Figs. 13–15, (a) is a 3/4-entrained oscillation, (b) is a 6/8-entrained oscillation that is generated by a period-doubling bifurcation of a 3/4-entrained oscillation of (a), (c) is a chaotic oscillation with an island structure both in the Poincaré sections and the return plots, which seems to emerge after accumulation of successive period-doubling bifurcations, and (d) is a more developed chaotic oscillation.

Figures 16(b)–16(d), 17(b)–17(d), and 18(b)–18(d) show another example of a similar route to chaos with period-doubling bifurcations. In the route to chaos with successive period-doubling bifurcations, we generally observe an n/m-entrained oscillation with mutually prime natural numbers m and n first, a $2n/2m$-entrained oscillation, a $4n/4m$-entrained oscillation, and so on successively, although it is quite difficult to observe higher-order period-doubling bifurcations probably due to nonstationarity and noise in squid giant axons.

Figures 16(e), 16(f), 17(e), 17(f), and 18(e), 18(f) show a different route to chaos with intermittency. In this case, the 4/5-entrained oscillation in (f) bifurcates to intermittent chaos in (e) by

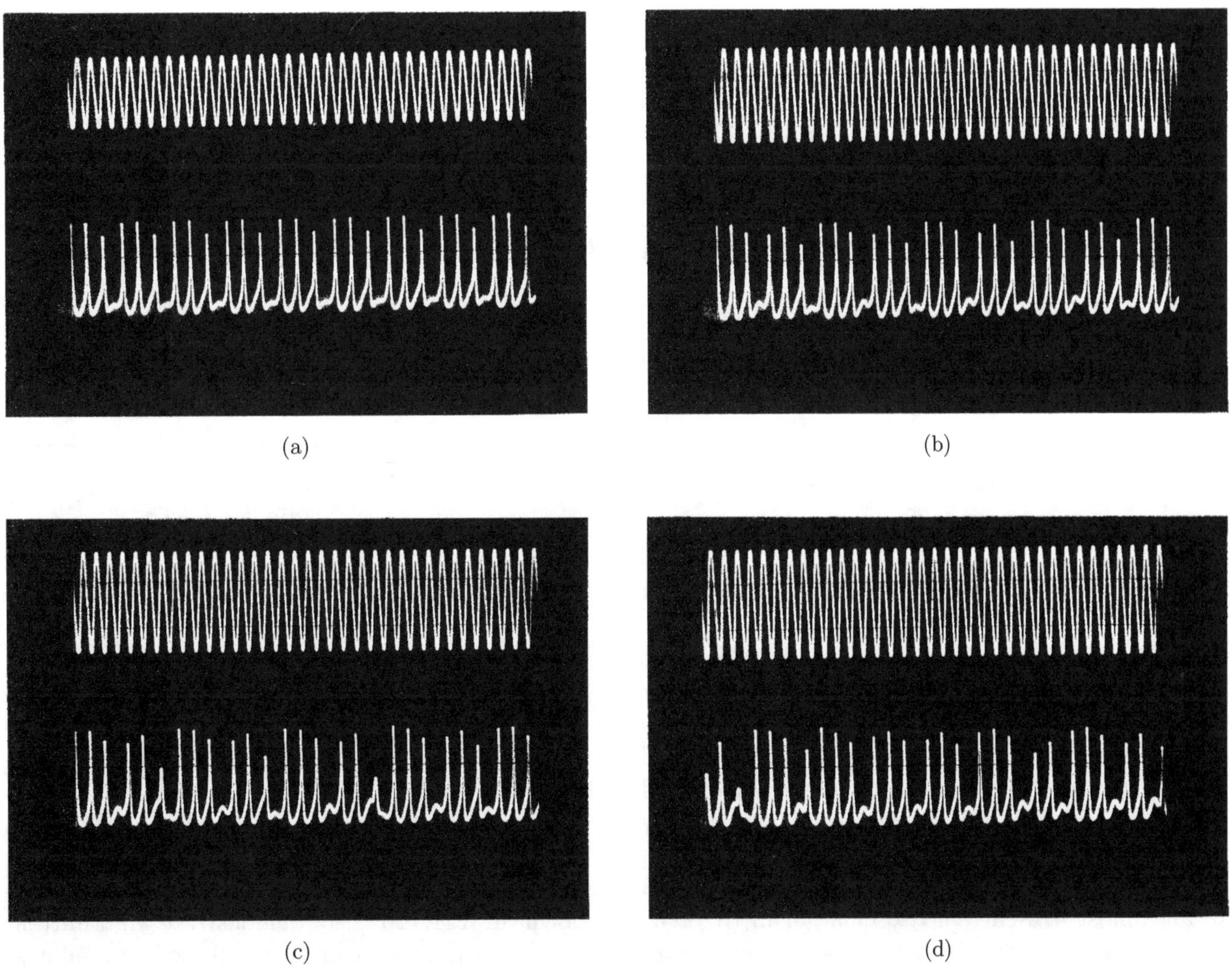

Fig. 13. A route to chaos through successive period-doubling bifurcations. Panels (a)–(d) show waveforms of (a) a 3/4-entrained oscillation, (b) a 6/8-entrained oscillation, (c) a chaotic oscillation just after accumulation of successive period-doubling bifurcations, and (d) a more developed chaotic oscillation. $f_N = 203\,\text{Hz}$, $f_F = 270\,\text{Hz}$, and (a) $A = 1.5\,\mu\text{A}$, (b) $A = 2.0\,\mu\text{A}$, (c) $A = 2.2\,\mu\text{A}$ and (d) $A = 2.32\,\mu\text{A}$.

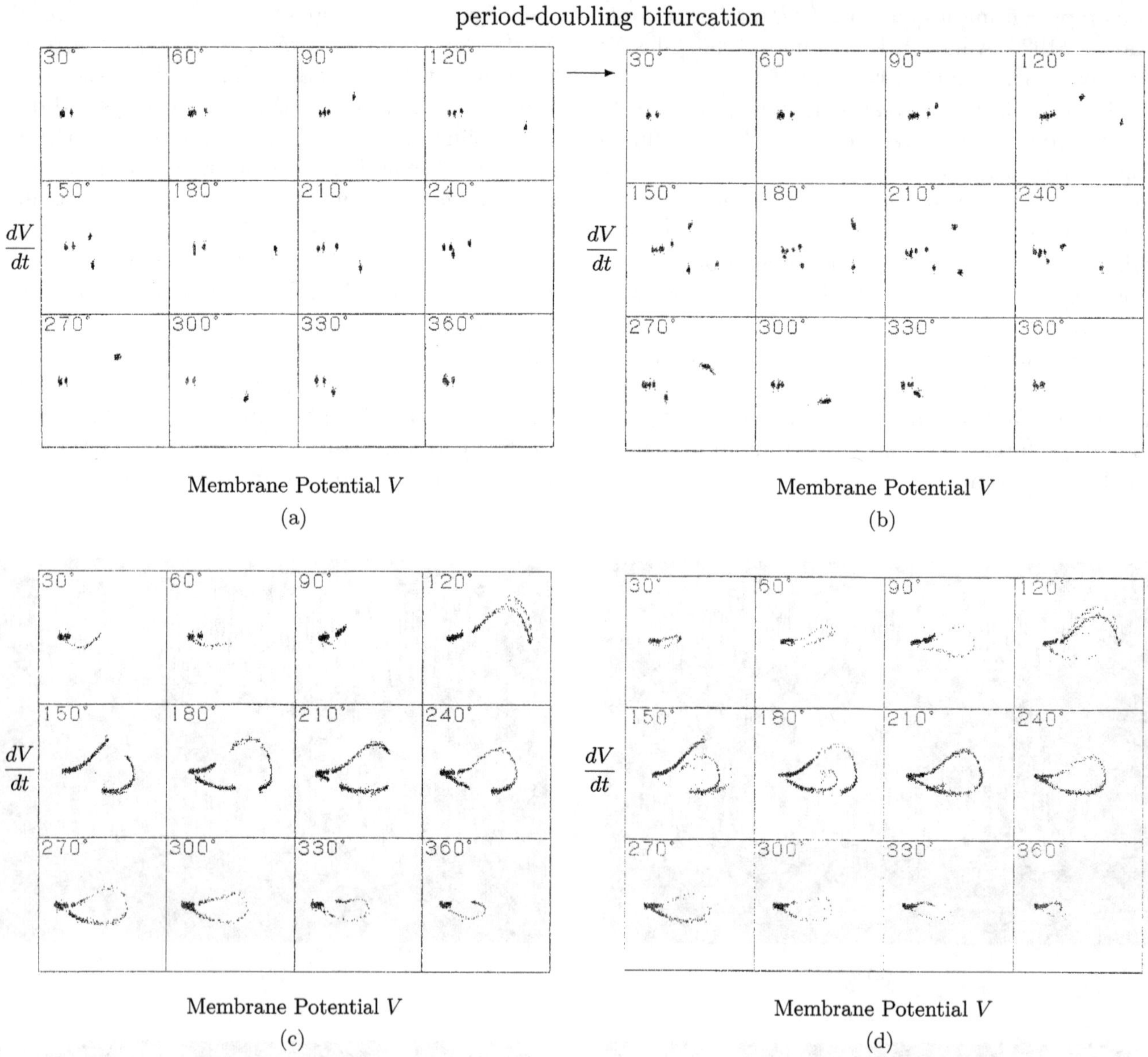

Fig. 14. A route to chaos through successive period-doubling bifurcations. Panels (a)–(d) show Poincaré sections corresponding to Figs. 13(a)–13(d), respectively. (*Source*: Modified from Fig. 3 in [Aihara, 1994] with permission.)

decreasing the forcing amplitude A from $2.58\,\mu$A to $2.45\,\mu$A while the forcing frequency f_F is fixed at 250 Hz. Figure 19 shows the long-term waveforms of this intermittent chaos in (e), which are observed for a longer duration where an almost 4/5-entrained oscillation in Fig. 19(a) transits to apparently chaotic oscillations in (b) and (c) with components like a 3/4-entrained oscillation, and again return to an almost 4/5-entrained oscillation in (d). The waveform of Fig. 18(e) is observed in a time window between Figs. 19(c) and 19(d), where transition to an almost 4/5-entrained oscillation is visible. The intermittent chaos of Figs. 16(e) and 17(e) repeats this transition between almost 4/5-entrained oscillations and chaotic oscillations, which may be understood to be generated by a saddle-node bifurcation of the 4/5-entrained oscillation in (f). Figures 16(a), 17(a), and 18(a) show another example of intermittent chaos that may bifurcate from a 3/4-entrained oscillation in Figs. 16(b), 17(b), and 18(b). The waveforms of this intermittent chaos shown in Fig. 20 also demonstrate intermittent transitions between chaotic oscillations and almost 3/4-entrained oscillations.

The Poincaré sections and the return plots of the intermittently chaotic oscillations in Figs. 16(a),

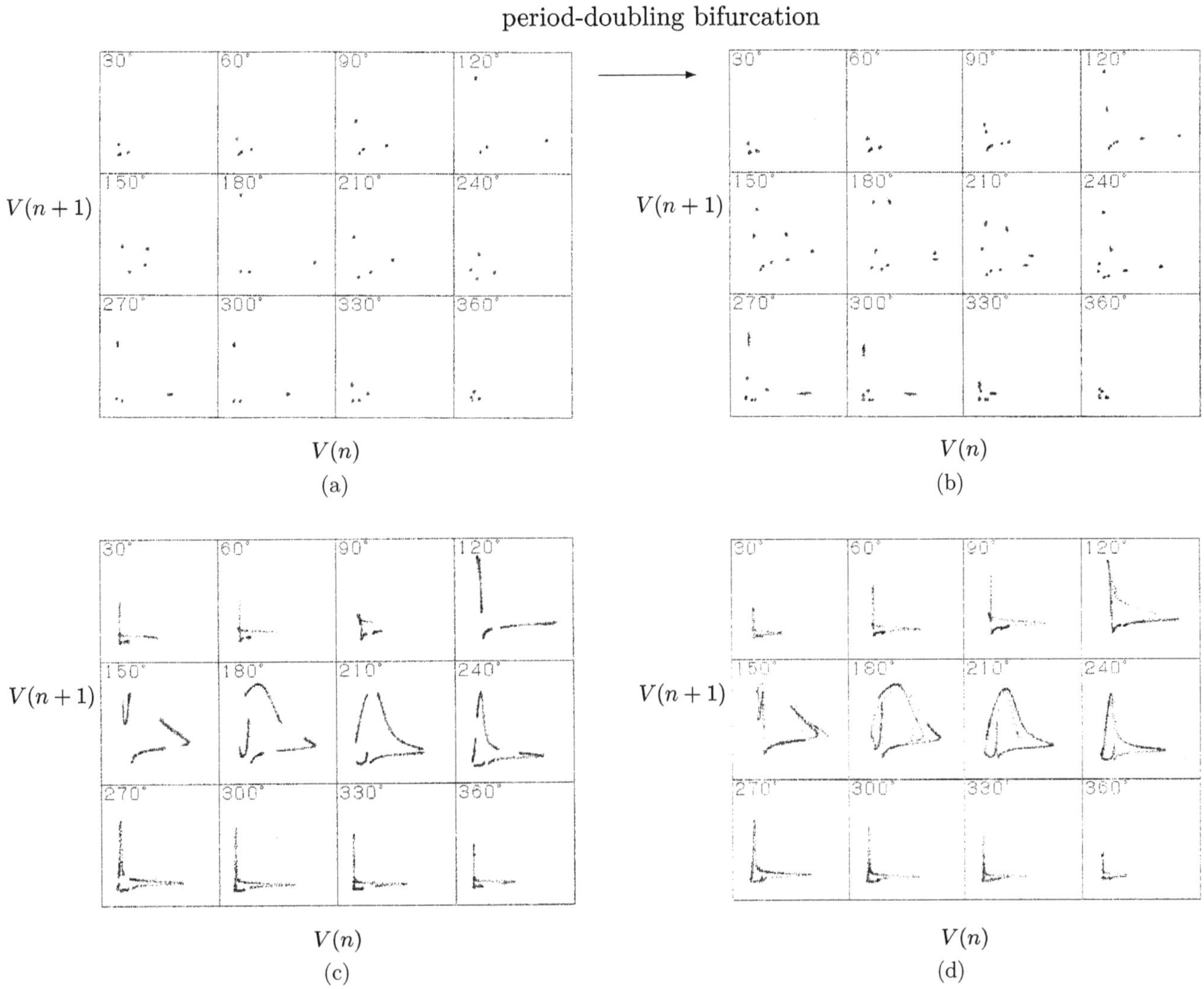

Fig. 15. A route to chaos through successive period-doubling bifurcations. Panels (a)–(d) show return plots corresponding to Figs. 13(a)–13(d), respectively. (*Source*: Modified from Fig. 3 in [Aihara, 1994] with permission.)

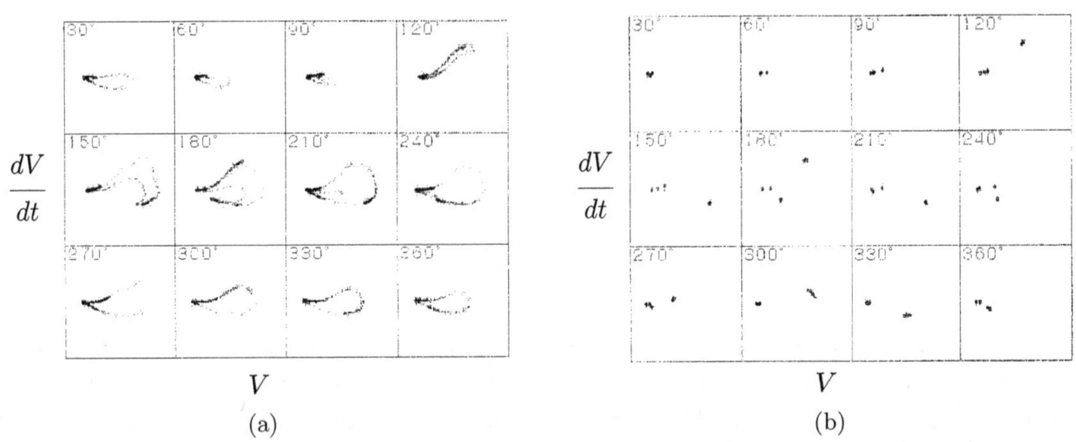

Fig. 16. Routes to chaos through successive period-doubling bifurcations and intermittency. Panels (a)–(h) show Poincaré sections of (a) a chaotic oscillation, (b) a 3/4-entrained oscillation, (c) a 6/8-entrained oscillation, (d) and (e) chaotic oscillations, (f) a 4/5-entrained oscillation, (g) a chaotic oscillation with island structure, and (h) a more developed chaotic oscillation, where $f_N = 173\,\text{Hz}$, $f_F = 250\,\text{Hz}$, and (a) $A = 1.63\,\mu\text{A}$, (b) $A = 1.7\,\mu\text{A}$, (c) $A = 2.28\,\mu\text{A}$, (d) $A = 2.34\,\mu\text{A}$, (e) $A = 2.45\,\mu\text{A}$, (f) $A = 2.58\,\mu\text{A}$, (g) $A = 2.70\,\mu\text{A}$, and (h) $A = 2.84\,\mu\text{A}$.

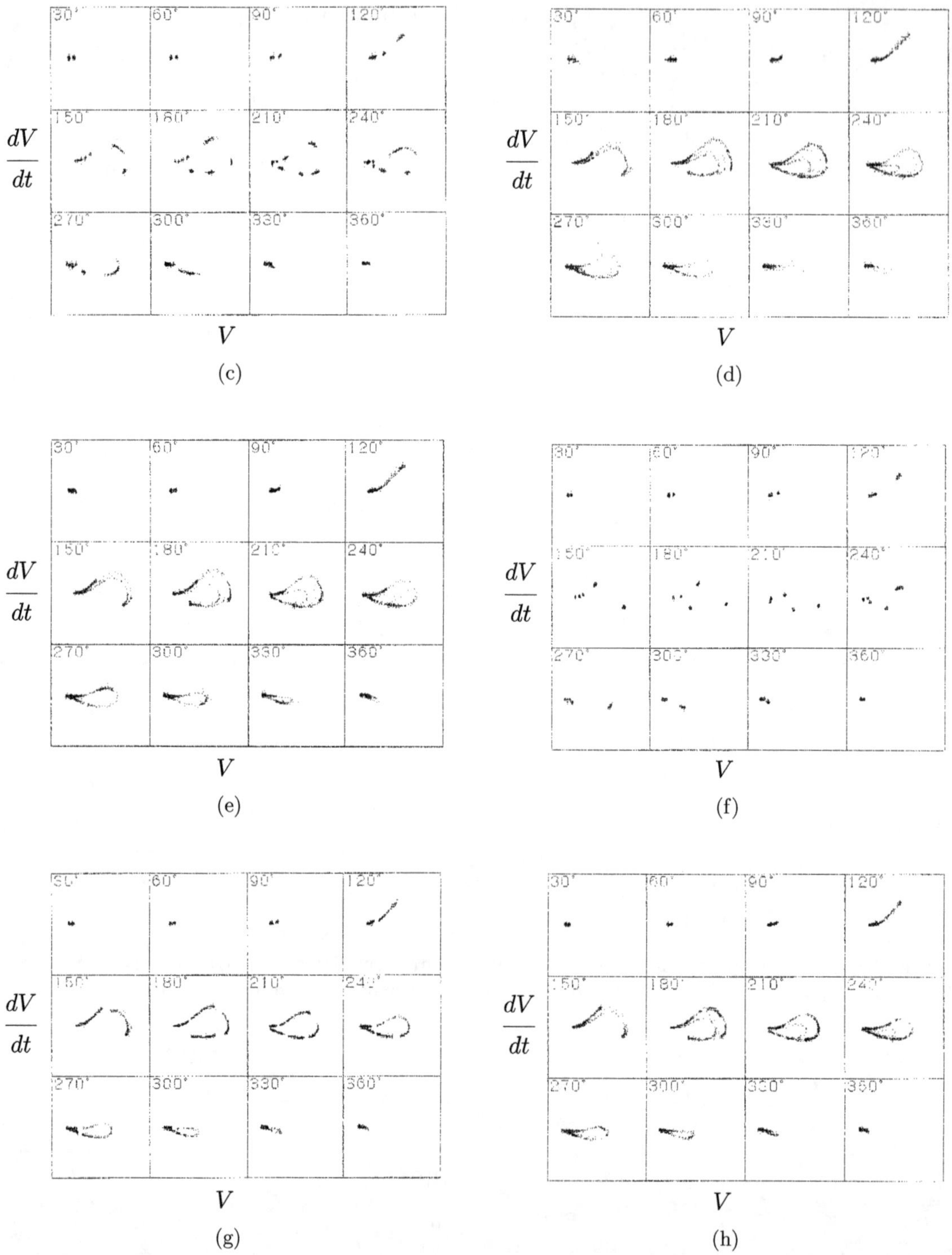

Fig. 16. (*Continued*)

16(e) and 17(a), 17(e) show higher densities of points appearing around the entrained oscillations in Figs. 16(b), 16(f) and 17(b), 17(f), respectively. Thus, saddle-node bifurcations seem to produce the intermittent chaos here. Figure 21 shows another example of intermittent chaos that probably emerges through a saddle-node bifurcation of a higher-order entrained oscillation, namely a 12/13-entrained oscillation in this case. Figure 22 shows transition of the waveforms in this intermittent chaos.

Generally speaking, a typical mechanism of intermittent transitions to chaos in a state space is composed of combination of destabilization of a

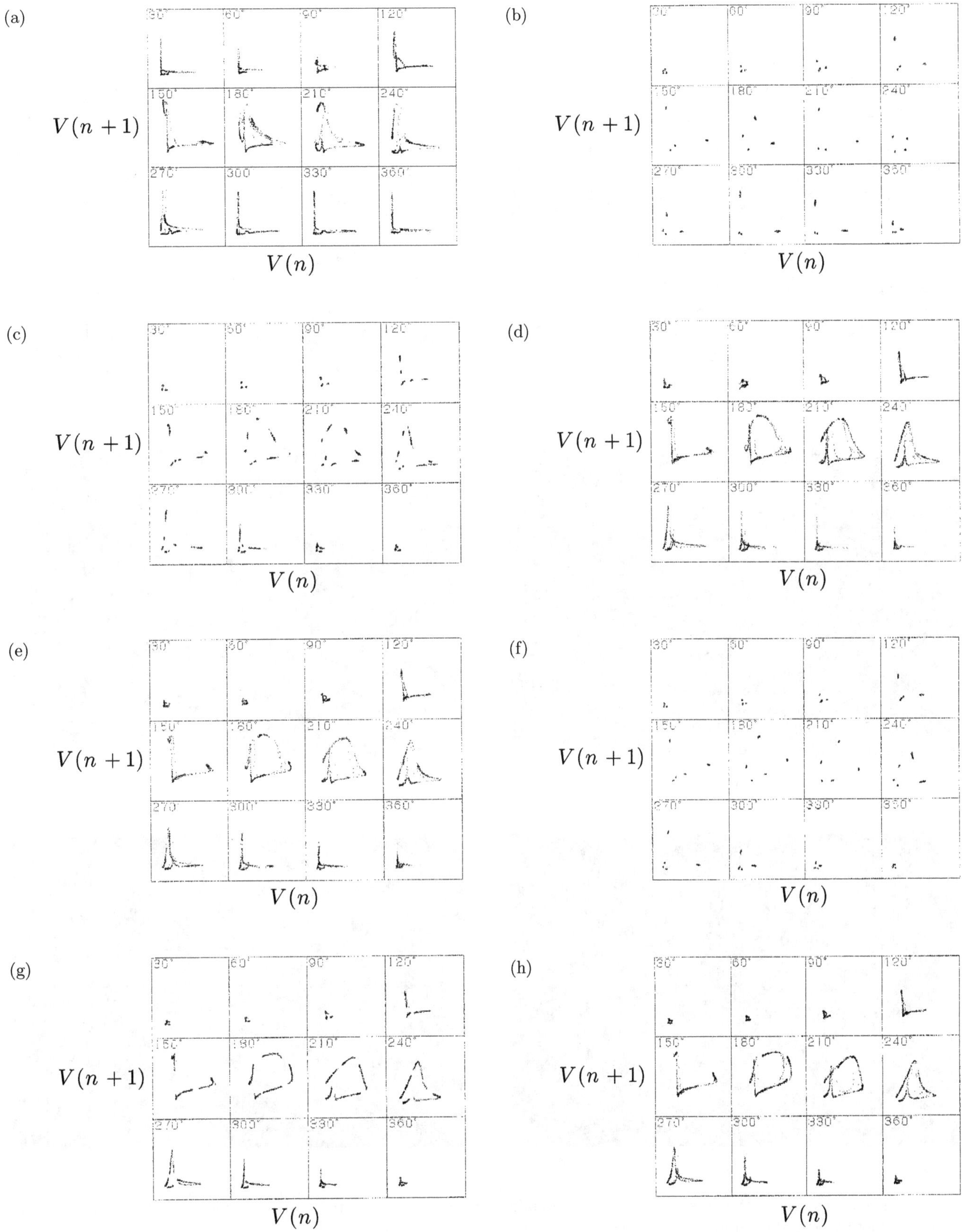

Fig. 17. Routes to chaos through successive period-doubling bifurcations and intermittency. Panels (a)–(h) show return plots corresponding to Figs. 16(a)–16(h), respectively.

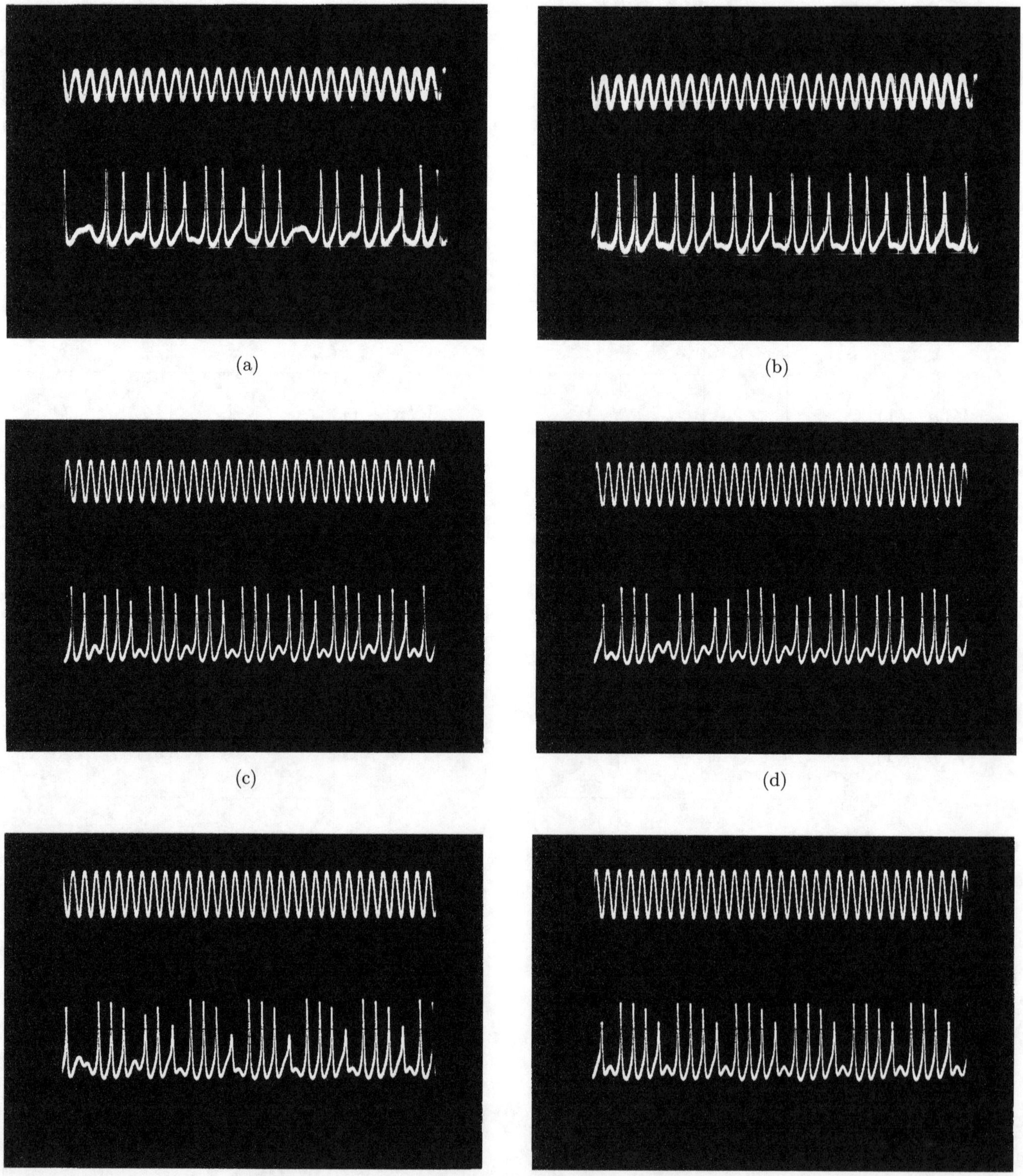

Fig. 18. Routes to chaos through successive period-doubling bifurcations and intermittency. Panels (a)–(f) show waveforms corresponding to Figs. 16(a)–16(f), respectively.

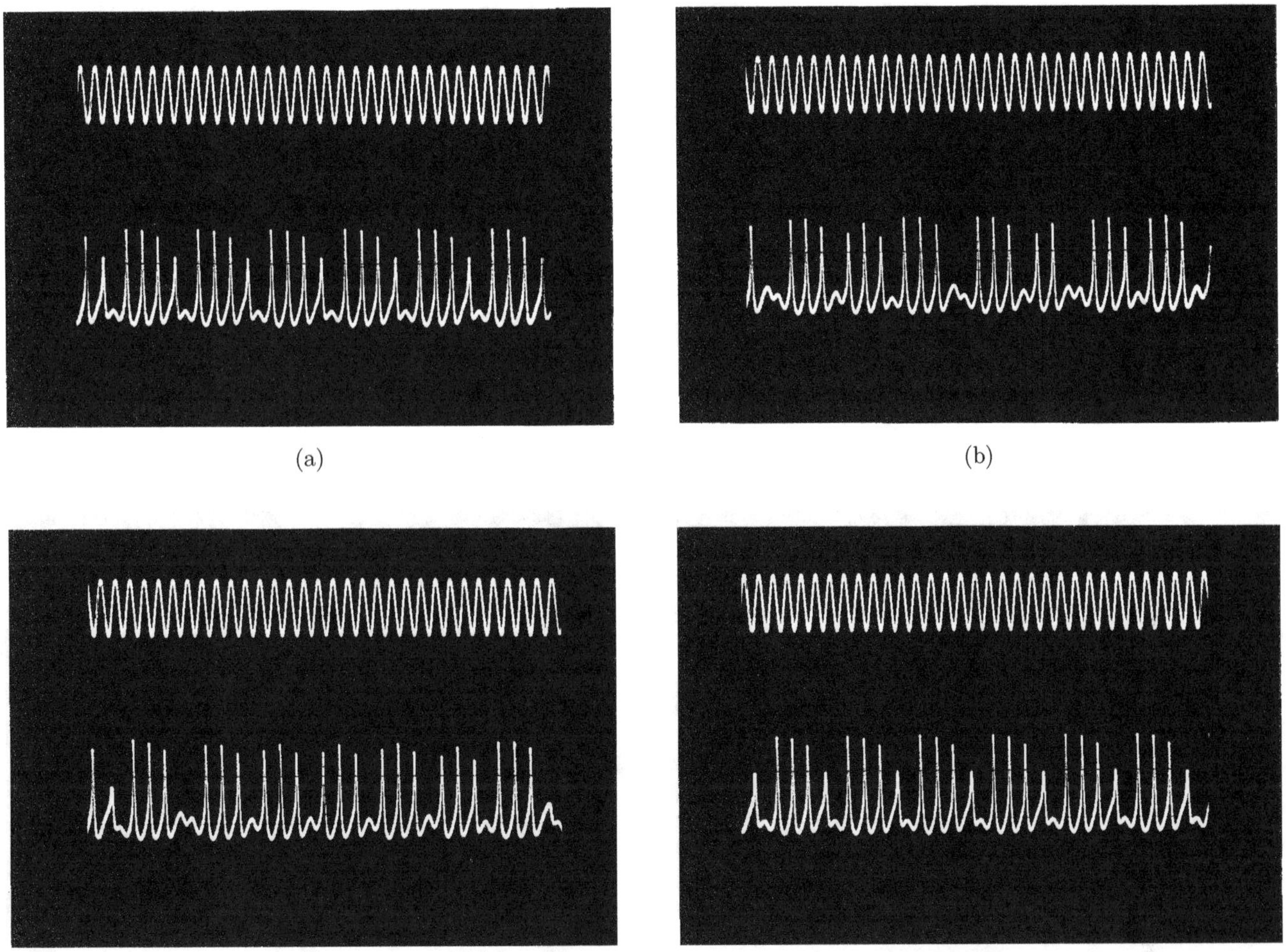

(a) (b) (c) (d)

Fig. 19. The long-term change of waveforms in the intermittent chaos in Figs. 16(e), 17(e), and 18(e) that is probably generated through a saddle-node bifurcation of the 4/5-entrained oscillation in Figs. 16(f), 17(f), and 18(f) by decreasing the forcing amplitude A from 2.58 μA to 2.45 μA while keeping the forcing frequency at $f_F = 250$ Hz.

limit cycle through a local bifurcation and reinjection of a trajectory into a neighborhood of a ruin of the destabilized limit cycle through global dynamics. The three types of intermittency are distinguished according to types of local bifurcations, namely type I with the saddle-node bifurcation, type II with the subcritical (or inverted) Neimark–Sacker bifurcation, and type III with the subcritical (or inverted) period-doubling bifurcation [Pomeau & Manneville, 1980]. Thus, the intermittent route that is explained above is type I, while type III intermittency is also observed in squid giant axons that are in the resting state and forced by periodic pulses [Matsumoto *et al.*, 1987]. Grebogi *et al.* further considered crisis-induced intermittency [Grebogi *et al.*, 1982; Grebogi *et al.*, 1987; Ott, 2002].

The saddle-node bifurcations also provide the mechanism of transition from quasi-periodic oscillations to entrained oscillations through frequency-locking or mode-locking. Figure 23 shows an example where a quasi-periodic oscillation (a) transits to an almost entrained oscillation in (b) through a locking probably produced by a saddle-node bifurcation on the two-dimensional torus of (a) when increasing the forcing amplitude A from 1.0 μA to 1.5 μA. Figure 24 shows the corresponding waveforms where Fig. 24(b) implies an almost 6/7-entrained oscillation.

The third route, or another route to chaos through quasi-periodicity has been also studied intensively. Especially in the early stage of studies on chaos, a genesis of strange attractors by perturbing low-dimensional tori was an important

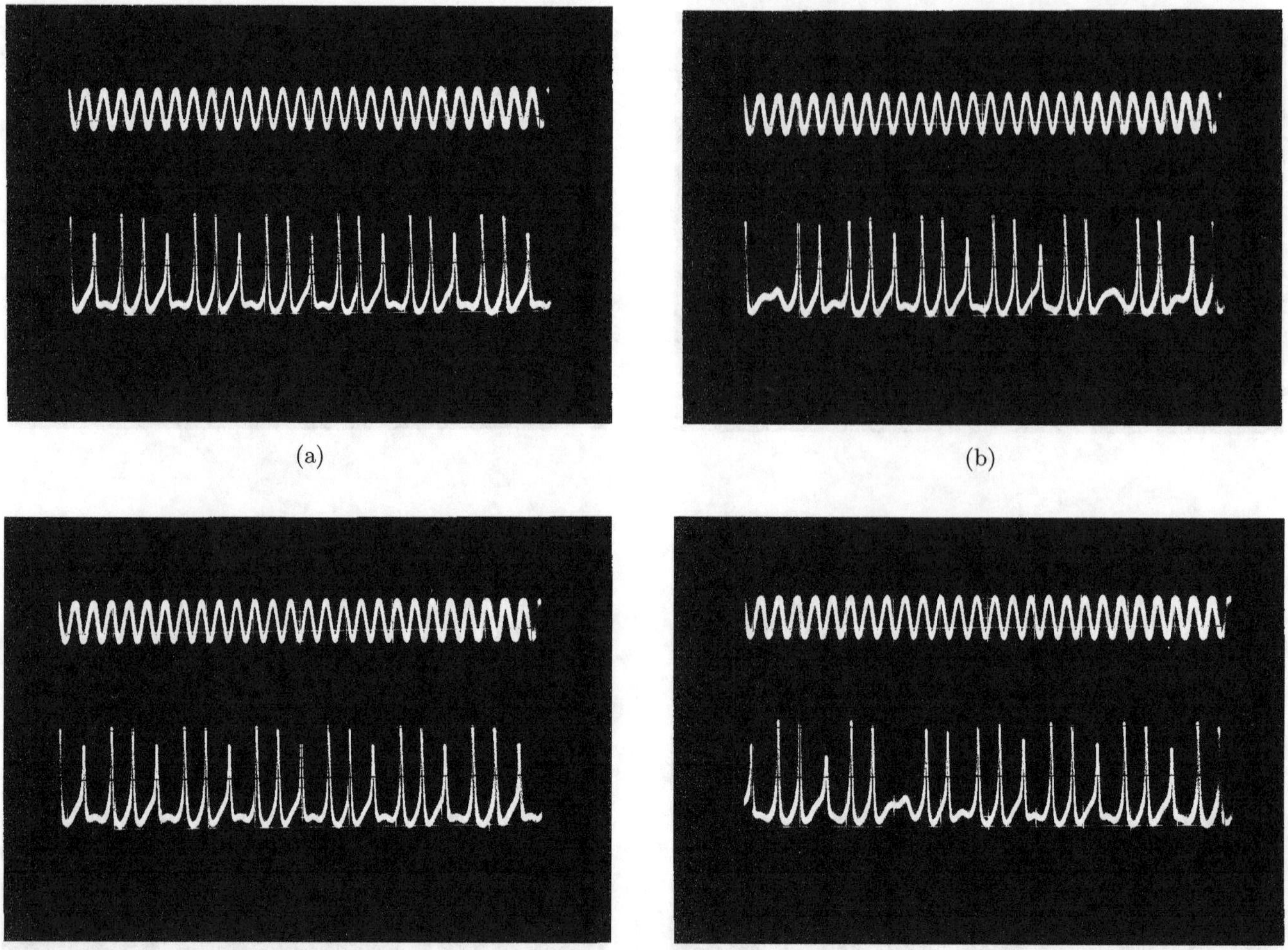

Fig. 20. The long-term change of waveforms in the intermittent chaos in Figs. 16(a), 17(a), and 18(a) that is probably generated through a saddle-node bifurcation of the 3/4-entrained oscillation in Figs. 16(b), 17(b), and 18(b) by decreasing the forcing amplitude A from $1.7\,\mu$A to $1.63\,\mu$A while keeping the forcing frequency at $f_F = 250\,$Hz.

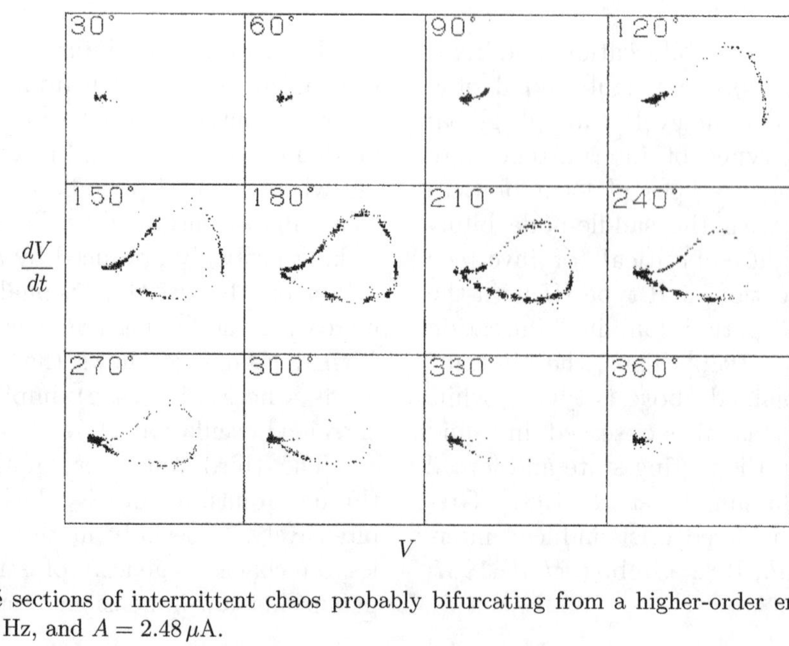

Fig. 21. The Poincaré sections of intermittent chaos probably bifurcating from a higher-order entrained oscillation where $f_N = 197\,$Hz, $f_F = 255\,$Hz, and $A = 2.48\,\mu$A.

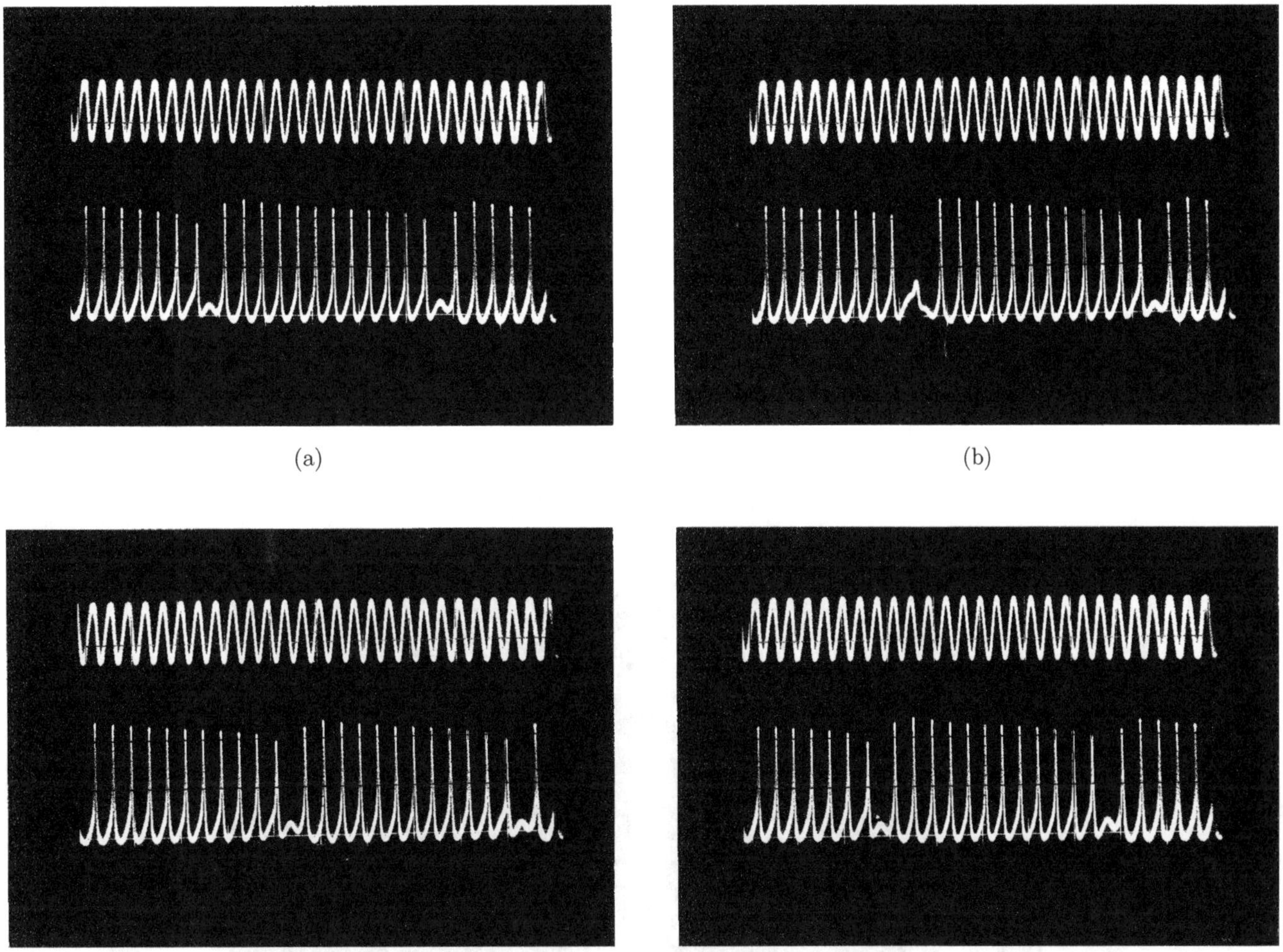

(a) (b)

(c) (d)

Fig. 22. The waveforms of the intermittent chaos of Fig. 21.

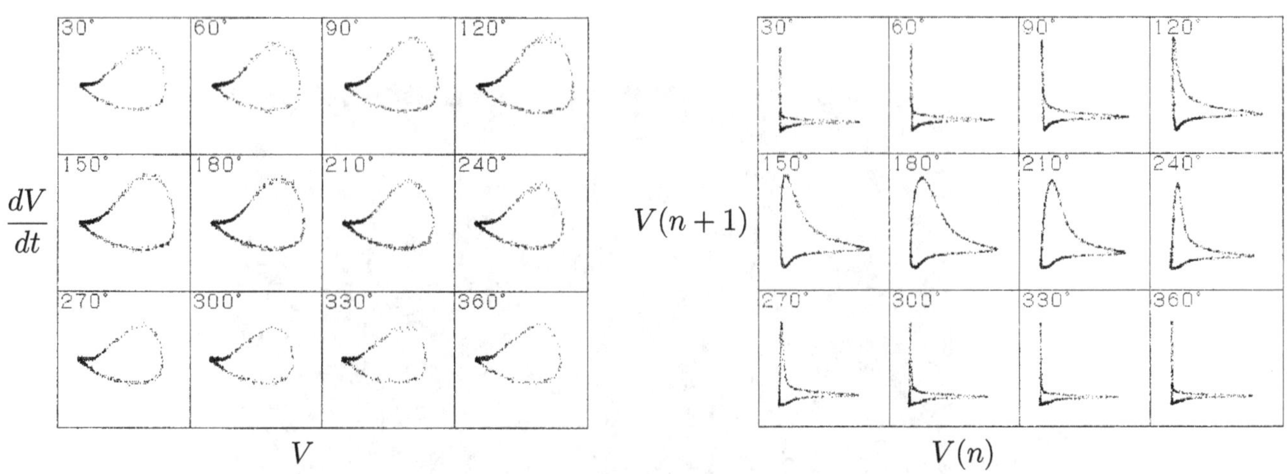

(a)

Fig. 23. Transition from (a) a quasi-periodic oscillation to (b) an almost entrained oscillation through locking by a saddle-node bifurcation. $f_N = 203\,\text{Hz}$, $f_F = 250\,\text{Hz}$, (a) $A = 1.0\,\mu\text{A}$ and (b) $A = 1.5\,\mu\text{A}$. The left pictures and right ones are the Poincaré sections and the return plots, respectively.

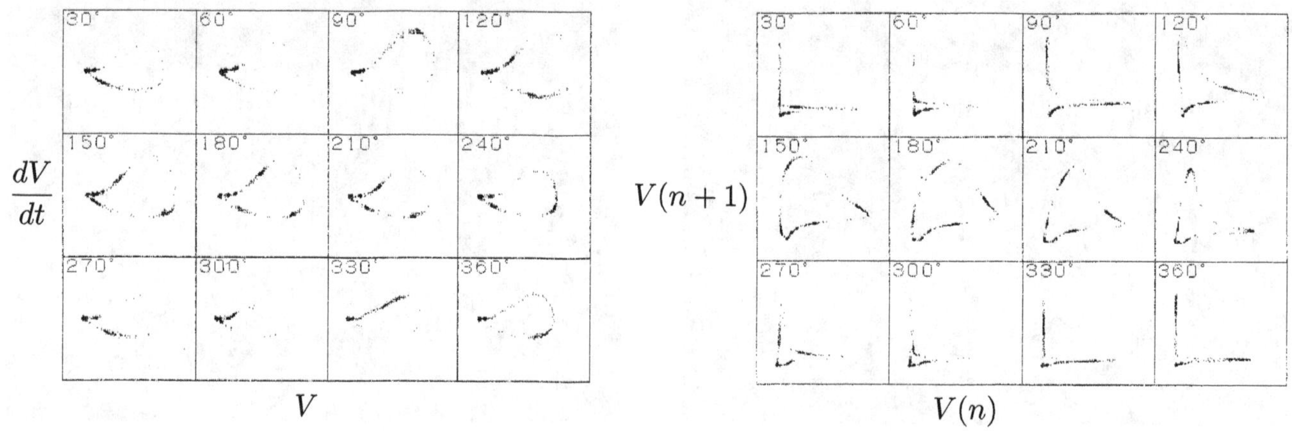

(b)

Fig. 23. (*Continued*)

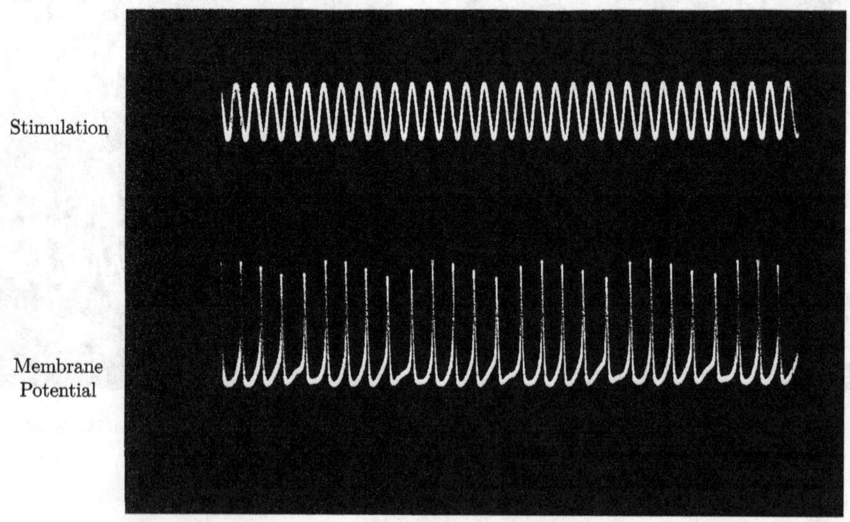

(a)

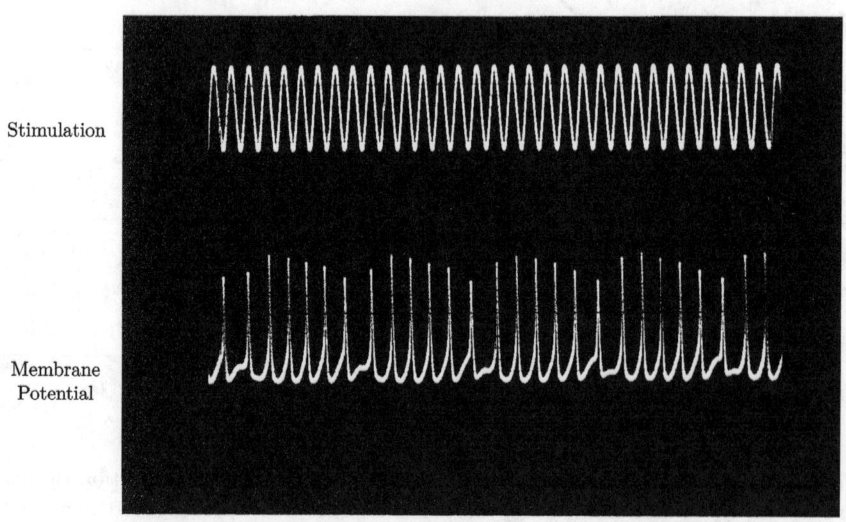

(b)

Fig. 24. The waveforms corresponding to Figs. 23(a) and 23(b).

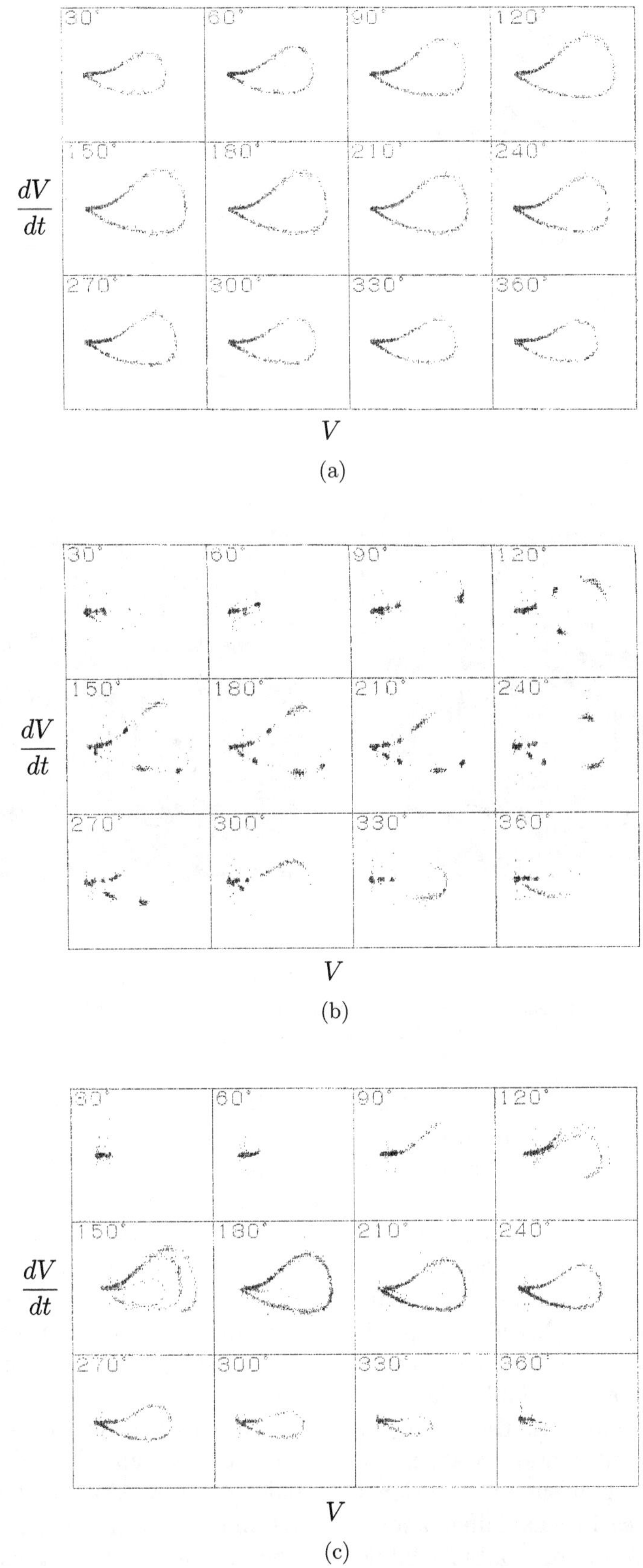

Fig. 25. A route to (c) chaos from (a) a quasi-periodic oscillation through (b) locking. $f_N = 208\,\text{Hz}$, $f_F = 256\,\text{Hz}$, (a) $A = 0.5\,\mu\text{A}$, (b) $A = 1.0\,\mu\text{A}$, and (c) $A = 1.5\,\mu\text{A}$.

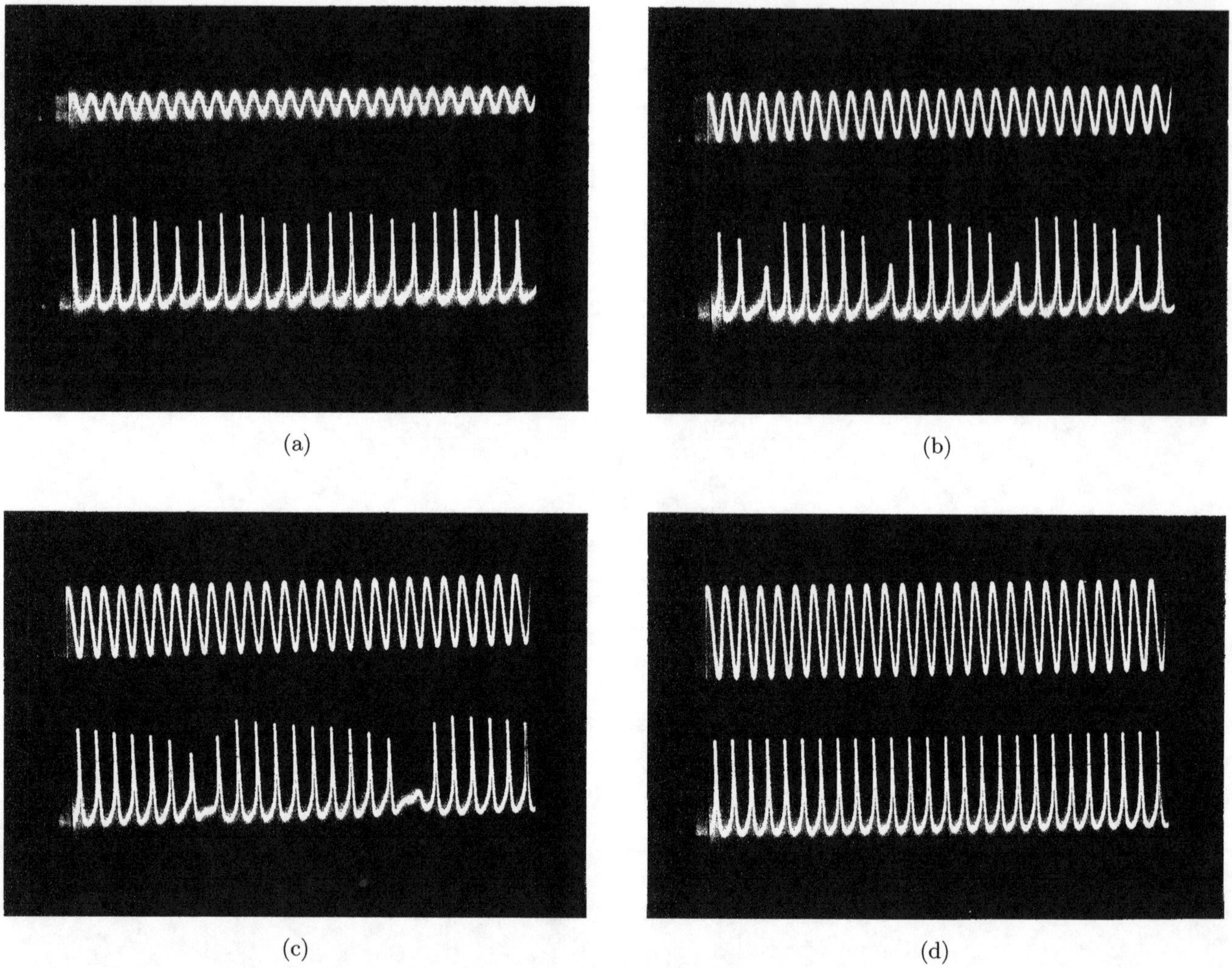

Fig. 26. A route from quasi-periodicity to chaos through collapse of a two-dimensional torus. Panels (a)–(d) show waveforms of (a) a quasi-periodic oscillation, (b) a quasi-periodic oscillation with a structure of an almost collapsing two-dimensional torus, (c) a chaotic oscillation observed after collapse of a two-dimensional torus, and (d) a 1/1-entrained oscillation. $f_N = 208\,\text{Hz}$, $f_F = 250\,\text{Hz}$, and (a) $A = 0.5\,\mu\text{A}$, (b) $A = 1\,\mu\text{A}$, (c) $A = 1.5\,\mu\text{A}$, (d) $A = 2\,\mu\text{A}$.

subject [Ruelle & Takens, 1971; Newhouse *et al.*, 1978] in contrast with Landau's scheme with an infinite sequence of Hopf bifurcations, each of which produces a new frequency incommensurate with the others [Ott, 2002]. However, a typical route to chaotic oscillations through quasi-periodicity that can be usually observed in experiments with squid giant axons is through locking. Namely, a quasi-periodic oscillation first transits to an entrained one when increasing the forcing amplitude A as suggested in Figs. 23 and 24 and the emergent entrained oscillation then bifurcates to a chaotic one with further increasing A through conventional local (codimension-1) bifurcations explained above, typically through successive period-doubling bifurcations. Figure 25 shows an example of such a route to chaos from quasi-periodicity through locking.

It should be noted, however, that if we carefully adjust values of two bifurcation parameters A and f_F, we can observe a direct route from quasi-periodicity to chaos through deformation and collapse of a two-dimensional torus representing a quasi-periodic oscillation as shown in Figs. 26 and 27. In particular, Figs. 26(b) and 27(b) show a quasi-periodic oscillation with a structure of an almost collapsing two-dimensional torus [Aihara, 1994].

Moreover, generally speaking, different kinds of routes to chaos through e.g. global bifurcations could be observed in the forced neural oscillator at least in principle, although their experimental verification is not so easy. For example, a transition from a chaotic oscillation with an island structure to a more developed chaotic one as shown in Figs. 14(c),

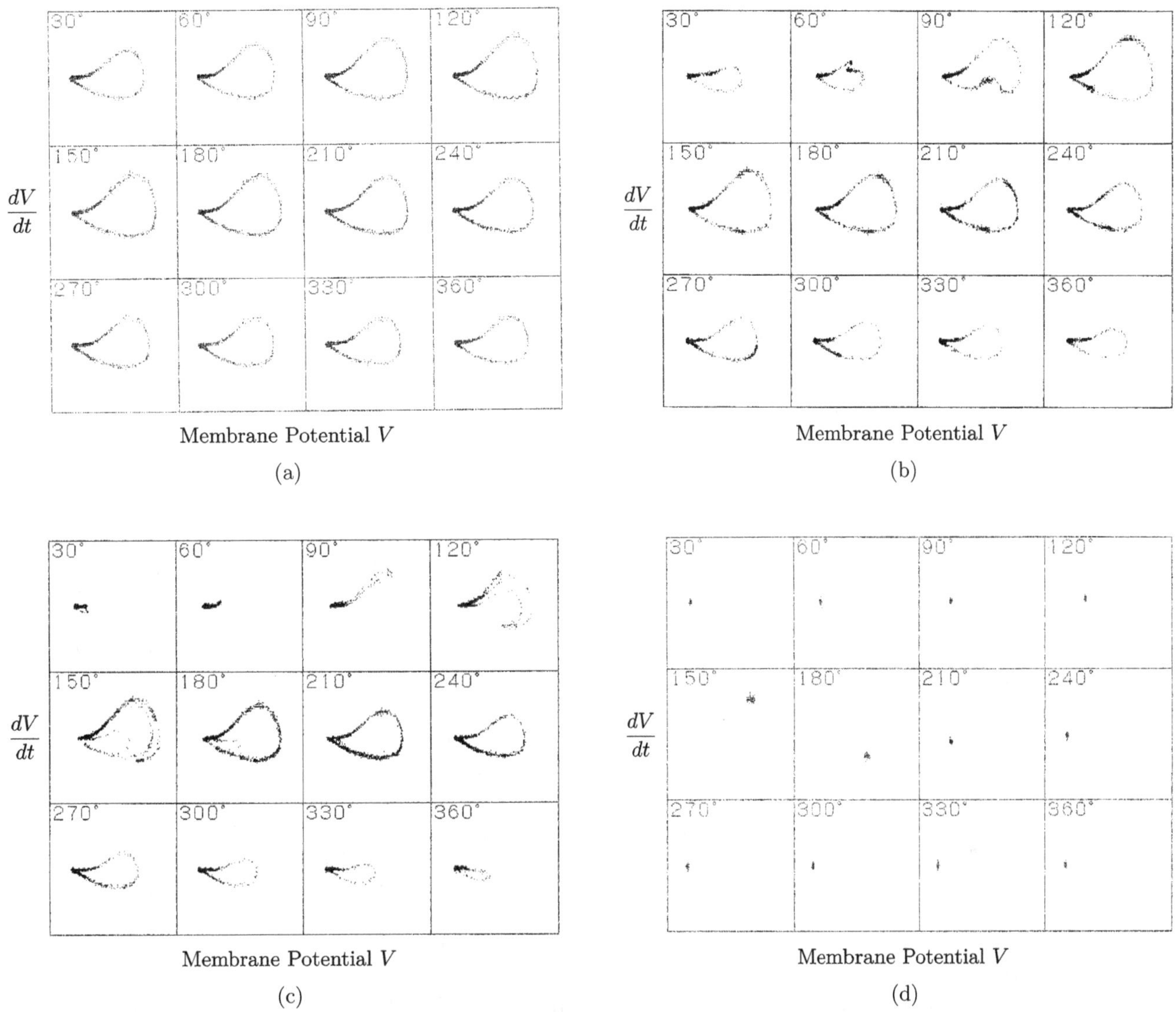

Fig. 27. A route from quasi-periodicity to chaos through the collapse of a two-dimensional torus. Panels (a)–(d) show Poincaré sections corresponding to Figs. 26(a)–26(d), respectively. (*Source*: Modified from Fig. 4 in [Aihara, 1994] with permission.)

14(d) and 16(g), 16(h) may be understood as a crisis [Grebogi *et al.*, 1982; Grebogi *et al.*, 1987; Ott, 2002].

When we change the values of the bifurcation parameter A and f_F in much wider ranges, we usually observe alternating periodic–chaotic sequences globally [Aihara *et al.*, 1985].

6. Discussion

I have reviewed attractors and bifurcations of forced oscillations experimentally observed in squid giant axons. These nonlinear phenomena can be theoretically well described by the Hodgkin–Huxley equations and modified versions [Hodgkin & Huxley, 1952; Aihara & Matsumoto, 1982, 1986, 1987; Aihara *et al.*, 1984; Aihara, 1994, 1995]. Chaos and bifurcations can be also observed both in squid giant axons and the Hodgkin–Huxley equations in the resting state stimulated by periodic pulses, that are more close to normal conditions of neurons [Matsumoto *et al.*, 1987; Aihara & Matsumoto, 1987; Takahashi *et al.*, 1990; Aihara, 1994, 1995, 2002]. Using these neurophysiological data, a mathematical model of chaotic neural networks was proposed [Aihara *et al.*, 1990] and its applications to combinatorial optimization and dynamical associative memory are intensively studied [Aihara, 1993, 2002; Horio *et al.*, 2012]. Further, information processing with chaotic dynamics is also generally studied [Kaneko & Tsuda, 1996] and hardware implementation of the chaotic neural networks is well developed [Horio *et al.*, 2003; Horio *et al.*, 2012]. Moreover, since the chaotic neuron model [Aihara *et al.*, 1990] can be easily implemented with the ReLU function that is widely used in the present AI [Tamura *et al.*, 2018], it is an important future problem to explore possible information processing of the chaotic neural networks for AI.

Acknowledgments

This study was made possible by the invaluable collaboration with the late Professor Gen Matsumoto, to whom I wish to express my deepest gratitude. This work was partially supported by International Research Center for Neurointelligence at The University of Tokyo Institutes for Advanced Study, and Institute for AI and Beyond at The University of Tokyo.

ORCID

Kazuyuki Aihara
https://orcid.org/0000-0002-4602-9816

References

Abraham, R. & Shaw, C. D. [1983] *Dynamics — The Geometry of Behavior, Part 2: Chaotic Behavior*, Visual Mathematics Library (Aerial Press).

Aihara, K. & Matsumoto, G. [1982] "Temporally coherent organization and instabilities in squid giant axons," *J. Theor. Biol.* **95**, 697–720.

Aihara, K., Matsumoto, G. & Ikegaya, Y. [1984] "Periodic and non-periodic responses of a periodically forced Hodgkin–Huxley oscillator," *J. Theor. Biol.* **109**, 249–269.

Aihara, K., Matsumoto, G. & Ichikawa, M. [1985] "An alternating periodic–chaotic sequence observed in neural oscillators," *Phys. Lett. A* **111**, 251–255.

Aihara, K. & Matsumoto, G. [1986] "Chaotic oscillations and bifurcations in squid giant axons," in *Chaos*, ed. Holden, A. V. (Manchester University Press, Manchester and Princeton University Press, Princeton, NJ), pp. 257–269.

Aihara, K., Numajiri, T., Matsumoto, G. & Kotani, M. [1986] "Structures of attractors in periodically forced neural oscillators," *Phys. Lett. A* **116**, 313–317.

Aihara, K. & Matsumoto, G. [1987] "Forced oscillations and routes to chaos in the Hodgkin–Huxley axons and squid giant axons," in *Chaos in Biological Systems*, eds. Degn, H., Holden, A. V. & Olsen, L. F. (Plenum Press, NY and London), pp. 121–131.

Aihara, K., Takabe, T. & Toyoda, M. [1990] "Chaotic neural networks," *Phys. Lett. A* **144**, 333–334.

Aihara, K. (ed.) [1993] *Chaos in Neural Systems* (Tokyo Denki University Press, Tokyo) (in Japanese).

Aihara, K. [1994] "Chaos in neural response and dynamical neural network models: Toward a new generation of analog computing," in *Towards the Harnessing of Chaos*, ed. Yamaguti, M. (Elsevier, Amsterdam), pp. 83–98.

Aihara, K. [1995] "Chaos in axons," in *The Handbook of Brain Theory and Neural Networks*, 1st edition, ed. Arbib, M. A. (The MIT Press, Cambridge, MA), pp. 183–185.

Aihara, K. [2002] "Chaos engineering and its application to parallel distributed processing with chaotic neural networks," *Proc. IEEE* **90**, 919–930.

Benardo, L. S. [1997] "Recruitment of GABAergic inhibition and synchronization of inhibitory interneurons in rat neocortex," *J. Neurophysiol.* **77**, 3134–3144.

Glass, L. & Mackey, M. C. [1988] *From Clocks to Chaos* (Princeton University Press, Princeton, NJ).

Glendinning, J. I. [1994] "Is the bitter rejection response always adaptive?" *Physiol. Behav.* **56**, 1217–1227.

Grebogi, C., Ott, E. & Yorke, J. A. [1982] "Chaotic attractors in crisis," *Phys. Rev. Lett.* **48**, 1507–1510.

Grebogi, C., Ott, E., Romeiras, F. & Yorke, J. A. [1987] "Critical exponents for crisis-induced intermittency," *Phys. Rev. A* **36**, 5365–5380.

Guevara, M. R., Glass, L. & Shrier, A. [1981] "Phase locking, period-doubling bifurcations, and irregular dynamics in periodically stimulated cardiac cells," *Science* **214**, 1350–1353.

Hayashi, H., Ishizuka, S., Ohta, M. & Hirakawa, K. [1982] "Chaotic behavior in the Onchidium giant neuron under sinusoidal stimulation," *Phys. Lett. A* **88**, 435–438.

Hilborn, R.C. [2000] *Chaos and Nonlinear Dynamics*, 2nd edition (Oxford University Press, NY).

Hodgkin, A. L. & Huxley, A. F. [1952] "A quantitative description of membrane current and its application to conduction and excitation in nerve," *J. Physiol. (Lond.)* **117**, 500–544.

Horio, Y., Aihara, K. & Yamamoto, O. [2003] "Neuron-synapse IC chip-set for large-scale chaotic neural networks," *IEEE Trans. Neural Netw.* **14**, 1393–1404.

Horio, Y., Adachi, M. & Ikeguchi, T. [2012] *Chaos Neuro-Computing* (KindaiKagaku-sha, Tokyo) (in Japanese).

Huxley, A. F. [1959] "Ion movements during nerve activity," *Ann. New York Acad. Sci.* **81**, 221–246.

Kaneko, K. [1986] *Collapse of Tori and Genesis of Chaos in Dissipative Systems* (World Scientific, Singapore).

Kaneko, K. & Tsuda, I. [1996] *Complex Systems: Chaos and Beyond* (Springer, Berlin).

Kapitaniak, T. & Bishop, S. R. [1999] *The Illustrated Dictionary of Nonlinear Dynamics and Chaos* (John Wiley & Sons Ltd., Chichester).

Matsumoto, G. [1981] "Long-range spatial interactions and a dissipative structure in squid giant axons and proposed physical model of nerve excitation," in *Nerve Membrane — Biochemistry and Function of Channel Proteins*, eds. Matsumoto, G. & Kotani, H. (University of Tokyo Press, Tokyo), pp. 203–220.

Matsumoto, G., Aihara, K., Ichikawa, M. & Tasaki, A. [1984] "Periodic and nonperiodic responses of membrane potentials in squid giant axons during sinusoidal current stimulation," *J. Theor. Neurobiol.* **3**, 1–14.

Matsumoto, G., Aihara, K., Hanyu, Y., Takahashi, N., Yoshizawa, S. & Nagumo, J. [1987] "Chaos and phase locking in normal squid axons," *Phys. Lett. A* **123**, 162–166.

Moser, J. [1969] "Perturbation theory of quasiperiodic solutions of differential equations," in *Bifurcation Theory and Nonlinear Eigenvalue Problems*, eds. Keller, J. B. & Antman, S. (Benjamin, NY), pp. 283–308.

Newhouse, S., Ruelle, D. & Takens, F. [1978] "Occurrence of strange Axiom$A$ attractors near quasi periodic flows on $T^m, m \geq 3$," *Commun. Math. Phys.* **64**, 35–40.

Ott, E., Sauer, T. & Yorke, J. A. (eds.) [1994] *Coping with Chaos: Analysis of Chaotic Data and the Exploration of Chaotic Systems* (John Wiley & Sons, NY).

Ott, E. [2002] *Chaos in Dynamical Systems* (Cambridge University Press, Cambridge).

Pomeau, Y. & Manneville, P. [1980] "Intermittent transition to turbulence in dissipative dynamical systems," *Commun. Math. Phys.* **74**, 189–197.

Ruelle, D. & Takens, F. [1971] "On the nature of turbulence," *Commun. Math. Phys.* **20**, 167–192.

Sell, G. R. [1981] "Hopf–Landau bifurcation near strange attractors," in *Chaos and Order in Nature*, ed. Haken, H. (Springer-Verlag, Berlin, Heidelberg, NY), pp. 84–91.

Shaw, R. [1981] "Strange attractors, chaotic behavior and information flow," *Z. Naturforsch. A* **36**, 80–112.

Takahashi, N., Hanyu, Y., Musha, T., Ryogo, K. & Matsumoto, G. [1990] "Global bifurcation structure in periodically stimulated giant axons of squid," *Physica D* **43**, 318–334.

Tamura, H., Kohno, T. & Aihara, K. [2018] "ReLU chaotic neuron model for deep chaotic neural networks," *Seisan Kenkyu* **70**, 45–47.

Thompson, J. M. T. & Stewart, H. B. [1984] "Folding and mixing in the Birkhoff–Shaw chaotic attractor," *Phys. Lett. A* **103**, 229–231.

Thompson, J. M. T. & Stewart, H. B. [1986] *Nonlinear Dynamics and Chaos* (John Wiley & Sons).

Ueda, Y. & Akamatsu, N. [1981] "Chaotically transitional phenomena in the forced negative-resistance oscillator," *IEEE Trans. Circuits Syst.* **28**, 217–224.

Zeeman, E. C. [1982] "Bifurcation, catastrophe, and turbulence," in *New Directions in Applied Mathematics*, eds. Hilton, P. J. and Young, G. S. (Springer-Verlag, NY), pp. 109–153.

https://doi.org/10.1142/9789819830633_0008

Chapter 8

Algebraic Limit Cycles[#]

Jaume Llibre
Departament de Matemàtiques, Universitat Autònoma de Barcelona, 08193 Bellaterra, Barcelona, Catalonia, Spain
jaume.llibre@uab.cat

In the qualitative theory of differential equations in the plane $\mathbb{R}^2$, one of the most difficult objects to study is the existence of limit cycles. Here, we summarize some results and open problems on the algebraic limit cycles of the planar polynomial differential systems. More precisely, (i) we study the algebraic limit cycles of the polynomial differential systems of degree 2; (ii) we provide the maximum number of generic algebraic limit cycles that the polynomial differential systems of degree n can exhibit; (iii) we show using the algebraic limit cycles that any finite configuration of limit cycles can be realized by some polynomial differential system; and (iv) we provide the maximum number of algebraic limit cycles formed by circles that a polynomial differential system of degree n can exhibit.

Keywords: Algebraic limit cycle; limit cycle; polynomial vector field; quadratic polynomial vector field.

1. Introduction

A *(real) planar polynomial differential system* is a differential system of the form

$$\frac{dx}{dt} = \dot{x} = P(x,y), \quad \frac{dy}{dt} = \dot{y} = Q(x,y), \qquad (1)$$

where P and Q are real polynomials in the variables x and y. The *degree* n of this polynomial differential system is the maximum of the degrees of the polynomials P and Q.

Associated to this polynomial differential system there is the *polynomial vector field*

$$\mathcal{X} = P(x,y)\frac{\partial}{\partial x} + Q(x,y)\frac{\partial}{\partial y},$$

or simply $\mathcal{X} = (P,Q)$. The *degree* of the polynomial differential system (1) or of its associated polynomial vector field $\mathcal{X}$ is the maximum of the degrees of the polynomials P and Q.

A *limit cycle* of the differential system (1) is an isolated periodic orbit in the set of all periodic orbits of system (1).

We denote by $\mathbb{R}[x,y]$ the ring of all the real polynomials in the variables x and y. Let $f(x,y)$ be a real polynomial irreducible in the ring $\mathbb{R}[x,y]$. Then, the zero set $f(x,y) = 0$ is an *algebraic curve*. An *algebraic limit cycle* is a limit cycle contained in an algebraic curve of the plane $\mathbb{R}^2$. The *degree* of an algebraic limit cycle is the degree of the irreducible polynomial $f(x,y)$ such that the algebraic curve $f(x,y) = 0$ contains the algebraic limit cycle.

It is well known that the orbits of a polynomial differential system (1) are contained in analytic curves, that in general are not algebraic curves.

It is not easy to distinguish if a limit cycle is algebraic or not. Thus, until the year 1995, it was not proved that the limit cycle exhibited in the famous van der Pol differential equation [van der Pol, 1926] is not algebraic, see [Odani, 1995]. The van der Pol differential system is a polynomial differential system of degree 3, but we do not know the explicit expression of the analytic curve that contains the nonalgebraic limit cycle of van der Pol

[#]To cite this article, please refer to its earlier version published in the *International Journal of Bifurcation and Chaos*, Vol. 35, No. 14 (2025) 2540008 (9 pages), DOI: 10.1142/S0218127425400085.

differential system. On the other hand, the explicit expression of an algebraic limit cycle is known, because we have the algebraic curve containing the limit cycle, otherwise we do not know that it is algebraic.

Let $f = f(x,y) \in \mathbb{R}[x,y]$ be. Then, the algebraic curve $f(x,y) = 0$ of $\mathbb{R}^2$ is an *invariant algebraic curve* of the vector field $\mathcal{X}$ if for some polynomial $K \in \mathbb{R}[x,y]$, we have that

$$\mathcal{X}f = P\frac{\partial f}{\partial x} + Q\frac{\partial f}{\partial y} = Kf. \tag{2}$$

The polynomial K is called the *cofactor* of the invariant algebraic curve $f = 0$. Since the polynomial differential system has degree m, a cofactor has degree at most $m - 1$.

On the points of the algebraic curve $f = 0$, the gradient $(\partial f/\partial x, \partial f/\partial y)$ of f is orthogonal to the vector field $\mathcal{X} = (P,Q)$, see (2). Hence, at every point of the curve $f(x,y) = 0$ the vector field $\mathcal{X}$ is tangent to the curve $f(x,y) = 0$, so, this curve is formed by trajectories of the vector field $\mathcal{X}$. This explains the name "invariant algebraic curve" because it is invariant under the flow defined by $\mathcal{X}$.

Darboux [1878] found connections between the invariant algebraic curves and the existence of first integrals in the polynomial vector fields. In fact, the invariant algebraic curves are the main ingredient in the Darboux theory of integrability, for more details, see [Dumortier *et al.*, 2006, Chapter 8]. Today, after more than one century of investigations, the theory of the invariant algebraic curves is still full of open questions.

The main objective of this paper is the study of the algebraic limit cycles of the polynomial vector fields. First, we need to study the invariant algebraic curves because the algebraic limit cycles are contained in them.

2. Invariant Algebraic Curves

When a polynomial vector field has or does not have invariant algebraic curves? The answer to this question is not easy, see the large section in the book of [Jouanolou, 1979], or the long paper of [Moulin Ollagnier *et al.*, 1995] devoted to show that one particular polynomial system has no invariant algebraic solutions.

One of the best results on the invariant algebraic curves of the polynomial differential systems is the following result due to [Jouanolou, 1979].

Theorem 1. *Let $\mathcal{X}$ be a polynomial vector field of degree n. Then, either $\mathcal{X}$ has less than $2+n(n+1)/2$ irreducible invariant algebraic curves, or $\mathcal{X}$ has a rational first integral, and then all the orbits of $\mathcal{X}$ live on invariant algebraic curves.*

For a shorter and simple proof of Theorem 1, see [Christopher & Llibre, 1999].

Theorem 1 shows that for a given polynomial vector field $\mathcal{X}$ of degree n, the maximum degree of its irreducible invariant algebraic curves is bounded, because either $\mathcal{X}$ has a finite number of invariant algebraic curves less than $2 + n(n+1)/2$, or $\mathcal{X}$ has a rational first integral $f(x,y)/g(x,y)$ (i.e. f and g are polynomials), and consequently all orbits of $\mathcal{X}$ are contained in invariant algebraic curves of the form $af(x,y) + bg(x,y) = 0$ for some $a, b \in \mathbb{R}$.

Thus, for each polynomial vector field there is a natural number N that bounds the degree of all its irreducible invariant algebraic curves. So, if we know N, then it is possible (at least in theory) to compute its invariant algebraic curves from its definition given in (2).

A natural question, going back to Poincaré, and for some people working on polynomial differential systems, known as the *Poincaré problem*, is: to give an effective procedure for finding N for a given polynomial vector field $\mathcal{X}$. There are only partial answers to this problem, see for instance [Campillo & Carnicer, 1997; Carnicer, 1994; Cerveau & Lins Neto, 1991; Walcher, 2000]. But for the class of polynomial vector fields with a fixed degree n having finitely many invariant algebraic curves (i.e. having no rational first integrals), there does not exist a uniform upper bound $N(n)$ for N, as it was shown in [Christopher & Llibre, 2002; Moulin Ollagnier, 2001]. These results imply that there are polynomial vector fields of degree n without having a rational first integral, but having an irreducible invariant algebraic curve of arbitrary large degree.

In summary, *a priori*, it is possible for polynomial vector fields with a fixed degree to have some algebraic limit cycle of arbitrary high degree.

3. Algebraic Limit Cycles of Quadratic Polynomial Vector Fields

Qin [1958] proved that after an affine change of variables and a rescaling of the time variable, the only

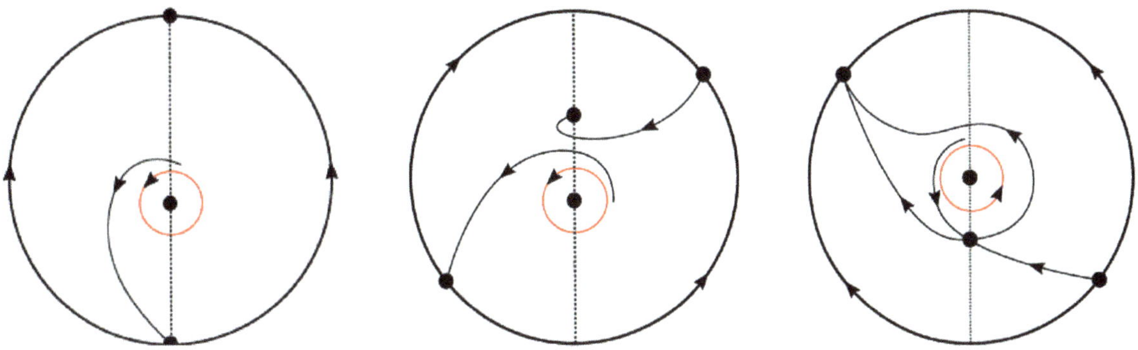

Fig. 1. The three kind of phase portraits in the Poincaré disc having an algebraic limit cycle of degree 2.

quadratic systems having an algebraic limit cycle of degree 2 can be written as

$$\begin{aligned}\dot{x} &= -y(ax+by+c) - (x^2+y^2-1),\\ \dot{y} &= x(ax+by+c),\end{aligned} \tag{3}$$

where $c^2+4(b+1)<0$ and $a^2+b^2<c^2$. These systems have the irreducible invariant algebraic curve $x^2+y^2-1=0$ of degree 2, that is an algebraic limit cycle. This algebraic limit cycle is the unique limit cycle of the quadratic polynomial differential systems (3). See the distinct phase portraits of systems (3) in the Poincaré disc in Fig. 1. In these phase portraits, the algebraic limit cycle is drawn in red color.

Roughly speaking, the *Poincaré disc* is a closed disc whose interior is diffeomorphic to the plane $\mathbb{R}^2$, and its boundary (the circle $\mathbb{S}^1$) is identified with the infinity of $\mathbb{R}^2$. Note that in the plane $\mathbb{R}^2$, the orbits can go or come from infinity in as many as directions as there are points on the circle $\mathbb{S}^1$. For more details on the Poincaré disc, see [Dumortier *et al.*, 2006, Chapter 5].

Evdokimenco [1970, 1974, 1979] proved that quadratic polynomial vector fields do not have algebraic limit cycles of degree 3. For two different and shorter proofs that quadratic polynomial vector fields do not have algebraic limit cycles of degree 3, see [Chavarriga *et al.*, 2004; Llibre, 2004].

Yablonskii [1996] found the first class of algebraic limit cycles of degree 4 inside the quadratic polynomial vector fields. This class is given by the quadratic polynomial differential systems

$$\begin{aligned}\dot{x} &= -4abcx - (a+b)y + 3(a+b)cx^2 + 4xy,\\ \dot{y} &= (a+b)abx - 4abcy\\ &\quad + \left(\frac{4abc^2-3(a+b)^2}{2} + 4ab\right)x^2\\ &\quad + 8(a+b)cxy + 8y^2,\end{aligned} \tag{4}$$

where

$$abc \neq 0, \quad a \neq b, \quad ab > 0 \quad \text{and}$$
$$4c^2(a-b)^2 + (3a-b)(a-3b) < 0.$$

Systems (4) have the irreducible invariant algebraic curve $(y+cx^2)^2 + x^2(x-a)(x-b) = 0$ of degree 4. This curve has two components, an isolated equilibrium point and an oval, that is the algebraic limit cycle, see Fig. 2. In the figure, the invariant algebraic curve of degree 4 is drawn in red color.

The second class of algebraic limit cycles of degree 4 was found in [Filiptsov, 1973]. This class is given by the quadratic polynomial differential systems

$$\dot{x} = 6(1+a)x + 2y - 6(2+a)x^2 + 12xy,$$

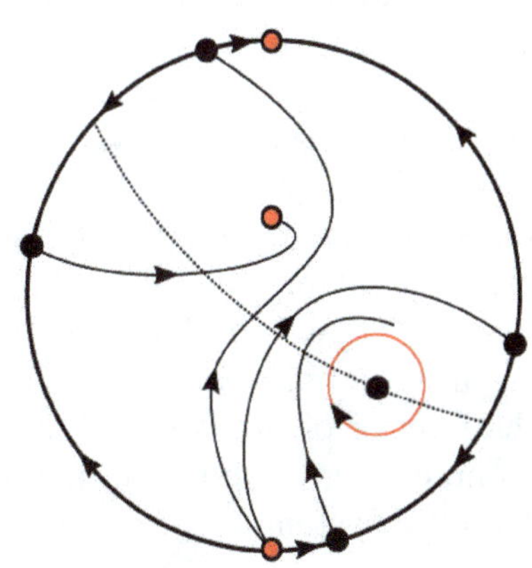

Fig. 2. Yablonskii algebraic limit cycle of degree 4 in the Poincaré disc.

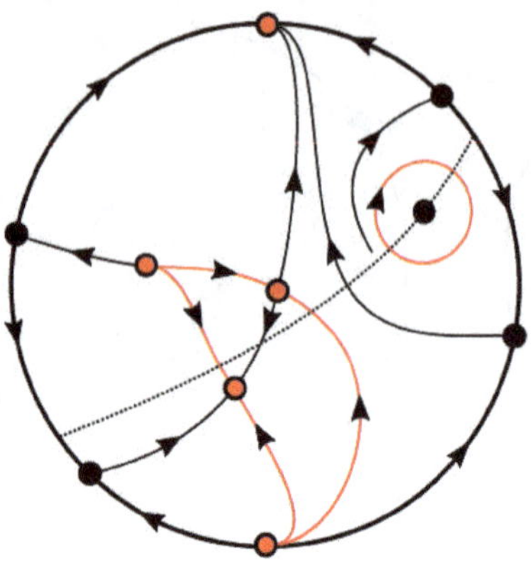

Fig. 3. Filiptsov algebraic limit cycle of degree 4 in the Poincaré disc.

$$\dot{y} = 15(1+a)y + 3a(1+a)x^2 - 2(9+5a)xy + 16y^2, \tag{5}$$

where $0 < a < 3/13$. These quadratic polynomial differential systems have the irreducible invariant algebraic curve $3(1+a)(ax^2+y)^2 + 2y^2(2y - 3(1+a)x) = 0$ of degree 4. This curve has two components, one homeomorphic to a straight line and the other to an oval, the algebraic limit cycle, see Fig. 3. The component homeomorphic to a straight line contains three equilibrium points of system (5).

More recently, two new classes of algebraic limit cycles of degree 4 have been found by [Chavarriga *et al.*, 2004]. The first class is given by the quadratic polynomial differential systems

$$\dot{x} = 5x + 6x^2 + 4(1+a)xy + ay^2,$$
$$\dot{y} = x + 2y + 4xy + (2+3a)y^2,$$

where $(-71 + 17\sqrt{17})/32 < a < 0$. This class has the irreducible invariant algebraic curve $x^2 + x^3 + x^2y + 2axy^2 + 2axy^3 + a^2y^4 = 0$ of degree 4 having three components, two are homeomorphic to a straight line and the third to an oval, the algebraic limit cycle, see Fig. 4(a). Each one of the components homeomorphic to a straight line contains one equilibrium point of the system. Note that on these components homeomorphic to a straight line, we only compute the finite equilibria, the ones contained in the interior of the Poincaré disc.

The second class is given by the quadratic polynomial differential systems

$$\dot{x} = 2(1 + 2x - 2ax^2 + 6xy),$$
$$\dot{y} = 8 - 3a - 14ax - 2axy - 8y^2,$$

where $0 < a < 1/4$. This class has the irreducible invariant algebraic curve $\frac{1}{4} + x - x^2 + ax^3 + xy + x^2y^2 = 0$ of degree 4 having three components, two are homeomorphic to a straight line and the third to an oval, the algebraic limit cycle, see Fig. 4(b). One of the components homeomorphic to a straight line contains two equilibrium points of the system, the other component homeomorphic to a straight line does not contain equilibrium points.

The authors of [Chavarriga *et al.*, 2004] also proved that these four classes of algebraic limit cycles of degree 4 are the unique classes of algebraic limit cycles of degree 4 for the quadratic polynomial vector fields. When such algebraic limit cycles of degree 4 exist, then they are the unique limit cycle of the quadratic polynomial differential system, see [Chavarriga *et al.*, 2001].

By doing convenient birational transformations of the plane to quadratic polynomial vector fields having algebraic limit cycles of degree 4, [Christopher *et al.*, 2005] obtained an algebraic limit cycle of degrees 5 and 6 for quadratic polynomial vector fields. More precisely, the quadratic polynomial differential systems

$$\dot{x} = 28x + 2(16-\alpha^2)(\alpha+12)x^2 + 6(3\alpha-4)xy - \frac{12}{\alpha+4}y^2,$$
$$\dot{y} = 2(16-\alpha^2)x + 8y + (16-\alpha^2)(\alpha+12)xy + 2(5\alpha-12)y^2,$$

where $\alpha \in (3\sqrt{7}/2, 4)$, have an algebraic limit cycle contained into the algebraic curve

$$f(x,y) = x^2 + (16-\alpha^2)x^3 + (\alpha-2)x^2y - \frac{2}{\alpha+4}xy^2 - \frac{1}{4}(4-\alpha)(\alpha+12)x^2y^2$$

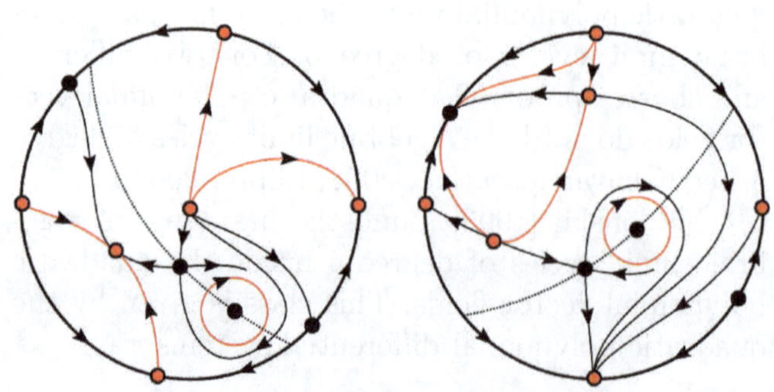

Fig. 4. The two additional algebraic limit cycles of degree 4 in the Poincaré disc.

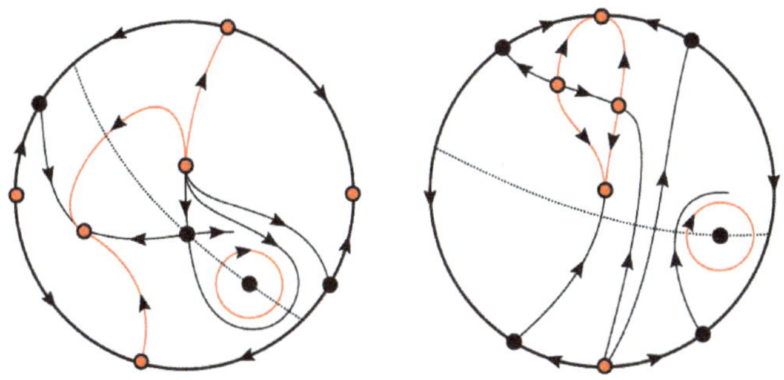

(a) (b)

Fig. 5. (a) The algebraic limit cycle of degree 5 in the Poincaré disc. (b) The algebraic limit cycle of degree 6 in the Poincaré disc.

$$+\frac{8-\alpha}{\alpha+4}xy^3+\frac{1}{(\alpha+4)^2}y^4+\frac{\alpha+12}{\alpha+4}xy^4$$
$$-\frac{6}{(\alpha+4)^2}y^5$$

of degree 5. This algebraic curve has two components, one is homeomorphic to a straight line, and the other to an oval, the algebraic limit cycle of degree 5, see Fig. 5(a). The component homeomorphic to a straight line contains two equilibrium points of the system.

The quadratic polynomial differential systems

$$\dot{x}=28\beta(\beta-30)x+y+168\beta^2x^2+3xy,$$
$$\dot{y}=224\beta^2(\beta-30)^2x+516\beta(\beta-30)y$$
$$+1344\beta^3(\beta-30)x^2+24\beta(17\beta-6)xy+6y^2,$$

where $\beta\in(3/2,2)$, have an algebraic limit cycle contained within the algebraic curve

$$f(x,y)=48\beta^3(\beta-30)^4x^2+24\beta^2(\beta-30)^3xy$$
$$+3\beta(\beta-30)^2y^2+64\beta^3(\beta-30)^3$$
$$\times(9\beta-4)x^3+24\beta^2(\beta-30)^2(9\beta-4)x^2y$$
$$+18\beta(\beta-30)(\beta-2)xy^2-7y^3+576\beta^3$$
$$\times(\beta-30)^2(\beta-2)^2x^4+144\beta^2(\beta-30)$$
$$\times(\beta-2)^2x^3y+27\beta(\beta-2)^2x^2y^2$$
$$-3456\beta^3(\beta-30)(\beta-2)^2(2\beta+3)x^5$$
$$-432\beta^2(\beta-2)^2(2\beta+3)x^4y$$
$$+3456\beta^3(\beta-2)^2(\beta+12)(2\beta+3)x^6,$$

of degree 6. This algebraic curve has two components; one is homeomorphic to a straight line and the other to an oval, the algebraic limit cycle of degree 6, see Fig. 5(b). The component homeomorphic to a straight line contains three equilibrium points of the system.

A new algebraic limit cycle of degree 5 was recently found by [Alberich-Carramiñana *et al.*, 2021]. The quadratic polynomial differential systems

$$\dot{x}=-8x+\frac{\gamma}{2}(\gamma-16)y-(5\gamma-64)x^2$$
$$+\frac{\gamma}{8}(\gamma^2-256)xy,$$
$$\dot{y}=-28y+\frac{24}{\gamma}x^2-3(3\gamma-32)xy$$
$$+\frac{\gamma}{4}(\gamma^2-256)y^2,$$

have the irreducible invariant algebraic curve

$$f(x,y)=\gamma y^2-4x^2y+\frac{\gamma}{2}(\gamma-12)xy^2$$
$$-\frac{\gamma^2}{4}(\gamma-16)y^3+\frac{4}{\gamma}x^4-(\gamma-24)x^3y$$
$$+\frac{\gamma}{16}(\gamma^2-256)x^2y^2-\frac{24}{\gamma}x^5$$
$$+(\gamma+16)x^4y,$$

of degree 5, where $\gamma\in(0,8-3\sqrt{7})$. This algebraic curve has two components; one is homeomorphic to a straight line and the other to an oval, the algebraic limit cycle of degree 5, see Fig. 6. The component homeomorphic to a straight line contains two equilibrium points of the system.

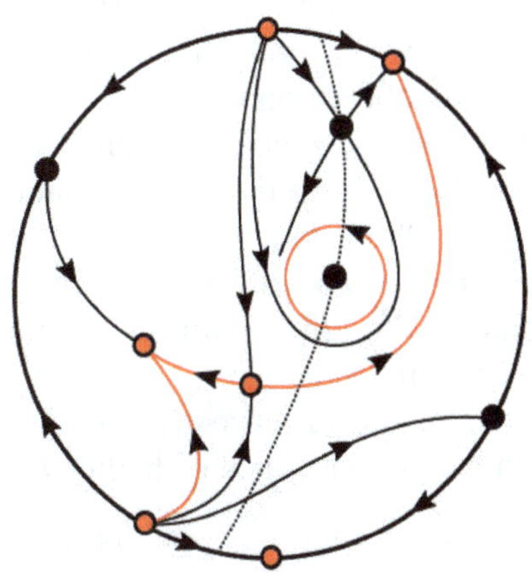

Fig. 6. The new algebraic limit cycle of degree 5.

Open problems

(i) *What is the maximum degree of an algebraic limit cycle of a quadratic polynomial vector field?*
(ii) *Is 1 the maximum number of algebraic limit cycles that a quadratic polynomial vector field can have?*

More open problems on algebraic limit cycles can be found in [Llibre, 2008].

4. 16th Hilbert Problem for Algebraic Limit Cycles

What can be said about the maximum number of limit cycles that planar polynomial vector fields of degree m can exhibit? What can be the distribution or configuration of their limit cycles? These two questions appear in the second part of Hilbert's 16th problem which became one of the most persistent problems in the famous Hilbert list of 1990, with the only exception of the Riemann ζ-function conjecture, see [Hilbert, 1900] and also [Ilyashenko, 2002; Li, 2003].

Consider the set Σ_n of all polynomial vector fields of degree n having irreducible invariant algebraic curves. We define the 16th Hilbert problem restricted to the algebraic limit cycles as: Does there exist a uniform bound for the maximum number of algebraic limit cycles that a polynomial vector field of Σ_n can exhibit?

We cannot solve this question for arbitrary irreducible algebraic invariant curves, but we have an answer for a class of generic invariant algebraic curves. We say that the set $f_j = 0$, $j = 1, \ldots, k$, of irreducible invariant algebraic curves is *generic* if it satisfies the following five conditions:

(i) There are no points of the curve $f_j = 0$ where its first derivatives vanish, (i.e. $f_j = 0$ is a nonsingular algebraic curve).
(ii) The homogeneous terms of higher degree of the polynomial f_j have no repeated factors.
(iii) If two curves intersect in a point of the affine plane, they intersect transversally.
(iv) A point of the affine plane never passes more than two curves.
(v) Two curves never have a common factor in their homogenous parts of higher degree.

All these conditions are open in the sense that if we do a sufficiently small perturbation to these algebraic curves, the conditions continue to be verified.

Theorem 2. *If all the invariant algebraic curves of a polynomial vector field $\mathcal{X}$ of degree n are generic, then the number of its algebraic limit cycles is at most $1 + (n-1)(n-2)/2$ if n is even, and at most $(n-1)(n-2)/2$ if n is odd. Moreover, these upper bounds are reached.*

Theorem 2 was proved by [Llibre *et al.*, 2010]. The same authors, one year later in [Llibre *et al.*, 2011], obtained the same result of Theorem 2 but only by assuming condition (i), and that the algebraic curves are irreducible being the sum of their degrees at most $n + 1$, of course n is the degree of the polynomial vector field.

Zhang [2011] obtained again the same result of Theorem 2 but now with the assumption that all invariant algebraic curves are nodal, i.e. an algebraic curve is *nodal* if all its singularities are intersections of exactly two transversal branches.

When the degree of the polynomial differential system is odd, the upper bound provided in Theorem 2 is not true if the assumptions of the papers [Llibre *et al.*, 2010, 2011; Zhang, 2011] are not satisfied. Indeed, for polynomial vector fields of degree 3, Theorem 2 says that one is the maximum number of algebraic limit cycles that such vector fields can have with only generic invariant algebraic curves. But we have examples of cubic polynomial vector fields having two algebraic limit cycles, of course the invariant algebraic curves of these cubic polynomial vector fields do not satisfy the assumptions of the papers [Llibre *et al.*, 2010, 2011; Zhang, 2011]. Indeed, the cubic polynomial vector field

$$\dot{x} = 2y(10 + xy),$$
$$\dot{y} = 20x + y - 20x^3 - 2x^2y + 4y^3,$$

has two algebraic limit cycles contained in the invariant algebraic curve $2x^4 - 4x^2 + 4y^2 + 1 = 0$.

All the results obtained until now about the algebraic limit cycles of the polynomial differential systems satisfy the following conjecture.

Conjecture. *The maximum number of algebraic limit cycles that a polynomial vector field of degree n can have is $1 + (n-1)(n-2)/2$.*

5. Configuration of Limit Cycles

We define a *configuration of limit cycles* as a finite set $C = \{C_1, \ldots, C_n\}$ of disjoint simple closed curves of the plane such that $C_i \cap C_j = \emptyset$ for all $i \neq j$.

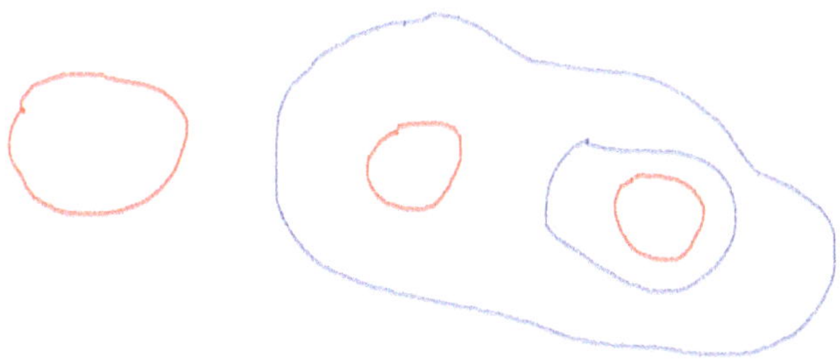

Fig. 7. A configuration of 5 limit cycles having 3 primary limit cycles, the red ones.

Two configurations of limit cycles $C = \{C_1, \ldots, C_n\}$ and $C' = \{C'_1, \ldots, C'_m\}$ are *(topologically) equivalent* if there is a homeomorphism $h : \mathbb{R}^2 \to \mathbb{R}^2$ such that $h(\bigcup_{i=1}^n C_i) = \bigcup_{i=1}^m C'_i$. Of course, for equivalent configurations of limit cycles C and C' we have that $n = m$.

Hilber [1900] also asked: *What are the classes of configurations of limit cycles realizable by the polynomial vector fields?*

Given a configuration of limit cycles $C = \{C_1, \ldots, C_n\}$, we say that the simple closed curve C_i is *primary* if there is no curve C_j of C contained within the bounded region limited by C_i, see an example of a configuration of limit cycles in Fig. 7.

The next result shows that any finite configuration of limit cycles is realizable by some polynomial vector fields. For a proof, see the article of [Llibre & Rodríguez, 2004].

Theorem 3. *Let $C = \{C_1, \ldots, C_n\}$ be an arbitrary configuration of limit cycles, and let r be its number of primary limit cycles. Then, the following statements hold.*

(a) *The configuration C is realizable by a polynomial differential system.*
(b) *The configuration C is realizable as algebraic limit cycles by a polynomial differential system of degree $\leq 2(n + r) - 1$.*

A key result in the proof of the previous theorem is the following.

Theorem 4. *Suppose that a C^1 differential system $\dot{x} = P(x, y)$, $\dot{y} = Q(x, y)$, satisfies*

$$\dot{x} = \frac{P(x,y)}{V(x,y)} = -\frac{\partial H}{\partial y}, \quad \dot{y} = \frac{Q(x,y)}{V(x,y)} = \frac{\partial H}{\partial x},$$

for some C^1 functions $V : \mathbb{R}^2 \to \mathbb{R}^2$ and $H : \mathbb{R}^2 \backslash \{V(x, y) = 0\} \to \mathbb{R}$. Then, if the differential system has a limit cycle, it is contained in the curve $V(x, y) = 0$.

Theorem 4 was proved for the first time in [Giacomini *et al.*, 1996], later on, an easier proof has been given in [Llibre & Rodríguez, 2004].

Later on and using ideas of the proof of the previous theorems, the same result that any finite configuration of limit cycles is realized by some polynomial differential was obtained for Liénard polynomial differential systems by [Coll *et al.*, 2013]. These authors worked with Liénard differential systems of the form

$$\dot{x} = y - F(x), \quad \dot{y} = g(x),$$

where F and g are polynomials.

6. Circular Algebraic Limit Cycles

Let U be an open and dense subset of $\mathbb{R}^2$. A C^1 function $H : U \to \mathbb{C}$ nonlocally constant is a *first integral* of the polynomial differential system (1) if it is constant on the orbits of system (1) contained in U. Of course, since the polynomial differential system (1) is real, the real and the imaginary parts of a first integral H are real first integrals of system (1) if they are nonzero.

We say that a polynomial differential system (1) is *Darboux integrable* if there exists a first integral of the form

$$H = H(x, y) = f_1^{\lambda_1} \cdots f_m^{\lambda_m} e^{g/h},$$

where f_k for $k = 1, \ldots, m$, g and h are complex polynomials in the variables x and y, and $\lambda_k \in \mathbb{C}$ for $k = 1, \ldots, m$.

Theorem 5. *A polynomial vector field $\mathcal{X}$ of degree n with n invariant circles is Darboux integrable, and the invariant circles are not limit cycles.*

Theorem 6. *For a polynomial vector field $\mathcal{X}$ of degree n, the maximum number of algebraic limit cycles given by circles is at most $n - 1$. Moreover, this upper bound is reached.*

Theorems 5 and 6 are proved by [Llibre *et al.*, 2016].

Acknowledgments

The author is partially supported by the Agencia Estatal de Investigación of Spain grant PID2022-136613NB-100, AGAUR (Generalitat de

Catalunya) grant 2021SGR00113, and by the Reial Acadèmia de Ciències i Arts de Barcelona.

ORCID

Jaume Llibre
https://orcid.org/0000-0002-9511-5999

References

Alberich-Carramiñana, M., Ferragut, A. & Llibre, J. [2021] "Quadratic planar differential systems with algebraic limit cycles via quadratic plane Cremona maps," *Adv. Math.* **389**, 107924.

Campillo, A. & Carnicer, M. M. [1997] "Proximity inequalities and bounds for the degree of invariant curves by foliations of $\mathbb{C}P^2$," *Trans. Am. Math. Soc.* **349**, 2211–2228.

Carnicer, M. M. [1994] "The Poincaré problem in the nondicritical case," *Ann. Math.* **140**, 289–294.

Cerveau, D. & Lins Neto, A. [1991] "Holomorphic foliations in $\mathbb{C}P(2)$ having an invariant algebraic curve," *Ann. Inst. Fourier* **41**, 883–903.

Chavarriga, J., Giacomini, H. & Llibre, J. [2001] "Uniqueness of algebraic limit cycles for quadratic systems," *J. Math. Anal. Appl.* **261**, 85–99.

Chavarriga, J., Llibre J. & Sorolla, J. [2004] "Algebraic limit cycles of degree 4 for quadratic systems," *J. Differ. Equ.* **200**, 206–244.

Christopher, C. & Llibre, J. [1999] "Algebraic aspects of integrability for polynomial systems," *Qual. Theory Dyn. Syst.* **1**, 71–95.

Christopher, C. & Llibre, J. [2002] "A family of quadratic polynomial differential systems with invariant algebraic curves of arbitrarily high degree without rational first integrals," *Proc. Am. Math. Soc.* **130**, 2025–2030.

Christopher, C., Llibre, J. & Świrszcs, G. [2005] "Invariant algebraic curves of large degree for quadratic system," *J. Math. Anal. Appl.* **303**, 450–461.

Coll, B., Dumortier, F. & Prohens, R. [2013] "Configurations of limit cycles in Liénard equations," *J. Differ. Equ.* **255**, 4169–4184.

Darboux, G. [1878] "Mémoire sur les équations différentielles algébriques du premier ordre et du premier degré (mélanges)," *Bull. Sci. Math.* **124A**, 60–96, 123–144, 151–200.

Dumortier, F., Llibre, J. & Artés, J. C. [2006] *Qualitative Theory of Planar Differential Systems* (Springer, NY).

Evdokimenco, R. M. [1970] "Construction of algebraic paths and the qualitative investigation in the large of the properties of integral curves of a system of differential equations," *Differ. Equ.* **6**, 1349–1358.

Evdokimenco, R. M. [1974] "Behavior of integral curves of a dynamic system," *Differ. Equ.* **9**, 1095–1103.

Evdokimenco, R. M. [1979] "Investigation in the large of a dynamic system with a given integral curve," *Differ. Equ.* **15**, 215–221.

Filiptsov, V. F. [1973] "Algebraic limit cycles," *Differ. Equ.* **9**, 983–986.

Giacomini, H., Llibre, J. & Viano, M. [1996] "On the non-existence, existence, and uniqueness of limit cycles," *Nonlinearity* **9**, 501–516.

Hilbert, D. [1900] "Mathematische Problem (lecture)," *Second Int. Congress Mathematics Paris, 1900, Nachrichten der Gesellschaft der Wissenschaften zu Göttingen, Mathematisch-Physikalische Klasse*, pp. 253–297.

Ilyashenko, Y. S. [2002] "Centennial history of Hilbert's 16th problem," *Bull. Am. Math. Soc.* **39**, 301–354.

Jouanolou, J. P. [1979] *Equations de Pfaff algébriques*, Lecture Notes in Mathematics, Vol. 708 (Springer, Berlin).

Li, J. [2003] "Hilbert's 16th problem and bifurcations of planar polynomial vector fields," *Int. J. Bifurcation and Chaos* **13**, 47–106.

Llibre, J. [2004] "Integrability of polynomial differential systems," in *Handbook of Differential Equations, Ordinary Differential Equations*, eds. Cañada, A., Drabek, P. & Fonda, A., Vol. 1 (Elsevier), pp. 437–533.

Llibre, J. [2008] "Open problems on the algebraic limit cycles of planar polynomial vector fields," *Bul. Acad. Stiinte Repub. Mold. Mat.* **1**, 19–26.

Llibre, J., Ramírez, R. & Sadovskaia, N. [2010] "On the 16th Hilbert problem for algebraic limit cycles," *J. Differ. Equ.* **248**, 1401–1409.

Llibre, J., Ramírez, R. & Sadovskaia, N. [2011] "On the 16th Hilbert problem for limit cycles on nonsingular algebraic curves," *J. Differ. Equ.* **250**, 983–999.

Llibre, J., Ramírez, R., Ramírez, V. & Sadovskaia, N. [2016] "The 16th Hilbert problem restricted to circular algebraic limit cycles," *J. Differ. Equ.* **260**, 5726–5760.

Llibre, J. & Rodríguez, G. [2004] "Configurations of limit cycles and planar polynomial vector fields," *J. Differ. Equ.* **198**, 374–380.

Moulin Ollagnier, J. [2001] "About a conjecture on quadratic vector fields," *J. Pure Appl. Algebra* **165**, 227–234.

Moulin Ollagnier, J., Nowicki, A. & Strelcyn, J. M. [1995] "On the non-existence of constants of derivations: The proof of a theorem of Jouanolou and its development," *Bull. Sci. Math.* **119**, 195–233.

Odani, K. [1995] "The limit cycle of the van der Pol equation is not algebraic," *J. Differ. Equ.* **115**, 146–152.

Qin, Y. [1958] "On the algebraic limit cycles of second degree of the differential equation $dy/dx = \sum_{0\le i+j\le 2} a_{ij}x^i y^j / \sum_{0\le i+j\le 2} b_{ij}x^i y^j$," *Acta Math. Sinica* **8**, 23–35.

van der Pol, B. [1926] "On relaxation-oscillations," *London Edinburgh Dublin Philos. Mag. J. Sci.* **2**, 978–992.

Walcher, S. [2000] "On the Poincaré problem," *J. Differ. Equ.* **166**, 51–78.

Yablonskii, A. I. [1966] "Limit cycles of a certain differential equations," *Differ. Equ.* **2**, 335–344.

Zhang, X. [2011] "The 16th Hilbert problem on algebraic limit cycles," *J. Differ. Equ.* **251**, 1778–1789.

https://doi.org/10.1142/9789819830633_0009

Chapter 9

Nonautonomous Dynamics: The Research Landscape#

Stephen Wiggins
School of Mathematics, University of Bristol, Fry Building, Woodland Road, Bristol BS8 1UG, UK
Department of Mathematics, United States Naval Academy, Chauvenet Hall, 572C Holloway Road, Annapolis, MD 21402-5002, USA
s.wiggins@bristol.ac.uk

We survey some aspects of nonautonomous dynamical systems theory through the lens of fundamental topics typically encountered in a first course on autonomous dynamical systems. Using concrete, analytically tractable examples, we highlight how classical concepts— such as flows, invariant sets, hyperbolicity, stability, stable and unstable manifolds of hyperbolic trajectories, Lyapunov functions, center manifolds, bifurcations, chaotic behavior, and Hamiltonian dynamics — are manifested in the nonautonomous setting. The discussion reflects on the historical trajectory of the field, contrasting its visibility and application with the autonomous theory, and identifies promising directions for future research, particularly in developing compelling nonautonomous analogs of familiar dynamical objects. Throughout, this paper underscores the importance of processes, pullback attraction, and time-dependent hyperbolicity in shaping a modern understanding of time-varying dynamical behavior. These developments open significant new opportunities for applied mathematicians, engineers, and physicists to model and analyze complex time-dependent phenomena — ranging from fluid dynamics to mechanical systems and data-driven modeling — where nonautonomous effects are intrinsic and essential.

Keywords: Nonautonomous; deterministic process; pullback attraction; exponential dichotomy; hyperbolic trajectory; time-dependent stable and unstable manifolds; time-dependent center manifold; time-dependent Lyapunov function; LaSalle invariance principle; nonautonomous chaos; nonautonomous Hamiltonian dynamics.

1. Introduction

In this paper, we discuss the nonautonomous manifestations of a collection of specifically chosen topics in dynamical systems theory through a series of concrete examples. We are not making a serious attempt to give a history of the subject of dynamical systems theory. The reader interested in that should consult [Diacu & Holmes, 1999; Abraham & Ueda, 2000; Holmes, 2005, 2007; Letellier *et al.*, 2021].

Applications of dynamical systems theory flowered in the late 1970's and 1980's. One significant reason for this was the wide availability of desktop computational resources coupled with visualization capabilities. That enabled scientists and engineers outside of the mathematics community to access and apply mathematical results in specific problems in applications as well as to discover new phenomena in dynamical systems, which subsequently led to new progress in mathematics. Three prime examples of this was the work of Lorenz on atmospheric dynamics that resulted in the celebrated Lorenz equations, and motivated literally a flood of work on that topic that continues to this day.

#To cite this article, please refer to its earlier version published in the *International Journal of Bifurcation and Chaos*, Vol. 35, No. 14 (2025) 2540009 (27 pages), DOI: 10.1142/S0218127425400097.

The periodically forced and damped Duffing equation ("two-well potential") has been a fundamental model in many areas of science and engineering for years, and the decades-long studies of [Ueda, 1991] have revealed the richness of chaotic dynamics and strange attractors in this system. The Hénon–Heiles Hamiltonian system [Hénon & Heiles, 1964] emerged in a study of galactic dynamics and serves as a paradigm for studies of the interplay of regular and chaotic dynamics as a function of the total energy of the system. We note that all of this work is in the context of autonomous dynamical systems.[1]

It is probably fair to say that the work of these investigators was somewhat outside the mainstream of their respective areas of investigation. Nevertheless, the accessibility of this mathematical work to nonmathematicians as well as the promise of opening up new areas of investigation and solving outstanding problems in traditional fields gave rise to an explosion of new research and activities along these lines, e.g. new scientific journals, meetings, organizations, and academic groupings, just to name a few. A particular history of these times can be found in [Gleick, 1987].

During these times, there was a great desire for information on this "new" approach to nonlinear dynamical systems, and two books, with very different emphases, emerged at about the same time: [Lichtenberg & Lieberman, 1983], which focused mostly on Hamiltonian systems and [Guckenheimer & Holmes, 1983], which focused more on general dynamical systems, surveyed many of the current results and topics at that time (e.g. the Lorenz equations and the damped, periodically forced Duffing equation) and served more as a "users guide". However, essentially all of the topics in these two books were concerned with autonomous dynamical systems (or in situations where the dynamics could be studied by a discrete time autonomous map).

In the midst of all of this "new mathematics", some mathematicians where a little annoyed since the mathematical analysis of Ordinary Differential Equations (ODEs) was a thriving topic with many relevant results, most of which remained out of the spotlight. In the West, this theory was represented by three excellent (graduate level) textbooks [Coddington & Levinson, 1955; Hartman, 2002; Hale, 1969]. These books contain a wealth of relevant dynamical systems theory, but it is presented in a way that is difficult for nonmathematicians to access, and they contain no significant applications.[2] This is in contrast to [Lichtenberg & Lieberman, 1983; Guckenheimer & Holmes, 1983] where, even though there is deep underlying mathematics, a variety of significant applications are discussed and the mathematics is directly utilized to give significant new insights.

The differential equations community generally treats autonomous and nonautonomous equations on an equal footing, which is true in [Coddington & Levinson, 1955; Hartman, 2002; Hale, 1969], and there has been considerable effort in developing a geometric theory of nonautonomous dynamical systems over the past 25 years or so (we will comment on aspects of this throughout our discussion), but this development has lacked compelling applications and examples and has failed to capture the imagination of a significant group of people working in applications in the same way that occurred during the development of applied dynamical systems in the 1970's and 1980's.

In this paper, we will discuss the situation by considering the typical topics one might encounter in a course in dynamical systems in the autonomous setting and consider how these are treated in the nonautonomous setting. Rather than discuss theorems covering each case, we will consider simple examples where the calculations can be carried out by hand. At the same time, we will discuss topics and directions for future research.

1.1. *Typical topics in a first course on dynamical systems (with a "geometrical and applied flavor")*

What is considered a "first course" in dynamical systems is very much subjective nowadays. Some aspects of dynamical systems theory and chaos are successfully taught even at the US high school level [Devaney, 2018]. This speaks to the widely accepted importance of this subject matter in the

[1]Of course, the periodically forced Duffing equation is time-dependent, but in the case of periodic forcing it can be studied by an autonomous Poincaré map.

[2]We note that one might view the book [Chicone, 2006] as a modern update of these three classic texts. It not only has the same classical rigor, but also a variety of applications with a dynamical systems flavor and point of view.

early education of those intending to pursue a career in not only mathematics, but also in a wide range of areas of science, engineering, the life sciences, social science, and economics. However, for this paper a "first course in dynamical systems" I will mean a course at the advanced undergraduate level or beginning postgraduate level with a background in basic linear algebra, multivariable calculus, and some exposure to differential equations (which will probably be typical types of first-order equations and linear, constant coefficient, second-order equations — a course that is generally in desperate need of updating). Such a course would generally be designed to run for 15 weeks, with 3 h of lectures per week. In this setting, a typical list of topics might be:

- How dynamics is generated — maps defined by trajectories.
- Invariance and invariant sets.
- Attraction and stability.
- Linearization and hyperbolicity.
- Lyapunov function methods.
- The LaSalle invariance principle.
- Stable and unstable manifolds of hyperbolic trajectories.
- Nonautonomous center manifold theory.
- Chaos (the Smale horseshoe, homoclinic orbits and heteroclinic cycles, and Melnikov's method).
- Bifurcation theory.
- Hamiltonian dynamics.

In the typical dynamical systems course following one of the standard textbooks, these ideas are developed for autonomous ODEs or maps (which might arise as the Poincaré map of a time-periodic, nonautonomous differential equation). In this paper, we will follow this outline for nonautonomous dynamical systems, and we will attempt to provide simple examples for each topic that can easily be calculated.

2. Generating the Dynamics — Flows and Processes

The geometrical point of view of dynamical systems begins by setting the stage for the consideration of dynamics generated by maps of points of phase space into itself. When the rules governing the dynamics are given by ODEs these maps are given by the solutions, or trajectories, of the differential equations.[3] Maps given by solutions of autonomous differential equations have a different dependence on the time parameters compared to the solutions of nonautonomous differential equations, and it is this what we now discuss.

We consider the following autonomous:

$$\dot{x} = f(x), \quad x \in \mathbb{R}^n, \tag{1}$$

and nonautonomous vector fields

$$\dot{x} = f(x, t), \quad x \in \mathbb{R}^n. \tag{2}$$

The space of dependent variables is referred to as the phase space [Hale, 1969], which we will take as $\mathbb{R}^n$ to simplify matters. Many of the geometrical objects that we will study for nonautonomous systems will vary in time. For a more complete study and understanding of this, we will find it insightful to append to the phase space time, the independent variable. It is standard to refer to this as the extended phase space, denoted by $\mathbb{R}^n \times \mathbb{R}$. We extend the vector field to the extended phase space in the usual manner by explicitly including time as a new dependent variable

$$\begin{aligned} \dot{x} &= f(x, t), \\ \dot{t} &= 1, \quad (x, t) \in \mathbb{R}^n \times \mathbb{R}, \end{aligned} \tag{3}$$

which results in an autonomous vector field on the extended phase space. We will see later that this seemingly trivial "trick" has some conceptual value for understanding the time-dependent nature of "phase space objects".

We will use the usual Euclidean metric as the measure of distance between points in both the phase space and the extended phase space, and we will denote it by $d(\cdot, \cdot)$ (whether we are considering it in the phase space or the extended phase space will be made clear in the context). We will also require the notion of the distance from a point to a set, which we define as follows:

$$\text{dist}(x, A) \equiv \inf_{y \in A} d(x, y), \tag{4}$$

as well as the distance between two sets (known as the Hausdorff distance)

$$\text{dist}(A, B) \equiv \sup_{x \in A} \text{dist}(x, B). \tag{5}$$

[3]In this paper, we will only consider continuous time dynamical systems, i.e. vector fields. For information on discrete time nonautonomous dynamical systems, see [Pötzsche, 2010].

We will assume that the vector fields are sufficiently regular so that existence and uniqueness of solutions on sufficiently long time intervals hold. For both autonomous and nonautonomous vector fields we will denote solutions, or "trajectories" as follows:

$$x(t, t_0, x_0), \quad x(t_0, t_0, x_0) = x_0. \tag{6}$$

In this notation, x_0 denotes the initial position, t denotes the current time, t_0 denotes the initial time (possibly the term "reference time" might be more useful) and $t - t_0$ is the time elapsed from the initial time, or just "elapsed time". What will distinguish the nature of the trajectories of autonomous and nonautonomous vector fields?

A key characteristic of the solutions of autonomous vector fields is that "time shifts of solutions are still solutions (but with a different initial condition), see [Wiggins, 2023]. This also means that the solutions of autonomous vector fields depend only on the elapsed time, $t - t_0$, rather than t and t_0 independently. Consequently, the trajectories of (1) can be expressed in the following form:

$$x(t - t_0, 0, x_0), \quad x(0, 0, x_0) = x_0.$$

As a consequence of this property, we often simplify the notation for the trajectories of autonomous vector fields by writing them as follows:

$$x(t, x_0), \quad x(0, x_0) = x_0.$$

It is instructive to see how this is manifested in a specific example. Consider the linear, autonomous one-dimensional vector field

$$\dot{x} = -2x, \quad x(t_0, t_0, x_0) = x_0, \quad x \in \mathbb{R}.$$

Trajectories are given by

$$x(t, t_0, x_0) = x_0 e^{-2(t - t_0)}.$$

Hence, we see that the trajectory depends only on the elapsed time $t - t_0$, and it is simple to check that if one adds a constant to t ("time shift") in this example, it is still a solution, but with a different initial condition.

In general, the time shift property of the trajectories of autonomous vector fields is expressed as follows:

$$x(t + s; x_0) = x(t; x(s; x_0)), \quad \forall\, s, t \in \mathbb{R}.$$

We define a family of maps (the *flow*) by

$$\varphi_t : \mathbb{R}^n \to \mathbb{R}^n, \quad \varphi_t(x_0) = x(t; x_0).$$

We can express the time shift property in terms of the family of maps

$$\begin{aligned}\varphi_{t+s}(x_0) &= x(t + s; x_0) = x(t; x(s; x_0)) \\ &= \varphi_t(\varphi_s(x_0)) = (\varphi_t \circ \varphi_s)(x_0).\end{aligned}$$

Thus, $\{\varphi_t\}_{t \in \mathbb{R}}$ satisfies:

(1) **Initial time identity.**

$$\varphi_0(x_0) = x_0,$$

or

$$\varphi_0 = \text{id}.$$

(2) **Composition rule.**

$$\varphi_{t+s} = \varphi_t \circ \varphi_s,$$

which is the group (flow) property.

This is the definition of a (continuous) *flow* on $\mathbb{R}^n$.

The trajectories of nonautonomous vector fields do not satisfy these properties. In particular, they do not depend on the elapsed time, $t - t_0$, but on t and t_0 independently, and time shifts of solutions are not solutions. This is easily seen in a simple example

$$\dot{x} = -2tx, \quad x(t_0) = x_0,$$

with solution given by

$$x(t, t_0, x_0) = x_0 e^{-(t^2 - t_0^2)}.$$

With a little algebra, it is straightforward to see that the solution cannot be manipulated to depend only on $t - t_0$, and also that time shifts of solutions are not solutions.

Denoting the trajectories of a nonautonomous vector field by

$$x(t, t_0, x_0), \quad x(t_0, t_0, x_0) = x_0,$$

we define a two-parameter family of maps, called a *process*, by

$$\phi_{t,t_0} : \mathbb{R}^n \to \mathbb{R}^n, \quad \phi_{t,t_0}(x_0) = x(t, t_0, x_0).$$

This map satisfies:

(1) **Initial time identity.**

$$\phi_{t_0,t_0}(x_0) = x_0.$$

(2) **Composition rule.** For all $t_0 \leq t_1 \leq t$

$$\phi_{t,t_1}(\phi_{t_1,t_0}(x_0)) = x(t; t_1, x(t_1; t_0, x_0)) = x(t; t_0, x_0),$$

or

$$\phi_{t,t_1} \circ \phi_{t_1,t_0} = \phi_{t,t_0}.$$

Sometimes, this is referred to as the *cocycle property*.

The collection of maps $\{\phi(t, t_0)\}_{t \geq t_0}$ forms a *process*, which might be viewed as the nonautonomous analog of a flow (but it is *not* a flow).

There has been considerable work on the process formulation of nonautonomous dynamical systems previously. For example, see [Miller, 1965; Sell, 1967a, 1967b; Dafermos, 1971; Hale, 1988], and more recently [Balibrea *et al.*, 2010; Kloeden & Rasmussen, 2011]. An alternative formulation of trajectory maps for nonautonomous ODEs is the skew product flow formulation. A description of this can be found in the above references (and see also [Meyer & Zhang, 1996]), but we will not consider this formulation in this paper.

3. Invariance and Invariant Sets

Invariant sets play an important role in dynamical systems. Roughly, a set is invariant if trajectories that start in the set remain in the set for all time.[4] Thus, the dynamics conspires in such a way that it constrains a collection of trajectories to be restricted to a subset in the phase space. Since invariant sets are composed of trajectories, no trajectories that start outside an invariant set may "cross" the invariant set. This is another way in which invariant sets constrain the dynamics and, as a result, have significant consequences on the dynamics. Consequently, the identification and characterization (e.g. their dimension and stability) of invariant sets are essential for understanding the global behavior and stability in dynamical systems.

Trajectories are trivial examples of invariant sets. Any point on a trajectory will always remain on that trajectory throughout its evolution (and as long as the trajectory exists, i.e. remains finite). In autonomous vector fields, hyperbolic equilibria and their stable and unstable manifolds are familiar examples of invariant sets. Periodic orbits and their stable and unstable manifolds, invariant tori and their stable and unstable manifolds and, more generally, normally hyperbolic invariant manifolds and their stable and unstable manifolds provide familiar examples of invariant sets. Finally, there are invariant sets associated with chaotic dynamics, e.g. the (hyperbolic) invariant set of the Smale horseshoe and chaotic attractors, whose "skeleton" is composed of a hierarchy of more simple invariant sets. There are many examples of how these types of invariant sets play the role of the essential building blocks for constructing the "global story" of the behavior of autonomous dynamical systems. A variety of such examples and situations are considered in [Wiggins, 1992].

We will consider the extent to which similar analyses can be carried out for nonautonomous vector fields. It should be evident that the invariant sets in nonautonomous systems may vary in time. So, we first need to define what we mean in that setting, and invariance will be with respect to a process, rather than a flow.

Following [Langa *et al.*, 2002; Balibrea *et al.*, 2010; Kloeden & Rasmussen, 2011], a time-varying family of sets

$$\{M(t)\}_{t \in \mathbb{R}},$$

is said to be invariant if

$$x(t, t_0, M(t_0)) = M(t), \quad \text{for all } t_0 \leq t.$$

If we know the invariant set, we would like some condition on the vector field that insures that the set is indeed invariant, since analytical expressions for the flow are typically not available. If the invariant set has the structure of a differentiable surface, or manifold:

For autonomous systems, the geometric condition that implies invariance is that the vector field is tangent to the invariant manifold, or equivalently, that the time derivative along trajectories is zero on the invariant manifold. We will now derive a computable condition to verify this for nonautonomous systems for a limited class of invariant manifolds in the extended phase space. We consider a differentiable scalar-valued function of the form $h(x, t)$, $(x, t) \in \mathbb{R}^n \times \mathbb{R}$. The zero-level set of

[4] In some applications remaining in the set for all time after the initial time (positive invariant set) is considered or remaining in the set for all time before the initial time (negative invariant set). In this paper, will only consider the case of "for all time" for simplicity.

this function defines a n-dimensional surface in the $n+1$-dimensional phase space, which we denote as follows:

$$\tilde{M} = \{(x,t) \mid h(x,t) = 0\} \subset \mathbb{R}^n \times \mathbb{R}.$$

We extend the vector field to the extended phase space by writing it as $(f(x,t), 1)$

$$\begin{aligned} \dot{x} &= f(x,t), \\ \dot{t} &= 1, \quad (x,t) \in \mathbb{R}^n \times \mathbb{R}. \end{aligned} \tag{7}$$

Then, the condition for this vector field to be tangent to $h(x,t) = 0$ in $\mathbb{R}^n \times \mathbb{R}$ is given by

$$\frac{\partial h}{\partial t}(x,t) + \nabla_x h(x,t) \cdot f(x,t) = 0. \tag{8}$$

We will give examples of how this is used to verify the existence of invariant sets in Secs. 7 and 8.

4. Attraction

The time-dependent nature of the dynamics motivates the development of a new type of attraction that we discuss in this section. We begin with an example that motivates the main issues.

Consider the following nonautonomous one-dimensional vector field considered in [Szeri *et al.*, 1991]:

$$\dot{x} = -x + t, \quad x \in \mathbb{R}, \ t \in \mathbb{R}. \tag{9}$$

This vector field generates the following process:

$$\begin{aligned} x(t, t_0, x_0) &= x(t_0)e^{-(t-t_0)} - (t_0 - 1)e^{-(t-t_0)} \\ &\quad + (t-1). \end{aligned} \tag{10}$$

It is clear from the form of (10) that for any x_0 and t_0, the solution is attracted to $M(t) = t - 1$ at an exponential rate. This is the standard type of *forward attraction* that we consider for autonomous dynamical systems. However, as time increases, $M(t)$ increases without bound. But for autonomous dynamical systems, the attracting sets do not "run away" — they are fixed regions in phase space. We could ask a different question regarding attraction. Rather than ask where initial conditions in the present evolve to in the future, we could ask which initial conditions in the past evolve to a specific set in the present. This is the notion of *pullback attraction*. More precisely, we fix t (the present time) and consider initial times t_0 in the far past, i.e. $t_0 \to -\infty$. Applying this idea to (10), we see that

$$\lim_{t_0 \to -\infty} x(t, t_0, x_0) = t - 1. \tag{11}$$

For autonomous vector fields, the notion of forward attraction and pullback attraction is the same. This is a consequence of the fact that the flow generated by an autonomous vector field depends on the elapsed time, $t - t_0$, rather than on t and t_0 separately. Hence, $\lim_{t\to\infty}(t - t_0) = \lim_{t_0 \to -\infty}(t - t_0)$ (with t_0 fixed in the left limit and t fixed in the right limit).

Now we define the notion of a general nonautonomous attractor. We consider a nonautonomous invariant set, i.e.

$$\mathcal{A} \equiv \{(x,t) \in \mathbb{R}^n \times \mathbb{R} \mid x \in \mathcal{A}(t)\}, \tag{12}$$

with the invariance condition

$$x(t, t_0, \mathcal{A}(t_0)) = \mathcal{A}(t), \quad \text{for all } t_0 \le t. \tag{13}$$

We say that $\mathcal{A}$ attracts a bounded subset $D \subset \mathbb{R}^n$ of initial conditions (x_0, t_0) if

$$\operatorname{dist}(x(t, t_0, D), \mathcal{A}(t)) \to 0 \quad \begin{cases} \text{as } t \to \infty \text{ with } t_0 \text{ fixed}, & \text{(forward attracting)}, \\ \text{as } t_0 \to -\infty \text{ with } t \text{ fixed}, & \text{(pullback attracting)}. \end{cases} \tag{14}$$

There is a significant literature on pullback attractors. In the context of random dynamical systems, see [Crauel & Flandoli, 1994; Arnold, 1998] and for deterministic nonautonomous systems, see [Temam, 1997; Chepyzhov & Vishik, 2002; Cheban *et al.*, 2002]. The textbooks [Kloeden & Rasmussen, 2011; Kloeden & Yang, 2021] provide an extensive development of this topic.

We note that for nonautonomous ODEs, probably, the two most common notions of stability of trajectories are Lyapunov stability ("start close and stay close forever) and asymptotic stability (start close and stay close forever, and asymptotically approach the trajectory). These ideas are developed in a pullback setting in [Langa *et al.*, 2002].

Finally, we note that even though there has been significant development of the theory of pullback attractors in nonautonomous systems, a broad collection of examples that have been studied in detail is lacking. This lack of examples is particularly notable when compared with the attractor literature for autonomous systems. See the recent paper [Wagemakers, 2025] for a remarkably extensive and useful collection of attractors from a diverse collection of applications. The Hénon map [Hénon, 1976] and the Lozi map [Lozi, 1978] are two seminal examples of two-dimensional maps with strange attractors. They have been generalized to the nonautonomous case in [Balibrea-Iniesta *et al.*, 2015; Lopesino *et al.*, 2015]. The focus of this work was on proving the existence of a chaotic invariant set in the autonomous case (we will discuss this in more detail in Sec. 9). However, a study of the nature of nonautonomous strange attractors would certainly be interesting and the paper of Lozi provides a roadmap of how one might approach this. We note that the work described in [Kominis & Bountis, 2010] provides a framework that one might use for constructing nonautonomous versions of a class of piecewise-linear maps. Finally, the paper of [Milnor, 1985] has played a fundamental role in how we think about attractors in autonomous systems. A similar paper for nonautonomous attractors would be timely and welcome.

5. Linearization and Hyperbolicity

Suppose $\hat{x}(t, t_0, x_0)$, $\hat{x}(t_0, t_0, x_0) = x_0$ is a trajectory of interest and we wish to understand the dynamics near this trajectory. To do this, we introduce a new coordinate system. $x = y + \hat{x}(t, t_0, x_0)$. Substituting this into (2), Taylor expanding in y about $y = 0$, and retaining only the terms linear in y give

$$y = D_x f(\hat{x}(t, t_0, x_0), t)y, \quad y \in \mathbb{R}^n. \tag{15}$$

This equation is referred to as the linearization of (2) about $\hat{x}(t, t_0, x_0)$. In general, the $n \times n$ matrix $D_x f(\hat{x}(t, t_0, x_0), t)$ depends on time.

If the vector field is autonomous (i.e. (1)) and $\hat{x}(t, t_0, x_0)$, $\hat{x}(t_0, t_0, x_0) = x_0$ is an equilibrium point, x_0, then $D_x f(x_0) \equiv A$ is a $n \times n$ matrix of real numbers and (15) takes the familiar form

$$\dot{y} = Ay, \quad y(t_0) = y_0. \tag{16}$$

The solution of (16) for any initial condition can be constructed using standard methods of linear algebra, see, e.g. [Hirsch & Smale, 1974], and it has the form

$$y(t) = e^{A(t-t_0)} y_0. \tag{17}$$

It is a simple matter to verify that (17) solves (16) and in the course of the computation of (17), it becomes evident that the spectral properties of A are related to the spectral properties of $e^{A(t-t_0)}$ and to the behavior of trajectories near $y = 0$. The spectral properties of A are used to characterize the nature of the stability of $y = 0$.

In particular, if all of the eigenvalues of A have negative real part, then the origin is asymptotically stable. If at least one eigenvalue of A has positive real part, then the origin is unstable. The case of eigenvalues with zero real parts leads to a variety of cases that are well understood, but we will not consider them here (complete details for the linear autonomous theory can be found in [Hirsch & Smale, 1974]).

The three types of eigenvalues classified according to the sign of their real parts (negative, positive, zero) are associated with invariant subspaces (invariant with respect to the flow $e^{A(t-t_0)}$). The span of the (generalized) eigenvectors of A corresponding to eigenvalues of A with negative real part is invariant under the dynamics generated by (16). It is referred to as the stable subspace, denoted by E^s, and corresponds to initial conditions of trajectories that approach the origin at an exponential rate as $t \to \infty$. The span of the (generalized) eigenvectors of A corresponding to eigenvalues of A with positive real part is invariant under the dynamics generated by (16). It is referred to as the unstable subspace, denoted by E^u, and corresponds to initial conditions of trajectories that approach the origin at an exponential rate as $t \to -\infty$. The span of the (generalized) eigenvectors of A corresponding to the eigenvalues of A with zero real part is also invariant under the dynamics generated by (16). It is referred to as the center subspace. We will not consider the behavior of trajectories restricted to the center subspace.

Now, we can introduce an important concept. The equilibrium point at the origin is said to be *hyperbolic* if none of the eigenvalues of A lie on the imaginary axis (i.e. have zero real part). It should be intuitively clear that this is a situation that is stable under appropriately defined perturbations. This is a different use of the word stability — rather than stability of a solution, we are discussing certain aspects of the stability of a particular dynamical system — in this case a linear autonomous system. This type of stability is referred to as *structural stability* and has, historically, played an important role in the development of autonomous dynamical systems. An elementary discussion of this can be found in [Wiggins, 2003], and a more advanced discussion is in [Robinson, 1974].

It is notable that the behavior of the solution (17) can be determined from the properties of the matrix A, i.e. the vector field (16). This is a property of linear autonomous systems and does not generally hold for linear nonautonomous systems, as we will show with an example.

Suppose that $D_x f(\hat{x}(t, t_0, x_0), t)$ or $D_x f(\hat{x}(t, t_0, x_0))$ is time-dependent, which for either case we denote the Jacobian matrix by $A(t)$, then the associated linearized equation has the form

$$\dot{y} = A(t)y, \quad y(t_0) = y_0. \tag{18}$$

It is tempting to hope that the eigenvalues of $A(t)$ provide information about the stability of $y = 0$ and that the solution of (18) might be related to $e^{A(t,t_0)}$, but this is not true, as the well-known Markus–Yamabe example [Markus & Yamabe, 1960] shows, which we now describe.

Example 5.1 (Markus–Yamabe Example). Consider the following two-dimensional linear nonautonomous ODE:

$$\begin{pmatrix} \dot{u}_1 \\ \dot{u}_2 \end{pmatrix} = A(t) \begin{pmatrix} u_1 \\ u_2 \end{pmatrix}, \tag{19}$$

where

$$A(t) = \begin{pmatrix} -1 + \frac{3}{2}\cos^2 t & 1 - \frac{3}{2}\sin t \cos t \\ -1 - \frac{3}{2}\sin t \cos t & -1 + \frac{3}{2}\sin^2 t \end{pmatrix}.$$

Despite the fact that $A(t)$ depends explicitly on time, its eigenvalues are constant and both eigenvalues have negative real parts

$$-\frac{1}{4} \pm i\frac{7}{4}.$$

Therefore, it is tempting to believe that the origin of (19) is asymptotically stable, which is not true. It can be shown that two linearly independent solutions of (19) are given by

$$e^{\frac{t}{2}} \begin{pmatrix} -\cos t \\ \sin t \end{pmatrix}, \quad e^{-t} \begin{pmatrix} \sin t \\ \cos t \end{pmatrix}. \tag{20}$$

Since the general solution of (19) can be written as a linear combination of these two linearly independent solutions, it is clear that the origin is unstable.

Therefore, for nonautonomous linear systems, we need a different approach for characterizing stability and developing the notion of hyperbolicity. There are several ways this might be done, but the method that we will discuss and use is the notion of an *exponential dichotomy*, which is generally attributed to [Perron, 1930], which we now define.

Consider the nonautonomous linear system

$$\dot{u}(t) = A(t)u(t), \quad u(t) \in \mathbb{R}^n. \tag{21}$$

Let $\Phi(t, t_0)$ be the fundamental solution matrix (or evolution operator) with $\Phi(t_0, t_0) = I$. The system is said to admit an *exponential dichotomy* on $\mathbb{R}$ if there exists:

- a projection matrix $P \in \mathbb{R}^{n \times n}$ $(P^2 = P)$,
- constants $K \geq 1$, $\alpha, \beta > 0$,

such that for all t, $t_0 \in \mathbb{R}$

$$\Phi(t, t_0)P = P\Phi(t, t_0),$$

and

$$\|\Phi(t, t_0)P\| \leq Ke^{\alpha(t-t_0)}, \quad \text{for } t \geq t_0,$$
$$\|\Phi(t, t_0)(I - P)\| \leq Ke^{-\beta(t-t_0)}, \quad \text{for } t \leq t_0.$$

This means the space splits into an invariant subspace of exponentially growing solutions (range of P) and a subspace of exponentially decaying solutions (range of $I - P$).[5]

Now, we compute an exponential dichotomy for a specific example. The setting we will consider is one that might arise in applications, i.e. we start with a nonlinear system, we identify a trajectory of interest, and then we want to determine if the trajectory is hyperbolic.

Example 5.2. We consider a two-dimensional, nonautonomous, nonlinear system

$$\begin{aligned} \dot{x} &= x, \\ \dot{y} &= -y + x^2 + t. \end{aligned} \tag{22}$$

The trajectory through the point (x_0, y_0) at $t = t_0$ is given by

$$x(t, t_0, x_0) = x_0 e^{(t-t_0)}, \tag{23}$$

[5] The norm used in the definition of the exponential dichotomy is the standard operator norm on $\mathbb{R}^n$.

$$y(t, t_0, x_0, y_0) = \left(y_0 - \frac{x_0^2}{3}\right) e^{-(t-t_0)} + \frac{x_0^2}{3} e^{2(t-t_0)} - (t_0 - 1)e^{-(t-t_0)} + t - 1. \quad (24)$$

It is easy to check that $(\hat{x}(t), \hat{y}(t)) = (0, t-1)$ is a solution of (22). We want to compute the linearization about this solution and show that the resulting linear system has an exponential dichotomy. This means that we have to linearize (22) about $(\hat{x}(t), \hat{y}(t))$ and verify that the associated linear system has an exponential dichotomy. In this way, we conclude that $(\hat{x}(t), \hat{y}(t)) = (0, t-1)$ is hyperbolic.

The linearization of (22) evaluated at $(\hat{x}(t), \hat{y}(t)) = (0, t-1)$ is

$$\begin{pmatrix} \dot{u}_1 \\ \dot{u}_2 \end{pmatrix} = \begin{pmatrix} 1 & 0 \\ 2e^t & -1 \end{pmatrix} \begin{pmatrix} u_1 \\ u_2 \end{pmatrix}. \quad (25)$$

Two linearly independent solutions of (25) are given by

$$\begin{pmatrix} e^t \\ \frac{2}{3}e^{2t} \end{pmatrix}, \quad \begin{pmatrix} 0 \\ e^{-t} \end{pmatrix}.$$

We form the invertible matrix

$$X(t) = \begin{pmatrix} e^t & 0 \\ \frac{2}{3}e^{2t} & e^{-t} \end{pmatrix}.$$

Then, the fundamental solution matrix is given by

$$\Phi(t, t_0) = X(t)X(t_0)^{-1} = \begin{pmatrix} e^{t-t_0} & 0 \\ \frac{2}{3}(e^{2t-t_0} - e^{-t+2t_0}) & e^{t_0-t} \end{pmatrix}.$$

We define the projection onto the unstable direction

$$P = \begin{pmatrix} 1 & 0 \\ 0 & 0 \end{pmatrix}.$$

Then, we have

$$\|\Phi(t, t_0)P\| \leq e^{t-t_0}, \quad \text{for all } t \geq t_0,$$

and

$$\|\Phi(t, t_0)(I - P)\| \leq e^{-(t-t_0)}, \quad \text{for all } t \leq t_0,$$

Thus, the system has an exponential dichotomy with

$$K = 1, \quad \alpha = \beta = 1.$$

We analyze another example that will be useful later when we consider the time-dependent center manifold theory.

Example 5.3. Consider the following two-dimensional, nonautonomous vector field:

$$\begin{aligned} \dot{x} &= xy, \\ \dot{y} &= -y + \sin t, \quad (x, y) \in \mathbb{R}^2. \end{aligned} \quad (26)$$

It is straightforward to verify that

$$(\hat{x}(t), \hat{y}(t)) = \left(0, \frac{1}{2}(\sin t - \cos t)\right), \quad (27)$$

is a solution of (26). We want to check whether or not this solution is hyperbolic, i.e. does it have an exponential dichotomy?

We compute the linearization of (26) about the solution (27)

$$\begin{pmatrix} \dot{u}_1 \\ \dot{u}_2 \end{pmatrix} = \begin{pmatrix} \frac{1}{2}(\sin t - \cos t) & 0 \\ 0 & -1 \end{pmatrix} \begin{pmatrix} u_1 \\ u_2 \end{pmatrix}. \quad (28)$$

The solution is given by

$$u_1(t) = u_{1,0} \exp\left(\frac{1}{2}(\cos t_0 + \sin t_0 - \cos t - \sin t)\right), \quad (29)$$

$$u_2(t) = u_{2,0}e^{-(t-t_0)}. \quad (30)$$

We see that the u_1 component is bounded and oscillatory in time. Therefore, the system does not have an exponential dichotomy.

We end this section with some final remarks.

- The study of exponential dichotomies is a large topic which we do not have the space to adequately review. The classic reference is [Coppel, 1978]. See also [Kloeden & Rasmussen, 2011].
- For linear autonomous systems "solving the system", i.e. constructing the fundamental solution matrix, determining stability, and constructing invariant subspaces are reduced to problems of linear algebra. However, for linear nonautonomous problems, you must solve for the fundamental solution matrix. In general, this cannot

be done analytically and accurate numerical integration schemes are required.

- For linear autonomous systems, the associated spectral theory is derived from the constant matrix A, and is just standard linear algebra. For linear nonautonomous systems, there are various spectral theories. For example, there is a spectral theory associated with exponential dichotomies and another with Lyapunov exponents. These spectral theories and the relation between the two are discussed in [Kloeden & Rasmussen, 2011]. Pioneering work in this area was carried out by [Miller, 1965; Sell, 1967a, 1967b; Sacker & Sell, 1978].
- We mentioned earlier the notion of structural stability for autonomous systems and its connection with the notion of hyperbolicity. Structural stability of exponential dichotomies is a topic of much research. In that setting, the term "roughness" is often used, as opposed to "structural stability". (Interestingly, English translations of early Russian work on structural stability also used the term "roughness".) For a discussion of structural stability in nonautonomous systems, see [Kryzhevich & Pliss, 2003], and also [Pliss, 1977, 1980a, 1980b, 1981].

6. Lyapunov Function Methods

Lyapunov function methods, or "Lyapunov's second method" or the "indirect method of Lyapunov" [Lyapunov, 1992] is a method for determining the nature of the stability of solutions. Excellent modern references that cover both the autonomous and nonautonomous cases are [Yoshizawa, 1966; LaSalle, 1976]. Lyapunov stability techniques have been particularly influential in the development of control theory [Khalil, 2009]. The method is used to determine stability of a given solution (what we mean by "given solution" will be discussed shortly) and it has several advantages over linearization, when it works. In particular, it may provide conclusions about stability when information from linearization is inconclusive (e.g. it applies in situations where the solution is nonhyperbolic) and it may provide an approach to estimating the size of the basin of attraction [Barreiro *et al.*, 2002; Giesl & Hafstein, 2015]. This method also applies in the nonautonomous setting, and we will provide an example of this. But first, for context, we will describe the autonomous setting.

The setup for Lyapunov's theorem is as follows. We consider a general C^r, $r \geq 1$ autonomous ODE

$$\dot{x} = f(x), \quad x \in \mathbb{R}^n, \tag{31}$$

having an equilibrium point, which without loss of generality can be translated to $x = 0$, i.e.

$$f(0) = 0. \tag{32}$$

For a scalar-valued C^r ($r \geq 1$) function defined on $\mathbb{R}^n$

$$\begin{aligned} V : \mathbb{R}^n &\to \mathbb{R}, \\ x &\mapsto V(x), \end{aligned} \tag{33}$$

we define the derivative of (33) along trajectories of (31) by

$$\begin{aligned} \frac{d}{dt}V(x) = \dot{V}(x) &= \nabla V(x) \cdot \dot{x} \\ &= \nabla V(x) \cdot f(x). \end{aligned} \tag{34}$$

We can now describe Lyapunov's theorem on stability of the equilibrium point $x = 0$.

Suppose V satisfies:

(1) $V(0) = 0$ and $V(x) > 0$ for x near 0.
(2) $\dot{V}(x) \leq 0$ for x near 0.

Then, $x = 0$ is Lyapunov stable. Moreover, if

$$\dot{V}(x) < 0, \quad \text{for } x \text{ near } 0,$$

then $x = 0$ is asymptotically stable.

The function $V(x)$ is referred to as a Lyapunov function.

There are multiple references that have examples of the method of Lyapunov functions for determining stability in the autonomous case, see, for example [Wiggins, 2023].

Now, we turn our attention to Lyapunov's method in the nonautonomous setting.

Consider the nonautonomous system

$$\dot{x} = f(x, t), \quad x \in \mathbb{R}^n, \ t \in \mathbb{R},$$

with $f(0, t) = 0$ for all t.

Let $V : \mathbb{R}^n \times \mathbb{R} \to \mathbb{R}$ be a continuously differentiable function and let $W_1(x)$, $W_2(x)$, and $W_3(x)$ be positive definite scalar-valued functions such that

(1)
$$W_1(x) \leq V(x, t) \leq W_2(x), \tag{35}$$
for all x near 0 and all $t \geq t_0$.

(2) The total derivative with respect to t satisfies
$$\dot{V}(x, t) = \frac{\partial V}{\partial x} f(x, t) + \frac{\partial V}{\partial t} \leq -W_3(x). \tag{36}$$

Then, the origin is stable. If $\dot{V}(x,t) < 0$ for $x \neq 0$, the origin is asymptotically stable.

We make some remarks on the "sandwich functions", $W_1(x)$, $W_2(x)$ and $W_3(x)$, which do not appear in the autonomous case. The idea behind the need for these functions in the nonautonomous case appears to be due to [Yoshizawa, 1966] and is a result of the time dependence of $V(x,t)$. If we did not bound $V(x,t)$ above and below by positive definite, time-independent functions of x, then the value of $V(x,t)$ could become arbitrarily small or large due to its time dependence alone, regardless of the actual distance of x from $x = 0$. The bound (36) ensures that any decrease in $V(x,t)$ along trajectories can be attributed to the system's stability properties, not just to the time variation of V itself.

Example 6.1. Now, we consider an example. Consider the following one-dimensional nonautonomous vector field:

$$\dot{x} = x \sin t - x^3, \quad x \in \mathbb{R}. \tag{37}$$

It is clear that $x = 0$ is a solution, and we want to examine its stability.

First, we try linearization. The linearization about $x = 0$ is given by

$$\dot{y} = y \sin t, \quad y(t_0) = y_0, \quad y \in \mathbb{R}. \tag{38}$$

This equation can be solved analytically and the solution is given by

$$y(t) = y_0 e^{(-\cos t + \cos t_0)}. \tag{39}$$

Hence, the solution is bounded and oscillatory for all time. Therefore, $x = 0$ is not hyperbolic (it does not have an exponential dichotomy) and we cannot conclude that the stability properties of the linearization carry over for the full nonlinear problem.

First, we construct a candidate for a Lyapunov function. We let

$$V(x,t) = \frac{1}{2}x^2 e^{\alpha(t)},$$

where $\alpha(t)$ is a sufficiently differentiable function that will be chosen to give the desired properties needed to apply Lyapunov's method.

We compute the total derivative of $V(x,t)$ along solutions of the system

$$\begin{aligned} \frac{d}{dt}V(x(t),t) &= \frac{\partial V}{\partial x}\dot{x} + \frac{\partial V}{\partial t} \\ &= x e^{\alpha(t)}\dot{x} + \frac{1}{2}x^2 \dot{\alpha}(t) e^{\alpha(t)}. \end{aligned}$$

Next, we substitute $\dot{x} = x \sin t - x^3$ into the expression

$$\begin{aligned} \dot{V} &= x e^{\alpha(t)}(x \sin t - x^3) + \frac{1}{2}x^2 \dot{\alpha}(t) e^{\alpha(t)} \\ &= \left(x^2 \sin t - x^4 + \frac{1}{2}x^2 \dot{\alpha}(t)\right) e^{\alpha(t)}. \end{aligned}$$

Now, to ensure $\dot{V} \leq 0$, we choose $\dot{\alpha}(t) = -2\sin t$, so that

$$\dot{V} = -x^4 e^{\alpha(t)} \leq 0.$$

Thus, we define

$$\alpha(t) = 2\cos t + C,$$

and we can ignore the constant C (absorbed into a scaling of V).

So, the Lyapunov function becomes

$$V(x,t) = \frac{1}{2}x^2 e^{2\cos t}.$$

This function satisfies:

- $V(x,t) > 0$ for all $x \neq 0$,
- $V(0,t) = 0$,
- $\dot{V}(x,t) = -x^4 e^{2\cos t} < 0$, for all $x \neq 0$.

Therefore, the origin $x = 0$ is asymptotically stable.

We remark that if we had used the Lyapunov function $V(x) = \frac{1}{2}x^2$, we could have proved Lyapunov stability, but not asymptotic stability, with this choice of Lyapunov function. This illustrates the fact that a particular choice of Lyapunov function may not yield optimal results.

6.1. *The LaSalle invariance principle*

There is another result due to [LaSalle, 1960] that is in the spirit of Lyapunov's method, which we will briefly describe.

The original proof was in the setting of autonomous systems, for which we will take the phase space to be $\mathbb{R}^n$. We first find a positive invariant compact set $\mathcal{M}$ and a function $V : \mathbb{R}^n \to \mathbb{R}$ having the property that the derivative of V along trajectories is less than or equal to zero on $\mathcal{M}$. i.e. $\dot{V} \leq 0$ on $\mathcal{M}$. With this setup, we determine the set of points in $\mathcal{M}$ for which $\dot{V} = 0$. We refer to this set as E. Finally, we seek the set of points in E that remain in E forever, i.e. the maximal invariant set in E (this is the "invariance" in the LaSalle invariance principle). We call this set M. Then, the LaSalle invariance principle says that given any point in $\mathcal{M}$, the trajectory through that point approaches M as $t \to \infty$.

We illustrate this with an example; the damped, unforced Duffing equation

$$\begin{aligned} \dot{x} &= y, \\ \dot{y} &= x - x^3 - \delta y, \quad \delta > 0, \quad (x, y) \in \mathbb{R}^2. \end{aligned} \tag{40}$$

This system has three equilibria: a hyperbolic saddle at $(0, 0)$, and two hyperbolic sinks at $(\pm 1, 0)$, all on the x-axis. Let $V(x, y) = \frac{y^2}{2} - \frac{x^2}{2} + \frac{x^4}{4}$. Then, the derivative of V along trajectories is $\dot{V} = -\delta y^2$. The area enclosed by the level set of V, $V(x, y) = c$, for c sufficiently large, defines a positive invariant set, $\mathcal{M}$, that contains the three equilibria. It is clear that $\dot{V} = 0$ on the x-axis, which contains the three equilibria, and it is not hard to show that the only trajectories that start on the x-axis and remain on the x-axis forever are the three equilibria. Hence, by LaSalle's invariance principle, every trajectory will asymptotically approach one of the three equilibria. This is a remarkable result. It is not a result about stability, necessarily (although there are situations where it can be used to infer stability), but a result about the limiting behavior of trajectories.

Now, we turn to the nonautonomous case. There is some confusion in the literature as to whether or not the LaSalle invariance principle applies to nonautonomous systems. It is certainly true that limiting behavior of trajectories in nonautonomous systems can be a bit more "tricky". Nevertheless, there is a rich literature on different generalizations of the LaSalle invariance principle to the nonautonomous setting. The paper of [Barkana, 2014] provides an excellent discussion of the issues, and with a thorough literature survey, and see also [Dafermos, 1971; LaSalle, 1971].

7. Stable and Unstable Manifolds of Hyperbolic Trajectories

Stable and unstable manifolds of hyperbolic trajectories play a foundational role in the geometric theory of dynamical systems. For example, they form the phase space "skeleton" on which the deterministic chaos reminiscent of Smale horseshoes is constructed and understood [Wiggins, 2003]. Stable manifolds of hyperbolic trajectories may act as the boundary between basins of attraction of distinct attractors [Wiggins, 1992]. More generally, stable and unstable manifolds constitute impenetrable barriers in phase space that constrain the regions where trajectories may evolve [Wiggins *et al.*, 2001]. Examples of all of these phenomena are not as common in nonautonomous systems (with the exception of systems that depend on time in a periodic manner). In short, stable and unstable manifolds of hyperbolic trajectories are basic building blocks in phase space that guide the way we "think" about dynamical systems.

We will discuss the nature of stable and unstable manifolds of hyperbolic trajectories in nonautonomous systems through examples. We will begin by considering an autonomous example, and build from the insights obtained in that example.

Consider the two-dimensional autonomous vector field

$$\begin{aligned} \dot{x} &= x, \\ \dot{y} &= -y + x^2, \quad (x, y) \in \mathbb{R}^2, \end{aligned} \tag{41}$$

whose solution through the point (x_0, y_0) at $t = t_0$ is given by

$$\begin{aligned} x(t, t_0, x_0) &= x_0 e^{(t-t_0)}, \\ y(t, t_0, x_0, y_0) &= \left(y_0 - \frac{x_0^2}{3} \right) e^{-(t-t_0)} \\ &\quad + \frac{x_0^2}{3} e^{2(t-t_0)}. \end{aligned} \tag{42}$$

We have purposefully written down the solution so that the exponentially growing and decaying terms are isolated in the expressions for the solution, with coefficients that are a function of the initial conditions. We will see the reason for that shortly, and how it relates to stable and unstable manifolds.

It is easy to see that $(x, y) = (0, 0)$ is an equilibrium point of (41) and the linearized vector field about this equilibrium point is given by

$$\begin{pmatrix} \dot{u}_1 \\ \dot{u}_2 \end{pmatrix} = \begin{pmatrix} 1 & 0 \\ 0 & -1 \end{pmatrix} \begin{pmatrix} u_1 \\ u_2 \end{pmatrix}. \tag{43}$$

It is clear that the origin is a hyperbolic saddle point (with eigenvalues ± 1) and that the x-axis (in the original coordinates) is the invariant unstable subspace and the y-axis is the invariant stable subspace.

Next, we turn our attention to the full nonlinear system. The stable manifold of the origin is a manifold corresponding to initial conditions of trajectories that approach the origin at an exponential rate as $t \to \infty$. The reason for presenting the solution as we did in (42) is that we can explicitly see the exponential growing and decaying parts of the solutions and their dependence on initial conditions. In particular, if we seek solutions that decay to zero as $t \to \infty$, then we must have $x_0 = 0$, i.e. the y-axis is the stable manifold of the origin. This can also be seen directly from the vector field (41). If we seek solutions that decay to zero as $t \to -\infty$, then we must have $y_0 = \frac{x_0^2}{3}$, which is the unstable manifold of the origin. We see that the stable and unstable manifolds of the origin, respectively, are tangent to the stable and unstable subspaces of the linearized vector field at the origin, respectively. Trajectories starting in the stable manifold exhibit the same exponential decay to the origin as $t \to \infty$ under the nonlinear dynamics as trajectories that start in the stable subspace of the origin under the linear dynamics. Trajectories starting in the unstable manifold exhibit the same exponential decay to the origin as $t \to -\infty$ under the nonlinear dynamics as trajectories that start in the unstable subspace of the origin under the linear dynamics.

Now, we apply these ideas to a nonautonomous vector field having a time-dependent hyperbolic trajectory and we compute the stable and unstable manifolds of the hyperbolic trajectory.

We consider the two-dimensional nonautonomous vector field that we considered in Example 5.2

$$\begin{aligned} \dot{x} &= x, \\ \dot{y} &= -y + x^2 + t, \quad (x, y) \in \mathbb{R}^2, \quad t \in \mathbb{R}, \end{aligned} \tag{44}$$

with the solution given by

$$x(t, t_0, x_0) = x_0 e^{(t-t_0)}, \tag{45}$$

$$\begin{aligned} y(t, t_0, x_0, y_0) &= \left(y_0 - \frac{x_0^2}{3} - (t_0 - 1) \right) e^{-(t-t_0)} \\ &\quad + \frac{x_0^2}{3} e^{2(t-t_0)} + t - 1. \end{aligned} \tag{46}$$

We showed in Example 5.2 that $(\hat{x}(t), \hat{y}(t)) = (0, t-1)$ is a hyperbolic trajectory, and we now compute its stable and unstable manifolds.

We know that $x_0 = 0$ is the stable manifold of $(0, t-1)$. It is easy to verify that $x = 0$ is invariant directly from (44), and we have previously shown in Sec. 4 that trajectories starting on the y-axis approach $t - 1$ at an exponential rate.

Next, we turn to the unstable manifold. This will be the set of initial conditions that approach $(0, t-1)$ at an exponential rate as $t \to -\infty$. This holds when the initial conditions lie on the curve

$$y_0 - \frac{x_0^2}{3} - (t_0 - 1) = 0. \tag{47}$$

It is interesting to think about why (47) is invariant. One answer is that it follows from construction. However, another check of invariance is that the vector field is tangent to the manifold. If we define

$$h(x, y, t) = y - \frac{x^2}{3} - (t - 1) = 0, \tag{48}$$

fits into the framework for the discussion of invariant sets in Sec. 3 and it is easy to verify that (48) satisfies the condition of invariance given in (8).

One useful characteristic of examples where the stable and unstable manifolds can be computed analytically is that they provide useful benchmarks for numerical methods. In this regard, it is interesting to contrast (41) and (44). The former is autonomous and has a hyperbolic equilibrium point at the origin which has an unstable manifold fixed in space given by $y = \frac{x^2}{3}$. The latter has a hyperbolic trajectory at $(\hat{x}(t), \hat{y}(t)) = (0, t-1)$ which has an unstable manifold given by $y = \frac{x^2}{3} + t - 1$. Algorithms for computing time-varying hyperbolic trajectories and their time-varying stable and unstable manifolds are more complicated and computationally intensive; see [Mancho *et al.*, 2003].

We conclude this discussion by considering general theorems for the existence of stable and unstable manifolds of trajectories in the nonautonomous setting. Such theorems have been around for some time. If you examine the proof of the stable and unstable manifold theorem in [Coddington & Levinson, 1955] you will see that it is valid for general time dependence (under reasonable regularity assumptions). In the classical ordinary differentiable equations literature, often the term "integral manifold" is used. This generally means that the extended phase space is used to study the manifold. The book of [Hale, 1969] contains general theorems on integral manifolds. A more recent exposition on stable, unstable (and center) manifolds of hyperbolic trajectories in nonautonomous systems is [Kloeden & Rasmussen, 2011]. We also note [Irwin, 1973; De Blasi & Schinas, 1973].

8. Nonautonomous Center Manifold Theory

Invariant center manifolds of nonhyperbolic equilibria for autonomous vector fields have been understood for some time in the mathematics community. Kelley [1967] gives a complete development of the topic, and a comprehensive history of center manifolds can be found in [Knobloch & Aulbach, 1982]. The book of [Carr, 1981] gives a concise, but complete, development of the theory for both finite and infinite dimensions, and many simple examples can be found in [Wiggins, 2023].

Center manifold theory began to become well known in the applied community in the late 1970's and early 1980's. The reasons for this are clear. It provided an algorithmic approach for dimensional reduction in a well-defined setting in a manner that captured the essential dynamics in the reduced-dimensional "center manifold equations". The paper [Holmes & Marsden, 1981] is probably one of the earliest examples of the application of this approach where it was shown that a low-dimensional center manifold could capture the dynamics of a fundamental instability in a structural mechanics system. Since that time, there have been literally thousands of applications of the center manifold methodology in science and engineering.

Essentially, all of this earlier work was carried out in the setting of autonomous vector fields (or autonomous maps in the discrete time setting). However, the setting for nonautonomous center manifold theory is sufficiently well developed for interesting applications, see [Kloeden & Rasmussen, 2011], but it has not seen widespread adoption in the applied community. In this section, we will describe a simple example of the application of nonautonomous center manifold theory. We will do this by first describing the essential elements for describing center manifold theory in the autonomous case, and then showing how they carry over to the nonautonomous setting

$$\begin{aligned}\dot{x} &= Ax + f(x, y),\\ \dot{y} &= By + g(x, y), \quad (x, y) \in \mathbb{R}^c \times \mathbb{R}^s,\end{aligned} \tag{49}$$

where A is a $c \times c$ matrix of real numbers having eigenvalues with zero real part, B is a $s \times s$ matrix of real numbers having eigenvalues with negative real parts, and $f(x, y)$ and $g(x, y)$ are $\mathcal{O}(2)$. (Often x denotes the center coordinates and y denotes the stable coordinates.) In this case, $(x, y) = (0, 0)$ is a nonhyperbolic equilibrium point, and therefore the linearization does not allow us to make conclusions about stability for the nonlinear system.[6]

This is the situation where center manifold theory can be applied through the systematic application of three theorems whose proofs are given in [Carr, 1981].

Existence of a Center Manifold. There exists a center manifold of (49) that has the form of a graph over the center coordinates

$$y = h(x). \tag{50}$$

The center manifold passes through the equilibrium point at the origin

$$h(0) = 0, \tag{51}$$

and it is tangent to the center subspace at the origin

$$Dh(0) = 0. \tag{52}$$

Approximation of the Center Manifold. The equation of invariance of $y = h(x)$ is given by

$$\begin{aligned}&Bh(x) + g(x, h(x))\\ &\quad = Dh(x)(Ax + f(x, h(x))).\end{aligned} \tag{53}$$

This can be solved using a power series representation of $h(x)$ and the error in the solution is the order of the power series approximation to the solution.

Dynamics Restricted to the Center Manifold. The dynamics of (49) restricted to the center manifold is given by

$$\dot{x} = Ax + f(x, h(x)). \tag{54}$$

Lyapunov stability (asymptotic stability) (instability) of $x = 0$ of (54) implies Lyapunov stability (asymptotic stability) (instability) of $(x, y) = (0, 0)$ of (49).

[6] If the system depends on external parameters, say $\mu \in \mathbb{R}^p$, then the parameters can be included directly into (49) as new dependent variables, $\dot{\mu} = 0$. This adds p new center variables to the system and the center manifold becomes a function of the parameters, $y = h(x, \mu)$. This is an important approach for bifurcation theory as the nonhyperbolic center variables play an essential role in bifurcation behavior and the invariant center manifold can capture the change in solution structure due to bifurcation.

8.1. *Time-dependent center manifold calculation*

Next, we turn to center manifold theory for nonautonomous systems and we illustrate the application of the time-dependent center manifold theory with the example from Sec. 5 for which we carried out much of the preliminary work required to set up the problem. The "three theorems" that we noted above for the autonomous case have analogs for the nonautonomous case. These can be found in [Kloeden & Rasmussen, 2011].

In Example 5.3, we considered the following nonautonomous vector field:

$$\begin{aligned} \dot{x} &= xy, \\ \dot{y} &= -y + \sin t. \end{aligned} \tag{55}$$

It is instructive to first consider the autonomous version of this vector field

$$\begin{aligned} \dot{x} &= xy, \\ \dot{y} &= -y. \end{aligned} \tag{56}$$

It is clear that $(0,0)$ is a nonhyperbolic equilibrium point and that $x = 0$ is the stable manifold of the origin and $y = 0$ is the center manifold of the origin. The dynamics restricted to the center manifold is given by $\dot{x} = 0$. Therefore, $(0,0)$ is Lyapunov stable.

Now, we turn our attention to the nonautonomous system (55). Clearly, this system has no equilibria. However, from our earlier discussion in Example 5.3, we know that (55) has a nonhyperbolic time-dependent solution of the following form:

$$\begin{aligned} (\hat{x}(t), \hat{y}(t)) &= (0, y_*(t)), \quad \text{where} \\ y_*(t) &= \frac{1}{2}(\sin t - \cos t). \end{aligned} \tag{57}$$

Therefore, this trajectory has a one-dimensional stable manifold and a one-dimensional center manifold. It is easy to verify that the stable manifold of this trajectory is given by $x = 0$, and we seek to compute an approximation to the center manifold of this trajectory and use it to determine stability of this trajectory for the full nonlinear equation.

We begin by setting up the problem and transforming this (time-dependent) trajectory to the origin by defining the new coordinates

$$x = x, \quad y = y_*(t) + \eta. \tag{58}$$

Then, (55) becomes

$$\begin{aligned} \dot{x} &= x(y_*(t) + \eta) = xy_*(t) + x\eta, \\ \dot{\eta} &= \dot{y} - \dot{y}_*(t) = (-y + \sin t) - \dot{y}_*(t). \end{aligned} \tag{59}$$

Now, note that

$$y_*(t) = \frac{1}{2}(\sin t - \cos t), \quad \dot{y}_*(t) = \frac{1}{2}(\cos t + \sin t),$$

and therefore

$$\sin t - y_*(t) - \dot{y}_*(t) = 0.$$

Thus, the second component of (59) becomes

$$\dot{\eta} = -\eta.$$

Hence, (55) in the coordinates (x, η) is

$$\begin{aligned} \dot{x} &= xy_*(t) + x\eta, \\ \dot{\eta} &= -\eta, \end{aligned} \tag{60}$$

for which we see that (x, η) has a one-dimensional stable manifold given by $x = 0$.

Existence of a Center Manifold. A center manifold exists and has the following form:

$$\begin{aligned} \eta &= h(x,t), \quad \text{with} \\ h(0,t) &= 0, \quad \frac{\partial h}{\partial x}(0,t) = 0. \end{aligned} \tag{61}$$

Approximation of the Center Manifold. Next, we will derive the invariance equation; the equation that (61) must satisfy in order to be invariant under (59) (cf. (53) and (8)). Differentiating $\eta = h(x,t)$ with respect to t and substituting in the relevant terms of (61) give

$$\dot{\eta} = \frac{\partial h}{\partial t} + \frac{\partial h}{\partial x}\dot{x} = \frac{\partial h}{\partial t} + \frac{\partial h}{\partial x}(xy_*(t) + xh(x,t)).$$

On the other hand, from the equation for η

$$\dot{\eta} = -\eta = -h(x,t).$$

So, the invariance equation is

$$\frac{\partial h}{\partial t} + \frac{\partial h}{\partial x}(xy_*(t) + xh(x,t)) = -h(x,t). \tag{62}$$

Assume a power series expansion for the center manifold of the form

$$h(x,t) = a_2(t)x^2 + a_3(t)x^3 + \cdots.$$

Then, we assemble the different pieces for the invariance equation

$$\frac{\partial h}{\partial t} = \dot{a}_2(t)x^2 + \dot{a}_3(t)x^3 + \cdots,$$

$$\frac{\partial h}{\partial x} = 2a_2(t)x + 3a_3(t)x^2 + \cdots,$$

$$\dot{x} = xy_*(t) + xh(x,t) = xy_*(t) + a_2(t)x^3 + \cdots.$$

Now,

$$\frac{\partial h}{\partial x}\dot{x} = (2a_2x + 3a_3x^2 + \cdots)(xy_*(t) + \cdots)$$
$$= 2a_2y_*(t)x^2 + 3a_3y_*(t)x^3 + \cdots .$$

Then, the left-hand side of the invariance equation is

$$[\dot{a}_2 + 2a_2y_*(t)]x^2 + [\dot{a}_3 + 3a_3y_*(t)]x^3 + \cdots ,$$

and the right-hand side is

$$-a_2x^2 - a_3x^3 + \cdots .$$

Matching powers of x

$$\dot{a}_2 + 2a_2y_*(t) = -a_2 \Rightarrow \dot{a}_2 = -a_2(1 + 2y_*(t)),$$
$$\dot{a}_3 + 3a_3y_*(t) = -a_3 \Rightarrow \dot{a}_3 = -a_3(1 + 3y_*(t)).$$

Recall

$$y_*(t) = \frac{1}{2}(\sin t - \cos t) \Rightarrow 1 + 2y_*(t)$$
$$= 1 + \sin t - \cos t.$$

So

$$\dot{a}_2 = -a_2(1 + \sin t - \cos t) \Rightarrow a_2(t)$$
$$= \frac{a_2(0)}{e}e^{-t+\cos t+\sin t}.$$

Therefore, the time-dependent center manifold is

$$\eta = h(x,t) = \frac{a_2(0)}{e}e^{-t+\cos t+\sin t}x^2 + \cdots ,$$

or, returning to the original variable y

$$y = \frac{1}{2}(\sin t - \cos t)$$
$$+ \frac{a_2(0)}{e}e^{-t+\cos t+\sin t}x^2 + \cdots .$$

Analysis of Stability of $x = 0$ on the Center Manifold. The reduced equation on the time-dependent center manifold is

$$\dot{x} = \frac{1}{2}(\sin t - \cos t)\,x + a_2(t)\,x^3 + \cdots , \qquad (63)$$

where $a_2(t) = \frac{a_2(0)}{e}e^{-t+\cos t+\sin t}$ is a smooth bounded function.

Method 1: Linearization

It is generally best to try the "easiest" approach first, which is usually linearization.

The linear part of (63) is

$$\dot{x} = \frac{1}{2}(\sin t - \cos t)\,x.$$

We can easily solve this linear nonautonomous ODE

$$x(t) = x_0 \exp\left(\int_{t_0}^{t}\frac{1}{2}(\sin s - \cos s)\,ds\right).$$

The integral evaluates to

$$\int_{t_0}^{t}\frac{1}{2}(\sin s - \cos s)\,ds$$
$$= \frac{1}{2}(-\cos t + \cos t_0 - \sin t + \sin t_0).$$

Thus,

$$x(t) = x_0 \exp\left(\frac{1}{2}[(-\sin t - \cos t) + (\sin t_0 + \cos t_0)]\right),$$

is bounded in time and therefore the origin is Lyapunov stable, in the linear approximation. Unfortunately, Lyapunov stability can be destroyed by the nonlinear terms, so, we need a more refined analysis.

Method 2 : Construct a Lyapunov function

As we noted earlier, Lyapunov's method may work even in the situation where the solution is nonhyperbolic, if we can construct a suitable Lyapunov function.

We could try the Lyapunov function $V(x) = \frac{1}{2}x^2$. But if you work through the analysis, you cannot do any better than conclude Lyapunov stability with this Lyapunov function. Instead, we will construct a time-dependent Lyapunov function in the spirit of Example 6.1.

We consider a time-dependent Lyapunov function of the form

$$V(x,t) = \frac{1}{2}p(t)x^2,$$

where $p(t) > 0$ is a function to be determined. Then, the time derivative of V along solutions is

$$\frac{dV}{dt} = \frac{1}{2}\dot{p}(t)x^2 + p(t)x\dot{x}$$
$$= \frac{1}{2}\dot{p}(t)x^2 + p(t)x\left(\frac{1}{2}(\sin t - \cos t)x + a_2(t)x^3\right)$$

$$= \left(\frac{1}{2}\dot{p}(t) + \frac{1}{2}p(t)(\sin t - \cos t)\right) x^2$$
$$+ p(t)a_2(t)x^4.$$

To eliminate the indefinite quadratic term, we require $p(t)$ to satisfy the differential equation

$$\dot{p}(t) + p(t)(\sin t - \cos t) = 0.$$

This is a linear first-order ODE, whose solution is

$$p(t) = C \exp\left(-\int (\sin t - \cos t)dt\right) = Ce^{\cos t + \sin t},$$

where $C > 0$ is arbitrary.

Thus, we define the Lyapunov function as follows:

$$V(x,t) = \frac{1}{2}e^{\cos t + \sin t}x^2.$$

The time derivative of V along solutions of the system is given by

$$\frac{dV}{dt} = e^{\cos t + \sin t}a_2(t)\, x^4.$$

From this, we can make conclusions about stability of $x = 0$.

- If $a_2(t) < 0$ for all t, then $\frac{dV}{dt} < 0$ for all $x \neq 0$. Thus, the origin $x = 0$ is asymptotically stable.
- If $a_2(t) \leq 0$, then $\frac{dV}{dt} \leq 0$, and the origin is Lyapunov stable.
- If $a_2(t)$ changes sign or is positive on some intervals, further analysis is required to determine stability.

9. Chaos

The discovery, analysis, and quantification of deterministic chaos have had a significant impact on a variety of diverse applications in science and engineering. Significant landmarks in the development of this topic have been Poincaré's discovery of the "homoclinic tangle" in the circular restricted three body problem [Barrow-Green, 1997], the work of Cartwright and Littlewood on the periodically forced van der Pol equation [Cartwright & Littlewood, 1945; McMurran & Tattersall, 1996; Guckenheimer, 2025] which later motivated Smale to develop his celebrated horseshoe map [Smale, 1967, 1980], and the work of Lorenz on a simplified model of Rayleigh–Bénard convection that spawned a multitude of studies on the resulting Lorenz equations and Lorenz attractor [Lorenz, 1963, 1993]. All of these studies were in the setting of autonomous vector fields, or autonomous maps for the case of the periodically forced van der Pol equation. In this section, we will review some work on chaos in the setting of nonautonomous dynamics. The approach we take will be to review related concepts in the autonomous setting and then consider how they can be generalized to the nonautonomous setting.

We begin with the definition of chaos. Phrases like "chaotic trajectory", "chaotic dynamical system", and "chaotic set" can be found throughout the literature. However, each phrase is a bit imprecise. In general, chaos is a property of the system that magnifies imprecise knowledge of the initial condition of a trajectory, resulting in the deviation of the trajectory from what would occur if one had precise knowledge of the initial condition. Therefore, a trajectory is not chaotic in isolation; chaos emerges from how nearby trajectories diverge and fold within the system.

There are many alternative ways of stating this notion of chaos which make the notion seem "almost evident" to the layman, based on experience. For example,

- Small differences in the present can lead to large differences in the future.
- The future is determined by the present, but uncertainty in the present leads to unpredictability in the future.
- Determinism does not imply predictability.

But we actually need a bit more since this description would apply to a linear autonomous system having a hyperbolic saddle equilibrium point. So, here is where a more precise mathematical definition is necessary. We require the chaos to occur in a compact invariant set of the dynamical system (this rules out the autonomous linear system with a hyperbolic saddle point). We require all trajectories in this invariant set to come arbitrarily close to all points in the invariant set during the course of their evolution. In the topological dynamics sense, we refer to this as *topological transitivity*, if we were working in the metric sense of ergodic theory, we refer to this as *ergodicity* (and we would also discount some sets of zero measure from our arguments). So, the standard definition of a compact, chaotic invariant set for autonomous systems has three elements:

(1) trajectories exhibit sensitive dependence on initial conditions,

(2) the dynamics on the invariant set is topologically transitive, and the following property is usually included
(3) periodic trajectories are dense in the invariant set.

These properties have been distilled from large amounts of numerical experimentation and analysis of benchmark examples. We note that one way to verify that each trajectory exhibits sensitive dependence on initial conditions is to verify that one of its Lyapunov exponents must be positive. This would be a sufficient condition for sensitive dependence on initial conditions. These properties are certainly not "written in stone" and could certainly be modified as new phenomena are discovered. In this regard, property 3 is probably not relevant to nonautonomous systems with aperiodic time dependence in situations where no periodic orbits exist.

The other notion to address is the meaning of the phrase "chaotic dynamical system". Typically one means that the system has one, or more, chaotic invariant sets. "Global chaos" is rare (and hard to prove). Many systems have trajectories that evolve to infinity (e.g. the Hénon map [Devaney & Nitecki, 1979] and the Hénon–Heiles Hamiltonian system [Churchill *et al.*, 2005]). But there are examples of systems with compact phase space where the dynamics is chaotic on the entire phase space, e.g. the Arnold cat map [Arnold & Avez, 1968]. But in general, typical nonlinear dynamical systems exhibit a mixture of chaotic regions and nonchaotic regions in their phase space, this statement is based on much experience and it has been proved rigorously for autonomous Hamiltonian systems by [Markus & Meyer, 1974]. So, with a definition of chaotic invariant set in hand, we now describe a way of understanding this type of dynamics and how it arises in specific autonomous dynamical systems.

Probably, the easiest way to understand the dynamical characteristics of chaos is through the topic of symbolic dynamics. In this setting, the phase space is the space of bi-infinite sequences of two symbols, say 0 and 1. A typical element in this phase space is denoted $\{\cdots s_{-n}\cdots s_{-1}\cdot s_0 s_1\cdots\}$ (the "dot" serves to separate the future from the past), and we denote the space of all such elements by Σ_2. Dynamics on Σ_2 is defined by a map of Σ_2 into itself, the shift map, denoted σ, where $\sigma(\{\cdots s_{-n}\cdots s_{-1}\cdot s_0 s_1\cdots\}) = \{\cdots s_{-n}\cdots s_{-1}s_0\cdot s_1\cdots\}$, i.e. σ moves the point one step forward in time, indicated by the "dot" moving one place to the right. The orbit structure can be easily deduced from the structure of the symbol sequences. Periodic sequences can be constructed by concatenating all sequences of length one, two, three, and so one. These sequences are contained in periodic orbits, hence, there are a countable infinity of periodic orbits. Thinking of the sequences of 0's and 1's as being binary expansions of numbers in the unit interval, we can deduce that there are an uncountable number of nonperiodic orbits. Σ_2 can be equipped with a metric, with which compactness can be deduced, as well as the fact that σ has sensitive dependence on initial conditions. An orbit that is dense in Σ_2 can also be constructed, and this implies topological transitivity. Therefore, the shift map acting on the space of sequences of two symbols is the model for what is accepted as the definition of deterministic chaos. (Proofs of all of these statements can be found in [Wiggins, 2003], and a history of symbolic dynamics can be found in [Coven & Nitecki, 2006].) However, this all seems rather abstract and far removed from the types of differential equations that arise in applications, but that is far from the case. Motivated by very practical work of [Cartwright & Littlewood, 1945] on the periodically forced van der Pol oscillator, Smale developed a simplified model to capture the essential dynamical features observed by Cartwright and Littlewood. The result was the two-dimensional "Smale horseshoe map" which contains a compact invariant set whose dynamics is topologically conjugate to the shift map acting on a space of symbol sequences [Smale, 1967, 1980]. In this way, all of the complexity described by symbolic dynamics goes over to the Smale horseshoe map. The resulting invariant set is often referred to as a "chaotic invariant set" or "chaotic saddle". But this raises the question of how do you recognize that, for a given differential equation, a horseshoe map construction can be carried out in a region of its phase space? The key here is the notion of homoclinic orbits or, in other words, the intersection of the stable and unstable manifolds of a hyperbolic invariant set. When the hyperbolic invariant set is a hyperbolic periodic orbit and its stable and unstable manifolds intersect transversely, then a Smale horseshoe map can be constructed in a neighborhood of the intersection point. Moser gave a proof of this [Moser, 1973], but he also gave general conditions for constructing a chaotic invariant set for two-dimensional maps, which are often referred to as the Conley–Moser conditions [Wiggins, 2003]. These conditions are very powerful (when they can be applied) and can be used to analyze "Smale horseshoe like chaos" over large regions of phase space and for large sets of parameter values. Examples where this has been carried are described in [Devaney & Nitecki, 1979; Holmes, 1982]. There is also a class of perturbation methods, referred to as Melnikov methods, that can be used to detect the type of homoclinic intersections that give rise to chaotic dynamics, see [Wiggins, 1988, 2003].

This story is for autonomous systems. It is natural to ask how much of it can be carried over for nonautonomous systems. Nonautonomous symbolic dynamics have been developed in [Stoffer, 1988a, 1988b; Wiggins, 1999]. Generalizations of the Conley–Moser conditions to the nonautonomous setting have been given in [Wiggins, 1999] and they have been applied to nonautonomous versions of the Hénon map [Balibrea-Iniesta *et al.*, 2015] and the Lozi map [Lopesino *et al.*, 2015]. This work demonstrated how the chaotic invariant sets constructed in this way varied in time. We note that [Lerman & Shilńikov, 1992] showed the existence of Smale horseshoe like dynamics associated with homoclinic orbits in nonautonomous ODEs.

Earlier, we mentioned a class of perturbation methods for detecting homoclinic orbits and characterizing some aspects of their geometry — the "Melnikov methods". Originally, they were developed for time-periodic perturbations [Melnikov, 1963], but were later extended to quasi-periodically time-dependent perturbations [Wiggins, 1988; Meyer & Sell, 1989; Beigie *et al.*, 1991; Beigie & Wiggins, 1992] as well as to time aperiodic perturbations [Malhotra & Wiggins, 1998; Balasuriya *et al.*, 2003], and to systems perturbed by noise [Frey & Simiu, 1993; Simiu, 2002]. This work established the existence of homoclinic orbits and it was then shown that these orbits could be used to conclude Smale horseshoe like dynamics for these particular types of time dependence.

While we have been mostly avoiding stochastic time dependence, we note that there has been very recent work on stochastically time-dependent forcing of the Duffing equation [Lu & Wang, 2010, 2011]. This involved a stochastic version of Melnikov's method and an associated construction of a "stochastic Smale horseshoe". Another approach to these problems using the skew product flow formulation of nonautonomous dynamics can be found in [Feng & Li, 2021].

10. Bifurcation Theory

In many areas of science and engineering where dynamical systems theory plays an important role in the modeling process, bifurcation theory is an important component. One might classify bifurcations as "local" (e.g. bifurcation of equilibria or periodic orbits) or "global" (e.g. bifurcations of homoclinic orbits, heteroclinic orbits, or general normally hyperbolic invariant manifolds). "Bifurcation" of any of these dynamical objects generally means a "loss of hyperbolicity". This occurs when the dynamical system depends on parameter(s) and a parameter value is reached where the object becomes nonhyperbolic.

In typical dynamical systems courses concerned with autonomous systems, bifurcation theory is usually developed in a hierarchical manner–bifurcations of equilibria are studied first, then periodic orbits (typically in the context of maps) and then more complicated settings, such as homoclinic and heteroclinic bifurcations.

Studying all of these settings in the context of both autonomous and nonautonomous dynamics is beyond the scope of this paper. Rather, we will start with a standard bifurcation of equilibria studied in the autonomous setting and then ask how it is manifested in the nonautonomous setting. This will serve to reveal a number of the issues that arise when considering nonautonomous bifurcation theory.

We begin by considering the following one-dimensional, one-parameter autonomous vector field:

$$\dot{x} = \mu x - x^3, \quad x(t_0) = x_0, \quad x \in \mathbb{R}, \ \mu \in \mathbb{R}. \tag{64}$$

This is the normal form for the well-known pitchfork bifurcation of equilibria in autonomous systems. It is trivial to understand. For $\mu < 0$, the system has one asymptotically stable equilibrium, at $\mu = 0$, the system has one stable nonhyperbolic equilibrium, and for $\mu > 0$, the system has three equilibria: $x = 0$ is unstable and $x = \pm\sqrt{\mu}$ are both asymptotically stable. Thus, $\mu = 0$ is the critical parameter value corresponding to the change of stability of $x = 0$ and the creation of two new stable equilibria. There are two features that define the pitchfork bifurcation: nonhyperbolicity at $\mu = 0$ (a necessary but not sufficient condition for the bifurcation) and the form of nonlinearity (e.g. if the nonlinear term were x^2 rather than x^3, we would have a transcritical bifurcation).

A nice feature of studying bifurcations of one-dimensional autonomous vector fields is that zeros of the right-hand side of the equation (i.e. equilibria) are actually solutions of the equations. These tend to organize the behavior of all of the solutions and there is no reason to actually solve for time-dependent solutions of the equation. This is not true in the nonautonomous setting, as we will see shortly. However, it is possible to solve (64) analytically, either by integrating using partial fractions or by recognizing that it is a Bernoulli differential equation. Choosing the latter approach, the solution is given by

$$x(t) = \frac{\pm 1}{\sqrt{\frac{1}{\mu} + \left(x_0^{-2} - \frac{1}{\mu}\right) e^{-2\mu(t-t_0)}}}. \tag{65}$$

From (64), it is easy to see that for $\mu < 0$ as $t \to \infty$, $x(t) \to 0$ and for $\mu > 0$ as $t \to \infty$, $x(t) \to \pm\sqrt{\mu}$.

Now, we turn our attention to understanding the nonautonomous manifestation of the pitchfork bifurcation. The understanding that we have described for the autonomous pitchfork bifurcation was derived from the simple form of (64). Consequently, it is natural to assume a similar normal form for the nonautonomous case. This approach was carried out in [Langa *et al.*, 2002], which we now describe.

Example 10.1. Their model for a system undergoing a nonautonomous pitchfork bifurcation has the form

$$\begin{aligned} \dot{x} &= \mu x - b(t)x^3, \\ x(t_0) &= x_0, \quad x \in \mathbb{R}, \ \mu \in \mathbb{R}, \end{aligned} \tag{66}$$

where $b(t)$ is a function satisfying $0 < b(t) \leq B$. It is clear that $x = 0$ is a solution that is asymptotically stable for $\mu < 0$, unstable for $\mu > 0$, and nonhyperbolic for $\mu = 0$. However, $x = \pm\sqrt{\frac{\mu}{b(t)}}$ are not solutions of (66), and therefore, the only way to understand possible bifurcation behavior for $\mu > 0$ is to analyze the time-dependent solutions of (66) as a function of μ. Fortunately, this equation is also a Bernoulli differential equation, as noted by [Langa *et al.*, 2002], and can be solved exactly. The solution is given explicitly by

$$x(t) = \pm\left[e^{-2\mu t}\left(\frac{e^{2\mu t_0}}{x_0^2} + 2\int_{t_0}^{t} b(s)e^{2\mu s}\,ds\right)\right]^{-1/2}.$$

We analyze the pullback behavior for $\mu < 0$, $\mu = 0$, and $\mu > 0$ by fixing t and taking $t_0 \to -\infty$ for each case. We denote pullback attractors by $\mathcal{A}(t)$.

Case $\boldsymbol{\mu < 0}$**.** As $t_0 \to -\infty$

$$\frac{e^{2\mu t_0}}{x_0^2} \to \infty, \quad \text{so } x(t)^2 \to 0.$$

Thus, the pullback attractor is

$$\mathcal{A}(t) = \{0\}.$$

Case $\boldsymbol{\mu = 0}$**.** In this case, the solution becomes

$$x(t) = \pm\left[\frac{1}{x_0^2} + 2\int_{t_0}^{t} b(s)\,ds\right]^{-1/2}.$$

Since $b(s) > 0$, the integral diverges as $t_0 \to -\infty$, and therefore

$$x(t) \to 0.$$

Therefore, the pullback attractor is again

$$\mathcal{A}(t) = \{0\}.$$

Case $\boldsymbol{\mu > 0}$**.** As $t_0 \to -\infty$

$$\frac{e^{2\mu t_0}}{x_0^2} \to 0,$$

so, the solution approaches

$$x(t) \to \pm\left[e^{-2\mu t}\, 2\int_{-\infty}^{t} b(s)e^{2\mu s}\,ds\right]^{-1/2}.$$

We define

$$\rho(t) := \left[e^{-2\mu t}\, 2\int_{-\infty}^{t} b(s)e^{2\mu s}\,ds\right]^{-1/2}.$$

Then, the pullback attractors are

$$\mathcal{A}(t) = \{-\rho(t), +\rho(t)\}.$$

We summarize these results in the following table:

Case	Pullback Attractor $\mathcal{A}(t)$	Description
$\mu < 0$	$\{0\}$	Global attractor at the origin
$\mu = 0$	$\{0\}$	Global attractor at the origin
$\mu > 0$	$\{-\rho(t), +\rho(t)\}$	Bistable time-dependent attractors

Example 10.2. In Example 10.1, the time dependence was in the nonlinearity. Now, we consider a similar example, but where the time dependence is in the linear part.

Consider the nonautonomous differential equation

$$\dot{x} = (\mu + \sin t)x - x^3, \qquad x(t_0) = x_0, \quad x \in \mathbb{R}, \ \mu \in \mathbb{R}. \tag{67}$$

This is also a Bernoulli differential equation and can therefore be solved analytically. The solution is given by

$$x(t) = \pm\left[e^{-2\mu t+2\cos t}\left(\frac{e^{2\mu t_0 - 2\cos t_0}}{x_0^2} + \int_{t_0}^{t} 2e^{2\mu s-2\cos s}\,ds\right)\right]^{-1/2}.$$

As in Example 10.2, we analyze the pullback behavior for $\mu < 0$, $\mu = 0$, and $\mu > 0$ by fixing t and taking $t_0 \to -\infty$ for each case, and we denote pullback attractors by $\mathcal{A}(t)$.

***Case* $\mu < 0$.** Note that

$$e^{2\mu t_0} \to \infty \quad \text{as } t_0 \to -\infty,$$

from which it follows that $x(t) \to 0$. Hence, the pullback attractor is given by

$$\mathcal{A}(t) = \{0\}.$$

***Case* $\mu = 0$.** In this case, the solution simplifies to

$$x(t) = \pm\left[e^{2\cos t}\left(\frac{e^{-2\cos t_0}}{x_0^2} + \int_{t_0}^{t} 2e^{-2\cos s}\,ds\right)\right]^{-1/2},$$

and as $t_0 \to -\infty$, the integral diverges. Thus, $x(t) \to 0$, and the pullback attractor is given by

$$\mathcal{A}(t) = \{0\}.$$

***Case* $\mu > 0$.** The following integral converges:

$$\int_{-\infty}^{t} 2e^{2\mu s-2\cos s}\,ds < \infty.$$

We define

$$\rho(t) := \left[e^{-2\mu t+2\cos t}\int_{-\infty}^{t} 2e^{2\mu s-2\cos s}\,ds\right]^{-1/2}.$$

Then,

$$x(t) \to \pm\rho(t) \quad \text{as } t_0 \to -\infty,$$

and therefore the pullback attractors are given by

$$\mathcal{A}(t) = \{-\rho(t), +\rho(t)\}.$$

We summarize the results in the following table:

μ	$\mathcal{A}(t)$	Description
$\mu < 0$	$\{0\}$	Global attractor at the origin
$\mu = 0$	$\{0\}$	Global attractor at the origin
$\mu > 0$	$\{-\rho(t), +\rho(t)\}$	Bistable time-dependent attractors

Both Examples 10.1 and 10.2 exhibit similar behavior (even though the nature of the loss of hyperbolicity is different in the two examples). The loss of stability is accompanied by the creation of two time-dependent pullback attractors.

The nonautonomous pitchfork bifurcation can be used to study the nonautonomous Hopf bifurcation. This is because the truncated two-dimensional normal form for the Hopf bifurcation can be written in polar coordinates, $x = r\cos\theta$, $y = r\sin\theta$ and the r component of this resulting vector field is independent of θ and has the form of the pitchfork bifurcation [Wiggins, 2003].

In an introductory dynamical systems course where bifurcation of equilibria in autonomous systems is considered, besides the pitchfork and Hopf bifurcations, the transcritical and saddle-node bifurcations are also considered. The transcritical bifurcation normal form, $\dot{x} = \mu x - x^2$, $x \in \mathbb{R}$, $\mu \in \mathbb{R}$ is also a Bernoulli differential equation and nonautonomous versions of that equation can be studied in the same way as for the pitchfork bifurcation. The saddle-node bifurcation normal form, $\dot{x} = \mu - x^2$, $x \in \mathbb{R}$, $\mu \in \mathbb{R}$ is not a Bernoulli differential equation (although it is a Riccati equation). Nevertheless, the nonautonomous saddle-node bifurcation is discussed in [Langa *et al.*, 2002; Kloeden & Rasmussen, 2011].

Finally, we remark that there is a significant literature on a type of time-dependent bifurcation associated with the slow variation of a parameter through a bifurcation value. Much of that work arose through the consideration of specific applications and a reconsideration of these results through the more recent nonautonomous dynamical systems framework might prove fruitful. See Haberman, 1979; Schecter, 1985; Mitropolśkii, 1965;

Leibovitz & Schaar, 1975, 1976; Neishtadt, 1987, 1988; Baer *et al.*, 1989; Erneux & Mandel, 1986; Leibovitz & Pesci, 1995; Raman & Bajaj, 1998; Arnold *et al.*, 1994; Baesens, 1991, 1995.

11. Hamiltonian Dynamics

In this section, we make some remarks on Hamiltonian systems. In particular, we would like to understand how much of the previous results on nonautonomous systems apply to nonautonomous Hamiltonian systems. We begin by defining the setting in which we pose this question.

We start with a scalar-valued function of $2n$ variables, $H(q,p)$, for the autonomous case, and $H(q,p,t)$, for the nonautonomous case. In both cases, we refer to H as the Hamiltonian function (or just "Hamiltonian") and from it we form the following Hamiltonian vector field:

$$\begin{aligned} \dot{q} &= \frac{\partial H}{\partial p}, \\ \dot{p} &= -\frac{\partial H}{\partial q}, \quad (q,p) \in \mathbb{R}^n \times \mathbb{R}^n. \end{aligned} \tag{68}$$

Both the autonomous and nonautonomous Hamiltonian vector fields are divergence-free (i.e. the divergence is zero at every point (q,p)). This means that phase space volumes are preserved under time evolution. This result is known as Liouville's theorem, and the proof for both autonomous and nonautonomous Hamiltonian systems can be found in [Kardar, 2007]. For autonomous systems, $H(q,p)$ is constant along trajectories ("energy conservations"). This is not true for nonautonomous systems.

Evidently, the property of volume preservation has implications for the stability of trajectories and the dimensions of stable and unstable manifolds of trajectories. It is tempting to state that the dimensions of the stable and unstable manifolds of (general) hyperbolic trajectories must be equal, for both autonomous and nonautonomous Hamiltonian systems. While there are specific examples of this statement, e.g. when the hyperbolic trajectory is an equilibrium point for autonomous systems, a periodic orbit, or quasi-periodic orbit that can be reduced to a specific "normal form", to our knowledge, these issues have not been studied in detail in the general nonautonomous systems.

The Hamiltonian structure and the constraints associated with phase space volume preservation have a significant influence on Hamiltonian bifurcation theory. In particular, for canonical Hamiltonian systems (even-dimensional, smallest dimension is two), the dimension of the normal form for the pitchfork, saddle-node, and transcritical bifurcations is two, and there is no analog of the two-dimensional Hopf bifurcation (we do not have attracting or repelling periodic orbits in Hamiltonian systems). There is a result known popularly as the Hamiltonian Hopf bifurcation, but the minimum dimension for that to occur is four and, curiously, E. Hopf had nothing to do with the result. It was proven by [Meyer & Schmidt, 1971]. An elementary exposition of these simple Hamiltonian bifurcations can be found in [Wiggins, 2023]. But the point we wish to make here is that there has been very little development of bifurcation theory for nonautonomous Hamiltonian systems (other than the time-periodic case, which can be reduced to the study of bifurcation of symplectic maps), and this seems to be a topic ripe for future development.

Hamiltonian systems, both autonomous and nonautonomous, have a rich geometric structure which is often pursued in its own right. For example, autonomous vector fields generate a flow. One of the basic results of autonomous Hamiltonian dynamics [Arnold, 1989] is that for fixed t, the resulting time t flow map is a symplectic map of the phase space to itself. An obvious question is whether or not such a result holds for maps of phase space from the process generated by a nonautonomous Hamiltonian vector field? Recent results [Ateşli *et al.*, 2025] show that the answer to this question is "no" and that a geometric structure different from the canonical symplectic structure arises in the study of the geometry of nonautonomous Hamiltonian systems.

Finally, we make some comments on the famous Kolmogorov–Arnold–Moser (KAM) theorem and the Nekhoroshev theorem. A very thorough discussion of these results can be found in [Dumas, 2014]. First, KAM theorems and Nekhoroshev theorems are perturbation theorems. The unperturbed system is assumed to be completely integrable, and based on some further assumptions related to the integrability, it can be represented in a special set of coordinates which are referred to as action–angle coordinates. The mathematics behind these statements are the content of the Liouville–Arnold theorem [Arnold, 1989] (but [Mineur, 1935, 1937] also has a claim to this result, and see also [Nekhoroshev, 1972]). In the canonical case ($2n$-dimensional phase space), there are n actions and n angles, and the actions are constant in time and the angles evolve linearly in time, giving rise to quasi-periodic motion. The precise nature of the quasi-periodic motion depends on the existence of resonance relations between the n angular motions. The KAM theorem is concerned with the persistence of nonresonant quasi-periodic motions when the perturbation is applied. The Nekhoroshev theorem is related to the time evolution of the action variables when the perturbation is applied.

In terms of nonautonomous near integrable Hamiltonian systems, the unperturbed part is generally considered to be time-independent and obeys the same integrability criteria as the autonomous case, i.e. the transformation to action–angle variables is possible. Explicit time dependence is introduced into the perturbation. There have been results for periodic and quasi-periodic time dependence for some time, but it is only recently that more general aperiodic time dependence has been considered. For example, see [Fortunati & Wiggins, 2014a, 2014b, 2015; Canadell & de la Llave, 2015; Fortunati & Wiggins, 2016a, 2016b, 2016c; Scarcella, 2024a, 2024b, 2024c, 2025].

12. Summary and Outlook

This paper has surveyed aspects of the theory of nonautonomous dynamical systems by systematically revisiting foundational topics from the autonomous setting — such as flows, invariant sets, stability, Lyapunov functions, center manifolds, exponential dichotomies, bifurcation theory, chaotic behavior, and Hamiltonian dynamics — and considering their manifestation through examples in the context of time-dependent dynamics. A distinctive feature of our exposition is the use of concrete, analytically tractable examples that are intended to illustrate essential differences (and similarities) between autonomous and nonautonomous systems, and also to expose geometric and analytic structures unique to the nonautonomous setting. We emphasize the role of processes as the natural generalization of flows, the relevance of pullback attraction, and the refinement of classical stability notions via tools like time-dependent Lyapunov functions and exponential dichotomies. Despite the considerable theoretical development of nonautonomous dynamics over recent decades, the field still lags behind its autonomous counterpart in terms of widely recognized applications and geometric constructs. We highlight promising directions for future research through developing nonautonomous analogs of well-known dynamical objects, such as invariant manifolds, attractors, and bifurcations. This area offers a rich arena of research and tools for applied mathematicians, engineers, and physicists. In many real-world systems — ranging from fluid flows to mechanical systems, biological networks, and control applications — the underlying dynamics may be inherently nonautonomous. Modeling such systems accurately requires the approach and tools presented here. Moreover, the rise of data-driven methods (e.g. Koopman operator theory and dynamic mode decomposition [Mauroy *et al.*, 2020; Budišić *et al.*, 2012]) creates further opportunities to connect theory and practice in new ways, especially when reformulated in terms of nonautonomous processes. We anticipate that continued integration of theoretical developments with computational and application-driven efforts will further establish nonautonomous dynamics as a central and indispensable component of the broader dynamical systems framework.

Acknowledgments

I am grateful for the support of the William R. Davis '68 Chair in the Department of Mathematics at the United States Naval Academy. The views expressed in this paper are those of the author and do not reflect the official policy or position of the U.S. Naval Academy, Department of the Navy, the Department of Defense, or the U.S. Government.

ORCID

Stephen Wiggins
https://orcid.org/0000-0002-0780-0911

References

Abraham, R. & Ueda, Y. [2000] *The Chaos Avant-garde: Memories of the Early Days of Chaos Theory*, Vol. 39 (World Scientific, Singapore).

Arnold, V. I. & Avez, A. [1968] *Ergodic Problems of Classical Mechanics* (Benjamin, NY), Translated from French; original title: "Problèmes Ergodiques de la Mécanique Classique".

Arnold, V. I. [1989] *Mathematical Methods of Classical Mechanics*, Graduate Texts in Mathematics, Vol. 60, 2nd edition (Springer, NY).

Arnold, V. I., Afrajmovich, V. S., Il'yashenko, Y. S. & Shil'nikov, L. P. [1994] *Dynamical Systems V: Bifurcation Theory and Catastrophe Theory*, Encyclopaedia of Mathematical Sciences, Vol. 5 (Springer, Berlin).

Arnold, L. [1998] *Random Dynamical Systems*, Springer Monographs in Mathematics (Springer, Berlin).

Ateşli, B., Esen, O., de León, M. & Sardón, C. [2025] "On non-autonomous Hamiltonian dynamics, dual spaces, and kinetic lifts," *Physica D* **472**, 134504.

Baer, S. M., Erneux, T. & Rinzel, J. [1989] "The slow passage through a Hopf bifurcation: Delay, memory effects, and resonance," *SIAM J. Appl. Math.* **49**, 55–71.

Baesens, C. [1991] "Slowly changing phase space structure in dynamical systems," *Physica D* **53**, 319–375.

Baesens, C. [1995] "Delayed bifurcations: Towards a general theory," *Physica D* **86**, 127–164.

Balasuriya, S., Mezić, I. & Jones, C. K. [2003] "Weak finite-time Melnikov theory and 3D viscous perturbations of Euler flows," *Physica D* **176**, 82–106.

Balibrea, F., Caraballo, T., Kloeden, P. E. & Valero, J. [2010] "Recent developments in dynamical systems: Three perspectives," *Int. J. Bifurcation and Chaos* **20**, 2591–2636.

Balibrea-Iniesta, F., Lopesino, C., Wiggins, S. & Mancho, A. M. [2015] "Chaotic dynamics in nonautonomous maps: Application to the nonautonomous Hénon map," *Int. J. Bifurcation and Chaos* **25**, 1550172.

Barkana, I. [2014] "Defending the beauty of the invariance principle," *Int. J. Control* **87**, 186–206.

Barreiro, A., Aracil, J. & Pagano, D. [2002] "Detection of attraction domains of non-linear systems using bifurcation analysis and Lyapunov functions," *Int. J. Control* **75**, 314–327.

Barrow-Green, J. [1997] *Poincaré and the Three-Body Problem*, History of Mathematics, Vol. 11 (American Mathematical Society/London Mathematical Society, Providence, RI/London).

Beigie, D., Leonard, A. & Wiggins, S. [1991] "Chaotic transport in the homoclinic and heteroclinic tangle regions of quasiperiodically forced two-dimensional dynamical systems," *Nonlinearity* **4**, 775.

Beigie, D. & Wiggins, S. [1992] "Dynamics associated with a quasiperiodically forced morse oscillator: Application to molecular dissociation," *Phys. Rev. A* **45**, 4803.

Budišić, M., Mohr, R. & Mezić, I. [2012] "Applied Koopmanism," *Chaos* **22**, 047510.

Canadell, M. & de la Llave, R. [2015] "KAM tori and whiskered invariant tori for non-autonomous systems," *Physica D* **310**, 104–113.

Carr, J. [1981] *Applications of Centre Manifold Theory*, Applied Mathematical Sciences, Vol. 35 (Springer, NY).

Cartwright, M. L. & Littlewood, J. E. [1945] "On non-linear differential equations of the second order: I. The equation $\ddot{y} - k(1-y^2)\dot{y} + y = b\lambda k\cos(\lambda t + a)$, k large," *J. London Math. Soc.* **20**, 180–189.

Cheban, D. N., Kloeden, P. E. & Schmalfuß, B. [2002] "The relationship between pullback, forwards and global attractors of nonautonomous dynamical systems," *Nonlin. Dyn. Syst. Theory* **2**, 125–144.

Chepyzhov, V. V. & Vishik, M. I. [2002] *Attractors for Equations of Mathematical Physics*, American Mathematical Society Colloquium Publications, Vol. 49 (American Mathematical Society).

Chicone, C. [2006] *Ordinary Differential Equations with Applications*, Texts in Applied Mathematics, Vol. 34 (Springer-Verlag, NY).

Churchill, R., Pecelli, G. & Rod, D. [2005] "A survey of the Hénon-Heiles Hamiltonian with applications to related examples," in *Stochastic Behavior in Classical and Quantum Hamiltonian Systems: Volta Memorial Conference, Como, 1977* (Springer), pp. 76–136.

Coddington, E. A. & Levinson, N. [1955] *Theory of Ordinary Differential Equations*, Vol. 158 (McGraw-Hill, NY).

Coppel, W. A. [1978] *Dichotomies in Stability Theory*, Lecture Notes in Mathematics, Vol. 629 (Springer, Berlin).

Coven, E. M. & Nitecki, Z. H. [2006] "On the genesis of symbolic dynamics as we know it," arXiv preprint arXiv:math/0611322.

Crauel, H. & Flandoli, F. [1994] "Attractors for random dynamical systems," *Probab. Theory Relat. Fields* **100**, 365–393.

Dafermos, C. M. [1971] "An invariance principle for compact processes," *J. Differ. Equ.* **9**, 239–252.

De Blasi, F. & Schinas, J. [1973] "On the stable manifold theorem for discrete time dependent processes in Banach spaces," *Bull. London Math. Soc.* **5**, 275–282.

Devaney, R. & Nitecki, Z. [1979] "Shift automorphisms in the Hénon mapping," *Commun. Math. Phys.* **67**, 137–146.

Devaney, R. L. [2018] *Discrete Dynamical Systems: A Pathway for Students to Become Enchanted with Mathematics* (Springer International Publishing, Cham), pp. 137–144.

Diacu, F. & Holmes, P. [1999] *Celestial Encounters: The Origins of Chaos and Stability*, Vol. 22 (Princeton University Press).

Dumas, H. S. [2014] *The KAM Story: A Friendly Introduction to the Content, History, and Significance of Classical Kolmogorov–Arnold–Moser Theory* (World Scientific, Singapore).

Erneux, T. & Mandel, P. [1986] "Slow passage through a steady bifurcation: Delay and memory effects," *Phys. Rev. A* **34**, 4110–4112.

Feng, J. & Li, Y. [2021] "The weak Smale horseshoe and mean hyperbolicity," *J. Differ. Equ.* **299**, 154–195.

Fortunati, A. & Wiggins, S. [2014a] "Normal form and Nekhoroshev stability for nearly integrable Hamiltonian systems with unconditionally slow aperiodic time dependence," *Regul. Chaotic Dyn.* **19**, 363–373.

Fortunati, A. & Wiggins, S. [2014b] "Persistence of Diophantine flows for quadratic nearly integrable Hamiltonians under slowly decaying aperiodic time dependence," *Regul. Chaotic Dyn.* **19**, 586–600.

Fortunati, A. & Wiggins, S. [2015] "A Kolmogorov theorem for nearly integrable Poisson systems with asymptotically decaying time-dependent perturbation," *Regul. Chaotic Dyn.* **20**, 476–485.

Fortunati, A. & Wiggins, S. [2016a] "Integrability and strong normal forms for non-autonomous systems in a neighbourhood of an equilibrium," *J. Math. Phys.* **57**, 092703.

Fortunati, A. & Wiggins, S. [2016b] "The Kolmogorov-Arnold-Moser (KAM) and Nekhoroshev theorems with arbitrary time dependence," in *Essays in Mathematics and Its Applications: In Honor of Vladimir Arnold* (Springer), pp. 89–99.

Fortunati, A. & Wiggins, S. [2016c] "Negligibility of small divisor effects in the normal form theory for nearly-integrable Hamiltonians with decaying non-autonomous perturbations," *Celest. Mech. Dyn. Astron.* **125**, 247–262.

Frey, M. & Simiu, E. [1993] "Noise-induced chaos and phase space flux," *Physica D* **63**, 321–340.

Giesl, P. & Hafstein, S. [2015] "Review on computational methods for Lyapunov functions," *Discrete Contin. Dyn. Syst. B* **20**, 2291–2331.

Gleick, J. [1987] *Chaos: Making a New Science* (Viking, NY).

Guckenheimer, J. & Holmes, P. [1983] *Nonlinear Oscillations, Dynamical Systems, and Bifurcations of Vector Fields*, Applied Mathematical Sciences, Vol. 42 (Springer-Verlag, NY).

Guckenheimer, J. [2025] "The legacy of the Cartwright-Littlewood collaboration," arXiv preprint arXiv:2506.06889.

Haberman, R. [1979] "Slowly varying jump and transition phenomena associated with algebraic bifurcation problems," *SIAM J. Appl. Math.* **37**, 69–106.

Hale, J. K. [1969] *Ordinary Differential Equations*, Pure and Applied Mathematics, Vol. 21 (John Wiley & Sons, NY).

Hale, J. K. [1988] *Asymptotic Behavior of Dissipative Systems*, Mathematical Surveys and Monographs, Vol. 25 (American Mathematical Society, Providence, RI).

Hartman, P. [2002] *Ordinary Differential Equations*, Classics in Applied Mathematics, Vol. 38, 2nd edition (Society for Industrial and Applied Mathematics, Philadelphia).

Hénon, M. & Heiles, C. [1964] "The applicability of the third integral of motion: Some numerical experiments," *Astron. J.* **69**, 73–79.

Hénon, M. [1976] "A two-dimensional mapping with a strange attractor," *Commun. Math. Phys.* **50**, 69–77.

Hirsch, M. W. & Smale, S. [1974] *Differential Equations, Dynamical Systems, and Linear Algebra*, Pure and Applied Mathematics, Vol. 60 (Academic Press, NY).

Holmes, P. & Marsden, J. [1981] "A partial differential equation with infinitely many periodic orbits: Chaotic oscillations of a forced beam," *Arch. Ration. Mech. Anal.* **76**, 135–165.

Holmes, P. J. [1982] "The dynamics of repeated impacts with a vibrating table," *J. Sound Vib.* **84**, 173–189.

Holmes, P. [2005] "Ninety plus thirty years of nonlinear dynamics: Less is more and more is different," *Int. J. Bifurcation and Chaos* **15**, 2703–2716.

Holmes, P. [2007] "A short history of dynamical systems theory: 1885–2007," in *Encyclopedia of Life Support Systems (EOLSS), History of Mathematics* (UNESCO), https://www.eolss.net/sample-chapters/c02/E6-132-21.pdf.

Irwin, M. [1973] "Hyperbolic time-dependent processes," *Bull. London Math. Soc.* **5**, 209–217.

Kardar, M. [2007] *Statistical Physics of Particles* (Cambridge University Press, Cambridge).

Kelley, A. [1967] "The stable, center-stable, center, center-unstable, unstable manifolds," *J. Differ. Equ.* **3**, 546–570.

Khalil, H. K. [2009] "Lyapunov stability," *Control Syst. Robot. Autom.* **12**, 115.

Kloeden, P. E. & Rasmussen, M. [2011] *Nonautonomous Dynamical Systems*, Mathematical Surveys and Monographs, Vol. 176 (American Mathematical Society, Providence, RI).

Kloeden, P. E. & Yang, M. [2021] "Nonautonomous attractors," in *Physics of Biological Oscillators: New Insights into Non-Equilibrium and Non-Autonomous Systems* (Springer), pp. 55–70.

Knobloch, H. W. & Aulbach, B. [1982] "The role of center manifolds in ordinary differential equations," *Proc. Fifth Czechoslovak Conf. Differential Equations and their Applications (Equadiff 5)* (BSB B.G. Teubner Verlagsgesellschaft), pp. 179–189.

Kominis, Y. & Bountis, T. [2010] "Analytical solutions of systems with piecewise linear dynamics," *Int. J. Bifurcation and Chaos* **20**, 509–533.

Kryzhevich, S. G. & Pliss, V. A. [2003] "Structural stability of nonautonomous systems," *Differ. Equ.* **39**, 1395–1403.

Langa, J. A., Robinson, J. C. & Suárez, A. [2002] "Stability, instability, and bifurcation phenomena in non-autonomous differential equations," *Nonlinearity* **15**, 887.

LaSalle, J. P. [1960] "Some extensions of Liapunov's second method," *IRE Trans. Circuit Theory* **7**, 520–527.

LaSalle, J. P. [1971] "Generalized invariance principles and theory of stability," in *Nonlinear Mechanics and Stability, Proceedings of the Instituto Nazionale di Alta Matematica* (Convegni del Consiglio Nazionale delle Ricerche, Rome, Italy, 1970). Published by the Instituto Nazionale di Alta Matematica (INdAM), Rome.

LaSalle, J. P. [1976] *The Stability of Dynamical Systems*, CBMS-NSF Regional Conference Series in Applied Mathematics, Vol. 25 (Society for Industrial and Applied Mathematics, Philadelphia).

Leibovitz, C. & Pesci, R. [1995] "Dynamic bifurcation in Hamiltonian systems with one degree of freedom," *SIAM J. Appl. Math.* **55**, 1233–1255.

Leibovitz, C. & Schaar, G. [1975] "Slow passage through bifurcation points," *SIAM J. Appl. Math.* **28**, 225–242.

Leibovitz, C. & Schaar, G. [1976] "Slow passage through bifurcation points: The nonlinear case," *SIAM J. Appl. Math.* **30**, 688–705.

Lerman, L. & Shilńikov, L. [1992] "Homoclinical structures in nonautonomous systems: Nonautonomous chaos," *Chaos* **2**, 447–454.

Letellier, C., Abraham, R., Shepelyansky, D. L., Rössler, O. E., Holmes, P., Lozi, R., Glass, L., Pikovsky, A., Olsen, L. F., Tsuda, I., Grebogi, C., Parlitz, U., Gilmore, R., Pecora, L. M. & Carroll, T. L. [2021] "Some elements for a history of the dynamical systems theory," *Chaos* **31**, 053110.

Lichtenberg, A. & Lieberman, M. [1983] *Regular and Stochastic Motion*, Applied Mathematical Sciences, Vol. 38 (Springer, NY).

Lopesino, C., Balibrea-Iniesta, F., Wiggins, S. & Mancho, A. M. [2015] "The chaotic saddle in the Lozi map, autonomous and nonautonomous versions," *Int. J. Bifurcation and Chaos* **25**, 1550184.

Lorenz, E. N. [1963] "Deterministic nonperiodic flow," *J. Atmos. Sci.* **20**, 130–141.

Lorenz, E. N. [1993] *The Essence of Chaos*, 1st edition (University of Washington Press, Seattle).

Lozi, R. [1978] "Un attracteur étrange du type attracteur de Hénon," *J. Phys.* **39**, 9–10.

Lu, K. & Wang, Q. [2010] "Chaos in differential equations driven by a nonautonomous force," *Nonlinearity* **23**, 2935.

Lu, K. & Wang, Q. [2011] "Chaotic behavior in differential equations driven by a Brownian motion," *J. Differ. Equ.* **251**, 2853–2895.

Lyapunov, A. M. [1992] *The General Problem of the Stability of Motion* (Taylor & Francis (CRC Press), London), English translation of the 1892 Russian original; includes biography by V. I. Smirnov and bibliography by J. F. Barrett.

Malhotra, N. & Wiggins, S. [1998] "Geometric structures, lobe dynamics, and Lagrangian transport in flows with aperiodic time-dependence, with applications to Rossby wave flow," *J. Nonlinear Sci.* **8**, 401–456.

Mancho, A. M., Small, D., Wiggins, S. & Ide, K. [2003] "Computation of stable and unstable manifolds of hyperbolic trajectories in two-dimensional, aperiodically time-dependent vector fields," *Physica D* **182**, 188–222.

Markus, L. & Yamabe, H. [1960] "Global stability criteria for differential systems," *Osaka Math. J.* **12**, 305–317.

Markus, L. & Meyer, K. R. [1974] *Generic Hamiltonian Dynamical Systems are Neither Integrable Nor Ergodic*, Memoirs of the American Mathematical Society, Vol. 144 (American Mathematical Society, Providence, RI).

Mauroy, A., Mezić, I. & Susuki, Y. (eds.) [2020] *The Koopman Operator in Systems and Control: Concepts, Methodologies, and Applications*, Lecture Notes in Control and Information Sciences, Vol. 484 (Springer, Cham, Switzerland).

McMurran, S. L. & Tattersall, J. J. [1996] "The mathematical collaboration of ML Cartwright and JE Littlewood," *Am. Math. Mon.* **103**, 833–845.

Melnikov, V. K. [1963] "On the stability of the center for time periodic perturbations," *Trans. Moscow Math. Soc.* **12**, 3–52.

Meyer, K. R. & Schmidt, D. S. [1971] "Periodic orbits near L_4 for mass ratios near the critical mass ratio of Routh," *Celest. Mech.* **4**, 99–107.

Meyer, K. R. & Sell, G. R. [1989] "Melnikov transforms, Bernoulli bundles, and almost periodic perturbations," *Trans. Am. Math. Soc.* **314**, 63–105.

Meyer, K. R. & Zhang, X. [1996] "Stability of skew dynamical systems," *J. Differ. Equ.* **132**, 66–86.

Miller, R. [1965] "Almost periodic differential equations as dynamical systems with applications to the existence of almost periodic solutions," *J. Differ. Equ.* **1**, 337–345.

Milnor, J. [1985] "On the concept of attractor," *Commun. Math. Phys.* **99**, 177–195.

Mineur, H. [1935] "Sur les systemes mécaniques admettant n intégrales premieres uniformes et l'extensiona ces systemes de la méthode de quantification de Sommerfeld," *C. R. Acad. Sci. Paris* **200**, 1571–1573.

Mineur, H. [1937] "Sur les systèmes mécaniques dans lesquels figurent des paramètres fonctions du temps. étude des systèmes admettant n intégrales premières uniformes en involution. Extension à ces systèmes des conditions de quantification de Bohr-Sommerfeld," *J. Ecole Polytechn.* **3**, 173–191.

Mitropolśkii, Y. A. [1965] *Problems of the Asymptotic Theory of Nonstationary Vibrations* (Daniel Davey and Co., NY), English translation, Israel Program for Scientific Translations, Jerusalem.

Moser, J. [1973] *Stable and Unstable Motions in Dynamical Systems* (Princeton University Press, Princeton, NJ).

Neishtadt, A. I. [1987] "Persistence of stability loss for dynamical bifurcations. I," *Differ. Equ.* **23**, 1385–1391.

Neishtadt, A. I. [1988] "Persistence of stability loss for dynamical bifurcations. II," *Differ. Equ.* **24**, 171–176.

Nekhoroshev, N. N. [1972] "Action-angle variables, and their generalizations," *Tr. Mosk. Mat. Obs.* **26**, 181–198.

Perron, O. [1930] "Die stabilitätsfrage bei differentialgleichungen," *Math. Z.* **32**, 703–728.

Pliss, V. A. [1977] "Uniformly bounded solutions of linear systems of differential equations," *Differ. Urav.* **13**, 883–891.

Pliss, V. A. [1980a] "Sets of linear systems of differential equations with uniformly bounded solutions," *Differ. Urav.* **16**, 1599–1616.

Pliss, V. A. [1980b] "Stability of an arbitrary system with respect to perturbations that are small in the sense of c^1," *Differ. Urav.* **16**, 1891–1892.

Pliss, V. A. [1981] "Connection between various conditions for structural stability," *Differ. Urav.* **17**, 828–835.

Pötzsche, C. [2010] *Geometric Theory of Discrete Nonautonomous Dynamical Systems*, Lecture Notes in Mathematics, Vol. 2002 (Springer, Berlin).

Raman, B. & Bajaj, Y. [1998] "On the non-stationary passage through bifurcations in resonantly driven nonlinear oscillators," *Int. J. Non-Linear Mech.* **33**, 251–269.

Robinson, C. [1974] "Structural stability of vector fields," *Ann. Math.* **99**, 154–175.

Sacker, R. J. & Sell, G. R. [1978] "A spectral theory for linear differential systems," *J. Differ. Equ.* **27**, 320–358.

Scarcella, D. [2024a] "Asymptotic motions converging to arbitrary dynamics for time-dependent Hamiltonians," *Nonlinear Anal.* **243**, 113528.

Scarcella, D. [2024b] "Asymptotically quasiperiodic solutions for time-dependent Hamiltonians," *Nonlinearity* **37**, 065005.

Scarcella, D. [2024c] "Biasymptotically quasi-periodic solutions for time-dependent Hamiltonians," *Regul. Chaotic Dyn.* **29**, 620–653.

Scarcella, D. [2025] "Weakly asymptotically quasiperiodic solutions for time-dependent Hamiltonians with a view to celestial mechanics," *J. Differ. Equ.* **431**, 113192.

Schecter, S. [1985] "Persistent unstable equilibria and closed orbits of a singularly perturbed equation," *J. Differ. Equ.* **60**, 131–141.

Sell, G. R. [1967a] "Nonautonomous differential equations and topological dynamics I. The basic theory," *Trans. Am. Math. Soc.* **127**, 241–262.

Sell, G. R. [1967b] "Nonautonomous differential equations and topological dynamics. II. Limiting equations," *Trans. Am. Math. Soc.* **127**, 263–283.

Simiu, E. [2002] *Chaotic Transitions in Deterministic and Stochastic Dynamical Systems: Applications of Melnikov Processes in Engineering, Physics, and Neuroscience* (Princeton University Press).

Smale, S. [1967] "Differentiable dynamical systems," *Bull. Am. Math. Soc.* **73**, 747–817.

Smale, S. [1980] *The Mathematics of Time: Essays on Dynamical Systems, Economic Processes, and Related Topics* (Springer, NY).

Stoffer, D. [1988a] "Transversal homoclinic points and hyperbolic sets for non-autonomous maps I," *Z. Angew. Math. Phys.* **39**, 518–549.

Stoffer, D. [1988b] "Transversal homoclinic points and hyperbolic sets for non-autonomous maps II," *Z. Angew. Math. Phys.* **39**, 783–812.

Szeri, A. J., Wiggins, S. & Leal, L. G. [1991] "On the dynamics of suspended microstructure in unsteady, spatially inhomogeneous, two-dimensional fluid flows," *J. Fluid Mech.* **228**, 207–241.

Temam, R. [1997] *Infinite-Dimensional Dynamical Systems in Mechanics and Physics*, Applied Mathematical Sciences, Vol. 68, 2nd edition (Springer-Verlag).

Ueda, Y. [1991] "Survey of regular and chaotic phenomena in the forced Duffing oscillator," *Chaos Solit. Fract.* **1**, 199–231.

Wagemakers, A. [2025] "The basins zoo," arXiv preprint arXiv:2504.01580.

Wiggins, S. [1988] *Global Bifurcations and Chaos: Analytical Methods*, Applied Mathematical Sciences, Vol. 73 (Springer-Verlag, NY).

Wiggins, S. [1992] *Chaotic Transport in Dynamical Systems* (Springer-Verlag, NY).

Wiggins, S. [1999] "Chaos in the dynamics generated by sequences of maps, with applications to chaotic advection in flows with aperiodic time dependence," *Z. Angew. Math. Phys.* **50**, 585–616.

Wiggins, S., Wiesenfeld, L., Jaffé, C. & Uzer, T. [2001] "Impenetrable barriers in phase-space," *Phys. Rev. Lett.* **86**, 5478–5481.

Wiggins, S. [2003] *Introduction to Applied Nonlinear Dynamical Systems and Chaos*, Texts in Applied Mathematics, Vol. 2, 2nd edition (Springer, NY).

Wiggins, S. [2023] *Ordinary Differential Equations: A Dynamical Point of View*, 1st edition (World Scientific, Singapore).

Yoshizawa, T. [1966] *Stability Theory by Liapunov's Second Method*, Publications of the Mathematical Society of Japan, Vol. 9 (The Mathematical Society of Japan, Tokyo).

https://doi.org/10.1142/9789819830633_0010

Chapter 10

Finding and Analyzing the Solutions of Singular Nonlinear Traveling Wave Equations: A Dynamical Systems Approach#

Jibin Li*,†,§ and Guanrong Chen‡,¶

*Department of Mathematics,
Zhejiang Normal University,
Jinhua, Zhejiang 321004, P. R. China

†School of Mathematical Science,
Huaqiao University,
Quanzhou, Fujian 362021, P. R. China

‡Department of Electrical Engineering,
City University of Hong Kong,
Hong Kong 999077, P. R. China

§lijb@zjnu.cn

¶eegchen@cityu.edu.hk

This overview paper reviews a dynamical systems approach to studying the singular nonlinear traveling wave equations. First, the notion of exact peakon, periodic peakon, pseudo-peakon, as well as compacton solutions for the generalized Camassa–Holm (CH) equation and the Degasperis–Procesi (DP) equation, is introduced. Based on the method of dynamical systems and the theory of singular traveling wave equations, the exact explicit parametric representations of the solutions of the above-mentioned equations are derived. These solutions show that peakon is a limit solution of a family of periodic peakons or a limit solution of a family of pseudo-peakons in different limit senses, whereas the pseudo-peakon and pseudo-periodic peakon families are smooth classical solutions in two different time scales. Second, nonlinear wave equation models are introduced to show that there exist various exact explicit peakon solutions, which are different from the peakon solution given by the generalized CH equation and the DP equation. Third, the so-called "peakon equations" discussed in some studies actually have no peakons. Corresponding to these "peakon equations", their traveling wave systems are singular traveling wave systems of the second kind, which cannot have peakon solutions. Finally, an application example is presented to illustrate our theory and methodology using a model of nonlinear elastodynamics of materials with strong ellipticity.

Keywords: Nonlinear wave equation; traveling wave solution; peakon; periodic peakon; pseudo-peakon; pseudo-periodic peakon; compacton; exact solution; bifurcation; planar dynamical system.

#To cite this article, please refer to its earlier version published in the *International Journal of Bifurcation and Chaos*, Vol. 35, No. 14 (2025) 2540010 (54 pages), DOI: 10.1142/S0218127425400103.

§Author for correspondence

1. Introduction

In the last three decades, nonlinear wave equations having nonsmooth solitary wave solution (the so-called peakon) have attracted considerable attention, leading to a large number of publications. Peakon was first proposed by [Cammasa & Holm, 1993; Cammasa *et al.*, 1994]; thereafter various peakon equations were developed (see [Degasperis & Procesi, 1999; Li & Liu, 2000; Degasperis *et al.*, 2002; Qiao, 2006, 2007; Li & Chen, 2007; Li & Dai, 2007; Novikov, 2009] and the cited references therein). Peakons are also called peaked solitons, including, for example, solitons with discontinuous first-order derivative at the peak point. Usually, the profile of a wave function is referred to as peakon if, at a continuous point, its left and right derivatives are finite with different signs (see [Fokas, 1995]).

It should be noted that (1+1)-dimensional nonlinear wave models are generally partial differential equations (PDEs), while their traveling wave systems are ordinary differential equations (ODEs). Peakon is a special traveling wave solution of a nonlinear model (a PDE). Therefore, to understand its existence and properties, it is very important to study the dynamical behavior of the corresponding traveling wave system (an ODE) of the original nonlinear model.

In [Li & Qiao, 2013a, 2013b], it is shown that there exists pseudo-peakon solution for nonlinear wave equations. The so-called "pseudo-peakon" means that the wave profile looks like a peakon, but the solution still has continuous first-order derivative. By using the dynamical systems theory, it was theoretically proved that if a boundary curve of a period annulus of a center contains a segment of the singular straight line of the traveling wave system, then near this straight line the parametric representation of every periodic orbit has two time scales. In fact, for a singular nonlinear traveling wave system of the first class, the following two results have been obtained (see [Li & Dai, 2007; Li & Chen, 2007; Li, 2013]).

Theorem A (The Rapid-Jump Property of the Derivative Near the Singular Straight Line). *Suppose that in a left (or right) neighborhood of a singular straight line, there exists a family of periodic orbits. Then, along a segment of every orbit near the straight line, the derivative of the wave function jumps down rapidly on a very short time interval.*

Theorem B (Existence of a Finite Time Interval of Solutions with Respect to the Wave Variable in the Positive or Negative Direction). *For a singular nonlinear traveling wave system of the first class with possible changes of the wave variable, if an orbit transversely intersects with a singular straight line at a point or if it approaches a singular straight line but the derivative tends to infinity, then it only takes a finite time interval of the wave variable to move the points of the orbit to arrive on the singular straight line.*

In order to understand rigorously the occurrence of "peaked" traveling wave solutions and the change of the wave profiles with the changes of the system parameters and level curves, it is naturally expected to obtain exact parametric representations of the traveling wave solutions for a given nonlinear partial differential system. Using exact solution formulas to draw the wave profiles, one can easily see how the wave profiles change. In Sec. 2, the following two nonlinear wave equations will be used as examples for study.

(1) The generalized Camassa–Holm (CH) equation with real parameters κ, α:

$$u_t + \kappa u_x - u_{xxt} + \alpha u u_x = 2u_x u_{xx} + u u_{xxx}. \tag{1}$$

Equation (1) with $\alpha = 3$ is exactly the standard CH equation (see [Cammasa & Holm, 1993; Cammasa *et al.*, 1994]), which is a shallow water model.

(2) The generalized Degasperis–Procesi (DP) equation with real parameters κ, α:

$$u_t + \kappa u_x - u_{xxt} + \alpha u u_x = 3u_x u_{xx} + u u_{xxx}. \tag{2}$$

Equation (2) with $\alpha = 4$ is exactly the standard DP equation (see [Degasperis & Procesi, 1999]), which is another shallow water model.

These two equations are shallow water wave models with moderate amplitude regimes (see [Li, 2018]).

Let $u(x,t) = \phi(x - ct) = \phi(\xi)$. Then, Eqs. (1) and (2) have the following traveling wave systems, respectively:

$$\frac{d\phi}{d\xi} = y, \quad \frac{dy}{d\xi} = \frac{-y^2 + \alpha\phi^2 + 2(\kappa - c)\phi}{2(\phi - c)}, \tag{3}$$

$$\frac{d\phi}{d\xi} = y, \quad \frac{dy}{d\xi} = \frac{-2y^2 + \alpha\phi^2 + 2(\kappa - c)\phi}{2(\phi - c)}, \tag{4}$$

where the integral constants are set to zero.

Systems (3) and (4) of Eqs. (1) and (2) are called the singular traveling systems of the first kind, defined by [Li & Chen, 2007; Li, 2013]. Note that the two systems define two piecewise smooth vector fields in two-half phase planes on the two sides of the singular straight line $\phi = c$. These two systems have the first integral, respectively, as follows:

$$H_{\rm ch}(\phi, y) = (\phi - c)y^2 - \left[(\kappa - c)\phi^2 + \frac{1}{3}\alpha\phi^3\right] = h; \tag{5}$$

$$H_{\rm dp}(\phi, y) = (\phi - c)^2 y^2 + \left[(\kappa - c)c\phi^2 + \frac{1}{3}(c\alpha + 2c - 2\kappa)\phi^3 - \frac{1}{4}\alpha\phi^4\right] = h. \tag{6}$$

Note also that the traveling wave systems (3) and (4) have a singular straight line $\phi = c$. Under some parameter conditions, there exists at least one family of periodic orbits such that the boundary curves of the period annulus are a homoclinic orbit, a curve triangle or a arch curve (see the phase portraits shown in Sec. 2). Applying the classical analysis method, one can obtain the parametric representations for these boundary curves. By considering these homoclinic orbits, arch curves or curve triangles as the limit curves of the periodic orbit family in the period annulus, these exact parametric representations provide very good understanding for the occurrence of peaked traveling wave solutions, namely, the homoclinic orbit gives rise to a solitary peakon-like wave solution (called pseudo-peakon), the arch curve gives rise to a periodic peakon, while the curve triangle gives rise to a peakon solution.

How to classify traveling wave solutions is interesting in studying traveling systems. Lenells [2005a, 2005b, 2006, 2007] investigated traveling wave solutions for some nonlinear shallow water wave models that admit smooth, peaked and cusped solutions, stumpons, as well as composite traveling waves with a fractal appearance. Lenells introduced the concept of distribution to describe peaked wave solutions. However, because the corresponding traveling wave equations of the considered models are integrable planar dynamical systems, these models cannot have the so-called composite traveling wave with a fractal appearance. Similar incorrect understanding can also be found from some other reports (for example [Geyer & Quirchmayr, 2018]), which claimed that a shallow water wave model has "Cantor wave with peak elements and cusp elements".

By adopting the method of dynamical systems with the above Theorems A and B, we can well understand the dynamical behavior of all traveling wave solutions to some integrable PDE models. Therefore, we know which orbit gives rise to what wave profiles and how the wave profiles are changed depending on the variation of the system parameters. In addition, by applying the first integrals of the integrable traveling wave systems, we are able to obtain exact explicit solutions for many mathematical physical equations and shallow water wave equations (see [Li, 2008a, 2008b, 2013, 2014a, 2014b, 2017, 2018, 2019; Li & Chen, 2005a, 2005b, 2007, 2010, 2012, 2016a, 2016b, 2019; Li & Dai, 2007, 2025; Li & Han, 2020; Li & Ke, 2025; Li & Kou, 2017; Li & Li, 2008; Li & Liu, 2000, 2002; Li & Qiao, 2010, 2012, 2013a, 2013b, 2015; Li & Shi, 2013; Li & Zhang, 2004, 2009, 2010, 2011a, 2011b; Li *et al.*, 2006, 2009a, 2009b, 2016, 2019, 2020a, 2020b, 2020c, 2020d, 2021, 2024; Shi & Li, 2021; Zhou *et al.*, 2022, 2023; Jin *et al.*, 2025; Yang & Li, 2025a, 2025b]).

Li [2014a] proved the following two assertions: (a) Under different parameter conditions, one nonlinear wave equation can have different exact one-peakon solutions. (b) Different nonlinear wave equations can have different explicit exact one-peakon solutions. In other words, there are various exact explicit one-peakon solutions, which are different from the peakon solutions given by Eqs. (1) and (2).

The rest of this paper is organized as follows. In Sec. 2, we first discuss the bifurcations of phase portraits and derive the exact solutions for the traveling wave systems of Eqs. (1) and (2), respectively, and then use the obtained results to study the classification of traveling wave solutions. In Sec. 3, we use three nonlinear wave equation models to show that there exist various exact explicit peakon solutions, which are different from the peakon solutions given by the generalized CH equation and the DP equation. In Sec. 4, we show that the so-called "peakon equations" claimed in some studies actually have no peakons. In fact, corresponding to these "peakon equations", their traveling systems are singular traveling systems of the second kind (see [Li, 2013]), which cannot have peakon solutions. In Sec. 5, we present an application example to demonstrate our theory and methodology to a model of nonlinear elastodynamics of materials with strong ellipticity.

2. Dynamics and Exact Solutions of the Singular Traveling Wave System of the First Kind

2.1. *Bifurcations of phase portraits of the traveling wave systems (3) and (4) of CH and DP equations*

We first discuss the bifurcations of phase portraits of system (3). Clearly, system (3) has two equilibrium points, $O(0,0)$ and $E_1(\frac{2(c-\kappa)}{\alpha},0)$, on the ϕ-axis.

Applying the transformation $d\xi = (\phi - c)d\zeta$ for $\phi \neq c$, system (3) becomes its associated regular system (see [Li, 2013]):

$$\frac{d\phi}{d\zeta} = 2y(\phi - c), \qquad \frac{dy}{d\zeta} = -y^2 + 2(\kappa - c)\phi + \alpha\phi^2. \tag{7}$$

Apparently, the straight line $\phi = c$ is a solution of system (7). On this straight line, system (7) has two equilibrium points, $S_{1,2}(c, \pm\sqrt{Y_0})$, of (7), when $Y_0 = \alpha c^2 + 2(\kappa - c)c > 0$.

Let

$$h_0 = H_{\rm ch}(0,0) = 0,$$

$$h_1 = H_{\rm ch}\left(\frac{2(c-\kappa)}{\alpha}, 0\right) = \frac{4(c-\kappa)^3}{3\alpha^2},$$

$$h_s = H_{\rm ch}(c, \pm\sqrt{Y_0}) = c^2\left[(c-\kappa) - \frac{1}{3}\alpha c\right].$$

It can be seen that, when $\alpha = -\frac{3}{c}(\kappa - c)$, one has $h_s = 0$; when $\alpha = \frac{1}{c}(\kappa - c)$ and $\alpha = -\frac{2}{c}(\kappa - c)$, one has $h_s = h_1$.

According to the theory of planar dynamical systems, for a fixed $c > 0$ and $\beta = \kappa - c$, by varying α, one obtains the bifurcations of phase portraits of system (3), as shown in Fig. 1.

Second, we consider the associated regular system of system (4):

$$\frac{d\phi}{d\zeta} = 2y(\phi - c), \quad \frac{dy}{d\zeta} = -2y^2 + 2(\kappa - c)\phi + \alpha\phi^2. \tag{8}$$

System (8) has two equilibrium points, $O(0,0)$ and $E_1(\frac{2(c-\kappa)}{\alpha},0)$, on the ϕ-axis. On this straight line, system (8) has two equilibrium points, $S_{1,2}(c, \pm\sqrt{Y_1})$ of (8), when

$$Y_1 = \frac{1}{2}(\alpha c^2 + 2(\kappa - c)c) > 0.$$

Let

$$h_0 = H_{\rm dp}(0,0) = 0,$$

$$h_1 = H_{\rm dp}\left(\frac{2(c-\kappa)}{\alpha}, 0\right) = \frac{4}{3\alpha^3}(\kappa - c)^3[(\alpha - 1)c + \kappa],$$

$$h_s = H_{\rm dp}(c, \pm\sqrt{Y_1}) = \frac{c^3}{12}[(\alpha - 4)c + 4\kappa].$$

Then, obviously, when $\alpha = -\frac{4}{c}(\kappa - c)$, one has $h_s = 0$; when $\alpha = \pm\frac{2}{c}(\kappa - c)$, one has $h_s = h_1$.

According to the theory of planar dynamical systems, for a fixed $c > 0$ and $\beta = \kappa - c$, by varying α, one obtains the bifurcations of phase portraits of system (4), which are similar to Fig. 1.

2.2. *Exact pseudo-peakon and pseudo-periodic peakon solutions of CH and DP equations*

To calculate some exact solutions of traveling wave systems, the following result is useful.

Let $X(R) = A + BR + CR^2$ and $\Delta = B^2 - 4AC$. Consider the integral

$$x = \int_R^{R_0} \frac{dR}{R\sqrt{X(R)}},$$

which solves the differential equation

$$\frac{dR}{dx} = R\sqrt{X(R)}.$$

Theorem C

(1) *When* $A > 0, \Delta > 0, BC \neq 0$ *and* $X(R_0) = 0$, *if* $R_0 = -\frac{B+\sqrt{\Delta}}{2C}$, *then*

$$R(x) = \frac{2A}{\sqrt{\Delta}\cosh(\sqrt{A}x) - B};$$

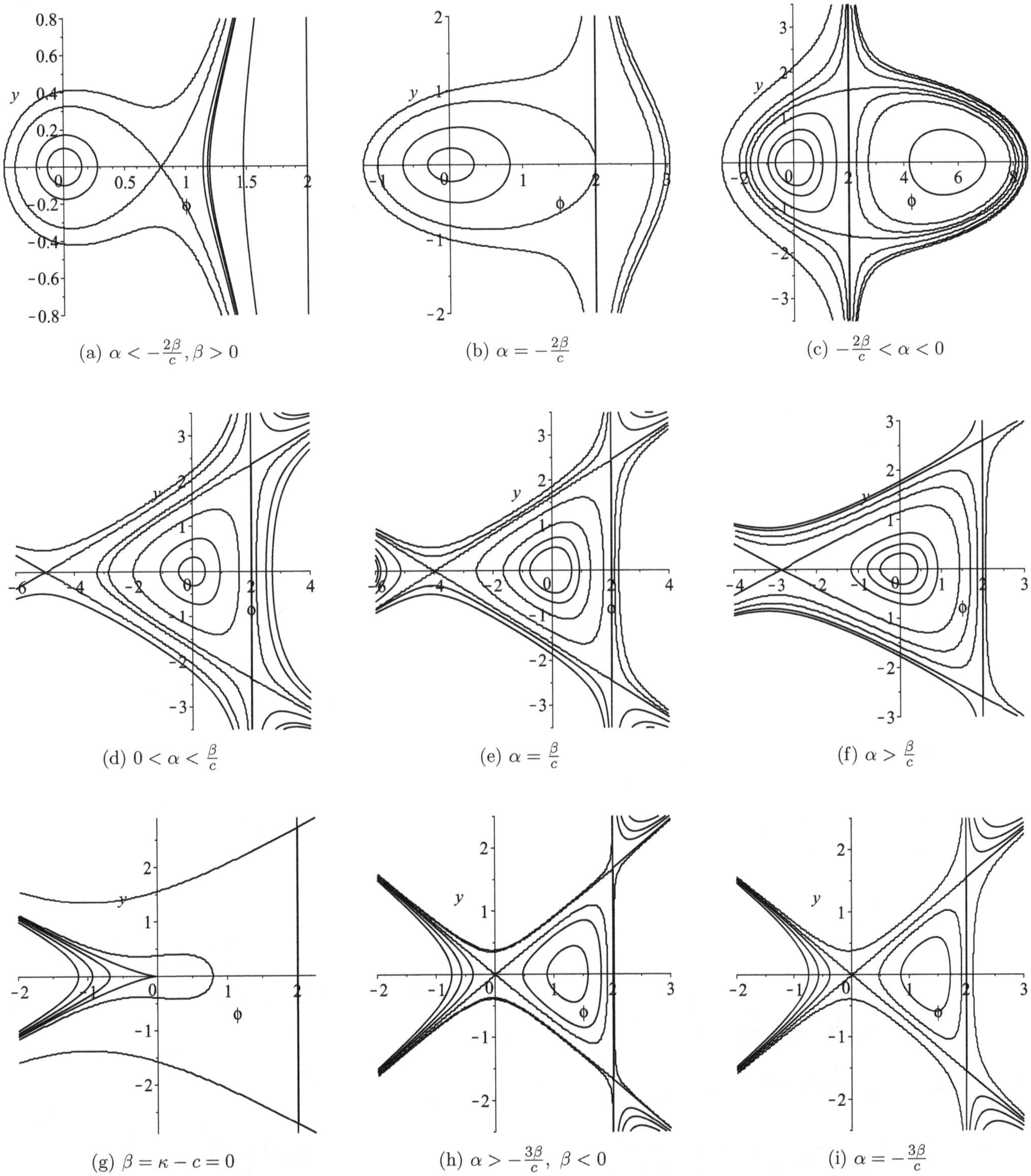

(a) $\alpha < -\frac{2\beta}{c}, \beta > 0$ (b) $\alpha = -\frac{2\beta}{c}$ (c) $-\frac{2\beta}{c} < \alpha < 0$

(d) $0 < \alpha < \frac{\beta}{c}$ (e) $\alpha = \frac{\beta}{c}$ (f) $\alpha > \frac{\beta}{c}$

(g) $\beta = \kappa - c = 0$ (h) $\alpha > -\frac{3\beta}{c},\ \beta < 0$ (i) $\alpha = -\frac{3\beta}{c}$

Fig. 1. Bifurcations of phase portraits of system (3) for $\alpha \neq 0, \beta = \kappa - c$.

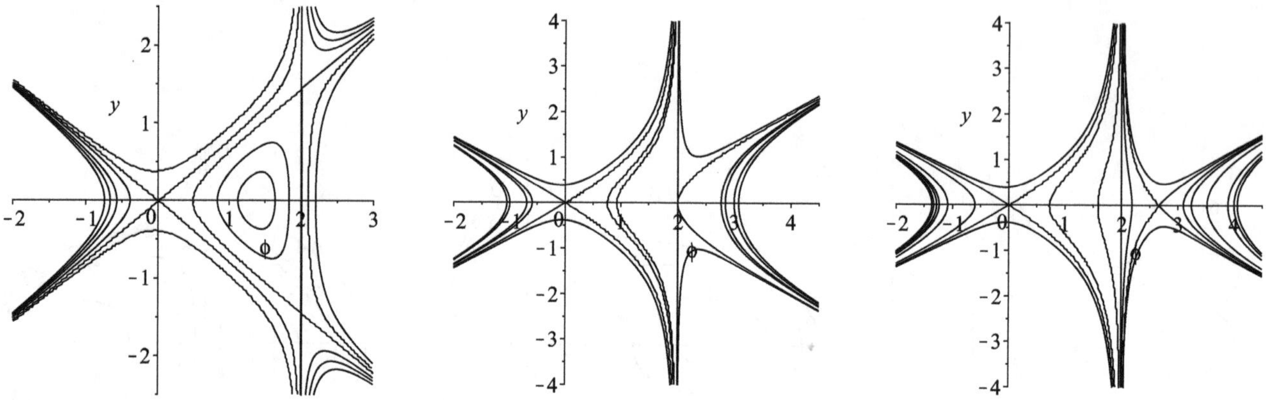

(j) $-\frac{2\beta}{c} < \alpha < -\frac{3\beta}{c}$ (k) $\alpha = -\frac{2\beta}{c}$ (l) $0 < \alpha < -\frac{2\beta}{c}$

Fig. 1. *(Continued)*

if $R_0 = \frac{-B+\sqrt{\Delta}}{2C}$, *then*

$$R(x) = -\frac{2A}{\sqrt{\Delta}\cosh(\sqrt{A}x) + B};$$

if $X(R_0) \neq 0$, *then*

$$R(x) = \frac{2A}{P\cosh_q(\sqrt{A}x) - B},$$

where

$$P = \frac{1}{R_0}(2\sqrt{AX(R_0)} + BR_0 + 2A),$$

$$q = \frac{\Delta}{P^2}, \quad \cosh_q(x) = \frac{1}{2}(e^x + qe^{-x}),$$

$$\frac{d\cosh_q(x)}{dx} = \sinh_q(x) = \frac{1}{2}(e^x - qe^{-x})$$

(see [Arab, 1991, 2001]).

(2) *When* $A = 0, BC \neq 0$,

$$R(x) = \frac{4B}{B^2(x - x_{01})^2 - C},$$

where

$$x_{01} = \frac{2\sqrt{BR_0 + CR_0^2}}{BR_0}.$$

(3) *When* $A < 0, \Delta > 0, BC \neq 0$,

$$R(x) = \frac{2|A|}{B - \sqrt{\Delta}\sin(\sqrt{|A|}(x + x_{02}))},$$

where

$$x_{02} = \frac{1}{\sqrt{|A|}}\sin^{-1}\left(\frac{BR_0 + 2A}{R_0\sqrt{\Delta}}\right).$$

(4) *When* $A > 0, \Delta < 0, BC \neq 0$,

$$R(x) = \frac{-2A}{B - \sqrt{-\Delta}\sinh(\sqrt{|A|}(x + x_{03}))},$$

where

$$x_{03} = \frac{1}{\sqrt{A}}\sinh^{-1}\left(\frac{BR_0 + 2A}{R_0\sqrt{-\Delta}}\right).$$

(5) *When* $A > 0, B = 0$, *if* $C > 0$, *then*

$$R(x) = \sqrt{\frac{A}{C}}\operatorname{cseh}(\sqrt{C}x + x_{04}),$$

where

$$x_{04} = \operatorname{cseh}^{-1}\left(\sqrt{\frac{C}{A}}R_0\right);$$

if $C < 0$, *then*

$$R(x) = \sqrt{\frac{A}{|C|}}\operatorname{sech}(\sqrt{|C|}x + x_{05}),$$

where

$$x_{05} = \operatorname{sech}^{-1}\left(\sqrt{\frac{|C|}{A}}R_0\right).$$

When $A < 0, C > 0$,

$$R(x) = \sqrt{\frac{|A|}{C}}\sec(\sqrt{|A|}x + x_0),$$

where

$$x_0 = \cos^{-1}\left(\sqrt{\frac{|A|}{C}}R_0^{-1}\right).$$

(6) *When* $C = 0, AB \neq 0$, *if* $A > 0$, *then*

$$R(x) = -\frac{A}{B}\operatorname{sech}^2\left(\frac{1}{2}\sqrt{A}x - x_{06}\right),$$

where

$$x_{06} = \tanh^{-1}\left(\frac{\sqrt{A + BR_0}}{\sqrt{A}}\right);$$

if $A < 0$, *then*

$$R(x) = \frac{|A|}{B}\sec^2\left(\frac{1}{2}\sqrt{|A|}x + x_{07}\right),$$

where

$$x_{06} = \tan^{-1}\left(\frac{\sqrt{A + BR_0}}{\sqrt{|A|}}\right).$$

(7) *When* $\Delta = 0$, *i.e. when* $C = \frac{B^2}{4A}$ *and* $A > 0$,

$$R(x) = \frac{2A}{B - e^{(\sqrt{A}x + x_{07})}},$$

where $x_{07} = \ln(\frac{2A+BR_0}{R_0})$.

One can see, from Figs. 1(h)–1(j), that for system (3), under the fixed parameter conditions of $\kappa < c, c > 0$, system (3) has three phase portraits, as shown in Figs. 2(a)–2(c). For system (4), it has the same phase portraits, as shown in Figs. 2(a)–2(c), but the parameter conditions become: (h) $\alpha > \frac{4(c-\kappa)}{c}$, (i) $\alpha = -\frac{4\beta}{c}$ and (j) $-\frac{2\beta}{c} < \alpha < -\frac{4\beta}{c}$.

Next, first consider the parametric representations of the bounded orbits given by Fig. 2(a).

(i) For the family of periodic orbits of Eq. (3) defined by $H_{\text{ch}}(\phi, y) = h,\ h \in (0, h_1)$ in (5), one can see that

$$y^2 = \frac{\alpha\left(-\frac{3h}{\alpha} + \frac{3}{\alpha}(c-\kappa)\phi^2 - \phi^3\right)}{3(c-\phi)} = \frac{\alpha(r_2 - \phi)(\phi - r_3)(\phi - r_4)}{3(c-\phi)}.$$

Thus, from the first equation of system (3) one has

$$\xi = \sqrt{\frac{3}{\alpha}}\int_{\phi}^{r_2}\frac{(c-\phi)d\phi}{\sqrt{(c-\phi)(r_2-\phi)(\phi-r_3)(\phi-r_4)}},$$

which leads to the parametric representation

$$\phi(\chi) = c - \frac{c - r_2}{1 - \hat{\alpha}_0^2\text{sn}^2(\chi, k)},$$

$$\xi(\chi) = \frac{2\sqrt{3}(c - r_2)}{\sqrt{\alpha(c-r_3)(r_2-r_4)}} \times \Pi(\arcsin(\text{sn}(\chi,k)), \hat{\alpha}_0^2, k), \tag{9}$$

where $k^2 = \frac{\hat{\alpha}_0^2(c-r_4)}{(r_2-r_4)}$, $\hat{\alpha}_0^2 = \frac{r_2-r_3}{c-r_3}$ and $\Pi(\cdot, \hat{\alpha}^2, k)$ is the elliptic integral of the third kind, $\text{sn}(u,k)$, $\text{cn}(u,k)$ and $\text{dn}(u,k)$ are the Jacobian elliptic functions (see [Byrd & Fridman, 1971]).

Note from Fig. 2(a) that, when parameter α is very close to $\frac{3(c-\kappa)}{c}$, one has $0 < |h_s| \ll 1$. It implies that a segment of the homoclinic orbit defined by a branch of the level curve $H_{\text{ch}}(\phi, y) = 0$ completely lies in a left neighborhood of the singular straight line $\phi = c$. According to Theorem A, the state coordinate y of the points in this segment rapidly jumps (as ξ varies) from a positive number to a negative number, consequently this homoclinic orbit gives rise to a solitary cusp wave, i.e. pseudo-peakon.

In addition, from (9), one can easily see that

$$\frac{d\phi}{d\xi} = \frac{\sqrt{3}\hat{\alpha}_0^2\text{sn}(\chi,k)\text{cn}(\chi,k)\text{dn}(\chi,k)}{\sqrt{\alpha}(1 - \hat{\alpha}_0^2\text{sn}^2(\chi,k))}. \tag{10}$$

When $|h - h_s| \ll 1$, i.e. $|1 - k| \ll 1$, the graph of $\frac{d\phi}{d\xi}$ is shown in Fig. 3(a). Clearly, when $\chi = 4nK(k)$, $n = 0, \pm 1, \pm 2, \ldots$ and $\frac{d\phi}{d\xi} = 0$, where $K(k)$ is the complete elliptic integral of the first kind modulo k. One can see from Fig. 3(a) that, when χ passes through $4nK(k)$, its sign changes from + to − and its value jumps rapidly from a positive maximum to a minus maximum. This implies that the wave profile of $\phi(\xi)$, determined by the periodic orbit closing to the homoclinic orbit, is a smooth periodic cusp wave [see Fig. 3(b)]. Finally, when $h \to 0$, $k \to 1$, the period of the periodic cusp wave solutions tends to ∞; thus, the homoclinic orbit shown in Fig. 2(a) gives rise to a smooth solitary cusp wave solution of Eq. (1), i.e. a pseudo-peakon [see Fig. 3(c)]. It is remarked that the word "pseudo" here means that, in the top peak part of the profile of every traveling wave solution $\phi(\xi)$, the first-order derivative $\phi'(\xi)$ is continuous.

For the family of periodic orbits of system (4) defined by $H_{\text{dp}}(\phi, y) = h,\ h \in (0, h_1)$ in (6), one can see that

$$y^2 = \frac{\alpha\left(-\frac{4h}{\alpha} + \frac{4c}{\alpha}(c-\kappa)\phi^2 - \frac{4}{3\alpha}(c\alpha + 2c - 2\kappa)\phi^3 + \phi^4\right)}{4(c-\phi)^2} = \frac{\alpha(r_1-\phi)(r_2-\phi)(\phi-r_3)(\phi-r_4)}{4(c-\phi)^2}.$$

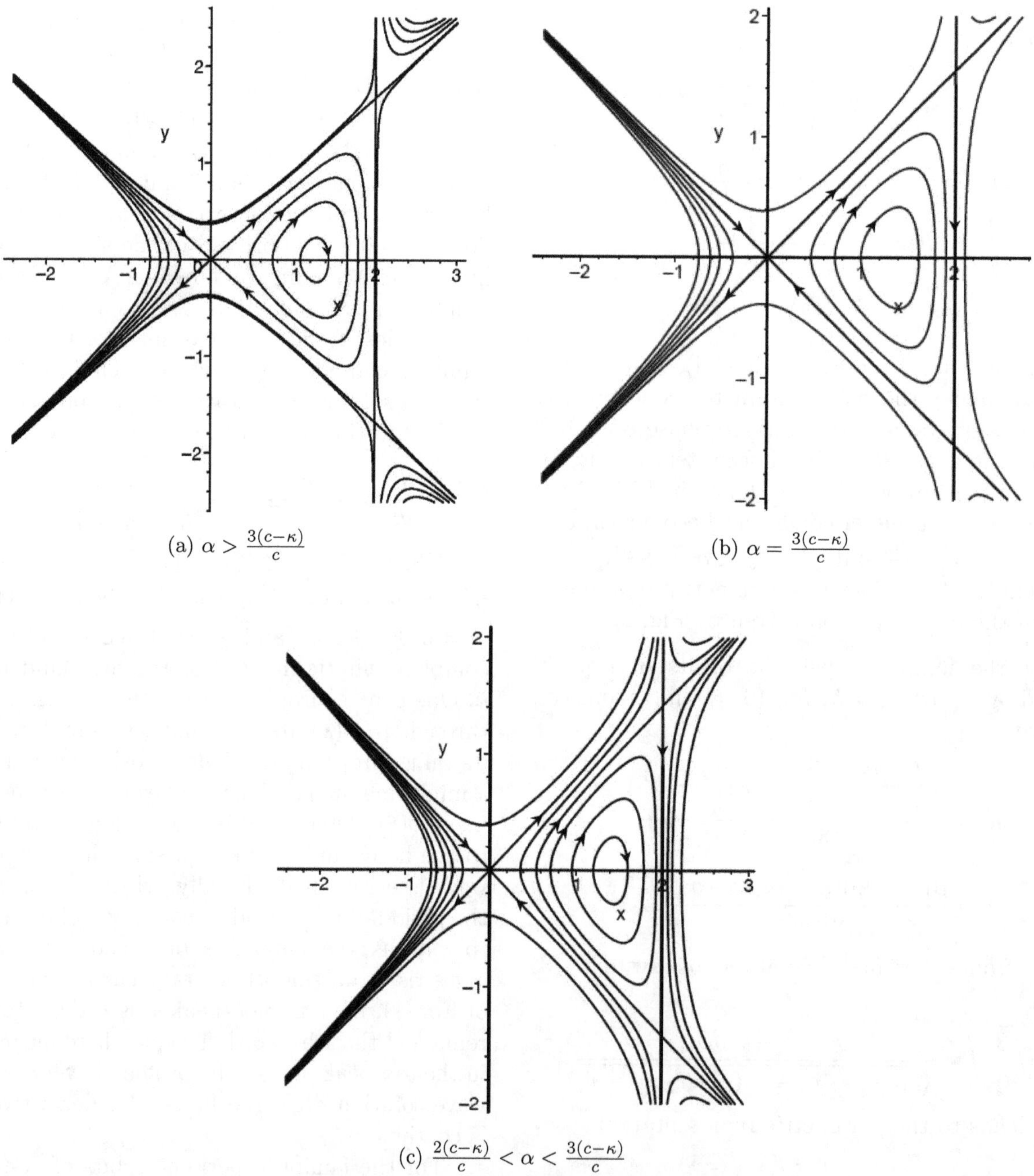

(a) $\alpha > \frac{3(c-\kappa)}{c}$

(b) $\alpha = \frac{3(c-\kappa)}{c}$

(c) $\frac{2(c-\kappa)}{c} < \alpha < \frac{3(c-\kappa)}{c}$

Fig. 2. The phase portraits of (3) for $\alpha > 0, \beta = \kappa - c < 0$.

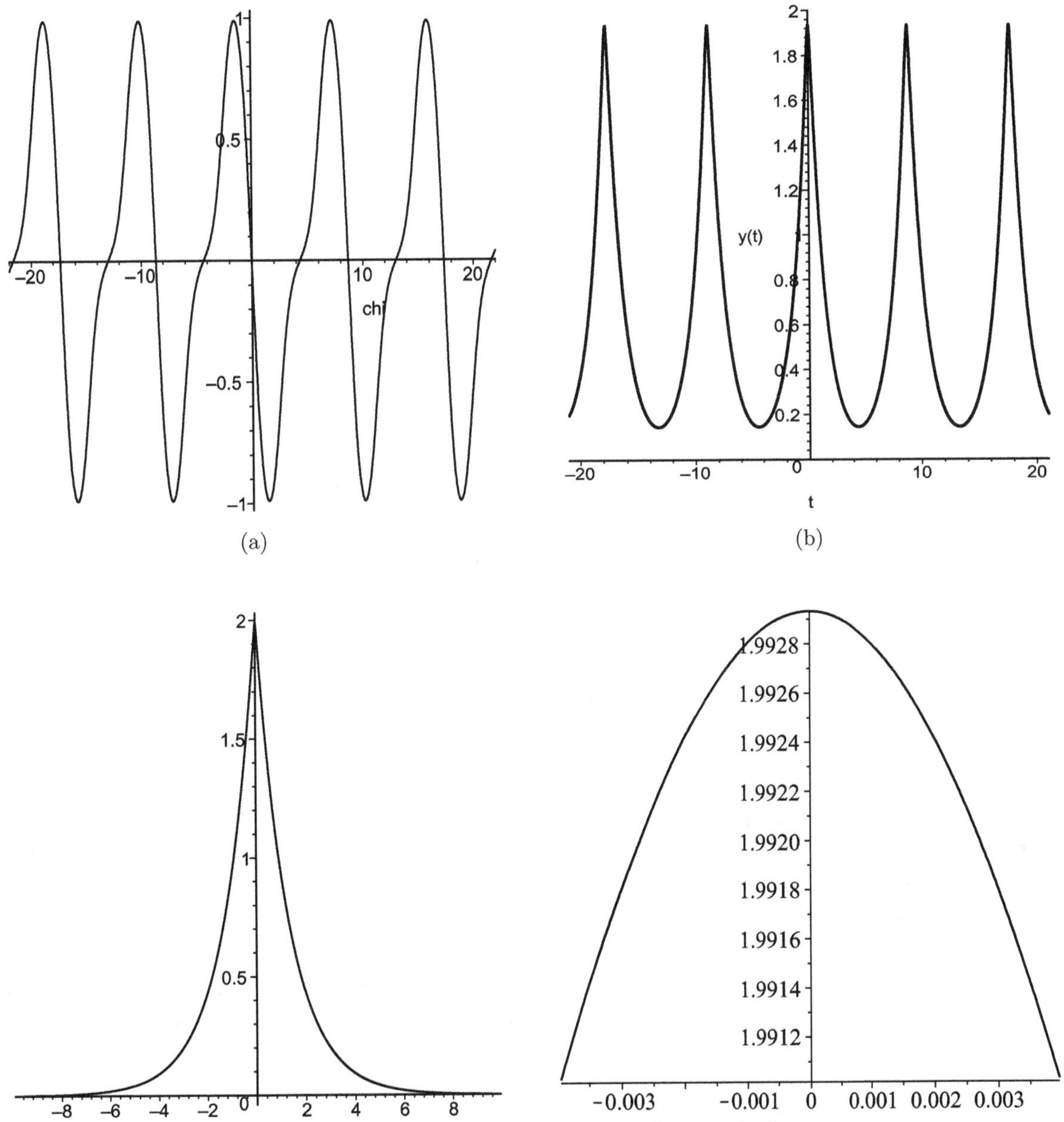

Fig. 3. The changes of wave profiles of $\phi(\xi)$. (a) The graph of $\frac{d\phi}{d\xi}$. (b) A pseudo-periodic peakon $\phi(\xi)$. (c) The pseudo-peakon $\phi(\xi)$. (d) The top part of pseudo-peakon.

Thus, from the first equation of system (4), one has

$$\xi = \sqrt{\frac{4}{\alpha}} \int_{\phi}^{r_2} \frac{(c-\phi)d\phi}{\sqrt{(r_1-\phi)(r_2-\phi)(\phi-r_3)(\phi-r_4)}},$$

which leads to the parametric representation

$$\phi(\chi) = r_1 - \frac{r_1-r_2}{1-\hat{\alpha}_1^2 \mathrm{sn}^2(\chi,k)},$$

$$\xi(\chi) = \frac{4(r_1-r_2)}{\sqrt{\alpha(r_1-r_3)(r_2-r_4)}} \times \Pi(\arcsin(\mathrm{sn}(\chi,k)), \hat{\alpha}_1^2, k), \tag{11}$$

where $k^2 = \frac{\hat{\alpha}_1^2(r_1-r_4)}{r_2-r_4}$, $\hat{\alpha}_1^2 = \frac{r_2-r_3}{r_1-r_3}$. Clearly, the periodic solutions defined by (11) have the same dynamical behavior as (9).

(ii) For the homoclinic orbit defined by $H_{\rm ch}(\phi,y) = 0$, let $P_M(\phi_M, 0)$ be the intersection point of the homoclinic orbit with the ϕ-axis. It is known that $\phi_M = \frac{3(c-\kappa)}{\alpha}$. When $h=0, H_{\rm ch}(\phi,y) = 0$ yields $y = \pm\sqrt{\frac{\alpha\phi^2(\phi_M-\phi)}{3(c-\phi)}}$. By using the first equation of system (3) and taking initial value $\phi(0) = \phi_M$, one obtains the parametric representation of the homoclinic orbit to the origin $E_0(0,0)$ defined by $H_{\rm ch}(\phi,y) = 0$, as follows:

$$\phi(\chi) = \frac{2c\phi_M}{(c-\phi_M)\cosh\left(\sqrt{c\phi_M}\chi\right) + (c+\phi_M)},$$

$$\xi(\chi) = \sqrt{\frac{3}{\alpha}}\left[c\chi \mp \left(\ln\left|\sqrt{(\phi_M-\phi)(c-\phi)} + \phi - \frac{1}{2}(c+\phi_M)\right| - \ln\left(\frac{1}{2}(c-\phi_M)\right)\right)\right], \tag{12}$$

for $\chi \in (-\infty, 0]$ and for $\chi \in [0,\infty)$, respectively.

When $\alpha - \frac{3(c-\kappa)}{c} \ll 0$, (12) gives rise to the pseudo-peakon solution of the CH equation [see Figs. 3(c) and 3(d)]. One can see, from Fig. 3(d), that in the top part of a pseudo-peakon solution, this solution is also smooth.

The level curves defined by $H_{\rm dp}(\phi,y) = 0$ contain a homoclinic orbit to the origin $O(0,0)$ and an open curve passing through the point $(\phi_L, 0)$ with $\phi_L > c$. Let $Q_M(\phi_M, 0)$ be the intersection point of the homoclinic orbit with the ϕ-axis. When $h = h_0 = 0$, $H_{\rm dp}(\phi,y) = 0$ gives

$$y^2 = \frac{\alpha\phi^2(a_0+b_0\phi+\phi^2)}{4(c-\phi)^2} = \frac{\alpha\phi^2(\phi_L-\phi)(\phi_M-\phi)}{4(c-\phi)^2},$$

where

$$a_0 = \frac{4c}{\alpha}(c-\kappa),$$

$$b_0 = -\frac{4}{3\alpha}[(\alpha-2)c-2\kappa],$$

$$\phi_L = \frac{1}{2}(-b_0+\sqrt{\Delta_1}),$$

$$\phi_M = \frac{1}{2}(-b_0-\sqrt{\Delta_1}), \quad \Delta_1 = b_0^2 - 4a_0.$$

By using the first equation of system (4) and taking initial value $\phi(0) = \phi_M$, one obtains the parametric representation of the homoclinic orbit to the origin $E_0(0,0)$, defined by $H_{\rm dp}(\phi,y) = 0$, as follows:

$$\phi(\chi) = \frac{2\phi_L\phi_M}{(\phi_L-\phi_M)\cosh(\sqrt{\phi_L\phi_M}\chi) + (\phi_L+\phi_M)},$$

$$\xi(\chi) = \frac{2}{\sqrt{\alpha}}\left[c\chi \mp \left(\ln\left|\sqrt{(\phi_M-\phi(\chi))(\phi_L-\phi(\chi))} + \phi(\chi) - \frac{1}{2}(\phi_L+\phi_M)\right| - \ln\left(\frac{1}{2}(\phi_L-\phi_M)\right)\right)\right], \tag{13}$$

for $\chi \in (-\infty, 0]$ and for $\chi \in [0,\infty)$, respectively.

When $\alpha - \frac{4(c-\kappa)}{c} \ll 0$, (13) gives rise to the pseudo-peakon solution of the DP equation [see Figs. 3(c) and 3(d)]. One can see, from Fig. 3(d), that in the top part of a pseudo-peakon solution, this solution is also smooth.

2.3. *Exact peakon solutions of the CH and DP equations*

Now, consider the parametric representations of the orbits given by Fig. 2(b). For system (3) and (4), one has

$$h_0 = h_s = 0, \quad \alpha = \frac{3(c-\kappa)}{c} \quad \text{and} \quad \alpha = \frac{4(c-\kappa)}{c},$$

respectively.

(i) For the family of periodic orbits of system (3) and (4) defined by $H_{\text{ch}}(\phi, y) = h$, $h \in (0, h_1)$ and $H_{\text{dp}}(\phi, y) = h, h \in (h_1, 0)$ in Fig. 2(b), one has the same parametric representation as (9) and (11).

(ii) The level curves defined by $H_{\text{ch}}(\phi, y) = 0$ and $H_{\text{dp}}(\phi y) = 0$ have a straight line triangle enclosing a family of periodic orbit family. It is a limit triangle of the family of the periodic orbit family enclosing the center $E_1(\phi_1, 0)$, as $h \to 0$.

For system (3), corresponding to two straight line orbits connecting the equilibrium points $O(0,0)$ and $S_1(c, \mp\sqrt{Y_0})$, defined by $H_{\text{ch}}(\phi, y) = 0$, one has $y^2 = \frac{1}{2}\alpha\phi^2$. By Theorem B, one can take initial value as $\phi(0) = c$. Then,

$$u(x,t) = \phi(\xi) = ce^{-\sqrt{\frac{\alpha}{3}}|\xi|}, \tag{14}$$

which is a real peakon solution to the CH equation (1) [see Fig. 4(b)].

For system (4), corresponding to two straight line orbits connecting the equilibrium points $O(0,0)$ and $S_1(c, \mp\sqrt{Y_1})$, defined by $H_{\text{dp}}(\phi, y) = 0$, one has $y^2 = \frac{1}{4}\alpha\phi^2$. By Theorem B, one can take initial value as $\phi(0) = c$. Then,

$$u(x,t) = \phi(\xi) = ce^{-\sqrt{\frac{\alpha}{4}}|\xi|}, \tag{15}$$

which is a real peakon solution to the DP equation (2) [see Fig. 4(b)]. Clearly, under fixed parameter condition, as h is varied and $h \to 0$, the above peakon solutions are the limit solutions of a family of pseudo-periodic peakons.

2.4. *Exact periodic peakons and compacton families of the CH and DP equations*

Here, the exact solutions for the orbits shown in Fig. 2(c) are discussed.

(i) For the family of periodic orbits of systems (3) and (4) defined by $H_{\text{ch}}(\phi, y) = h$ and $H_{\text{dp}}\phi, y) = h$, $h \in (h_s, h_1)$, respectively, one has the same parametric representations as (9) and (11).

(ii) For the arch orbit defined by $H_{\text{ch}}(\phi, y) = h_s$ and $H_{\text{dp}}(\phi, y) = h_s$, one has $y^2 = \frac{\alpha}{3}(cm_0 + m_0\phi + \phi^2)$, where $m_0 = c - \frac{3}{\alpha}(c-\kappa)$ and $y^2 = \frac{\alpha}{4}(cm_1 + m_1\phi + \phi^2)$, with $m_1 = \frac{1}{3\alpha}[(\alpha-4)c + 4\kappa]$, respectively. Hence, one obtains the parametric representation of the periodic peakon wave solution of the CH equation, as follows [see Fig. 5(a)]:

$$\begin{aligned} &\phi(\xi) = \frac{1}{2}\left[\sqrt{m_0(m_0-4c)}\cosh\left(\sqrt{\frac{\alpha}{3}}\xi\right) - m_0\right], \\ &0 \le |\xi| \le \tilde{\xi}_0 = \sqrt{\frac{3}{\alpha}}\cosh^{-1} \\ &\quad \times \left(\frac{m_0 + 2c}{\sqrt{m_0(m_0-4c)}}\right), \end{aligned} \tag{16}$$

and the periodic peakon solution of the DP equation (2), as follows:

$$\begin{aligned} &\phi(\xi) = \sqrt{m_1(m_1-c)}\cosh \\ &\quad \times \left(\frac{\sqrt{\alpha}}{2}\xi\right) - m_1, \\ &0 \le |\xi| \le \hat{\xi}_0 = \frac{2}{\sqrt{\alpha}}\cosh^{-1} \\ &\quad \times \left(\frac{m_1 + c}{\sqrt{m_1(m_1-c)}}\right). \end{aligned} \tag{17}$$

Note that (16) and (17) only give the wave profiles of $\phi(\xi)$ in one period. When extending $\phi(\xi)$ to $\xi \in (-\infty, \infty)$, because $\phi'(\xi)$ at $\pm\tilde{\xi}_0$ and $\pm\hat{\xi}_0$ have opposite signs, these solutions have discontinuous first-order derivatives at the peak points. Thus, (16) and (17) give rise to two periodic peakons.

Note also that solutions (16) and (17) hold for all α satisfying $\frac{2(c-\kappa)}{c} < \alpha < \frac{3(c-\kappa)}{c}$ and $\frac{2(c-\kappa)}{c} < \alpha < \frac{4(c-\kappa)}{c}$, respectively. Therefore, when $\alpha \to \frac{3(c-\kappa)}{c}$ and $\alpha \to \frac{4(c-\kappa)}{c}$, respectively, the periodic peakons defined by (16) and (17) tend to the peakon solutions given by (14) and (15), respectively.

(iii) The level curves defined by $H_{\text{ch}}(\phi, y) = h$, $h \in (h_s, 0)$, contain three open curve families, $\{\Gamma_j\}$, $j = 1, 2, 3$, passing through the ϕ-axis at the points $(r_1, 0), (r_2, 0), (r_3, 0), r_3 < 0 < r_2 < \phi_1 < c < r_1$. The level curve family $\{\Gamma_2\}$ passes through the ϕ-axis at the point $(r_2, 0)$ and tends to the singular straight line $\phi = c$, as $|y| \to \infty$. For this family of

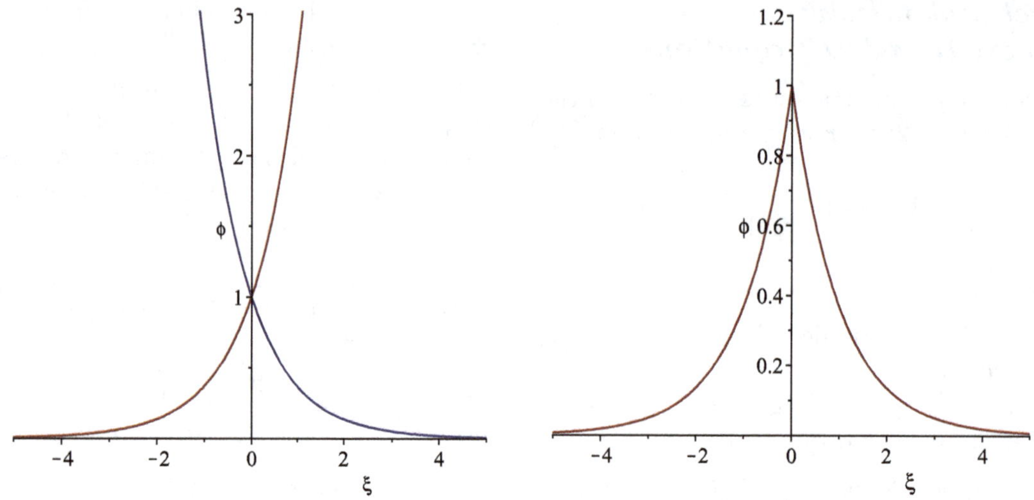

(a) (b)

Fig. 4. The peakon of the CH equation. (a) Stable and unstable manifolds of the origin O. (b) Peakon solution of the CH equation.

open orbits, one can see that

$$y^2 = \frac{\alpha\left[-\dfrac{3h}{\alpha} + \dfrac{3}{\alpha}(c-\kappa)\phi^2 - \phi^3\right]}{3(c-\phi)} = \frac{\alpha((r_1-\phi)(\phi-r_3)(\phi-r_4))}{3(c-\phi)}.$$

Thus, from the first equation of system (3), one has

$$\xi = \sqrt{\frac{3}{\alpha}}\int_\phi^c \frac{(c-\phi)d\phi}{\sqrt{(r_1-\phi)(c-\phi)(\phi-r_3)(\phi-r_4)}},$$

which leads to the following parametric representation [see Fig. 5(b)]:

$$\begin{aligned} \phi(\chi) &= r_1 - \frac{r_1-c}{1-\hat{\alpha}_2^2\mathrm{sn}^2(\chi,k)}, \\ &\qquad \chi \in (-K(k),0],\ [0,K(k)), \\ \xi(\chi) &= \frac{2\sqrt{3}(r_1-c)}{\sqrt{\alpha(r_1-r_3)(c-r_4)}} \\ &\qquad \times \Pi(\arcsin(\mathrm{sn}(\chi,k)),\hat{\alpha}_2^2,k), \end{aligned} \tag{18}$$

where

$$k^2 = \frac{\hat{\alpha}_2^2(r_1-r_4)}{c-r_4}, \quad \hat{\alpha}_2^2 = \frac{c-r_3}{r_1-r_3}.$$

The level curves defined by $H_{\mathrm{dp}}(\phi,y) = h$, $h \in (h_s,0)$, contain four open curve families, for which two families $\{\Gamma_j\}$, $j=1,2$, passing through the ϕ-axis at the points $(r_1,0), (r_2,0), r_2 < 0 < r_1 < \phi_1 < c$. The level curve family $\{\Gamma_1\}$ passes through the ϕ-axis at the point $(r_1,0)$ and tends to the singular straight line $\phi = c$, as $|y| \to \infty$. For this family of open orbits, one can see that

$$y^2 = \frac{\alpha(r_1-\phi)(\phi-r_2)[(\phi-b_1)^2+a_1^2]}{4(c-\phi)}.$$

Thus, from the first equation of system (4), one has

$$\xi = \sqrt{\frac{4}{\alpha}}\int_{r_1}^\phi \frac{(c-\phi)d\phi}{\sqrt{(r_1-\phi)(\phi-r_2)[(\phi-b_1)^2+a_1^2]}},$$

which leads to the following parametric representation:

$$\begin{aligned} \phi(\chi) &= \hat{A}_0 + \frac{\hat{B}_0}{1+\hat{\alpha}_3\mathrm{cn}(\chi,k)}, \quad \chi \in [-\hat{\chi}_0,0],[0,\hat{\chi}_0], \\ \xi(\chi) &= \frac{2}{\sqrt{\alpha A_1 B_1}}\left[\left(c + \frac{r_1B_1 - r_2B_2}{B_1-A_1}\right)\chi \pm \left(\frac{A_1+B_1}{2A_1} + \frac{2A_1(r_1B_1+r_2A_1)}{(A_1-B_1)(B_1r_1-r_2A_1)}\right)\right. \\ &\qquad \left.\times\left[\Pi\left(\arccos(\mathrm{cn}(\chi,k)), \frac{\hat{\alpha}_3^2}{\hat{\alpha}_3^2-1}, k\right) + \hat{\alpha}_3 f_1(\chi)\right]\right., \end{aligned} \tag{19}$$

where

$$A_1 = (r_1 - b_1)^2 + a_1^2, \quad B_1 = (r_2 - b_1)^2 + a_1^2,$$

$$\hat{A}_0 = \frac{r_1 B_1 + r_2 A_1}{A_1 + B_1}, \quad \hat{B}_0 = \frac{2A_1 B_1 (r_1 - r_2)}{B_1^2 - A_1^2},$$

$$\hat{\alpha}_3 = \frac{A_1 + B_1}{A_1 - B_1}, \quad k^2 = \frac{(A_1 + B_1)^2 - (r_1 - r_2)^2}{4A_1 B_1},$$

$$\hat{\chi}_0 = \mathrm{cn}^{-1}\left(\frac{1}{\hat{\alpha}_3}\left(1 - \frac{\hat{B}_0}{c - \hat{B}_0}\right)\right),$$

the function $f_1(\chi)$ can be found from [Byrd & Fridman, 1971, formula 361.54].

Equations (18) and (19) give rise to, respectively, a family of compacton solutions of the CH equation (1) and the DP equation (2). One can see that the compacton family is a solution family of a singular system for which all solutions $\phi(\xi)$ have finite support sets (see [Olver & Rosenau, 1996]), i.e. the defined region of every $\phi(\xi)$ with respect to ξ is finite and the value region of function ϕ is bounded [see Fig. 5(b)]. These solutions are also called "breaking wave solutions", because $\xi \notin (-\infty, \infty)$.

(iv) The level curves defined by $H_{\mathrm{ch}}(\phi, y) = 0$ contain the stable and unstable manifolds of the saddle point $O(0,0)$ and a open curve passing through $(\frac{3}{\alpha}(c-\kappa), 0)$. For the right stable and unstable manifolds, one has

$$y^2 = \frac{\alpha\left(\frac{3}{\alpha}(c-\kappa) - \phi\right)\phi^2}{3(c-\phi)} \equiv \frac{\alpha(\beta_1 - \phi)\phi^2}{3(c-\phi)}.$$

On the basis of Theorem B, one can take initial value $\phi(0) = c$. Using the first equation of (3), one obtains the following two exact solutions [see Figs. 5(c) and 5(d)] of the CH equation (1):

$$\begin{aligned} \phi(\chi) &= \frac{2\beta_1 c}{(\beta_1 - c)\cosh(\sqrt{c\beta_1}\chi) + (\beta_1 + c)}, \quad \chi \in (-\infty, 0] \ \text{ and } \ \chi \in [0, \infty), \ \text{ respectively}, \\ \xi(\chi) &= \sqrt{\frac{3}{\alpha}}\left[\sqrt{\frac{c}{\beta_1}}\chi \mp \left(\ln\left|\sqrt{(\beta_1 - \phi(\chi))(c - \phi(\chi))} + \phi(\chi) - \frac{1}{2}(\beta_1 + c)\right| - \ln\left(\frac{1}{2}\kappa\right)\right)\right]. \end{aligned} \tag{20}$$

The level curves defined by $H_{\mathrm{dp}}(\phi, y) = 0$ contain the stable and unstable manifolds of the saddle point $O(0,0)$ and two open curves which have no intersection point with ϕ-axis. For the right stable and unstable manifolds, similar to Sec. 2.2, one has $y^2 = \frac{\alpha\phi^2(a_0 + b_0\phi + \phi^2)}{4(c-\phi)}$, now $\Delta_1 < 0$. Using the first equation of (4) and taking the initial value $\phi(0) = c$, one obtains the following two exact solutions of the DP equation (2):

$$\begin{aligned} \phi(\chi) &= \frac{2a_0}{\sqrt{|\Delta_1|}\sinh(\sqrt{a_0}\chi + \tilde{\chi}_0) - b_0}, \quad \chi \in (-\infty, 0] \ \text{ and } \ \chi \in ([0, \infty), \ \text{ respectively}, \\ \xi(\chi) &= \frac{2}{\alpha}\left[c\chi \pm \sinh^{-1}\left(\frac{2\phi(\chi) + b_0}{\sqrt{|\Delta_1|}}\right) \mp \sinh^{-1}\left(\frac{2c + b_0}{\sqrt{|\Delta_1|}}\right)\right], \end{aligned} \tag{21}$$

where

$$\tilde{\chi}_0 = \sinh^{-1}\left(\frac{2a_0 + cb_0}{c\sqrt{|\Delta_1|}}\right).$$

Note that, in some existing publications, the two solutions of stable manifold and unstable manifold are composed together as a *cuspon solution*. However, as can be seen from Figs. 5(c) and 5(d), actually the CH equation and the DP equation both *have no such cuspon solutions*.

In summary, we have the following result.

Theorem 1. *For the generalized CH equation (1) and the DP equation (2), considering their traveling wave systems (3) and (4) with a fixed wave velocity parameter $c > 0$, the following conclusions hold:*

(1) *When the parameter group (α, κ) of systems (3) and (4) satisfy the condition that $\alpha = \frac{3(c-\kappa)}{c}$ and $\alpha = \frac{4(c-\kappa)}{c}$ with $\kappa < c$, respectively, there exists a straight line triangle of systems (3) and (4), respectively, defined by $\phi = c$ and the level curves $H_{\mathrm{ch}}(\phi, y) = 0$ and $H_{\mathrm{dp}}(\phi, y) = 0$, respectively, which gives rise to a peakon solutions given by (14) and (15), respectively. When h varies from h_1 to 0, periodic wave*

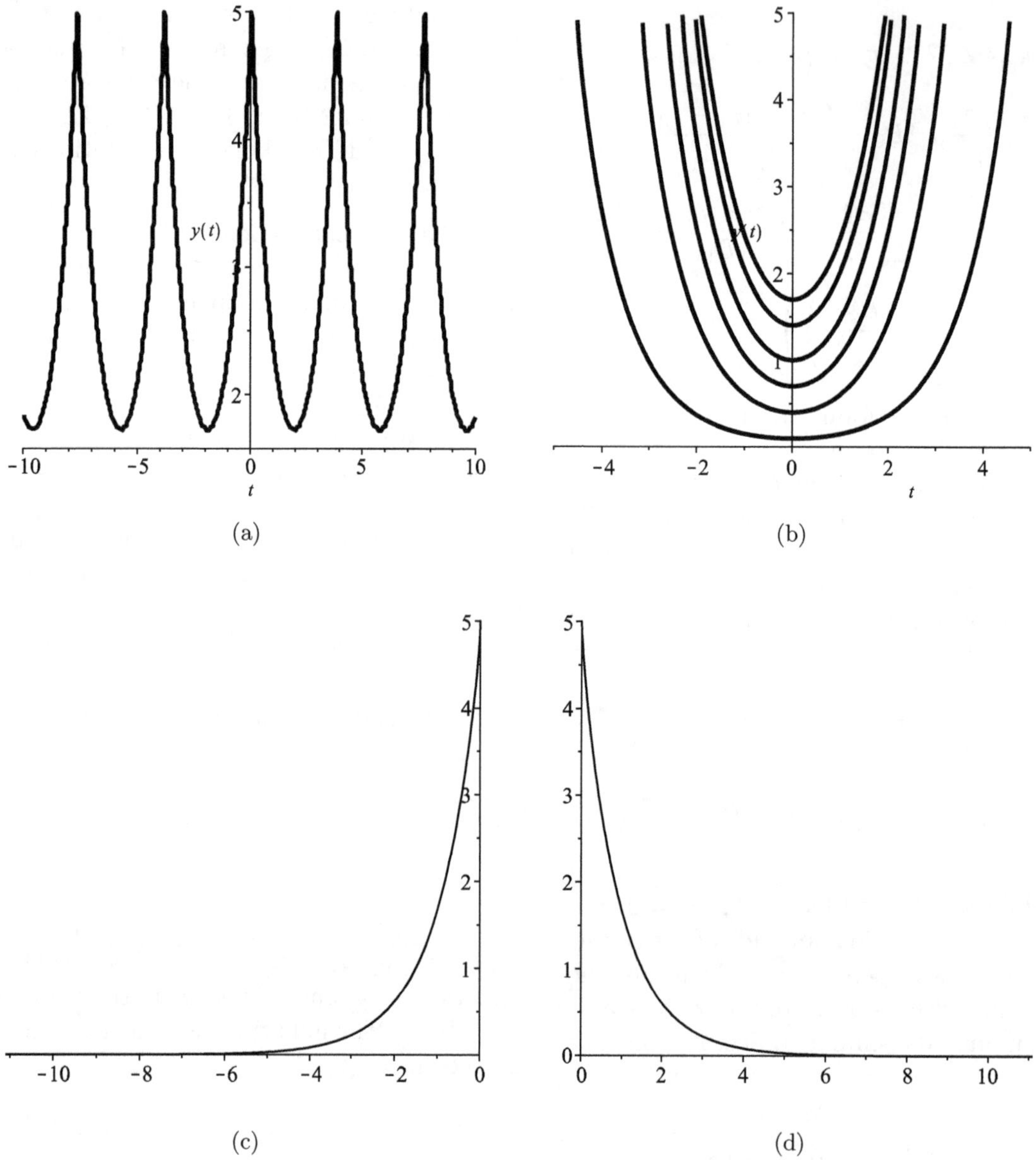

Fig. 5. The different wave profiles of $\phi(\xi)$. (a) Periodic peakon. (b) Compacton family. (c) Unstable manifold. (d) Stable manifold.

solutions of systems (3) and (4), defined by (9) and (11), will gradually become peaked periodic waves, which evolve from nonpeaked periodic waves to smooth pseudo-periodic peakon waves and finally converge to peakon solutions. Thus, peakon is a limit solution of a family of pseudo-periodic peakons given by (9) and (11), respectively.

(2) *When the parameter group (α, κ) of systems (3) and (4) satisfies the conditions that $\alpha > \frac{3(c-\kappa)}{c}$ and $\alpha > \frac{4(c-\kappa)}{c}$ with $\kappa < c$, respectively, there exist homoclinic orbits of systems (3) and (4) given by a branch of the level curves $H_{\text{ch}}(\phi, y) = 0$ and $H_{\text{dp}}(\phi, y) = 0$, respectively. These homoclinic orbits have the exact parametric representations given by (12) and (13), respectively. When $\alpha - \frac{3(c-\kappa)}{c} \ll 1$ and $\alpha - \frac{4(c-\kappa)}{c} \ll 1$, respectively, i.e. $0 < |h_s| \ll 1$, as limit curves of a family of periodic orbits of system (3) and (4) defined by the closed branch of the curves $H_{\text{ch}}(\phi, y) = h, h \in (h_1, 0)$ and $H_{\text{dp}}(\phi, y) = h$, $h \in (h_1, 0)$, respectively, the homoclinic orbits give rise to smooth solitary cusp wave solutions, which are referred to as pseudo-peakons of Eqs. (1) and (2).*

(3) *When the parameter group (α, κ) of systems (3) and (4) satisfies the conditions that $\frac{2(c-\kappa)}{c} < \alpha < \frac{3(c-\kappa)}{c}$ and $\frac{2(c-\kappa)}{c} < \alpha < \frac{4(c-\kappa)}{c}$ with $\kappa < c$, respectively, there exist arch orbits of systems (3) and (4), given by a branch of the level curves $H_{\text{ch}}(\phi, y) = h_s$ and $H_{\text{dp}}(\phi, y) = h_s$, respectively. These arch orbits have the exact parametric representations given by (16) and (17), respectively. A limit curve of a family of periodic orbits of systems (3) and (4) defined by the closed branch of the level curves, $H_{\text{ch}}(\phi, y) = h$ and $H_{\text{dp}}(\phi, y) = h$, $h \in (h_1, h_s)$, respectively, gives rise to periodic peakon solutions of Eqs. (1) and (2), respectively. As parameter α varies such that $h_s \to 0$, the arch orbit becomes the curve triangle. Thus, a peakon is a limit solution of a family of periodic peakons.*

Remark 2.1. One can see, from Figs. 1(d)–1(f), that when $\beta = \kappa - c > 0$ and $\alpha > 0$, there exist three similar phase portraits, as shown in Fig. 2. By moving the origin $O(0,0)$ in Fig. 2 to $E_1(\phi_1, 0)$, one obtains the same phase portraits as those shown in Figs. 1(d)–1(f). Hence, by shifting $-\phi_1$ in the above exact solution formulas, one correspondingly obtains new exact solutions.

Remark 2.2. Generally, consider the planar dynamical system

$$\frac{d\phi}{d\xi} = y, \quad \frac{dy}{d\xi} = \frac{\delta y^2 + \alpha\phi^2 + 2(\kappa - c)\phi}{2(\phi - c)}, \tag{22}$$

with $\delta \neq 1, 2$, system (22) has the first integral

$$H_\delta(\phi, y) = (c-\phi)^{-\delta} y^2 + \frac{(c-\phi)^{-\delta}}{\delta(\delta-1)(\delta-2)} \times [\alpha\delta(\delta-1)\phi^2 - 2p\delta\phi + 2cp], \tag{23}$$

where $p = c\alpha - (\delta - 2)(\kappa - c)$. When $\delta = -1, -2$, one obtains $H_{\text{ch}}(\phi, y)$ and $H_{\text{dp}}(\phi, y)$, respectively.

2.5. *Classification of wave profiles: Using phase orbits to identify the wave profiles of a nonlinear wave equation*

In concluding the above discussions, corresponding to different phase orbits of traveling wave systems, the following classification of wave profiles are obtained.

(i) Pseudo-periodic peakon

If the boundary curves of a period annulus of a center contain a segment of the singular straight line of a traveling wave system, then near this straight line, the parametric representation of every periodic orbit has two time scales. These periodic orbits give rise to a family of pseudo-periodic peakons.

Proposition 1. *A pseudo-periodic peakon is a smooth classical solution, it is not a weak solution in the sense of the distribution.*

(ii) Peakon

Peakon is a special traveling wave solution; at the cusp point, the left derivative and right derivative of the wave function are finite with have different signs. In the phase plane of traveling wave system, as the bounded curves of a periodic annulus of a center, a curve triangle consisting of three segments of the stable and unstable manifolds of a saddle point and a singular straight line gives rise to a peakon solution.

Proposition 2. *Peakon has a determined geometrical property if and only if, in the phase plane of traveling wave system, there exists a periodic annulus of a center such that it has a curve triangle as its bounded curves.*

(iii) Periodic peakon

Periodic peakon is also a special traveling wave, at many cusp points, the left derivative and right derivative of the wave function are finite with different signs.

Proposition 3. *Periodic peakon has a determined geometrical property if and only if, in the phase plane of traveling wave system, there exists a periodic annulus of a center such that it has an arch curve as its bounded curves.*

(iv) The family of pseudo-peakons

An example is the homoclinic orbit shown in Fig. 2(a) where, as the parameter α tends to $\frac{3(c-\kappa)}{c}$, one has $0 < |h_s| \ll 1$. This implies that the homoclinic orbit given by $H(\phi, y) = 0$ has a segment lying in a vertical strip very near the straight line $\phi = c$. Thus, as a limit solution of a family of pseudo-periodic peakons, this homoclinic orbit gives rise a pseudo-peakon solution. Depending on the change of the parameter $\alpha \to \frac{3(c-\kappa)}{c}$, one has a family of homoclinic orbits, which gives rise to a family of pseudo-peakon solutions.

Proposition 4. *A pseudo-peakon is a smooth classical solution, which is not a weak solution in the sense of distribution.*

(v) Compacton family

Definition. Depending on the change of the wave variable ξ, if the "existence time interval" of the wave function $\phi(\xi)$ with respect to the wave variable ξ in the positive or negative direction is always finite, $\phi(\xi)$ is called a breaking wave solution.

As an example, in Fig. 2(c), between the stable and unstable manifolds of the origin and the arch curve defined by $H(\phi, y) = h_s$, there exists a family of open curves given by $H(\phi, y) = h$, $h \in (0, h_s)$. It follows from Theorem 2 that these open curves give rise to uncountably infinitely many bounded breaking wave solutions. These solutions have compact support sets (define area) and bounded value areas, which are called a compacton family.

Proposition 5. *If a traveling wave system has an open orbit family tending to a singular strait line as $y \to \pm\infty$, then this system has a compacton solution family.*

How to identify the profiles of the traveling wave solutions by using known phase orbits? For ease of understanding, Table 1 is used to show the wave profiles determined by different phase orbits of a singular traveling system of the first kind.

In Table 1, (1) Homoclinic orbit to left equilibrium; (2) Homoclinic orbit to right equilibrium; (3) Curve triangle in the left side of $\phi = \phi_s$; (4) Curve triangle in the right side of $\phi = \phi_s$; (1a) Solitary wave of peak type; (2a) Solitary wave of valley type; (3a) Peakon; (4a) Anti-peakon; (5) Arch curve in the left side of $\phi = \phi_s$; (6) Arch curve in the right side of $\phi = \phi_s$; (7) An oval contacting to $\phi = \phi_s$; (8) Two heteroclinic orbits; (5a) Periodic peakon; (6a) Anti-periodic peakon; (7a) Smooth periodic wave; (8a) Kink wave; (9) Homoclinic orbit near the singular straight line; (10) Open orbit families approaching to singular straight line when $|y| \to \infty$; (11) Infinitely many homoclinic orbits to a high-order equilibrium point; (12) The periodic family between two singular straight lines; (9a) Pseudo-peakon; (10a) Compacton solution families; (11a) Infinitely many solitary wave solutions; (12a) Saw-tooth cusp wave.

2.6. *Smooth traveling wave solutions: In the case of existing two node points of the associated regular system in the singular straight line*

An interesting question is as follows: For a singular traveling wave system of the first class, if there exist two node points of the associated regular system in the singular straight line, what dynamics of the solutions of the singular system will emerge?

This section uses a representative model to show that, if there exist two node points of the associated regular system in the singular straight line, then the singular system has no peakon, periodic peakon and compacton solutions. It will be shown that, in this case, the traveling wave system has no curve triangle surrounding a periodic annulus of a center. Instead, there exist smooth periodic wave, solitary wave and kink wave solutions of the singular system because the singular system and its associated regular system define different vector fields, respectively.

As shown by Fig. 6, on the left-hand side of the singular straight line $\phi = \phi_s$, the direction of the orbits of the vector field defined by the singular system has the inverse direction of the orbits of the vector field defined by its associated regular system (see [Li *et al.*, 2009a, 2009b; Li, 2013]).

The following Green–Naghdi (GN) equations (see [Green *et al.*, 1974; Green & Naghdi, 1976; Deng *et al.*, 2011, 2013]) are an approximate system for the full water wave problem that models surface wave propagations through an irrotational, inviscid and incompressible gravity flow, as follows:

$$\begin{aligned} &\eta_t + (u\eta)_x = 0, \\ &u_t + uu_x + \eta_x = \frac{1}{3\eta}[\eta^3(u_{xt} + uu_{xxx} - (u_x)^2)]_x. \end{aligned} \tag{24}$$

To consider the traveling wave solutions of system (24), by letting $\eta(x,t) = \phi(\xi)$, $u(x,t) = u(\xi)$, $\xi = x - ct$, where c is the wave speed, substituting them to system (24), and then integrating the resulting system once, one obtains

$$\begin{aligned} &u = c - \frac{a}{\phi}, \\ &a^2\phi\phi_{\xi\xi} = a^2(\phi_\xi)^2 - \frac{3}{2}\phi^3 + 3b\phi - 3a^2, \end{aligned} \tag{25}$$

Table 1. The wave profiles determined by the phase orbits of a singular system.

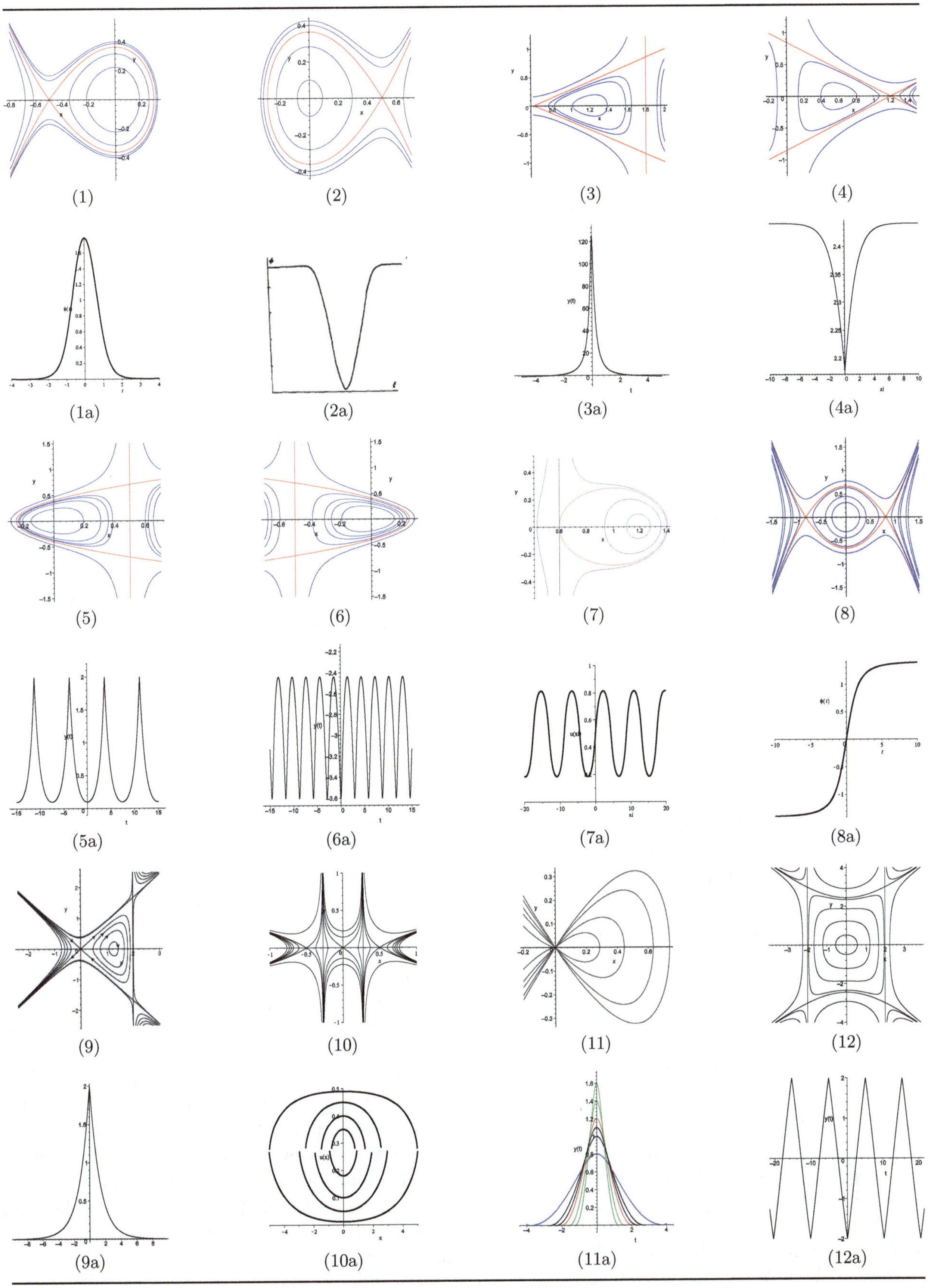

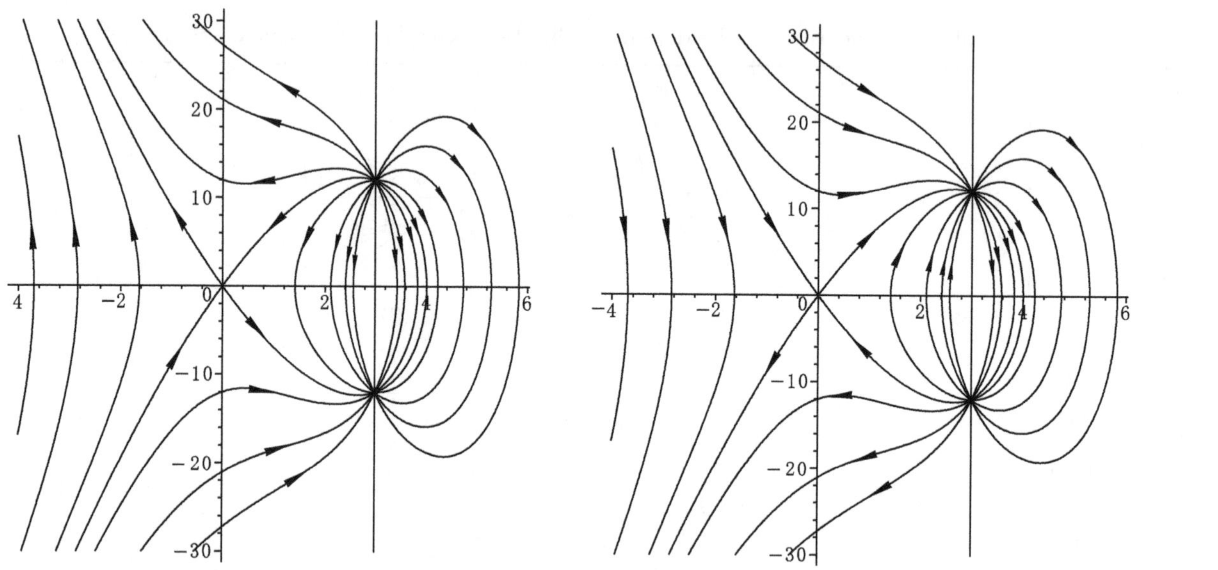

(a) (b)

Fig. 6. Two different vector fields. (a) Vector field defined by the regular system. (b) Vector field defined by the singular system.

where a and b are two integral constants. Assume that $a \neq 0$. The second equation is equivalent to the following two-parameter planar dynamical system:

$$\frac{d\phi}{d\xi} = y, \qquad \frac{dy}{d\xi} = \frac{2a^2y^2 - 3(\phi^3 - 2b\phi + 2a^2)}{2a^2\phi}. \tag{26}$$

System (26) is an integrable system with the first integral

$$H(\phi, y) = \frac{y^2}{\phi^2} - \frac{3}{\phi^2} + \frac{6b}{a^2\phi} + \frac{3}{a^2}\phi = h. \tag{27}$$

Clearly, system (26) is a singular nonlinear traveling wave system of the first kind, defined in [Li & Chen, 2007; Li, 2013], with the singular straight line $\phi = 0$.

Now, consider the associated regular system of system (26):

$$\frac{d\phi}{d\zeta} = 2a^2\phi y, \qquad \frac{dy}{d\zeta} = 2a^2y^2 - 3(\phi^3 - 2b\phi + 2a^2), \tag{28}$$

where $d\xi = 2a^2\phi d\zeta$, with $\phi \neq 0$.

Write $f(\phi) = \phi^3 - 2b\phi + 2a^2$, and note that $f(0) = 2a^2 > 0$ and $f'(\phi)$ has two real zeros at $\phi = \pm\sqrt{\frac{2}{3}}b$ with $b > 0$. It can be easily verified that, when $b > \frac{3}{2}a^{\frac{4}{3}}$, $f(\phi)$ has three real zeros: ϕ_j, $j = 1, 2, 3$, with $\phi_1 < 0 < \phi_2 < \phi_3$; when $b = \frac{3}{2}a^{\frac{4}{3}}$, $f(\phi)$ has one simple real zero and double real zeros with $\phi_1 < 0 < \phi_2 = \phi_3$; when $b < \frac{3}{2}a^{\frac{4}{3}}$, $f(\phi)$ has only one real zero with $\phi_1 < 0$. It implies that, along the ϕ-axis, system (28) has at most three equilibrium points. In the singular straight line $\phi = 0$, there exist two equilibrium points, $N_\pm(0, \pm\sqrt{3})$, of system (28).

Next, let $M(\phi_j, y_j)$ be the coefficient matrix of the linearized system of system (28) at the equilibrium point $E_j(\phi_j, y_j)$, and denote $J(\phi_j, y_j) = \det M(\phi_j, y_j)$. Then, one has

$$J(\phi_j, 0) = 6a^2\phi_j f'(\phi_j),$$

$$J(0, \pm\sqrt{3}) = 24a^4 > 0,$$

$$(\text{trace}M(0, \pm\sqrt{3}))^2 - 4J(0, \pm\sqrt{3}) = 12a^4 > 0.$$

By the theory of planar dynamical systems (see [Li, 2013]), for an equilibrium point of a planar integrable system, if $J < 0$ then the equilibrium point is a saddle point; if $J > 0$ and $(\text{Trice}M)^2 - 4J < 0$ (>0) then it is a center point (a node point); if $J = 0$ and the Poincaré index of the equilibrium point is 0, then this equilibrium point is a cusp.

From the above discussion, one can see that the equilibrium points $N_\pm$ are node points of system (28), $E_1(\phi_1, 0)$ and $E_2(\phi_2, 0)$ are two saddle points, and $E_3(\phi_3, 0)$ is a center point. Write $h_j = H(\phi_j, 0)$ and obtain the bifurcations of phase portraits of system (28), as shown in Figs. 7(a)–7(c).

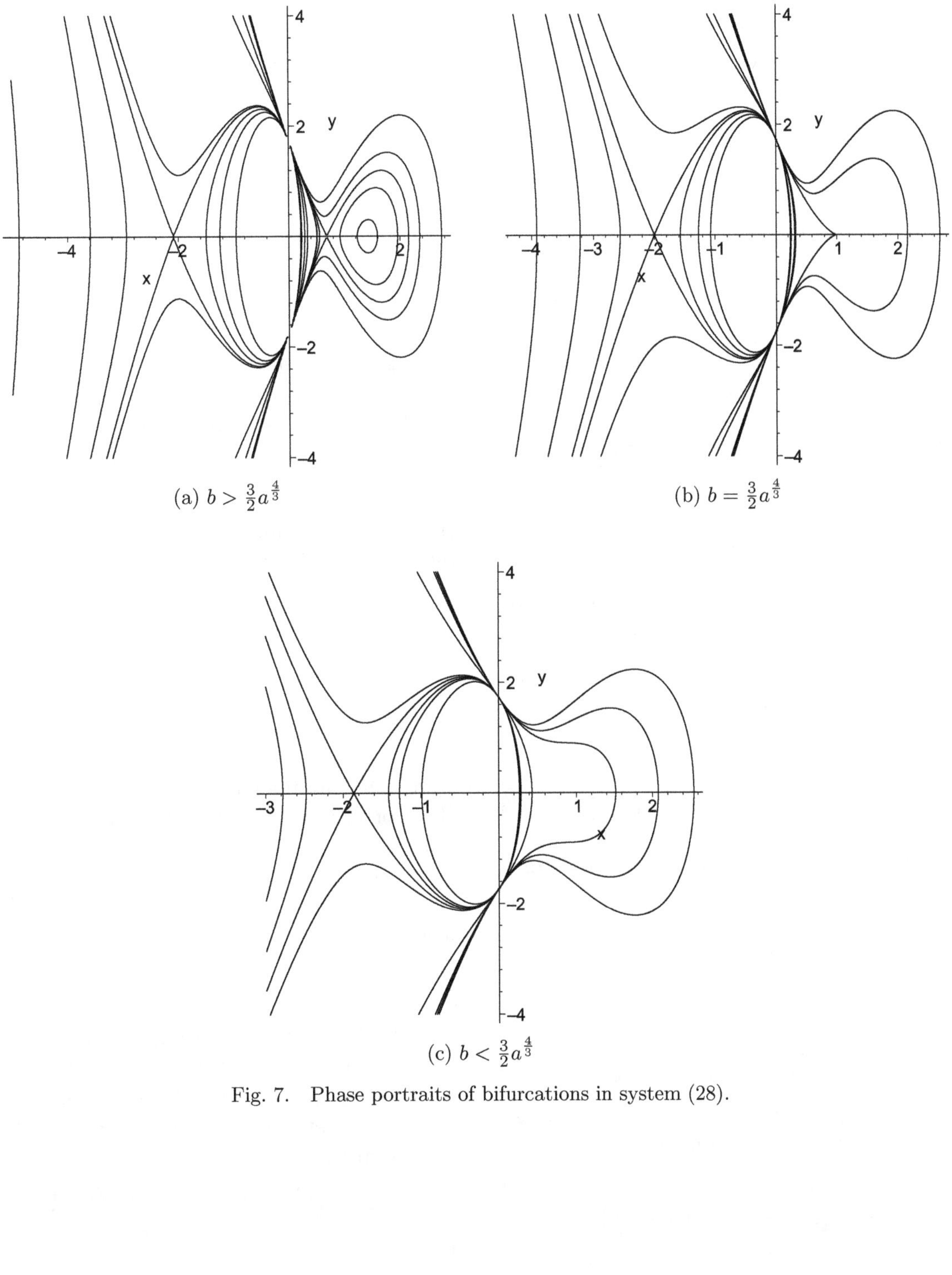

(a) $b > \frac{3}{2}a^{\frac{4}{3}}$

(b) $b = \frac{3}{2}a^{\frac{4}{3}}$

(c) $b < \frac{3}{2}a^{\frac{4}{3}}$

Fig. 7. Phase portraits of bifurcations in system (28).

It is noted that, when the phase points pass through the singular straight line, the direction of the vector field defined by system (26) is changed to the inverse direction defined by system (28). Thus, when there exist two node points of system (28), as shown in the straight line $\phi = 0$ in Fig. 7, the singular system (26) may have a family of periodic orbits and homoclinic orbits.

(a) The case of $b > \frac{3}{2}a^{\frac{4}{3}}$ [see Fig. 7(a)].

One can easily see that, when h is varied, the changes of the level curves defined by $H(\phi, y) = h$ in (27) are as shown in Figs. 8(a)–8(f).

Next, consider the exact parametric representations of the orbits of system (26).

(i) Corresponding to the periodic orbits defined by $H(\phi, y) = h, h \in (-\infty, h_1)$ [see Fig. 8(a)], the Hamiltonian can be written as

$$y^2 = \frac{3}{a^2}\left[a^2 - 2b\phi + \frac{a^2 h}{3}\phi^2 - \phi^3\right].$$

Thus, it follows from the first equation in system (26) that

$$\frac{\sqrt{3}}{a}\xi = \int_{r_2}^{\phi} \frac{d\phi}{\sqrt{(r_1 - \phi)(\phi - r_2)(\phi - r_3)}}.$$

It leads to the following parametric representation of the family of periodic orbits of system (26):

$$\phi(\xi) = r_3 + \frac{r_2 - r_3}{\mathrm{dn}^2(\Omega_1 \xi, k)}, \tag{29}$$

where

$$\Omega_1 = \frac{1}{2a}\sqrt{\frac{3}{r_1 - r_3}} \quad \text{and} \quad k^2 = \frac{r_1 - r_2}{r_1 - r_3}.$$

(ii) Corresponding to the homoclinic orbit defined by $H(\phi, y) = h_1$ in Fig. 8(b), one has

$$y^2 = \frac{3}{a^2}(\phi - \phi_1)^2(\phi_{M1} - \phi)$$

and

$$\frac{\sqrt{3}}{a}\xi = \int_{\phi}^{\phi_{M1}} \frac{d\phi}{(\phi - \phi_1)\sqrt{\phi_{M1} - \phi}}.$$

Thus, the parametric representation of the homoclinic orbit to the equilibrium point $E_1(\phi_1, 0)$ of system (26) is

$$\phi(\xi) = \phi_1 + (\phi_{M1} - \phi_1)\,\mathrm{sech}^2(\omega_1 \xi), \tag{30}$$

where

$$\omega_1 = \frac{1}{2a}\sqrt{3(\phi_{M1} - \phi_1)}.$$

On the other hand, corresponding to the stable manifold of the saddle $E_1(\phi_1, 0)$ defined by $H(\phi, y) = h_1$ in the upper half-phase plane shown in Fig. 8(b), one has

$$\frac{\sqrt{3}}{a}\xi = \int_{\phi_0}^{\phi} \frac{d\phi}{(\phi_1 - \phi)\sqrt{\phi_{M1} - \phi}}, \qquad \phi_0 < \phi_1.$$

Hence, the parametric representation of the stable manifold of the equilibrium point $E_1(\phi_1, 0)$ of system (26) is

$$\phi(\xi) = \phi_1 - (\phi_{M1} - \phi_1)\mathrm{csch}^2(\omega_1 \xi - \xi_0), \tag{31}$$

where

$$\xi_0 = \mathrm{ctnh}^{-1}\frac{\sqrt{\phi_{M1} - \phi_0}}{\phi_{M1} - \phi_1}.$$

Obviously, (31) is an unbounded solution of system (26).

(iii) Corresponding to the periodic orbits defined by $H(\phi, y) = h$, $h \in (h_3, h_2)$, as shown in Fig. 8(d), one has the same parametric representation of the family of periodic orbits of system (26), as shown in (29).

(iv) Corresponding to the homoclinic orbit defined by $H(\phi, y) = h_2$ in Fig. 8(e), one has

$$y^2 = \frac{3}{a^2}(\phi - \phi_2)^2(\phi_{M2} - \phi)$$

and

$$\frac{\sqrt{3}}{a}\xi = \int_{\phi}^{\phi_{M2}} \frac{d\phi}{(\phi - \phi_2)\sqrt{\phi_{M2} - \phi}}.$$

Thus, the parametric representation of the homoclinic orbit to the equilibrium point $E_2(\phi_2, 0)$ of system (26) is

$$\phi(\xi) = \phi_2 + (\phi_{M2} - \phi_2)\,\mathrm{sech}^2(\omega_2 \xi), \tag{32}$$

where

$$\omega_2 = \frac{1}{2a}\sqrt{3(\phi_{M2} - \phi_2)}.$$

Similarly, corresponding to the stable manifold of the saddle $E_2(\phi_2, 0)$ defined by $H(\phi, y) = h_2$ in the upper-half phase plane as shown in Fig. 8(e), the parametric representation of an unbounded solution of system (26) is

$$\phi(\xi) = \phi_2 - (\phi_{M1} - \phi_2)\mathrm{csch}^2(\omega_2 \xi - \xi_0), \tag{33}$$

where

$$\xi_0 = \mathrm{ctnh}^{-1}\frac{\sqrt{\phi_{M2} - \phi_0}}{\phi_{M2} - \phi_2}, \qquad \phi_0 < \phi_2, \ \phi_0 \neq 0.$$

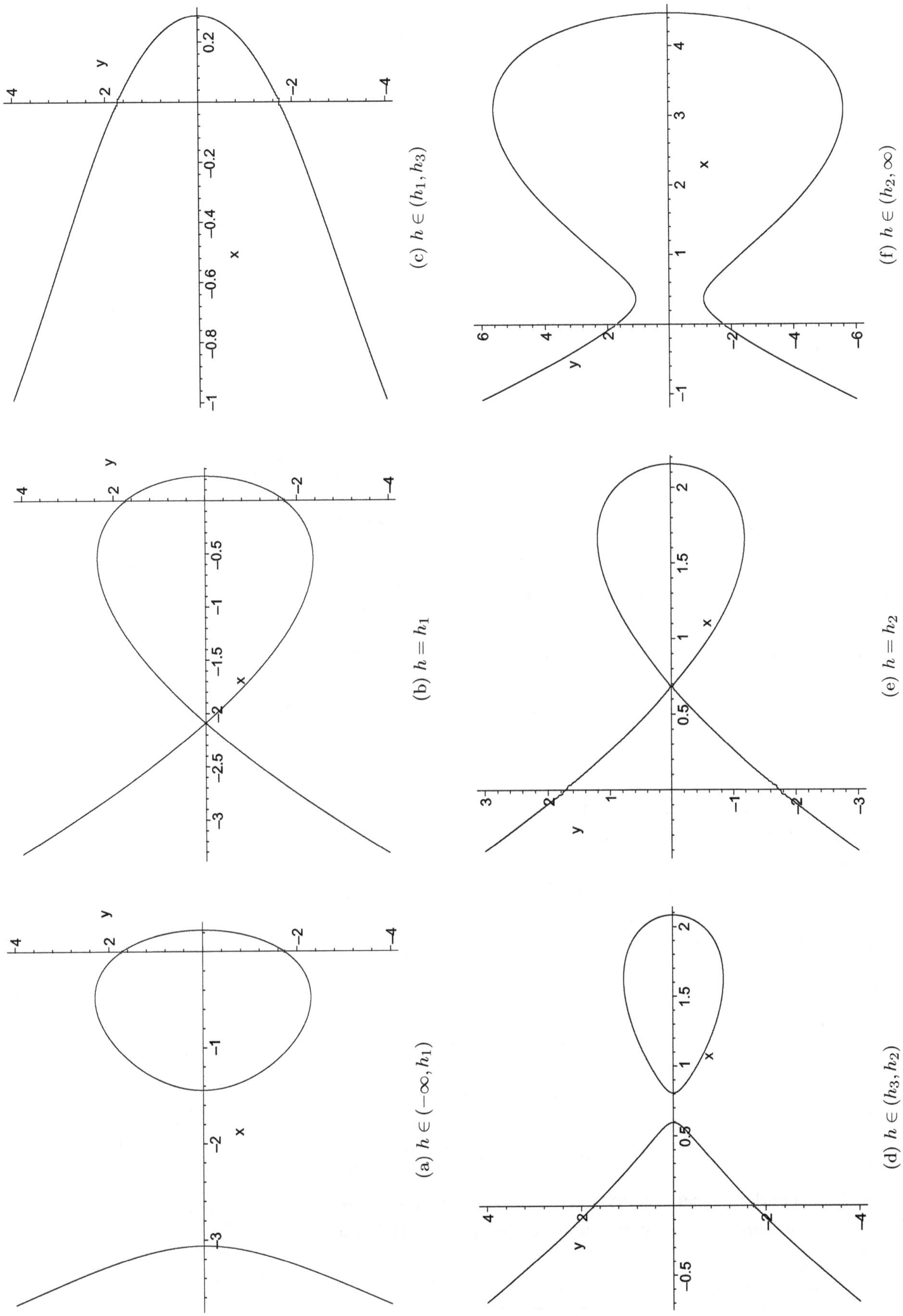

(a) $h \in (-\infty, h_1)$ (b) $h = h_1$ (c) $h \in (h_1, h_3)$

(d) $h \in (h_3, h_2)$ (e) $h = h_2$ (f) $h \in (h_2, \infty)$

Fig. 8. Changes of the level curves defined by $H(\phi, y) = h$ with $b > \frac{3}{2}a^{\frac{4}{3}}$.

(b) The case of $b \le \frac{3}{2}a^{\frac{4}{3}}$ [see Figs. 7(b) and 7(c)].

When $b = \frac{3}{2}a^{\frac{4}{3}}$ or $b < \frac{3}{2}a^{\frac{4}{3}}$, for $h \in (-\infty, h_1)$ and $h = h_1$, the level curves defined by $H(\phi, y) = h$ are the same as the case of $b > \frac{3}{2}a^{\frac{4}{3}}$. Therefore, in these two cases, system (26) has a periodic family and a homoclinic orbit given by (29) and (30), respectively.

When $b = \frac{3}{2}a^{\frac{4}{3}}$, one has $\phi_1 = -2a^{\frac{2}{3}}$, $\phi_2 = \phi_3 = a^{\frac{2}{3}}$ and $h_1 = -\frac{45}{4}a^{-\frac{4}{3}}$, $h_2 = 9a^{\frac{4}{3}}$. For $h = h_2$, the level curves defined by $H(\phi, y) = h_2$ are the stable manifold and the unstable manifold to the double equilibrium point (cusp point) $E_2(\phi_2, 0)$, respectively. In this case, $y^2 = \frac{3}{a^2}(a^{\frac{2}{3}} - \phi)^3$. Thus, by using the first equation of system (26), one obtains the parametric representation of the stable manifold of the equilibrium point $E_2(\phi_2, 0)$, as follows:

$$\phi(\xi) = a^{\frac{2}{3}} - \frac{4}{\left(2a^{-\frac{1}{3}} + \frac{\sqrt{3}}{a}\xi\right)^2}. \tag{34}$$

Obviously, when $\xi \to -\frac{2}{\sqrt{3}}a^{\frac{2}{3}}$, $\phi(\xi) \to \infty$, i.e. (34) gives rise to an unbounded solution of system (26).

Summarizing the above analysis gives the following result.

Theorem 2. *For the traveling wave system (26) of the GN equation:*

(1) *For any parameter pair $(a, b) \ne (0, 0)$, system (26) has a smooth periodic solution family given by (29) and a smooth solitary wave solution given by (30).*
(2) *Furthermore, if $b > \frac{3}{2}a^{\frac{4}{3}}$, then corresponding to two families of periodic orbits and two homoclinic orbits of system (26), there exist two families of periodic solutions given by (29) and two smooth solitary wave solutions given by (30) and (32), respectively.*
(3) *Corresponding to the stable manifolds of the equilibrium points of system (26), there exist unbounded solutions of system (26) given by (31), (33) and (34).*

3. Variform Exact Peakon Solutions for Some Nonlinear Wave Models

This section discusses the exact peakon solutions for three nonlinear wave models. It will be shown that there are various exact explicit one-peakon solutions, which are different from the peakon solution given by Eqs. (1) and (2). It will be seen that, under different parameter conditions, one nonlinear wave equation can have different exact one-peakon solutions and different nonlinear wave equations can have different explicit exact one-peakon solutions.

3.1. *Exact peakons of modified Serre shallow water wave equations*

Consider the following modified Serre equation (see [Carter & Cienfuegos, 2011; Clamond *et al.*, 2017; Li *et al.*, 2019]):

$$\begin{aligned} &h_t + (hu)_x = 0, \\ &(hu)_t + \left(hu^2 + \frac{1}{2}h^2\right)_x \\ &\quad = \mu^2\left(\left[\frac{1}{3}[u_{xt} + uu_{xx} - u_x^2]h^3\right]_x \right. \\ &\quad\quad \left. -B\left[\frac{1}{2}h_x^2 - hh_{xx}\right]_x\right). \end{aligned} \tag{35}$$

Let $h(x, t) = h(\xi)$, $u(x, t) = \phi(\xi)$, $\xi = x - ct$. It can be seen from the first equation of (24) that $h(\xi) = \frac{g_1}{\phi - c}$, where $g_1 \ne 0$ is an integral constant. Substituting it into the second equations of (35) and then integrating the resulting equations once, one obtains

$$\phi'' = \frac{\frac{1}{3}\mu^2 g_1^2\left[(\phi - c) - \frac{9B}{2g_1}\right](\phi')^2 + (\phi - c)^2\left[\frac{1}{2}g_1 + (\phi - c)^2\left(\phi - \frac{g_2}{g_1}\right)\right]}{\frac{1}{3}\mu^2 g_1^2(\phi - c)\left[(\phi - c) - \frac{3B}{g_1}\right]}, \tag{36}$$

where g_2 is the second integral constant. By introducing new parameters $a = \frac{1}{3}\mu^2 g_1$, $b = \frac{3B}{g_1}$, it can be verified that Eq. (36) are equivalent to the following planar dynamical system:

$$\frac{d\phi}{d\xi} = y, \quad \frac{dy}{d\xi} = \frac{a\left(\phi - c - \frac{3}{2}b\right)y^2 + (\phi - c)^2\left[\frac{1}{2}g_1 + (\phi - c)^2\left(\phi - \frac{g_2}{g_1}\right)\right]}{a(\phi - c)(\phi - c - b)}. \tag{37}$$

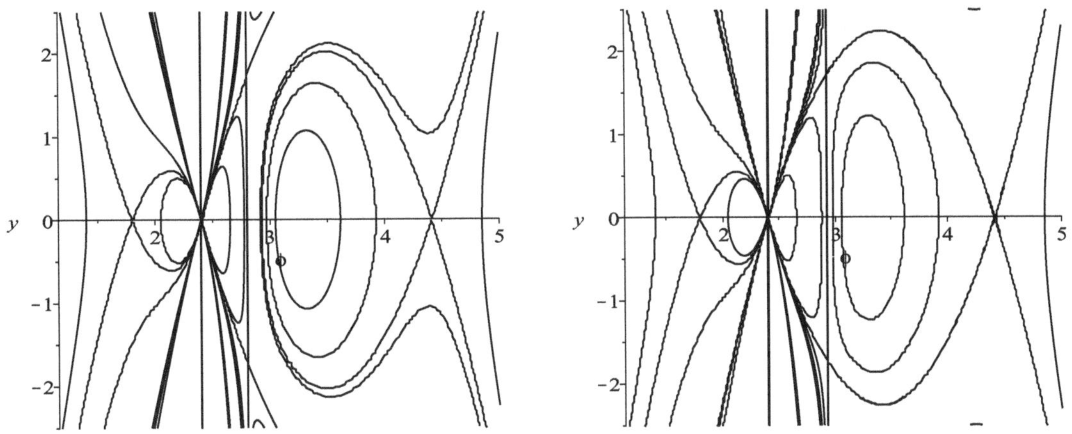

(a) $h_3 < h_s$ (b) $h_3 = h_s$

Fig. 9. Two phase portraits of system (37).

System (37) has the following first integral:

$$H_{se}(\phi, y) = \frac{y^2(\phi - c - b)}{(\phi - c)^3} - \frac{\phi\left(\phi - \frac{2g_2}{g_1}\right)(\phi - c) - g_1}{a(\phi - c)} = h. \tag{38}$$

Clearly, for fixed $c > 0, a > 0$, system (37) has a double equilibrium point $E_c(c, 0)$. Assume that the parameter group (g_1, g_2) makes system (37) have three simple equilibrium points $E_j(\phi_j, 0)$, $j = 1, 2, 3$, satisfying $0 < \phi_1 < c < c + b < \phi_2 < \phi_3$. In the straight line $\phi = c+b$, there exist two equilibrium points, $S_{\mp}(c+b, \mp y_s)$, of the associated regular system of (37), where $y_s^2 = \frac{2b}{a}[\frac{1}{2}g_1 + b^2(c + b - \frac{g_2}{g_1})]$. Write $h_j = H_{se}(\phi_j, 0)$, $h_s = H_{se}(c + b, y_s)$. Then, system (37) has the two phase portraits shown in Figs. 9(a) and 9(b).

Next, consider the exact parametric representations of peakon solutions of (37), which correspond to the curve triangles shown in Figs. 9(a) and 9(b). One can see from (38) that

$$y^2 = \frac{(\phi - c)^2\left[ah(\phi - c) + \phi(\phi - c)\left(\phi - \frac{2g_2}{g_1}\right) - g_1\right]}{a(\phi - c - b)} \equiv \frac{(\phi - c)^2 P_3(\phi)}{a(\phi - c - b)}.$$

Note that, when $h = h_s$, one has

$$P_3(\phi) = (\phi - c - b)\left[\phi^2 + \left(b - \frac{2g_2}{g_1}\right)\phi - \left(c^2 + bc - \frac{2cg_2}{g_1} - \frac{g_1}{b}\right)\right] \equiv (\phi - c - b)(\phi^2 + b_1\phi + a_1).$$

Hence, one obtains $y^2 = \frac{1}{a}(\phi - c)^2(\phi^2 + b_1\phi + a_1)$. By using the first equation of (37), generally one has

$$\frac{1}{\sqrt{a}}\xi = \int_{\phi_0}^{\phi} \frac{|\phi - c - b|d\phi}{|\phi - c|\sqrt{|\phi - c - b|P_3(\phi)}}.$$

When $h = h_s$, one obtains

$$\frac{1}{\sqrt{a}}\xi = \int_{\phi_0}^{\phi} \frac{d\phi}{|\phi - c|\sqrt{\phi^2 + b_1\phi + a_1}}.$$

(i) First, consider the curve triangle shown in Fig. 9(a), which connects the three equilibrium points $E_c(c, 0)$ and $S_{\mp}(c + b, \mp y_s)$ of system (37) defined by $H_{se}(\phi, y) = h_s$. Corresponding to the curve segment connecting $E_c(c, 0)$ and $S_+(c+b, y_s)$, the above integral becomes

$$\frac{1}{\sqrt{a}}\xi = \int_{\phi}^{c+b} \frac{d\phi}{(\phi - c)\sqrt{a_1 + b_1\phi + \phi^2}} = \int_{\psi}^{b} \frac{d\psi}{\psi\sqrt{a_0 + b_0\psi + \psi^2}},$$

where

$$\psi = \phi - c, \quad a_0 = \frac{g_1}{b},$$

$$b_0 = 2c + b - \frac{2g_2}{g_1} < 0.$$

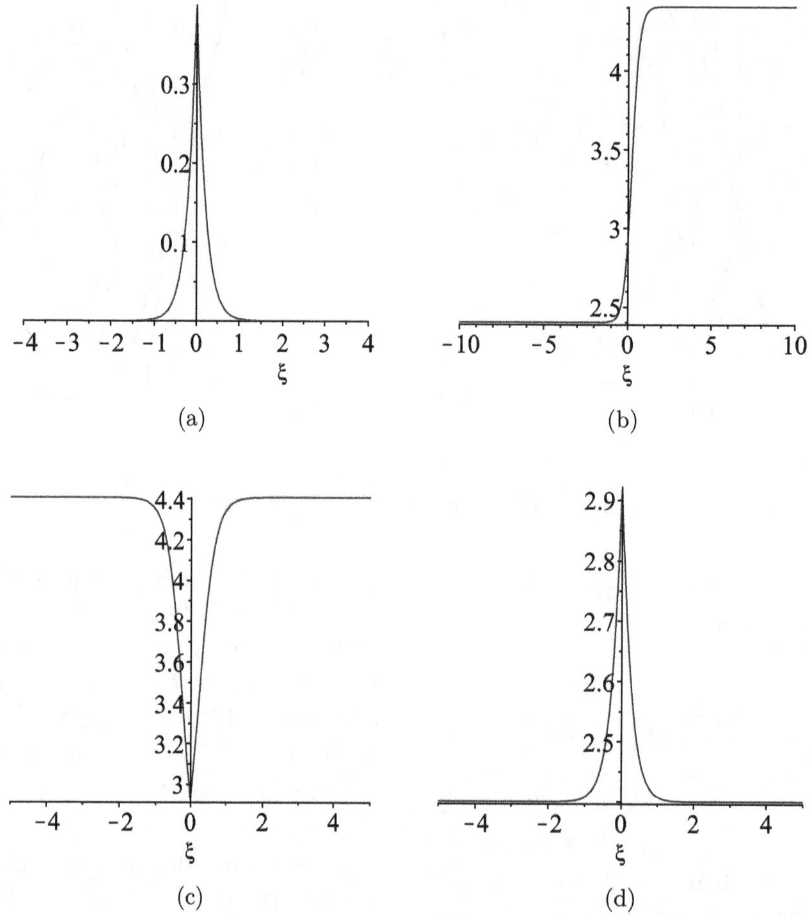

Fig. 10. Three wave profiles of the u-component of system (35). (a) Peakon given by (39). (b) Kink wave given by (40). (c) Anti-peakon. (d) Peakon given by (42)

It then follows the peakon solution of the u-component of system (35) [see Fig. 10(a)]:

$$\phi(\xi) = c + \frac{2a_0}{\sqrt{-\Delta_1}\sinh\left(\sqrt{\frac{g_1}{ab}}|\xi| + \xi_{01}\right) + |b_0|},$$

$$\xi \in (-\infty, \infty), \tag{39}$$

where

$$\Delta_1 = b_0^2 - 4a_0 < 0 \quad \text{and}$$

$$\xi_{01} = \sinh^{-1}\left(\frac{2a_0 + bb_0}{b\sqrt{-\Delta_1}}\right).$$

(ii) Second, consider the two curve triangles shown in Fig. 9(b), which are defined by the level curves $H_2(\phi, y) = h_s = h_3$.

For two heteroclinic orbits connecting the equilibrium points $E_c(c, 0)$ and $E_3(\phi_3, 0)$, passing through the straight line $\phi = c + b$, one has

$$\frac{1}{\sqrt{a}}\xi = \int_{c+b}^{\phi} \frac{d\phi}{(\phi - c)(\phi_3 - \phi)}.$$

Thus, one obtains the following kink wave and anti-kink wave solutions [see Fig. 10(b)]:

$$\phi(\xi) = \phi_3 + \frac{c - \phi_3}{1 + \gamma_0 e^{\pm\omega_0 \xi}},$$

$$\xi \in (-\infty, \infty), \tag{40}$$

where

$$\omega_0 = \frac{\phi_3 - c}{\sqrt{a}} \quad \text{and} \quad \gamma_0 = \frac{b}{\phi_3 - c - b}.$$

Corresponding to the left curve triangle $E_c S_+ S_-$, as a limit solution of a family of homoclinic orbits, as $h \to h_s$, it gives rise to an anti-peakon, which has the following parametric representation [see Fig. 10(c)]:

$$\phi(\xi) = \phi_3 + \frac{c - \phi_3}{1 + \gamma_0 e^{\omega_0 |\xi|}}, \quad \xi \in (-\infty, \infty). \tag{41}$$

Corresponding to the right curve triangle $S_+ E_3 S_-$, as a limit solution of a family of periodic orbits enclosing the center $E_2(\phi_2, 0)$, as $h \to h_s$, it gives rise to a peakon, which has the following parametric representation [see Fig. 10(d)]:

$$\phi(\xi) = \phi_3 + \frac{c - \phi_3}{1 + \gamma_0 e^{-\omega_0|\xi|}}, \quad \xi \in (-\infty, \infty). \tag{42}$$

3.2. *Peakons of the multicomponent Korteweg–de Vries equation with dispersion*

Kupershmidt [1985] proposed the following equations:

$$\begin{aligned} u_t &= -u_{xxx} + 6uu_x + 2v^T v_x + C^T v_{xx}, \\ v_t &= (2uv)_x - u_{xx}C, \end{aligned} \tag{43}$$

where $v = (v_1, v_2, \ldots, v_n)^T$ and $C = (c_1, c_2, \ldots, c_n)^T$ is a constant (column) vector. System (43) is a bi-Hamiltonian system with an infinite number of conservation laws.

To investigate the traveling wave solutions of system (43), let

$$\begin{aligned} u(x,t) &= \phi(x - ct) = \phi(\xi), \\ v_i(x,t) &= v_i(x - ct) = v_i(\xi), \quad \xi = x - ct. \end{aligned} \tag{44}$$

Substituting (44) into the second equation of system (43) and integrating the obtained equation once, it yields

$$v_i = \frac{c_i\phi_\xi}{c + 2\phi}, \quad (v_i)_\xi = \frac{c_i\phi_{\xi\xi}(c + 2\phi) - 2c_i\phi_\xi^2}{(c + 2\phi)^2}, \quad i = 1, 2, \ldots, n, \tag{45}$$

where the integral constant is taken as zero.

Substituting (45) into the first equation of system (43) and integrating the obtained result once, it gives the following planar dynamical system (see [Li & Han, 2020]):

$$\begin{aligned} \frac{d\phi}{d\xi} &= y, \\ \frac{dy}{d\xi} &= \frac{-a^2y^2 + (c + 2\phi)^2(3\phi^2 + c\phi + g)}{(c + 2\phi)(c + 2\phi - a^2)}, \end{aligned} \tag{46}$$

which has the first integral

$$H_{ku}(\phi, y) = \frac{(c + 2\phi - a^2)y^2}{c + 2\phi} - (2g\phi + c\phi^2 + 2\phi^3) = h, \tag{47}$$

where h and g are constants.

Consider the associated regular system of system (46), given as follows:

$$\begin{aligned} \frac{d\phi}{d\zeta} &= y(c + 2\phi)(c + 2\phi - a^2), \\ \frac{dy}{d\zeta} &= -a^2y^2 + (c + 2\phi)^2(3\phi^2 + c\phi + g), \end{aligned} \tag{48}$$

where $d\xi = (c + 2\phi)(c + 2\phi - a^2)d\zeta$, for $\phi \neq \phi_{s1} = -\frac{1}{2}c$ and $\phi \neq \phi_{s2} = \frac{a^2-c}{2}$. Systems (43) and (48) have the same first integral, but in the phase plane, these two systems define different vector fields in the two sides of the singular straight lines.

Obviously, system (48) has the equilibrium points $E_1(\phi_1, 0)$, $E_2(\phi_2, 0)$ and $E_s(\phi_{s1}, 0)$, where $\phi_{1,2} = \frac{1}{6}(-c \mp \sqrt{\Delta})$, when $\Delta = c^2 - 12g > 0$. When $F_s = 3a^4 - 4ca^2 + c^2 + 4g > 0$, in the singular line $\phi = \phi_{s2} = \frac{a^2-c}{2}$, there exist two equilibrium points, $S_{\mp}(\phi_{s2}, \mp y_s)$, where $y_s = \frac{a}{2}\sqrt{F_s}$. Clearly, when $\Delta < 0$, one has $F_s > 0$. The point $E_s(-\frac{1}{2}c, 0)$ is a double equilibrium point of system (48).

For the first integral given by (47), write

$$h_1 = H_{ku}(\phi_1, 0) = \frac{1}{54}(-\Delta^{\frac{3}{2}} - c + 18cg),$$

$$h_2 = H_{ku}(\phi_2, 0) = \frac{1}{54}(\Delta^{\frac{3}{2}} - c + 18cg)$$

and

$$h_s = H_{ku}(\phi_{s2}, \mp y_s) = -\frac{1}{4}(a^2 - c)(a^4 - ca^2 + 4g).$$

For a fixed pair (a^2, c) with $c > 0$, when $g = g_s = \frac{1}{4}ca^2 - \frac{3}{16}a^4$, one has $h_2 = h_s$. Taking $g = g_s$, one obtains the phase portrait of system (46) shown in Fig. 11(a). The level curves defined by $H_{ku}(\phi, y) = h_s$ are shown in Fig. 11(b).

It follows from (47) that

$$y^2 = \frac{(h + 2g\phi + c\phi^2 + 2\phi^3)(c + 2\phi)}{(c + 2\phi - a^2)}.$$

By using the first equation of (46), one obtains

$$\sqrt{2}\xi = \int_{\phi_0}^{\phi} \frac{(\phi_{s2} - \phi)d\phi}{\sqrt{\left(\frac{1}{2}h + g\phi + \frac{1}{2}c\phi^2 + \phi^3\right)(\phi_{s2} - \phi)(\phi - \phi_{s1})}}.$$

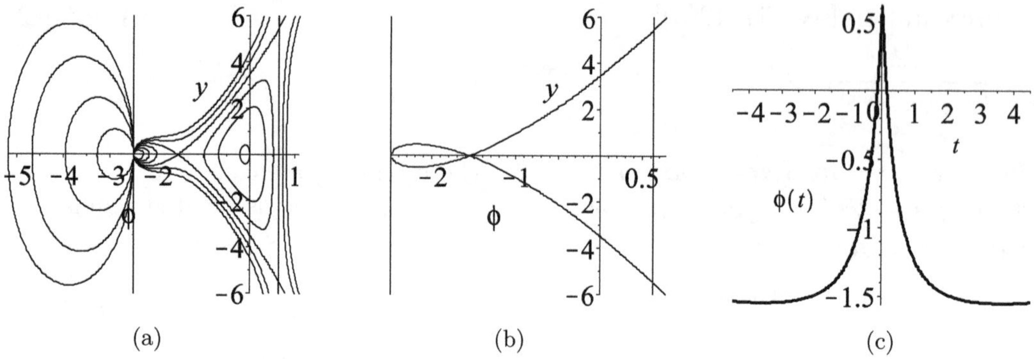

Fig. 11. The curve triangle defined by $H_{ku}(\phi, y) = h_s$ of system (43). (a) Phase portrait when $g = g_s$. (b) Level curves defined by $H_{ku}(\phi, y) = h_s$. (c) Peakon.

When $h = h_s$, for the curve triangle, this integral becomes

$$\sqrt{2}\xi = \int_{\phi}^{\phi_{s2}} \frac{d\phi}{(\phi_1 - \phi)\sqrt{\phi - \phi_{s1}}}.$$

Thus, one obtains the following peakon solution [see Fig. 11(c)] along with its parametric representation:

$$\begin{aligned}\phi(\xi) &= \phi_{s1} + (\phi_1 - \phi_{s1}) \left(\frac{e^{\omega_1|\xi|} + q_0 e^{-\omega_1|\xi|}}{e^{\omega_1|\xi|} - q_0 e^{-\omega_1|\xi|}} \right)^2 \\ &\equiv \phi_{s1} + \frac{\phi_1 - \phi_{s1}}{\tanh_{q0}^2(\omega_1|\xi|)} \\ &= \phi_{s1} + (\phi_1 - \phi_{s1})\mathrm{ctnh}_{q0}^2(\omega_1|\xi|), \end{aligned} \tag{49}$$

where

$$q_0 = \frac{\sqrt{\phi_{s2} - \phi_{s1}} - \sqrt{\phi_1 - \phi_{s1}}}{\sqrt{\phi_{s2} - \phi_{s1}} + \sqrt{\phi_1 - \phi_{s1}}},$$

$$\omega_1 = \sqrt{\frac{\phi_1 - \phi_{s1}}{2}},$$

and $\tanh_{q0}(\xi), \mathrm{ctnh}_{q0}(\xi)$ are the Arab q-deformed functions (see [Arab, 1991, 2001]).

3.3. *Peakons of the rotation-two-component Camassa–Holm system*

Consider the rotation-two-component CH system (see [Fan *et al.*, 2016; Fan & Yuen, 2019]):

$$\begin{aligned} &u_t - u_{xxt} - Au_x + 3uu_x - \sigma(2u_x u_{xx} + uu_{xxx}) \\ &\quad + \mu u_{xxx} + (1 - 2\Omega A)\rho\rho_x - 2\Omega\rho(u\rho)_x = 0, \\ &\rho_t + (\rho u)_x = 0. \end{aligned} \tag{50}$$

To investigate the traveling wave solutions of system (50), by letting

$$u(x,t) = \phi(x - ct) = \phi(\xi),$$
$$\rho(x,t) = v(x - ct) = v(\xi),$$

where c is the wave speed, one can see from the second equation of system (50) that $v(\xi) = \frac{\beta}{\phi - c}$, where β is an integration constant and $\beta \neq 0$. By the first equation of system (50), one has

$$(\sigma\phi - c - \mu)\phi'' = -\frac{1}{2}\sigma(\phi')^2 - (A + c)\phi + \frac{3}{2}\phi^2 + \frac{(1 - 2A\Omega + 2c\Omega)\beta^2}{2(\phi - c)^2} - \frac{1}{2}g,$$

where $\frac{1}{2}g$ is the second integration constant. Write $\alpha = 1 - 2A\Omega + 2c\Omega$. Then, the above equation is equivalent to the following two-dimensional system:

$$\frac{d\phi}{d\xi} = y, \quad \frac{dy}{d\xi} = \frac{-\sigma y^2(\phi - c)^2 + (\phi - c)^2[3\phi^2 - 2(A + c)\phi - g] + \alpha\beta^2}{2(\phi - c)^2(\sigma\phi - c - \mu)}, \tag{51}$$

which has the following first integral:

$$H_{\mathrm{rch}}(\phi, y) = y^2(\sigma\phi - c - \mu) - \phi^3 + (A + c)\phi^2 + g\phi + \frac{\alpha\beta^2}{(\phi - c)} = h. \tag{52}$$

Assume that $A > 0, \alpha > 0$. System (51) has two singular lines $\phi = c$ and $\phi = \frac{c+\mu}{\sigma}$.

To see the equilibrium points of system (51), write

$$f(\phi) = (\phi - c)^2(3\phi^2 - 2(A+c)\phi - g) + \alpha\beta^2,$$

$$f'(\phi) = 2(\phi - c)[6\phi^2 - 3(A+2c)\phi + c(A+c) - g],$$

$$f''(\phi) = 2(18\phi^2 - 6(A+4c)\phi + c(4A+7c) - 2g.$$

Apparently, $f'(\phi)$ has one zero at $\phi = \phi_{s1} = c$. When

$$\Delta = 9A^2 + 12Ac + 12c^2 + 24g > 0, \quad f'(\phi)$$

has two real zeros, at

$$\phi = \tilde{\phi}_{1,2} = \frac{1}{12}[3(A+2c) \mp \sqrt{\Delta}].$$

Thus,

$$f(c) = \alpha\beta^2, f'(c) = 0 \quad \text{and}$$

$$f''(c) = 2(c^2 - 2cA - g), \quad f(0) = \alpha\beta^2 - gc^2.$$

On the ϕ-axis, the equilibrium points $E_j(\phi_j, 0)$ of system (51) satisfy $f(\phi_j) = 0$. Obviously, system (51) has at most four equilibrium points at $E_j(\phi_j, 0)$, $j = 1, 2, 3, 4$. On the straight line $\phi = c$, there is no equilibrium point of the associated regular system of (51) because $\beta \neq 0$. On the straight line $\phi = \frac{c+\mu}{\sigma}$, there exist two equilibrium points,

$$S_{\mp}\left(\frac{c+\mu}{\sigma}, \mp Y_s\right)$$

of the associated regular system of (51) with

$$Y_s = \sqrt{\frac{f\left(\frac{c+\mu}{\sigma}\right)}{\sigma\left(\frac{c+\mu}{\sigma} - c\right)^2}}, \quad \text{if } \sigma f\left(\frac{c+\mu}{\sigma}\right) > 0.$$

Note that $f(c) = \alpha\beta^2 > 0$. For a given wave speed $c + \mu > 0$, assume that one of the following two conditions holds:

(a_1) $g > 0, c < A + \sqrt{A^2 + g}$. For given A and g, $f(\tilde{\phi}_1) < 0,\ f(\tilde{\phi}_2) < 0$.

(a_2) $g < 0,\ A^2 + 4g > 0,\ A - \sqrt{A^2+g} < c < A + \sqrt{A^2+g}$. For given A and g, $f(\tilde{\phi}_1) < 0,\ f(\tilde{\phi}_2) < 0$.

Then, system (51) has four simple equilibrium points, $E_j(\phi_j, 0)$, $j = 1, 2, 3, 4$, satisfying $\phi_1 < \tilde{\phi}_1 < \phi_2 < c < \phi_3 < \tilde{\phi}_2 < \phi_4$. The existence of ϕ_j does not depend on the parameter σ.

Write $h_i = H_{\text{rch}}(\phi_i, 0)$ and $h_s = H_{\text{rch}}(\frac{c+\mu}{\sigma}, \mp Y_s)$, where H_{rch} is given by (52).

Suppose that $\sigma \neq 0$. For a fixed $c + \mu$, by increasing σ from $\sigma < 1$ to $\sigma \geq 1$, i.e. letting the singular line $\phi = \frac{c+\mu}{\sigma}$ move from right to left on the (ϕ, y)-phase plane, one obtains different topological phase portraits of system (51). Especially, two phase portraits contain curve triangles, as shown in Figs. 12(a) and 12(b), where the corresponding values of $H_{\text{rch}}(\phi_j, 0)$ and parameter conditions are given.

Next, consider the exact parametric representations of peakon and anti-peakon solutions of system (50) corresponding to two curve triangles in Figs. 12(a) and 12(b). As can be seen from (52), for a fixed integral constant h, one has

$$y^2 = \frac{\phi^4 - (A+2c)\phi^3 + (c^2 + Ac - g)\phi^2 + (h + cg)\phi - (ch + \alpha\beta^2)}{(\phi - c)(\sigma\phi - c - \mu)} \equiv \frac{G(\phi)}{(\phi - c)(\sigma\phi - c - \mu)}.$$

By using the first equation of system (51) and taking integration along a branch of the level curve defined by $H_{\text{rch}}(\phi, y) = h$ with initial value $\phi(\xi_0) = \phi_0$, one obtains

$$\xi - \xi_0 = \int_{\phi_0}^{\phi} \sqrt{\frac{(\phi - c)(\sigma\phi - c - \mu)}{G(\phi)}} d\phi. \quad (53)$$

(i) Corresponding to the orbit triangle [see Fig. 12(c)] connecting the equilibrium points $E_3(\phi_3, 0)$ and $S_{\mp}$ of the associated regular system of (51) and enclosing the center $E_4(\phi_4, 0)$ in Fig. 12(c), which is the level curve defined by $H_{\text{rch}}(\phi, y) = h_s = h_3$, one has $G(\phi) = (\frac{c+\mu}{\sigma} - \phi)(\phi - \phi_3)^2(\phi - \phi_l)$. Hence, taking integrals along the curves E_3S_+ and S_-E_3, it yields from (53) that

$$\pm\frac{\xi}{\sqrt{\sigma}} = \int_{\phi_4}^{\phi} \frac{d\phi}{\sqrt{(\phi - c)(\phi - \phi_l)}} + (\phi_3 - c)\int_{\phi_4}^{\phi} \frac{d\phi}{(\phi - \phi_3)\sqrt{(\phi - c)(\phi - \phi_l)}}. \quad (54)$$

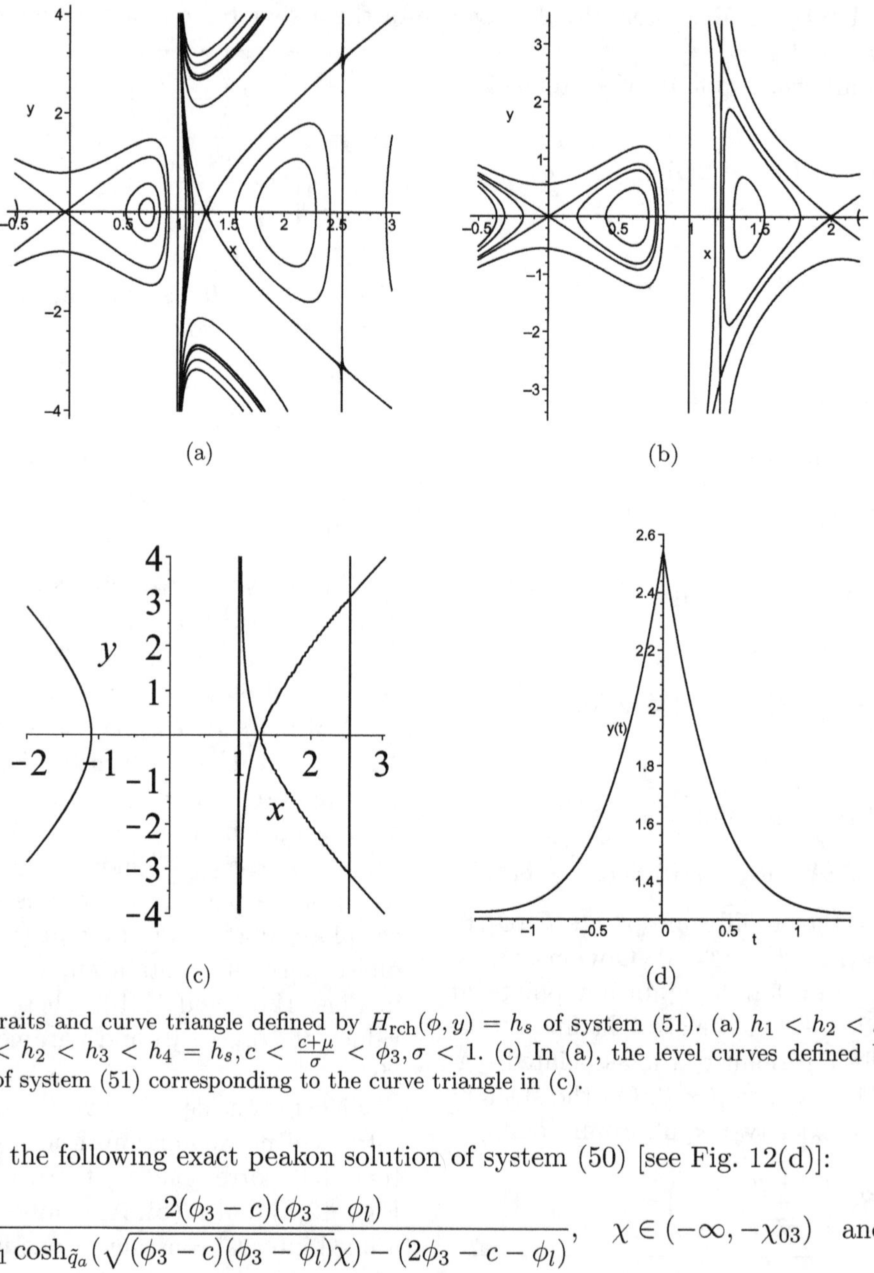

Fig. 12. Phase portraits and curve triangle defined by $H_{\text{rch}}(\phi, y) = h_s$ of system (51). (a) $h_1 < h_2 < h_s = h_3 < h_4, \phi_4 < \frac{c+\mu}{\sigma}, \sigma < 1$. (b) $h_1 < h_2 < h_3 < h_4 = h_s, c < \frac{c+\mu}{\sigma} < \phi_3, \sigma < 1$. (c) In (a), the level curves defined by $H_{\text{rch}}(\phi, y) = h_s$. (d) Peakon solution of system (51) corresponding to the curve triangle in (c).

Thus, one obtains the following exact peakon solution of system (50) [see Fig. 12(d)]:

$$\begin{aligned}
\phi(\chi) &= \phi_3 + \frac{2(\phi_3 - c)(\phi_3 - \phi_l)}{\tilde{P}_1 \cosh_{\tilde{q}_a}(\sqrt{(\phi_3 - c)(\phi_3 - \phi_l)}\chi) - (2\phi_3 - c - \phi_l)}, \quad \chi \in (-\infty, -\chi_{03}) \quad \text{and} \quad (\chi_{03}, \infty), \\
\xi(\chi) &= \mp\sqrt{\sigma}\left[(\phi_3 - c)\chi + \ln\left(\sqrt{(\phi(\chi) - c)(\phi(\chi) - \phi_l)} + \phi(\chi) - \frac{1}{2}(c + \phi_l)\right) - \xi_{01}\right],
\end{aligned} \tag{55}$$

where

$$\chi_{03} = \frac{1}{\sqrt{(\phi_3 - c)(\phi_3 - \phi_l)}} \cosh_{\tilde{q}_a}^{-1}\left(\frac{1}{\tilde{P}_1}\left(\frac{2(\phi_3 - c)(\phi_3 - \phi_l)}{\frac{c+\mu}{\sigma} - \phi_3} + (2\phi_3 - c - \phi_l)\right)\right),$$

$$\tilde{P}_1 = \frac{1}{\phi_4}\left[2\sqrt{(\phi_3 - c)(\phi_3 - \phi_l)(\phi_4 - c)(\phi_4 - \phi_l)} + (2\phi_3 - c - \phi_3)\phi_4 + 2(\phi_3 - c)(\phi_3 - \phi_l)\right],$$

$$\tilde{q}_a = \frac{(c - \phi_l)^2}{(\tilde{P}_1)^2}, \quad \xi_{01} = \ln\left(\sqrt{(\phi_4 - c)(\phi_4 - \phi_l)} + \phi_4 - \frac{1}{2}(c + \phi_l)\right).$$

(ii) Corresponding to the orbit triangle [see Fig. 12(b)] connecting the equilibrium points $E_4(\phi_4, 0)$ and $S_\mp$ of system (51) and enclosing the center $E_3(\phi_3, 0)$ in Fig. 12(b), which is the level curve defined by $H_{\mathrm{rch}}(\phi, y) = h_s = h_4$, one has $G(\phi) = (\phi_4 - \phi)^2(\phi - \frac{c+\mu}{\sigma})(\phi - \phi_l)$.

Then, taking integrals along the curves S_+E_4 and E_4S_-, it yields from (53) that

$$\pm\frac{\xi}{\sqrt{\sigma}} = -\int_{\phi_3}^{\phi} \frac{d\phi}{\sqrt{(\phi - c)(\phi - \phi_l)}} + (\phi_4 - c)\int_{\phi_3}^{\phi} \frac{d\phi}{(\phi_4 - \phi)\sqrt{(\phi - c)(\phi - \phi_l)}}.$$

Thus, one obtains the following exact anti-peakon solution of system (50):

$$\begin{aligned} \phi(\chi) &= \phi_4 - \frac{2(\phi_4 - c)(\phi_4 - \phi_l)}{\tilde{P}_2 \cosh_{\tilde{q}_b}(\sqrt{(\phi_4 - c)(\phi_4 - \phi_l)}\chi) + (2\phi_4 - c - \phi_l)}, \quad \chi \in (-\infty, -\chi_{04}) \text{ and } (\chi_{04}, \infty), \\ \xi(\chi) &= \mp\sqrt{\sigma}\left[(\phi_3 - c)\chi - \ln\left(\sqrt{(\phi(\chi) - c)(\phi(\chi) - \phi_l)} + \phi(\chi) - \frac{1}{2}(c + \phi_l)\right) + \xi_{02}\right], \end{aligned} \tag{56}$$

where

$$\begin{aligned} \chi_{04} &= \frac{1}{\sqrt{(\phi_4 - c)(\phi_4 - \phi_l)}} \cosh_{\tilde{q}_b}^{-1}\left(\frac{1}{\tilde{P}_2}\left(\frac{2(\phi_4 - c)(\phi_4 - \phi_l)}{\phi_4 - \dfrac{c+\mu}{\sigma}} - (2\phi_4 - c - \phi_l)\right)\right), \\ \tilde{P}_2 &= \frac{1}{\phi_4}\left[2\sqrt{(\phi_4 - c)(\phi_4 - \phi_l)(\phi_3 - c)(\phi_3 - \phi_l)} - (2\phi_4 - c - \phi_3)\phi_3 + 2(\phi_4 - c)(\phi_4 - \phi_l)\right], \\ \tilde{q}_b &= \frac{(c - \phi_l)^2}{(\tilde{P}_2)^2}, \quad \xi_{02} = \ln\left(\sqrt{(\phi_3 - c)(\phi_3 - \phi_l)} + \phi_3 - \frac{1}{2}(c + \phi_l)\right). \end{aligned}$$

4. Dynamics of the Singular Traveling Systems of the Second Kind — The So-Called "Peakon Equations" Have No Peakons

Anco and Recio [2019] introduced a general family of peakon equations involving two arbitrary functions of the wave amplitude and the wave gradient, in which one-parameter subfamilies of the CH–mCH Hamiltonian family are explored. It was suggested to study the following "family of Hamiltonian multi-peakon equations":

$$\begin{aligned} m_t + u_x f_1(u^2 - u_x^2)m + [uf_1(u^2 - u_x^2) \\ + g_1(u^2 - u_x^2)m]_x = 0, \end{aligned} \tag{57}$$

which involves two arbitrary functions, f_1 and g_1 of $u^2 - u_x^2$.

Two subfamilies represent nonlinear generalizations of the mCH equation given by

$$m_t + ((u^2 - u_x^2)^p m)_x = 0, \quad m = u - u_{xx}, \quad p \geq 1, \tag{58}$$

and

$$\begin{aligned} m_t + ((u^2 - u_x^2)^{p-1})u_x m \\ + [u((u^2 - u_x^2)^{p-1})m]_x = 0, \quad p \geq 1. \end{aligned} \tag{59}$$

A unified generalization of these two subfamilies is

$$\begin{aligned} m_t + au_x(u^2 - u_x^2)^{\frac{1}{2}k}m + [au(u^2 - u_x^2)^{\frac{1}{2}k}m \\ + b(u^2 - u_x^2)^{\frac{1}{2}(k+1)}m]_x = 0, \quad k \geq 0. \end{aligned} \tag{60}$$

By putting $k = 2p - 2, b = 0, a = 1$ in the three-parameter family (60), one obtains the one-parameter family of the generalized CH equation (58). Taking $a = 0, b = 1, k = 2p - 1$, Eq. (60) becomes Eq. (59).

Anco and Recio [2019] stated that "in the case of the $p = 2$ generalized mCH equation (58), the peakon and anti-peakon can form a bound pair which has a maximum finite separation in the asymptotic past and future". In [Anco & Kraus, 2018; Anco & Recio, 2018a, 2018b, 2019], other peakon equations were studied. Unfortunately, the corresponding traveling wave systems of Eqs. (58)–(60) were not derived.

This section considers the bifurcations problems of the solutions of the corresponding traveling wave systems of Eqs. (58)–(60), depending on the parameters of systems (see [Fu & Li, 2021]).

To study the traveling wave solutions of Eqs. (58)–(60), setting $u(x,t) = u(x-ct) \equiv \phi(\xi)$, where $\xi = x - ct$ and c is the wave speed. Substituting it into (58)–(60), and then integrating the resulting equations once, one obtains

$$[(\phi^2 - \phi_\xi^2)^p - c]\phi_{\xi\xi} = \phi(\phi^2 - \phi_\xi^2)^p - c\phi + g, \tag{61}$$

and

$$[\phi(\phi^2 - \phi_\xi^2)^{p-1} - c]\phi_{\xi\xi} = \frac{1}{2p}(\phi^2 - \phi_\xi^2)^p + \phi^2(\phi^2 - \phi_\xi^2)^{p-1} - c\phi + g, \tag{62}$$

$$\begin{aligned}&[a\phi(\phi^2 - \phi_\xi^2)^{\frac{1}{2}k} + b(\phi^2 - \phi_\xi^2)^{\frac{1}{2}(k+1)} - c]\phi_{\xi\xi}\\ &\quad = g - c\phi + \frac{a}{k+2}(\phi^2 - \phi_\xi^2)^{\frac{1}{2}k+1} + a\phi^2(\phi^2 - \phi_\xi^2)^{\frac{1}{2}k} + b\phi(\phi^2 - \phi_\xi^2)^{\frac{1}{2}(k+1)},\end{aligned} \tag{63}$$

where g is an integral constant. For Eq. (61), assume that $g \neq 0$; otherwise, (61) becomes a linear equation. Equations (61)–(63) are, respectively, equivalent to the following planar dynamical systems:

$$\frac{d\phi}{d\xi} = y, \quad \frac{dy}{d\xi} = \frac{[c - (\phi^2 - y^2)^p]\phi - g}{c - (\phi^2 - y^2)^p}, \tag{64}$$

$$\frac{d\phi}{d\xi} = y, \quad \frac{dy}{d\xi} = \frac{-\frac{1}{2p}(\phi^2 - y^2)^p - \phi^2(\phi^2 - y^2)^{p-1} + c\phi - g}{c - \phi(\phi^2 - y^2)^{p-1}}, \tag{65}$$

and

$$\begin{aligned}\frac{d\phi}{d\xi} &= y,\\ \frac{dy}{d\xi} &= \frac{g - c\phi + \frac{a}{k+2}(\phi^2 - y^2)^{\frac{1}{2}k+1} + a\phi^2(\phi^2 - y^2)^{\frac{1}{2}k} + b\phi(\phi^2 - y^2)^{\frac{1}{2}(k+1)}}{a\phi(\phi^2 - y^2)^{\frac{1}{2}k} + b(\phi^2 - y^2)^{\frac{1}{2}(k+1)} - c}.\end{aligned} \tag{66}$$

Systems (64)–(66) have the following first integrals, respectively:

$$H_{p1}(\phi, y) = \frac{1}{p+1}(\phi^2 - y^2)^{p+1} + cy^2 - c\phi^2 + 2g\phi = h, \tag{67}$$

$$H_{p2}(\phi, y) = \frac{1}{p}\phi(\phi^2 - y^2)^p + cy^2 - c\phi^2 + 2g\phi = h, \tag{68}$$

and

$$\begin{aligned}H_k(\phi, y) &= \frac{2a}{k+2}\phi(\phi^2 - y^2)^{\frac{1}{2}k+1} + \frac{2b}{k+3}(\phi^2 - y^2)^{\frac{1}{2}(k+3)} + cy^2\\ &\quad + \frac{a}{k+2}\int(\phi^2 - y^2)^{\frac{1}{2}k+1}d\phi - \frac{1}{2}c\phi^2 + g\phi = h,\end{aligned} \tag{69}$$

where the calculation of integral $\int(\phi^2 - y^2)^{\frac{1}{2}k+1}d\phi$ can use the recursive formula as follows:

$$\int(\phi^2 - y^2)^{\frac{1}{2}k+1}d\phi = \frac{1}{k+3}\left[\phi(\phi^2 - y^2)^{\frac{1}{2}k+1} - (k+2)y^2\int(\phi^2 - y^2)^{\frac{1}{2}k}d\phi\right]. \tag{70}$$

Clearly, for $c > 0$, on the curves $c - (\phi^2 - y^2)^p = 0$, $c - \phi(\phi^2 - y^2)^{p-1} = 0$ and $a\phi(\phi^2 - y^2)^{\frac{1}{2}k} + b(\phi^2 - y^2)^{\frac{1}{2}(k+1)} - c = 0$, respectively, systems (64)–(66) are discontinuous. Such systems are called singular traveling wave systems of *the second kind* defined by [Li *et al.*, 2009a, 2009b].

Especially, when $p = 1$, system (64) becomes

$$\frac{d\phi}{d\xi} = y, \quad \frac{dy}{d\xi} = \frac{[c - \phi^2 + y^2]\phi - g}{c - \phi^2 + y^2} \qquad (64_{p=1})$$

with the first integral (see [Li & Zhang, 2009])

$$H_{11}(\phi, y) = \frac{1}{2}(c - \phi^2 + y^2)^2 + 2g\phi = h. \quad (67_{p=1})$$

When $p = 1$, system (65) becomes the traveling wave system of CH equation as follows:

$$\frac{d\phi}{d\xi} = y, \quad \frac{dy}{d\xi} = \frac{-y^2 + 3\phi^2 - 2c\phi + 2g}{2(\phi - c)}, \quad (65_{p=1})$$

with the first integral

$$H_{12}(\phi, y) = (\phi - c)y^2 - \phi^3 + c\phi^2 - 2g\phi = h. \qquad (68_{p=1})$$

The bifurcations and exact solutions of system $(65_{p=1})$ can be seen in the following section. Therefore, for Eq. (65), only the case of $p \geq 2$ is discussed.

4.1. *Bifurcations of phase portraits of systems (64)–(66)*

Always assume that $c > 0$, $g > 0$ and $p(\geq 1)$ is a positive integer.

(1) First, consider all possible phase portraits of system (64). It is known that system (64) has the same invariant curve solutions as the following associated regular system:

$$\begin{aligned} \frac{d\phi}{d\zeta} &= y[c - (\phi^2 - y^2)^p], \\ \frac{dy}{d\zeta} &= \phi[c - (\phi^2 - y^2)^p] - g, \end{aligned} \qquad (71)$$

where $d\xi = [c - (\phi^2 - y^2)^p]d\zeta$, for $c - (\phi^2 - y^2)^p \neq 0$.

Denote $f_1(\phi) = \phi^{2p+1} - c\phi + g$, $f_1'(\phi) = (2p+1)\phi^{2p} - c$. Thus, when $\phi = \mp\tilde{\phi}_0 = \mp(\frac{c}{2p+1})^{\frac{1}{2p}}$, one has $f_1'(\mp\tilde{\phi}_0) = 0$.

The equilibrium points $E_j(z_j, 0)$ of system (71) satisfy $f_1(z_j) = 0$. Write

$$f_a = f_1(-\tilde{\phi}_0) = g + \frac{2pc}{2p+1}\tilde{\phi}_0,$$

$$f_b = f_1(\tilde{\phi}_0) = g - \frac{2pc}{2p+1}\tilde{\phi}_0.$$

It is easy to show that, for a fixed $c > 0$, the following holds.

(1) When $g > \frac{2pc}{2p+1}\tilde{\phi}_0 = 2p(\frac{c}{2p+1})^{1+\frac{1}{2p}}$, $f_1(\phi)$ has only one negative zero $z_1 < -c^{\frac{1}{2p}} < -\tilde{\phi}_0 < 0$.

(2) When $g = \frac{2pc}{2p+1}\tilde{\phi}_0$, $f_1(\phi)$ has one simple zero $z_1 < -c^{\frac{1}{2p}} < -\tilde{\phi}_0 < 0$ and a double zero $z_{23} = \tilde{\phi}_0$.

(3) When $0 < g < \frac{2pc}{2p+1}\tilde{\phi}_0$, $f_1(\phi)$ has three simple zeros: $z_1 < -c^{\frac{1}{2p}} < -\tilde{\phi}_0 < 0 < z_2 < \tilde{\phi}_0 < z_3 < c^{\frac{1}{2p}}$.

Let $M(z_j, 0)$ be the coefficient matrix of the linearized system of (71) at equilibrium point $E_j(z_j, 0)$. One has

$$J(z_j, 0) = \det M(z_j, 0) = -f_1'(z_j)(z_j^{2p} - c).$$

By the theory of planar dynamical systems, for an equilibrium point of a planar integral system, if $J < 0$, then the equilibrium point is a saddle point; if $J > 0$, then it is a center point; if $J = 0$ and the Poincaré index of the equilibrium point is 0, then this equilibrium point is a cusp.

Write $h_j = H_{p1}(z_j, 0)$, $j = 1, 2, 3$, where H_{p1} is given by (67).

Obviously, when $g > \frac{2pc}{2p+1}\tilde{\phi}_0$, the unique equilibrium point $E_1(z_1, 0)$ of system (71) is a saddle point. When $0 < g < \frac{2pc}{2p+1}\tilde{\phi}_0$, the point $E_3(z_3, 0)$ is a center, while $E_1(z_1, 0)$ and $E_2(z_2, 0)$ are saddle points. When $g = g^* = \frac{1}{2}(h_2 + \frac{p}{p+1}c^{1+\frac{1}{p}})c^{-\frac{1}{2p}}$, the homoclinic orbit defined by $H_{p1}(\phi, y) = h_2$ to the saddle point $E_2(z_2, 0)$ passes through the point $E_s(c^{\frac{1}{2p}}, 0)$.

By using the above results to perform qualitative analysis, with the change of the parameter g, one obtains the bifurcations of phase portraits of (71), as shown in Fig. 13. Also, the graph of the hyperbola $(\phi^2 - y^2)^p - c = 0$ is drawn in every phase portrait, in order to show the position of the singular curve.

When $g < 0$, the bifurcations of phase portraits of system (64) are just the reflections of Figs. 13(a)–13(e) with respect to the y-axis. Therefore, only the case of $g > 0$ is discussed.

(2) Consider all possible phase portraits of system (65). System (65) has the same invariant curve solutions as the following associated regular system:

$$\begin{aligned} \frac{d\phi}{d\zeta} &= y[c - \phi(\phi^2 - y^2)^{p-1}], \\ \frac{dy}{d\zeta} &= -\frac{1}{2p}(\phi^2 - y^2)^p - \phi^2(\phi^2 - y^2)^{p-1} + c\phi - g, \end{aligned} \qquad (72)$$

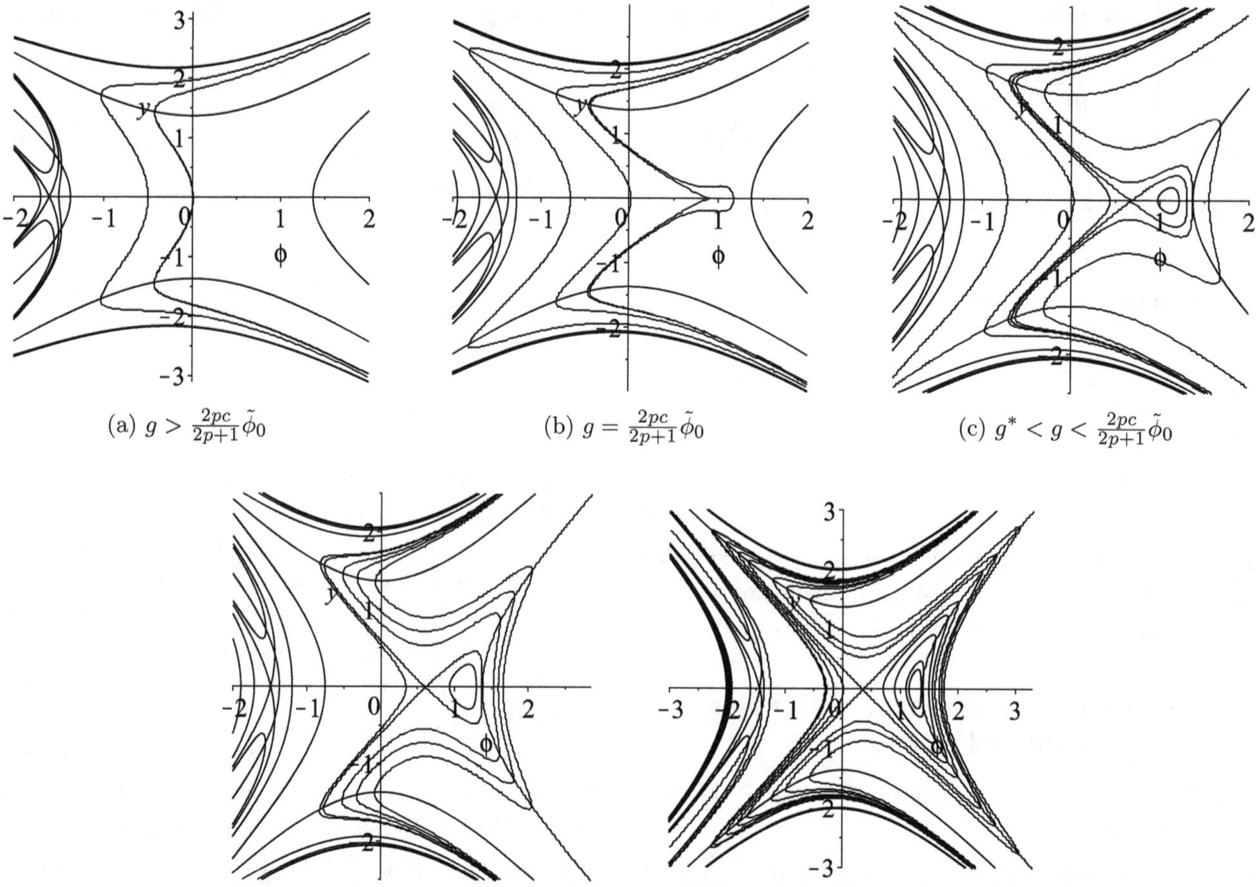

(a) $g > \frac{2pc}{2p+1}\tilde{\phi}_0$ (b) $g = \frac{2pc}{2p+1}\tilde{\phi}_0$ (c) $g^* < g < \frac{2pc}{2p+1}\tilde{\phi}_0$

(d) $g = g^*$ (e) $0 < g < g^*$

Fig. 13. The bifurcations of phase portraits of system (64), for fixed $c > 0$, $p \geq 1$.

where $d\xi = [c - \phi(\phi^2 - y^2)^{p-1}]d\zeta$, for $c - \phi(\phi^2 - y^2)^{p-1} \neq 0$.

Write

$$f_2(\phi) = \left(1 + \frac{1}{2p}\right)\phi^{2p} - c\phi + g,$$

$$f_2'(\phi) = (2p+1)\phi^{2p-1} - c.$$

Thus, when

$$\phi = \hat{\phi}_0 = \left(\frac{c}{2p+1}\right)^{\frac{1}{2p-1}}, \quad f_2'(\hat{\phi}_0) = 0,$$

$$f_2(\hat{\phi}_0) = g - \left(1 - \frac{1}{2p}\right)c\hat{\phi}_0.$$

The equilibrium points $E_j(z_j, 0)$ of system (72) satisfy $f_2(z_j) = 0$. It is easy to show that, for a fixed $c > 0$, the following holds:

(1) When $g > (1 - \frac{1}{2p})c\hat{\phi}_0 = (2p - \frac{1}{2p})(\frac{c}{2p+1})^{1+\frac{1}{2p-1}}$, $f_2(\phi)$ has no real zeros.
(2) When $g = (1 - \frac{1}{2p})c\hat{\phi}_0$, $f_2(\phi)$ has a double zero $z_{12} = \hat{\phi}_0$.
(3) When $0 \leq g < (1 - \frac{1}{2p})c\hat{\phi}_0$, $f_2(\phi)$ has two simple zeros: $0 \leq z_1 < \hat{\phi}_0 < z_2 < c^{\frac{1}{2p-1}}$.

Let $M(z_j, 0)$ be the coefficient matrix of the linearized system of (72) at equilibrium point $E_j(z_j, 0)$. Then, one has

$$J(z_j, 0) = \det M(z_j, 0) = -f_2'(z_j)(z_j^{2p-1} - c).$$

Write $h_j = H_{p2}(z_j, 0)$, $j = 1, 2$, where H_{p2} is given by (68).

It can be seen that, when $0 < g < (1 - \frac{1}{2p})c\hat{\phi}_0$, the point $E_1(z_1, 0)$ is a saddle point; $E_2(z_2, 0)$ is a center point. When

$$g = g^{**} = \frac{1}{2}\left[h_1 + \left(1 - \frac{1}{p}\right)c^{1+\frac{2}{2p-1}}\right]c^{-\frac{1}{2p-1}},$$

the homoclinic orbit defined by $H_{p2}(\phi, y) = h_1$ to the saddle point $E_1(z_1, 0)$ passes through the point $E_s(c^{\frac{1}{2p-1}}, 0)$.

By using the above results to perform qualitative analysis, with the change of the parameter g,

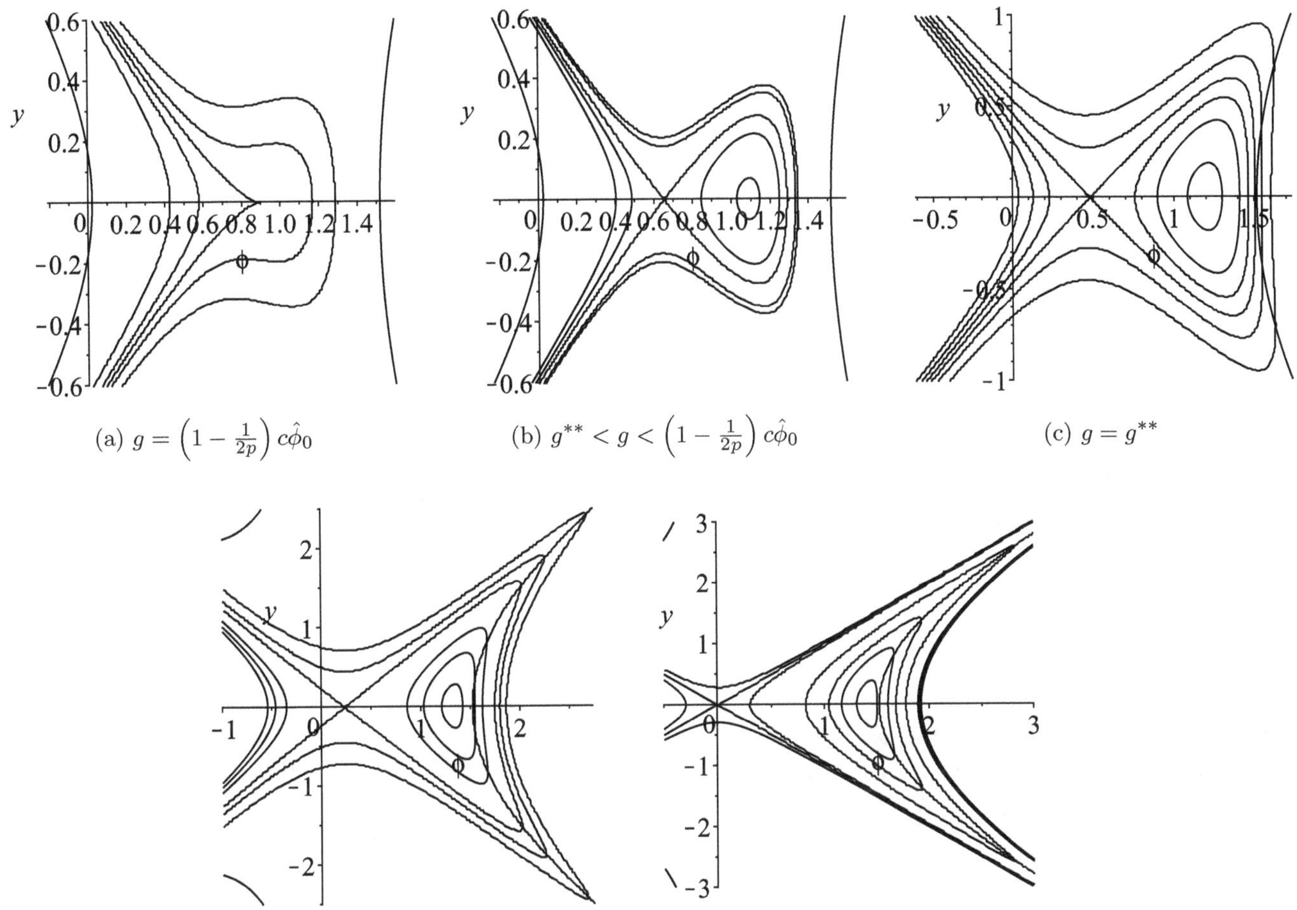

(a) $g = \left(1 - \frac{1}{2p}\right) c\hat{\phi}_0$ (b) $g^{**} < g < \left(1 - \frac{1}{2p}\right) c\hat{\phi}_0$ (c) $g = g^{**}$

(d) $0 < g < g^{**}$ (e) $g = 0$

Fig. 14. The bifurcations of phase portraits of system (65) for fixed $c > 0$, $p \geq 2$.

one obtains the bifurcations of phase portraits of system (65), as shown in Fig. 14. Also, the graph of the hyperbola $\phi(\phi^2 - y^2)^{p-1} - c = 0$ is drawn in every phase portrait, in order to show the position of the singular curve.

(3) Consider all possible phase portraits of system (66). System (66) has the same invariant curve solutions as the following associated regular system:

$$\begin{aligned}
\frac{d\phi}{d\zeta} &= y[a\phi(\phi^2 - y^2)^{\frac{1}{2}k} + b(\phi^2 - y^2)^{\frac{1}{2}(k+1)} - c], \\
\frac{dy}{d\zeta} &= g_1 - c\phi + \frac{a}{k+2}(\phi^2 - y^2)^{\frac{1}{2}k+1} \\
&\quad + a\phi^2(\phi^2 - y^2)^{\frac{1}{2}k} + b\phi(\phi^2 - y^2)^{\frac{1}{2}(k+1)},
\end{aligned} \tag{73}$$

where

$$d\xi = [a\phi(\phi^2 - y^2)^{\frac{1}{2}k} + b(\phi^2 - y^2)^{\frac{1}{2}(k+1)} - c]d\zeta,$$

for

$$a\phi(\phi^2 - y^2)^{\frac{1}{2}k} + b(\phi^2 - y^2)^{\frac{1}{2}(k+1)} - c \neq 0.$$

Write

$$f_3(\phi) = \left(\frac{a}{k+2} + a + b\right)\phi^{k+2} - c\phi + g,$$

$$f_3'(\phi) = (a + (k+2)(a+b))\phi^{k+1} - c.$$

Thus, when $a + (k+2)(a+b) > 0$, if

$$\phi = \bar{\phi}_0 = \left(\frac{c}{a + (k+2)(a+b)}\right)^{\frac{1}{k+1}},$$

$$f_3'(\bar{\phi}_0) = 0, \quad f_3(\bar{\phi}_0) = g - \left(1 - \frac{1}{k+2}\right)c\bar{\phi}_0.$$

The equilibrium points $E_j(z_j, 0)$ of system (64) satisfy $f_3(z_j) = 0$. It is easy to show that, for a fixed $c > 0$, the following hold.

(1) When $g > (1 - \frac{1}{k+2})c\bar{\phi}_0$, $f_3(\phi)$ has no positive real zeros.
(2) When $g = (1 - \frac{1}{k+2})c\bar{\phi}_0$, $f_3(\phi)$ has a positive double real zero $z_{23} = \bar{\phi}_0$.

(3) When $0 \leq g < (1-\frac{1}{k+2})c\bar{\phi}_0$, $f_3(\phi)$ has two positive simple zeros: $0 < z_2 < \bar{\phi}_0 < z_3 < (\frac{c}{a+b})^{\frac{1}{k+1}}$.

Let $M(z_j, 0)$ be the coefficient matrix of the linearized system of (73) at equilibrium point $E_j(z_j, 0)$. Then, one has

$$J(z_j, 0) = \det M(z_j, 0) = -f_3'(z_j)[(a+b)z_j^{k+1} - c].$$

Write $h_j = H_k(z_j, 0)$, $j = 2, 3$, where H_k is given by (69).

By using the above result to perform qualitative analysis, with the change of the parameter g, one obtains the bifurcations of phase portraits of system (66) similar to Fig. 13, when k is an even number; when k is a odd number, the bifurcations of phase portraits of system (66) are similar to Fig. 12.

Note that, when $g = g^{***}$, the homoclinic orbit defined by $H_k(\phi, y) = h_2$ contacts a branch of singular curve $a\phi(\phi^2-y^2)^{\frac{1}{2}k} + b(\phi^2-y^2)^{\frac{1}{2}(k+1)} - c = 0$, where g^{***} is a solution of $H_k(\bar{\phi}_0, 0) = h_2$.

4.2. *Solitary wave solutions determined by the homoclinic orbits of systems (64)–(66)*

It is well known that a smooth homoclinic orbit of a traveling system gives rise to a solitary wave solution of the corresponding nonlinear wave equation. In the following, always assume that $c > 0$ and it is fixed.

It is known that, for systems (64)–(66), when conditions

$$g^* < g < 2p\left(\frac{c}{2p+1}\right)^{1+\frac{1}{2p}},$$

$$g^{**} < g < \left(1 - \frac{1}{2p}\right)c\hat{\phi}_0 \quad \text{and}$$

$$g^{***} < g < \left(1 - \frac{1}{k+2}\right)c\bar{\phi}_0$$

hold, respectively, the homoclinic orbits defined by $H_{p1}(\phi, y) = h_2$, $H_{p2}(\phi, y) = h_1$ and $H_k(\phi, y) = h_2$ have no intersection points with the one of three singular curves, respectively [see Figs. 12(c) and 13(b)]. Thus, the following conclusions follow.

Theorem 3. *Assume that $c > 0$ is fixed.*

(i) *When $g^* < g < 2p(\frac{c}{2p+1})^{1+\frac{1}{2p}}$, Eq. (58) has a smooth solitary wave solution given by the homoclinic orbit of system (64) defined by $H_{p1}(\phi, y) = h_2$. In addition, Eq. (58) has a family of smooth periodic wave solutions defined by the closed branch of $H_{p1}(\phi, y) = h$, $h \in (h_3, h_2)$ [see Fig. 12(c)].*

(ii) *When $g^{**} < g < (1 - \frac{1}{2p})c\hat{\phi}_0$, Eq. (59) has a smooth solitary wave solution given by the homoclinic orbit of system (65) defined by $H_{p2}(\phi, y) = h_1$. In addition, Eq. (48) has a family of smooth periodic wave solutions defined by the closed branch of $H_{p2}(\phi, y) = h$, $h \in (h_2, h_1)$ [see Fig. 13(b)].*

(iii) *When $g^{***} < g < (1 - \frac{1}{k+2})c\bar{\phi}_0$, Eq. (60) has a smooth solitary wave solution given by the homoclinic orbit of system (66) defined by $H_k(\phi, y) = h_2$. In addition, Eq. (60) has a family of smooth periodic wave solutions defined by the closed branch of $H_k(\phi, y) = h$, $h \in (h_3, h_2)$ [see Fig. 13(b)].*

For $p = 1$, to find the exact explicit parametric representation of the solitary wave solution, it follows from $(67_{p=1})$ that

$$y^2 = \phi^2 \pm \sqrt{h - 4g\phi} - c. \tag{74}$$

The signs $\pm$ before the term $\sqrt{h - 4g\phi}$ are dependent on the interval of ϕ. Let $\psi^2 = h - 4g\phi$, i.e. $\phi = \frac{1}{4g}(h - \psi^2)$. Then, one obtains

$$y^2 = \frac{1}{16g^2}[\psi^4 - 2h\psi^2 \pm 16g^2\psi + (h^2 - 16g^2c)]. \tag{75}$$

Under the condition of $g^* < g < \frac{2c}{3}\sqrt{\frac{c}{3}}$, for $\phi \in (\phi_2, \sqrt{c})$, one needs to take $+$ before the term $16g^2\psi$. By the first equation of $(64_{p=1})$, corresponding to the homoclinic orbit defined by $H_{11}(\phi, y) = h_2$, one has

$$\frac{\psi d\psi}{\sqrt{\psi^4 - 2h_2\psi^2 + 16g^2\psi + (h_2^2 - 16g^2c)}} = -\frac{1}{2}d\xi. \tag{76}$$

Denote $(c-\phi^2)^2 + 4g\phi - h_2 = (\phi-\phi_2)^2(\phi-\phi_M)(\phi-\phi_3)$, where $\phi_3 < \phi_2 < \phi_M$. The point $(\phi_M, 0)$ is the intersection point of the homoclinic orbit to $(\phi_2, 0)$ of system $(64_{p=1})$ on the positive ϕ-axis. Thus, one has

$$\begin{aligned}\psi^4 - 2h_2\psi^2 + 16g^2\psi + (h_2^2 - 16g^2c) \\ = (\psi - \psi_1)^2(\psi - \psi_m)(\psi - \psi_3),\end{aligned}$$

where

$$\psi_1 = \sqrt{h_2 - 4g\phi_2}, \quad \psi_m = \sqrt{h_2 - 4g\phi_M},$$

$$\psi_3 = -\sqrt{h_2 - 4g\phi_3}.$$

By introducing a parametric variable χ and integrating (76), one obtains

$$\psi(\chi) = \psi_1 - \frac{2(\psi_1 - \psi_m)(\psi_1 - \psi_3)}{(\psi_m - \psi_3)\cosh(\sqrt{(\psi_1 - \psi_m)(\psi_1 - \psi_3)}\chi) - (2\psi_1 - \psi_m - \psi_3)},$$
$$\xi(\chi) = x - ct = -2\left[\psi_1\chi + \ln\left(\frac{2\sqrt{(\psi - \psi_m)(\psi - \psi_3)} + 2\psi - (\psi_m - \psi_3)}{\psi_m - \psi_3}\right)\right]. \quad (77)$$

Thus, one has the following exact explicit parametric representations of smooth solitary solution of system $(64_{p=1})$ (see [Li & Zhang, 2009]):

$$\phi(\chi) = \frac{1}{4g}(1 - \psi^2(\chi)),$$
$$\xi(\chi) = x - ct = -2\left[\psi_1\chi + \ln\left(\frac{2\sqrt{(\psi - \psi_m)(\psi - \psi_3)} + 2\psi - (\psi_m - \psi_3)}{\psi_m - \psi_3}\right)\right]. \quad (78)$$

For $p \geq 2$, because one cannot explicitly solve the exact y from $H_{p1}(\phi, y) = h_2$, it is not possible to give the exact solitary wave solutions of Eq. (64).

4.3. *M-shape traveling wave solutions determined by the homoclinic orbit of systems (64)–(66)*

In this section, assume that $0 < g < g^*$, $0 < g < g^{**}$ and $0 < g < g^{***}$, respectively. Note that, unlike the case with the existence of a singular straight line discussed in Sec. 2, for systems (64), (65) with $p \geq 2$ and (66), the curves $(\phi^2 - y^2)^p - c = 0$, $\phi(\phi^2 - y^2)^{p-1} - c = 0$ and $a\phi(\phi^2 - y^2)^{\frac{1}{2}k} + b(\phi^2 - y^2)^{\frac{1}{2}(k+1)} - c = 0$ are not solutions of systems (64)–(66), respectively. For every fixed $c > 0$, when $0 < g < g^*$, $0 < g < g^{**}$ and $0 < g < g^{***}$ hold, respectively, there exist uncountably infinitely many periodic orbits and a homoclinic orbit of systems (64)–(66), which are transversely intersecting with an open branch in each of the above three singular curves [see Figs. 13(e) and 14(d)]. The phase portraits, Fig. 13(e) of system (71) and Fig. 14(d) of system (72), are redrawn on the left of Fig. 15, while the vector fields defined by systems (64) and (65) have been represented on the right of Fig. 15. Obviously, the vector fields defined by systems (71) and (64) [or systems (72) and (65)] are different (see Fig. 15).

The curve $c - (\phi^2 - y^2)^p = 0$ is an infinite isocline of the vector field of system (71). When ζ is varied along the homoclinic orbit defined by $H_{p1}(\phi, y) = h_2$, the vector field of system (64) has the same direction, as shown on the left-hand side of Fig. 15. On the left-hand side of a branch of the

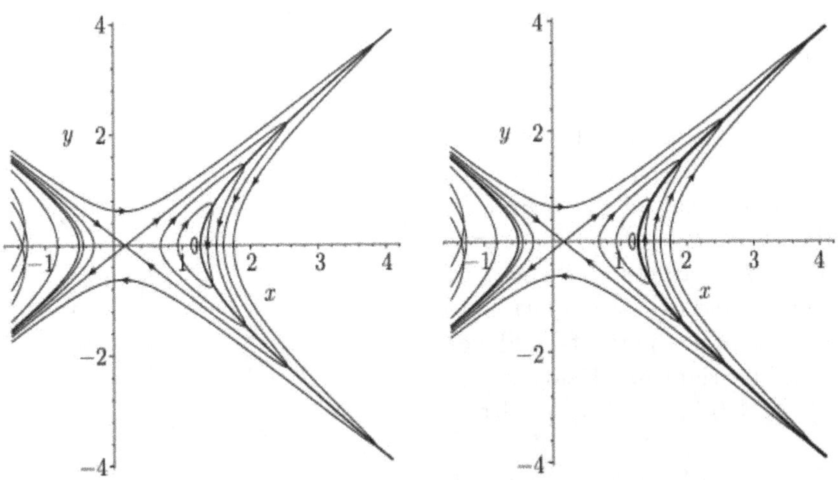

Fig. 15. The graph of $(\phi^2 - y^2)^p - c = 0$ and the vector fields defined by (71) and (64).

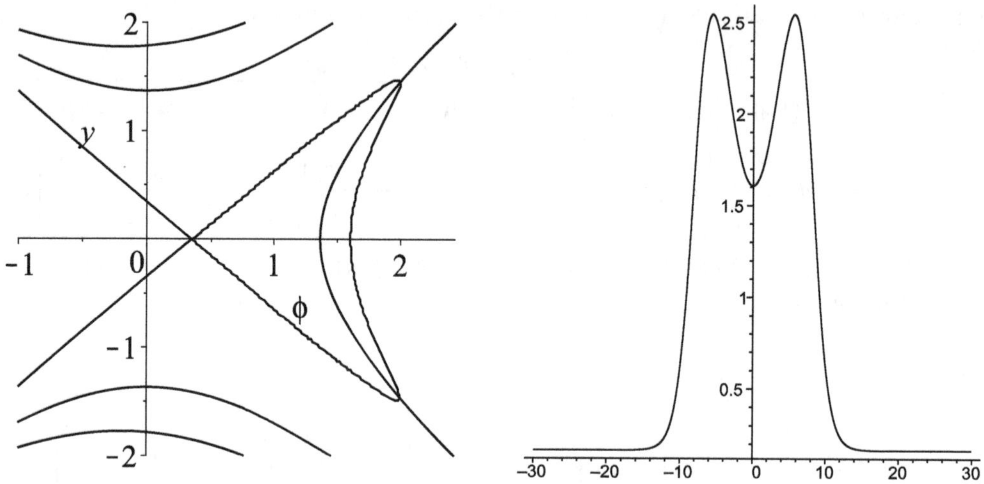

(a) (b)

Fig. 16. M-shape wave solution of system (71) on the (ζ, ϕ)-plane when $0 < g < g^*$. (a) Homoclinic orbit of system (71). (b) The M-shape solitary wave of system (64).

curve $c - (\phi^2 + y^2)^p = 0$ in the first quadrant, one has

$$\frac{d\phi}{d\zeta} = y[c - (\phi^2 - y^2)^p] > 0,$$

$$\frac{dy}{d\zeta} = -g + \phi[c - (\phi^2 + y^2)^p] > 0.$$

On the right-hand side of the hyperbola $(\phi^2 - y^2)^p - c = 0$ in the first quadrant, one has

$$\frac{d\phi}{d\zeta} = y[c - (\phi^2 - y^2)^p] < 0,$$

$$\frac{dy}{d\zeta} = -g + \phi[c - (\phi^2 + y^2)^p] < 0.$$

Corresponding to the homoclinic orbit of system (71) [see Fig. 16(a)], one has the M-shape wave profile, as shown in Fig. 16(b).

Similarly, the vector fields for systems (72) and (65) can be discussed.

Differing from system (71), for system (64), the hyperbola $(\phi^2 - y^2)^p - c = 0$ is a singular curve of the vector field of the system. Here, consider the case of $\phi > 0$. Clearly, on both the left-hand and the right-hand sides of the hyperbola $(\phi^2 - y^2)^p - c = 0$, when ξ is varied along the loop orbit defined by $H_{p1}(\phi, y) = h_2$, the vector field of system (64) has a different direction, as shown on the right-hand side of Fig. 15. In fact, on the right-hand side of the hyperbola $(\phi^2 - y^2)^p - c = 0$ in the first quadrant, one has

$$\frac{d\phi}{d\xi} = y > 0, \quad \frac{dy}{d\xi} = \frac{-g + \phi[c - (\phi^2 - y^2)]}{c - (\phi^2 - y^2)^p} > 0.$$

This implies that the loop orbit of system (64) defined by $H_{p1}(\phi, y) = h_2$ consists of three breaking solutions of system (64) (see Fig. 17).

The above statement also holds for systems (72) and (65).

When $p = 1$ in system (64), the exact parametric representation of the M-shape wave solution was given in [Li & Zhang, 2009].

In summary, the following conclusions can be drawn.

Theorem 4. *Assume that $c > 0$ is fixed.*

(i) *When $0 < g < g^*$, Eq. (64) has an M-shape wave solution defined by a branch of the level curves $H_{p1}(\phi, y) = h_2$, which consists of three breaking wave solutions.*

(ii) *When $0 < g < g^{**}$, Eq. (65) has an M-shape wave solution defined by a branch of the level*

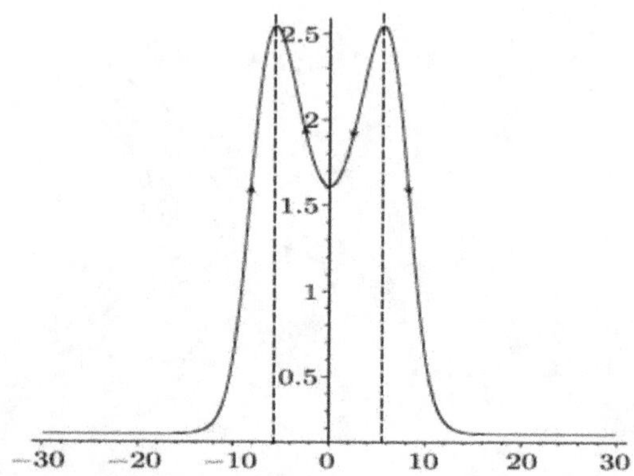

Fig. 17. M-shape wave consisting of three breaking waves of system (53).

curves $H_{p2}(\phi, y) = h_1$, which consists of three breaking wave solutions.

(iii) *When $0 < g < g^{***}$, Eq. (66) has an M-shape wave solution defined by a branch of the level curves $H_k(\phi, y) = h_2$, which consists of three breaking wave solutions.*

It follows from the discussions in this section that, for the singular traveling wave systems of the second kind, the dynamics of the orbits are different from that of the singular traveling wave systems of the first kind. Because every singular curve intersects an orbit of the singular traveling wave systems of the second kind only in a few isolated points, in systems (64) and (65), the variables ξ and ζ have the same "time scale". The singular curves make the direction of vector field defined by the singular traveling wave systems of the second class swiftly change. It leads to new breaking wave solutions. Equations (64), (65) with $p \geq 2$ and (66) have no peakon solutions in the sense of peakon in the CH equation. Therefore, for the many new peakon models proposed in other studies, the traveling wave systems there need to be revisited carefully.

5. Application Example: A Model of Nonlinear Elastodynamics of Materials with Strong Ellipticity

Rogers *et al.* [2014, 2015] studied the propagation of transverse waves in a class of Mooney–Rivlin incompressible materials, subject to a physical force corresponding to a nonlinear isotropic substrate potential. A class of generalized nonlinear Schrödinger equation is derived, and integrable Hamiltonian subsystems are separated for a wide range of deformations. On this basis, a class of incompressible elastomers with infinite sizes is studied, and the Cartesian coordinate system is introduced to describe the incompressible elastomers in Rogers *et al.* [2015], which has extended the results of Rogers *et al.* [2014] under the framework of the Klein–Gordon system and some Carroll-type solutions are studied. The model equation is

$$\Psi_{tt} - [(c^2 + c_{nl}^2|\Psi_z|^2)\Psi_z]_z + g(|\Psi|^2)\Psi = 0, \quad (79)$$

where

$$c^2 = \frac{\mu}{\rho} > 0, \quad c_{nl}^2 = \frac{2}{\rho}(C_{20} + 2C_{11} + C_{02}) > 0$$

are material constants, $g(s)$ is a given function, $g(0) \neq 0$, and $G(s) = \int_0^s g(s)ds$ with $G(0) = 0$.

To consider the *compact wave solutions* of Eq. (79), letting $\Psi = \Psi(\xi) = \phi(\xi)e^{i\hat{\theta}(\xi)}$, $\xi = x - vt$, and substituting it into Eq. (79), by separation of the real and imaginary parts and performing integration once for the imaginary part equation, Rogers *et al.* [2015] obtained the following two-order ODE:

$$[(v^2 - c^2 - c_{nl}^2(\phi')^2)\phi']' + g(\phi^2)\phi = 0, \quad (80)$$

where $\phi' = \frac{d\phi}{d\xi}$. Note that the phase $\hat{\theta}(\xi)$ must be constant. Write $\alpha_0 = v^2 - c^2$. Then, Eq. (80) is equivalent to the following planar dynamical system:

$$\frac{d\phi}{d\xi} = y, \quad \frac{dy}{d\xi} = \frac{-g(\phi^2)\phi}{\alpha_0 - 3c_{nl}^2 y^2}. \quad (81)$$

When $\alpha_0 > 0$, system (81) is a singular traveling wave system of the second kind, which has the following first integral:

$$H_1(\phi, y) = \alpha_0 y^2 - \frac{3}{2}c_{nl}^2 y^4 + G(\phi^2) = h. \quad (82)$$

Rogers *et al.* [2015] assumed that $g(\phi) = \alpha_1 + \alpha_2\phi^p$, $G(\phi^2) = \alpha_1\phi^2 + \frac{\alpha_2}{p+1}\phi^{2(p+1)}$ and investigate a few exact solutions. However, the bifurcations depending on the parameters and dynamics of solutions of system (81) have not been studied. Here, we discuss these problems for system (81).

Consider *the motions with oscillatory spatial dependence* in the form of

$$\Psi = \phi(t)e^{i[\kappa z + \tilde{\theta}(t)]}, \quad k \in (-\infty, \infty). \quad (83)$$

Substituting (83) into Eq. (79), Rogers *et al.* [2015] obtained the following equations:

$$\begin{aligned} \phi_{tt} - \frac{a^2}{\phi^3} + \kappa^2(c^2 + c_{nl}^2\kappa^2\phi^2)\phi + g(\phi^2)\phi = 0, \\ \tilde{\theta}_t = \frac{a}{\phi^2}, \end{aligned} \quad (84)$$

where a is an integral constant. The first equation of Eq. (84) is equivalent to the following planar dynamical system:

$$\begin{aligned} \frac{d\phi}{dt} &= y, \\ \frac{dy}{dt} &= \frac{a^2 - \phi^3(\kappa^2(c^2 + c_{nl}^2\kappa^2\phi^2)\phi + g(\phi^2)\phi)}{\phi^3}. \end{aligned} \quad (85)$$

Assume $g(\phi) = \alpha_1 + \alpha_2\phi$ and write $\beta_1 = -(c^2\kappa^2 + \alpha_1)$, $\beta_2 = -(c_{nl}^2\kappa^4 + \alpha_2)$. Then, system (85) becomes

$$\frac{d\phi}{dt} = y, \quad \frac{dy}{dt} = \frac{\beta_2\phi^6 + \beta_1\phi^4 + a^2}{\phi^3} = \frac{F(\phi)}{\phi^3}. \quad (86)$$

System (86) is a singular traveling wave system of the first kind, which has the following first integral:

$$H_2(\phi, y) = \frac{1}{2}y^2 + \frac{a^2}{2\phi^2} - \frac{1}{2}\beta_1\phi^2 - \frac{1}{4}\beta_2\phi^4 = h. \tag{87}$$

Because [Rogers *et al.*, 2015] did not study the bifurcations and dynamical behavior of systems (86), we studied these problems for system (86) in [Zhou & Li, 2023].

5.1. *Compact wave solutions of Eq. (79)*

(1) Bifurcations of the phase portraits of system (81) when $g(\phi) = \alpha_1 + \alpha_2\phi^p$.

Consider the following associated regular system of system (81):

$$\frac{d\phi}{d\zeta} = y(\alpha_0 - 3c_{nl}y^2), \quad \frac{dy}{d\zeta} = -\phi(\alpha_1 + \alpha_2\phi^{2p}), \tag{88}$$

where $d\xi = (\alpha_0 - 3c_{nl}y^2)d\zeta$.

Clearly, when $\alpha_0 < 0$ and $\alpha_1\alpha_2 > 0$, the origin $O(0,0)$ is the only equilibrium point of system (88).

When $\alpha_0 < 0$ and $\alpha_1\alpha_2 < 0$, there exist three equilibrium points of system (88), at $O(0,0)$ and $E_{1\mp}(\mp\phi_1, 0)$, where $\phi_1 = (\frac{-\alpha_1}{\alpha_2})^{\frac{1}{2p}}$.

When $\alpha_0 > 0$ and $\alpha_1\alpha_2 > 0$, system (88) has three equilibrium points, $O(0,0)$ and $E_{0\mp}(0, \mp y_s)$, where $y_s = \sqrt{\frac{\alpha_0}{3c_{nl}^2}}$.

When $\alpha_0 > 0$ and $\alpha_1\alpha_2 < 0$, system (88) has nine equilibrium points, at $O(0,0)$, $E_{0\mp}(0, \mp y_s)$, $E_{1\mp}(\mp\phi_1, 0)$ and $E_{s\mp}(\mp\phi_1, \mp y_s)$.

Now, let $M(\phi_j, 0)$ be the coefficient matrix of the linearized system of system (10) at the equilibrium point (ϕ_j, y_j). Then, one has

$$J(0,0) = \det M(0,0) = \alpha_1\alpha_0,$$

$$J(\phi_1, 0) = \det M(\phi_1, 0) = -2p\alpha_1\alpha_0,$$

$$J(0, y_s) = \det M(0, y_s) = -3\alpha_1\alpha_0,$$

$$J(\phi_1, y_s) = \det M(\phi_1, y_s) = 4p\alpha_1\alpha_0.$$

By the theory of planar dynamical systems, for an equilibrium point of a planar integrable system, if $J < 0$, then the equilibrium point is a saddle point; if $J > 0$ and $(\text{Trace}M)^2 - 4J < 0 (> 0)$, then it is a center point (a node point); if $J = 0$ and the Poincaré index of the equilibrium point is 0, then this equilibrium point is a cusp.

Write

$$h_0 = H_1(0,0) = 0,$$

$$h_1 = H_1(\phi_1, 0) = \frac{p\alpha_1\phi_1^2}{p+1},$$

$$h_s = H_1(0, y_s) = \frac{\alpha_0^2}{6c_{nl}^2},$$

$$h_2 = H_1(\phi_1, y_s) = h_1 + h_s.$$

Based on the above qualitative analysis, one obtains the bifurcations of the phase portraits of system (81), as presented in Figs. 18–20.

It can be seen from Fig. 20 that, because there exist two singular straight lines $y = \mp y_s$ of system (81), on the upper-hand side of the singular straight lines $y = \mp y_s$ of system (81), the direction of the orbits of the vector field defined by the singular system (81) is the inverse direction of the orbits of the vector field defined by its associated regular system (88). Therefore, corresponding to a family of closed orbits defined by $H_1(\phi, y) = h$, passing through the singular straight lines $y = \mp y_s$, there exist two families of breaking waves of the singular system (81).

(2) Exact solutions and dynamics of system (81) when $g(\phi) = \alpha_1 + \alpha_2\phi$, $\alpha_0 \leq 0$.

Consider the following system:

$$\frac{d\phi}{d\xi} = y, \quad \frac{dy}{d\xi} = \frac{-\phi(\alpha_1 + \alpha_2\phi^2)}{\alpha_0 - 3c_{nl}^2y^2}, \tag{89}$$

and its associated regular system

$$\frac{d\phi}{d\zeta} = y(\alpha_0 - 3c_{nl}^2y^2), \quad \frac{dy}{d\zeta} = -\phi(\alpha_1 + \alpha_2\phi^2), \tag{90}$$

where $d\xi = (\alpha_0 - 3c_{nl}y^2)d\zeta$. These two systems have the same first integral:

$$H_1(\phi, y) = \alpha_0 y^2 - \frac{3}{2}c_{nl}^2y^4 + \alpha_1\phi^2 + \frac{1}{2}\alpha_2\phi^4 = h. \tag{91}$$

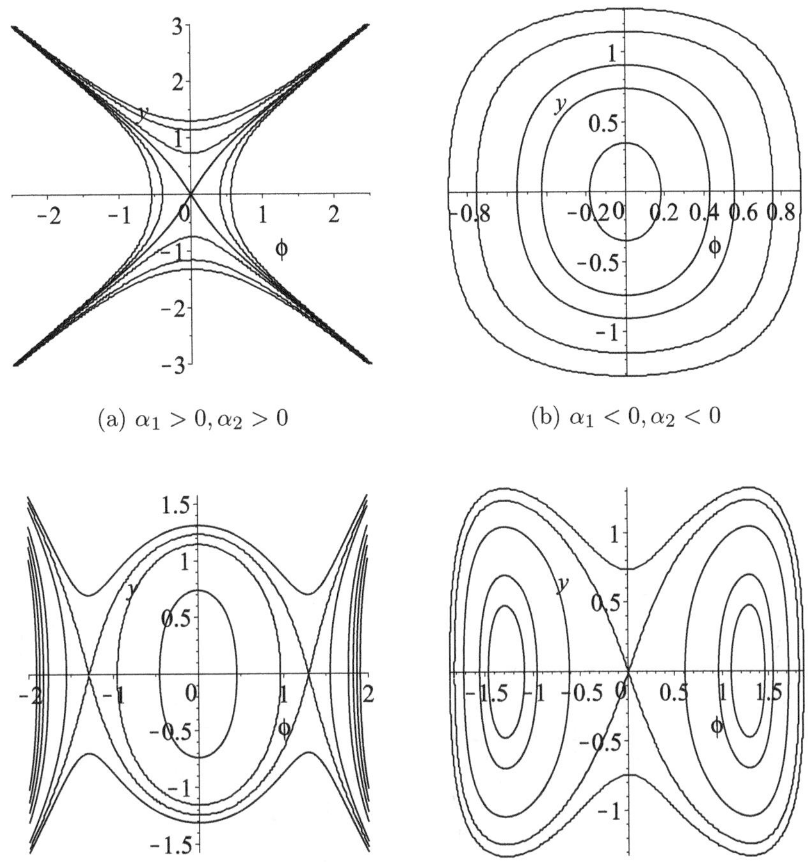

(a) $\alpha_1 > 0, \alpha_2 > 0$ (b) $\alpha_1 < 0, \alpha_2 < 0$

(c) $\alpha_1 < 0, \alpha_2 > 0$ (d) $\alpha_1 > 0, \alpha_2 < 0$

Fig. 18. Bifurcations of phase portraits of system (81) when $\alpha_0 < 0$.

In order to obtain the parametric representations of the level curves defined by (91), polar coordinates $\phi = r\cos\theta, y = r\sin\theta$ are used. Then, (91) can be written as

$$H_1(r,\theta) = \left(\frac{1}{2}\alpha_2\cos^4\theta - \frac{3}{2}c_{nl}^2\sin^4\theta\right)r^4 + r^2(\alpha_1\cos^2\theta + \alpha_0\sin^2\theta) \equiv r^4u_1(\theta) + r^2u_2(\theta) = h. \tag{92}$$

Thus, (92) yields

$$r^2(\theta) = \frac{-u_2(\theta) \mp \sqrt{u_2^2(\theta) + 4hu_1(\theta)}}{2u_1(\theta)}. \tag{93}$$

Consequently, (90) becomes

$$\frac{dr}{d\zeta} = r\cos\theta\sin\theta[(\alpha_0 - \alpha_1) - r^2(\alpha_2\cos^2\theta + 3c_{nl}^2\sin^2\theta)],$$

$$\frac{d\theta}{d\zeta} = -(\alpha_1\cos^2\theta + \alpha_0\sin^2\theta) - r^2(\alpha_2\cos^4\theta - 3c_{nl}^2\sin^4\theta) = -u_2(\theta) - 2r^2u_1(\theta). \tag{94}$$

It follows from (93) and (94) that

$$\frac{d\theta}{d\zeta} = \mp\sqrt{u_2^2(\theta) + 4hu_1(\theta)}, \qquad \zeta = \mp\int\frac{d\theta}{\sqrt{u_2^2(\theta) + 4hu_1(\theta)}}. \tag{95}$$

By using (95), for a given h, one can calculate the exact solutions for system (90). Obviously, if one knows $\tan\theta = \psi(\zeta)$, then from

$$r^2(\zeta) = \frac{-u_2(\theta(\zeta)) \mp \dfrac{d\theta(\zeta)}{d\zeta}}{2u_1(\theta(\zeta))} = \frac{[-(\alpha_1 + \alpha_0\psi^2(\zeta)) \pm \psi'(\zeta)](1 + \psi^2(\zeta))}{\alpha_2 - 3c_{nl}^2\psi^4(\zeta)}, \tag{96}$$

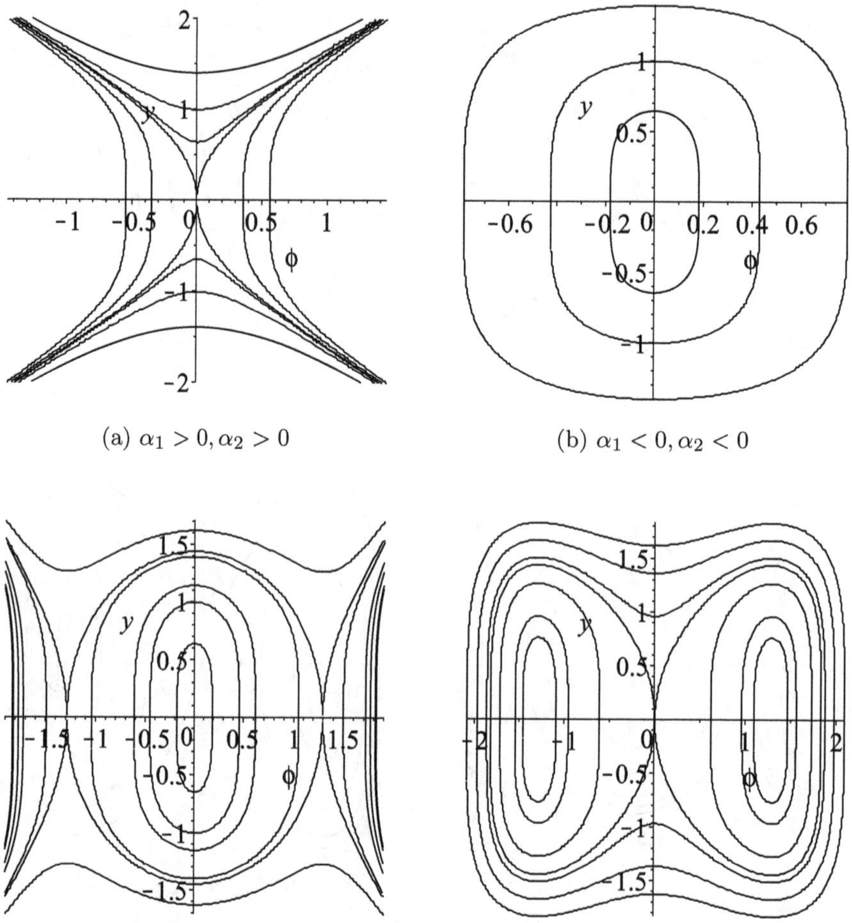

(a) $\alpha_1 > 0, \alpha_2 > 0$ (b) $\alpha_1 < 0, \alpha_2 < 0$

(c) $\alpha_1 < 0, \alpha_2 > 0$ (d) $\alpha_1 > 0, \alpha_2 < 0$

Fig. 19. Bifurcations of phase portraits of system (81) when $\alpha_0 = 0$.

one can obtain the parametric representation of the orbit given by level curve $H_1(\phi, y) = h$ of system (90), as follows:

$$\begin{aligned} &\phi(\zeta) = r(\zeta)\cos\theta(\zeta), \quad y(\zeta) = r(\zeta)\sin\theta(\zeta), \\ &\xi(\zeta) = \int (\alpha_0 - 3c_{nl}^2 y^2(\zeta))d\zeta. \end{aligned} \tag{97}$$

Note that

$$\begin{aligned} u_1(\theta) &= \frac{1}{2}\cos^4\theta(\alpha_2 - 3c_{nl}^2\tan^4\theta) \\ &= \frac{1}{8}[(\alpha_2 - 3c_{nl}^2)\cos^2 2\theta \\ &\quad + 2(\alpha_2 + 3c_{nl}^2)\cos 2\theta \\ &\quad + (\alpha_2 - 3c_{nl}^2)], \\ u_2(\theta) &= \cos^2\theta(\alpha_1 + \alpha_0\tan^2\theta) \\ &= \frac{1}{2}[(\alpha_1 - \alpha_0)\cos 2\theta + (\alpha_1 + \alpha_0)]. \end{aligned}$$

Hence,

$$\begin{aligned} &u_2^2(\theta) + 4hu_1(\theta) \\ &\quad = \left[\frac{1}{2}h(\alpha_2 - 3c_{nl}^2) + \frac{1}{4}(\alpha_1 - \alpha_0)^2\right]\cos^2 2\theta \\ &\quad\quad + \left[h(\alpha_2 + 3c_{nl}^2) + \frac{1}{2}(\alpha_1^2 - \alpha_0^2)\right]\cos 2\theta \\ &\quad\quad + \frac{1}{2}h(\alpha_2 - 3c_{nl}^2) + (\alpha_1 + \alpha_0)^2. \end{aligned} \tag{98}$$

Let $u = \tan\theta$. Then, one has

$$\zeta = \int \frac{d\theta}{\sqrt{u_2^2(\theta) + 4hu_1(\theta)}} = \int \frac{du}{\sqrt{F(u)}}, \tag{99}$$

where

$$\begin{aligned} F(u) = (\alpha_0^2 - 6hc_{nl}^2)u^4 + 2\alpha_0\alpha_1 u^2 \\ + (2\alpha_2 h + \alpha_1^2). \end{aligned}$$

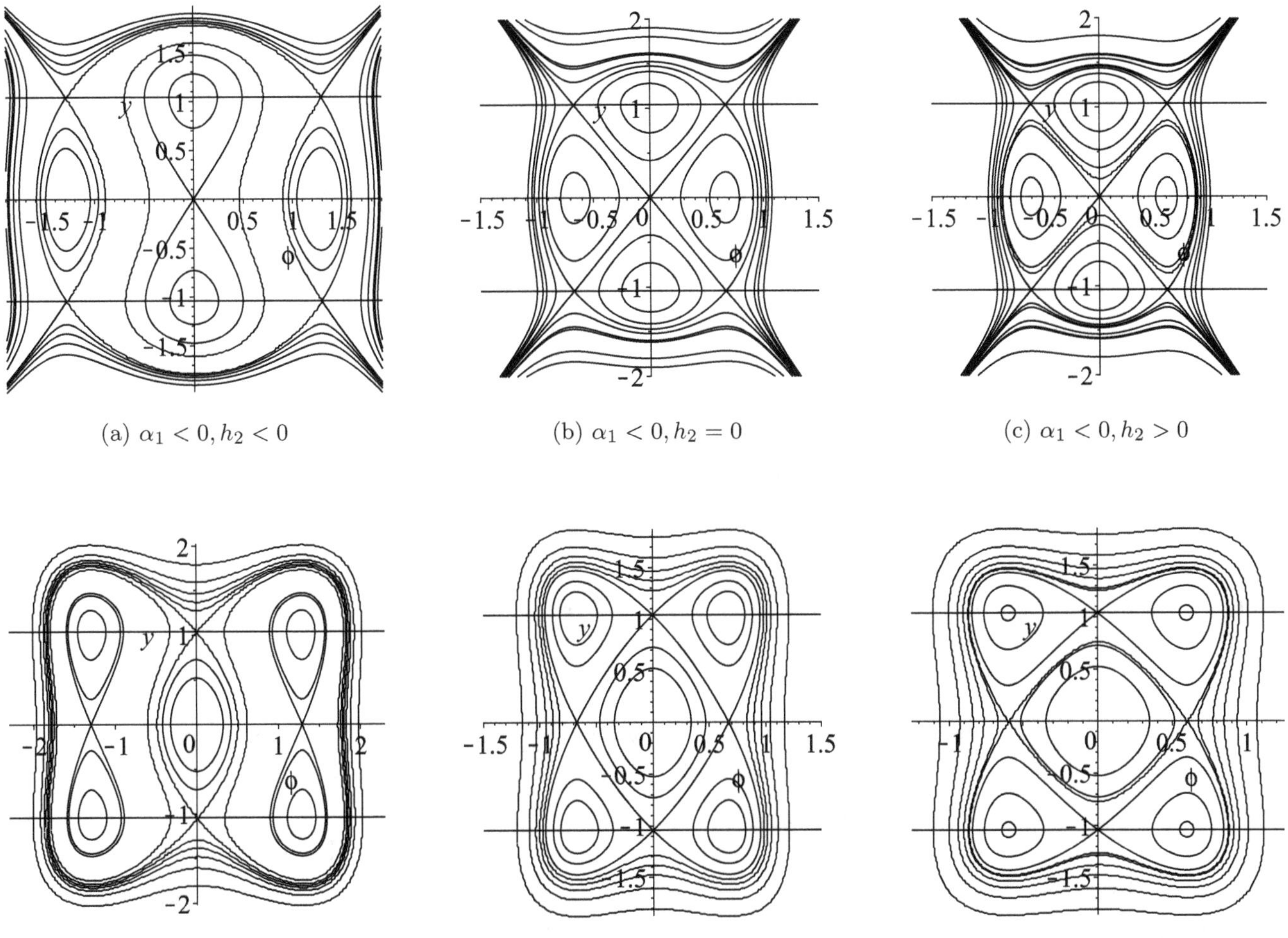

(a) $\alpha_1 < 0, h_2 < 0$ (b) $\alpha_1 < 0, h_2 = 0$ (c) $\alpha_1 < 0, h_2 > 0$

(d) $\alpha_1 > 0, h_1 > h_s$ (e) $\alpha_1 > 0, h_1 = h_s$ (f) $\alpha_1 > 0, h_1 < h_s$

Fig. 20. Bifurcations of phase portraits of system (81) when $\alpha_0 > 0$, $\alpha_1\alpha_2 < 0$.

By calculating (99) and using (97), one obtains

$$\begin{aligned}\phi^2(\zeta) &= r^2(\zeta)\cos^2\theta(\zeta) = \frac{r^2(\zeta)}{1+\psi^2(\zeta)},\\ y^2(\zeta) &= r^2(\zeta)\sin^2\theta(\zeta) = \frac{r^2(\zeta)\psi^2(\zeta)}{1+\psi^2(\zeta)}\\ &= \phi^2(\zeta)\psi^2(\zeta),\\ \xi(\zeta) &= \alpha_0\zeta - 3c_{nl}^2\int\phi^2(\zeta)\psi^2(\zeta)d\zeta,\end{aligned} \tag{100}$$

where r^2 is given by (96).

It can be seen from (100) that one needs to first calculate the parametric representations for the bounded phase orbits of the associated regular system (90), as shown in Figs. 18–20. Then, by coupling the solutions $\xi(\zeta) = \alpha_0\zeta - 3c_{nl}^2\int\phi^2(\zeta)\psi^2(\zeta)d\zeta$, one can obtain the parametric representations of the singular system (89).

(A) The case of $\alpha_0 = v^2 - c^2 \leq 0$, $\alpha_1 < 0$, $\alpha_2 < 0$.

When $\alpha_0 < 0$, corresponding to the closed orbit family defined by $H_1(\phi, y) = h$, $h \in (-\infty, 0)$ in Fig. 18(b), (99) can be written as

$$\sqrt{\alpha_0^2 + 6c_{nl}^2|h|}\zeta = \int_u^\infty \frac{du}{\sqrt{(u^2+\rho^2)(u^2+\bar{\rho}^2)}},$$

where

$$(\rho\bar{\rho})^2 = \frac{\alpha_1^2 + 2|\alpha_2 h|}{\alpha_0^2 + 6c_{nl}^2|h|},\quad \rho^2 + \bar{\rho}^2 = \frac{2\alpha_0\alpha_1}{\alpha_0^2 + 6c_{nl}^2|h|}.$$

It then follows that

$$\begin{aligned}\tan^2\theta(\zeta) = u^2 &= \frac{\rho\bar{\rho}(1+\mathrm{cn}(2\Omega_1\zeta,k))}{(1-\mathrm{cn}(2\Omega_1\zeta,k))}\\ &= \frac{\rho\bar{\rho}}{\mathrm{tn}^2(\Omega_1\zeta,k)\mathrm{dn}^2(\Omega_1\zeta,k)}\\ &= \rho\bar{\rho}\,\mathrm{cs}^2(\Omega_1\zeta,k)\mathrm{nd}^2(\Omega_1\zeta,k),\end{aligned} \tag{101}$$

where

$$\Omega_1 = \sqrt{\rho\bar{\rho}(\alpha_0^2 + 6c_{nl}^2|h|)}, \quad k^2 = -\frac{(\rho - \bar{\rho})^2}{4\rho\bar{\rho}}.$$

Note that

$$\frac{d\theta}{d\zeta} = \frac{\Omega_1\sqrt{\rho\bar{\rho}}(k^2\mathrm{cd}^2(\Omega_1\zeta, k) - \mathrm{ns}^2(\Omega_1\zeta, k))}{1 + \rho\bar{\rho}\mathrm{cs}^2(\Omega_1\zeta, k)\mathrm{nd}^2(\Omega_1\zeta, k)}. \quad (102)$$

Then, it follows from (98) that

$$\phi^2(\zeta) = \frac{-(|\alpha_1| + |\alpha_0|\rho\bar{\rho}\mathrm{cs}^2(\Omega_1\zeta, k)\mathrm{nd}^2(\Omega_1\zeta, k)) + \Omega_1\rho\bar{\rho}(k^2\mathrm{cd}^2(\Omega_1\zeta, k) - \mathrm{ns}^2(\Omega_1\zeta, k))}{|\alpha_2| + 3c_{nl}^2\rho^2\bar{\rho}^2\mathrm{cs}^4(\Omega_1\zeta, k)\mathrm{nd}^4(\Omega_1\zeta, k)}. \quad (103)$$

Therefore, $(\phi(\zeta), \xi(\zeta))$ given by (103) and (100) give rise to a family of periodic solutions of system (89).

(B) The case of $\alpha_0 \le 0$, $\alpha_1 < 0$, $\alpha_2 > 0$.

(a) For the closed orbit family defined by $H_1(\phi, y) = h$, $h \in (h_1, 0)$ in Fig. 18(c), enclosing the origin $O(0,0)$, where $h_1 = -\frac{\alpha_1^2}{2\alpha_2} < 0$, it has the same parametric representation as (103).

(b) Corresponding to the heteroclinic orbits defined by $H_1(\phi, y) = h_1$ in Fig. 18(c), connecting to the singular points $E_{1-}(-\phi_1, 0)$ and $E_{1+}(\phi_1, 0)$, (99) can be written as

$$\sqrt{\alpha_0^2 + \frac{3c_{nl}^2\alpha_1^2}{\alpha_2}}\zeta = \int_0^u \frac{du}{u\sqrt{u^2 + \rho_0^2}} = \frac{1}{\rho_0}\,\mathrm{cseh}^{-1}\left(\frac{u}{\rho_0}\right),$$

where $\rho_0^2 = \frac{2\alpha_0\alpha_1}{\alpha_0^2 + \frac{3c_{nl}^2\alpha_1^2}{\alpha_2}}$. Thus, one has $u = \tan\theta = \frac{\rho_0}{\sinh(\sqrt{2\alpha_0\alpha_1}\zeta)}$. It implies that

$$\phi^2(\zeta) = \frac{[|\alpha_1|\sinh^2(\sqrt{2\alpha_0\alpha_1}\zeta) + |\alpha_0|\rho_0^2 - \rho_0\sqrt{2\alpha_0\alpha_1}\cosh(\sqrt{2\alpha_0\alpha_1}\zeta)]\sinh^2(\sqrt{2\alpha_0\alpha_1}\zeta)}{\alpha_2\sinh^4(\sqrt{2\alpha_0\alpha_1}\zeta) - 3c_{nl}^2\rho_0^4}. \quad (104)$$

Hence, $(\phi(\zeta), \xi(\zeta))$ given by (104) and (100) give rise to a kink wave solution and an anti-kink wave solution of system (89).

(C) The case of $\alpha_0 \le 0$, $\alpha_1 > 0, \alpha_2 < 0$.

(i) When $\alpha_0 < 0$, for the homoclinic orbits defined by $H_1(\phi, y) = 0$, enclosing the equilibrium points $E_{1\mp}(\mp\phi_1, 0)$, respectively, in Fig. 18(d), one has

$$\zeta = \int_\theta^0 \frac{d\theta}{u_2(\theta)} = \int_\theta^0 \frac{d(2\theta)}{(\alpha_1 - \alpha_0)\cos 2\theta + (\alpha_1 + \alpha_0)}.$$

Thus, one obtains $\tan\theta = \sqrt{\frac{\alpha_1}{|\alpha_0|}}\tanh(\sqrt{|\alpha_0|\alpha_1}\zeta)$. It follows that

$$r^2(\zeta) = -\frac{u_2(\theta)}{u_1(\theta)} = \frac{2\alpha_1\,\mathrm{sech}^2(\sqrt{|\alpha_0|\alpha_1}\zeta)\left(1 + \frac{\alpha_1}{|\alpha_0|}\tanh^2(\sqrt{|\alpha_0|\alpha_1}\zeta)\right)}{|\alpha_2| + \frac{3c_{nl}^2\alpha_1^2}{\alpha_0^2}\tanh^4(\sqrt{|\alpha_0|\alpha_1}\zeta)}.$$

Therefore, it follows that

$$\phi^2(\zeta) = \frac{2\alpha_1\,\mathrm{sech}^2(\sqrt{|\alpha_0|\alpha_1}\zeta)}{|\alpha_2| + \frac{3c_{nl}^2\alpha_1^2}{\alpha_0^2}\tanh^4(\sqrt{|\alpha_0|\alpha_1}\zeta)}. \quad (105)$$

Thus, $(\phi(\zeta), \xi(\zeta))$ given by (105) and (100) give rise to two solitary wave solutions of system (89).

When $\alpha_0 = 0$, for the homoclinic orbits defined by $H_1(\phi, y) = 0$, enclosing the equilibrium points $E_{1\mp}(\mp\phi_1, 0)$, respectively, in Fig. 19(d), one has

$$\zeta = \int_\theta^0 \frac{d\theta}{u_2(\theta)} = \int_\theta^0 \frac{d(2\theta)}{\alpha_1(1 + \cos 2\theta)} = \frac{1}{\alpha_1}\tan\theta,$$

i.e. $\tan\theta = \alpha_1\zeta$. Therefore, one obtains

$$\phi^2(\zeta) = \frac{2\alpha_1}{(|\alpha_2| + 3c_{nl}^2(\alpha_1\zeta)^4)(1 + (\alpha_1\zeta)^2)},$$
$$\xi(\zeta) = -3c_{nl}^2\int \phi^2(\zeta)(\alpha_1\xi)^2 d\zeta. \quad (106)$$

The integral formula of $\xi(\zeta)$ is very long, so it is omitted here.

(ii) When $\alpha_0 \le 0$, for the two families of periodic orbits defined by $H_1(\phi, y) = h$, $h \in (0, h_1)$, enclosing the equilibrium points $E_{1\mp}(\mp\phi_1, 0)$, respectively, in Fig. 18(d), one has

$$\sqrt{6hc_{nl}^2-\alpha_0^2}\zeta=\int_u^{\rho_b}\frac{du}{\sqrt{(\rho_a^2+u^2)(\rho_b^2-u^2)}},$$

where

$$(\rho_a\rho_b)^2=\frac{2h\alpha_2+\alpha_1^2}{6hc_{nl}^2-\alpha_0^2},\quad \rho_a^2-\rho_b^2=\frac{2|\alpha_0|\alpha_1}{6hc_{nl}^2-\alpha_0^2}.$$

It gives rise to

$$u=\tan\theta=\rho_b\mathrm{cn}(\Omega_2\zeta,k),$$

where

$$k^2=\frac{\rho_b^2}{\rho_a^2+\rho_b^2},$$

$$\Omega_2=\sqrt{(\rho_a^2+\rho_b^2)(6hc_{nl}^2-\alpha_0^2)}.$$

Thus, one obtains

$$\phi^2(\zeta)=\frac{\alpha_1-|\alpha_0|\rho_b^2\mathrm{cn}^2(\Omega_2\zeta,k)+\rho_b\Omega_2\mathrm{sn}(\Omega_2\zeta,k)\mathrm{dn}(\Omega_2\zeta,k)}{|\alpha_2|+3c_{nl}^2\rho_b^4\mathrm{cn}^4(\Omega_2\zeta,k)}. \tag{107}$$

Consequently, $(\phi(\zeta),\xi(\zeta))$ given by (107) and (100) give rise to two families of periodic wave solutions of system (89).

When $\alpha_0=0$, one has $\rho_a=\rho_b, k^2=\frac{1}{2}$ in (107).

(iii) When $\alpha_0\le 0$, for the global family of periodic orbits defined by $H_1(\phi,y)=h$, $h\in(h_1,\infty)$, enclosing three equilibrium points $O(0,0)$ and $E_{1\mp}(\mp\phi_1,0)$, it yields the same parametric representation as (103).

In summary, the following conclusions can be drawn.

Theorem 5. *Assume that $\alpha_0=v^2-c^2\le 0$, $p=1$. Then, the following conclusions hold*:

(1) *System (89) has the bifurcations of phase portraits shown in Figs. 18 and 19.*
(2) *When $\alpha_1<0$, $\alpha_2<0$, system (89) has a global periodic solution family $(\phi(\zeta),\xi(\zeta))$, where $\phi(\zeta)$ given by (103), $\xi(\zeta)$ defined by (100).*
(3) *When $\alpha_1<0$, $\alpha_2>0$, system (89) has a kink wave solution and an anti-kink wave solution $(\phi(\zeta),\xi(\zeta))$, where $\phi(\zeta)$ is given by (104), and $\xi(\zeta)$ is defined by (100).*
(4) *When $\alpha_1>0,\alpha_2<0$, system (89) has two solitary wave solutions $(\phi(\zeta),\xi(\zeta))$, where $\phi(\zeta)$ is given by (105). In addition, system (89) has two families of periodic wave solutions $(\phi(\zeta),\xi(\zeta))$, where $\phi(\zeta)$ is given by (107) and $\xi(\zeta)$ is defined by (100).*

(3) Exact solutions and dynamics of system (81) when $g(\phi)=\alpha_1+\alpha_2\phi$, $\alpha_0>0$.

When $\alpha_0=v^2-c^2>0$, $\alpha_1\alpha_2<0$, one can see from Fig. 20 that there exist two singular straight lines $y=\mp y_s$ of system (89). Thus, generally, one only needs to consider the parametric representations for all orbits that do not pass through the two singular straight lines shown in Fig. 20.

(A) The case of $\alpha_0>0$, $\alpha_1<0$, $\alpha_2>0$.

First, it is noted that, for the level curves $H_1(\phi,y)=h_2$, the direction of the vector field of defined by system (89) is shown in Fig. 21. These curves consist of an ellipse and a hyperbola (or two straight lines). The ellipse gives rise to a smooth periodic solution, while the hyperbola gives rise to two unbounded solutions of system (89).

(i) $h_1<h_2<0<h_s$ [see Fig. 20(a)]

In this case, as h is varied, the changes of the level curves of system (89) defined by $H_1(\phi,y)=h$ are shown in Fig. 22.

(a) The level curves defined by $H_1(\phi,y)=h$, $h\in(h_1,h_2)$ are two families of periodic orbits enclosing the centers $E_{1\mp}(\mp\phi_1,0)$, respectively [see Fig. 22(b)]. Now, (99) can be written as

$$\sqrt{\alpha_0^2+6|h|c_{nl}^2}\zeta=\int_\infty^0\frac{du}{\sqrt{(u^2-\rho_a^2)(u^2-\rho_b^2)}},$$

where

$$\rho_a^2\rho_b^2=\frac{2\alpha_2h+\alpha_1^2}{\alpha_0+6|h|c_{nl}^2},$$

$$\rho_a^2+\rho_b^2=\frac{2\alpha_0|\alpha_1|}{\alpha_0+6|h|c_{nl}^2}.$$

It follows that

$$u=\tan\theta=\rho_a\mathrm{ns}(\Omega_3\zeta,k),$$

$$k^2=\frac{\rho_b^2}{\rho_a^2},$$

$$\Omega_3=\rho_a\sqrt{\alpha_0^2+6|h|c_{nl}^2}.$$

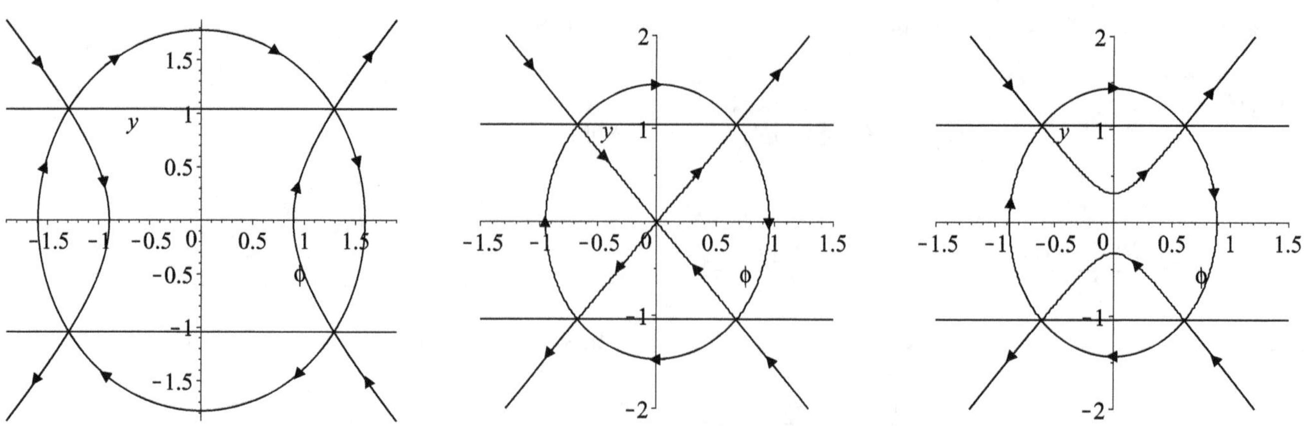

(a) $h_2 < 0$ (b) $h_2 = 0$ (c) $h_2 > 0$

Fig. 21. The directions of the level curves defined by $H_1(\phi, y) = h_2$ of system (89).

Hence, one obtains

$$\phi^2(\zeta) = \frac{|\alpha_1| - \alpha_0\rho_a^2 \mathrm{ns}^2(\Omega_3\zeta, k) + \rho_a\Omega_3 \mathrm{cs}(\Omega_3\zeta, k)\mathrm{ds}(\Omega_3\zeta, k)}{\alpha_2 - 3c_{nl}^2\rho_a^4 \mathrm{ns}^4(\Omega_3\zeta, k)}. \tag{108}$$

Consequently, $(\phi(\zeta), \xi(\zeta))$ given by (108) and (100) give rise to two families of periodic wave solutions of system (89).

(b) The level curves defined by $H_1(\phi, y) = h_2$ are six heteroclinic orbits of the associated regular system (90), connecting the singular points $E_{1\mp}(\mp\phi_1, \mp y_s)$ [see Fig. 22(c)], whereas four heteroclinic orbits enclose the centers $E_{1\mp}(\mp\phi_1, 0)$, respectively; and two heteroclinic orbits connect $E_{1-}(-\phi_1, -y_s)$ and $E_{1+}(\phi_1, -y_s)$, $E_{1+}(-\phi_1, y_s)$ and $E_{1+}(\phi_1, y_s)$, respectively.

For the four heteroclinic orbits, (99) can be written as

$$\sqrt{\frac{3c_{nl}^2\alpha_1^2}{\alpha_2}}\zeta = \int_0^u \frac{du}{\rho_1^2 - u^2} = \frac{1}{\rho_1}\tanh^{-1}\left(\frac{u}{\rho_1}\right)$$

where $\rho_1^2 = \frac{\alpha_0\alpha_2}{3c_{nl}^2|\alpha_1|}$. It gives rise to $u = \tan\theta = \rho_1 \tanh(\sqrt{\alpha_0|\alpha_1|}\zeta)$. Thus, one has

$$\phi^2(\zeta) = \frac{|\alpha_1| - \alpha_0\rho_1^2 \tanh^2(\sqrt{\alpha_0|\alpha_1|}\zeta) \pm \rho_1\sqrt{\alpha_0|\alpha_1|}\,\mathrm{sech}^2(\sqrt{\alpha_0|\alpha_1|}\zeta)}{\alpha_2 - 3c_{nl}^2\rho_1^4 \tanh^4(\sqrt{\alpha_0|\alpha_1|}\zeta)}. \tag{109}$$

Clearly, $\phi^2(\zeta)$ given by (109) gives rise to the four heteroclinic solutions of the associated regular system (90). However, it is not the periodic solution of system (89).

For the two heteroclinic orbits of the associated regular system (90), connecting $E_{1-}(-\phi_1, -y_s)$ and $E_{1+}(\phi_1, -y_s)$, $E_{1+}(-\phi_1, y_s)$ and $E_{1+}(\phi_1, y_s)$, respectively, the integral (99) can be written as

$$\left(\frac{\alpha_0\alpha_2 + 3c_{nl}^2|\alpha_1|}{\sqrt{3\alpha_2 c_{nl}^2}}\right)\zeta = \int_\pi^\theta \frac{d\theta}{\rho_2 + \cos 2\theta} = \frac{2}{\sqrt{1-\rho_2^2}}\tanh\left(\sqrt{\frac{1-\rho_2}{1+\rho_2}}\tan\theta\right),$$

where

$$\rho_2 = \frac{\alpha_0\alpha_2 - 3c_{nl}^2|\alpha_1|}{\alpha_0\alpha_2 + 3c_{nl}^2|\alpha_1|}.$$

It follows that

$$\tan\theta = \sqrt{\frac{1+\rho_2}{1-\rho_2}}\tanh(\sqrt{\alpha_0|\alpha_1|}\zeta).$$

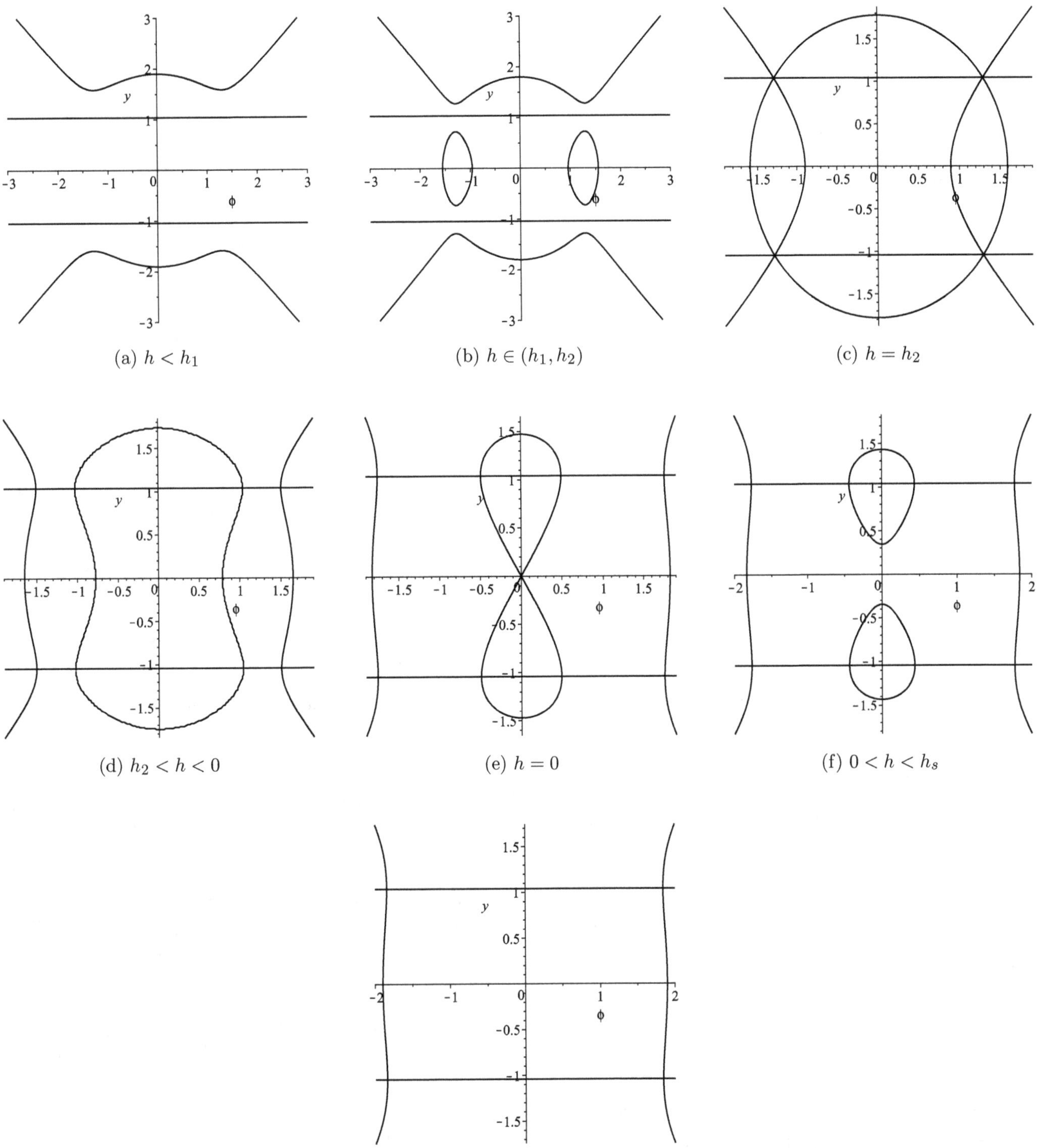

(a) $h < h_1$ (b) $h \in (h_1, h_2)$ (c) $h = h_2$

(d) $h_2 < h < 0$ (e) $h = 0$ (f) $0 < h < h_s$

(g) $h \geq h_s$

Fig. 22. The changes of the level curves of system (89) defined by $H_1(\phi, y) = h$.

Therefore, one has

$$\phi^2(\zeta) = \frac{|\alpha_1| - \alpha_0\left(\frac{1+\rho_2}{1-\rho_2}\right)\tanh^2(\sqrt{\alpha_0|\alpha_1|}\zeta) - \sqrt{\frac{1+\rho_2}{1-\rho_2}}\sqrt{\alpha_0|\alpha_1|}\,\mathrm{sech}^2(\sqrt{\alpha_0|\alpha_1|}\zeta)}{\alpha_2 - 3c_{nl}^2\left(\frac{1+\rho_2}{1-\rho_2}\right)^2\tanh^4(\sqrt{\alpha_0|\alpha_1|}\zeta)}. \tag{110}$$

Obviously, $\phi^2(\zeta)$ given by (110) gives rise to a kink wave solution and an anti-kink wave solution of the associated regular system (90). However, it is not the periodic solution of system (89).

Note that $\xi(\zeta) = \int(\alpha_0 - 3c_{nl}^2 y^2(\zeta))d\zeta$. Thus, corresponding to the ellipse in Fig. 21(a), $(\phi(\zeta), \xi(\zeta))$ give rise to a smooth periodic solution of system (89), where $\phi(\xi)$ is given by (109) and (110).

(ii) $h_1 < h_2 = 0 < h_s$ [see Fig. 20(b)]

When $\alpha_2 = \frac{3c_{nl}^2\alpha_1^2}{\alpha_0^2}$, one has $h = H_1(\phi, y) = h_2 = 0$. Under this parameter condition, $H_1(\phi, y) = h_2$ given by (91) can be written as

$$\begin{aligned} H_1(\phi, y) &= (\alpha_0 y^2 - |\alpha_1|\phi^2) \\ &\quad \times \left(1 - \frac{3c_{nl}^2 y^2}{2\alpha_0} - \frac{3c_{nl}^2|\alpha_1|\phi^2}{2\alpha_0^2}\right) \\ &= 0. \end{aligned} \tag{111}$$

(a) Corresponding to the ellipse orbit $\frac{3c_{nl}^2 y^2}{2\alpha_0} + \frac{3c_{nl}^2|\alpha_1|\phi^2}{2\alpha_0^2} = 1$ shown in Fig. 21(b), one has

$$\sqrt{\frac{2\alpha_0}{3c_{nl}^2}}\xi = \int_0^\phi \frac{d\phi}{\sqrt{1 - \left(\frac{\phi}{\rho_3}\right)^2}},$$

where $\rho_3 = \sqrt{\frac{2\alpha_0^2}{3c_{nl}^2|\alpha_1|}}$. Thus, one obtains the following periodic solution of system (89):

$$\phi(\xi) = \rho_3 \sin\left(\sqrt{\frac{|\alpha_1|}{\alpha_0}}\xi\right). \tag{112}$$

(b) Corresponding to the four half-straight lines $y = \mp\sqrt{\frac{|\alpha_1|}{\alpha_b}}\phi$ shown in Fig. 21(b), one has

$$\phi(\xi) = \mp\sqrt{\frac{|\alpha_1|}{\alpha_2}}e^{\mp\sqrt{\frac{|\alpha_1|}{\alpha_0}}\xi}. \tag{113}$$

These are four unbounded solutions of system (89).

(c) The level curves defined by $H_1(\phi, y) = h$, $h \in (h_1, 0)$ are two families of periodic orbits enclosing the centers $E_{1\mp}(\mp\phi_1, 0)$, respectively. They have the same parametric representation as (108) with $\alpha_2 = \frac{3c_{nl}^2\alpha_1^2}{\alpha_0^2}$.

(iii) $h_1 < 0 < h_2 < h_s$ [see Fig. 20(c)]

In this case, as h is varied, the changes of the level curves of system (89) defined by $H_1(\phi, y) = h$ are shown in Fig. 23.

(a) For the homoclinic orbits defined by $H_1(\phi, y) = 0$, enclosing the equilibrium points $E_{1\mp}(\mp\phi_1, 0)$, respectively, in Fig. 23(c), one has

$$\zeta = \int_\theta^0 \frac{d\theta}{u_2(\theta)} = \int_\theta^0 \frac{d(2\theta)}{(\alpha_1 - \alpha_0)\cos 2\theta + (\alpha_1 + \alpha_0)}.$$

Thus, one obtains $\tan\theta = \sqrt{\frac{|\alpha_1|}{\alpha_0}}\tanh(\sqrt{\alpha_0|\alpha_1|}\zeta)$. It follows that

$$\phi^2(\zeta) = \frac{2|\alpha_1|\,\mathrm{sech}^2(\sqrt{\alpha_0|\alpha_1|}\zeta)}{\alpha_2 - \frac{3c_{nl}^2\alpha_1^2}{\alpha_0^2}\tanh^4(\sqrt{\alpha_0|\alpha_1|}\zeta)}. \tag{114}$$

Consequently, $(\phi(\zeta), \xi(\zeta))$ given by (114) and (100) give rise to two solitary wave solutions of system (89).

(b) For the two families of periodic orbits defined by $H_1(\phi, y) = h$, $h \in (h_1, 0)$, enclosing the equilibrium points $E_{1\mp}(\mp\phi_1, 0)$, respectively, in Fig. 23(b), one has

$$\sqrt{6|h|c_{nl}^2 + \alpha_0^2}\zeta = \int_0^u \frac{du}{\sqrt{(\rho_a^2 - u^2)(\rho_b^2 - u^2)}},$$

where

$$(\rho_a\rho_b)^2 = \frac{\alpha_1^2 - 2|h|\alpha_2}{\alpha_0^2 + 6|h|c_{nl}^2}, \quad \rho_a^2 + \rho_b^2 = \frac{2\alpha_0|\alpha_1|}{\alpha_0^2 + 6|h|c_{nl}^2}.$$

It gives rise to $u = \tan\theta = \rho_b \mathrm{sn}(\Omega_4\zeta, k)$, where

$$k^2 = \frac{\rho_b^2}{\rho_a^2}, \quad \Omega_4 = \sqrt{\rho_a^2(\alpha_0^2 + 6|h|c_{nl}^2)}.$$

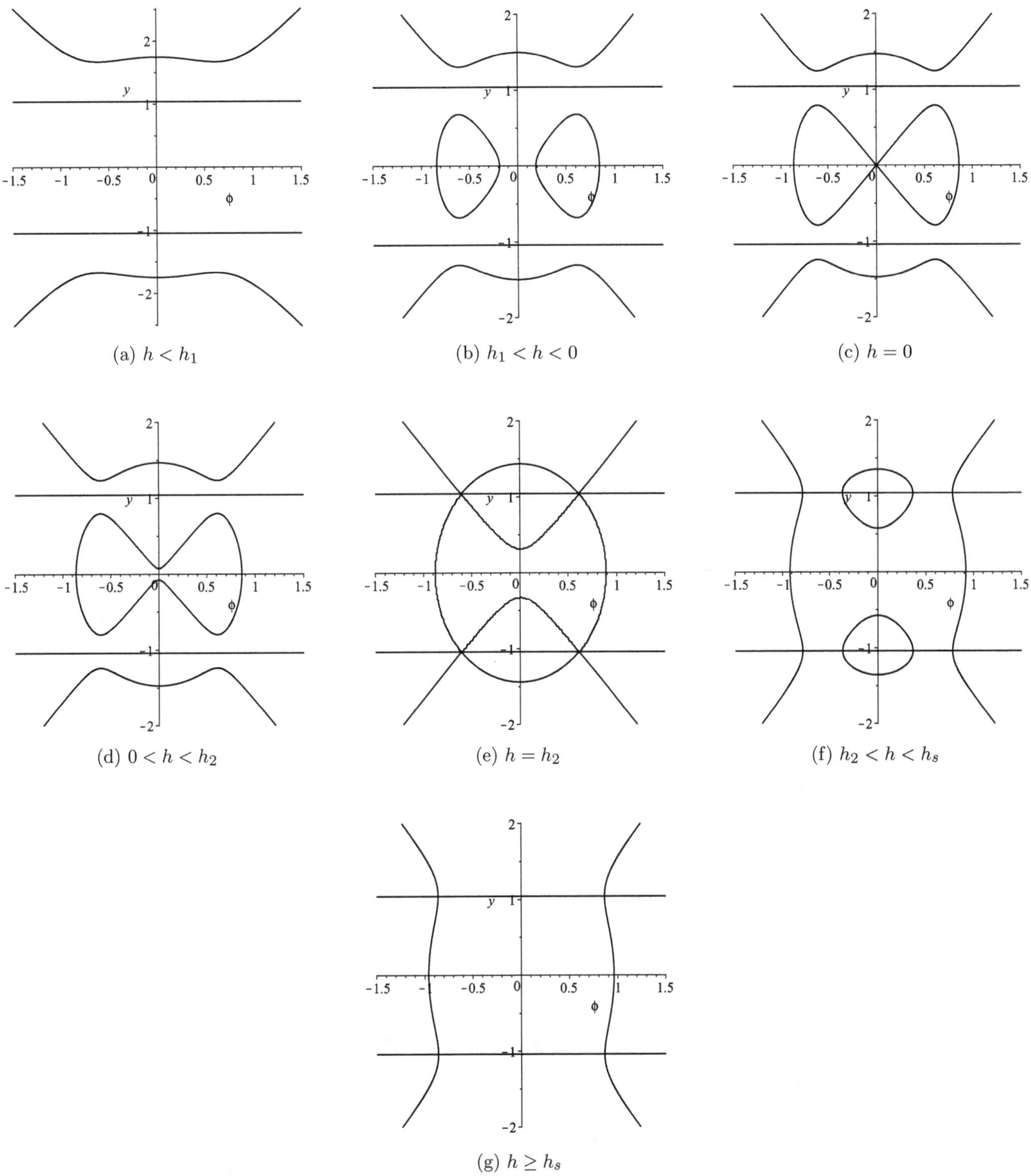

(a) $h < h_1$ (b) $h_1 < h < 0$ (c) $h = 0$

(d) $0 < h < h_2$ (e) $h = h_2$ (f) $h_2 < h < h_s$

(g) $h \geq h_s$

Fig. 23. The changes of the level curves of system (89) defined by $H_1(\phi, y) = h$.

Thus, one obtains

$$\phi^2(\zeta) = \frac{|\alpha_1| - \alpha_0\rho_b^2\mathrm{sn}^2(\Omega_4\zeta,k) + \rho_b\Omega_4\mathrm{cn}(\Omega_4\zeta,k)\mathrm{dn}(\Omega_4\zeta,k)}{\alpha_2 - 3c_{nl}^2\rho_b^4\mathrm{sn}^4(\Omega_4\zeta,k)}. \tag{115}$$

Consequently, $(\phi(\zeta), \xi(\zeta))$ given by (115) and (100) give rise to two families of periodic wave solutions of system (89).

(c) For the family of periodic orbits defined by $H_1(\phi, y) = h$, $h \in (0, h_2)$, enclosing three equilibrium points $O(0,0)$ and $E_{1\mp}(\mp\phi_1, 0)$ in Fig. 23(d), (99) can be written as

$$\sqrt{\alpha_0^2 - 6hc_{nl}^2}\zeta = \int_u^\infty \frac{du}{\sqrt{(u^2+\rho^2)(u^2+\bar{\rho}^2)}},$$

where

$$(\rho\bar{\rho})^2 = \frac{\alpha_1^2 + 2\alpha_2 h}{\alpha_0^2 - 6c_{nl}^2 h},$$

$$\rho^2 + \bar{\rho}^2 = \frac{2\alpha_0\alpha_1}{\alpha_0^2 - 6c_{nl}^2 h}.$$

It follows that

$$\tan^2\theta(\zeta) = u^2 = \frac{\rho\bar{\rho}(1 + \mathrm{cn}(2\Omega_5\zeta,k))}{1 - \mathrm{cn}(2\Omega_5\zeta,k)}$$
$$= \frac{\rho\bar{\rho}}{\mathrm{tn}^2(\Omega_5\zeta,k)\mathrm{dn}^2(\Omega_5\zeta,k)}$$
$$= \rho\bar{\rho}\mathrm{cs}^2(\Omega_5\zeta,k)\mathrm{nd}^2(\Omega_5\zeta,k),$$

where

$$\Omega_5 = \sqrt{\rho\bar{\rho}(\alpha_0^2 - 6c_{nl}^2 h)}, \quad k^2 = -\frac{(\rho-\bar{\rho})^2}{4\rho\bar{\rho}}.$$

Note that

$$\frac{d\theta}{d\zeta} = \frac{\Omega_5\sqrt{\rho\bar{\rho}}(k^2\mathrm{cd}^2(\Omega_5\zeta,k) - \mathrm{ns}^2(\Omega_5\zeta,k))}{1 + \rho\bar{\rho}\mathrm{cs}^2(\Omega_5\zeta,k)\mathrm{nd}^2(\Omega_5\zeta,k)}.$$

Thus, one obtains

$$\phi^2(\zeta) = \frac{|\alpha_1| - \alpha_0\rho\bar{\rho}\mathrm{cs}^2(\Omega_5\zeta,k)\mathrm{nd}^2(\Omega_5\zeta,k) + \Omega_5\rho\bar{\rho}(k^2\mathrm{cd}^2(\Omega_5\zeta,k) - \mathrm{ns}^2(\Omega_5\zeta,k))}{\alpha_2 - 3c_{nl}^2\rho^2\bar{\rho}^2\mathrm{cs}^4(\Omega_5\zeta,k)\mathrm{nd}^4(\Omega_5\zeta,k)}. \tag{116}$$

Consequently, $(\phi(\zeta), \xi(\zeta))$ given by (116) and (100) give rise to a family of periodic solutions of system (89).

Similarly to (A)(ii), one can obtain the parameter representations of the ellipse periodic orbit shown in Fig. 23(e), as given in (112) and (100).

(B) The case of $\alpha_0 > 0$, $\alpha_1 > 0, \alpha_2 < 0$.

(i) $0 < h_s < h_1 < h_2$ [see Fig. 20(d)]

(1) For the two heteroclinic orbits of the associated regular system (90), connecting $E_{s-}(0, -y_s)$ and $E_{s+}(0, y_s)$, enclosing the center $O(0,0)$, the integral (99) can be written as

$$\sqrt{2\alpha_0\alpha_1}\zeta = \int_0^u \frac{du}{\sqrt{u^2+\rho_3^2}} = \sinh^{-1}\left(\frac{u}{\rho_3}\right),$$

where $\rho_3^2 = \frac{\alpha_0\alpha_2}{6c_{nl}^2\alpha_1} + \frac{\alpha_1}{2\alpha_0}$. It follows that $\tan\theta = u = \rho_3\sinh(\sqrt{2\alpha_0\alpha_1}\zeta)$.

$$\phi^2(\zeta) = \frac{\alpha_1 + \alpha_0\rho_2\sinh^2(\sqrt{2\alpha_0\alpha_1}\zeta) \mp \rho_3\sqrt{2\alpha_0\alpha_1}\cosh(\sqrt{2\alpha_0\alpha_1}\zeta)}{|\alpha_2| + 3c_{nl}^2\rho_3^4\sinh^4(\sqrt{2\alpha_0\alpha_1}\zeta)}. \tag{117}$$

(2) For the family of periodic orbits of the associated regular system (90) defined by $H_1(\phi, y) = h$, $h \in (0, h_s)$, enclosing the center $O(0,0)$, the integral (99) can be written as

$$\sqrt{\alpha_0^2 - 6c_{nl}^2 h}\zeta$$
$$= \int_0^u \frac{du}{\sqrt{(u^2+\rho_a^2)(u^2+\rho_b^2}} = \frac{1}{\rho_a}\mathrm{tn}^{-1}\left(\frac{u}{\rho_b}\right),$$

where

$$(\rho_a\rho_b)^2 = \frac{\alpha_1^2 + 2\alpha_2 h}{\alpha_0^2 - 6c_{nl}^2 h}, \quad \rho_a^2 + \rho_b^2 = \frac{2\alpha_0\alpha_1}{\alpha_0^2 - 6c_{nl}^2 h}.$$

Hence, $u = \tan\theta = \rho_b\mathrm{tn}(\Omega_6\zeta, k)$, where

$$\Omega_6 = \rho_a\sqrt{\alpha_0^2 - 6c_{nl}^2 h}, \quad k^2 = \frac{\rho_a^2 - \rho_b^2}{\rho_a^2}.$$

Thus, one has

$$\phi^2(\zeta) = \frac{[\alpha_1 \mathrm{cn}^2(\omega_6\zeta, k) + \alpha_0\rho_b^2 \mathrm{sn}^2(\omega_6\zeta, k) + \rho_b\Omega_6 \mathrm{dn}^2(\omega_6\zeta, k)]\mathrm{cn}^2(\omega_6\zeta, k)}{|\alpha_2|\mathrm{cn}^4(\omega_6\zeta, k) + 3c_{nl}^2\rho_b^4 \mathrm{sn}^4(\omega_6\zeta, k)}. \tag{118}$$

(ii) $0 < h_1 = h_s < h_2$ [see Fig. 20(d)]

(1) For the four heteroclinic orbits of the associated regular system (90), connecting $E_{s\pm}(0, \pm y_s)$ and $E_{1\mp}(\mp\phi_1, 0)$, enclosing the center $O(0,0)$, the integral (99) can be written as

$$\mp\sqrt{2\alpha_0\alpha_1}\zeta = \int_{u_0}^{u} \frac{du}{u} = \ln\left(\frac{u}{u_0}\right),$$

where u_0 is a constant. Thus, $u = \tan\theta = \mp e^{\mp\sqrt{2\alpha_0\alpha_1}\zeta}$. It gives rise to the following solutions:

$$\phi^2(\zeta) = \frac{\alpha_1 + \alpha_0 u_0^2 e^{\mp 2\sqrt{2\alpha_0\alpha_1}\zeta} \pm u_0\sqrt{2\alpha_0\alpha_1} e^{\mp\sqrt{2\alpha_0\alpha_1}\zeta}}{|\alpha_2| + 3c_{nl}^2 u_0^4 e^{\mp 4\sqrt{2\alpha_0\alpha_1}\zeta}}. \tag{119}$$

(2) For the family of periodic orbits of the associated regular system (90) defined by $H_1(\phi, y) = h$, $h \in (0, h_1)$, $h_1 = h_s$, enclosing the center $O(0,0)$, it has the same parametric representation as (118).

(iii) $0 < h_1 < h_s < h_2$ [see Fig. 20(e)]

For the two heteroclinic orbits of the associated regular system (90), connecting $E_{1-}(-\phi_1, 0)$ and $E_{1+}(\phi_1, 0)$, enclosing the center $O(0,0)$, similar to (26), it has the following parametric representations:

$$\phi^2(\zeta) = \frac{[-\alpha_1 \sinh^2(\sqrt{2\alpha_0\alpha_1}\zeta) - \alpha_0\rho_0^2 + \rho_0\sqrt{2\alpha_0\alpha_1}\cosh(\sqrt{2\alpha_0\alpha_1}\zeta)]\sinh^2(\sqrt{2\alpha_0\alpha_1}\zeta)}{|\alpha_2|\sinh^4(\sqrt{2\alpha_0\alpha_1}\zeta) + 3c_{nl}^2\rho_0^4}, \tag{120}$$

where

$$\rho_0^2 = \frac{2\alpha_0\alpha_1}{\alpha_0^2 + \dfrac{3c_{nl}^2\alpha_1^2}{\alpha_2}}.$$

For the family of periodic orbits of the associated regular system (90) defined by $H_1(\phi, y) = h$, $h \in (0, h_1)$, enclosing the center $O(0,0)$, it has the same parametric representation as (118).

In summary, the following results are obtained.

Theorem 6. *Assume that $\alpha_0 = v^2 - c^2 > 0$, $p = 1$. Then, the following conclusions hold:*

(1) *System (89) has the bifurcations of phase portraits shown in Fig. 20.*

(2) *Suppose that $\alpha_1 < 0, \alpha_2 > 0$.*

   (i) *If $h_1 < h_2 < 0 < h_s$, system (89) has two periodic solution families, $(\phi(\zeta), \xi(\zeta))$, where $\phi(\zeta)$ is given by (108) and $\xi(\zeta)$ is defined by (100). In addition, corresponding to the ellipse level curve $H_1(\phi, y) = h_2$, the system has a periodic solution $(\phi(\zeta), \xi(\zeta))$, where $\phi(\zeta)$ is given by (109) and (110), and $\xi(\zeta)$ is defined by (100).*

   (ii) *In particular, if $h_1 < h_2 = 0 < h_s$, then system (89) has a periodic solution given by $\phi(\xi) = \rho_3 \sin(\sqrt{\frac{|\alpha_1|}{\alpha_0}}\xi)$.*

   (iii) *If $h_1 < 0 < h_2 < h_s$, then system (89) has two solitary wave solutions, $(\phi(\zeta), \xi(\zeta))$, where $\phi(\zeta)$ is given by (114) and $\xi(\zeta)$ is defined by (100). In addition, system (89) has two periodic wave solution families, $(\phi(\zeta), \xi(\zeta))$, where $\phi(\zeta)$ is given by (115) and $\xi(\zeta)$ is defined by (100).*

(3) *Suppose that $\alpha_1 > 0$, $\alpha_2 < 0$.*

   (i) *If $0 < h_s < h_1 < h_2$, then system (89) has two bounded wave solutions, $(\phi(\zeta), \xi(\zeta))$, where $\phi(\zeta)$ is given by (117) and $\xi(\zeta)$ is defined by (100). In addition, system (89) has a family of periodic solutions, $(\phi(\zeta), \xi(\zeta))$, where $\phi(\zeta)$ is given by (118) and $\xi(\zeta)$ is defined by (100).*

   (ii) *If $0 < h_1 = h_s < h_2$, then system (89) has four bounded wave solutions, $(\phi(\zeta), \xi(\zeta))$, where $\phi(\zeta)$ is given by (120) and $\xi(\zeta)$ is defined by (100). In addition, system (89) has a family of periodic solutions, $(\phi(\zeta), \xi(\zeta))$, where $\phi(\zeta)$ is given by (118) and $\xi(\zeta)$ is defined by (100).*

   (iii) *If $0 < h_1 < h_s < h_2$, then system (89) has a kink wave solution and an anti-kink wave solution, $(\phi(\zeta), \xi(\zeta))$, where $\phi(\zeta)$ is given by (120) and $\xi(\zeta)$ is defined by (100).*

5.2. Motions with oscillatory spatial dependence

(1) Bifurcations of the phase portraits and exact solutions of system (86) when $g(\phi) = \alpha_1 + \alpha_2\phi$.

Consider the following associated regular system of system (86):

$$\frac{d\phi}{d\zeta} = y\phi^3, \quad \frac{dy}{d\zeta} = \beta_2\phi^6 + \beta_1\phi^4 + a^2 = F(\phi), \tag{121}$$

where $dt = \phi^3 d\zeta$, for $\phi \neq 0$.

Clearly, $F'(\phi) = \phi^3(6\beta_2\phi^2 + 4\beta_1)$. When $\beta_1\beta_2 < 0$, $F'(\phi)$ has two zeros at $\phi = \mp\phi_0 = \mp\sqrt{-\frac{2\beta_1}{3\beta_2}}$ and $F(\mp\phi_0) = \frac{27a^2\beta_2^2+4\beta_1^3}{27*\beta 2^2}$. It is easy to verify that, for a fixed $a > 0$, if and only if $0 < \beta_2 < \frac{2}{3a\sqrt{3}}\beta_1^{\frac{3}{2}}$, the $F(\phi)$ has four zeros, namely, system (121) has four equilibrium points.

Let $M(\phi_j, 0)$ be the coefficient matrix of the linearized system of system (121) at the equilibrium point $(\phi_j, 0)$. Then, one has

$$J(\phi_j, 0) = \det M(0,0) = -2\phi_j^6(2\beta_1 + 3\beta_3\phi_j^2).$$

Let $h_j = H_2(\phi_j, 0)$. For a fixed $a > 0$, Fig. 24(a) shows the partition of the (β_1, β_2)-parameter plane. When $(\beta_1, \beta_2) \in (I)$, system (121) has no equilibrium points. When $(\beta_1, \beta_2) \in (III)$, system (121) has two equilibrium points. When $(\beta_1, \beta_2) \in (II)$, system (121) has four equilibrium points. Figures 24(b)–24(d) show the bifurcations of the phase portraits of system (121).

(2) The exact solutions when $0 < \beta_2 < \frac{2}{3a\sqrt{3}}\beta_1^{\frac{3}{2}}$.

It can be seen from (87) that $y^2 = 2h - \frac{a^2}{\phi^2} - \beta_1\phi^2 + \frac{1}{2}\beta_2\phi^4$. Thus, by using the first equation of system (121), one obtains

$$\sqrt{2\beta_2}t = \int_{\phi_0}^{\phi} = \frac{d\phi^2}{\sqrt{-\frac{2a^2}{\beta_2} - \frac{4h}{\beta_2}\phi^2 + \frac{2\beta_1}{\beta_2}\phi^4 + \phi^6}}. \tag{122}$$

(i) First, consider the two homoclinic orbits defined by $H_2(\phi, y) = h_2$ shown in Fig. 5(c). Now, (122) can be written as

$$\sqrt{2\beta_2}t = \int_{\psi_m}^{\psi} \frac{d\psi}{(\psi_2 - \psi)\sqrt{(\psi - \psi_m)}},$$

where $\psi = \phi^2$. It follows that

$$\begin{aligned}\phi^2(t) &= \phi_2^2 - (\phi_2^2 - \phi_m^2)\,\mathrm{sech}^2(\bar{\omega}_1 t)\\ &= \phi_m^2 + (\phi_2^2 - \phi_m^2)\tanh^2(\bar{\omega}_2 t),\end{aligned} \tag{123}$$

where $\bar{\omega}_1 = \frac{1}{2}\sqrt{2\beta_2(\phi_2^2 - \phi_m^2)}$.

By (84), one has

$$\begin{aligned}\tilde{\theta}(t) &= \int \frac{a^2}{\phi^2(t)}\\ &= a^2 \int \frac{dt}{\phi_m^2 + (\phi_2^2 - \phi_m^2)\tanh^2(\bar{\omega}_2 t)}\\ &= \frac{a^2}{2\bar{\omega}_1\phi_2^2}\left[\ln\left(\frac{1 + \tanh(\bar{\omega}_1 t)}{-1 + \tanh(\bar{\omega}_1 t)}\right) + \frac{2\sqrt{\phi_2^2 - \phi_m^2}}{\phi_m}\arctan\left(\frac{1}{\phi_m}\sqrt{\phi_2^2 - \phi_m^2}\tanh(\bar{\omega}_1 t)\right)\right].\end{aligned} \tag{124}$$

(ii) For the two families of periodic orbits, enclosing the centers $E_{1\mp}(\mp\phi_1, 0)$, defined by $H_2(\phi, y) = h$, $h \in (h_1, h_2)$, (122) can be written as

$$\sqrt{2\beta_2}t = \int_{\phi_c^2}^{\phi^2} \frac{d\phi^2}{\sqrt{(\phi_a^2 - \phi^2)(\psi_b^2 - \phi^2)(\phi^2 - \phi_c^2)}},$$

where $0 < \phi_c < \phi_1 < \phi_b < \phi_2 < \phi_a$. It gives rise to the following periodic solutions:

$$\phi^2(t) = \phi_c^2 + (\phi_b^2 - \phi_c^2)\mathrm{sn}^2(\bar{\Omega}_1 t, k), \tag{125}$$

where

$$k^2 = \frac{\phi_b^2 - \phi_c^2}{\phi_a^2 - \phi_c^2}, \quad \bar{\Omega}_1 = \frac{1}{2}\sqrt{2\beta_2(\phi_a^2 - \phi_c^2)}.$$

Therefore, one has

$$\begin{aligned}\tilde{\theta}(t) &= \int \frac{a^2}{\phi^2(t)}\\ &= a^2 \int \frac{dt}{\phi_c^2 + (\phi_b^2 - \phi_c^2)\mathrm{sn}^2(\bar{\Omega}_1 t)}\end{aligned}$$

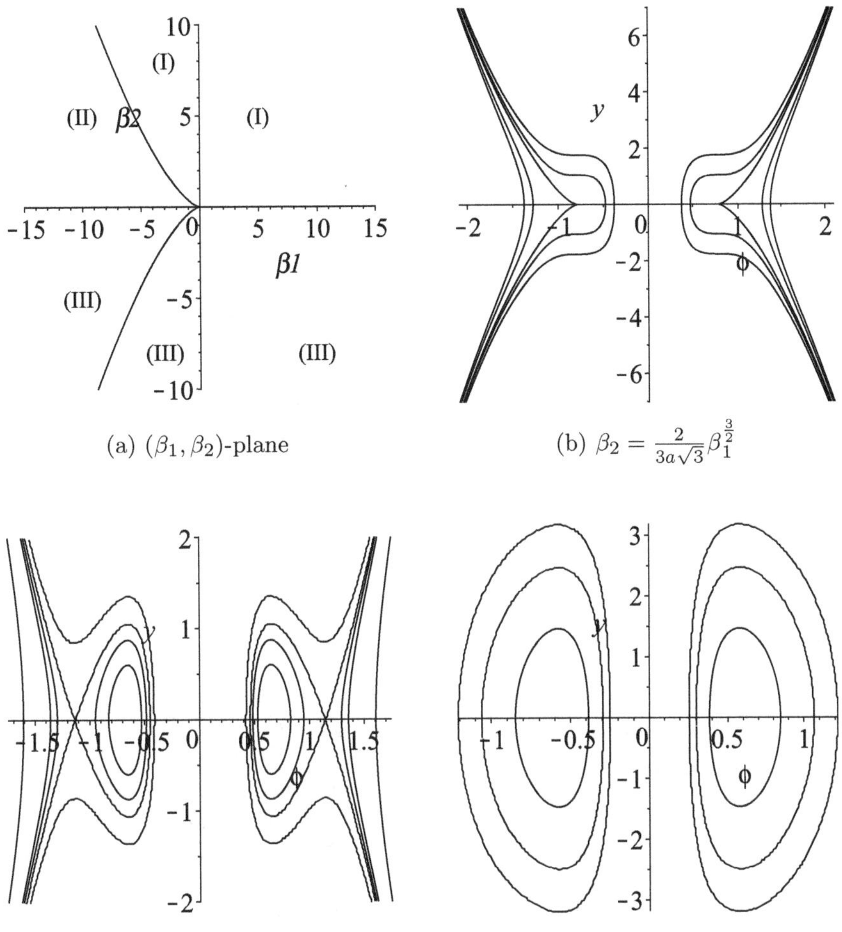

(a) (β_1, β_2)-plane (b) $\beta_2 = \frac{2}{3a\sqrt{3}}\beta_1^{\frac{3}{2}}$

(c) $(\beta_1, \beta_2) \in (II)$ (d) $(\beta_1, \beta_2) \in (III)$

Fig. 24. Partition of the (β_1, β_2)-parameter plane and bifurcations of the phase portraits.

$$
\begin{aligned}
&= \frac{a^2}{\phi_c^2}\int \frac{dt}{1+\left(\frac{\phi_b^2}{\phi_c^2}-1\right)\operatorname{sn}^2(\overline{\Omega}_1 t, k)} \\
&= \frac{a^2}{\overline{\Omega}_1\phi_c^2}\Pi\left(\arcsin(\operatorname{sn}(\overline{\Omega}_1 t, k)), \left(\frac{\phi_b^2}{\phi_c^2}-1\right), k\right).
\end{aligned} \tag{126}
$$

(3) The exact solutions when $(\beta_1, \beta_2) \in (III)$. Corresponding to the two global families of periodic orbits defined by $H_2(\phi, y) = h$, $h \in (h_1, \infty)$, enclosing the centers $E_{1\mp}(\mp\phi_1, 0)$, respectively, (122) can be written as

$$
\sqrt{2\beta_2}t = \int_{\phi_b^2}^{\phi^2} \frac{d\phi^2}{\sqrt{(\phi_a^2-\phi^2)(\psi_b^2-\phi^2)(\phi^2+\phi_c^2)}},
$$

where $\phi_c^2 < 0 < \phi_b^2 < \phi_1^2 < \phi_a^2$. Thus, one has

$$
\phi^2(t) = -\phi_c^2 + \frac{\phi_b^2+\phi_c^2}{\operatorname{dn}(\overline{\Omega}_2 t, k)}, \tag{127}
$$

where

$$
k^2 = \frac{\phi_a^2-\phi_b^2}{\phi_a^2+\phi_c^2}, \quad \overline{\Omega}_2 = \frac{1}{2}\sqrt{2\beta_2(\phi_a^2+\phi_c^2)}.
$$

Hence, it follows from (84) that

$$
\begin{aligned}
\tilde{\theta}(t) &= \int \frac{a^2}{\phi^2(t)} \\
&= a^2\int \frac{dt}{-\phi_c^2+(\phi_b^2+\phi_c^2)\operatorname{nd}^2(\overline{\Omega}_2 t, k)} \\
&= \frac{a^2}{\phi_b^2}\int \frac{\operatorname{dn}^2(\overline{\Omega}_2 t, k)dt}{1+\left(\frac{k^2\phi_c^2}{\phi_b^2}\right)\operatorname{sn}^2(\overline{\Omega}_2 t, k)} \\
&= \frac{a^2}{\overline{\Omega}_2\phi_b^2\tilde{\alpha}_0^2}[k^2\overline{\Omega}_2 t \\
&\quad + \Pi(\arcsin(\operatorname{sn}(\overline{\Omega}_2 t, k)), \tilde{\alpha}_0^2, k)],
\end{aligned} \tag{128}
$$

where $\tilde{\alpha}_0^2 = \frac{k^2\phi_c^2}{\phi_b^2}$.

Finally, the following results are obtained.

Theorem 7. *For Eq. (79), it has solutions of the form* $\Psi = \phi(t)e^{i[\kappa z+\tilde{\theta}(t)]}$ *with oscillatory spatial dependence. Assume that* $g(\phi) = \alpha_1 + \alpha_2\phi$ *and write* $\beta_1 = -(c^2\kappa^2 + \alpha_1)$, $\beta_2 = -(c_{nl}^2\kappa^4 + \alpha_2)$. *Then,* $\phi(t)$ *is a solution of system (86) and* $\tilde{\theta}_t = \frac{a}{\phi^2}$.

(1) *When* $0 < \beta_2 < \frac{2}{3a\sqrt{3}}\beta_1^{\frac{3}{2}}$, $\phi(t)$ *has homoclinic solutions given by (123), and* $\tilde{\theta}(t)$ *given by (124). In addition,* $\phi(t)$ *has two periodic solution families given by (125), and* $\tilde{\theta}(t)$ *given by (126).*
(2) *When* $(\beta_1, \beta_2) \in (III)$ *in Fig. 24(a),* $\phi(t)$ *has two global periodic solution families given by (127), and* $\tilde{\theta}(t)$ *given by (128).*

Acknowledgment

This research was partially supported by the National Natural Science Foundation of China (Nos. 11871231, 12071162 and 11701191).

ORCID

Jibin Li
https://orcid.org/0000-0003-4839-5956
Guanrong Chen
https://orcid.org/0000-0003-1381-7418

References

Anco, S. & Kraus, D. [2018] "Hamiltonian structure of peakons as weak solutions for the modified Camassa–Holm equation," *Discrete Contin. Dyn. Syst. Ser. A* **38**, 4449–4465.

Anco, S. & Recio, E. [2018a] "Accelerating dynamical peakons and their behaviour," *Discrete Contin. Dyn. Syst. Ser. A* **39**, 6131–6148.

Anco, S. & Recio, E. [2018b] "Conserved norms and related conservation laws for multi-peakon equations," *J. Phys. A: Math. Theor.* **51**, 065203.

Anco, S. & Recio, E. [2019] "A general family of multi-peakon equations and their properties," *J. Phys. A: Math. Theor.* **52**, 125203.

Arab, A. [1991] "Exactly solvable supersymmetric quantum mechanics," *J. Math. Anal. Appl.* **158**, 63–79.

Arab, A. [2001] "Exact solutions of multi-component nonlinear Schrödinger and Klein–Gordon equations in two-dimensional space-time," *J. Phys. A: Math. Gen* **34**, 4281.

Byrd, B. & Fridman, M. [1971] *Handbook of Elliptic Integrals for Engineers and Scientists* (Springer, Berlin).

Cammasa, R. & Holm, D. [1993] "An integrable shallow water equation with peaked solution," *Phys. Rev. Lett.* **71**, 1161–1164.

Cammasa, R., Holm, D. & Hyman, M. [1994] "A new integrable shallow water equation," *Adv. Appl. Mech.* **31**, 1–33.

Carter, D. & Cienfuegos, R. [2011] "The kinematics and stability of solitary and cnoidal wave solutions of the Serre equations," *Eur. J. Mech. B* **30**, 259–268.

Clamond, D., Dutykh, D. & Mitsotakis, D. [2017] "Conservative modified Serre–Green–Naghdi equations with improved dispersion characteristics," *Commun. Nonlin. Sci. Numer. Simul.* **45**, 245–257.

Degasperis, A., Holm, D. & Hone, A. [2002] "A new integrable equation with peakon solutions," *Theor. Math. Phys.* **133**, 1463–1474.

Degasperis, A. & Procesi, M. [1999] "Asymptotic integrability," in *Symmetry and Perturbation Theory*, eds. Degasperis, A. & Gaeta, G. (World Scientific, Singapore), pp. 23–27.

Deng, S., Guo, B. & Wang, T. [2011] "Some traveling wave solutions of the Green–Naghdi system," *Int. J. Bifurcation and Chaos* **21**, 575–585.

Deng, S., Guo, B. & Wang, T. [2013] "Traveling wave solutions of the Green–Naghdi system," *Int. J. Bifurcation and Chaos* **23**, 1350087.

Fan, L., Gao, J. & Liu, Y. [2016] "On the rotation-two-component Camassa–Holm system modelling the equatorial water waves," *Adv. Math.* **291**, 59–79.

Fan, N. & Yuen, M. [2019] "Peakon weak solutions for the rotation-two-component Camassa–Holm system," *Appl. Math. Lett.* **97**, 53–59.

Fokas, S. [1995] "On class of physically important integrable equations," *Physica D* **87**, 145–150.

Fu, Y. & Li, J. [2021] "Bifurcations of travelling wave solutions in three modified Camassa–Holm equations," *J. Appl. Anal. Comput.* **11**, 1051–1061.

Geyer, A. & Quirchmayr, R. [2018] "Traveling wave solutions of a highly nonlinear shallow water equation," *Discrete Contin. Dyn. Syst.* **38**, 1567–1604.

Green, A., Laws, N. & Naghdi, P. [1974] "On the theory of water waves," *Proc. R. Soc. Lond. A* **338**, 43–55.

Green, A. & Naghdi, P. [1976] "A derivation of equations for wave propagation in water of variable depth," *J. Fluid Mech.* **78**, 237–246.

Jin, X., Chen, G. & Li, J. [2025] "Bifurcation and exact solutions of a concatenation physical model," *Int. J. Bifurcation and Chaos* **35**, 2550011.

Kupershmidt, A. [1985] "A coupled Korteweg–de Vries equation with dispersion," *J. Phys. A: Math. Gen.* **18**, L571–L573.

Lenells, J. [2005a] "Traveling wave solutions of the Camassa–Holm equation," *J. Differ. Equ.* **217**, 393–430.

Lenells, J. [2005b] "Traveling wave solutions of the Degasperis–Procesi equation," *J. Math. Anal. Appl.* **306**, 72–82.

Lenells, J. [2006] "Classification of traveling waves for a class of nonlinear wave equations," *J. Dyn. Differ. Equ.* **18**, 381–391.

Lenells, L. [2007] "Classification of all travelling-wave solutions for some nonlinear dispersive equations," *Philos. Trans. R. Soc. A* **365**, 2291–2298.

Li, J. & Liu, Z. [2000] "Smooth and non-smooth travelling waves in a nonlinearly dispersive equation," *Appl. Math. Model.* **25**, 41–56.

Li, J. & Liu, Z. [2002] "Traveling wave solutions for a class of nonlinear dispersive equations," *Chin. Ann. Math. Ser. B* **23**, 1–22.

Li, J. & Zhang, J. [2004] "Bifurcations of travelling wave solutions in generalization form of the modified KdV equation," *Chaos Solit. Fract.* **20**, 899–913.

Li, J. & Chen, G. [2005a] "Bifurcations of traveling wave and breather solutions of a general class of nonlinear wave equations," *Int. J. Bifurcation and Chaos* **15**, 2913–2926.

Li, J. & Chen, G. [2005b] "Bifurcations of traveling wave solutions for four classes of nonlinear wave equations," *Int. J. Bifurcation and Chaos* **15**, 3973–3998.

Li, J., Wu, J. & Zhu, H. [2006] "Traveling waves for an integrable higher order KdV type wave equations," *Int. J. Bifurcation and Chaos* **16**, 2235–2260.

Li, J. & Chen, G. [2007] "On a class of singular nonlinear traveling wave equations," *Int. J. Bifurcation and Chaos* **17**, 4049–4065.

Li, J. & Dai, H. [2007] *On the Study of Singular Nonlinear Travelling Wave Equations: Dynamical Approach* (Science Press, Beijing).

Li, J. & Li, Y. [2008] "Bifurcations of travelling wave solutions for a two-component Camassa–Holm equation," *Acta Math. Sin. Engl. Ser.* **24**, 1319–1330.

Li, J. [2008a] "Bifurcations of travelling wave solutions for two generalized Boussinesq systems," *Sci. China Math.* **51**, 1577–1592.

Li, J. [2008b] "Exact explicit peakon and periodic cusp wave solutions for several nonlinear wave equations," *J. Dyn. Differ. Equ.* **20**, 909–922.

Li, J. & Zhang, Y. [2009] "Exact M/W-shape wave solutions determined by a singular travelling wave equation," *Nonlin. Anal.: Real World Appl.* **10**, 1797–1802.

Li, J., Zhang, Y. & Zhao, X. [2009a] "On a class of singular nonlinear traveling wave equations (II): An example of GCKdV equations," *Int. J. Bifurcation and Chaos* **19**, 1955–2007.

Li, J., Zhao, X. & Chen, G. [2009b] "On the Breaking property for the second class of singular nonlinear traveling wave equations," *Int. J. Bifurcation and Chaos* **19**, 1289–1306.

Li, J. & Chen, G. [2010] "On nonlinear wave equations with breaking loop-solutions," *Int. J. Bifurcation and Chaos* **20**, 519–537.

Li, J. & Qiao, Z. [2010] "Bifurcations of traveling wave solutions for an integrable equation," *J. Math. Phys.* **51**, 042703.

Li, J. & Zhang, Y. [2010] "Exact solitary wave and quasi-periodic wave solutions for four fifth-order nonlinear wave equations," *Discrete Contin. Dyn. Syst. Ser. B* **13**, 623–631.

Li, J. & Zhang, Y. [2011a] "Solitary wave and chaotic behavior of traveling wave solutions for the coupled KdV equations," *Appl. Math. Comput.* **218**, 1794–1797.

Li, J. & Zhang, Y. [2011b] "Homoclinic manifolds, center manifolds and exact solutions of four-dimensional traveling wave systems for two classes of nonlinear wave equations," *Int. J. Bifurcation and Chaos* **21**, 527–543.

Li, J. & Chen, G. [2012] "Exact traveling wave solutions and their bifurcations for the Kudryashov–Sinelshchikov equation," *Int. J. Bifurcation and Chaos* **22**, 1250118.

Li, J. & Qiao, Z. [2012] "Bifurcations and exact traveling wave solutions of the generalized two-component Camassa–Holm equation," *Int. J. Bifurcation and Chaos* **22**, 1250305.

Li, J. [2013] *Singular Nonlinear Traveling Wave Equations: Bifurcations and Exact Solutions* (Science Press, Beijing).

Li, J. & Shi, J. [2013] "Traveling wave solutions for a class of nonlinear diffusion–convection–reaction models," *Int. J. Bifurcation and Chaos* **23**, 1350168.

Li, J. & Qiao, Z. [2013a] "Peakon, pseudo-peakon, and cuspon solutions for two generalized Camassa–Holm equations," *J. Math. Phys.* **54**, 123501.

Li, J. & Qiao, Z. [2013b] "Bifurcations and exact traveling wave solutions for a generalized Camassa–Holm equation," *Int. J. Bifurcation and Chaos* **23**, 1350057.

Li, J. [2014a] "Variform exact one-peakon solutions for some singular nonlinear traveling wave equations of the first kind," *Int. J. Bifurcation and Chaos* **24**, 1450160.

Li, J. [2014b] "Bifurcations and exact traveling wave solutions of the generalized two-component Hunter–Saxton system," *Discrete Contin. Dyn. Syst. Ser. B* **19**, 1719–1729.

Li, J. & Qiao, Z. [2015] "Bifurcation and traveling wave solutions for the Fokas equation," *Int. J. Bifurcation and Chaos* **25**, 1550136.

Li, J., Zhu, W. & Chen, G. [2016] "Understanding peakons, periodic peakons and compactons via a shallow water wave equation," *Int. J. Bifurcation and Chaos* **26**, 1650207.

Li, J. & Chen, F. [2016a] "Bifurcations of traveling wave solutions of a nonlinear wave model created by diffraction in periodic media," *Int. J. Bifurcation and Chaos* **26**, 1650032.

Li, J. & Chen, F. [2016b] "Bifurcations and exact traveling wave solutions of degenerate coupled multi-KdV equations," *Int. J. Bifurcation and Chaos* **26**, 1650045.

Li, J. [2017] "Exact modulated wave solutions in a nonlinear inhomogeneous elasticity material model," *Sci. Sin. Math.* **27**, 147–154.

Li, J. & Kou, K. [2017] "Dynamics of traveling wave solutions to a new highly nonlinear shallow water wave equation," *Int. J. Bifurcation and Chaos* **27**, 1750044.

Li, J. [2018] "Periodic peakons, pseudo-peakons and compactons of ion-acoustic wave model in electronegative plasmas with electrons featuring Tsallis distribution," *Int. J. Bifurcation and Chaos* **28**, 1850015.

Li, J. [2019] *Bifurcation and Exact Solutions in Invariant Manifolds for Singular Nonlinear Traveling Wave Equations* (Science Press, Beijing).

Li, J. & Chen, G. [2019] "More on bifurcations and dynamics of traveling wave solutions for a higher-order shallow water wave equation," *Int. J. Bifurcation and Chaos* **29**, 1950014.

Li, J., Chen, G. & Song, J. [2019] "Exact traveling wave solutions and bifurcations of classical and modified Serre shallow water wave equations," *Int. J. Bifurcation and Chaos* **29**, 1950153.

Li, J. & Han, M. [2020] "Exact peakon solutions given by the generalized hyperbolic functions for some nonlinear wave equations," *J. Appl. Anal. Comput.* **10**, 1708–1719.

Li, J., Chen, G. & Song, J. [2020a] "Bifurcations of traveling wave solutions for fully nonlinear water waves with surface tension in the generalized Serre–Green–Naghdi equations," *Int. J. Bifurcation and Chaos* **30**, 2050019.

Li, J., Chen, G. & Song, J. [2020b] "Completing the study of traveling wave solutions for three two-component shallow water wave models," *Int. J. Bifurcation and Chaos* **30**, 2050036.

Li, J., Chen, G. & Song, J. [2020c] "Bifurcations and dynamics of traveling wave solutions for the regularized Saint-Venant equation," *Int. J. Bifurcation and Chaos* **30**, 2050109.

Li, J., Chen, G. & Zhou, Y. [2020d] "Exact peakon, periodic peakon and pseudo-peakon solutions of the rotation-two-component Camassa–Holm system," *Int. J. Bifurcation and Chaos* **30**, 2050139.

Li, J., Chen, G. & Zhou, Y. [2021] "Bifurcations and exact traveling wave solutions of two shallow water two-component systems," *Int. J. Bifurcation and Chaos* **31**, 2150001.

Li, J., Han, M. & Ke, A. [2024] "Bifurcations and exact traveling wave solutions of the Khorbatly's geophysical Boussinesq system," *J. Math. Anal. Appl.* **537**, 128263.

Li, J. & Dai, Y. [2025] "Bifurcations and exact traveling wave solutions for the model of slightly dispersive quasi-incompressible hyperelastic materials," *Qual. Theory Dyn. Syst.* **24**, 10.

Li, J. & Ke, A. [2025] "Dynamical behavior and exact explicit traveling wave solutions of a generalized Vakhnenko–Parkes equation," *Qual. Theory Dyn. Syst.* **24**, 119.

Novikov, V. [2009] "Generalizations of the Camassa–Holm equation," *J. Phys. A: Math. Theor.* **42**, 342002.

Olver, J. & Rosenau, P. [1996] "Tri-Hamiltonian duality between solitons and solitary-wave solutions having compact support," *Phys. Rev. E* **53**, 1900.

Qiao, Z. [2006] "A new integrable equation with cuspons and W/M-shape-peaks solitons," *J. Math. Phys.* **47**, 112701.

Qiao, Z. [2007] "New integrable hierarchy, parametric solutions, cuspons, one-peak solitons, and M/W-shape peak solutions," *J. Math. Phys.* **48**, 082701.

Rogers, C., Saccomandi, G. & Vergori, L. [2014] "Carroll-type deformations in nonlinear elastodynamics," *J. Phys. A: Math. Theor.* **47**, 205204.

Rogers, C., Saccomandi, G. & Vergori, L. [2015] "Nonlinear elastodynamics of materials with strong ellipticity condition: Carroll-type solutions," *Wave Motion* **56**, 147–164.

Shi, J. & Li, J. [2021] "Bifurcations and exact solitary wave, compacton and pseudo-peakon solutions in a modified generalized KdV equation," *Int. J. Bifurcation and Chaos* **31**, 2150215.

Yang, S. & Li, J. [2025a] "Bifurcations and exact solutions of generalized nonlinear Schrödinger equation with nonlinear chromatic dispersion and quintuple power law of refractive index in optical fibers," *Int. J. Bifurcation and Chaos* **35**, 2550113.

Yang, J. & Li, J. [2025b] "Bifurcations and exact solutions of the generalized Radhakrishnan–Kundu–Lakshmanan equation with four powers of nonlinearity," *Int. J. Bifurcation and Chaos* **35**, 2550147.

Zhou, Y., Chen, G. & Li, J. [2022] "Bifurcations, exact peakon, periodic peakons and solitary wave solutions of the modified Camassa–Holm equation," *Int. J. Bifurcation and Chaos* **32**, 2250076.

Zhou, Y. & Li, J. [2023] "Bifurcations and exact solutions in the model of nonlinear elastodynamics of materials with strong ellipticity condition," *Nonlin. Dyn.* **112**, 1–13.

Zhou, Y., Chen, G. & Li, J. [2023] "Bifurcations, exact peakon, periodic peakons and solitary wave solutions of the cubic Camassa–Holm type equation," *Int. J. Bifurcation and Chaos* **33**, 2350014.

https://doi.org/10.1142/9789819830633_0011

Chapter 11

Semi-Analytical Periodic Motions and Bifurcations in a Periodically Impulsive, Damped Pendulum#

Albert C. J. Luo* and Tianji Ma
Department of Mechanical and Mechatronics Engineering,
Southern Illinois University Edwardsville,
Edwardsville, IL 62026-1805, USA
**aluo@siue.edu*

In this paper, a computational method is developed for the bifurcation dynamics of periodic motions in impulsive nonlinear dynamical systems. Such a method based on the implicit mapping method is to discretize the nonlinear dynamical systems in the time interval of two impulses. The impulsive pendulum is studied as an example herein. Using specific mapping structures, periodic motions in the impulsive dynamical systems are determined, and the corresponding stability and bifurcation analysis of the periodic motions are carried out. The impulsive homoclinic orbits related to the different periodic motions are achieved, and the homoclinic bifurcations of periodic motions are determined. The sampled periodic motions and impulsive homoclinic orbits are illustrated for a better understanding of the bifurcation dynamics of impulsive nonlinear systems. Such a method can be applied to other impulsive nonlinear systems.

Keywords: Impulsive nonlinear system; impulsive periodic motion; impulsive homoclinic orbit; homoclinic bifurcation; saddle-node bifurcation; period-doubling bifurcation.

1. Introduction

Impulsive dynamical systems are used as the mathematical modeling of practical processes and controls, which possess the characteristics of continuous and discrete dynamics [e.g. Lakshmikantham *et al.*, 1989; van der Schaft & Schumacher, 2000]. Such hybrid dynamical systems exist extensively in control systems, such as mechanical impact systems, and control systems with controllers. A combination of discrete and continuous dynamics of the impulsive systems leads to a higher complexity in solution behaviors compared with a purely discrete or continuous system. Thus, such solution properties cause difficulty in studying stability and robustness of dynamical systems, especially for impulsive nonlinear systems. As in [Lakshmikantham *et al.*, 1989; van der Schaft & Schumacher, 2000], the impulsive systems were systematically discussed. One often focused on the stability and control of the impulsive systems near equilibriums through the Lyapunov method. The impulsive system possesses discontinuous trajectories. Lakshmikanthem and Liu [1989] discussed the stability criteria for impulsive differential equations with two measures. Liu [1994] discussed the stability of impulsive differential systems and applied it to the population growth model. Liu and Willms [1996] presented the impulsive controllability conditions of linear dynamical systems, which also were used for such impulsive controls in the maneuver of spacecraft. Lakshmikanthem and Liu [1998] discussed impulsive hybrid systems and stability theory.

Zheng and Fu [2013a, 2013b] discussed the chatter dynamics on the impulsive surface and

---

#To cite this article, please refer to its earlier version published in the *International Journal of Bifurcation and Chaos*, Vol. 35, No. 14 (2025) 2540011 (15 pages), DOI: 10.1142/S0218127425400115.

*Author for correspondence

pulse phenomena of the impulsive systems, and the discontinuous dynamical system theory was used. Akhmet and Cag [2018] presented the Tikhonov theorem for differential equations with singular impulses. Further, Akhmet *et al.* [2022] gave the unpredictable solutions of impulsive quasi-linear systems. Akhmet and Erim [2023] developed the almost periodic solutions of recurrent neural networks with state-dependent and structured impulses. In fact, the state-dependent and structured impulses are quite similar to the impact motions of dynamical systems. Recently, one discussed the stability and control of fractional dynamical systems with impulses. Ramkumar *et al.* [2021] discussed the approximate controllability for time-dependent impulsive neutral stochastic partial differential equations with fractional Brownian motion and Poisson jumps. Benjemaa *et al.* [2023] gave the stability analysis of nonlinear systems with impulsive perturbations through the Lyapunov function method, and such results were applied to the stability analysis of the Hopfield neural networks. Tamilarasi and Radhakrishnan [2023] discussed the existence and controllability for higher-order Caputo-fractional impulsive damping systems.

For the traditional analytical and numerical methods, it is very difficult to handle the discontinuity for the switching of dynamical systems. Luo [2011, 2012] systematically presented the general theory for impulsive systems and showed how to determine periodic motions of switching systems with general transport laws. Luo and Wang [2009] studied the periodic motions of switching systems with multiple linear oscillators. With periodic and random piecewise forcing, the solutions of dynamical oscillation were presented. For linear subsystems, the periodic motions of the switching system with impulses (transport laws) in each time interval can be expressed analytically. However, for nonlinear switching systems with impulses, no analytical solutions have been provided. Even for the nonlinear continuous systems, it is very difficult to get periodic motions until now.

An implicit mapping method for nonlinear dynamical systems was developed in [Luo, 2015a, 2015b] through the discretization of differential equations of nonlinear systems. From the implicit mapping method, the periodic motions of nonlinear dynamical systems are determined with specific mapping structures. Luo and Guo [2015] used the implicit mapping method to obtain the periodic motions in a periodically forced Duffing oscillator. Xing and Luo [2017, 2018] applied the implicit mapping method for the bifurcation trees of periodic motions in a time-delayed twin-well Duffing oscillator. Luo and Zhu [2024] discussed the bifurcation dynamics of an impact Duffing oscillator. To study the bifurcation dynamics of periodic motions in the impact Duffing oscillator, during two impacts, the Duffing oscillator in such a time interval was discretized through the implicit mapping method. The bifurcation dynamics of the impact Duffing oscillator with two displacement boundaries were also presented in [Zhu & Luo, 2025].

In this paper, the bifurcation dynamics of periodic motions in impulsive nonlinear dynamical systems will be studied through the implicit mapping method. The nonlinear subsystems during two impulses exist in the impulsive nonlinear system, and the implicit mapping method should be employed to discretize the nonlinear dynamical systems in the time interval of two impulses. The specific mapping structures will be used for periodic motions in the impulsive dynamical systems, and the corresponding stability and bifurcation analysis of periodic motions will be carried out. The impulsive homoclinic orbits related to the different periodic motions are obtained. The numerical illustrations of periodic motions and impulsive homoclinic orbits will be given for a better understanding of the bifurcation dynamics of impulsive nonlinear systems.

2. An Impulsive Pendulum System

Consider a simple pendulum forced by a periodic impulse, as shown in Fig. 1. The pendulum has a particle mass of m, connected with a massless bar of length l. Another end of the massless bar is fixed on the wall. The angular displacement of the pendulum

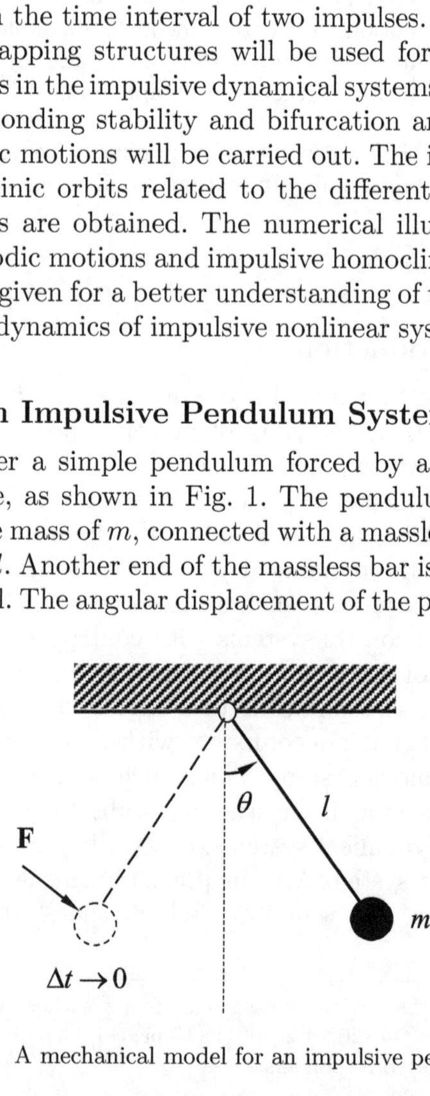

Fig. 1. A mechanical model for an impulsive pendulum.

is θ. Such an impulsive system has periodic motions under periodically impulsive force $\mathbf{F}$ at the time interval of $\Delta t \to 0$. Using the Lagrange equation gives the equation of motion for the pendulum system.

The mass moves in a circular track; the position and velocity vectors are

$$\begin{aligned}\mathbf{r} &= l\sin\theta\mathbf{i} - l\cos\theta\mathbf{j},\\ \mathbf{v} &= \dot{\mathbf{r}} = l\dot{\theta}(\cos\theta\mathbf{i} + \sin\theta\mathbf{j}).\end{aligned} \tag{1}$$

The kinetic energy is

$$T = \frac{1}{2}m\mathbf{v}\cdot\mathbf{v} = \frac{1}{2}ml^2\dot{\theta}^2. \tag{2}$$

Taking the zero-potential energy at the lowest point, the potential energy is

$$V = mgl(1 - \cos\theta). \tag{3}$$

The Lagrange function is defined as

$$L = T - V = \frac{1}{2}ml^2\dot{\theta}^2 - mgl(1 - \cos\theta). \tag{4}$$

The air damping force is expressed by

$$\mathbf{F}_d = -d\dot{\theta}l(\cos\theta\mathbf{i} + \sin\theta\mathbf{j}), \tag{5}$$

at the location of

$$\mathbf{r} = l(\sin\theta\mathbf{i} - \cos\theta\mathbf{j}). \tag{6}$$

Application of the Lagrange's equation

$$\frac{d}{dt}\left(\frac{\partial L}{\partial\dot{\theta}}\right) - \frac{\partial L}{\partial\theta} = Q_\theta^{\text{non}}, \tag{7}$$

where

$$Q_\theta^{\text{non}} = \frac{\partial\mathbf{r}}{\partial\theta}\cdot\mathbf{F}_d = -d\dot{\theta}l^2 \tag{8}$$

and

$$\begin{aligned}\frac{d}{dt}\left(\frac{\partial L}{\partial\dot{\theta}}\right) &= \frac{d}{dt}(ml^2\dot{\theta}) = ml^2\ddot{\theta},\\ \frac{\partial L}{\partial\theta} &= -mgl\sin\theta.\end{aligned} \tag{9}$$

Thus, Eq. (7) gives

$$ml^2\ddot{\theta} + d\dot{\theta}l^2 + mgl\sin(\theta) = 0. \tag{10}$$

The deformation of the foregoing equation is

$$\ddot{\theta} + \delta\dot{\theta} + \alpha\sin(\theta) = 0, \tag{11}$$

where

$$\delta = \frac{d}{ml^2}, \quad \alpha = \frac{g}{l}. \tag{12}$$

Let $x = \theta$. In state space, the equation of motion is

$$\begin{aligned}y &= \dot{x},\\ \dot{y} &= -\delta y - \alpha\sin x.\end{aligned} \tag{13}$$

Introduce two vectors as

$$\mathbf{f} = (f_1, f_2)^{\text{T}}, \quad \mathbf{x} = (x, \dot{x})^{\text{T}}, \tag{14}$$

where

$$\begin{aligned}f_1 &= y,\\ f_2 &= -\delta\dot{x} - \alpha\sin x.\end{aligned} \tag{15}$$

Thus, the dynamical subsystem for pendulum is

$$\dot{\mathbf{x}} = \mathbf{f}(\mathbf{x}). \tag{16}$$

Consider the system under an impulsive force $\mathbf{F}$ at $t = t_k$ $(k = 1, 2, \ldots)$, which can be expressed through a velocity change from the impulse theorem as

$$\begin{aligned}x(t_k^+) &= x(t_k^-),\\ y(t_k^+) &= y(t_k^-) + \Delta y,\end{aligned} \quad \text{for } k = 1, 2, \ldots. \tag{17}$$

where Δy is the velocity change caused by the impulsive force at $\Delta t \to 0$.

The time history of impulsive responses of the impulsive pendulum is represented in Fig. 2. The impulsive points are labeled through $\mathbf{x}_{N_{k-1}}$, $\mathbf{x}_{N_k}$, $\mathbf{x}_{N_{k+1}}$ for time of t_{k-1}, t_k, t_{k+1}. Consider a nonlinear subsystem S_σ for the interval of

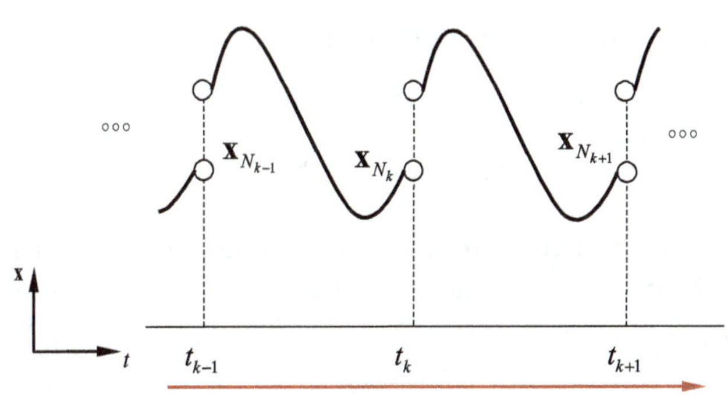

Fig. 2. Impulsive responses for an impulsive nonlinear system.

$t \in [t_{N_{k-1}}, t_{N_k}]$ partitioned to $N_k - N_{k-1}$ time segments. Thus,

$$[t_{N_{k-1}}, t_{N_k}] = \bigcup_{l_k=N_{k-1}+1}^{N_k} [t_{l_k-1}, t_{l_k}]. \tag{18}$$

In the time interval $t \in [t_{l_k-1}, t_{l_k}]$ ($l_k = N_{k-1}+1, N_{k-1}+2, \ldots, N_k$), the nonlinear subsystem can be discretized as

$$\begin{aligned} x_{l_k} &= x_{l_k-1} + h\left(\frac{y_{l_k-1}+y_{l_k}}{2}\right), \\ y_{l_k} &= y_{l_k-1} \\ &\quad + h\left(-\alpha \sin\frac{x_{l_k-1}+x_{l_k}}{2} - \delta\frac{y_{l_k-1}+y_{l_k}}{2}\right), \\ &\qquad \text{for } l_k = N_{k-1}+1, N_{k-1}+2, \ldots, N_k. \end{aligned} \tag{19}$$

Define a vector as

$$\mathbf{g} = (g_1, g_2)^{\mathrm{T}}. \tag{20}$$

The equations in Eq. (16) become

$$\mathbf{g}(\mathbf{x}_{l_k-1}, \mathbf{x}_{l_k}) = \mathbf{0}, \tag{21}$$

where

$$\begin{aligned} g_1(\mathbf{x}_k, \mathbf{x}_{k+1}) &= x_{l_k-1} + h\left(\frac{y_{l_k-1}+y_{l_k}}{2}\right) - x_{l_k}, \\ g_2(\mathbf{x}_k, \mathbf{x}_{k+1}) &= y_{l_k-1} + h\left(-\alpha \sin\frac{x_{l_k-1}+x_{l_k}}{2} \right. \\ &\quad \left. - \delta\frac{y_{l_k-1}+y_{l_k}}{2}\right) - y_{l_k}. \end{aligned} \tag{22}$$

3. Mapping Structures

A switching mapping is defined as

$$P_k : \mathbf{x}_{N_{k-1}^+} \to \mathbf{x}_{N_k^-}. \tag{23}$$

Between two impulses (switching points) with $t \in [t_{N_{k-1}}, t_{N_k}]$, there are many implicit submaps as

$$\begin{aligned} &p_{l_k} : \mathbf{x}_{l_k-1} \to \mathbf{x}_{l_k}, \\ &\qquad \text{for } l_k = N_{k-1}^+ + 1, N_{k-1}^+ + 2, \ldots, N_k^-. \end{aligned} \tag{24}$$

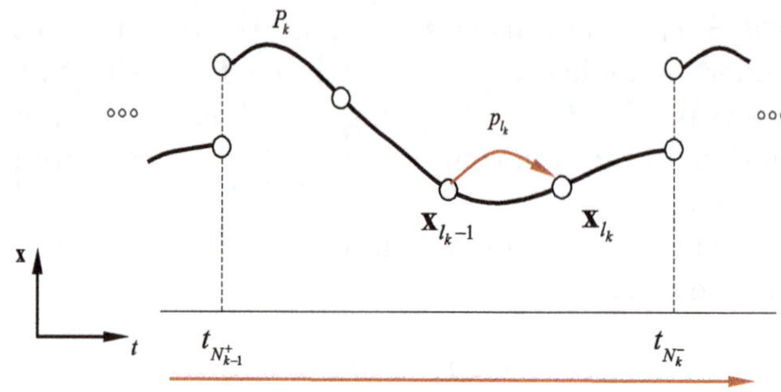

Fig. 3. Implicit submaps for a nonlinear system.

Such a submapping $p_{l_k} : \mathbf{x}_{l_k-1} \to \mathbf{x}_{l_k}$ in the switching mapping is sketched in Fig. 3. Thus,

$$P_k : \mathbf{x}_{N_{k-1}^+} \to \mathbf{x}_{N_k^-}, \tag{25}$$

where

$$P_k = p_{N_k^-} \circ \cdots \circ p_{N_{k-1}^+ + 1}. \tag{26}$$

The corresponding algebraic equations are

$$\mathbf{g}_{l_k}(\mathbf{x}_{l_k-1}, \mathbf{x}_{l_k}) = \mathbf{0}. \tag{27}$$

where

$$\mathbf{g}_{l_k} = (g_{l_k 1}, g_{l_k 2})^{\mathrm{T}}. \tag{28}$$

For $l_k = N_{k-1}^+ + 1, N_{k-1}^+ + 2, \ldots, N_k^-$, the corresponding algebraic functions of $g_{l_k 1}$ and $g_{l_k 2}$ are

$$\begin{aligned} g_{l_k 1} &= x_{l_k-1} + h\left(\frac{y_{l_k-1}+y_{l_k}}{2}\right) - x_{l_k}, \\ g_{l_k 2} &= y_{l_k-1} + h\left(-\alpha \sin\frac{x_{l_k-1}+x_{l_k}}{2} \right. \\ &\quad \left. - \delta\frac{y_{l_k-1}+y_{l_k}}{2}\right) - y_{l_k}. \end{aligned} \tag{29}$$

The impulsive mapping $P_{I_k} : \mathbf{x}_{N_k^-} \to \mathbf{x}_{N_k^+}$ is given by

$$\left.\begin{aligned} x_{N_k^+} &= x_{N_k^-}, \\ y_{N_k^+} &= y_{N_k^-} + \Delta y, \end{aligned}\right\} \text{for } k = 1, 2, \ldots. \tag{30}$$

4. Period-m Motions

In general, a period-m motion in the impulsive pendulum oscillator can be discussed through a discrete mapping structure as

$$P : \mathbf{x}_0 \to \mathbf{x}_{N_m^+}, \tag{31}$$

where

$$P = P_{I_m} \circ P_m \circ \cdots \circ P_{I_k} \circ P_k \circ \cdots \circ P_{I_2} \circ P_2 \circ P_{I_1} \circ P_1, \quad \text{for } k = 1, 2, \ldots, m. \tag{32}$$

For each discrete mapping, the submapping structure is given by

$$P_k = p_{N_k^-} \circ \cdots \circ p_{l_k} \circ \cdots \circ p_{N_{k-1}^+ + 1} \quad (k = 1, \ldots, m) \tag{33}$$

and

$$l_k = N_{k-1}^+ + 1, N_{k-1}^+ + 2, \ldots, N_k^-, \tag{34}$$

for $k = 1, 2, \ldots, m$.

When the system is forced by an impulse, the system has an impulsive map $P_{I_k} : \mathbf{x}_{N_k^-} \to \mathbf{x}_{N_k^+}$. From Eq. (27), define a vector as

$$\mathbf{g}_{I_k} = (g_{I_k 3}, g_{I_k 4})^{\mathrm{T}}, \tag{35}$$

thus

$$\mathbf{g}_{I_k}(\mathbf{x}_{N_k^-}, \mathbf{x}_{N_k^+}) = 0. \tag{36}$$

The corresponding algebraic equations are

$$\begin{aligned} g_{I_k 3}(\mathbf{x}_{N_k^-}, \mathbf{x}_{N_k^+}) &= x_{N_k^-} - x_{N_k^+}, \\ g_{I_k 4}(\mathbf{x}_{N_k^-}, \mathbf{x}_{N_k^+}) &= y_{N_k^-} + \Delta y - y_{N_k^+}, \\ &\text{for } k = 1, 2, \ldots, m. \end{aligned} \tag{37}$$

The periodicity conditions with an impulsive mapping require

$$\mathbf{g}_{I_m}(\mathbf{x}_{N_m^-}, \mathbf{x}_{N_m^+}) = 0, \quad \mathbf{x}_0 = \mathbf{x}_{N_m^+}. \tag{38}$$

Further,

$$\mathbf{g}_{I_m}(\mathbf{x}_{N_m^-}, \mathbf{x}_0) = \mathbf{0}. \tag{39}$$

From the mappings defined before, a set of algebraic equations for the periodic motion with a mapping structure of $P = P_{I_k} \circ P_k \circ \cdots \circ P_{I_2} \circ P_2 \circ P_{I_1} \circ P_1$ is obtained as

$$\mathbf{G} \equiv \begin{Bmatrix} \mathbf{G}_1(\mathbf{x}_0, \ldots, \mathbf{x}_{N_1^+}) \\ \vdots \\ \mathbf{G}_k(\mathbf{x}_{N_{k-1}^+}, \ldots, \mathbf{x}_{N_k^+}) \\ \vdots \\ \mathbf{G}_m(\mathbf{x}_{N_{m-1}^+}, \ldots, \mathbf{x}_{N_m^+}) \\ \mathbf{x}_{N_m^+} - \mathbf{x}_0 \end{Bmatrix} = \mathbf{0}, \tag{40}$$

where

$$\mathbf{G}_k(\mathbf{x}_{N_{k-1}^+}, \ldots, \mathbf{x}_{N_k^+}) \equiv \begin{Bmatrix} \mathbf{g}_{N_{k-1}^+ + 1}(\mathbf{x}_{N_{k-1}^+}, \mathbf{x}_{N_{k-1}^+ + 1}) \\ \vdots \\ \mathbf{g}_{l_k}(\mathbf{x}_{l_k}, \mathbf{x}_{l_k + 1}) \\ \vdots \\ \mathbf{g}_{N_k^-}(\mathbf{x}_{N_k^- - 1}, \mathbf{x}_{N_k^-}) \\ \mathbf{g}_{I_k}(\mathbf{x}_{N_k^-}, \mathbf{x}_{N_k^+}) \end{Bmatrix}, \tag{41}$$

for $k = 1, 2, \ldots, m$. Using the Newton–Raphson method, the discretized points on the impulsive periodic motions are determined.

5. Stability and Bifurcation Analysis

From the mapping structure of Eq. (31), through the perturbation, the variational equation is

$$\Delta \mathbf{x}_{N_m}^+ = DP \Delta \mathbf{x}_0, \tag{42}$$

where

$$DP = DP_{I_m} \cdot DP_m \cdot \ldots \cdot DP_{I_k} \cdot DP_k \cdot \ldots \cdot DP_{I_1} \cdot DP_1 \tag{43}$$

and

$$DP_k = Dp_{N_k^- - 1} \cdot \ldots \cdot Dp_{l_k} \cdot \ldots \cdot Dp_{N_{k-1}^+}, \tag{44}$$

with $l_k = N_{k-1}^+ + 1, N_{k-1}^+ + 2, \ldots N_k^-$ and $k = 1, 2, \ldots, m$. For a mapping p_{l_k}, the variational equation of $\mathbf{g}_{l_k}(\mathbf{x}_{l_k - 1}, \mathbf{x}_{l_k}) = \mathbf{0}$ in vicinity of $(\mathbf{x}_{l_k-1}^*, \mathbf{x}_{l_k}^*)$ gives

$$\left[\frac{\partial \mathbf{g}_{l_k}}{\partial \mathbf{x}_{l_k - 1}}\right]\Bigg|_{(\mathbf{x}_{l_k-1}^*, \mathbf{x}_{l_k}^*)} \Delta \mathbf{x}_{l_k - 1} + \left[\frac{\partial \mathbf{g}_{l_k}}{\partial \mathbf{x}_{l_k}}\right]\Bigg|_{(\mathbf{x}_{l_k-1}^*, \mathbf{x}_{l_k}^*)} \Delta \mathbf{x}_{l_k} = 0. \tag{45}$$

Thus,

$$Dp_{l_k} \equiv \left[\frac{\partial \mathbf{x}_{l_k}}{\partial \mathbf{x}_{l_k - 1}}\right]\Bigg|_{(\mathbf{x}_{l_k-1}^*, \mathbf{x}_{l_k}^*)} = -\left[\frac{\partial \mathbf{g}_{l_k}}{\partial \mathbf{x}_{l_k}}\right]^{-1} \cdot \left[\frac{\partial \mathbf{g}_{l_k}}{\partial \mathbf{x}_{l_k - 1}}\right]\Bigg|_{(\mathbf{x}_{l_k-1}^*, \mathbf{x}_{l_k}^*)}. \tag{46}$$

where

$$Dp_{l_k} \equiv \left[\frac{\partial \mathbf{x}_{l_k}}{\partial \mathbf{x}_{l_k-1}}\right]\Bigg|_{(\mathbf{x}^*_{l_k-1},\mathbf{x}^*_{l_k})} = \begin{bmatrix} \dfrac{\partial x_{l_k}}{\partial x_{l_k-1}} & \dfrac{\partial x_{l_k}}{\partial y_{l_k-1}} \\ \dfrac{\partial y_{l_k}}{\partial x_{l_k-1}} & \dfrac{\partial y_{l_k}}{\partial y_{l_k-1}} \end{bmatrix}_{(\mathbf{x}^*_{l_k-1},\mathbf{x}^*_{l_k})} \quad (47)$$

and

$$\left[\frac{\partial \mathbf{g}_{l_k}}{\partial \mathbf{x}_{l_k-1}}\right] = \begin{bmatrix} \dfrac{\partial g_{l_k 1}}{\partial x_{l_k-1}} & \dfrac{\partial g_{l_k 1}}{\partial y_{l_k-1}} \\ \dfrac{\partial g_{l_k 2}}{\partial x_{l_k-1}} & \dfrac{\partial g_{l_k 2}}{\partial y_{l_k-1}} \end{bmatrix}, \quad \left[\frac{\partial \mathbf{g}_{l_k}}{\partial \mathbf{x}_{l_k}}\right] = \begin{bmatrix} \dfrac{\partial g_{l_k 1}}{\partial x_{l_k}} & \dfrac{\partial g_{l_k 1}}{\partial y_{l_k}} \\ \dfrac{\partial g_{l_k 2}}{\partial x_{l_k}} & \dfrac{\partial g_{l_k 2}}{\partial y_{l_k}} \end{bmatrix}. \quad (48)$$

The matrix components are

$$\begin{aligned} &\frac{\partial g_{l_k 1}}{\partial x_{l_k-1}} = 1, \\ &\frac{\partial g_{l_k 1}}{\partial y_{l_k-1}} = \frac{h}{2}, \\ &\frac{\partial g_{l_k 2}}{\partial x_{l_k-1}} = -\frac{h}{2}\alpha \cos\frac{x_{l_k-1}+x_{l_k}}{2}, \\ &\frac{\partial g_{l_k 2}}{\partial y_{l_k-1}} = 1 - \frac{h}{2}\delta; \end{aligned} \quad (49)$$

$$\begin{aligned} &\frac{\partial g_{l_k 1}}{\partial x_{l_k}} = -1, \\ &\frac{\partial g_{l_k 1}}{\partial y_{l_k}} = \frac{h}{2}, \\ &\frac{\partial g_{l_k 2}}{\partial x_{l_k}} = -\frac{h}{2}\alpha \cos\frac{x_{l_k-1}+x_{l_k}}{2}, \\ &\frac{\partial g_{l_k 2}}{\partial y_{l_k}} = -\frac{h}{2}\delta - 1. \end{aligned} \quad (50)$$

For the impulsive mapping P_{I_k}, the variational equation of $\mathbf{g}_{I_k}(\mathbf{x}_{N_k^-},\mathbf{x}_{N_k^+}) = 0$ in vicinity of $(\mathbf{x}^*_{N_k^-},\mathbf{x}^*_{N_k^+})$ in Eq. (35) gives

$$\left[\frac{\partial \mathbf{g}_{I_k}}{\partial \mathbf{x}_{N_k^-}}\right]_{(\mathbf{x}^*_{N_k^-},\mathbf{x}^*_{N_k^+})} \Delta\mathbf{x}_{N_k^-} + \left[\frac{\partial \mathbf{g}_{I_k}}{\partial \mathbf{x}_{N_k^+}}\right]_{(\mathbf{x}^*_{N_k^-},\mathbf{x}^*_{N_k^+})} \Delta\mathbf{x}_{N_k^+} = 0. \quad (51)$$

Thus,

$$DP_{I_k} \equiv \left[\frac{\partial \mathbf{x}_{N_k^+}}{\partial \mathbf{x}_{N_k^-}}\right]\Bigg|_{(\mathbf{x}^*_{N_k^-},\mathbf{x}^*_{N_k^+})} = -\left[\frac{\partial \mathbf{g}_{I_k}}{\partial \mathbf{x}_{N_k^+}}\right]^{-1} \cdot \left[\frac{\partial \mathbf{g}_{I_k}}{\partial \mathbf{x}_{N_k^-}}\right]_{(\mathbf{x}^*_{N_k^-},\mathbf{x}^*_{N_k^+})}. \quad (52)$$

where

$$DP_{I_k} \equiv \left[\frac{\partial \mathbf{x}_{N_k^+}}{\partial \mathbf{x}_{N_k^-}}\right]\Bigg|_{(\mathbf{x}^*_{N_k^-},\mathbf{x}^*_{N_k^+})} = \begin{bmatrix} \dfrac{\partial x_{N_k^+}}{\partial x_{N_k^-}} & \dfrac{\partial x_{N_k^+}}{\partial y_{N_k^-}} \\ \dfrac{\partial y_{N_k^+}}{\partial x_{N_k^-}} & \dfrac{\partial y_{N_k^+}}{\partial y_{N_k^-}} \end{bmatrix}_{(\mathbf{x}^*_{N_k^-},\mathbf{x}^*_{N_k^+})} \quad (53)$$

and

$$\frac{\partial \mathbf{g}_{I_k}}{\partial \mathbf{x}_{N_k^-}} = \begin{bmatrix} \dfrac{\partial g_{I_k 1}}{\partial x_{N_k^-}} & \dfrac{\partial g_{I_k 1}}{\partial y_{N_k^-}} \\ \dfrac{\partial g_{I_k 2}}{\partial x_{N_k^-}} & \dfrac{\partial g_{I_k 2}}{\partial y_{N_k^-}} \end{bmatrix}, \quad \frac{\partial \mathbf{g}_{I_k}}{\partial \mathbf{x}_{N_k^+}} = \begin{bmatrix} \dfrac{\partial g_{I_k 1}}{\partial x_{N_k^+}} & \dfrac{\partial g_{I_k 1}}{\partial y_{N_k^+}} \\ \dfrac{\partial g_{I_k 2}}{\partial x_{N_k^+}} & \dfrac{\partial g_{I_k 2}}{\partial y_{N_k^+}} \end{bmatrix}. \quad (54)$$

For the impulsive map P_{I_k}, the matrix components are

$$\begin{aligned} &\frac{\partial g_{I_k 1}}{\partial x_{N_k^-}} = 1, \quad \frac{\partial g_{I_k 1}}{\partial y_{N_k^-}} = 0, \\ &\frac{\partial g_{I_k 2}}{\partial x_{N_k^-}} = \frac{\partial \Delta y}{\partial x_{N_k^-}}, \quad \frac{\partial g_{I_k 2}}{\partial y_{N_k^-}} = 1 + \frac{\partial \Delta y}{\partial y_{N_k^-}}; \end{aligned} \quad (55)$$

$$\frac{\partial g_{I_N 1}}{\partial x_{N_k^+}} = -1, \quad \frac{\partial g_{I_k 1}}{\partial y_{N_k^+}} = 0,$$

$$\frac{\partial g_{I_k 2}}{\partial x_{N_k^+}} = \frac{\partial \Delta y}{\partial x_{N_k^+}}, \tag{56}$$

$$\frac{\partial g_{I_k 2}}{\partial y_{N_k^+}} = -1 + \frac{\partial \Delta y}{\partial y_{N_k^+}}.$$

From Eq. (38), for the impulsive map P_{I_m}, the corresponding matrix components are

$$\frac{\partial g_{I_m 1}}{\partial x_{N_m^-}} = 1, \quad \frac{\partial g_{I_m 1}}{\partial y_{N_m^-}} = 0,$$

$$\frac{\partial g_{I_m 2}}{\partial x_{N_m^-}} = \frac{\partial \Delta y}{\partial x_{N_m^-}}, \tag{57}$$

$$\frac{\partial g_{I_m 2}}{\partial y_{N_m^-}} = 1 + \frac{\partial \Delta y}{\partial y_{N_m^-}};$$

$$\frac{\partial g_{I_m 1}}{\partial x_0} = -1, \quad \frac{\partial g_{I_m 1}}{\partial y_0} = 0,$$

$$\frac{\partial g_{I_m 2}}{\partial x_0} = \frac{\partial \Delta y}{\partial x_0}, \tag{58}$$

$$\frac{\partial g_{I_m 2}}{\partial y_0} = -1 + \frac{\partial \Delta y}{\partial y_0}.$$

Let

$$\Delta \mathbf{x}_{N_m^+} = \lambda \Delta \mathbf{x}_0,$$

the eigenvalues are determined by

$$|DP - \lambda I_{2\times 2}| = 0. \tag{59}$$

For stability analysis, the following conditions are satisfied.

(i) If $|\lambda_i| < 1$ for $(i = 1, 2)$, the periodic motion is stable.
(ii) If $|\lambda_i| > 1$ for $(i \in \{1, 2\})$, the periodic motion is unstable.

For the bifurcation of the periodic motion, the following conditions are satisfied.

(i) If $\lambda_i = -1$ and $|\lambda_j| < 1$ for $(i, j \in \{1, 2\}$ and $i \neq j)$, the period-doubling bifurcation (PD) of periodic motion occurs.
(ii) If $\lambda_i = 1$ and $|\lambda_j| < 1$ for $(i, j \in \{1, 2\}$ and $i \neq j)$, the saddle-node bifurcation (SN) of periodic motion occurs.
(iii) If $|\lambda_{i,j}| = 1$ and $\lambda_i = \alpha + \mathbf{i}\beta$, $\lambda_j = \alpha - i\beta$ for $(i, j \in \{1, 2\}$ and $i \neq j)$, the Neimark bifurcation occurs.

6. Bifurcation Trees of Periodic Motions

In this section, bifurcation trees of periodic motions to homoclinic orbits in the periodically impulsive pendulum are presented. Consider a set of system parameters as

$$\delta = 0.1, \quad \Delta y = 2, \quad l = 2. \tag{60}$$

The bifurcation trees of period-1 to period-4 librational motions with period-doubling bifurcations are predicted analytically through the implicit mapping method. The acronyms "SN", "PD", and "HO" are for saddle-node, period-doubling, and homoclinic bifurcations, respectively.

Such a bifurcation tree for impulsive node displacement and velocity is presented in Fig. 4. In Figs. 4(a) and 4(b), the global bifurcation trees are presented. The zoomed views of the bifurcation tree are presented in Figs. 4(c)–4(f). For $\Omega \in (0.4, 15.4023885)$ and $(0.34364, 1.64282)$, the period-1 motion is unstable, and for $\Omega \in (1.64282, 15.4023885)$, the period-1 motion is stable. At $\Omega \approx 0.34364, 0.4$, there are two approximate period-1 homoclinic orbits. Such two points are called the homoclinic bifurcations of the period-1 motions. At $\Omega \approx 15.4023885$, the saddle-node bifurcation of the period-1 motions occurs. At $\Omega \approx 1.64282$, the period-doubling bifurcation of period-1 motion occurs. For $\Omega \in (1.627085, 1.64282)$, the period-2 motion is stable, and for $\Omega \in (0.586, 1.627085)$, the period-2 motion is unstable. At $\Omega \approx 0.586$, the approximate period-2 homoclinic orbit exists. The homoclinic bifurcation of the period-2 motions occurs. At $\Omega \approx 1.64282$, there is a saddle-node bifurcation of the period-2 motion. At $\Omega \approx 1.627085$, the period-doubling bifurcation of the period-2 motion occurs. For $\Omega \in (1.6257, 1.627085)$, the period-4 motion is stable, and for $\Omega \in (0.872, 1.62757)$, the period-4 motion is unstable. At $\Omega \approx 0.872$, the homoclinic bifurcation of the period-4 motions occurs. At $\Omega \approx 1.627085$, there is a saddle-node bifurcation of the period-4 motion. At $\Omega \approx 1.62757$, the period-doubling bifurcation of the period-4 motion occurs. The corresponding results are tabulated in Table 1, and the corresponding eigenvalue analysis is presented in Fig. 5.

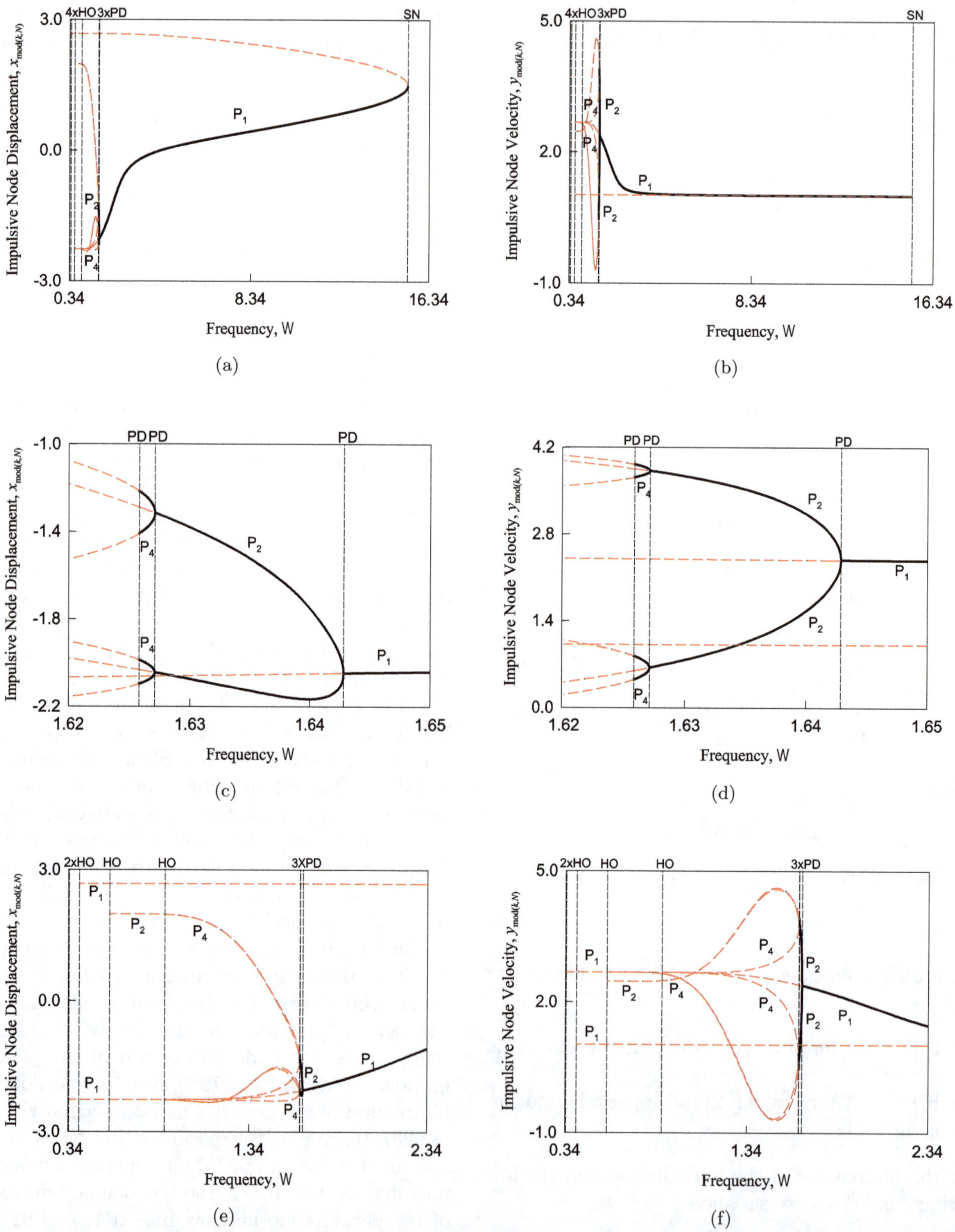

Fig. 4. Bifurcation tree of period-1 to period-4 motions varying with excitation frequency. (a) Impulsive displacement $x_{\text{mod }(k,N)}$, (b) impulsive velocity $y_{\text{mod }(k,N)}$. Zoomed view with $\Omega \in (1.62, 1.65)$: (c) impulsive displacement $x_{\text{mod }(k,N)}$, (d) impulsive velocity $y_{\text{mod }(k,N)}$. Zoomed view with $\Omega \in (0.34, 2.34)$: (e) impulsive displacement $x_{\text{mod }(k,N)}$, (f) impulsive velocity $y_{\text{mod }(k,N)}$ (parameters: $\delta = 0.1, \Delta y = 2, l = 2$).

Table 1. Impulsive period-1 to period-4 motions and bifurcations ($\delta = 0.1, \Delta y = 2, l = 2$).

Period-m	Ω	Ω_{cr}	Bifurcation	Stability
P_1	$0.4 \to 15.4023885$	0.4	HO	Unstable
	$15.4023885 \to 1.64282$	15.4023885	SN	Stable
	$1.64282 \to 0.34364$	1.64282	PD	Unstable
		0.34364	HO	
P_2	$1.627085 \to 1.64282$	1.627085	PD	Stable
	$0.568 \to 1.627085$	0.568	HO	Unstable
P_4	$1.6257 \to 1.627085$	1.6257	PD	Stable
	$0.872 \to 1.6257$	0.872	HO	Unstable

Note: SN: saddle-node bifurcation, PD: period-doubling bifurcation, HO: homoclinic orbit or homoclinic bifurcation.

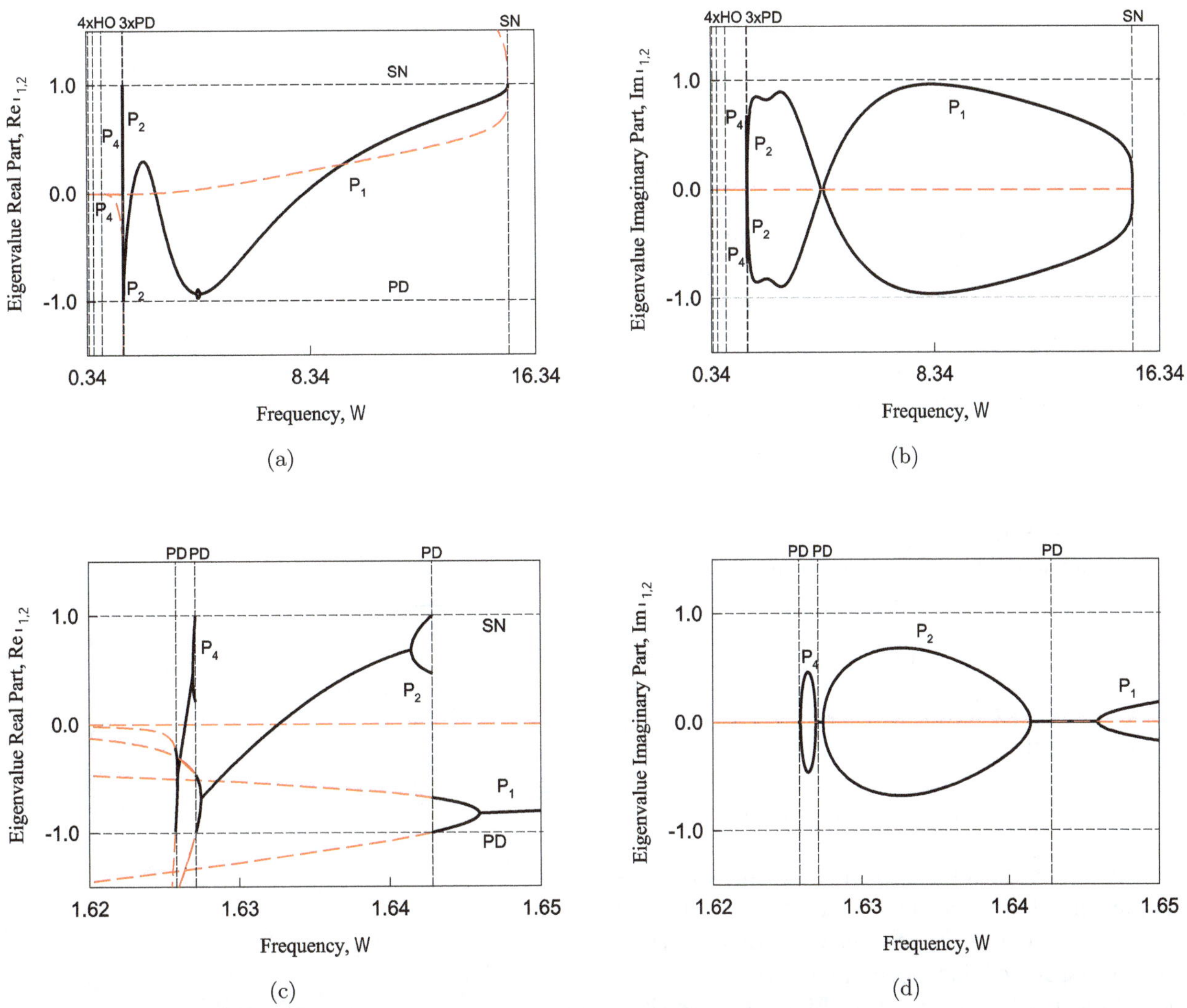

Fig. 5. Eigenvalues varying with excitation frequency for stability and bifurcation of period-1 to period-4 impulsive motions. A global view: (a) real parts of eigenvalues, (b) imaginary parts of eigenvalues; a zoomed view with $\Omega \in (1.62, 1.65)$: (c) real parts of eigenvalues, (d) imaginary parts of eigenvalues (parameters: $\delta = 0.1, \Delta y = 2, l = 2$).

7. Illustrations of Periodic Motions

To demonstrate periodic impulsive motions in the periodically impulsive pendulum, numerical results will be completed through the midpoint method, and the analytical and numerical results are compared. The initial conditions for numerical results are based on the analytical predictions of periodic motion. The acronyms "IC" and "IP" are for initial point and impulsive points, respectively.

The period-1 to period-4 motions in the periodically impulsive pendulum will be presented with different initial conditions under the same parameters from the method presented before. The mapping structure of an impulsive period-1 motion is

$$P = P_{I_1} \circ P_1. \tag{61}$$

For $\Omega = 2.0$, the impulsive period-1 motion with the initial condition is

$$(x_0, y_0) = (-0.84199715, 1.30732315). \tag{62}$$

From the above initial condition, the time histories of displacement and velocity and phase trajectory are presented in Fig. 6, respectively. In Fig. 6(a), the time history of displacement for the impulsive period-1 motion is very simple. In Fig. 6(b), the time history of velocity for the impulsive period-1 motion is presented. One impulse is observed to form the impulsive period-1 motion. In Fig. 6(c), the phase trajectory of the impulsive period-1 motion is presented.

The mapping structure of an impulsive period-2 motion is

$$P = (P_{I_2} \circ P_2) \circ (P_{I_1} \circ P_1). \tag{63}$$

For $\Omega = 1.628$, the impulsive period-2 motion with the initial condition is

$$(x_0, y_0) = (-1.33402452, 3.79507603). \tag{64}$$

From the above initial condition, the time histories of displacement and velocity and phase trajectory are presented in Fig. 7, respectively. In Fig. 7(a), the time history of displacement for the impulsive period-2 motion is two waves. In Fig. 7(b), the time history of velocity for the impulsive period-2 motion is presented with two impulses to form the impulsive period-2 motion. In Fig. 7(c), the phase trajectory of the impulsive period-2 motion is presented with two cycles and two impulses.

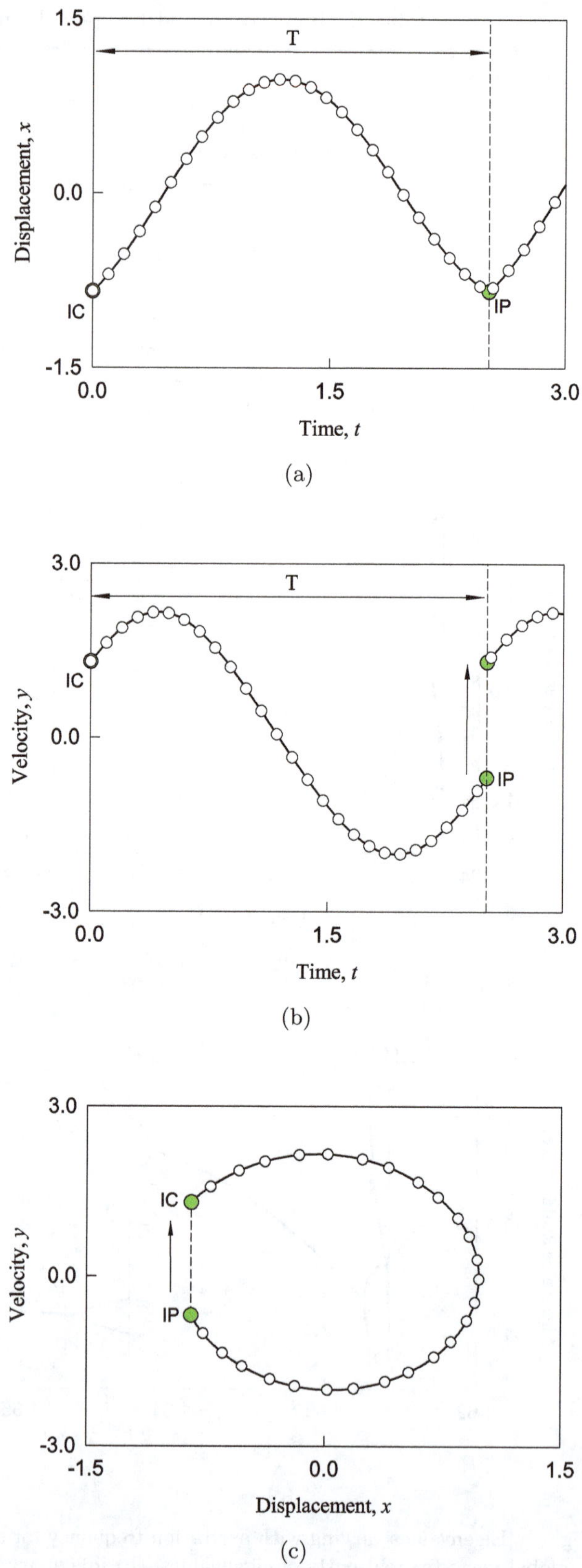

Fig. 6. An impulsive period-1 motion ($\Omega = 2.5$): (a) displacement, (b) velocity and (c) trajectory. Initial condition: $(x_0, y_0) = (-0.84199715, 1.30732315)$ (parameters: $\delta = 0.1$, $\Delta y = 2$, $l = 2$).

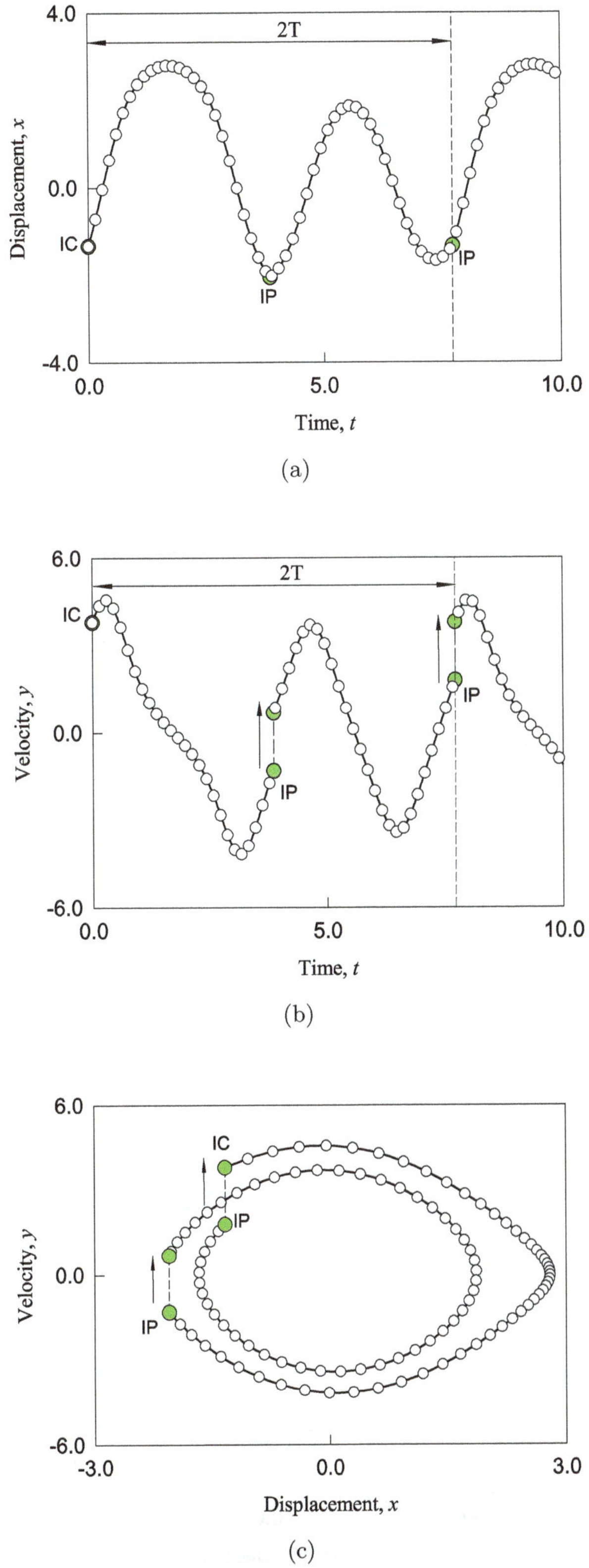

Fig. 7. A period-1 motion ($\Omega = 1.628$): (a) displacement, (b) velocity and (c) trajectory. Initial condition: $(x_0, y_0) = (-1.33402452, 3.79507603)$ (parameters: $\delta = 0.1$, $\Delta y = 2$, $l = 2$).

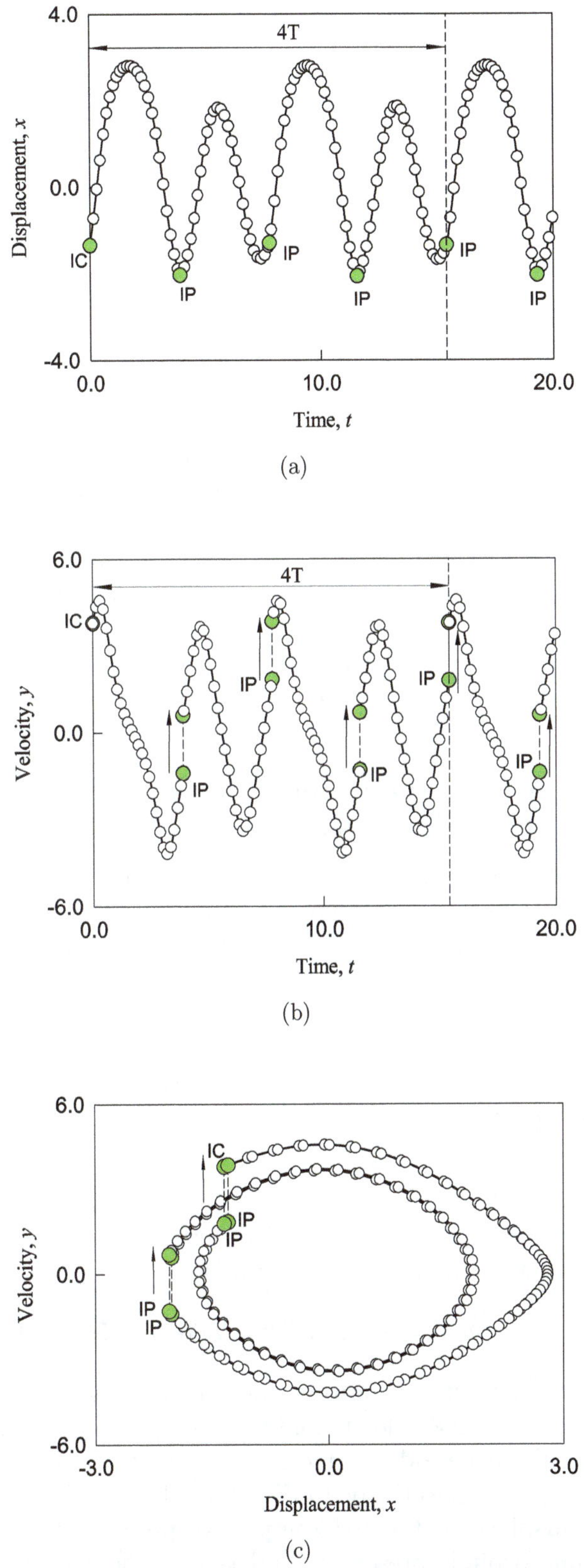

Fig. 8. A period-4 motion in the impulsive pendulum ($\Omega = 1.627$): (a) displacement, (b) velocity and (c) trajectory. Initial condition: $(x_0, y_0) = (-1.33880322, 3.79360711)$ (parameters: $\delta = 0.1$, $\Delta y = 2$, $l = 2$).

The mapping structure of an impulsive period-4 motion is

$$P = (P_{I_4} \circ P_4 \circ P_{I_3} \circ P_3) \circ (P_{I_2} \circ P_2 \circ P_{I_1} \circ P_1). \tag{65}$$

For $\Omega = 1.627$, the impulsive period-4 motion with the initial condition is

$$(x_0, y_0) = (-1.33880322, 3.79360711). \tag{66}$$

From the above initial condition, the time histories of displacement and velocity and phase trajectory are presented in Fig. 8, respectively. In Fig. 8(a), the time history of displacement for the impulsive period-4 motion is two waves. In Fig. 8(b), the time history of velocity for the impulsive period-4 motion is presented with four impulses to form the impulsive period-4 motion. In Fig. 8(c), the phase trajectory of the impulsive period-4 motion is presented with four cycles and four impulses.

8. Approximate Impulsive Homoclinic Orbits

To demonstrate the approximate impulsive homoclinic orbits in the impulsive pendulum, the impulsive homoclinic orbits will be presented. The period-1 to period-4 homoclinic orbits in the impulsive pendulum will be presented with different initial conditions under the same parameters.

The mapping structure of a period-1 impulsive homoclinic orbit is

$$P = P_{I_1} \circ P_1. \tag{67}$$

For $\Omega = 0.346$, the impulsive period-1 homoclinic orbit with the initial condition is

$$(x_0, y_0) = (-2.25676395, 2.68409595). \tag{68}$$

From the initial condition, the time histories of displacement and velocity and phase trajectory for the impulsive period-1 homoclinic orbit are presented in Fig. 9, respectively. In Fig. 9(a), the time history of displacement for the impulsive period-1 homoclinic orbit is presented, and the homoclinic orbit approaches the equilibrium $x_h = \pi$. In Fig. 9(b), the time history of velocity for the impulsive period-1 motion is presented, and the velocity is very close to zero. One impulse is observed to form the impulsive period-1 homoclinic orbit. In Fig. 9(c), the

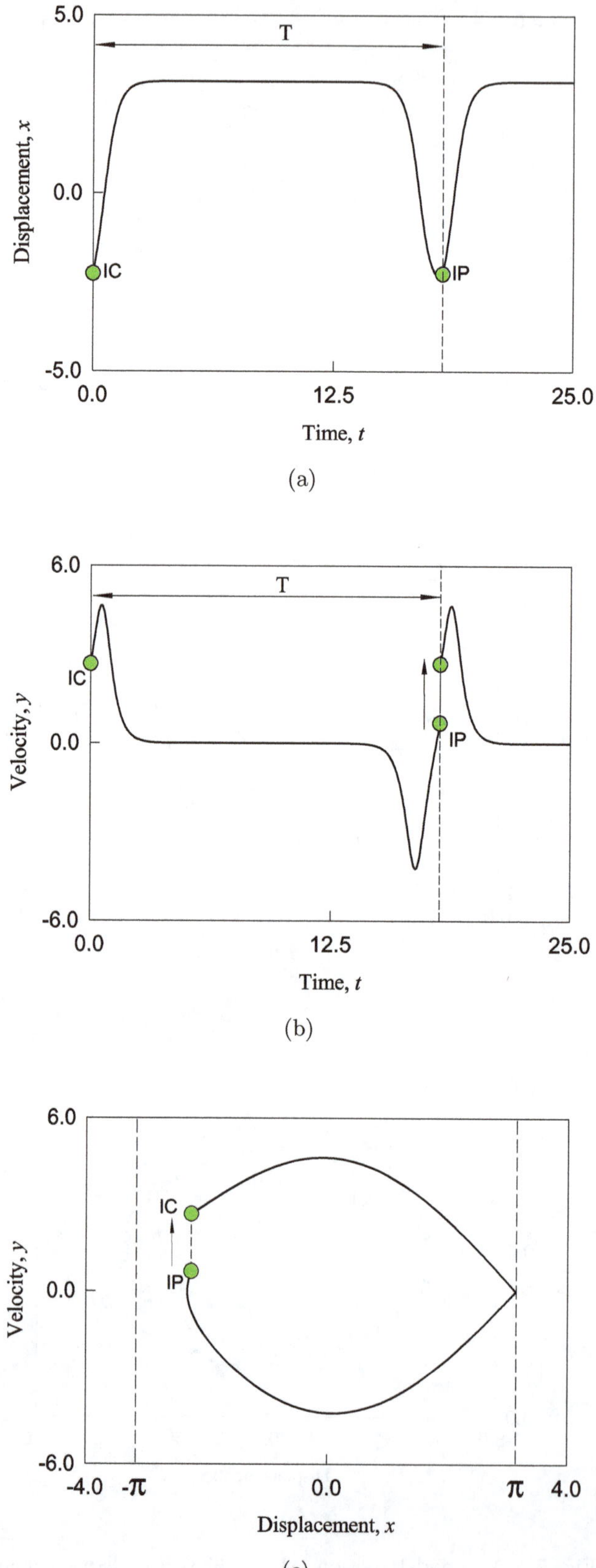

Fig. 9. An impulsive period-1 homoclinic orbit ($\Omega = 0.346$): (a) displacement, (b) velocity and (c) trajectory. Initial condition: $(x_0, y_0) = (-2.25676395, 2.68409595)$ (parameters: $\delta = 0.1$, $\Delta y = 2$, $l = 2$).

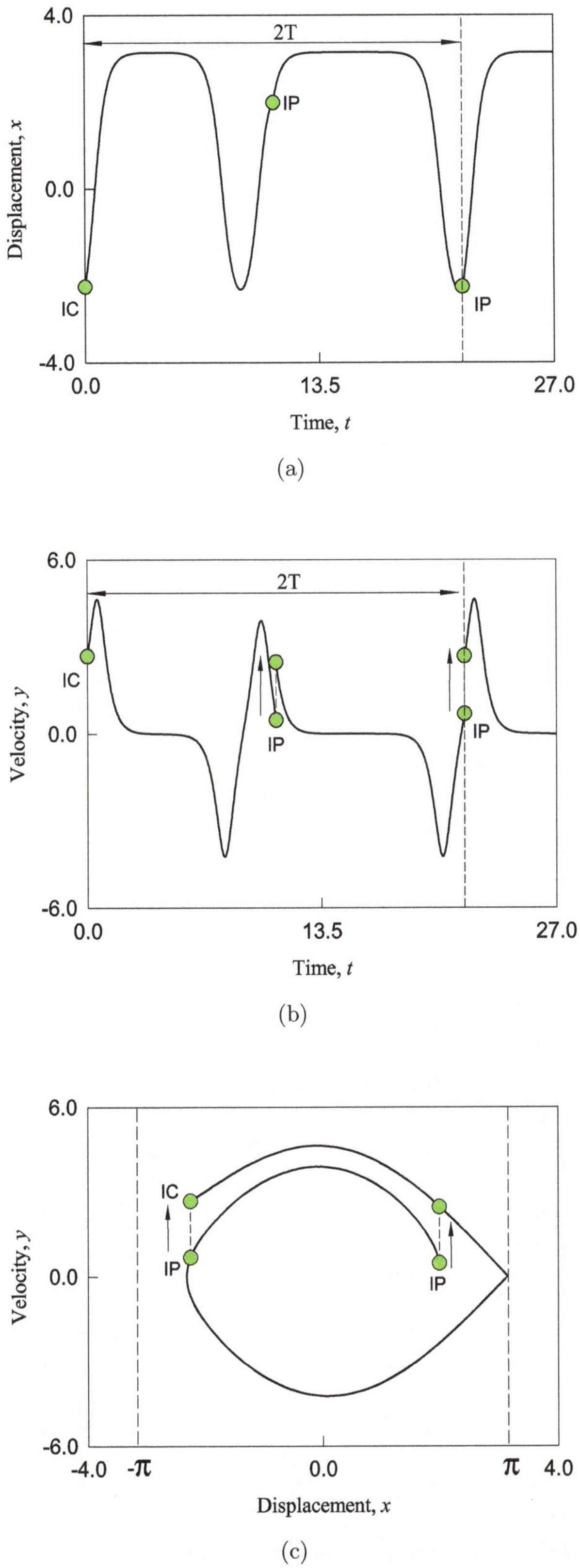

Fig. 10. An impulsive period-2 homoclinic orbit ($\Omega = 0.568$): (a) displacement, (b) velocity and (c) trajectory. Initial condition: $(x_0, y_0) = (-2.25641267, 2.68481425)$ (parameters: $\delta = 0.1$, $\Delta y = 2$, $l = 2$).

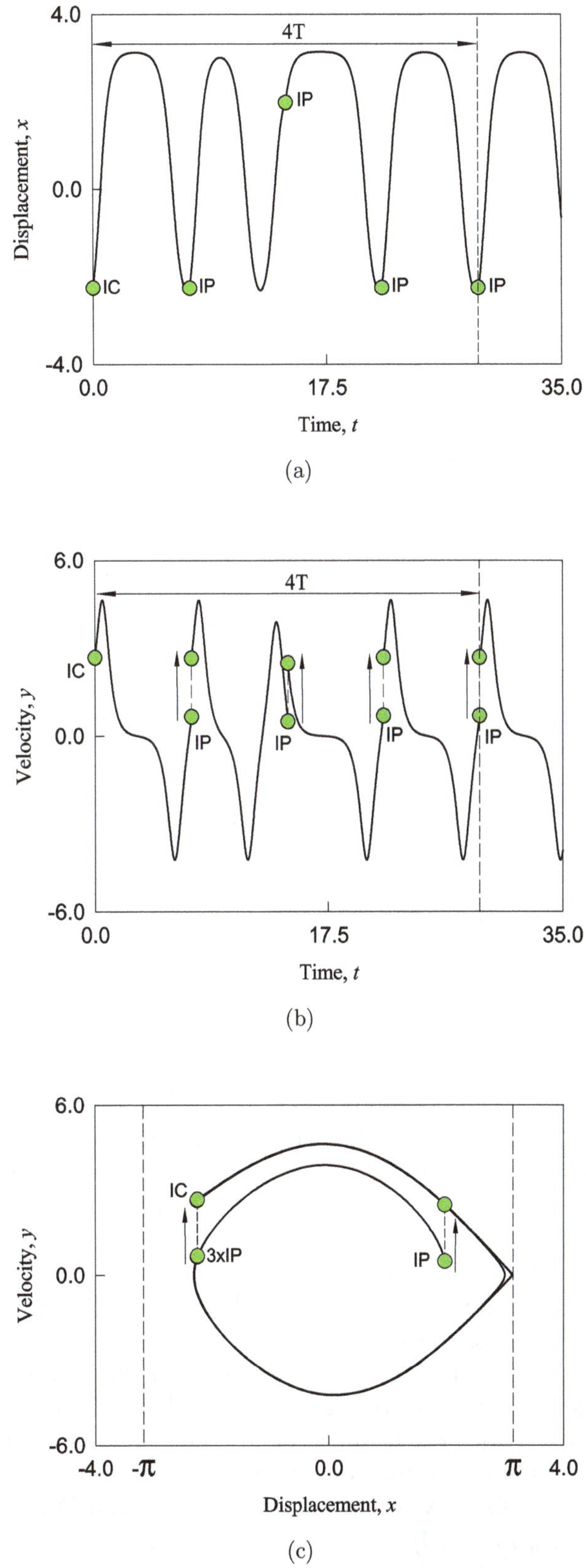

Fig. 11. An impulsive period-4 homoclinic orbit ($\Omega = 0.972$): (a) displacement, (b) velocity and (c) trajectory. Initial condition: $(x_0, y_0) = (-2.25618794, 2.68465860)$ (parameters: $\delta = 0.1$, $\Delta y = 2$, $l = 2$).

phase trajectory of the impulsive period-1 homoclinic orbit is presented, and the homoclinic orbit approaches the equilibrium of $(x_h, y_h) = (\pi, 0)$.

The mapping structure of an impulsive period-2 homoclinic orbit is

$$P = (P_{I_2} \circ P_2) \circ (P_{I_1} \circ P_1). \tag{69}$$

For $\Omega = 0.568$, the impulsive period-2 homoclinic orbit with the initial condition is

$$(x_0, y_0) = (-2.25641267, 2.68481425). \tag{70}$$

From the initial condition, the time histories of displacement and velocity and phase trajectory are presented in Fig. 10, respectively. In Fig. 10(a), the time history of displacement for the impulsive period-2 homoclinic orbit is two waves approaching $x_h = \pi$. In Fig. 10(b), the time history of velocity for the impulsive period-2 homoclinic orbit is presented with two impulses, and the two slowing movement velocities approach zero. In Fig. 10(c), the phase trajectory of the impulsive period-2 motion is presented with two cycles and two impulses. One of the two cycles approaches the equilibrium of $(x_h, y_h) = (\pi, 0)$.

The mapping structure of an impulsive period-4 homoclinic orbit is

$$P = (P_{I_4} \circ P_4 \circ P_{I_3} \circ P_3) \circ (P_{I_2} \circ P_2 \circ P_{I_1} \circ P_1). \tag{71}$$

For $\Omega = 1.627$, the impulsive period-4 motion with the initial condition is

$$(x_0, y_0) = (-1.33880322, 3.79360711). \tag{72}$$

From the initial condition, the time histories of displacement and velocity and phase trajectory are presented in Fig. 11, respectively. In Fig. 11(a), the time history of displacement for the impulsive period-4 homoclinic orbit is four waves approaching $x_h = \pi$. In Fig. 11(b), the time history of velocity for the impulsive period-4 homoclinic orbit is presented with four impulses, and the four slowing movement velocities approach zero. In Fig. 11(c), the phase trajectory of the impulsive period-4 motion is presented with four cycles and four impulses. One of the four cycles approaches the equilibrium of $(x_h, y_h) = (\pi, 0)$.

9. Conclusions

In this paper, the computational method was presented for the bifurcation dynamics of periodic motions in impulsive nonlinear dynamical systems. Such a method was based on the implicit mapping method, which was used to discretize the nonlinear dynamical subsystems in the time interval of two impulses. The specific mapping structures were adopted for periodic motions in the impulsive dynamical systems, and the stability and bifurcations of the corresponding periodic motions were completed. The impulsive homoclinic orbits related to the different periodic motions were obtained for the corresponding homoclinic bifurcations. The numerical illustrations of periodic motions and impulsive homoclinic orbits were for a better understanding of the bifurcation dynamics of impulsive nonlinear systems. From such a method, one can do similar analysis for other impulsive nonlinear systems.

ORCID

Albert C. J. Luo
https://orcid.org/0000-0001-8208-6108
Tianji Ma
https://orcid.org/0009-0000-4399-7777

References

Akhmet, M. & Cag, S. [2018] "Tikhonov theorem for differential equations with singular impulses," *Discontinuity Nonlinearity Complex.* **7**, 291–303.

Akhmet, M., Tleubergenova, M. & Nugayeva, Z. [2022] "Unpredictable solutions of impulsive quasi-linear systems," *Discontinuity Nonlinearity Complex.* **11**, 73–89.

Akhmet, M. & Erim, G. [2023] "Almost periodic Solutions of recurrent neural networks with state-dependent and structured impulses," *Discontinuity Nonlinearity Complex.* **12**, 141–165.

Benjemaa, M., Gouadri, W. & Hammami, M. A. [2023] "Stability analysis of nonlinear systems with impulsive perturbations: Application to Hopfield neural networks," *Discontinuity Nonlinearity Complex.* **12**, 789–801.

Lakshmikanthem, V. & Liu, Z. [1989] "Stability criteria for impulsive differential equations in terms of two measures," *J. Math. Anal. Appl.* **137**, 591–604.

Lakshmikantham, V., Bainov, D. D. & Simeonov, P. S. [1989] *Theory of Impulsive Differential Equations*, Series in Modern Applied Mathematics, Vol. 6 (World Scientific, Singapore).

Lakshmikanthem, V. & Liu, Z. [1998] "Impulsive hybrid systems and stability," *Dyn. Syst. Appl.* **7**, 1–10.

Liu, X. [1994] "Stability results of impulsive differential systems with applications to population growth models," *Dyn. Stab. Syst.* **9**, 163–174.

Liu, X. & Willms, A. R. [1996] "Impulsive controllability of linear dynamical systems with applications to maneuvers of spacecraft," *Math. Probl. Eng.* **2**, 277–299.

Luo, A. C. J. & Wang, Y. [2009] "Switching dynamics of multiple linear oscillators," *Commun. Nonlin. Sci. Numer. Simul.* **14**, 3472–3485.

Luo, A. C. J. [2011] *Regularity and Complexity in Dynamical Systems* (Springer, NY).

Luo, A. C. J. [2012] *Discrete and Switching Dynamical Systems* (Higher Education Press/L&H Scientific Publishing LLC, Beijing/Glen Carbon).

Luo, A. C. J. [2015a] *Discretization and Implicit Mapping Dynamics* (Springer Berlin Heidelberg).

Luo, A. C. J. [2015b] "Periodic flows to chaos based on discrete implicit mappings of continuous nonlinear systems," *Int. J. Bifurcation and Chaos* **25**, 1550044.

Luo, A. C. J. & Guo, Y. [2015] "A semi-analytical prediction of periodic motions in Duffing oscillator through mapping structures," *Discontinuity Nonlinearity Complex.* **4**, 13–44.

Luo, A. C. J. & Zhu, Y. [2024] "Periodic motions with impact chatters in an impact Duffing oscillator," *Chaos* **34**, 053124.

Ramkumar, K., Ravikumar, K. & Anguraj, A. [2021] "Approximate controllability for time-dependent impulsive neutral stochastic partial differential equations with fractional Brownian motion and Poisson jumps," *Discontinuity Nonlinearity Complex.* **10**, 227–235.

Tamilarasi, M. & Radhakrishnan, B. [2023] "Existence and controllability results for higher order Caputo fractional damping system with impulses," *J. Appl. Nonlinear Dyn.* **12**, 609–630.

van der Schaft, A. & Schumacher, H. [2000] *An Introduction to Hybrid Dynamical Systems*, Lecture Notes in Control and Information Sciences, Vol. 251 (Springer, London).

Xing, S. Y. & Luo, A. C. J. [2017] "Towards infinite bifurcation trees of period-1 motions to chaos in a time-delayed, twin-well Duffing oscillator," *J. Vib. Test. Syst. Dyn.* **1**, 353–392.

Xing, S. Y. & Luo, A. C. J. [2018] "On possible infinite bifurcation trees of period-3 motions to chaos in a time-delayed, twin-well Duffing oscillator," *Int. J. Dyn. Control* **6**, 1429–1464.

Zheng, S. & Fu, X. [2013a] "Pulse phenomena for impulsive dynamical systems," *Discontinuity Nonlinearity Complex.* **2**, 225–245.

Zheng, S. & Fu, X. [2013b] "Chatter dynamics on impulse surfaces in impulsive differential systems," *J. Appl. Nonlin. Dyn.* **2**, 373–396.

Zhu, Y. & Luo, A. C. J. [2025] "Bifurcation dynamics of periodic motions in an impact Duffing oscillator with two boundaries," *Int. J. Bifurcation and Chaos* **35**, 2550089.

https://doi.org/10.1142/9789819830633_0012

Chapter 12

Doubling Bifurcations of Invariant Closed Curves in 3D Maps#

Sayanho Biswas*[*] and Soumitro Banerjee[†]
Indian Institute of Science Education & Research, Kolkata, India
**sb21ms164@iiserkol.ac.in*
†soumitro@iiserkol.ac.in

Viktor Avrutin
University of Stuttgart, Germany
avrutin.viktor@gmail.com

Iryna Sushko
Institute of Mathematics, National Academy of Sciences of Ukraine, Kyiv, Ukraine
sushko@imath.kiev.ua

Numerous studies have reported two types of doubling of Invariant Closed Curves (ICCs) in dynamical systems: (a) the creation of two disjoint ICCs such that iterations flip between them; and (b) the creation of a single ICC of double the length of the original ICC. Various methods have been proposed to analyze such bifurcations. We show that the methods proposed so far are not always adequate for analyzing and predicting these bifurcations. We propose a new method to overcome the difficulties.

Keywords: Nonlinear dynamics; quasi-periodicity; torus; closed invariant curve; doubling bifurcation.

1. Introduction

An Invariant Closed Curve (ICC) in a Poincaré section corresponds to a two-frequency torus in the phase space of a continuous-time dynamical system. If the asymptotic dynamics on the ICC is periodic, it is called a mode-locked ICC or a resonant torus. In the simplest case, a stable mode-locked ICC contains a pair of cycles, namely a stable cycle and a saddle, along with the unstable manifold of the latter converging to the former. Multiple such pairs of cycles may also be present in a resonant ICC. If the asymptotic dynamics is quasi-periodic, the curve is called a quasi-periodic ICC or an ergodic torus, and sometimes also a drift ring. In this case, the ICC is defined as the closure of any of the quasi-periodic orbits it contains.

An ICC (also called a loop) has been reported to undergo doubling in two ways. In the first type, two disjoint loops develop, and the iterates flip between the two loops. Such bifurcations have been reported in the Mira map [Karatetskaia *et al.*, 2021; Gonchenko *et al.*, 2021], mechanical vibro-impacting systems [Ding *et al.*, 2004; Luo, 2006], in experiments on molten gallium [McKell *et al.*, 1990],

#To cite this article, please refer to its earlier version published in the *International Journal of Bifurcation and Chaos*, Vol. 35, No. 14 (2025) 2540012 (16 pages), DOI: 10.1142/S0218127425400127.
†Author for correspondence

and in coupled Chua's circuits [Zhong *et al.*, 1998]. We shall call this kind of doubling bifurcation a "loop doubling".

The second type of ICC doubling bifurcation results in a single loop of double length. Such bifurcations have been reported in a map with quadratic nonlinearity [Arneodo *et al.*, 1983], in experiments on vibrating strings [Molteno & Tufillaro, 1990], a circuit with hysteresis element [Sekikawa *et al.*, 2001], a 3D Lotka–Volterra model [Bischi & Tramontana, 2010], in a resonant circuit with two saturable inductors [Schilder *et al.*, 2005], in a bimode laser system [Letellier *et al.*, 2007], and in the food chain models used in ecological studies [Osipenko, 2007]. This kind of doubling bifurcation is called "length doubling" (a.k.a. "double covering").

A few approaches have been proposed to analyze the bifurcations of ICCs. In Sec. 2, we briefly describe these approaches and show that none of the available procedures is effective in predicting which kind of doubling is going to occur in a system. Moreover, each available approach is applicable to quasi-periodic or mode-locked ICCs, but not both.

In this paper, we propose a new approach based on the topological structure of the tangent space in the doubling direction. Our objective is to develop a general method for predicting the kind of doubling bifurcation that would apply to both resonant and ergodic tori.

2. State of the Art

2.1. *The second Poincaré section*

To analyze bifurcations of quasi-periodic ICCs, Banerjee *et al.* [2012] proposed placing a "second Poincaré section" transverse to the ICC and analyzing the local dynamics around the intersection of the ICC with that plane. Although this approach aptly captures other bifurcations of tori, it cannot distinguish between the two types of ICC doubling because, in the second Poincaré section, both are seen as period doubling bifurcation of a fixed point.

2.2. *The sign of the third eigenvalue*

This approach proposed by [Gardini & Sushko, 2012] is applicable to mode-locked ICCs. At each point of a stable p-cycle belonging to the ICC, there are three eigendirections associated with the eigenvalues of the Jacobian matrix for the pth iterate. In the cited work, the eigenvalue 1 is assumed to be

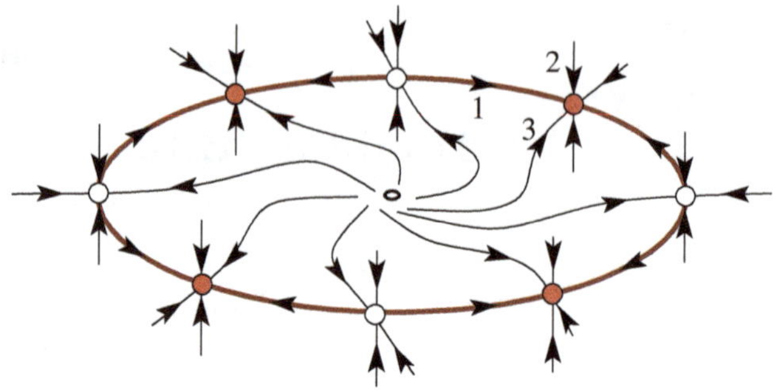

Fig. 1. Schematic representation of a resonant ICC, showing the eigendirections. The red circles represent the stable cycle, and the white circles represent the saddle cycle. Eigendirection 1 lies along the ICC, eigendirection 2 is the one that bifurcates, i.e. its eigenvalue approaches -1. Eigendirection 3 is along the radial direction.

positive (an assumption certainly valid sufficiently close to the Neimark–Sacker bifurcation of a fixed point leading to the appearance of the considered ICC, see Fig. 1). In case of a period-doubling of the stable cycle, eigenvalue 2 will necessarily be negative as it goes through -1. In the cited work, it is mentioned that after bifurcation, the manifold that includes ICCs has the form of a normal strip (i.e. a cylinder) in case of loop doubling and a Möbius strip in case of length doubling. To distinguish between these two cases, the authors conjectured that the type of doubling depends on the sign of the third eigenvalue: If it is positive, two disjoint ICCs result from the bifurcation, and if it is negative, a single ICC of double length appears.

2.3. *The "Dominant Lyapunov Bundle" method*

This method applies to quasi-periodic ICCs. For a quasi-periodic orbit, the largest Lyapunov exponent is zero, and hence vital information about the orbit is contained in the second-largest Lyapunov exponent, called the "dominant" Lyapunov exponent. The Lyapunov exponents give the rates of exponential divergence or convergence in the direction of the Lyapunov vectors. The collection of Lyapunov vectors corresponding to the dominant Lyapunov exponent computed on the quasi-periodic ICC is called its *dominant Lyapunov bundle.*

The method for obtaining Lyapunov bundles is given in [Kamiyama *et al.*, 2014]. The authors showed that, for a loop doubling bifurcation, the topological structure of the dominant Lyapunov bundle is cylindrical [orientation-reversing annulus

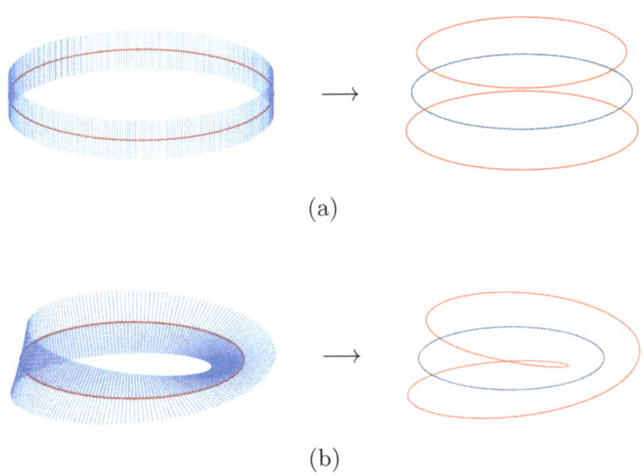

Fig. 2. The topological structures of the dominant Lyapunov bundle close to torus doubling. (a) If it is a cylinder, a loop-doubling bifurcation occurs; (b) if it is a Möbius strip, a length-doubling bifurcation occurs. The stable orbit is shown in red, and Lyapunov vectors are shown in blue.

in their language, see Fig. 2(a)]. For a length-doubling bifurcation, it takes a Möbius strip structure [orientation-preserving annulus in their language, see Fig. 2(b)].

Note that the Lyapunov bundle method [Kamiyama *et al.*, 2014] is not applicable to resonant ICCs, as the orbits used to compute the Lyapunov bundles converge to the stable cycle on the ICC and therefore do not give information on global structures in phase space.

2.4. *The center manifold*

In the work by [Gonchenko *et al.*, 2021], a more formal foundation for the analysis of bifurcations of quasi-periodic ICCs has been presented. With the assumption that ICCs must be intersection-free (which is certainly true for invertible maps), they showed that a doubling bifurcation of such a curve is associated with a 2D center manifold and the type of doubling is determined by the topology of this manifold. Specifically, this center manifold has a cylindrical or a Möbius strip structure, leading to loop and length-doubling bifurcations, respectively.

2.5. *Motivation of this work*

- The existing approaches consider resonant and quasi-periodic ICCs separately. It is reasonable to assume that the issue can be treated in a unified way, particularly because quasi-periodic ICCs can often be seen as limiting cases of periodic ICCs (e.g. in period-adding structures).
- The approaches mentioned in Secs. 2.1–2.3 have some issues. The secondary Poincaré section approach cannot distinguish between loop doubling and length doubling. Later in this paper, we present counterexamples that violate predictions of the "sign of the third eigenvalue" approach. The computation of Lyapunov bundles is a complicated and computation-intensive procedure, and the method lacks theoretical grounding.
- The theoretical foundation provided by Gonchenko *et al.* [2021] does not provide a way to apply it in the analysis of concrete examples.
- There has been no investigation on the doubling bifurcation of resonant ICCs involving a saddle–focus connection. This situation needs to be covered under the unified approach.

In this paper, we are interested in developing a practical method to predict the type of doubling bifurcation applicable both to resonant and quasi-periodic ICCs.

3. The Doubling Tangent Space

In this section, we propose a new method for analyzing the doubling of ICCs.

3.1. *Quasi-periodic ICCs*

As discussed in [Gonchenko *et al.*, 2021], the type of doubling of a quasi-periodic ICC can be predicted based on the topology of the center manifold associated with the eigenvalue -1. Unfortunately, this approach is not useful in practical situations, as there is no straightforward method to compute this 2D center manifold.

However, for our purposes, we need neither an analytic representation of this manifold nor its precise approximation, but only the topology. Evidently, near the ICC, this topology is reflected in the topology of a suitable linear tangent space along the ICC, which can be seen as a linear approximation of the relevant 2D center manifold in proximity to the ICC. This tangent space can easily be calculated by evaluating the eigenvectors of the Jacobian matrices for a suitable iterate at the points of the ICC, and it is natural to expect that it preserves the topology of the center manifold.

To do this, we first need to calculate the rotation number ρ of the ICC. If the ICC is quasi-periodic, ρ is an irrational number. We choose a sufficiently good rational approximation (one could use a Diophantine approximation) $\rho \approx q/p$. We evaluate the eigenvalues of the Jacobian matrix J_{f^p} of the pth composition of the map f at the points of the ICC. Since we consider the ICC sufficiently close to a doubling bifurcation, each of these matrices has an eigenvalue close to -1. The union of the eigenvectors corresponding to these eigenvalues computed at points on the ICC approximates the tangent eigenspace. This tangent eigenspace reflects the topology of the 2D center manifold, and hence can be used instead of the center manifold to predict the type of doubling bifurcation.

Next, note that the concept of a center manifold is applicable only at the bifurcation point. However, for a smooth system sufficiently close to the bifurcation, there exists a slow manifold of the same topology. The structure of this manifold changes to a small extent for small changes of a parameter, and, like the center manifold at the bifurcation point, it can be approximated by the corresponding linear eigenspace. Therefore, the topology of the tangent eigenspace before the bifurcation predicts the topology of the tangent eigenspace at the bifurcation which, in turn, reflects the topology of the center manifold at the bifurcation point. Hence, the topology of the tangent eigenspace before the bifurcation predicts the kind of doubling bifurcation the ICC would eventually undergo.

3.2. *Resonant ICCs*

Setup

Let us first fix our setup. We consider a resonant ICC in a 3D map f, consisting of a stable period-p node, a period-p saddle, and the unstable manifold of the saddle converging to the node. To describe the dynamics on the ICC, we consider the eigenvalues and eigenvectors of the Jacobian J_{f^p} of the pth iterate of the map, for which the cycle points are fixed points.

Each cycle has three eigendirections determined by J_{f^p}, as illustrated in Fig. 1. For simplicity, we denote these directions as radial, tangential, and transversal. This notation is borrowed from the configuration near a Neimark–Sacker bifurcation, where the ICC can be approximated by a circle in a plane. In this context, the radial direction corresponds to the radius of the circle, the tangential direction aligns with the tangent to the circle in the plane, and the transverse direction is orthogonal to this plane.

Note that in publications related to resonant ICCs, several mistakes have been made by mixing the eigenvalues of the Jacobians of f and f^p. Throughout this paper, we use the Jacobian of f^p.

Assumptions

Regarding the eigenvalues of J_{f^p} associated with the three directions described above, we make the following assumptions.

(A1) The eigenvalue associated with the radial direction is positive and less than one for both cycles (consistent with dynamics soon after a Neimark–Sacker bifurcation).

(A2) The eigenvalue associated with the tangential direction is positive for both cycles: it is less than one for the stable node and greater than one for the saddle (as expected after a fold bifurcation determining the boundaries of the related periodicity tongue issuing from a Neimark–Sacker bifurcation boundary).

(A3) The eigenvalue associated with the transverse direction is negative for both cycles. Initially, it lies between -1 and 0, but under parameter variation, these eigenvalues approach and eventually cross -1. Consequently, the period-doubling of the ICC occurs in the transversal direction.

The saddle and node belonging to the resonant ICC generally do not undergo doubling bifurcations at the same parameter value. Therefore, the doubling of a resonant ICC occurs not through a single bifurcation but via a sequence of two bifurcations. Thus, two distinct doubling scenarios arise, depending on whether the saddle or the node doubles first.

The two scenarios are independent of the type of doubling exhibited by the ICC (loop or length doubling). Moreover, the doubling bifurcation of one cycle does not imply that the other cycle must also undergo a doubling bifurcation. Therefore, we introduce an additional assumption:

(A4) The doubling bifurcations of the saddle and node belonging to the resonant ICC occur at parameter values sufficiently close to each other, and no other bifurcations involving these cycles arise between these values.

Main result

Under assumptions (A1)–(A4), we develop a way of predicting the type of doubling exhibited by a resonant ICC. Note that at the point of bifurcation of the saddle or node, the center manifolds are 1D curves that are transversal to the ICC. These 1D center manifolds do not yield conclusive information regarding the type of doubling.

Recall that, in the case of a quasi-periodic ICC, the type of doubling it undergoes can be inferred from the topology of a 2D center manifold. We proceed to define a similar 2D attracting manifold $\mathcal{M}$ with the following properties:

(1) The ICC is contained within $\mathcal{M}$.
(2) At the parameter values corresponding to the doubling bifurcations, the 1D center manifolds are also contained within $\mathcal{M}$.
(3) At other points along the ICC, $\mathcal{M}$ is tangent to the eigenvectors that, at the points of the stable cycle and at the bifurcation parameter value, correspond to the eigenvalue -1.

We claim that such an attracting surface exists because of the strong contraction in the radial direction (see Fig. 3). If assumption (A4) is satisfied, $\mathcal{M}$ would persist throughout the parameter interval between the doubling bifurcations of both cycles. The key statement of our theory is as follows.

If the 2D manifold $\mathcal{M}$ exists, its topology determines the type of doubling: either cylindrical or Möbius-strip-like — corresponding respectively to loop- or length-doubling of the resonant ICC.

Computation

To determine the type of doubling a resonant ICC undergoes, we must determine the topology of $\mathcal{M}$. As with quasi-periodic ICCs, this topology is reflected — ultimately via the Hartman–Grobman theorem — in the topology of the corresponding tangent space. When analyzing resonant ICCs, we apply the same computational procedure used for quasi-periodic ICCs. We compute the eigenvectors of the matrices J_{f^p} associated with the doubling

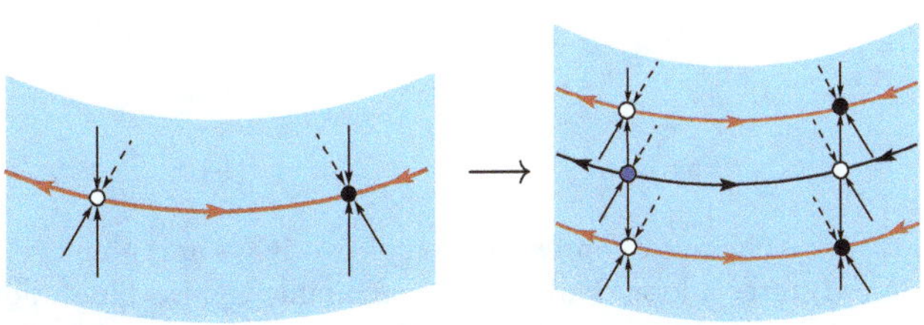

Fig. 3. Schematic picture of the attracting manifold $\mathcal{M}$, showing the local structure before and after the doubling of the ICC.

(transversal) direction at points along the ICC — both the cycle points and the points on the saddle's unstable manifold. In contrast to the quasi-periodic case, we do not need to approximate the rotation number ρ, as the period p is known *a priori*.

Dynamics inside $\mathcal{M}$

We now return to the two doubling scenarios, which depend on whether the saddle or the node doubles first. Due to the contracting nature of the radial direction, these scenarios unfold within $\mathcal{M}$.

If the saddle doubles first:

The process starts with an ICC with a saddle–node connection; see Fig. 4(a).

In the first step, the original saddle becomes a repelling node. In its vicinity, a period-$2p$ saddle emerges. As a result, an attracting bilayered ICC arises, consisting of a period-p node, a period-$2p$ saddle, and its unstable manifold [see Fig. 4(b)]. At this stage, the repelling node has a 2D unstable manifold (a subset of $\mathcal{M}$ including repelling ICC associated with the corresponding eigenvector), while the stable node possesses a stable 2D manifold (a subset of $\mathcal{M}$ as well).

In the next stage, the original node becomes a saddle. In its vicinity, a period-$2p$ node appears. At the bifurcation, the branches of the unstable manifold of the period-$2p$ saddle are tangent to the eigenvectors associated with the eigenvalue -1 [Fig. 4(b)].

After the bifurcation, a period-$2p$ ICC is present, consisting of a stable period-$2p$ node, a period-$2p$ saddle, and its unstable manifold [Fig. 4(e)]. Note that, immediately after bifurcation, the manifold branches approach the saddle from the same side [see Fig. 4(c)]. The unstable period-p ICC is composed of the unstable period-p node, a period-p saddle and its stable manifold.

If the node doubles first:

The sequence of events starts again with a stable period-p ICC, as shown in Fig. 4(a).

In the first stage, the original node becomes a saddle. In its vicinity, a stable period-$2p$ node emerges [see Fig. 4(d)]. As a result, a repelling period-p ICC is formed, containing two period-p saddles — one with an attracting transversal direction and the other with a repelling one. The unstable manifold of the first coincides with the stable manifold of the second, forming the ICC. At

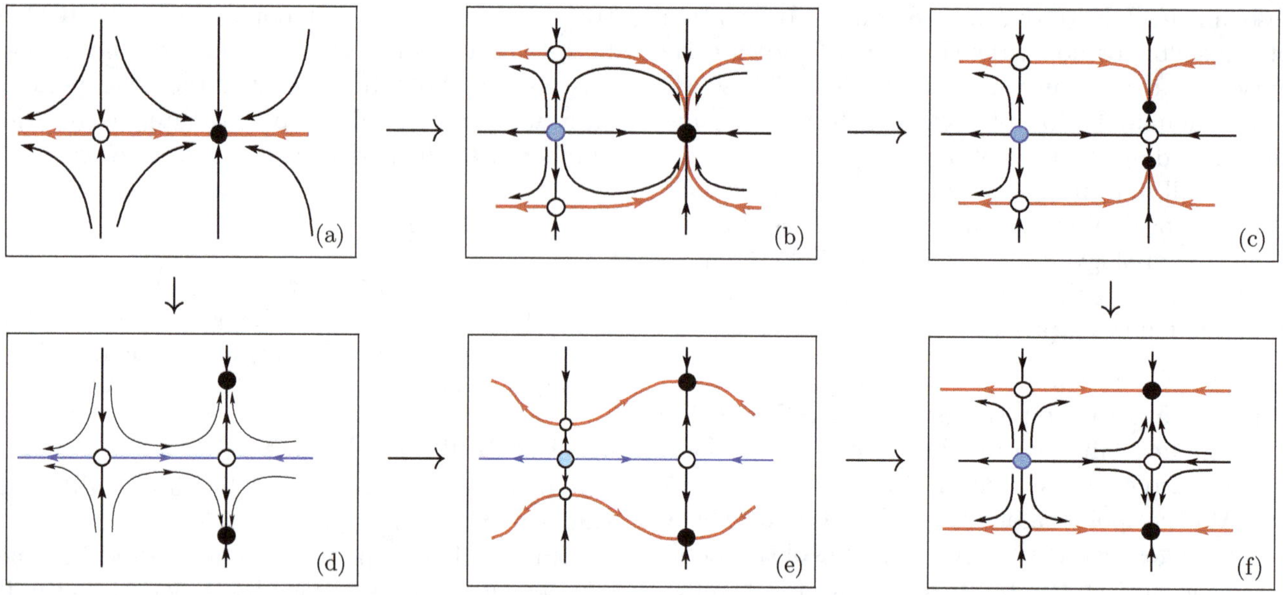

Fig. 4. Schematic picture of dynamics on $\mathcal{M}$. (a) The saddle–node connection; (b) The saddle has doubled but the node has not; (c) The node has just doubled; (d) The node has doubled but the saddle has not; (e) The saddle has also doubled; (f) The final situation. The system takes either (a) → (b) → (c) → (f) route, or (a) → (d) → (e) → (f) route.

this stage, the period-$2p$ attracting node is not yet involved in any ICC.

In the second stage, the original saddle — previously with an attracting transverse direction — becomes a repelling node. In its vicinity, a period-$2p$ saddle emerges. After bifurcation, an attracting period-$2p$ ICC is established [Fig. 4(f)]. The unstable period-p ICC exists as well [see Fig. 4(e)], which includes the period-p repelling node, the period-p saddle, and an unstable manifold of the former connecting to the latter.

3.3. *A note on cycles of even period*

As explained above, the type of doubling can be predicted based on the topology of the tangent doubling eigenspace. However, it is straightforward to show that for a cycle of even period p, the only possible topology of this eigenspace is that of a Möbius strip. To demonstrate this, suppose that the doubling eigenspace has a cylindrical topology. In this case, the second iterate f^2 of the map after the doubling bifurcation must have two separate ICCs, each with an attracting p-cycle, together forming an attracting $2p$-cycle for f. This means that the iterates of a point x_0 of this $2p$-cycle under f jump from one ICC to the other at each step. However, this is not possible, because iterates that jump in this manner return to the point x_0 after p steps (i.e. $x_0 = f^p(x_0)$), rather than $2p$ [see Fig. 5(a) where $p = 4$]. This implies that the topology of the tangent eigenspace cannot be cylindrical and must instead be a Möbius strip. For odd periods, this problem does not occur and the cylindrical topology is possible [see Fig. 5(b) with $p = 5$].

3.4. *Practical issues*

At this stage, some practical issues are worth mentioning.

- Note that computing the eigenvalues of J_{f^p} for large p may be an ill-conditioned task.
- The topology (cylinder or Möbius) of the tangent eigenspace may not always be clear from its appearance, since the eigenspace can be arbitrarily twisted. Still, determining this topology is a

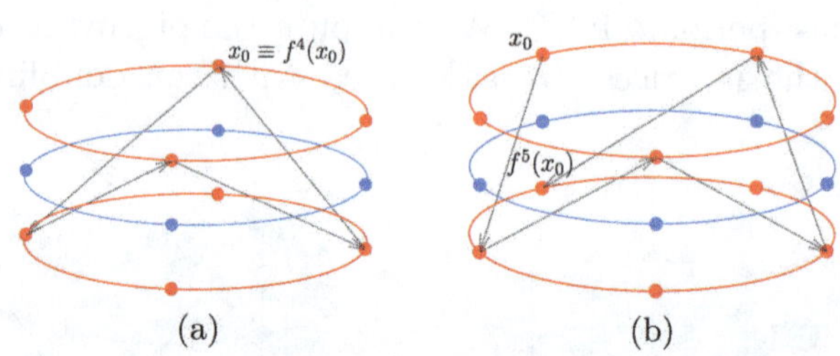

Fig. 5. Schematic representation of two ICCs and a $2p$-cycle (in red) after a loop doubling of a resonant ICC (in blue). For each curve, only one of the two cycles is shown to avoid overloading the figure. Here, p is (a) even (not possible); (b) odd (possible). The iterates of a cycle point x_0 are indicated by gray arrows.

simple task, as one can add the angles between the eigenvectors at the neighboring points of the ICC and the sum (mod 2π) converges either to zero or to π. If the sum yields zero, it has a cylindrical topology and if it yields π, it is Möbius.
- For quasi-periodic ICCs, both Lyapunov bundles and the tangent eigenspaces are approximations of the 2D center manifold. However, the tangent eigenspace is computationally much easier to obtain than the dominant Lyapunov bundles.
- For resonant ICCs, the computation procedure is more complicated than for quasi-periodic ones, as it includes computation of the ICCs itself, which, in this case, cannot be obtained by forward iterations starting from the points of the p-cycle. One has to detect the saddle cycle and then compute the unstable manifold of the saddle. When dealing with resonant ICCs of a sufficiently high period, this step might be unnecessary if the topology of the tangent eigenspace is evident from the union of doubling eigenvectors at the points of the stable cycle.

4. The Considered Systems

In this section, we introduce the systems used to illustrate the above theoretical results.

4.1. *The Mira map*

We consider the Mira map [Mira *et al.*, 1996], defined by

$$\begin{aligned} x_{n+1} &= y_n, \\ y_{n+1} &= z_n, \\ z_{n+1} &= bx_n + cy_n + az_n - x_n^2. \end{aligned} \tag{1}$$

The map has a trivial fixed point at $x = y = z = 0$. The trace, determinant, and second trace of the Jacobian, evaluated at this fixed point, are $\tau = a$, $\delta = b$, and $\sigma = -c$, respectively. According to the generic Neimark–Sacker bifurcation condition for fixed points in 3D maps, namely $\delta^2 - \tau\delta + \sigma = 1$, we conclude that this fixed point undergoes a Neimark–Sacker bifurcation at the set

$$\phi_{\text{NS}} = \{(a, b, c) \mid b^2 - ab - c = 1\}. \tag{2}$$

4.2. *The Kamiyama maps*

Next, we consider two systems used in [Kamiyama *et al.*, 2014], namely

$$\begin{aligned} x_{n+1} &= R\cos\theta \cdot x_n - R\sin\theta \cdot y_n \\ &\quad - (R - 0.9) \cdot (x_n^2 + y_n^2) + E \cdot z_n, \\ y_{n+1} &= R\sin\theta \cdot x_n - R\cos\theta \cdot y_n \\ &\quad - (R - 0.9) \cdot (x_n^2 + y_n^2), \\ z_{n+1} &= x_n \end{aligned} \tag{3}$$

and

$$\begin{aligned} x_{n+1} &= R\cos\theta \cdot x_n - R\sin\theta \cdot y_n \\ &\quad - (R - 0.9) \cdot (x_n^2 + y_n^2) + E \cdot z_n, \\ y_{n+1} &= R\sin\theta \cdot x_n + R\cos\theta \cdot y_n \\ &\quad - (R - 0.9) \cdot (x_n^2 + y_n^2), \\ z_{n+1} &= C\tanh z_n + D, \end{aligned} \tag{4}$$

where θ is an angle in degrees.

5. Results

5.1. *Loop doubling of a resonant torus*

Figure 6(a) shows a 2D bifurcation diagram for the Mira map (1) in the (b, c) parameter plane with a fixed at -2.5, which shows the bifurcation curve ϕ_{NS}. Several periodicity tongues originate from this bifurcation curve, including the period-5 tongue magnified in Fig. 6(b). The figure also shows two parameter paths marked by ψ_1 (inside the resonant tongue) and ψ_2 (outside the tongue), corresponding to the doubling of a resonant ICC and a quasi-periodic ICC. The corresponding one-parameter bifurcation diagrams are shown in Figs. 6(c) and 6(d), respectively.

As can be seen in Fig. 6(c), in this case, the node doubles first. The stable period-5 cycle undergoes a flip bifurcation, leading to the appearance of a stable 10-cycle. According to the Neimark–Sacker theorem, we can expect that, sufficiently close to its origin, the considered period-5 tongue contains a resonant ICC.

This is confirmed by Fig. 7(a), which shows the structure of the phase space corresponding to the point marked A in Fig. 6(c). It shows an attracting ICC just before the doubling bifurcation: a union of the stable 5-node P_5^{sss}, the 5-saddle Q_5^{uss}, and its 1D unstable manifold. In the notation used here, the three symbols in the superscript indicate the stability of the cycles's three multipliers, while the

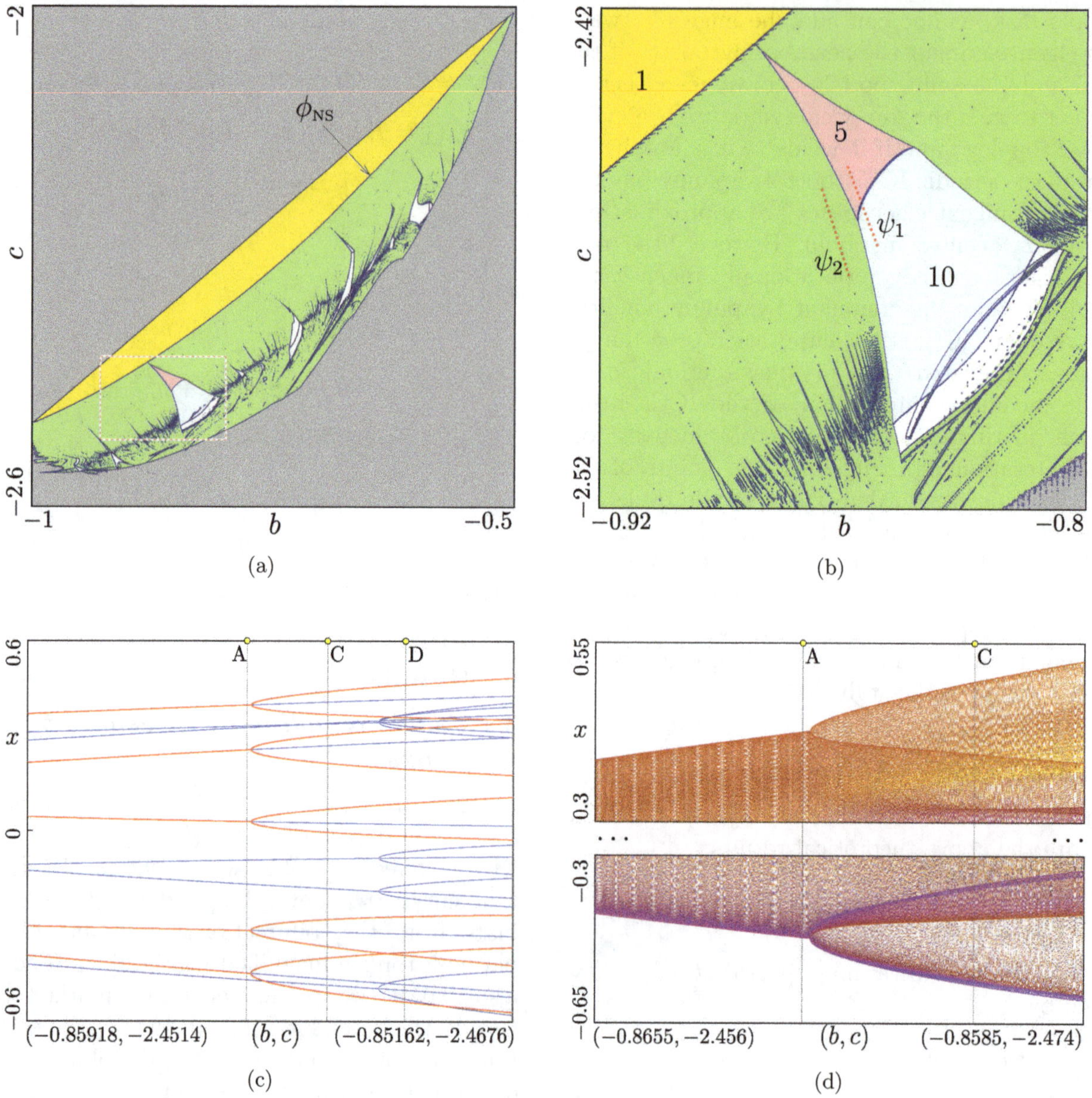

Fig. 6. (a) Parameter plane (b, c) of the Mira map with $a = -2.5$, showing the overall bifurcation structure. Gray indicates divergence, yellow indicates a fixed point, and green indicates no detected periodicity. (b) Magnification of the rectangle marked in (a), showing the region near the 5-tongue. Parameter paths ψ_1 and ψ_2 exhibit the doubling of a resonant and a quasi-periodic invariant curves, as shown in (c) and (d), respectively. Parameter values corresponding to the configurations shown in Figs. 7 and 9 are marked. The points A, C, and D in (c) refer to Figs. 7(a), 7(c), and 7(d), respectively. The points A and C in (d) refer to Figs. 9(a) and 9(c), respectively.

subscript refers to its period. To determine the type of doubling bifurcation for this invariant curve, we compute its doubling tangent eigenspace by approximating it with the eigenvectors of the Jacobian J_{f^5} corresponding to eigenvalues approaching -1. The eigenvectors have been obtained at a large number of points on the ICC and their tips have been joined by a green line, and different points on each arrow have been joined by lines of different color. In this way, the structure of the doubling tangent space becomes evident. As illustrated in Fig. 7(b), this tangent space has a cylindrical topology. This implies that a loop-doubling bifurcation must occur.

The loop-doubling process is illustrated in Figs. 7(c) and 7(d). First, a flip bifurcation transforms the stable 5-node P_5^{sss} into a 5-saddle P_5^{uss}, leading to the appearance of a stable 10-node, P_{10}^{sss}. The 5-saddle Q_5^{uss} remains unaffected by this bifurcation. As a result, we obtain a repelling ICC composed of the 5-saddles P_5^{uss} and Q_5^{uss}, along

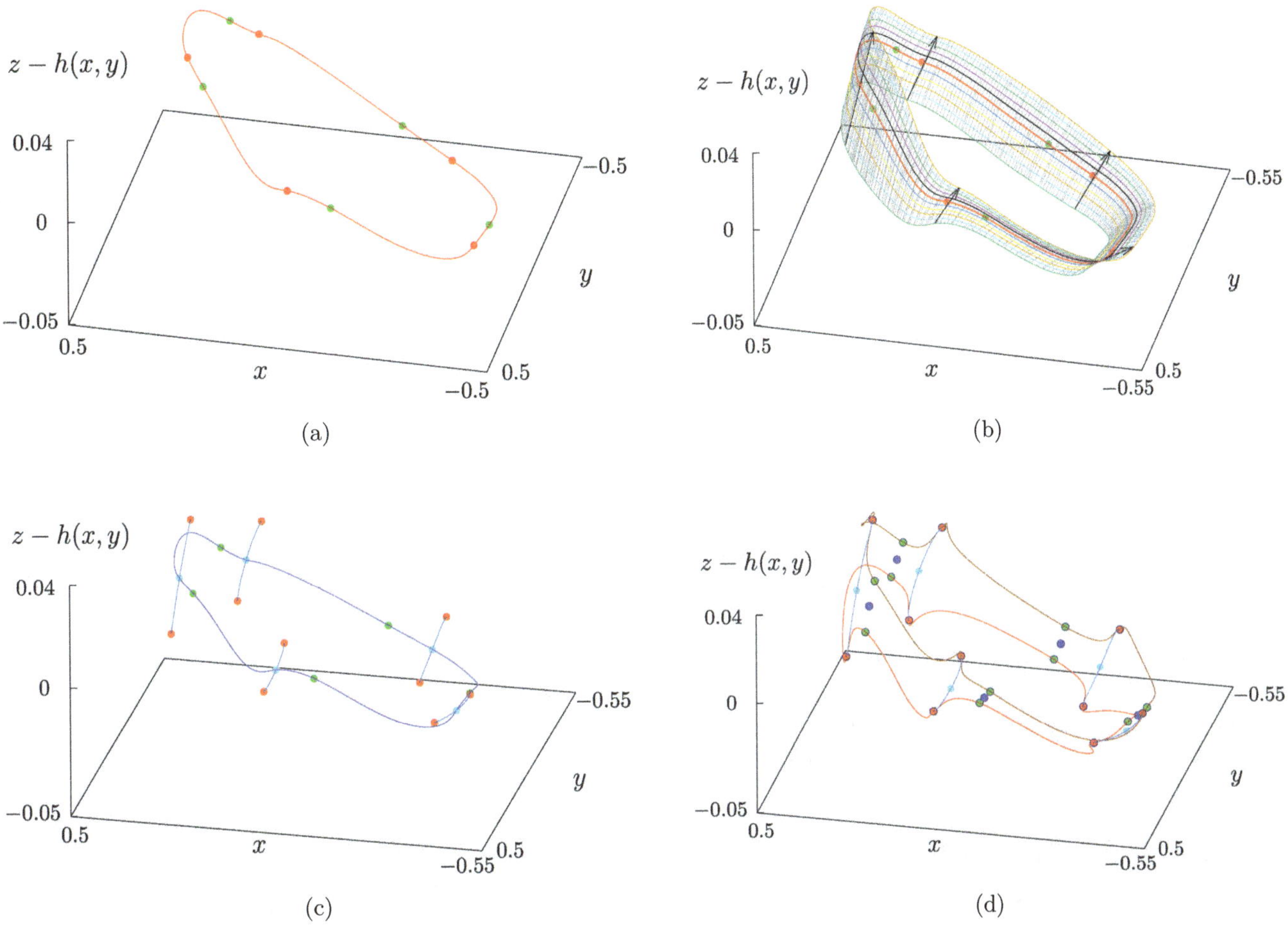

Fig. 7. Loop-doubling of a resonant ICC in the Mira map. (a) Attracting ICC before the doubling; (b) Doubling tangent space at the same parameter values as in (a); (c) Repelling ICC after the doubling of the node but before the doubling of the saddle; (d) Attracting ICC consisting of two loops after both the node and the saddle have bifurcated. Additionally, in (c) and (d), unstable manifolds of the saddle P_5^{uss} are also shown. Parameter values: $a = -2.5$; (a), (b) $b = -0.85578$, $c = -2.45869$; (c) $b = -0.8545207$, $c = -2.4613842$; (d) $b = -0.8533$, $c = -2.464$. The corresponding points are indicated in Fig. 6(c) by A, B, and C, respectively. To improve visibility, the ICCs and the doubling tangent space are shown relative to the surface $h(x, y) = a_1x^2 + a_2y^2 + a_3xy + a_4x + a_5y + a_6$ with the coefficients a_1–a_6 obtained by least-square fitting of the ICC.

with the 1D unstable manifold of the latter [see Fig. 7(c), which corresponds to the schematic illustration in Fig. 4(d)]. The 1D unstable manifold of P_5^{uss} approaches the stable 10-node P_{10}^{sss} that is, at this stage, not yet involved in any ICC.

Next, the 5-saddle Q_5^{uss} undergoes a flip bifurcation, transforming into a 5-saddle Q_5^{uus}, while a 10-saddle Q_{10}^{uss} appears around it. The unstable manifold of this saddle converges to the stable 10-node P_{10}^{sss}, forming an ICC. As predicted, the resulting ICC consists of two loops invariant under the second iterate f^2. The sequence of events follows the route shown in Fig. 4, (a) → (d) → (e) → (f).

The eigenvalues of J_{f^5} corresponding to the parameter values of Fig. 7(a) corresponding to the doubling, tangent and third directions are −0.99767964, 0.86129688, and 0.53415424, respectively. The third eigenvalue is positive, which conforms to the Gardini–Sushko conjecture mentioned in Sec. 2.2.

However, for the Kamiyama map (4), we get a different result. For this system, the loop doubling phenomenon is shown in Fig. 8. In this case also, the tangent eigenspace [Fig. 8(c)] shows a cylindrical structure. For the ICC in Fig. 8(a), the eigenvalues of J_{f^5} corresponding to the doubling, tangent and third directions evaluated at the points of the stable cycle are $-0.99501, 0.72672, -0.06028$, respectively. Note that the third eigenvalue is negative, which violates the Gardini–Sushko conjecture.

5.2. *Loop doubling of an ergodic torus*

To illustrate the doubling of a quasi-periodic ICC,[1] consider the bifurcation diagram for the Mira map, shown in Fig. 6(d), which corresponds to the parameter path marked in Fig. 6(b) as ψ_2. In Fig. 6(d), only the upper and lower parts of the bifurcation diagram are shown (with the middle part omitted), making the doubling clearly visible. The sufficiently quasi-periodic ICC at the parameter point marked as A in Fig. 6(d) immediately before the doubling bifurcation is shown in Fig. 9(a). To determine the type of doubling the ICC undergoes, we need to choose the iterate f^p to compute the tangent space along the ICC. To this end, we numerically estimate the rotation number ω of the ICC, obtaining $\omega \approx 0.39860585$. Then, we consider $\frac{2}{5}$ as a sufficiently close approximation of ω and calculate the tangent eigenspace in the doubling direction using the fifth iterate f^5. Note that the same procedure, namely a rational approximation of the numerically computed rotation number, can also be used to compute the eigenvalues in the second Poincaré section discussed in [Banerjee *et al.*, 2012].

The doubling tangent space, approximated using the doubling eigenvectors of J_{f^5} at points along the quasi-periodic ICC, is shown in Fig. 9(b). As clearly visible, it has a cylindrical structure. Therefore, the doubling bifurcation that the considered ICC undergoes is a loop doubling. This is confirmed by Fig. 9(c), which shows the ICC after the doubling.

5.3. *Length doubling of a resonant torus*

For the map (3), Fig. 10(a) shows a 2D bifurcation diagram showing several periodicity tongues, and a zoom on to the period-5 tongue is shown in Fig. 10(b). It shows two paths ψ_1 and ψ_2 along which parameters are varied.

Figure 11 shows the ICCs before and after a length-doubling bifurcation as the parameter is varied along ψ_1. Figure 11(a) shows the ICC before the bifurcation begins, corresponding to the parameter marked A in Fig. 10(c). Applying our proposed method, we obtain the set of eigenvectors along the ICC. The tangent eigenspace related to the

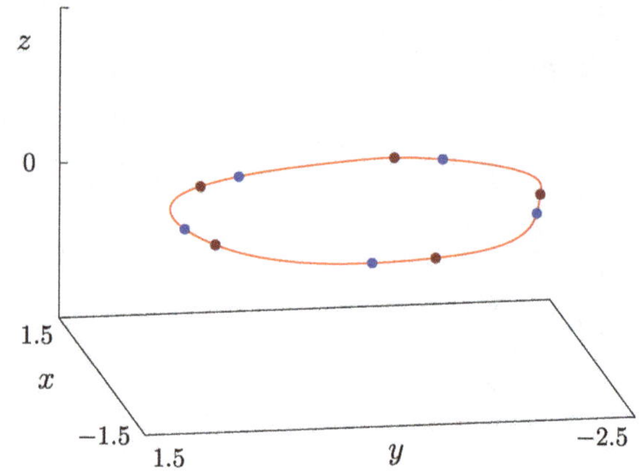

(a)

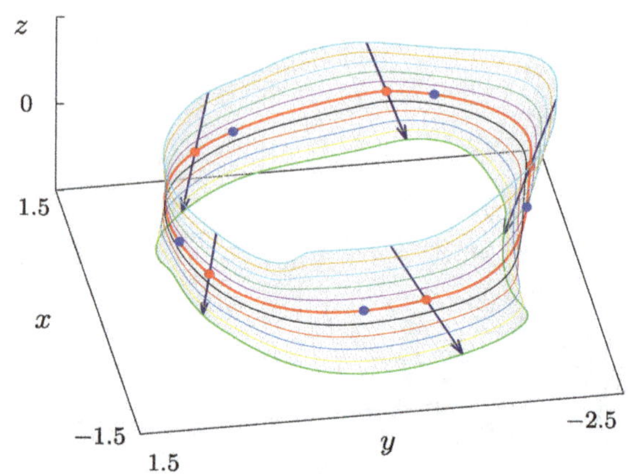

(b)

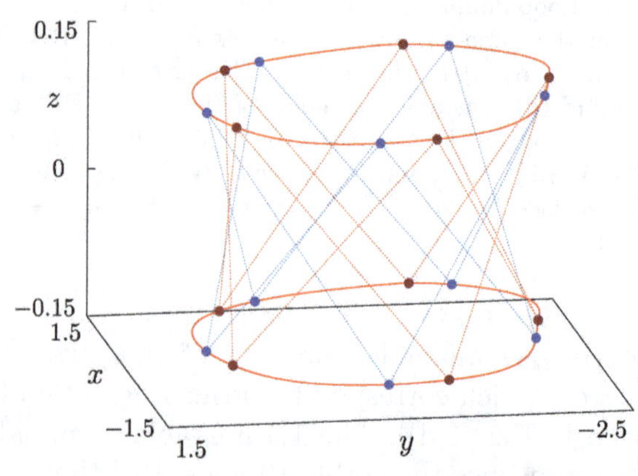

(c)

Fig. 8. Loop doubling in map (4). (a) Before the bifurcation; (b) The tangent eigenspace showing the eigenvectors of J_{f^5} corresponding to the doubling eigenvalue evaluated at the nodes; (c) The ICC after the bifurcation. The red dots represent a period-5 stable cycle and the blue dots represent a period-5 saddle. The saddle–node connection via the unstable manifolds of the saddle forms the ICC, shown in red. Parameters: $R = 1.15$, $E = 0.1$, $\theta = 82$, $D = 0$, (a), (b) $C = -0.999$, (c) $C = -1.03$.

[1] One should note that a finite-precision computer cannot identify a truly irrational frequency ratio. So, when we say an orbit is quasi-periodic, one should take it in the sense of being "sufficiently quasi-periodic".

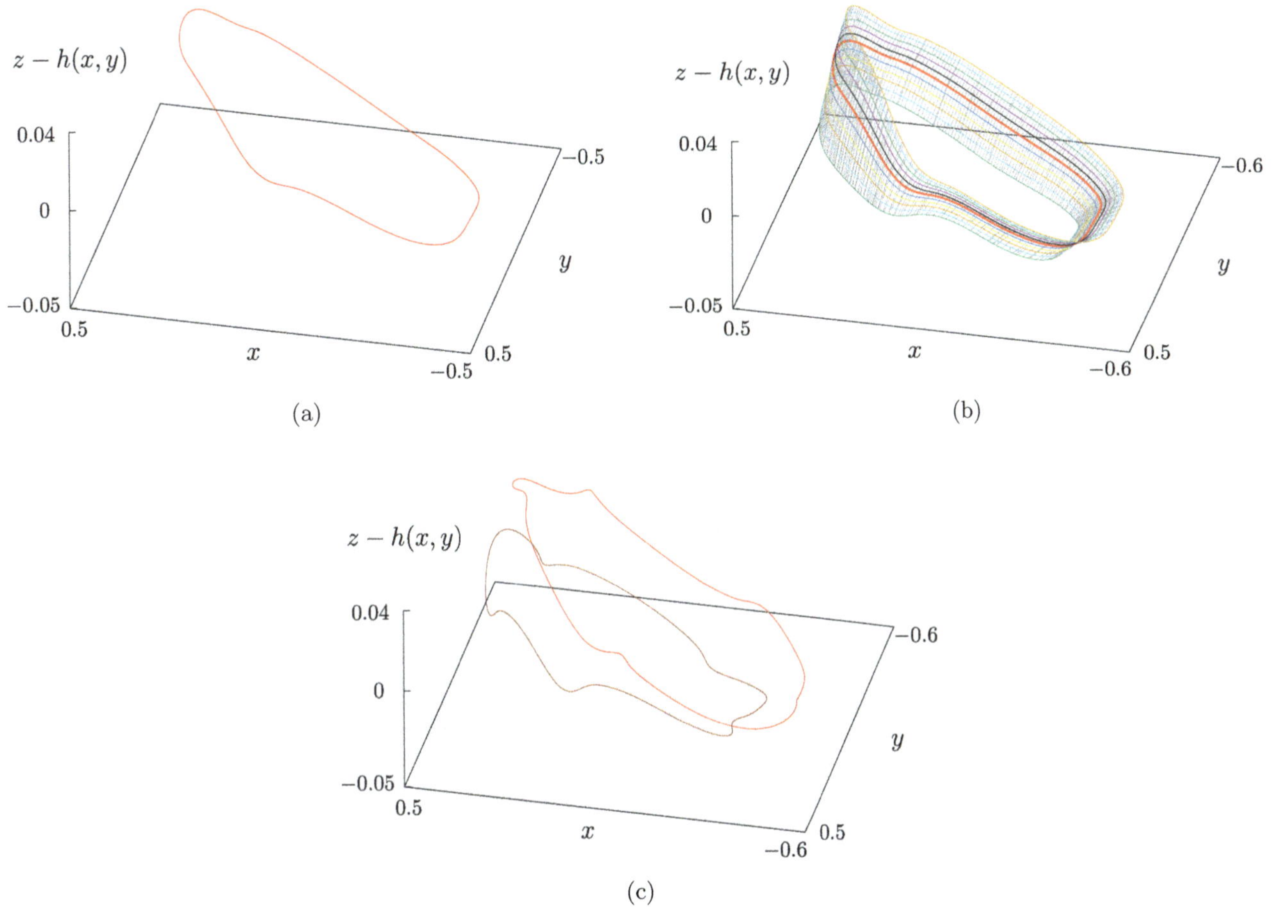

Fig. 9. Loop-doubling of a sufficiently quasi-periodic ICC in the Mira map. (a) Attracting ICC before the doubling; (b) Doubling tangent space at the same parameter values as in (a); (c) Attracting ICC consisting of two loops after the doubling. Parameter values: $a = -2.5$; (a), (b) $b = -0.862501$, $c = -2.463746$; (c) $b = -0.86$, $c = -2.47$. The corresponding points are indicated in Fig. 6(d) as A and C, respectively.

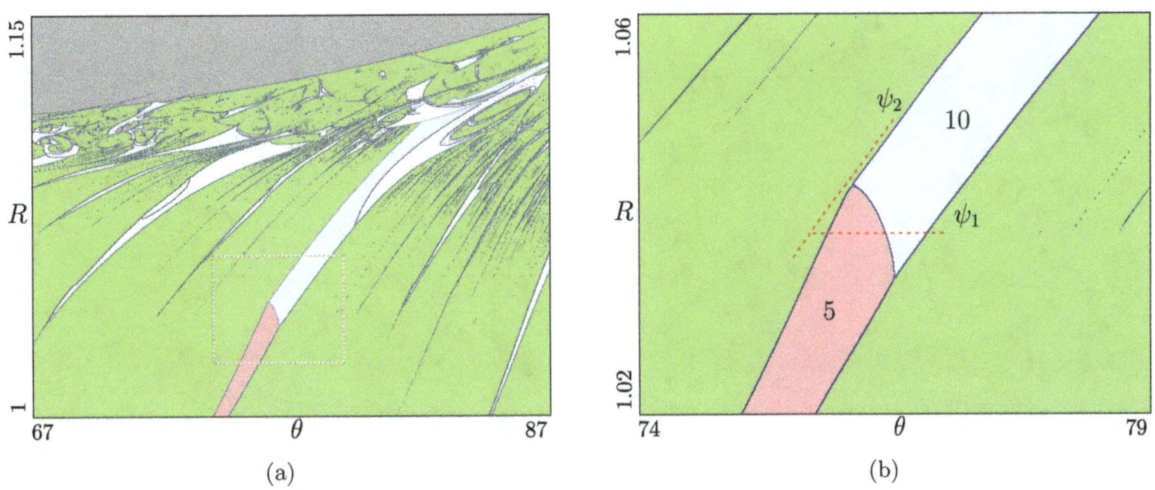

Fig. 10. (a) Parameter plane (b, c) of the Kamiyama map (3) with $E = -0.3$, showing the overall bifurcation structure. Gray indicates divergence, blue indicates period 5 and green indicates period 10. (b) Magnification of the rectangle marked in (a), showing the region near the 5-tongue. Parameter paths ψ_1 and ψ_2 exhibit the doubling of resonant and quasi-periodic invariant curves, as shown in (c) and (d), respectively. Parameter values corresponding to the configurations shown in Figs. 11 and 12 are marked. The points A, C, and D in (c) refer to Figs. 11(a), 11(c), and 11(d), respectively. The points A and C in (d) refer to Figs. 12(a) and 12(c), respectively.

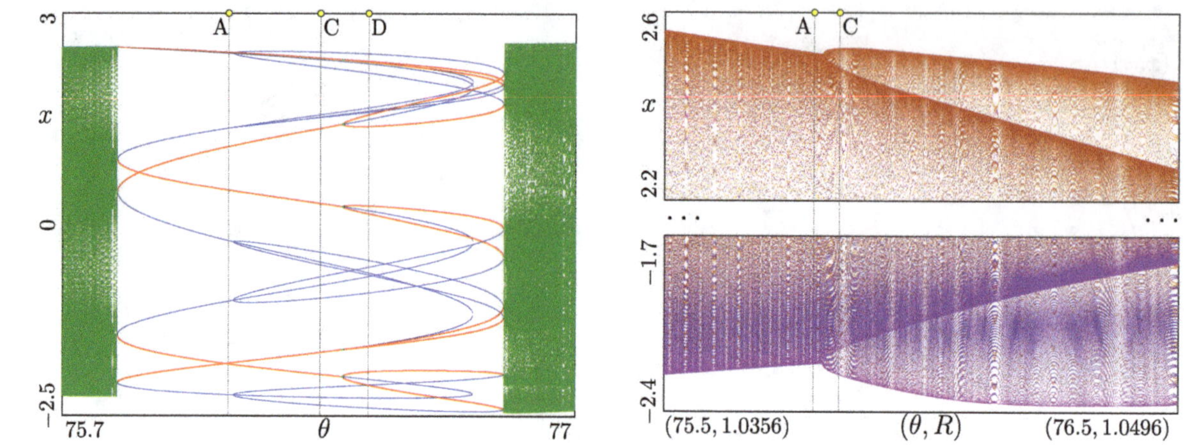

(c) (d)

Fig. 10. (*Continued*)

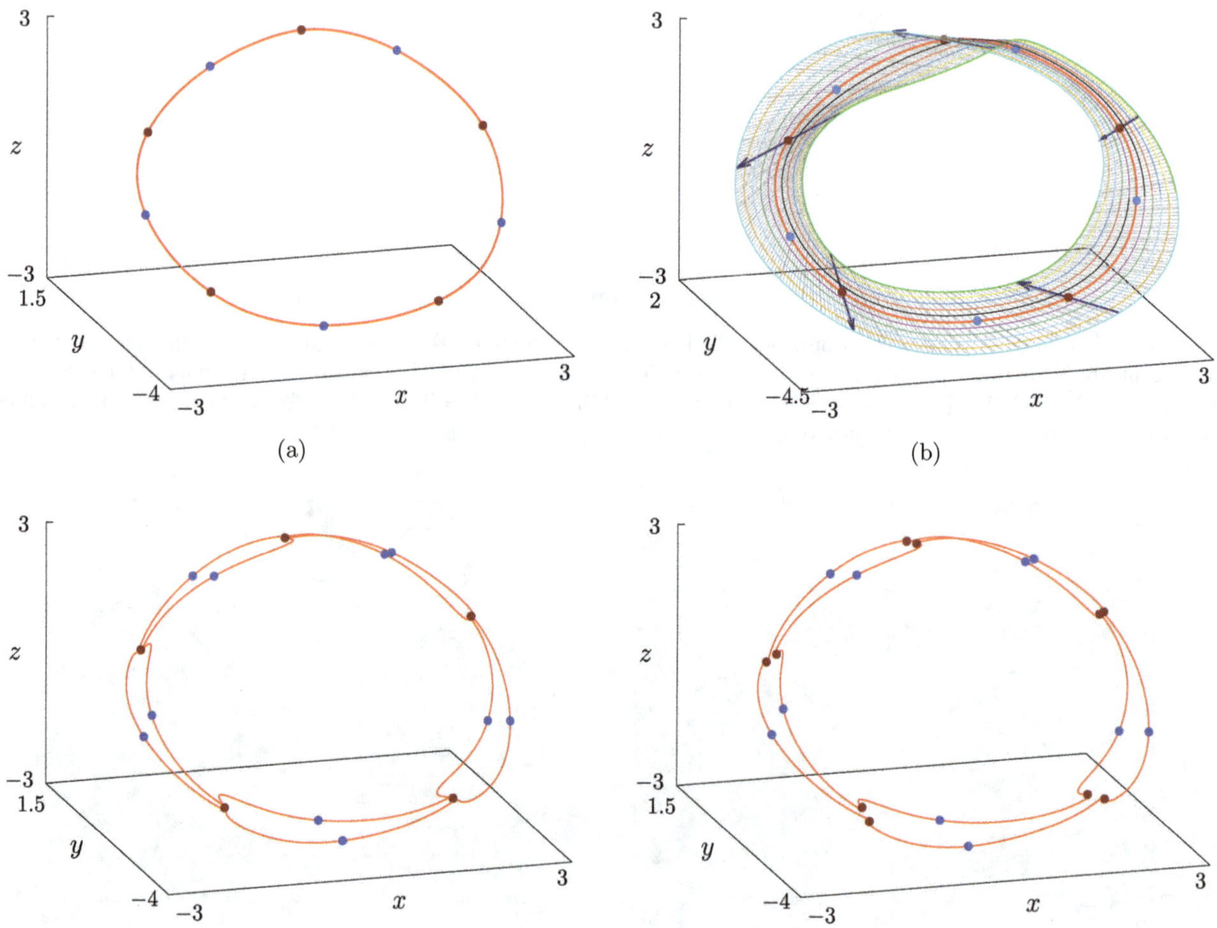

(a) (b) (c) (d)

Fig. 11. Length doubling in map (3). (a) The ICC before the bifurcation; (b) The tangent space obtained by the eigenvectors of J_{f^5} corresponding to the doubling eigenvalue evaluated at the points of the ICC. The eigenvectors at the nodes are also shown. (c) Bi-layered ICC after doubling bifurcation of the saddle but before doubling bifurcation of the node; (d) Period-10 ICC after doubling bifurcation of the node. Figures (a), (c), and (d) refer to parameters marked as A, B, and C in Fig. 10(c), respectively. Parameters: $R = 1.038$, $E = -0.3$, (a), (b) $\theta = 76.12$, (c) $\theta = 76.35$, (d) $\theta = 76.47$.

doubling eigenvalue before the bifurcation is shown in Fig. 11(b). It is clear that this tangent eigenspace has a Möbius strip structure. According to our theory, there should be a length-doubling bifurcation as the parameter is varied.

Figure 10(c) shows that, in this case, the saddle doubles first. The situation after the saddle doubles but before the node doubles is shown in Fig. 11(c), for the parameter value marked B in Fig. 10(c). In the next stage, the node also undergoes a period-doubling bifurcation, resulting in the length-doubled ICC [Fig. 11(d)] for the parameter value marked C in Fig. 10(c). Therefore, this ICC doubling occurs following the sequence illustrated in Fig. 4, (a) → (b) → (c) → (f).

The eigenvalues of J_{f^5} corresponding to the doubling, tangent, and third directions just before the bifurcation, evaluated at the points of the stable cycle are $-0.95979, 0.85489, 3.58612 \times 10^{-6}$, respectively. We notice that the third eigenvalue is positive, which contradicts the Gardini–Sushko conjecture.

5.4. *Length doubling of a quasi-periodic torus*

Figure 12 shows the appearance of a double-length quasi-periodic ICC for the map (3) as the parameter is varied along the line ψ_2 shown in Fig. 10(b). Figure 12(a) shows the quasi-periodic ICC before bifurcation. The tangent eigenspace [Fig. 12(b)] computed following the procedure outlined above shows a Möbius strip structure. Finally, Fig. 12(c) shows the resulting length-doubled ICC.

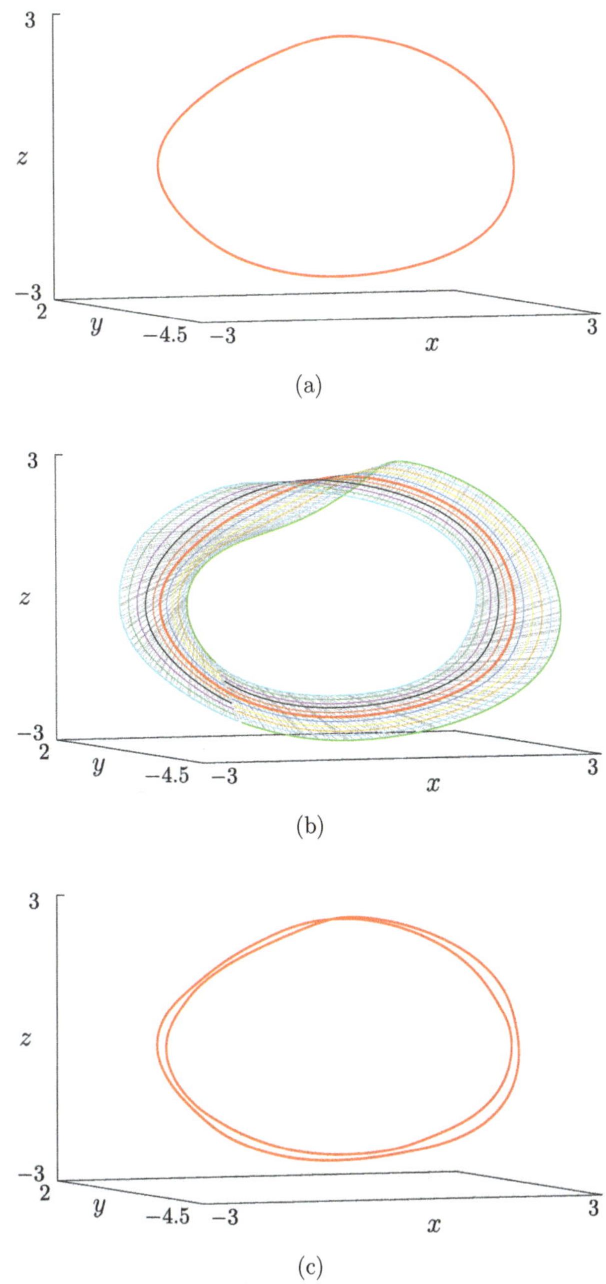

Fig. 12. Length doubling of a sufficiently quasi-periodic ICC in map (3). (a) The ICC before the bifurcation; (b) The tangent eigenspace; (c) The orbit after the bifurcation. The parameter values for (a) and (b) are $E = -0.3$, $R = 1.0397$, $\theta = 75.79$ [marked as A in Fig. 10(d)] and those for (c) are $E = -0.3$, $R = 1.0404$ $\theta = 75.8398$ [marked as C in Fig. 10(d)]. Rotation number ≈ 0.199220.

5.5. *ICC with saddle–focus connection*

The phenomena discussed in this paper occur in maps of dimension 3 or higher. If we consider a 3D system, there are three eigenvalues of the Jacobian matrix. All three eigenvalues can be real, or one can be real and the other two complex conjugate. In the latter case, the ICC could be formed in two possible ways:

(1) The ICC is formed by the unstable manifolds of the saddle cycle that are tangent to the real eigenvectors;
(2) The unstable manifolds of the saddles spiral onto the stable nodes (the saddle–focus connection), forming an ICC of infinite length.

Figure 13 shows the two types of structures schematically.

Since a torus-doubling bifurcation involves a period doubling of a stable cycle, the real-valued eigenvector (along which the doubling occurs) cannot be a part of the ICC. Therefore, the ICC structure must be formed by a saddle–focus connection.

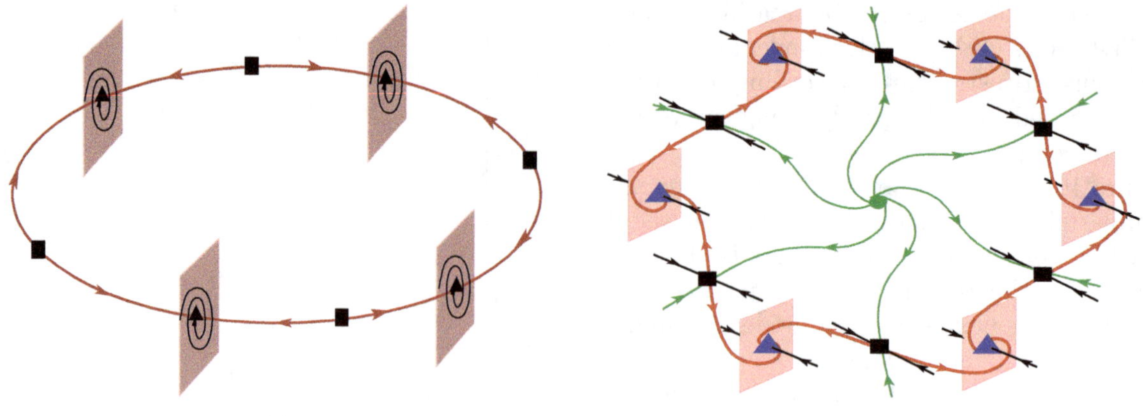

Fig. 13. Schematic diagram showing two types of ICCs connecting saddle cycles and focus-type nodes.

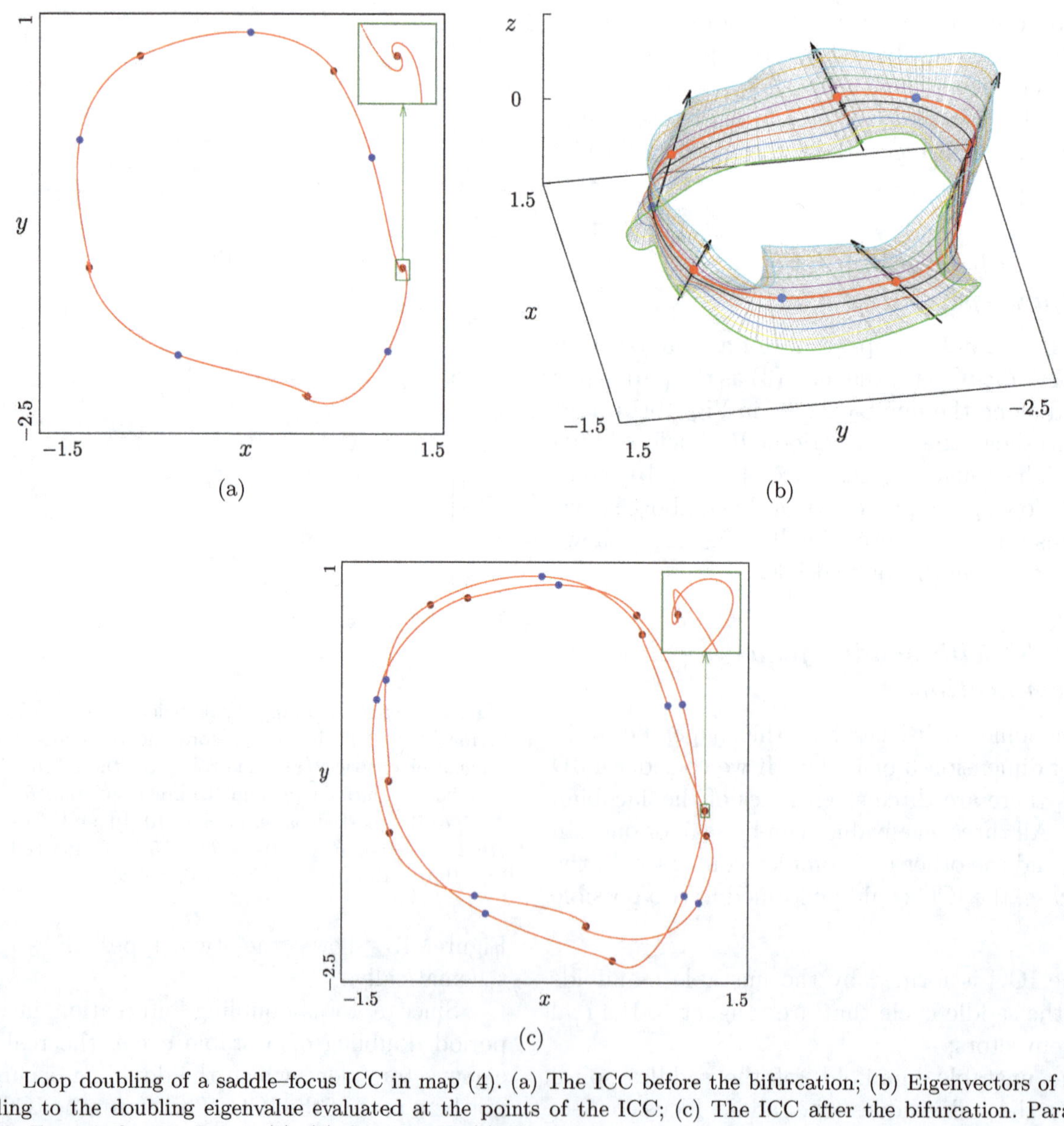

Fig. 14. Loop doubling of a saddle–focus ICC in map (4). (a) The ICC before the bifurcation; (b) Eigenvectors of J_{f^5} corresponding to the doubling eigenvalue evaluated at the points of the ICC; (c) The ICC after the bifurcation. Parameters: $R = 1.19$, $E = 0.1$, $\theta = 85$, $D = 0$, (a), (b) $C = -0.999$, (c) $C = -1.03$.

For Fig. 13(a), since the real eigendirections are along the ICC, it cannot be a period-doubling direction. Therefore, a torus doubling cannot occur for a structure like this.

Figure 14 shows an ICC with a saddle–focus connection for system (4) that bifurcates to two disjoint loops. The only real eigenvalue of J_{f^5} evaluated at the points of the stable cycle before the bifurcation is ≈ -0.99501; the other two eigenvalues are complex. The set of eigenvectors corresponding to the doubling eigenvalue is shown in Fig. 14(c). Although the direction of the vector changes at different points in the loop, the structure is topologically a cylinder.

6. Conclusions

In this work, we have developed a unified and practical framework for predicting the type of doubling bifurcation (loop doubling or length doubling) exhibited by ICCs in 3D maps. Our approach resolves the critical limitations of existing methods by leveraging the topology of the tangent eigenspace (the eigenvectors associated with the eigenvalue approaching -1 along the ICC). Specifically, a cylindrical topology predicts loop doubling (two disjoint ICCs with iterates flipping between them), while a Möbius strip topology predicts length doubling (a single ICC of double length). Unlike earlier methods, this technique provides a single, universally applicable framework for both quasi-periodic and resonant ICCs. Furthermore, we prove that resonant ICCs with even-period cycles can only undergo length doubling due to topological constraints, which explains why loop doubling is observed less frequently than length doubling.

Acknowledgments

S. Banerjee acknowledges financial support from J. C. Bose Grant of the Anusandhan National Research Foundation, Government of India, No. JBR/2020/000049. The work of V. Avrutin and Sushko were supported by the German Research Foundation within the scope of the project "Global bifurcation phenomena in discontinuous piecewise-smooth maps in theory and applications for power converter systems", project number AV 111/3-1. S. Biswas acknowledges financial support from the WISE program of DAAD (No. 57698568) to carry out a summer internship in Germany during which some of the work was done.

ORCID

Sayanho Biswas
https://orcid.org/0000-0003-0752-5400
Soumitro Banerjee
https://orcid.org/0000-0003-3576-0846
Viktor Avrutin
https://orcid.org/0000-0001-7931-8844
Iryna Sushko
https://orcid.org/0000-0001-5879-0699

References

Arneodo, A., Coullet, P. H. & Spiegel, E. A. [1983] "Cascade of period doublings of tori," *Phys. Lett. A* **94**, 1–6.

Banerjee, S., Giaouris, D., Missailidis, P. & Imrayed, O. [2012] "Local bifurcations of a quasiperiodic orbit," *Int. J. Bifurcation and Chaos* **22**, 1250289.

Bischi, G. & Tramontana, F. [2010] "Three-dimensional discrete-time Lotka–Volterra models with an application to industrial clusters," *Commun. Nonlin. Sci. Numer. Simul.* **15**, 3000–3014.

Ding, W.-C., Xie, J. H. & Sun, Q. G. [2004] "Interaction of Hopf and period doubling bifurcations of a vibro-impact system," *J. Sound Vib.* **275**, 27–45.

Gardini, L. & Sushko, I. [2012] "Doubling bifurcation of a closed invariant curve in 3D maps," *ESAIM: Proc.*, Vol. 36 (EDP Sciences), pp. 180–188.

Gonchenko, A., Gonchenko, S. & Turaev, D. [2021] "Doubling of invariant curves and chaos in three-dimensional diffeomorphisms," *Chaos* **31**, 113130.

Kamiyama, K., Komuro, M., Endo, T. & Aihara, K. [2014] "Classification of bifurcations of quasi-periodic solutions using Lyapunov bundles," *Int. J. Bifurcation and Chaos* **24**, 1430034.

Karatetskaia, E., Shykhmamedov, A. & Kazakov, A. [2021] "Shilnikov attractors in three-dimensional orientation-reversing maps," *Chaos* **31**, 011102.

Letellier, C., Bennoud, M. & Martel, G. [2007] "Intermittency and period-doubling cascade on tori in a bimode laser model," *Chaos Solit. Fract.* **33**, 782–794.

Luo, G. W. [2006] "Hopf-flip bifurcations of vibratory systems with impacts," *Nonlinear Anal.: Real World Appl.* **7**, 1029–1041.

McKell, K. E., Broomhead, D. S., Jones, R. & Hurle, D. T. J. [1990] "Torus doubling in convecting molten gallium," *Europhys. Lett.* **12**, 513–518.

Mira, C., Gardini, L., Barugola, A. & Cathala, J.-C. [1996] *Chaotic Dynamics in Two-dimensional Non-invertible Maps*, World Scientific Series on Nonlinear Science, Vol. 20 (World Scientific, New Jersey).

Molteno, T. C. A. & Tufillaro, N. B. [1990] "Torus doubling and chaotic string vibrations: Experimental results," *J. Sound Vib.* **137**, 327–330.

Osipenko, G. [2007] *Dynamical Systems, Graphs, and Algorithms* (Springer-Verlag, Berlin).

Schilder, F., Osinga, H. M. & Vogt, W. [2005] "Continuation of quasi-periodic invariant tori," *SIAM J. Appl. Dyn. Syst.* **4**, 459–488.

Sekikawa, M., Miyoshi, T. & Inaba, N. [2001] "Successive torus doubling," *IEEE Trans. Circuits Syst. I: Fundam. Theory Appl.* **48**, 28–34.

Zhong, G. Q., Wu, C. W. & Chua, L. O. [1998] "Torus-doubling bifurcations in four mutually coupled Chua's circuits," *IEEE Trans. Circuits Syst. I: Fundam. Theory Appl.* **45**, 186–193.

www.ingramcontent.com/pod-product-compliance
Lightning Source LLC
LaVergne TN
LVHW081947200926
848484LV00009B/2519